Volumes already published

Volume 1: Trees I (1986)
Volume 2: Crops I (1986)
Volume 3: Potato (1987)
Volume 4: Medicinal and Aromatic Plants I (1988)
Volume 5: Trees II (1989)
Volume 6: Crops II (1988)
Volume 7: Medicinal and Aromatic Plants II (1989)
Volume 8: Plant Protoplasts and Genetic Engineering I (1989)
Volume 9: Plant Protoplasts and Genetic Engineering II (1989)
Volume 10: Legumes and Oilseed Crops I (1990)
Volume 11: Somaclonal Variation in Crop Improvement I (1990)
Volume 12: Haploids in Crop Improvement I (1990)

Volumes in preparation

Volume 13: Wheat (1990)
Volume 14: Rice (1991)
Volume 15: Medicinal and Aromatic Plants III (1991)
Volume 16: Trees III (1991)

Biotechnology in Agriculture and Forestry 11

Somaclonal Variation in Crop Improvement I

Edited by Y. P. S. Bajaj

With 168 Figures

Springer-Verlag Berlin Heidelberg New York
London Paris Tokyo Hong Kong Barcelona

Professor Dr. Y.P.S. BAJAJ
A-137
New Friends Colony
New Delhi 110065, India

ISBN 978-3-642-08077-7

Library of Congress Cataloging-in-Publication Data. Somaclonal variation in crop improvement / edited by Y.P.S. Bajaj. p. cm. -- (Biotechnology in agriculture and forestry ; 11-Includes bibliographical references. 1. Plant propagation--In vitro. 2. Plant breeding. 3. Botany--Variation. 4. Crops--Breeding. I. Bajaj, Y.P.S., 1936- . II. Series. SB123.6.S65 1990 631.5'23--dc20 89-26245

Printed in Germany

2131/3145(3011)-543210 – Printed on acid-free paper

Dedicated to my Mother
Mrs. Kundan Kaur Bajaj

Preface

Genetic erosions in plant cell cultures, especially in chromosome number and ploidy level, have now been known for over 25 years. Until the mid –1970s such changes were considered undesirable and therefore discarded because the main emphasis was on clonal propagation and genetic stability of cultures. However, since the publication on somaclonal variation by Larkin and Scowcroft (1981) there has been a renewed interest to utilize these in vitro obtained variations for crop improvement. Studies conducted during the last decade have shown that callus cultures, especially on peridical subculturing over an extended period of time, undergo morphological and genetic changes, i.e. polyploidy, aneuploidy, chromosome breakage, deletions, translocations, gene amplification, inversions, mutations, etc. In addition, there are changes at the molecular and biochemical levels including changes in the DNA, enzymes, proteins, etc. Such changes are now intentionally induced, and useful variants are selected. For instance in agricultural crops such as potato, tomato, tobacco, maize, rice and sugarcane, plants showing tolerance to a number of diseases, viruses, herbicides and salinity, have been isolated in cell cultures. Likewise induction of male sterility in rice, and wheat showing various levels of fertility and gliadin, have been developed in vitro. These academic excercises open new avenues for plant breeders and pathologists.

Another area of tremendous commercial importance in the pharmaceutical industry is the selection of cell lines showing high levels of medicinal and industrial compounds. Already high shikonin containing somaclones in *Lithospermum* are being used commercially. Somaclonal variation also enables the addition or intensification of only one feature to an established variety possessing a combination of most of the useful agronomic traits. Although desirable material generated in vitro would help to increase the much needed genetic variability in crops, one has to be cautious with respect to field aspects which are the most important criteria for crop improvement. A judicious rather than an overenthusiastic approach is to be commended.

Taking the above mentioned developments into consideration the present book on SOMACLONAL VARIATION IN CROP IMPROVEMENT I has been compiled. It comprises 29 chapters contributed by international experts on various aspects of somaclonal variation, i.e. the genetic and molecular basis of variability, gene amplification, mosaics and chimeras, role of environments, and the variability of tolerance to salinity, nematodes, diseases, etc. in plant cell cultures. Eighteen individual chapters deal exclusively with the induction of somaclonal variation in cereals, vegetables, fruits and ornamentals, to including such crops as rice, maize, barley, potato, tomato, eggplant, cucurbits, sugarbeet, chicory, strawberry, peach, geraniums, fuchsia, carnations, *Haworthia, Weigela,* alfalfa, *Nicotiana* spp. The merits and demerits of somaclonal variation and their implications in plant breeding are discussed.

Since somaclonal variation in plant cell cultures has been the focus of interest recently for the induction of much needed genetic variability in crops, this book will be of great assistance to research workers, teachers and advanced students of plant biotechnology, tissue culture, genetics, horticulture, plant pathology, and especially plant breeding.

New Delhi, March 1990

Professor Y.P.S. Bajaj
Series Editor

Contents

Section I Somaclonal Variation; Chromosomal, Genetical and Molecular Variability; Gene Amplification; Mosaics and Chimeras; Variability for Tolerance to Salinity and Nematodes

Section II Cereals (For Wheat and Triticale see Vol. 13)

II. 1 Somaclonal Variation in Cereals
C. TONELLI

II. 2 Somaclonal Variation in Rice
Z-X SUN and K-L ZHENG (With 1 Figure)

II. 3 Somaclonal Variation in Maize
E.D. EARLE and A.R. KUEHNLE (With 2 Figures)

II. 4 Somaclonal Variation in Barley (*Hordeum vulgare* L.)
A. BREIMAN and D. ROTEM-ABARBANELL (With 4 Figures)

Section III Vegetables and Fruits

List of Contributors

ALICCHIO, R., Department of Evolutionary and Experimental Biology, University of Bologna, Via F. Selmi 1, 40126 Bologna, Italy

ANORUO, A.O., Yale School of Forestry and Environmental Studies, New Haven, CT 06511, USA

BAJAJ, Y.P.S., Former Professor of Tissue Culture, Punjab Agricultural University, Ludhiana, India, (Present address: A-137 New Friends Colony, New Delhi 110065, India)

BALL, S.G., Chaire de Génétique et d'Amélioration des Plantes, Faculté des Sciences Agronomiques de l'Etat, 5800 Gembloux, Belgium

BECK, R.C., Yale School of Forestry and Environmental Studies, New Haven, CT 06511, USA

BENNICI, A., Plant Biology Department, Faculty of Agriculture, University of Florence, Italy (Address for correspondence: Dipartimento di Biologia Vegetale, P.le delle Cascine 28, 50144 Firenze, Italy)

BERLYN, G.P., Yale School of Forestry and Environmental Studies, New Haven, CT 06511, USA

BOUHARMONT, J., Laboratoire de Cytogénétique, Université Catholique Louvain, Place Croix du Sud 4, 1348 Louvain-la-Neuve, Belgium

BREIMAN, A., Department of Botany, The George S. Wise Faculty of Life Science, Tel Aviv University, Tel Aviv 69978, Israel

BUIATTI, M., Department of Animal Biology and Genetics, Via Romana 17, Florence, Italy

CULLIS, C.A., Department of Biology, Case Western Reserve University, Cleveland, OH 44106, USA

DABIN, P., Laboratoire de Cytogénétique, Université Catholique Louvain, Place Croix du Sud 4, 1348 Louvain-la-Neuve, Belgium

DE PAEPE, R., Laboratoire de Génétique Moléculaire des Plantes, UA 115, Université Paris-Sud, 91405 Orsay Cedex, France

DECOURTYE, L., Station d'Amélioration des Expéces Fruitères et Ornementales, Laboratoire des Arbustes d'Ornement, INRA, 49000 Angers, France

DOLEY, W.P., Department of Crop and Soil Sciences, Michigan State University, East Lansing, MI 481824, USA

DURON, M., Station d'Amélioration des Expéces Fruitères et Ornementales, Laboratoire des Arbustes d'Ornement, INRA, 49000 Angers, France

EARLE, E.D., Department of Plant Breeding, Cornell University, 516 Bradfield Hall, Ithaca, NY 14853-1902, USA

FASSULIOTIS, G., US Department of Agriculture, Agriculture Research Service, US Vegetable Laboratory, 2875 Savannah Highway, Charleston, SC 29414, USA

GRAZIA, S., Department of Evolutionary Experimental Biology, University of Bologna, Via Belmeloro 8, 40126 Bologna, Italy

HAMMERSCHLAG, F., Tissue Culture and Molecular Biology Laboratory, USDA/ARS/BA, Beltsville, MD 20705, USA

JOHNSON, L.B., Department of Plant Pathology, Throckmorton Hall, Kansas State University, Manhatten, KS 66506, USA

KARP, A., Biochemistry Department, Rothamsted Experimental Station, Harpenden, Hertshire AL5 2JQ, United Kingdom

KUEHNLE, A.R., Department of Horticulture, 102 St. John Lap., University of Hawaii, Honolulu, HI 96822, USA

LESHEM, B., Department of Plant Genetics and Breeding, ARO, The Volcani Center, P.O. Box 6, Bet Dagan 50250, Israel

LI, X.Q., University of Beijing, Beijing, China

MARCOTRIGIANO, M., Department of Plant and Soil Sciences, University of Massachusetts, Amherst, MA 01003, USA

MORENO, V., Plant Tissue Culture Laboratory, Department of Biotechnology, Escuela Técnica Superior de Ingenieros Agrónomos, Universidad Politecnica de Valencia, C/Camino de Vera 14, 46022 Valencia, Spain

MORPURGO, R., Department of Physiology, International Potato Center (CIP), P.O. Box 5696, Lima, Peru

NAGL, W., Division of Cell Biology, Department of Biology and Biotechnology Program, The University of Kaiserslautern, 6750 Kaiserslautern, Federal Republic of Germany

NIEMIROWICZ-SZCZYTT, K., Department of Genetics and Horticultural Plant Breeding, ul Nowoursynowska 166, 02-766 Warsaw, Poland

OGIHARA, Y., Section of Cytogenetics, Kihara Institute of Biological Research, Yokohama City University, Yokohama 232, Japan

OGURA, H., Department of Agricultural Sciences, Ishikawa College of Agriculture, Nonoichi, Ishikawa 921, Japan

PIERAGOSTINI, E., Department of Animal Production, University of Bari, Via Amendola 165 A, 70100 Bari, Italy

PRAT, D., Laboratoire de Génétique et Physiologie du Dévelopement des Plantes, CNRS, 91190 Gif-sur-Yvette, France, (Present address: Laboratoire de Génétique des Populations d'Arbes Forestiers, ENGREF, 54042 Nancy Cedex, France)

ROCCHETTA, G., Department of Evolutionary Experimental Biology, University of Bologna, Via Belmeloro 8, 40126 Bologna, Italy

ROIG, L.A., Plant Tissue Culture Laboratory, Department of Biotechnology, Escuela Técnica Superior de Ingenieros Agrónomos, Universidad Politecnica de Valencia, C/Camino de Vera 14, 46022 Valencia, Spain

ROTEM-ABARBANELL, D., Department of Botany, The George S. Wise Faculty of Life Science, Tel Aviv University, Tel Aviv 69978, Israel

SAUNDERS, J.W., US Department of Agriculture, Agricultural Research Service, Crops and Entomology Research Unit, Box 1633, East Lansing, MI 48826-6633, USA

SIBI, M., E.N.S.A.I.A 2, Avenue de la Forêt de Haye, 545050 Vandoeuvre, France

SUN, Z-X, Plant Genetics and Breeding Department, China National Rice Research Institute, Hangzhou, China 310006

TAL, M., Department of Biology, Ben-Gurion University of the Negev, P.O.Box 653, Beer-Sheva 84105, Israel

THEURER, J.C., US Department of Agriculture, Agricultural Research Service, Crops and Entomology Research Unit, Box 1633, East Lansing, MI 48826-6633, USA

THOMAS, M.R., Department of Plant Pathology, Throckmorton Hall, Kansas State University, Manhatten, KS 66506, USA

TONELLI, C., Dipartimento di Genetica e di Biologia dei Microorganismi, Universitá degli Studi di Milano, Via Celoria 26, 20133 Milano, Italy

YU, M.H., US Department of Agriculture, Agricultural Research Service, Sugarbeet Production Unit, 1636 East Alisal St. Salinas, CA 93905, USA

ZHENG, K-L, Biotechnology Department, China National Rice Institute, Hangzhou, China 310006

Section I Somaclonal Variation; Chromosomal, Genetical and Molecular Variability; Gene Amplification; Mosaics and Chimeras; Variability for Tolerance to Salinity and Nematodes

I.1 Somaclonal Variation – Origin, Induction, Cryopreservation, and Implications in Plant Breeding

Y.P.S. Bajaj[1]

1 Introduction

It has been known for over 25 years now that plant cell and tissue cultures undergo genetic erosions and show changes of various types, especially in chromosome numbers and ploidy level (Partanen 1963; D'Amato 1965; Murashige and Nakano 1966). However, till the mid 1970's such changes were considered undesirable and were therefore discarded because the main emphasis was on clonal propagation and genetic stability of the cell cultures. Extensive studies conducted during the last decade have shown that the cell cultures, especially on periodical subculturing, undergo various morphological and genetic changes, i.e., polyploidy, aneuploidy, chromosome breakage, deletions, translocations, gene amplification, inversions, mutations, etc. (see Nagl 1972; Meins 1983; D'Amato 1985). In addition, there are changes at the molecular and biochemical levels (Day and Ellis 1984; Landsman and Uhrig 1985; Ball and Seilleur 1986) including changes in the DNA (Berlyn 1982; Cullis 1983), enzymes (Brettell et al. 1986b), gliadin (Cooper et al. 1986), etc. Since the publication of Larkin and Scowcroft (1981) there has been an upsurge of interest, and the attitude toward these changes has shifted to using the somaclonal variations for plant improvement; they are intentionally induced and much sought after.

The success of any crop improvement program depends on the extent of genetic variability in the base population. However, there is a lack of existing genetic variability in most of the agricultural crops, which hampers their improvement through conventional plant breeding methods. In this context, in vitro technology is a powerful tool for the induction of much-needed genetic variability. Already plants showing tolerance/resistance to phytotoxins (Gengenbach et al. 1977), herbicides (Chaleff and Ray 1984), nematodes and viruses (Wenzel and Uhrig 1981), salt (Nabors et al. 1980; Bajaj and Gupta 1987), that are high-yielding (Ogura et al. 1988), rich in protein (Schaeffer and Sharpe 1987) and sugar contents (Liu and Chen 1976) and male sterile (Ling et al. 1987), have been obtained, and are being incorporated in crop improvement programs. Important agricultural crops such as wheat, rice, maize, potato, sugarcane, brassica, etc. have already yielded positive results to the extent of new cultivars being released.

[1] Former Professor of Tissue Culture, Punjab Agricultural University, Ludhiana, India. Present address: A-137 New Friends Colony, New Delhi 110065, India

Biotechnology in Agriculture and Forestry, Vol. 11
Somaclonal Variation in Crop Improvement I (ed. by Y.P.S. Bajaj)

In my opinion, any genetic variability brought about by in vitro culture may be included in the term somaclonal variation. Therefore in this chapter the term is used in a broad sense to include the variability not only brought about in cell and callus cultures (as is generally believed), but also the changes induced by other in vitro systems and techniques. In this chapter literature on somaclonal variation is compiled (Table 1), dealing with the present state of the art, origin of variation, and induction of somaclones in various crops, and the implications in plant breeding are discussed. Our work on crops such as rice, wheat, potato, peanut, chickpea, beans pigeonpea, cotton, pearl-millet, panicum, and napier grass is summarized.

Table 1. Some references on somaclonal variation/in vitro chromosomal variability in various plant species

Cereals	
Wheat	Day and Ellis (1984); Karp and Maddock (1984); Larkin et al. (1984); Ahloowalia and Sherington (1985); Bajaj (1986a); Hu Han (1986); Maddock and Semple (1986); Cooper et al. (1986); Bajaj et al. (1986a); Breiman et al. (1987a); Rode et al. (1987); Ryan et al. (1987); Metakovsky et al. (1987); Lazar et al. (1988); Bajaj (1989a)
Rice	Bajaj and Bidani (1980, 1986); Sun et al. (1983); Schaeffer et al. (1984); Oono (1985); Chu et al. (1985); Loo and Xu (1986); Sun et al. (1986); Wong et al. (1986); Ling (1987); Ling et al. (1987); Terada et al. (1987a); Ogura et al. (1987, 1988); Toriyama et al. (1988); Zhang et al. (1988); (see also Chap. II.2, this Vol.)
Maize	Gengenbach and Green (1975); Brettell et al. (1980, 1986b); Gengenbach and Connelly (1981); Edallo et al. (1981); Chourey and Kemble (1982); Hibberd and Green (1982); McNay et al. (1984); Tuberosa and Phillips (1986); Dennis et al. (1987); Earle et al. (1987); Zehr et al. (1987); Rhodes et al. (1988)
Barley	Mix et al. (1978); Orton (1980b); Ruiz and Vazquez (1982); Foroughi and Friedt (1984); Powell et al. (1984, 1986); Kott et al. (1986); Chawla and Wenzel (1987); Breiman et al. (1987b); Karp et al. (1987); Snappe et al. (1988)
Triticale	Jordan and Larter (1985); Brettell et al. (1986a)
Rye	Linacero and Vazquez (1986)
Oat	McCoy et al. (1982); Nabors et al. (1982); Rines (1986)
Sorghum	Smith et al. (1983)
Pearl Millet	Bajaj and Gupta (1986, 1987)
Grasses	
Panicum	Bajaj et al. (1981a)
Napier grass	Bajaj and Gupta (1986, 1987)
Ryegrass	Ahloowalia (1983)
Red festuca	Torello et al. (1984)
Sugarcane	Heinz and Mee (1971); Krishnamurthi and Tlaskal (1974); Heinz et al. (1977); Liu et al. (1982); Lourens and Martin (1987)

Table 1. *(Continued)*

Vegetables, fruits, root and tuber crops	
Potato	Shepard et al. (1980); Wenzel and Uhrig (1981); Binding et al. (1982); Sree Ramulu et al. (1983); Gressel et al. (1984); Landsmann and Uhrig (1985); Bajaj (1986b); van Swaaij et al. (1986); Weller et al. (1987); Matern and Strobel (1987); Rietveld et al. (1987); Meulmans et al. (1987). For more references see Bajaj (1987a)
Colocasia	Nyman et al. (1983)
Chicory	Caffaro et al. (1982); Grazia et al. (1985)
Sugarbeet	Saunders and Doley (1986); Pua and Thorpe (1986); Steen et al. (1986); Girod and Zryd (1987)
Tomato	Meredith (1978); Sibi (1980); Evans and Sharp (1983); Handa et al. (1983); Sibi et al. (1984); Toyoda et al. (1984a,b); Buiati et al. (1984, 1985, 1987); O'Connell et al. (1986); Scheller et al. (1986); Shahin and Spivey (1986); Shepherd and Sohndal (1986); Gavazzi et al. (1987); Korneef et al. (1987)
Pepper	Dix et al. (1975)
Brassica spp.	Wenzel et al. (1977); Gleba and Hoffmann (1980); Schenck and Röbbelen (1982); Pelletier et al. (1983); Newell et al. (1984); Mohapatra and Bajaj (1984, 1987); Chetrit et al. (1985); Spangenberg and Schweiger (1986); Sundberg and Glimelius (1986); Bajaj et al. (1986b); Guerche et al. (1987); Robertson et al. (1987); Terada et al. (1987b); Toriyama et al. (1987)
Cucurbits	Moreno and Roig (1989)
Lettuce	Sibi (1976); Engler and Grogan (1984)
Eggplant	Alicchio et al. (1982, 1984); Gleddie et al. (1986)
Celery	Williams and Collin (1976); Browers and Orton (1982); Murata and Orton (1984)
Strawberry	Niemrowicz-Szczytt et al. (1984)
Vitis	Lebrun et al. (1985)
Legumes	
Alfalfa	Reisch and Bingham (1981); Croughan et al. (1981); Latunde-Dada and Lucas (1983); Hartman et al. (1984); Johnson et al. (1984); Durante et al. (1984); Donn et al. (1984); Bingham and McCoy (1986); Groose and Bingham (1984)
Clover, Lucerne	Phillips et al. (1982); Smith and McComb (1983); Tan and Stern (1985); see also Williams et al. (1989)
Bean	Bajaj and Saettler (1970)
Peanut	Mroginski and Fernandes (1979); Bajaj et al. (1981b, 1982); Bajaj (1985a, 1987b)
Pea	Gosal and Bajaj (1988)
Chickpea	Bajaj and Gosal (1987); Gosal and Bajaj (1988)
Pigeon pea	Bajaj et al. (1980a)
Lotus	Pezzotti et al. (1984)

Table 1. *(Continued)*

Vicia	Negruk et al. (1986)
Greengram, blackgram	Bajaj and Singh (1980); Gosal and Bajaj (1983, 1988)
Medicinal and aromatic plants	
Atropa belladonna	Bajaj et al. (1978a); Kamada et al. (1986)
Anisodus acutangulus	Zheng et al. (1982)
Foeniculum vulgare	Hunault and Desmarest (1989)
Gossypium spp.	Bajaj and Gill (1985, 1989); Bajaj et al. (1986a)
Hyoscyamus muticus	Oksman-Caldentey and Strauss (1986)
Lithospermum erythrorhizon	Fujita et al. (1985)
Linum usitatissimum	Cullis and Cleary (1986)
Nicotiana tabacum	Carlson (1973); Aviv and Galun (1977); Radin and Carlson (1978); Bajaj et al. (1978a); Ogino et al. (1978); Miller and Hughes (1980); Chaleff and Keil (1981); Thanutong et al. (1983); Barbier and Dulieu (1983); Sjodin and Glimelius (1986)
Ornamental plants	
Weigela	Duron and Decourtype (1986)
Chrysanthemums	Broertjes et al. (1976); Sutter and Langhans (1981)
Fuchsia	Bourharmont and Dabin (1986)
Pelargonium	Bennici et al. (1968); Bennici (1974); Skirvin and Janick (1976); Tokumasu and Kato (1979)
Carnation	Kakehi (1972); Leshem (1986)
Haworthia	Yamabe and Yamada (1973)
Trees	
Poplar	Ettinger et al. (1986)
Loblolly Pine	Renfroe and Berlyn (1985)
Caribbean Pine	Berlyn et al. (1987)
Peach	Hammerschlag (1986)

2 Origin and Molecular Basis of Variation

Various types of changes in cell cultures, i.e., phenotypic, genetic/epigenetic, karyotypic, physiological, biochemical, or at the molecular level, may (1) be derived from changes already existing in the explant from which the culture is initiated, or (2) be induced in vitro by the culture media and environments. The genetic variation that is created in tissue culture plants according to Berlyn et al. (see Chap. I.7, this Vol.) "may be benign, i.e., a consequence of dedifferentiation, rejuvenation, or redifferentiation, may be transient, or may be persistent". They are not readily reversible, are stable, and can be enhanced by mutagens or tissue culture environments. Physiological changes are induced in cultures, and they may become habituated or even heritable (Berlyn et al. 1986). Factors promoting instability in cultures are prolonged culture time, high level of growth regulators, and environment. For detailed discussion and information on these aspects refer to Meins 1983; D'Amato 1985; Kudirka et al. 1986, and also see Chaps. I.2 and I.8 this Vol.).

To understand the origin of the intriguing phenomenon of somaclonal variation, knowledge of the molecular/biochemical basis will guarantee better control of its manifestation. Recently much interest has arisen in this area (see Chap. I.5, this Vol.), and observations indicate that changes in the organelle DNA, and proteins including enzymes can be correlated to the occurrence of variation. This has been demonstrated in wheat, rice, potato, maize, barley, flax, etc. For instance, in potato, changes in organelle DNA were observed in protoplast-derived plants (Kemble and Shepard 1984). Landsman and Uhrig (1985) examined 36 plants from 12 somaclones of potato and analyzed them by Southern hybridization with 26 tester DNA clones. Two of the somaclones showed progenies with a reduced copy number of sequence homologous to the probe 25S rDNA. Ball and Seilleur (1986) observed copy number reduction of repetitive DNA sequence in two somaclones with a probe, rDNA. In these studies, the five fold reductions were in the relative intensities of two hybridization signals.

Cullis (1983), working on flax cell cultures, observed about 15% decrease in total DNA quantity, and reported a dramatic effect for a satellite DNA sequence in the twofold range that led to sequence deamplification. Great diversity was also noted in the circular mitochondrial DNA of whole plants and cell cultures of *Vicia faba* by Negruk et al. (1986).

Considerable work has been conducted on molecular aspects of somaclones of cereals. For instance, in wheat, variation in gliadin (Maddock et al. 1985; Cooper et al. 1986), alterations in Nor (nucleolus organizer) loci (Rode et al. 1987), qualitative and quantitative differences in rDNA spacer length (Breiman et al. 1987a), and chloroplast DNA changes (Day and Ellis 1984) have been reported. In maize, alcohol dehydrogenase zymograms were studied in 645 plants from embryo culture (Brettell et al. 1986b), and one stable mutant was detected at the ADH1 locus. Likewise, methylation (Schwartz and Dennis 1986) and mDNA variation (Chourey and Kemble 1982) were also observed in cell cultures.

In *Triticale* different types of somaclones have been observed (Jordan and Larter 1985), and Brettell et al. (1986a) described alterations in Nor loci.

Recently, rice and barley have been extensively studied. In barley some variation was observed in hordeins in the callus-derived plants (Breiman et al. 1987b), and also several rDNA spacer TaqI fragments disappeared in some plants. Some changes in the cDNA have been reported in the anther-derived plants of rice and correlated with the albinos. Sun et al. (1983) presented evidence for cytoplasmic inheritance of the albino character. For more information on various aspects of the molecular basis of somaclonal variation see Chap I.5, this Vol.

Although at present most of the molecular changes cannot be pinpointed with regard to their significance in the origin or induction of variation, such studies are nevertheless very important, and with the accumulation of data, will give insight into understanding this phenomenon.

3 Induction of Variability

3.1 Disease-Resistant Plants from Phytotoxin-Treated Cells

During the past decade it has been established that disease resistance can be induced at a cellular level and, eventually, plants that are resistant to various pathogens can be obtained (Table 2). The isolated cells and protoplasts, like the microbes, can also be cultured in large numbers in a single Petri dish, and exposed to mutagens (both radiation and chemicals). Thus the induction of mutations for resistance to various pathotoxins holds a promise and will facilitate the selection of disease-resistant plants.

The following four approaches were previously discussed (see Bajaj 1981) to obtain disease-resistant plants. (1) production of disease-resistant plants from pathotoxin-resistant cells, (2) regeneration of disease-resistant plants from cell cultures of genetic mosaics, (3) somatic hybridization between a disease-resistant and a susceptible plant, (4) uptake of a selective genome or exogenous DNA by isolated protoplasts and pollen. Considerable literature has now accumulated on the first aspect (see Table 2), which is summarized here.

The study of the phytotoxins not only helps to understand their mode and site of action at cellular and subcellular levels, but also gives an insight into various physiological and biochemical aspects associated with susceptibility, tolerance, and resistance to diseases (Schaeffer and Yodder 1972).

Our work on the effect of pathotoxins on tissue cultures was initiated in 1968 (Bajaj and Saettler) with a view to understanding the mechanism of action of toxins and to selecting toxin-tolerant/resistant cell lines. Since then significant progress has been made in this field (Table 2).

The *Phaseolus vulgaris* callus cultures showed differential growth response when cultured on a host-specific toxin filtrate from the halo-blight bacterium, *Pseudomonas phaseolicola* (Bajaj and Saettler 1970). There was inhibition in growth to the extent of 77% (Table 3); however, some islets of cells resumed growth (Fig. 1), showing differential tolerance to the toxin filtrate. There were also similarities in certain physiological effects and the ultrastructure of callus and green leaves (Bajaj et al. 1969), particularly the dramatic increase, up to 55-fold, of ornithine in callus cultures (Table 4). This work was extended to another bacterial disease, the wildfire of tobacco caused by *Pseudomonas tabaci* and the regeneration of tobacco plants from protoplast-derived callus, which showed resistance to the toxin produced by *P. tabaci* (Carlson 1973).

Later, the differential response of protoplasts isolated from resistant and susceptible plants of maize to the host-specific race T toxin produced by *Helminthosporium maydis* was demonstrated (Pelcher et al. 1975; Earle et al. 1978). It was further observed (Gengenbach and Green 1975) that callus from susceptible plants (male sterile) showed sensitivity to the T toxin medium, whereas callus from resistant cell lines resulted in the regeneration of complete plants which were resistant to *Helminthosporium* (Gengenbach et al. 1977). Likewise, toxic filtrates from other pathogens have given a similar response. For instance, from potato cell suspensions treated with culture filtrates of *Phytophthora infestans*, plants showing resistance to late blight were obtained (Behnke 1979).

Table 2. Effect of various toxins/pathogens on plant protoplast, cell, tissue and organ cultures

Host	Pathogen/toxin	Culture material	Response	References
Phaseolus vulgaris (Kidney bean) cv. Manitou, Redkote	*Pseudomonas phaseolicola* (halo-blight)	Excised roots, stem callus, and cell suspension	Differential growth inhibition up to 77%; increase in abnormal cells; 55 fold increase in ornithine	Bajaj and Saettler (1970)
Nicotiana tabacum (tobacco)	Methionine sulfoximine (wildfire toxin from *Pseudomonas tabaci* is an analog of methionine)	Haploid protoplasts	Methionine sulfoximine-resistant plants	Carlson (1973)
Zea mays (maize) Various inbred lines	*Helminthosporium maydis* race T toxin	Embryo callus	Toxin-resistant cells and plants	Gengenbach and Green (1975) Gengenbach et al. (1977)
"	"	Leaf protoplasts	Differential response of susceptible and resistant protoplasts	Pelcher et al. (1975)
Saccharum officinarum	*Helminthosporium sacchari* (crude extract)	Cell suspension	Disease-resistant clones	Heinz et al. (1977)
Solanum tuberosum (potato) cv. Kennebec	*Phytophthora infestans* (cell wall extract, called Eiicitor)	Leaf protoplasts	Rapid agglutination and death of protoplasts	Peters et al. (1978)
Solanum tuberosum (cv. Russet Burbank)	*Alternaria solani*	Leaf protoplasts	Partial resistance	Matern et al. (1978)
Solanum tuberosum	*Phytophthora infestans* (Late blight)	Callus cells	Resistant plants	Behnke (1979)
Pennisetum americanum (pearl millet) cv. PHB 10, 12,14	*Claviceps fusiformis* (ergot) sclerotia extract	Embryos, excised roots, mesocotyl and callus	Differential growth inhibition; tolerant tissues	Bajaj et al. (1980b)

Table 2. *(Continued)*

Host	Pathogen/toxin	Culture material	Response	References
Tobacco	*Pseudomonas syringae*, and *Alternaria alternata*	Protoplast-derived calli	Resistant plants	Thanutong et al. (1983)
	Phoma lingam	Protoplasts subjected to purified sirodesmin PL	Increased resistance	Sjodin and Glimelius (1986)
Tomato	*Alternaria solani* (early blight)	Callus-regenerated plants	Resistant plants	Shepherd and Sohndal (1986)
Tomato	*Fusarium oxysporum* fusaric acid	Protoplasts	Resistant plants	Shahin and Spivey (1986)
Prunus persica (peach)	*Xanthomonas campestris* pv. pruni (leaf spot)	Callus from immature embryo	Two clones were resistant	Hammerschalg (1986)
Hybrid *Populus*	*Septoria musiva*	Callus challenged with conidia	Resistance expressed in plants	Ettinger et al. (1986)
Rice	*Xanthomonas oryzae*	Callus	Resistant plants	Sun et al. (1986)
Barley	*Fusarium* fusaric acid	Microspore callus	4 out 1000 calli showed resistance	Foroughi-Wehr et al. (1986)
Oat	*Helminthosporium victoriae* toxin victorin	Callus	Resistance heritable in regenerated plants	Rines (1986)
Potato	*Phytophthora infestans*	2500 calliclones inoculated with sporangia	68 resistant calliclones	Meulemans et al. (1987)
	Alternaria solani	Protoplasts, callus	Resistant plants	Matern and Strobel (1987)
Allium spp.	*Pyrenochaeta terrestris* filtrate	Callus	Differential response of susceptible and resistant genotypes	Gourd et al. (1988)

Table 3. The effect of control and culture filtrates (10% v/v) of *Pseudomonas phaseolicola* (*Pp*), *P. morsprunorum* (*Pmp*), and *P. syringae* (*Ps*), on the growth of Manitou bean callus tissue after 35 days[a] (Bajaj and Saettler 1970)

Treatment	Callus growth mg dry wt/tube
Control filtrate	41.4 ± 4.0
Pp filtrate	13.5 ± 6.2
Pmp filtrate	37.1 ± 6.4
Ps filtrate	24.5 ± 4.1

[a] Each treatment was replicated eight times.

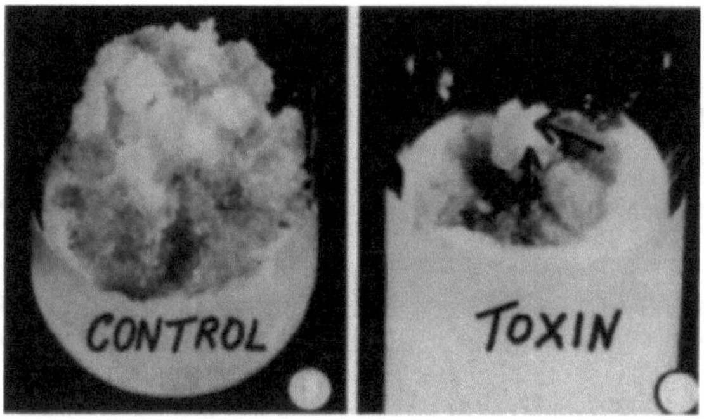

Fig. 1A,B. Effect of toxin filtrate of *Pseudomonas phaseolicola* on bean callus cultures. **A** Control. **B** Grown on toxin filtrate for 35 days; note the growth inhibition and survival of few tolerant cells (*arrow*). (Bajaj and Saettler 1970)

Table 4. The effect of control and culture filtrates (10% v/v) of *Pseudomonas phaseolicola* (*Pp*) on the levels of ammonia and certain free amino acids in Manitou bean callus tissue grown for 14 days (Bajaj and Saettler 1970)

Amino compound	Control filtrate	*Pp* filtrate
	μmol/g fresh wt	
Ammonia	1.7215	1.4039
Alanine	0.1434	0.3214
Ethanolamine	0.1827	0.2031
Histidine	0.0318	0.0797
Lysine	0.0607	0.1312
Methionine	0.0358	0.0564
Ornithine	0.0051	0.2814
Serine	0.1894	0.2527
Valine	0.5274	0.6197

Our work (Bajaj et al. 1980b) on the effect of the extract of ergot (*Claviceps fusiformis*) on the growth and development of embryos, excised segments of seedings, and the callus cultures of pearl millet (Table 5) has demonstrated the differential tolerance. At higher concentrations the extract completely stopped growth; however, certain cells survived (Fig. 2) and in one instance the shoot appeared. Since then a number of reports have appeared on the selection of somaclones showing resistance to various pathogens (see Table 2). Among them the regeneration of maize plants showing resistance to *Helminthosporium* (Gengenbach et al. 1977), tomato resistant to *Fusarium* (Shahin and Spivey 1986), potato resistant to early and late blight (Behnke 1979; Matern and Strobel 1987; Meulemans et al. 1987), and rice showing resistance to *Xanthomonas oryzae* (Sun et al. 1986) is noteworthy. These academic exercises open up new avenues for plant pathologists to incorporate somaclones into their research programs.

Table 5. Growth response of embryos/seeds of pearl-millet on MS medium containing extract from ergot (*Claviceps fusiformis*) infected earheads (Bajaj et al. 1980b)

Medium[a]	Percent germination	Root	Shoot	Callus	General remarks
MS (control)	66	Normal	Normal	Profuse callusing	Germination is not affected. Extract at higher concentrations stops the growth of roots and callus. Occasionally a few cells survive and continue to divide
MS + Seed extract	60	Slightly inhibited	Not affected	Slight inhibition	
MS + Ergot extract (24 g/l)	65	Inhibited	Not affected	Inhibition	
MS + Ergot extract (48 g/l)	66	Completely stopped	Not affected	Completely stopped	

[a] For callus experiments MS medium was supplemented with 5 mg/l of 2,4-D.

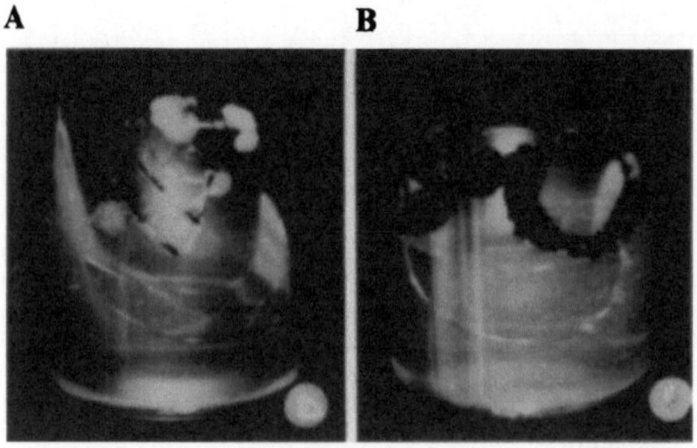

Fig. 2A,B. Effect of extract from ergot (*Claviceps fusiformis*) on excised root cultures of *Pennisetum americanum*. **A,B** 6-week-old cultures of excised roots on control and ergot extract medium respectively; note the active callus formation in **A**, and death of callus in **B**, but survival of few (*white*) cells on the right hand corner. (Bajaj et al. 1980b)

3.2 Resistance to Viruses and Nematodes

The production of virus-free plants by the culture of excised meristems is now an established technique and numerous plant species belonging to diverse genera have been freed from viruses (see Quak 1977; Boxus and Druart 1986). However, a number of precautions are required to keep them free of viruses. Therefore it is highly pertinent to devise and adopt new and unconventional means to complement the traditional methods of breeding for disease resistance.

A high level of resistance to Fiji virus disease was reported in somaclones of a susceptible sugarcane cultivar Pindar (Krishnamurthi and Tlaskal 1974). Later, Shepard (1975) regenerated plants from protoplasts of PVX-infected tobacco leaves, and obtained 7.5% regenerants free of virus.

Significant work has been done on potato to induce virus-resistant somaclones. Wenzel and Uhrig (1981) experimented with anther-derived clones for the study of resistance to potato virus X (PVX), potato virus Y (PVY), and potato leaf role virus (PLRV). In the case of PVY they used 46 microspore-derived plants and 15 different anther donor clones were analyzed after PVY inoculation, resistance being observed in some cases. They also tested resistance to PVY, PVX, and PLRV in the field. The resistance to PVY and PLRV was maintained. Likewise, Murakishi et al. (1984) found five of 300 protoplast-derived clones of Russet Burbank that had a low PVY titer. For more examples and details see Uhrig and Wenzel (1987) and Murakishi and Harris (1987).

Nematodes cause tremendous loss to crop yield. Although breeders have been relatively successful in transferring resistance to nematodes into some crops, the selection of somaclones clearly offers more opportunities. In this regard useful information has been obtained in potato and eggplant.

In conventional potato breeding, *Solanum vernei* normally serves as one of the sources of resistance to potato cyst nematode (*Globodera pallida*). However, transfer of resistance is hampered by low fertility of *S. vernei* and also the low seed set in a cross between *S. vernei* and *S. tuberosum* (Uhrig and Wenzel 1981). This has partly been overcome by the use of anther culture. Wenzel and Uhrig (1981) demonstrated that in potato, doubled monohaploids can be produced which possess resistance to potato cyst nematode in the homozygous condition. Resistance was transferred to the next generation with very high efficiency.

Another approach for incorporating resistance to nematodes is through somatic hybridization. Gleddie et al. (1985, 1986) fused protoplasts of *Solanum sisymbriifolium* (resistant to root knot nematode, *Meloidogyne incognita*) and *Solanum tuberosum* (susceptible), and regenerated 26 hybrid plants. These plants, inoculated with eggs of *M. incognita*, did not support nematode reproduction and displayed a type of reaction similar to that of the resistant parent (Gleddie et al. 1985). Several galls found on the hybrid plants contained vacuolated, sexually undeveloped juveniles, which suggests that the resistance was transferred from the wild species to the hybrid via protoplast fusion.

Still another possible approach for the induction of nematode resistance would be to regenerate plants from cell lines and excised roots showing resistance to nematodes (Tanda et al. 1980).

3.3 Salt-Tolerant Plants

Salt tolerance is a widespread agricultural problem (Epstein 1972; Waisel 1972) and the determination of "absolute tolerance" under in vivo conditions has been impossible because of both the complex interactions between the plant system and the environmental factors. However, recent developments in in vitro technology offer a meaningful tool for determining the tolerance and also for screening and developing salt-tolerant genotypes. Like most other agronomic traits, salt tolerance seems to be under genetic control (Rush and Epstein 1976; Iyengar et al. 1977; Meredith 1978; Orton 1980a). Thus efficient in vitro screening even at the differentiated stage could result in plant types capable of withstanding saline environments. With the first report on the in vitro regeneration of tobacco plants showing tolerance to sodium chloride (Nabors et al. 1980), this area of research attracted the attention of plant breeders and soil scientists, and now a number of reports are available on cell lines and plants showing tolerance to salinity. Some notable examples are *Capsicum* (Dix and Street 1975), *Lycopersicon* (Meredith 1978; Tal and Katz 1980), *Medicago* (Croughan et al. 1978; Smith and McComb 1983), *Nicotiana* (Nabors et al. 1980), rice (Croughan et al. 1981), oats (Nabors et al. 1982), *Sorghum* (Smith et al. 1983), *Colocasia* (Nyman et al. 1983), *Cicer*, *Vigna*, and *Pisum* (Gosal and Bajaj 1984), *Linum* (McHughen and Swartz 1984), *Citrus* (Spiegel-Roy and Ben-Hayyim 1985), *Vitis* (Lebrun et al. 1985), potato van Swaaij et al. 1986), pearl-millet and napier grass (Bajaj and Gupta 1986, 1987). For more examples and details see Chap. I.9, this Volume.

In our work (Gosal and Bajaj 1984) on various food legumes, *Cicer arietinum*, *Pisum sativum* and *Vigna radiata* (Table 6, Fig. 3), in vitro-induced NaCl-tolerant cell lines were isolated by exposing the cultures to increasing levels of NaCl (0.5, 1,2,3%). Actively growing cell suspension treated with EMS (0.25%) for 2–4 h increased the efficiency of salt-resistant colonies. The growth in all genotypes was drastically inhibited at 2% NaCl when added to normal MS medium containing 2 mg/l 2,4-D. However, gradual increase in the salt concentration in the growth medium led to the development of salt-resistant lines capable of growing well, even on 2.5% NaCl. The callus cultures of *Cicer* and *Vigna* thus isolated underwent rhizogenesis upon transferring to the basal medium containing NAA (0.5 mg/l).

Table 6. Number of colonies and fresh weight of largest colony as observed after 7 weeks of incubation on basal MS medium containing 0.05% (control) and 2% NaCl. (Gosal and Bajaj 1984)

	Inoculum from cell suspension growing in presence of NaCl				Inoculum from mutagen (EMS) treated suspension			
	Control		2% NaCl		Control		2% NaCl	
Species	No. of colonies	Wt (mg)	No. of colonies	Wt (mg)	No. of colonies	Wt (mg)	No. of colonies	Wt (mg)
Vigna radiata	34.7	219	14.0	87	29.2	301	11.0	103
Cicer arietinum	21.5	179	10.2	67	23.5	159	9.5	71

Colony number represent the mean number as counted with naked eye in three petriplates, and the values in parenthesis indicate the fresh weight (mg) of largest colony formed in each treatment

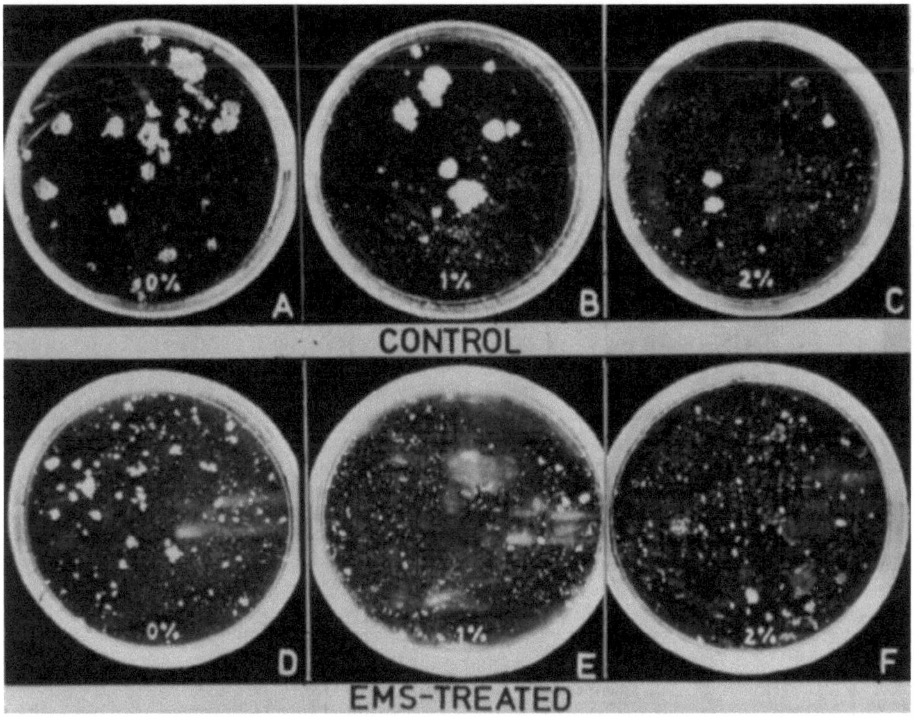

Fig. 3A-F. Screening of variation in cell suspensions of *Vigna radiata* for tolerance of NaCl. **A-C** Comparative development of colonies from cell suspensions plated on MS-2.4-D (2 mg/l) with 0, 1, and 2% NaCl respectively as observed after 7 weeks of incubation. Note highly reduced number and size of the resistant colonies developing on 2% salt. **D-F** Development of colonies from EMS-treated cell suspensions plated on MS-2.4-D (2 mg/l) with 0, 1, and 2% salt respectively as observed after 7 weeks of incubation. (Gosal and Bajaj 1984)

Furthermore, in *Cicer*, the comparative effects of varying doses of NaCl on in vivo seed germination, shoot and root growth, and on the in vitro culture of callus, hypocotyl, and epicotyl were also observed.

In another study on *Pennisetum purpureum* and *P. americanum* (Bajaj and Gupta 1986, 1987; Table 7, Fig. 4), the callus cultures showed different tolerance to NaCl (0.1–2%). *P. purpureum* was more tolerant and showed a higher growth rate. There was a sudden and drastic decrease in fresh and dry weights in both the species at a concentration of 0.2% salt. Whereas in *P. americanum* complete growth inhibition occurred at 1% salt and death at 2% salt, in *P. purpureum* at 2% salt some white patches of callus were still observed in an otherwise dark brown callus. In media containing auxin and kinetin, complete plants were regenerated from *P. purpureum* callus tolerant to 0.5% salt.

Although in most of the plant species changes seem to be physiological and epigenetic, nevertheless, continued efforts to study the heritability might be rewarding.

Table 7. Percentage of cultures from excised segments of inflorescences of *Pennisetum americanum* and *P. purpureum* showing callus formation on media containing various levels of salts (0.1–1%). (Bajaj and Gupta 1986)

Species	NaCl conc. (%)	Time interval (weeks)											
		1			2			3			4		
		+	++	+++	+	++	+++	+	++	+++	+	++	+++
P. americanum	0.1	12	0	0	14	4	0	4	12	25	10	27	43
	0.2	12	0	0	12	24	0	18	25	4	0	27	34
	0.3	10	0	0	12	0	0	8	0	0	4	20	0
	0.4	10	0	0	10	0	0	10	0	0	0	0	0
	0.5	0	0	0	0	0	0	0	0	0	0	0	0
	1.0	0	0	0	0	0	0	0	0	0	0	0	0
P. purpureum	0.1	48	40	0	52	36	4	44	40	4	32	52	12
	0.2	48	22	0	44	32	0	52	32	0	48	32	4
	0.3	12	4	0	20	8	0	20	8	0	16	12	0
	0.4	4	0	0	20	4	0	16	4	0	12	8	0
	0.5	0	0	0	8	4	0	16	4	0	12	4	0
	1.0	0	0	0	8	0	0	8	0	0	8	0	0

Callus growth rate: + low, + + medium, + + + high.

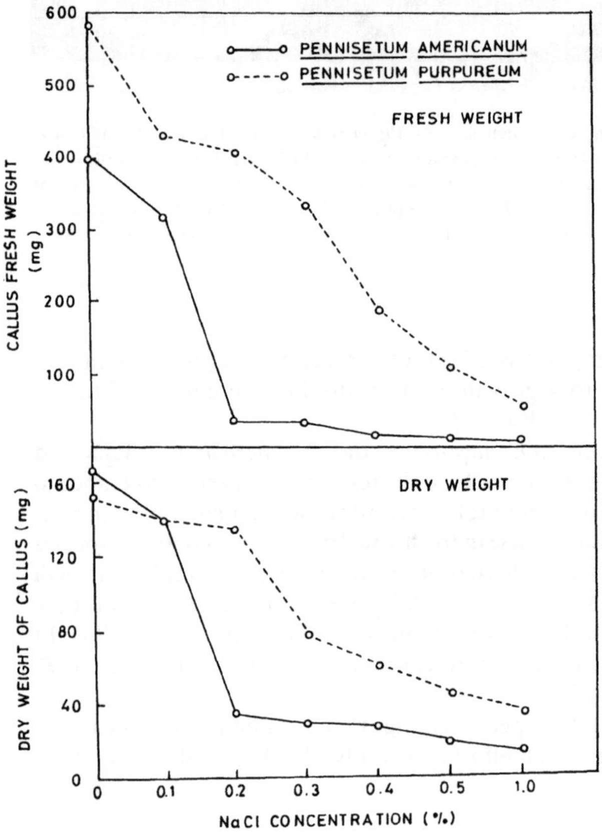

Fig. 4. Effect of increasing concentrations of sodium chloride (0.1–1%) on the fresh and dry weight of callus cultures of *Pennisetum americanum* and *P. purpureum* grown for 7 weeks. (Bajaj and Gupta 1986)

3.4 Herbicide-Tolerant Plants

Weeds are generally controlled by herbicides; however, these have to be applied repeatedly and are expensive. Moreover, they tend to have a short soil-residual life, and some crops are susceptible to them (Freeman 1982). To overcome some of these problems, three in vitro approaches have been followed to obtain plants showing resistance to herbicides, i.e., induction of mutants in cell cultures (Chaleff and Ray 1984), fusion of protoplasts (Gressel et al. 1984), and recombinant DNA techniques (Shah et al. 1986). The subject has been recently discussed (Weller et al. 1987).

The protoplasts and cell cultures subjected to various herbicides may undergo mutations to yield tolerant cell lines which subsequently regenerate into plants. In earlier studies, herbicide-resistant plants were regenerated from cell lines of *Nicotiana tabacum* resistant to isopropyl N-carbamate (Aviv and Galun 1977), picloram (Chaleff and Parsons 1978), and paraquat (Miller and Hughes 1980). Some of the picloram-resistant plants also showed resistance to hydroxyurea (Chaleff and Keil 1981). Likewise, cell lines of *Cordyllis* (Amrhein et al. 1983) and alfalfa (Donn et al. 1984) showing resistance to glyphosate and glufosinate respectively were obtained. Later, tobacco plants showing resistance to chlorsulfuron (Chaleff and Ray 1984) and corn plants resistant to imazaquin were regenerated (Anderson et al. 1984) and the trait is being bred into commercial cultivars. Now a number of other examples of herbicide tolerance/resistance are available (see Weller et al. 1987).

The fusion between protoplasts of *Solanum tuberosum* (sensitive) and *S. nigrum* (atrazine-resistant) resulted in many regenerants with intermediate morphological characters, which eventually segregated into resistant or susceptible plants (Binding et al. 1982). The only putative cybrid was a triazine-susceptible strain of *S. nigrum.* Later the triazine resistance was transferred from *S. nigrum* into potato (Gressel et al. 1984).

Shah et al. (1986) followed a different approach following the observation that herbicide glyphosate is a potent inhibitor of enzyme 5-enolpyru-vylshikimate-3-phosphate (EPSP) synthase in higher plants. They isolated a complementary DNA (cDNA) clone encoding EPSP synthase from a complementary DNA library of a glyphosate-tolerant *Petunia hybrida* cell line (MP4-G) that overproduces the enzyme. This line overproduces EPSP synthase messenger RNA as a result of a 20-fold amplification of the gene. A chimeric EPSP synthase gene was constructed with the use of the cauliflower mosaic virus 35S promoter to attain high level expression of EPSP synthase, and introduced into *Petunia* cells. They observed that the transformed plants were tolerant to glyphosate (Shah et al. 1986). These results represent a major step toward establishing selective herbicide tolerance in crop plants.

3.5 Somaclones for Nutrition Improvement

The essential amino acids which limit the nutritional value of most plant proteins are lysine, methionine, and tryptophane. The vegetable proteins could acquire a good balance if the quantity of essential amino acids can be increased to the desired level as compared to egg protein or to standard human amino acid requirements. Various workers have reported increase in tryptophan, proline, tyrosine, and phenylalanine contents in plant cell cultures (see Table 8). It is evident from these studies that there is great scope for nutritional quality improvement of limiting amino acids through resistance and alterations in synthetic pathways in somaclones.

3.6 Production of Male Sterile Lines

Male-sterile plants are of great importance in the utilization of heterosis for the production of hybrid seeds in many crops, such as maize, rice, brassica etc. Male sterility has been induced by protoplast fusion in brassica (Pelletier et al. 1983) and tobacco (Kumashiro and Kubo 1984). Induction of somaclones through callus cultures is still another approach. Recently, Ling et al. (1987) have successfully obtained the regeneration of male sterile plants from callus cultures of indica rice. Out of 157 plants regenerated, three were male sterile in the first generation (R_1). In the second generation (R_2), one line segregated male sterile and fertile plants. The segregation ratio of fertile:sterile in both R_2 of line 91 and the F_2 of male sterile plant IR24 fitted the formula 15/16 (fertile): 1/16 (sterile) which fits the expected value of Mendel's law 9:3:3:1. This work has far-reaching implications in rice breeding, as it will facilitate the production of hybrid rice. Exploration of translocation induced by somaclonal variation would also offer the possibility of breeding for seedless fruit crops (Ling 1987).

3.7 Environmentally Induced Heritable Changes

Plant genome is variable and is in a state of flux. The frequency of changes is influenced by a number of factors, including external events. The changed plants/lines are termed genotrophs (Durrants 1962). In this regard flax (*Linum usitatissimum*) cv. Stormont Cirrur has been extensively studied by Cullis (1977, 1983).

Cullis and Cleary (1986) compared the DNA's from leaves and callus of a series of flax genotrophs. The extent of variation observed between leaf DNA and callus DNA from a single genotroph was greater than that observed between the genotypes in vivo. The DNA's from the progeny of a number of regenerated plants were also compared. They sometimes differed both from the callus from which the plants were regenerated and the original line from which the callus line was derived. Individual progeny from a single inbred regenerated plant also differed. Thus it is possible that the phenotypic changes are a consequence of alterations in gene control brought about by changes in chromosome structure that are mediated by these changes in blocks of tandem arrays – blocks of heterochromatic (Cullis 1986).

3.8 Somaclonal Variation in Protoplast-Derived Plants

Isolated protoplasts, although once thought to be a genetically stable system vs. cell suspensions, have nevertheless yielded much chromosomal variability in the regenerated plants. Bajaj et al. (1978a) regenerated variable plants from mesophyll

Table 8. In vitro increase in essential amino acids in various plant systems (After S. Bajaj 1987)

Plant system	Amino acid	Inhibitor	Reference
Potato cell cultures	Tryptophan	5-MT	Widholm (1974). Carlson and Widholm (1978)
Potato shoot cultures	Proline	HYP	van Swaaij et al. (1984. 1985)
Potato cell suspension	Tyrosine. phenylalanine	5-MT	Jacobsen et al. (1986)
Tobacco haploid cells	Methionine	MSO	Carlson (1973)
Haploid *Nicotiana sylvestris* suspension culture	Methionine	ETH	Zenk (1974)
Tobacco. carrot. soyabean	Tryptophan	5-MT	Widholm (1974)
Rice callus	Lysine	AEC	Chaleff and Carlson (1975)
Tobacco suspension cultures	Lysine	AEC + δHYL	Widholm (1976)
Carrot suspension cultures	Methionine	ETH	Widholm (1976. 1978)
Carrot. tobacco cell suspension	Phenylalanine	PFP	Gathercole and Street (1976)
Carrot suspension culture	Proline	A2C	Widholm (1976)
Carrot cell suspension	Lysine	AEC	Widholm (1978)
Carrot cell suspension	Lysine	AEC	Mathews et al. (1980)
Maize callus	Lysine. threonine. isoleucine. methionine	Lysine	Hibberd et al. (1980) Hibberd and Green (1982)
Rice anther callus	Lysine	AEC	Schaeffer and Sharpe (1981. 1987)
Barley embryos	Proline	HYP	Kueh and Bright (1981. 1982)
Alfalfa suspension cultures	Methionine. cysteine	ETH	Reisch et al. (1981)
Rice callus	Tryptophan	5-MT	Wakasa and Widholm (1982. 1987)
Barley embryos. carrot embryoids. *N. sylvestris* leaf protoplasts	Threonine	AEC. lysine. threonine. ETH	Cattoir-Reynaerts et al. (1981); Jacobs et al. (1982)
Barley embryo cultures	Lysine. threonine	Lysine + threonine	Bright et al. (1982)
Tobacco callus	Lysine	4-Oxalysine	Ho et al. (1982)
Spirulina platensis and carrot	Proline	A2C	Riccardi et al. (1983)
N. sylvestris	Lysine	AEC	Negrutiu et al. (1984)
Tobacco cells	Methionine. threonine. lysine. isoleucine	ETH	Gonzales et al. (1984)
Maize cell and tissue culture	Tryptophan	5-MT	Sun (1985)
Pearl millet callus	Lysine	AEC	Boyes and Vasil (1987)
Maize callus	Tryptophan	AEC	Miao et al. (1988)

Abbreviations: PFP = phenylalanine analog; A2C = proline analog; 5-MT = 5-methyl tryptophan; MSO = methyl sulphoxide; ETH = D-L ethionine; AEC = S-(2-amino ethyl cysteine); δHYL = delta hydroxylysine; HYP = hydroxyproline.

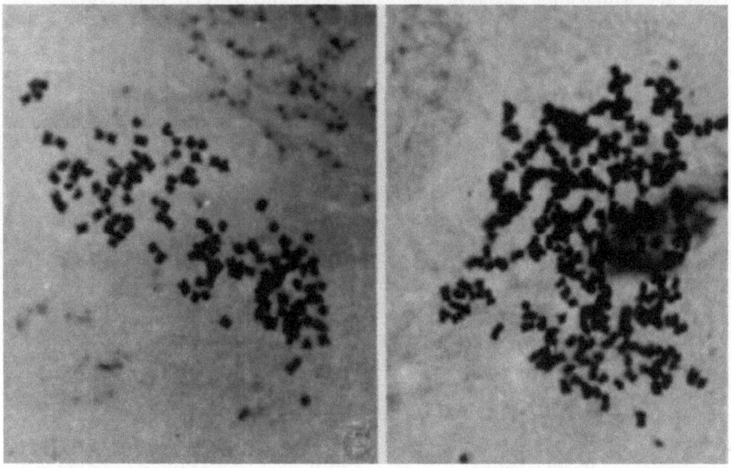

A B

Fig. 5A,B. Root-tip squashes from protoplast-regenerated plants of *Atropa belladonna* (the protoplasts were obtained from anther-derived haploid plants, n = 36); note 132 chromosomes (3n = 108) in **A**, and about 200 chromosomes (3n = 216) in **B**. (Bajaj et al. 1978a)

protoplasts of *Nicotiana tabacum* and *Atropa belladonna*. The tobacco plants were predominantly (91%) haploids with occasional formation of diploids, while *Atropa* showed more genetic unstability and plants with ploidy level ranging from haploids, diploids, triploids, and hexaploid along with aneuploids of various types were obtained (Fig. 5A,B). Since then, a number of examples of protoplast-derived plants are available which show variation of different types (see D'Amato 1985).

This variation seems to occur in the protoplasts at an early stage, as demonstrated in tobacco by Barbier and Dulieu (1983). They concluded that the exact time of occurrence of variants could be either before protoplast isolation, as lesions repaired by a recombination process (if unrepaired, they would yield deletions), during aging of resting mesophyll cells, or induced when cell division is resumed after isolation. Likewise, in *Nicotiana sylvestris*, it was observed that two plants regenerated from the same protoplast-derived callus produced different mutations (Prat 1983).

Regarding various types of variation, the potato protoplast system has been extensively studied by various workers (Shepard et al. 1980; Karp et al. 1982; Jacobsen et al. 1983; Kemble and Shepard 1984; Sree Ramulu et al. 1984; Bajaj 1986b; see also Bajaj 1987a for more details). The variations are change in chromosome numbers and structure, phenotype of the plant, tuber color, shape and size, yield, etc. In a vegetatively propagated crops such as potato, such variation can be of great help to reduce the time of release of new varieties.

Recently, a wide range of variability and differences in the crop yield of rice protoplast-derived plants has been observed (Ogura et al. 1987, 1988).

4 Somaclonal Variation in Various Crops

4.1 Somaclonal Variability in Cereals and Grasses

There is great potential for in vitro generation of genetic variability and also the alien genetic transfer for the improvement of cereals. Wide hybridization has been promising in *Triticum* crossed with *Hordeum* (Fedak 1980; Snappe et al. 1988), *Agropyron* (Mujeeb-Kazi et al. 1984), *Secale* (Bajaj et al. 1978b), *Triticale* (May 1983) and amongst various species of rice (Jana and Khush 1984). Anther culture has generated a wide range of genetic variability which has resulted in the release of varieties of wheat and rice (Hu 1986; Loo and Xu 1986; de Buyser et al. 1987). Attempts to obtain somatic hybrids in rice (*Oryza sativa* × *Echinochloa oryziola*) have recently yielded positive results (Terada et al. 1987a). The transformation studies on cereals have greatly advanced, and the recent success of the regeneration of complete maize and rice plants from transformed protoplasts (Rhodes et al. 1988; Toriyama et al. 1988; Zhang et al. 1988) is a clear indication that the same can be applicable to other cereals within a short span of time.

The callus-derived plants of various cereals and grasses have proved to be a rich source of somaclones, and some of the examples are cited in Table 1. In wheat, immature embryo-derived callus on transfer to a hormone-free medium underwent regeneration. On transfer to field plots the plants continued further growth, matured, and set seed (Fig. 6; Bajaj 1986a). The plants showed a range of morphological variation, especially in their height, size and shape of the leaves, length of the awns, fertility of the spikes, and size, shape, and color of the seeds. In a few instances two or three spikes were formed on one culm (Bajaj 1986a; Bajaj et al. 1986a). In addition to the morphological changes, the somaclonal variants show changes in various enzymes and proteins (Maddock et al. 1985; Davies et al. 1986; Ryan et al. 1987; Metakovsky et al. 1987). The field assessment of somaclonal variants (Maddock and Semple 1986) and heritability, especially for gliadin protein, was also studied (Cooper et al. 1986); however, the practical utility of these somaclones has yet to be realized.

Rice is one of the systems which has been extensively studied, and somaclones have been derived from cells, protoplasts, anthers, embryos, etc. In our work genetically variable plants were regenerated from embryo- and seedling-derived callus (Bajaj and Bidani 1980, 1986). The endosperm-derived plants were triploid, and the callus showed a wide variability (Fig. 7). The triploid plants produced from the endosperm (Bajaj et al. 1980c) showed broader leaves, faster rate of growth, and greater tillering than the embryo-derived plants. The triploid and hexaploid plants are more vigorous than diploids (Morinaga and Fukushima 1935) and their foremost use is in hybridization programs, especially for the augmentation of germplasm reservoirs. Hybrid selection can then be made for the desirable agronomic traits. Sun et al. (1983) studied inheritance and variation in more than 2000 somatic cell-derived plants and reported that variation phenotyped in the second generation proved to breed true. Later, some mutants (Oono 1985) and plants resistant to *Xanthomonas oryzae* (Sun et al. 1986) and sodium chloride (Wong et al. 1986), male sterile plants (Ling et al. 1987) high protein (Schaeffer et al. 1984) were selected. Recently, Ogura et al. (1987, 1988) induced somaclonal variants in

A C B

Fig. 6A-C. Induction of variation in plants regenerated from embryo-derived callus of *Triticum aestivum*. **A** Callus obtained from immature embryo on transfer to IAA + BAP medium showing differentiation of shoots and plantlets. **B** The same transferred to field. **C** Variation in the spikes obtained from **B**. (Bajaj 1986a)

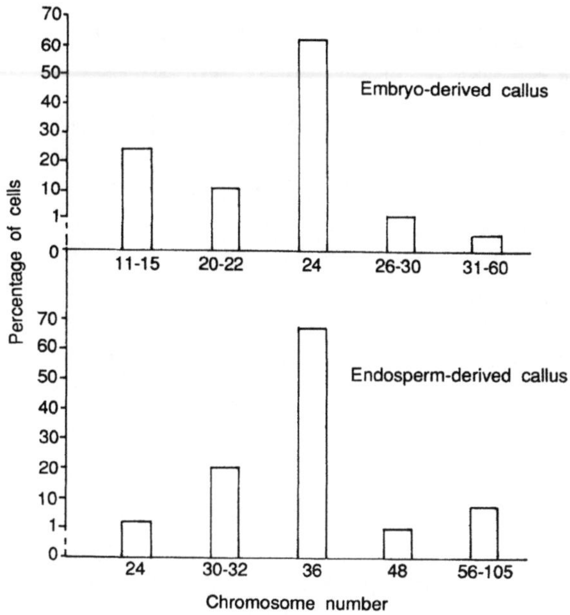

Fig. 7. Embryo and endosperm-derived callus of *Oryza sativa*. (Bajaj and Bidani 1986)

protoplast-derived plants and tested them under field conditions for various agronomic traits. For a detailed account of somaclonal variation in rice see Chapter II.2, this Volume. Amongst the grasses, *Lolium* was one of the first to be studied (Ahloowalia 1983).

Plants regenerated from inflorescence-derived callus of three species of *Panicum* showed genetic diversity, and some plants were albinos. (Fig. 8). The callus cells and root tip squashes showed a predominantly diploid chromosome number (2n = 36); however, aneuploids were very common. The chromosome number ranged from 29–36 in young cultures, and polyploids in older cultures. The callus-derived plants showed variation in morphological characters, i.e., size, leaf shape, and tillering (Bajaj et al. 1981a). Callus cultures of *Pennisetum americanum* and *P. purpureum* showed varied degrees of tolerance to sodium chloride (Bajaj and Gupta 1986, 1987).

4.2 Somaclonal Variation in Legumes

Legumes are the most important source of plant proteins and forages, and are cultivated throughout the world in wide areas. The major objectives in their breeding are to increase grain yield, methionine, cystine, and protein digestibility, and to decrease antienzymes and seed hardness or cooking time. There has been no significant increase in the production of pulses during the last decade because of their low static yields and susceptibility to various diseases. For qualitative and quantitative improvement, the breeder has to rely upon the extent of genetic

A B

Fig. 8A,B. Regeneration of plants from tissue cultures of *Panicum*. **A** Differentiation of callus 5 weeks after transfer to a 2,4-D-free medium. **B** A callus-derived plant of *P. maximum*, 7 weeks after transfer, showing variation in the size and shape of spikes. (Bajaj et al. 1981a)

variability present in the base population. The lack of genetic variability has been considered to be a major limiting factor for the slow progress made in their improvement. The desired goals can be achieved by incorporating additional genetic variability in the existing germplasm (see Bajaj and Gosal 1982). In this regard interspecific hybrids have been obtained in a number of incompatible crosses using embryo rescue techniques, notable among these being *Phaseolus* (Mok et al. 1978), *Glycine* (Broué et al. 1982), *Arachis* (Bajaj et al. 1982), *Vigna* Gosal and Bajaj (1983), etc. (for more examples see Bajaj 1990a). Protoplast fusion (Bajaj and Gosal 1988) and somatic hybridization (Hammatt and Davey 1989) have been reported. Likewise, genetic transformation has been successful in soybean (Baldes et al. 1987) *Medicago* (Deak et al. 1986; Reich et al. 1986) and other forage legumes (Webb 1986). *Medicago, Trifolium*, and *Lotus* are the forage legumes in which extensive work has been done on somaclonal variation (Reisch and Bingham 1981; Groose and Bingham 1984; Pezzotti et al. 1985; Williams et al. 1990).

In grain legumes, mostly changes in chromosome numbers in vitro have been reported. Salt-tolerant cell lines were obtained in *Cicer, Pisum*, and *Vigna* (Gosal and Bajaj 1988), and bean callus showing tolerance to halo-blight toxin filtrate were reported (Bajaj and Saettler 1970). The callus cultures of *Cicer arietnum* (Fig. 9; Bajaj and Gosal 1987), *Cajanus cajan* (Fig. 10; Bajaj et al. 1980a), and *Arachis* spp. (Fig. 11; Bajaj et al. 1981b) showed a wide range of chromosomal variability. Recently, Graybosch et al. (1987) reported somaclonal variation in soybean plants.

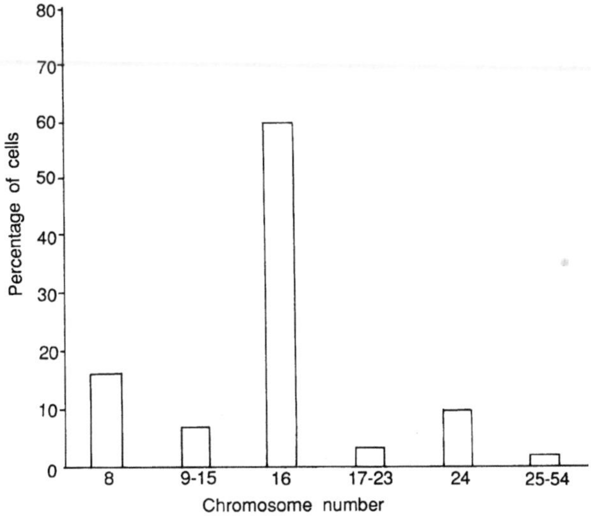

Fig. 9. Anther-derived callus of *Cicer arietinum*. (Gosal and Bajaj 1988)

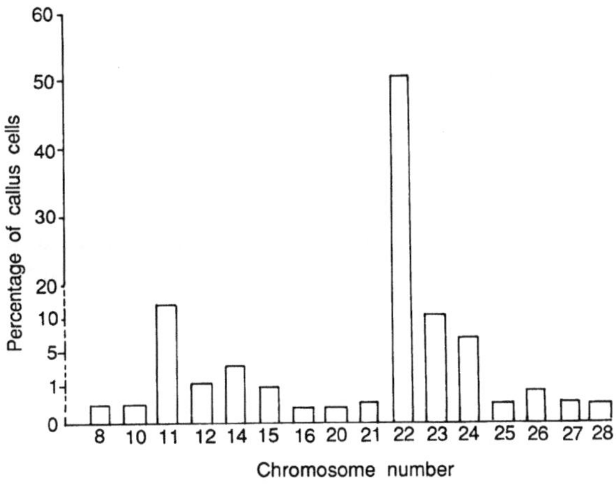

Fig. 10. Anther-derived callus of *Cajanus cajan*. (Bajaj et al. 1980a)

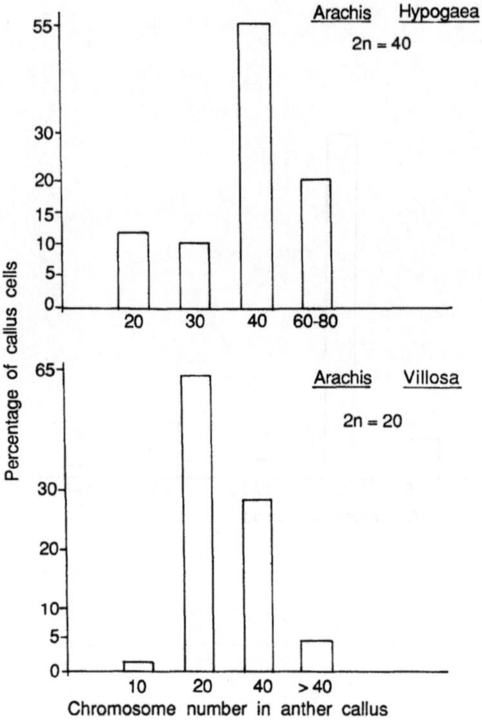

Fig. 11. Anther-derived callus of *Arachis hypogaea* and *Arachis villosa*. (Bajaj et al. 1981b)

4.3 Induced Variability in Potato

Potato is perhaps the best example for the study of in vitro-induced variation, and extensive work has been done and numerous publications have appeared on the subject (see Bajaj 1987c). Different types of variation from plants regenerated from isolated and fused protoplasts and callus cultures (Shepard et al. 1980; Melchers et al. 1978; Sree Ramulu et al. 1983, 1985; Bajaj 1986b), tuber discs (Rietveld et al. 1987), and anthers (Sopory and Bajaj 1987) have been obtained and genetic transformation induced (Ooms et al. 1985). The plants show differences in height, canopy traits, leaf size and shape, color, skin texture, yield, size and shape of the tubers, etc. (Fig. 12; Bajaj 1986b). The plants show tolerance/resistance to early blight (Matern et al. 1978), late blight (Behnke 1980a), *Fusarium* (Behnke 1980b), common scab (Gunn 1982), herbicides (Gressel et al. 1984) and nematodes and viruses (Wenzel and Uhrig 1981; Austin et al. 1985).

4.4 Variability in *Brassica* Crops

Extensive work has been done with a view to inducing genetic variability in various species of *Brassica*, both oilseed crops and vegetables, and plants derived in vitro

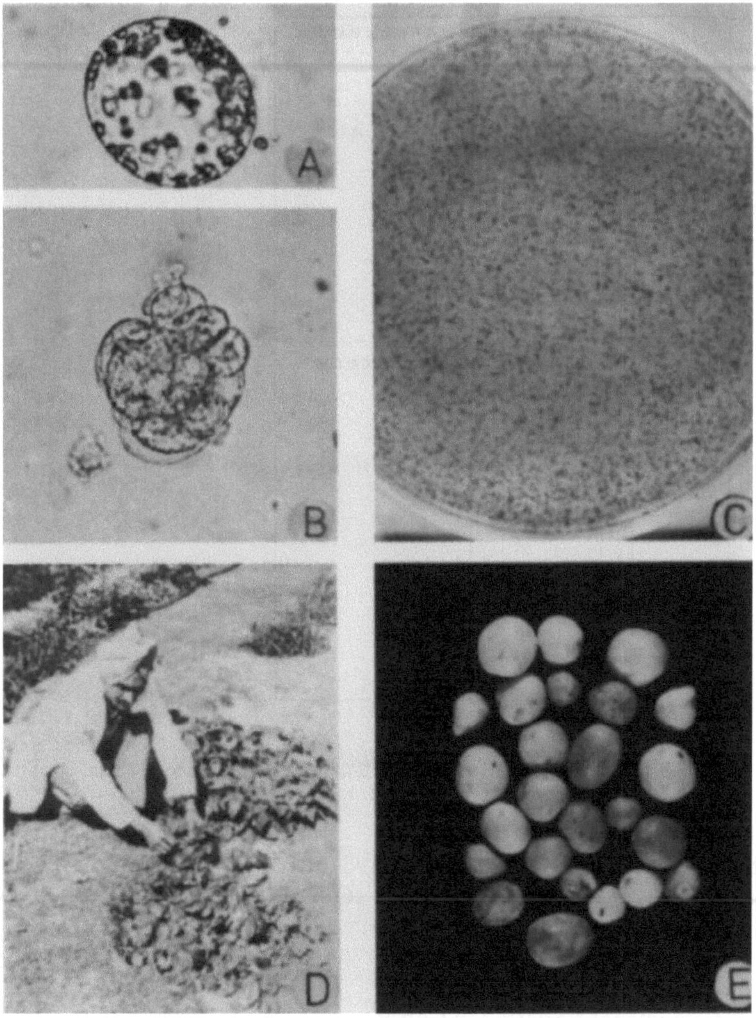

Fig. 12A-E. Somaclonal variation in potato plants regenerated from mesophyll protoplasts. **A** Protoplast isolated from mesophyll cells of young leaves. **B** A micro-colony of cells formed from a protoplast. **C** Protoplasts (now cell colonies) 4 weeks after plating in agar medium (0.6%). **D** Protoplast-derived plants transferred to field. **E** Large variation in size and shape (and color) of tubers obtained from plants in **D**. (Bajaj 1986b)

show variability of different types. For instance, interspecific hybrids obtained through embryo/ovule culture (Bajaj et al. 1986b; Mohapatra and Bajaj 1984, 1987), callus culture (Horák et al. 1971; Bajaj et al. 1986a; Fig. 13), anther-derived plants (Keller and Armstrong 1981), protoplast culture (Newell et al. 1984), somatic hybridization (Gleba and Hoffmann 1980; Schenck and Röbbelen 1982; Chetrit et al. 1985; Sundberg and Glimelius 1986; Spangenberg and Schweiger 1986;

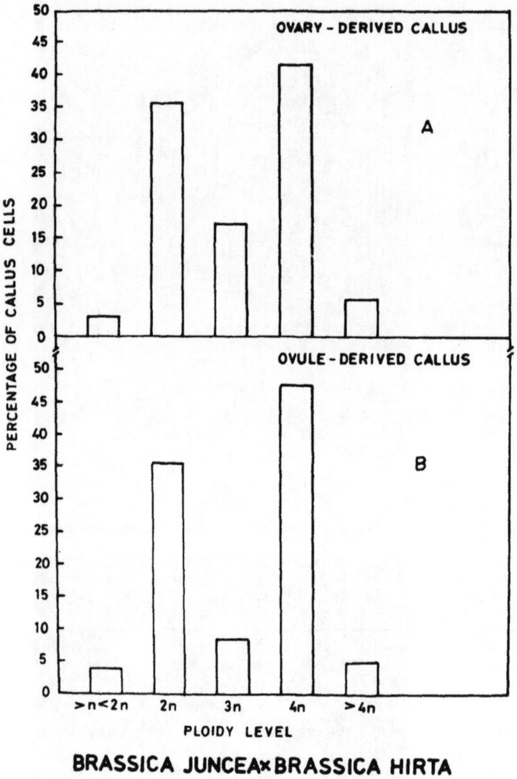

Fig. 13A,B. Ovary-, and ovule-derived callus from a cross, *Brassica juncea* × *Brassica*. (Bajaj et. al. 1986b)

Robertson et al. 1987; Terada et al. 1987b, Toriyama et al. 1987), cybrids (Pelletier et al. 1983), genetic transformants (Guerche et al. 1987), etc. showed various morphological and genetically inherited traits.

4.5 Somaclonal Variation in Cotton

Cotton is the most important source of textile fiber and has 35 wild and four cultivated species, both diploid (2n = 26) and tetraploid (2n = 52). Although such desirable traits as quality of fiber and resistance to various diseases and pests are present in the wild species, due to incompatibility barriers they have not been successfully incorporated into cultivated cotton. Various in vitro methods have been employed to generate genetic variability (Bajaj et al. 1986a; Bajaj and Gill 1987), and much chromosomal variation in the callus obtained from excised hypocotyl, embryos, ovules, anthers, and hybrid callus was obtained (Bajaj and Gill 1985; Stelly et al. 1989).

Hypocotyl-Derived Callus. The ploidy level of callus cells varied from haploid to more than hexaploid (6n = 78). The frequency of haploid cells was quite low

(0.74%). Most of the cells had a diploid (45.2%) or tetraploid (40.5%) number. Triploid (4.6%), pentaploid (2.6%), hexaploid (4.4%), or even a higher ploidy than hexaploid (2.9%) cells were also observed (Fig. 14).

Ovule-Derived Callus. The range of variability was greater in the ovule-derived callus of *G. herbaceum* than in those of *G. arboreum*. In *G. herbaceum* callus, haploid, triploid, pentaploid, hexaploid, and higher than hexaploid were present in higher frequency, whereas *G. arboreum* callus had a higher number of diploid and tetraploid cells. In general, ovule-derived callus of both species contained cells with a more variable number of chromosomes than the hypocotyl-, or anther-derived callus (Fig. 14).

Anther-Derived Callus. The callus cells obtained from young anthers showed a wide range of chromosome numbers, varying from haploid to hexaploid. In most of the cells (73.1%), the number was close to the diploid (38.3%) or tetraploid (34.8%) level (Fig. 14, Bajaj and Gill 1989).

Hybrid Embryo-Derived Callus. The embryos of a cross *G. hirsutum* × *G. arboreum*, proliferated profusely to form callus which showed chromosome numbers varying from less than 26 to more than 45 (Fig. 15). However, the majority of the cells belonged to the category 26–41. The cells with less than 26 and more than 45 chromosomes were rare, i.e., 1.4% and 0.84%, respectively. The occurrence of cells with less chromosomes may be due to laggards at anaphase. The very low frequency of cells with more than 39 chromosomes can also be accounted for by laggards, rather than by endomitosis or other types of polyploidization. It has been speculated

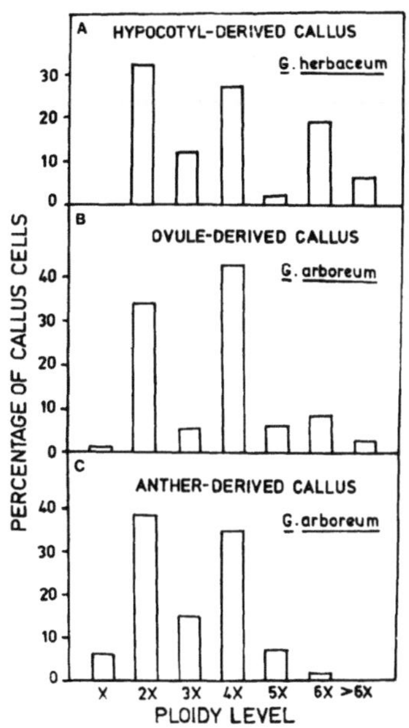

Fig. 14A-C. Hypocotyl-, ovule-, and anther-derived callus of *Gossypium* species. (Bajaj and Gill 1985)

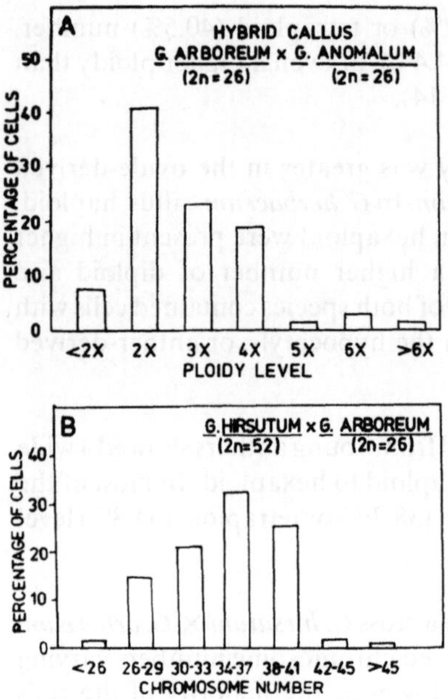

Fig. 15A,B. Interspecific (embryo-derived) hybrid callus of cotton. (Bajaj and Gill 1985)

that tissue culture may generate environments for enhancing chromosome breakage and reunion events, and thus a tissue culture cycle of the hybrid material may provide the means for obtaining the genetic exchange needed between two genomes in the interspecific hybrid. The hybrid callus may enhance the frequency of requisite exchange (Larkin and Scowcroft 1981). Hybrid callus, which is a rich source of genetic diversity, needs to be exploited for crop improvement.

In addition to the callus cultures, wide hybridization through embryo/ovule culture (Gill and Bajaj 1984, 1987) and *Agrobacterium*-mediated transformation (Firoozabady 1989) have yielded good results, and somatic hybridization is another avenue which is yet to be explored.

4.6 High-Yielding Somaclones of Medicinal Plants

Variation in morphology and chromosome numbers of callus-derived medicinal plants, such as *Atropa belladonna*, is a common observation (Bajaj et al. 1978a). However, selection of somaclonal variants for higher production of medicinal/industrial compounds from these cell cultures is an area of tremendous commercial importance, and has great potential, especially in the pharmaceutical industry. Already a number of elite cell lines of various plants have been isolated. Some examples are cell cultures of *Nicotiana* for nicotine (Ogino et al. 1978), *Anisodus* for scopolamine (Zheng et al. 1982), *Catharanthus* for tryptophan (Sasse et al. 1983), *Coptis* for berberine (Sato and Yamada 1984), *Lithospermum* for shikinon (Fujita et al. 1985), *Hyoscyamus* for scopolamine (Oksman-Caldentey and Strauss

1986), *Atropa* for tropanes (Kamada et al. 1986), *Foeniculum* for anethole (Hunault and Desmarest 1989), etc. These studies are likely to be compounded soon.

4.7 Somaclonal Variation in Ornamentals

Micropropagation of ornamentals and foliage has been commercially exploited for the past more than 25 years, primarily based on the need for genetic stability and clonal propagation of elite, rare, and valuable plants. However, now with the available knowledge on the potential of somaclones, in vitro induction of variants for flower color, shape, and other esthetic traits has also been accepted as a method for creating "new types". Some examples of ornamentals are geraniums (Bennici 1974; Skirvin and Janick 1976), carnations (Kakehi 1972; Leshem 1986), *Haworthia* (Yamabe and Yamada 1973); *Fuchsia* (Bourharmont and Dabin 1986), weigela (Duron and Decourtye 1986), and chrysanthemum (Sutter and Langhans 1981).

5 Cryopreservation of Somaclones

Cell cultures on periodical subculturing undergo genetic erosions resulting in variations of various types. Most of these variations may not be of any significance and are therefore discarded. Nevertheless, rare, elite, and potentially useful cell lines, mutants and high-yielding somaclones, especially those of the medicinal plants, need to be preserved (Bajaj 1988a). At present there is no method that prevents genetic deterioration, or enables long-term storage without transferring to fresh media, in this respect cryopreservation holds promise (Bajaj 1979, 1983, 1986c). Successful attempts have been made to cryopreserve isolated protoplasts, cells, tissues and organs of a number of plant species (Bajaj 1986c, 1988a, 1990b). In the author's work, cell suspensions of tobacco and carrot (Bajaj 1976), and isolated protoplasts of *Atropa, Nicotiana,* and *Datura* (Bajaj 1988b) regenerated plants after storage in liquid nitrogen (−196 °C) for various lengths of time. Likewise, protoplast-derived callus colonies of potato survived freezing. The retrieved cell colonies lay quiescent for about a week, then resumed growth and underwent morphogenesis to form shoots (Fig. 16; Bajaj 1986b). This is somewhat similar to the potato and cassava plants obtained from meristems cryopreserved for up to 4 years, that underwent normal tuberization (Bajaj 1985b).

The technology of cryopreservation has been refined and it enables the storage of in vitro cultures for the conservation of germplasm and somaclones of medicinal plants. The retention of biosynthetic potential of the retrieved cultures amply demonstrates the use of this technology for the storage of high alkaloid/secondary metabolites/medicines producing cell lines for pharmaceutical purposes. The successful cryopreservation of cell cultures of *Nicotiana, Atropa, Datura, Catharanthus, Panax, Dioscorea, Anisodus,* etc. (Table 9) without any deterioration, the experience gained, and the potential benefits expected warrant the extension of this

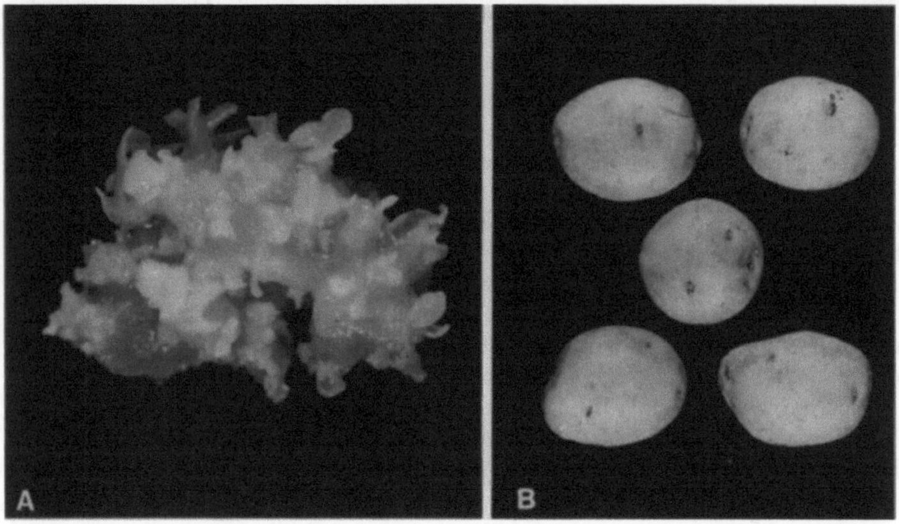

Fig. 16A,B. Normal tuberization in plants regenerated from cryopreserved protoplast-derived cell colonies. **A** A mass of callus (obtained from retrieved colony undergoing differentiation to form buds and shoots. **B** Tubers from plants derived from cultures, note the uniformity in the size and shape of the tubers. (Bajaj 1986b)

work to other plant species. This subject has been recently reviewed, for more details see Bajaj 1988a. Such studies are of special significance for the pharmaceutical industry.

6 Implications in Plant Breeding

As mentioned earlier, the success of a crop improvement program depends on the extent of genetic variability in the base population. However, there is continual depletion and shrinkage of the naturally occurring reservoirs of germplasm, which has caused great concern. There is a lack of genetic variability in most agricultural crops. This situation is further aggravated by the clearing of forest land, which results in the loss of native plants. Therefore, efforts are being made to resort to methods other than conventional for the induction as well as for the conservation of genetic variability. In this regard, in vitro technology has come of age and is quite competent to meet the challenge.

Cell cultures undergo genetic changes like polyploidy, aneuploidy, mutations, gene amplification, chromosome breakage, deletions, inversions, translocations etc., and they can be enhanced through the use of mutagens. Some of the useful ones can then be selected and stabilized in a single generation. One of the main advantages of in vitro selection is that it is faster and requires less effort and space. Each cubic centimeter of callus is believed to contain the equivalent of a thousand

Table 9. Cryopreservation of in vitro cultures of some medicinal and alkaloid-producing plants. (Bajaj 1988a)

Species	Culture	Response	Reference
Anisodus acutangulus	Cell suspension	Hyoscyamine and scopolamine synthesis	Zheng et al. (1983)
Atropa belladonna	Cell-suspension	Callus	Nag and Street (1975)
A. belladonna	Pollen embryos	Haploid plants	Bajaj 1977, 1978)
A. belladonna	Protoplasts	Plants	Bajaj (1988b)
Catharanthus roseus	Cell suspension	Callus	Chen et al. (1984)
Cinchona ledgeriana	Callus	Callus	Hunter (1986)
Citrus spp.	Ovule/nucellar embryo	Callus, plants	Bajaj (1984)
Datura stramonium	Cell suspension	40% Cell survival	Bajaj (1976)
D. innoxia	Cell suspension	Callus	Weber et al. (1983)
D. innoxia	Protoplasts	Callus	Bajaj (1988b)
Digitalis lanata	Cell suspension	Callus, biotransformation of β-methyldigitoxin to β-methyldigoxin	Diettrich et al. (1982); Seitz et al. (1983)
Dioscorea deltoidea	Cell suspension	Diosgenine synthesis	Butenko et al. (1984)
Lavandula vera	Cell suspension	Metabolic potential retained	Watanabe et al. (1983)
Nicotiana tabacum	Cell suspension	Haploid plants Callus	Bajaj (1976) Hauptmann and Widholm (1982); Maddox et al. (1983)
N. tabacum	Pollen embryos	Plants	Bajaj (1977)
N. tabacum	Protoplasts	Plants	Bajaj (1988b)
N. plumbaginifolia *N. sylvestris*	Cell suspension	Callus	Maddox et al. (1983)
Panax ginseng	Cell suspension	Retention of biosynthetic capacity	Butenko et al. (1984)

shoot apices for whole plant (Nabors 1982). Thus, even if only one cell among millions in one Petri dish is a mutant and transfers the desirable trait to the regenerated plant, it is much more efficient than the field experiment. Moreover, in somaclones, plants regenerate from single cells, so the frequency of mosaic is very low in contrast to mutation breeding.

Nature has taken, through evolution, millions of years to bring the plants to the present stage of development. Recent advances in plant cell cultures and genetic manipulation (see Bajaj 1989b) have demonstrated that such a process can be enhanced in a test tube, and novel plants and genetic variants, which do not exist in nature, can be obtained in a matter of a couple of years.

Conventional methods of plant breeding, although very successful in the past, seem not to be able to cope with the situation. The incorporation of biotechnology in plant breeding is imminent, and for this in vitro methods are the most important. Some of the newer methods of biotechnology, especially recombinant DNA technology, uptake and incorporation of foreign DNA, and somatic hybridization, are still in the process of being developed or refined, and it will be quite some time before they are available for routine use. However, in vitro changes brought about in in vitro cell cultures are a rich source of variability. The frequency of genetic changes in somaclones is much higher than the spontaneous genetic changes brought about in the entire plant (Prat 1983). Such genetic changes can be effectively used, provided entire plants can be regenerated from them – it is heartening to note that in most of the important agricultural crops, including some cereals like rice and maize, this can now be achieved.

It has been speculated (Larkin and Scowcroft 1981) that "tissue culture may generate the environment for enhancing chromosome breakage and reunion events, and thus a tissue culture cycle of the hybrid material may provide the means for obtaining genetic exchange needed between two genomes in the interspecific hybrid. The hybrid callus may enhance the frequency of requisite exchange". Hybrid callus, whether obtained through embryo culture or protoplast fusion, is a rich source of variation (Bajaj and Gill 1985) and needs to be exploited for crop improvement.

Somaclonal variation also makes it possible, to add or to intensify only one feature to an established variety, having combined most of the useful agronomic traits. As pointed out by Evans and Sharp (1986), "As single gene mutations and organelle gene mutations have been produced by somaclonal variation, one obvious strategy is to introduce the best available varieties into cell culture to select for incremental improvements of existing varieties by somaclonal variation. Hence, somaclonal variation could be used to uncover new variants that retain all the favorable qualities of an existing variety while adding one additional trait, such as disease resistance or herbicide resistance".

Plants showing tolerance/resistance to phytotoxins, salinity, herbicides, bacteria, nematodes, viruses, etc. have now been regenerated from cell cultures, under experimental conditions, in a number of species of economic importance, especially potato, tomato, sugarcane, maize, and rice (see Sects. 3 and 4). The induction of male sterility is another example which might play an important role in hybrid seed production. Likewise, the obtaining of high yielding protein/amino acid-rich plants is being pursued. Another area of tremendous commercial importance is the selection of cell lines showing high levels of secondary products. The induction and commercial use of high shikonin cell lines is a reality. Such cell lines can also be cryopreserved – avoiding periodical subculturing – which has far-reaching implications in the pharmaceutical industry. The applications and limitations of somaclonal variation have already been discussed (see Semal 1986; Evans and Sharp 1986), and need not be repeated here. However, it must be said that there is a mixed reaction from plant breeders regarding the utility of these variants, and it is argued that rigorous selection and field trials for various agronomic traits would be necessary before such variants are accepted. Although desirable in vitro-generated material would help to increase much needed genetic variability in crops, nevertheless one has to be cautious about field performance, which is the most important criterion for crop improvement. A judicious approach is more desirable than being over-optimistic.

References

Ahloowalia BS (1983) Spectrum of variation in somaclones of triploid ryegrass. Crop Sci 23:1141–1147

Ahloowalia BS, Sherington J (1985) Transmission of somaclonal variation in wheat. Euphytica 34:525–537

Alicchio R, Del Grosso E, Cavina R, Palenzona DL (1982) Effects of filtrates of *Verticillium dahliae* on the growth of *Solanum melongena* callus tissue. Genet Agrar 36:156

Alicchio R, Antonioli C, Palenzona DL (1984) Karyotypic variability in plants of *Solanum melongena* regenerated from callus grown in presence of culture filtrate of *Verticillium dahliae*. Theor Appl Genet 67:267–271

Amrhein N, Johanning D, Schab J, Schulz A (1983) Biochemical basis for glyphosate-tolerance in a bacterium and a plant tissue culture. FEBS Lett 157:191–196

Anderson PC, Georgeson MA, Hibberd KA (1984) Cell culture selection of herbicide-resistant corn. Crop Sci Soc Am Meet (Abstr) p 56

Austin S, Baer MA, Helgeson JP (1985) Transfer of resistance from potato leaf roll virus from *Solanum brevidens* into *Solanum tuberosum* by somatic fusion. Plant Sci 39:75–82

Aviv D, Galun E (1977) Isolation of tobacco protoplasts in the presence of isopropyl-N-phenylcarbamate and their culture and regeneration into plants. Z Pflanzenphysiol 83:267–273

Baertlein DA, McDaniel RG (1987) Molecular divergence of alfalfa somaclones. Theor Appl Genet 73:575–580

Bajaj S (1987) Biotechnology of nutritional improvement of potato. In: Bajaj YPS (ed) Biotechnology in agriculture and forestry, vol 3. Potato. Springer, Berlin Heidelberg New York Tokyo, pp 136–154

Bajaj YPS (1976) Regeneration of plants from cell suspensions frozen at –20°, –70° and –196°C. Physiol Plant 37:263–268

Bajaj YPS (1977) Survival of *Atropa* and *Nicotiana* pollen embryos frozen at –196°C. Curr Sci 46:305–307

Bajaj YPS (1978) Effect of superlow temperature on excised anthers and pollen embryos of *Atropa*, *Nicotiana* and *Petunia*. Phytomorphology 28:71–176

Bajaj YPS (1979) Technology and prospects of cryopreservation of germplasm. Euphytica 28:267–285

Bajaj YPS (1981) Production of disease-resistant plants through cell culture – a novel approach. J Nucl Agric Biol 10:1–5

Bajaj YPS (1983) Cryopreservation and international exchange of germplasm. In: Sen SK, Giles KL (eds) Plant cell culture in crop improvement. Plenum, New York, pp 19–41

Bajaj YPS (1984) Induction of growth in frozen embryos of coconut and ovules of citrus. Curr Sci 53:215–1216

Bajaj YPS (1985a) In vitro induction of genetic variability in groundnut. Proc Int Workshop Cytogenetics of *Arachis*. ICRISAT, Patancheru, 165 pp

Bajaj YPS (1985b) Cryopreservation of germplasm of potato (*Solanum tuberosum* L.) and cassava (*Manihot esculenta* Crantz): Viability of excised meristems cryopreserved up to 4 years. Indian J Exp Biol 23:285–287

Bajaj YPS (1986a) In vitro regeneration of diverse plants and the cryopreservation of germplasm in wheat (*Triticum aestivum* L.) Cereal Res Commun 14:305–311

Bajaj YPS (1986b) Cryopreservation of potato somaclones. In: Semal J (ed) Somaclonal variations and crop improvement. Nijhoff, Dordrecht, pp 244–250

Bajaj YPS (1986c) In vitro preservation of genetic resources. Proc Int Symp Nuclear techniques and in vitro culture for plant improvement. IAEA, Vienna, pp 43–57

Bajaj YPS (1987a) Biotechnology in agriculture and forestry, vol 3. Potato. Springer, Berlin Heidelberg New York Tokyo

Bajaj YPS (1987b) Induction and cryopreservation of somaclonal variation in peanut (*Arachis hypogaea* L.). Int Congr Plant Tissue Culture – Tropical species. Bogota, Colombia

Bajaj YPS (1987c) Biotechnology and 21st century potato. In: Bajaj YPS (ed) Biotechnology in agriculture and forestry, vol 3. Potato. Springer, Berlin Heidelberg New York Tokyo, pp 3–22

Bajaj YPS (1988a) Cryopreservation and the retention of biosynthetic potential in cell cultures of medicinal and alkaloid-producing plants. In: Bajaj YPS (ed) Biotechnology in agriculture and forestry, vol 4. Medicinal and aromatic plants I. Springer, Berlin Heidelberg New York Tokyo, pp 169–187

Bajaj YPS (1988b) Regeneration of plants from frozen (-196°C) protoplasts of *Atropa belladonna, Datura innoxia* and *Nicotiana tabacum.* Indian J Exp Biol 26:289–292
Bajaj YPS (1989a) Induction and cryopreservation of somaclonal variation in wheat and rice. Int Symp Genetic manipulation in plants. CIMMYT Mexico, pp 195–203
Bajaj YPS (ed) (1989b) Biotechnology in agriculture and forestry, vol 8, 9. Plant protoplasts and genetic engineering I, II. Springer, Berlin Heidelberg New York Tokyo
Bajaj YPS (1990a) Wide hybridization in legumes and oilseed crops through embryo, ovule and ovary culture. In: Bajaj YPS (ed) Biotechnology in agriculture and forestry, vol 10. Legumes and oilseed crops I. Springer, Berlin Heidelberg New York Tokyo, pp 1–37
Bajaj YPS (1990b) Cryopreservation of legumes and oilseed crops. In: Bajaj YPS (ed) Biotechnology in agriculture and forestry, vol 10. Legumes and oilseed crops I. Springer, Berlin Heidelberg New York Tokyo (in press)
Bajaj YPS, Bidani M (1980) Differentiation of genetically variable plants from embryo-derived callus of rice. Phytomorphology 30:290–294
Bajaj YPS, Bidani M (1986) In vitro induction of genetic variability in rice (*Oryza sativa* L.). In: Siddiqui KA, Faruqui AM (eds) New genetical approaches to crop improvement. PIDC, Karachi, pp 63–74
Bajaj YPS, Gill MS (1985) In vitro induction of genetic variability in cotton (*Gossypium* spp.). Theor Appl Genet 70:363–368
Bajaj YPS, Gill MS (1987) Biotechnology of cotton improvement. In: Crocomo OJ, Tavares FCA, Evans DA, Sharp WR, Bravo JE, Paddock EF (eds) Biotechnology of plants and microorganisms. Ohio State Univ Press, Columbus, pp 118–151
Bajaj YPS, Gill MS (1989) Pollen-embryogenesis and chromosomal variation in anther culture of a diploid cotton (*Gossypium arboreum* L.). SABRAO J 21:57–63
Bajaj YPS, Gosal SS (1982) Induction of genetic variability in grain legumes through tissue culture. In: Rao AN (ed) Tissue culture of economically important plants. Nat Univ, Singapore, COSTED, pp 25–41
Bajaj YPS, Gosal SS (1987) Pollen embryogenesis and chromosomal variation in cultured anthers of chickpea. Int Chickpea Newslett 17:12–13
Bajaj YPS, Gosal SS (1988) Isolation and fusion of protoplasts of *Arachis hypogaea* and *Arachis villosa.* Int Arachis Newslett 3:13–14
Bajaj YPS, Gupta RK (1986) Different tolerance of callus cultures of *Pennisetum americanum* L. and *Pennisetum purpureum* Schum. to sodium chloride. J Plant Physiol 125:491–495
Bajaj YPS, Gupta RK (1987) Plants from salt tolerant cell lines of napier grass *Pennisetum purpureum* Schum. Indian J Exp Biol 25:58–60
Bajaj YPS, Saettler AW (1968) Effect of culture filtrates of *Pseudomonas phaseolicola* on the growth of excised roots and callus tissue cultures of bean. Phytopathology 58:1041–1042
Bajaj YPS, Saettler AW (1970) Effect of halo toxin-containing filtrates of *Pseudomonas phaseolicola* on the growth of bean callus tissue. Phytopathology 60:1065–1067
Bajaj YPS, Spink GC, Saettler AW (1969) Ultrastructure of chloroplasts in bean leaves infected by *Pseudomonas phaseolicola.* Phytopathology 59:1017
Bajaj YPS, Gosch G, Ottma M, Weber A, Grobler A (1978a) Production of polyploid and aneuploid plants from anthers and mesophyll protoplasts of *Atropa belladonna* and *Nicotiana tabacum.* Indian J Exp Biol 16:947–953
Bajaj YPS, Gill KS, Sandha GS (1978b) Some factors enhancing the in vitro production of hexaploid triticale (*Triticum durum* × *Secale cereal*) Crop Improv 5:62–72
Bajaj YPS, Singh H, Gosal SS (1980a) Haploid embryogenesis in anther cultures of pigeon-pea (*Cajanus cajan*). Theor Appl Genet 58:157–159
Bajaj YPS, Phul PS, Sharma SK (1980b) Differential tolerance of tissue cultures of pearl-millet to ergot extract. Indian J Exp Biol 18:429–432
Bajaj YPS, Saini SS, Bidani M (1980c) Production of triploid plants from the immature and mature endosperm cultures of rice. Theor Appl Genet 58:17–18
Bajaj YPS, Sidhu BS, Dubey VK (1981a) Regeneration of genetically diverse plants from tissue cultures of forage grass – *Panicum* spp. Euphytica 30:135–140
Bajaj YPS, Ram AK, Labana KS, Singh H (1981b) Regeneration of genetically variable plants from the anther-derived callus of *Arachis hypogaea* and *A. villosa.* Plant Sci Lett 23:35–39

Bajaj YPS, Kumar P, Singh MM, Labana KS (1982) Interspecific hybridization in the genus *Arachis* through embryo culture. Euphytica 31:365–370
Bajaj YPS, Gill MS, Mohapatra D (1986a) Somaclonal and gametoclonal variation in wheat, cotton and brassica. In: Semal J (ed) Somaclonal variations and crop improvement. Nijhoff, Dordrecht, pp 160–169
Bajaj YPS, Mahajan SK, Labana KS (1986b) Interspecific hybridization of *Brassica napus* and *Brassica juncea* through ovary, ovule and embryo culture. Euphytica 35:103–109
Baldes R, Moos K, Geider K (1987) Transformation of soybean protoplasts from permanent suspension cultures by co-cultivation with cells of *Agrobacterium tumefaciens*. Plant Mol Biol 9:135–145
Ball SB, Seilleur P (1986) Characterization of somaclonal variations in potato: a biochemical approach. In: Semal J (ed) Somaclonal variations and crop improvement. Nijhoff, Dordrecht, pp 229–235
Barbier M, Dulieu H (1983) Early occurrence of genetic variants in protoplast cultures. Plant Sci Lett 29:201–206
Behnke M (1979) Selection of potato callus for resistance to culture filtrates of *Phytophthora infestans* and regeneration of plants. Theor Appl Genet 55:69–71
Behnke M (1980a) General resistance to late blight of *Solanum tuberosum* plants regenerated from callus resistant to culture filtrates of *Phytophthora infestans*. Theor Appl Genet 56:151–156
Behnke M (1980b) Selection of dihaploid potato callus for resistance to culture filtrate of *Fusarium oxysporum*. Z Pflanzenzücht 85:254–258
Bennici A (1974) Cytological analysis of roots, shoots and plants regenerated from suspension and solid in vitro cultures of haploid *Pelargonium*. Z Pflanzenzücht 72:199–205
Bennici A, Buiatti M, D'Amato F (1968) Nuclear conditions in haploid *Pelargonium* in vivo and in vitro. Chromosoma (Berlin) 24:194–201
Berlyn MB (1982) Variation in nuclear DNA content of isonicotinic acid hydrazide-resistant cell lines and mutant plants of *Nicotiana tabacum*. Theor Appl Genet 63:57–63
Berlyn GP, Beck RC, Renfoe MH (1986) Tissue culture and the genetic improvement of conifers: problems and possibilities. Tree Physiol 1:224–240
Berlyn GP, Anoruo AO, Beck RC, Cheng J (1987) DNA content polymorphism and tissue culture regeneration in Caribbean pine. Can J Bot 65:954–961
Binding H, Jain SM, Finger J, Mordhorst G, Nehls R, Gressel J (1982) Somatic hybridization of an atrazine-resistant biotype of *Solanum nigrum* with *Solanum tuberosum* Part I. Clonal variation in morphology and atrazine sensitivity. Theor Appl Genet 63:273–277
Bingham ET, McCoy TJ (1986) Somaclonal variation in alfalfa. In: Janick J (ed) Plant Breeding Rev, vol 4. AVI, Westport, pp 123–152
Blundy KS, Cullis CA, Hepburn A (1987) Ribosomal DNA methylation in a flax genotroph and a crown gall tumour. Plant Mol Biol 8:217–225
Bouharmont J, Dabin P (1986) Application des cultures in vitro à l'amélioration du fuchsia par mutation. Proc Int Symp. Nuclear techniques and in vitro culture for plant improvement. IAEA Vienna, pp 339–347
Boyes CJ, Vasil IK (1987) In vitro selection for tolerance to S-(2-aminoethyl)-L-cysteine and over-production of lysine by embryogenic calli and regenerated plants of *Pennisetum americanum* (L.) K Schum Plant Sci 50:195–203
Boxus P, Druart P (1986) Virus-free trees through tissue culture. In: Bajaj YPS (ed) Biotechnology in agriculture and forestry, vol 1. Trees I. Springer, Berlin Heidelberg New York Tokyo, pp 24–30
Breiman A (1985) Plant regeneration from *Hordeum spontaneum* and *Hordeum bulbosum* immature embryo derived calli. Plant Cell Rep 4:70–73
Breiman A, Felsenburg T, Galun E (1987a) Nor loci analysis in progenies of plants regenerated from the scutellar callus of bread-wheat. Theor Appl Genet 73:827–831
Breiman A, Rotem-Abarbanell D, Karp A, Shaskin H (1987b) Heritable somaclonal variation in wild barley (*Hordeum spontaneum*). Theor Appl Genet 74:104–112
Brettell RIS, Thomas E, Ingram DS (1980) Reversion of Texas male-sterile cytoplasm maize in culture to give fertile, T-toxin-resistant plants. Theor Appl Genet 58:55–58
Brettell RIS, Pallotta MA, Gustafson JP, Appels R (1986a) Variation at the Nor loci in triticale derived from tissue culture. Theor Appl Genet 71:637–643
Brettell RIS, Dennis ES, Scowcroft WR, Peacock JW (1986b) Molecular analysis of a somaclonal mutant of maize alcohol dehydrogenase. Mol Gen Genet 202:235–239

Bright SWJ, Miflin BJ, Rognes SE (1982) Threonine accumulation in the seeds of a barley mutant with an altered aspartate kinase. Biochem Genet 20:229–243
Broertjes C, Roest S, Bokelman GS (1976) Mutation breeding of *Chrysantheum morifolium* Ram., using in vivo and in vitro adventitious bud techniques. Euphytica 25:11–19
Broué P, Douglass J, Grace JP, Marshal DR (1982) Interspecific hybridization of soybeans and perennial *Glycine* species indigenous to Australia via embryo culture. Euphytica 31:715–724
Browers MA, Orton TJ (1982) Transmission of gross variability from suspension cultures into regenerated plants of celery. J Hered 73:159–162
Buiatti M, Tognoni F, Lipucci M, Marcheschi G, Pellegrini G, Grenci FC, Scala A, Bettini P, Bogani P, Betti L (1984) Genetic variability in tomato (*Lycopersicon esculentum*) plants regenerated from in vitro cultured cotyledons. Genet Agrar 38:325
Buiatti M, Marcheschi G, Tognoni F, Collina Grenci F, Martini G (1985) Genetic variability induced by tissue culture in the tomato (*Lycopersicon esculentum*). J Plant Breed 94:162–165
Buiatti M, Simeti C, Vannini S, Marcheschi G, Scala A, Bettini P, Bogani P, Pellegrini MG (1987) Isolation of tomato cell lines with altered response to *Fusarium* cell wall components. Theor Appl Genet 75:37–40
Butenko RG, Popov AS, Volkova LA, Chernyak ND, Nosov AM (1984) Recovery of cell cultures and their biosynthetic capacity after storage of *Dioscorea deltoidea* and *Panax ginseng* cells in liquid nitrogen. Plant Sci Lett 33:285–292
Caffaro L, Dameri RM, Profumo P, Bennici A (1982) Callus induction and plantlet regeneration in *Cichorium intybus* L.: 1. A cytological study. Protoplasma 111 (2):107–112
Carlson JE, Widholm JM (1978) Separation of two forms of anthralinate synthetase from 5-methyltryptophane-susceptible and -resistant cultured *Solanum tuberosum* cells. Physiol Plant 44:251–255
Carlson PS (1973) Methionine sulfoximine-resistant mutants of tobacco. Science 180:1366–1368
Cattoir-Reynaerts A, Degryse E, Jacobs M (1981) Selection and analysis of mutants over-producing amino acids of the aspartate family in barley, *Arabidopsis* and carrot. In: Induced mutations – a tool in plant breeding. IEAE, Vienna, pp 353–361
Chaleff RS, Carlson PS (1975) Higher plant cells as experimental organisms. In: Markham R, Davies DR, Hopwood DA, Horne RW (eds) Modification of the information content of plant cells. North-Holland, New York, pp 197–214
Chaleff RS, Keil RL (1981) Genetic and physiological variability among cultured cells and regenerated plants of *Nicotiana tabacum*. Mol Gen Genet 131:254–258
Chaleff RS, Parsons MF (1978) Direct selection in vitro for herbicide-resistant mutants of *Nicotiana tabacum*. Proc Natl Acad Sci USA 75:5104–5107
Chaleff RS, Ray TB (1984) Herbicide-resistant mutants from tobacco cell cultures. Science 223:1148–1151
Chawla HS, Wenzel G (1987) In vitro selection of barley and wheat for resistance against *Helminthosporium sativum*. Theor Appl Genet 74:841–845
Chen THH, Kartha KK, Leung NL, Kurz WGW, Chatson KB, Constabel F (1984) Cryopreservation of alkaloid-producing cell cultures of periwinkle (*Catharanthus roseus*). Plant Physiol 75:726–731
Chetrit P, Mathieu C, Vedel F, Pelletier G, Primard C (1985) Mitochondrial DNA polymorphism induced by protoplast fusion in Cruciferae. Theor Appl Genet 69:361–366
Chourey PS, Kemble RJ (1982) Transpositionment in tissue cultured cells of maize. In: Fujiwara A (ed) Plant tissue culture 1982. Maruzen, Tokyo, pp 425–426
Chu QR, Zhang Z-H, Cao Y-H (1985) Cytogenetical analysis on aneuploids obtained from pollen clones of rice (*Oryza sativa* L.). Theor Appl Genet 71:506–512
Cooper DB, Sears RG, Lookhart GL, Jones BL (1986) Heritable somaclonal variation in gliadin proteins of wheat plants derived from immature embryo callus culture. Theor Appl Genet 71: 784–790
Croughan TP, Stavarek SJ, Rains DW (1978) Selection of NaCl-tolerant line of cultured alfalfa cells. Crop Sci 18:959–963
Croughan TP, Stavarek SJ, Rains DW (1981) In vitro development of salt-resistant plants. Environ Exp Bot 21:317–324
Cullis CA (1977) Molecular aspects of the environmental induction of heritable changes in flax. Heredity 42:237–246
Cullis CA (1983) Environmentally induced DNA changes in plants. CRC Crit Rev Plant Sci 1:117–129

Cullis CA (1986) Phenotypic consequences of environmentally induced changes in plant DNA. Trends Genet 2:307–309
Cullis CA, Cleary W (1986) DNA variation in flax tissue culture. Can J Genet Cytol 28:247–251
D'Amato F (1965) Endopolyploidy as a factor in plant tissue development. In: White PR, Grove RA (eds) Proc Int Conf Plant tissue culture. McCutchan, Berkeley, pp 449–462
D'Amato F (1977) Cytogenetics of differentiation in tissue and cell culture. In: Reinert J, Bajaj YPS (eds) Applied and fundamental aspects of plant cell, tissue and organ culture. Springer, Berlin Heidelberg New York, pp 343–357
D'Amato F (1985) Cytogenetics of plant cell and tissue cultures and their regenerates. CRC Critical Rev Plant Sci 3:73–112
Davies PA, Pallotta MA, Ryan SA, Scowcroft WR, Larkin PJ (1986) Somaclonal variation in wheat: genetic and cytogenetic characterization of alcohol dehydrogenase I mutants. Theor Appl Genet 72:644–653
Day A, Ellis THN (1984) Chloroplast DNA deletions associated with wheat plants regenerated from pollen: possible basis for maternal inheritance of chloroplasts. Cell 39:359–368
Deak M, Kiss GB, Koncz C, Dudits D (1986) Transformation of *Medicago* by *Agrobacterium*-mediated gene transfer. Plant Cell Rep 5:97–100
de Buyser J, Henry Y, Lonnet P, Hertzoh R, Hespel R, Hespel A (1987) Florin: A doubled haploid wheat variety developed by the anther culture method. Plant Breed 98:53–56
Dennis ES, Brettell RIS, Peacock WJ (1987) A tissue culture induced *Adh*1 null mutant of maize results from a single base change. Mol Gen Genet (in press)
Diettrich B, Popov AS, Pfeiffer B, Neumann D, Butenko R, Luckner M (1982) Cryopreservation of *Digitalis lanata* cell cultures. Planta Med 46:82–87
Dix PJ, Street HE (1975) Sodium chloride-resistant cultured cell lines from *Nicotiana sylvestris* and *Capsicum annuum*. Plant Sci Lett 5:231–237
Donn G, Tischer E, Smith JA, Goodman HM (1984) Herbicide-resistant alfalfa cells: An example of gene amplification in plants. J Mol Appl Genet 2:621–635
Durante M, Geri C, Grisvard JA, Goodman HM (1984) Herbicide-resistant alfalfa cells: an example of gene amplification in plants. J Mol Appl Genet 2:621–635
Duron M, Decourtye L (1986) Effets biologiques des rayons Gamma appliqués à des plantes de *Weigela* cv. Bristol Ruby cultivés in vitro. Proc Int Symp Nuclear techniques and in vitro culture for plant improvement. IAEA, Vienna, pp 103–111
Durrant A (1962) The environmental induction of heritable changes in *Linum*. Heredity 17:27–61
Earle ED, Gracen VE, Best VM, Batts LA, Smith ME (1987) Fertile revertants from S-type male-sterile maize grown in vitro. Theor Appl Genet 74:601–609
Earle ED, Gracen VE, Yodder OC, Gemill KP (1978) Plant Physiol 61:420
Edallo S, Zuccinali C, Perenzin M, Salamini F (1981) Chromosomal variation and frequency of spontaneous mutation associated with in vitro culture and plant regeneration in maize. Maydica 26:39–56
Engler DE, Grogan RG (1984) Variation in lettuce plants regenerated from protoplasts. J Hered 75:426–430
Epstein E (1972) Mineral nutrition of plants – principles and perspectives. Wiley, New York
Ettinger TL, Ostry ME, Hackett WP, Read PE, Skilling DD (1986) Utilization of somaclonal variation in the development of *Septoria musiva*-resistant hybrid *Populus*. Abstr VI Int Congr Plant tissue and cell culture, Minnesota, p 214
Evans DA, Sharp WR (1983) Single gene mutations in tomato plants regenerated from tissue culture. Science 221:949–951
Evans DA, Sharp WR (1986) Applications of somaclonal variation. Biotechnology 4:528–532
Fedak G (1980) Production, morphology and meiosis of reciprocal barley-wheat hybrids. Can J Genet Cytol 22:117–123
Firoozabady E (1989) Genetic engineering in cotton. In: Bajaj YPS (ed) Biotechnology in agriculture and forestry, vol 9. Plant protoplasts and genetic engineering II. Springer, Berlin Heidelberg New York Tokyo, pp 140–154
Foroughi-Wehr B, Friedt W (1984) Rapid production of recombinant barley yellow mosaic virus-resistant *Hordeum vulgare* lines by anther culture. Theor Appl Genet 67:377–382
Foroughi-Wehr B, Friedt W, Schuchmann R, Kohler F, Wenzel G (1986) In vitro selection for resistance. In: Semal J (ed) Somaclonal variations and crop improvement. Nijhoff, Dordrecht, pp 35–44

Freeman JA (1982) The influence of weather on the response of potato cultivars to metribuzin. J Am Soc Hort Sci 107:189–194
Fujita Y, Takahashi S, Yamada Y (1985) Selection of cell lines with high productivity of shikonin derivatives by protoplast culture of *Lithospermum erythrorhizon*. Agric Biol Chem 49:1755–1759
Gathercole RWE, Street HE (1976) Isolation, stability and biochemistry of a parafluorophenylalanine-resistant cell line of *Acer pseudoplatanus* L. New Phytol 77:29–41
Gavazzi G, Tonelli C, Todesco G, Arreghini E, Raffaldi F, Vecchio F, Barbuzzi G, Biasini MG, Sala F (1987) Somaclonal variation versus chemically induced mutagenesis in tomato (*Lycopersicon esculentum* L.). Theor Appl Genet 74:733–738
Gengenbach BG, Connelly JA (1981) Mitochondrial DNA variation in maize plants regenerated during tissue culture selection. Theor Appl Genet 59:161–167
Gengenbach BG, Green CE (1975) Selection of T-cytoplasm maize callus cultures resistant to *Helminthosporium maydis* race T pathotoxin. Crop Sci 15:645–649
Gengenbach BG, Green CE, Donovan CM (1977) Inheritance of selected pathotoxin-resistance in maize plants regenerated from cell cultures. Proc Natl Acad Sci USA 74:5113–5117
Gill MS, Bajaj YPS (1984) Interspecific hybridization in the genus *Gossypium* through embryo culture. Euphytica 33:305–311
Gill MS, Bajaj YPS (1987) Hybridization between diploid (*Gossypium arboreum*) and tetraploid (*Gossypium hirsutum*) cotton through ovule culture. Euphytica 36:625–630
Girod P-A, Zryd J-P (1987) Clonal variability and light induction of betalain synthesis in red beet cell cultures. Plant Cell Rep 6:27–30
Gleba YY, Hoffmann F (1980) *Arabidobrassica*: A novel plant obtained by protoplast fusion. Planta 149:112–117
Gleddie S, Fassuliotis G, Keller WA, Setterfield G (1985) Somatic hybridization as a potential method of transferring nematode and mite resistance into eggplant. Z Pflanzenzücht 94:348–351
Gleddie S, Keller WA, Setterfield G (1986) Production and characterization of somatic hybrids between *Solanum melongena* L. and *S. sisymbriifolium* Lam. Theor Appl Genet 71:613–621
Gonzales RA, Das PK, Widholm JM (1984) Characterization of cultured tobacco cell lines selected for resistance to a methionine analog, ethionine. Plant Physiol 74:640–644
Gosal SS, Bajaj YPS (1983) Interspecific hybridization between *Vigna mungo* and *Vigna radiata* through embryo culture. Euphytica 32:129–137
Gosal SS, Bajaj YPS (1984) Isolation of sodium chloride-resistant cell lines in some grain-legumes. Indian J Exp Biol 22:209–214
Gosal SS, Bajaj YPS (1988) Pollen embryogenesis and chromosomal variation in anther culture of three food legumes — *Cicer arietinum*, *Pisum sativum* and *Vigna mungo*. SABRAO J 20
Gourd JM, Morris Southward G, Phillips GC (1988) Response of *Allium* tissue cultures to filtrates of *Pyrenochaeta terrestris*. Hort Sci 23:766–768
Graybosch RA, Edge ME, Delannay X (1987) Somaclonal variation in soybean plants regenerated from the cotyledonary node system. Crop Sci 27:803–806
Grazia S, Rocchetta G, Pieragostini E (1985) Somatic variation in micropropagated clones of *Cichorium intybus*. Experientia 41:393–395
Gressel J, Cohen J, Binding H (1984) Somatic hybridization of an atrazine-resistant biotype of *Solanum nigrum* with *Solanum tuberosum*. 2. Segregation of plastomes. Theor Appl Genet 66:131–134
Groose RW, Bingham ET (1984) Variation in plants regenerated from tissue culture of tetraploid alfalfa heterozygous for several traits. Crop Sci 24:655–658
Guerche P, Jouanin L, Tepfer D, Pelletier G (1987) Genetic transformation of oilseed rape (*Brassica napus*) by the Ri T-DNA of *Agrobacterium rhizogenes* and analysis of inheritance of the transformed phenotype. Mol Gen Genet 206:382–386
Gunn RE (1982) Breeding new potato varieties through protoplasts. 8th Long Ashton Symp Improv Veg Propagated Crops, Bristol, England
Hammatt N, Davey MR (1989) Soybean: Isolation, culture and fusion of protoplasts. In: Bajaj YPS (ed) Biotechnology in agriculture and forestry, vol 10. Legumes and oilseed crops I. Springer, Berlin Heidelberg New York Tokyo, pp 149–169
Hammerschlag F (1986) In vitro selection of peach cells for insensitivity to a toxin produced by *Xanthomonas campestris* pv. *pruni*. In: 6th Int Congr plant tissue and cell culture. Abstr Univ Minnesota, Minneapolis, p 74

Handa AK, Bressan RA, Handa S, Hasegawa PM (1983) Clonal variation for tolerance to polyethylene glycol-induced water stress in cultured tomato cells. Plant Physiol 72:645–653

Hartman CL, McCoy TJ, Knous TR (1984) Selection of alfalfa (*Medicago sativa*) cell lines and regeneration of plants resistant to the toxin(s) produced by *Fusarium oxysporum* f. sp. *medicaginis*. Plant Sci Lett 34:183–194

Hauptmann RM, Widholm JM (1982) Cryostorage of cloned amino acid analog-resistant carrot and tobacco suspension culture. Plant Physiol 70:30–37

Heinz DJ, Mee GWP (1971) Morphologic, cytogenetic, and enzymatic variation in *Saccharum* species hybrid clones derived from callus tissue. Am J Bot 58:257–262

Heinz DJ, Krishnamurthi M, Nickell LG, Maretzki A (1977) Cell, tissue and organ culture in sugarcane improvement. In: Reinert J, Bajaj YPS (eds) Applied and fundamental aspects of plant cell, tissue, and organ culture. Springer, Berlin Heidelberg New York, pp 3–17

Hibberd KA, Green CE (1982) Inheritance and expression of lysine plus threonine resistance selected in maize tissue culture. Proc Natl Acad Sci USA 79:559–563

Hibberd KA, Walter T, Green CE, Gengenbach BG (1980) Selection and characterization of a feedback-insensitive tissue culture of maize. Planta 148:183–187

Ho C, Loo S, Xu Z, Xu S, Li W (1982) Experiments on the regenerated callus tissue from 4-oxalysine-resistant mutant plant cell line. In: Fujiwara A (ed) Plant tissue culture 1982. Maruzen, Tokyo, pp 469–470

Holwerda BC, Jana S, Crosby WL (1986) Chloroplast and mitochondrial DNA variation in *Hordeum vulgare* and *Hordeum spontaneum*. Genetics 114:1271–1291

Horák J, Landa Z, Luštinek J (1971) Production of polyploid plants from tissue cultures of *Brassica oleracea* L. Phyton 28:7–10

Hu Han (1986) Wheat: Improvement through anther culture. In: Bajaj YPS (ed) Biotechnology in agriculture and forestry, vol 2. Crops I. Springer, Berlin Heidelberg New York Tokyo, pp 55–72

Hunault G, Desmarest P (1989) *Foeniculum vulgare* Miller (Fennel): Cell culture, regeneration and the production of anethole. In: Bajaj YPS (ed) Biotechnology in agriculture and forestry, vol 7. Medicinal and aromatic plants II. Springer, Berlin Heidelberg New York Tokyo, pp 185–212

Hunter CS (1986) In vitro propagation and germplasm storage of *Cinchona*. In: Withers LA, Alderson P (eds) Plant tissue culture and its agricultural applications. Butterworth, London, pp 291–301

Iyengar ERR, Patobia JS, Kurian TZ (1977) Varietal differences in barley to salinity. Z Pflanzenphysiol 84:355–361

Jana KK, Khush GS (1984) Embryo rescue of interspecific hybrid and its scope in rice improvement. Int Rice Genet Newslett 1:133–134

Jacobs M, Cattoir-Reynaerts A, Negrutiu I, Verbruggen I, Degryse E (1982) Comparison of selection schemes for the isolation of resistant mutants to aspartate-derived amino acids and S-(2-aminoethyl) cysteine in various cell culture and "whole plant" systems. In: Fujiwara A (ed) Plant tissue culture 1982. Maruzen, Tokyo, pp 453–454

Jacobsen E, Tempelaar MJ, Bijmolt EW (1983) Ploidy levels in leaf callus and regenerated plants of *Solanum tuberosum* determined by cytophotometric measurements of protoplasts. Theor Appl Genet 65:113–118

Jacobsen E, Visser RGF, Wijbrandi J (1986) Phenylalanine and tyrosine accumulating cell lines of a dihaploid potato selected by resistance to 5-methyl tryptophan. Plant Cell Rep 4:151–154

Johnson LB, Stuteville DL, Schlarbaum SE, Skinner DZ (1984) Variation in phenotype and chromosome number in alfalfa protoclones regenerated from nonmutagenized calli. Crop Sci 24:948–951

Jordan MC, Larter EN (1985) Somaclonal variation in triticale (x *Triticosecale* Wittmack) cv. Carman. Can J Genet Cytol 27:151–157

Kakehi M (1972) Studies on the tissue culture of carnation. II. Cytological studies on cultured cells. J Jpn Soc Hort Sci 41:72–75

Kamada H, Okamura N, Satake M, Harada H, Shimomura K (1986) Alkaloid production by hairy root cultures in *Atropa belladonna*. Plant Cell Rep 5:239–242

Karp A, Maddock SE (1984) Chromosome variation in wheat plants regenerated from cultured immature embryos. Theor Appl Genet 67:249–256

Karp A, Nelson RS, Thomas E, Bright SWJ (1982) Chromosome variation in protoplast-derived potato plants. Theor Appl Genet 63:265–272

Karp A, Steele SH, Parmar S, Jones MGK, Shewry PR, Breiman A (1987) Relative stability among barley plants regenerated from cultured immature embryos. Genome 29:405–412

Keller WA. Armstrong KC (1981) Production of anther-derived dihaploid plants in autotetraploid marrowstem kale (*Brassica oleracea* var. *acephala*). Can J Genet Cytol 23:259-265
Kemble RJ. Flavell RB (1982) Mitochondrial DNA analyses of fertile and sterile maize plants derived from tissue culture with the Texas male sterile cytoplasm. Theor Appl Genet 69:211-216
Kemble RJ. Shepard JF (1984) Cytoplasmic DNA variation in a protoclonal population. Theor Appl Genet 69:211-216
Korneef M. Hanhart CJ. Martinelli L (1987) A genetic analysis of cell culture traits in tomato. Theor Appl Genet 74:633-641
Kott LS. Flack S. Kasha KJ (1986) A comparative study of initiation and development of embryogenic callus from haploid embryos of several barley cultivars. II Cytophotometry of embryos and callus. Can J Bot 64:2107-2112
Krishnamurthi M. Tlaskal J (1974) Fiji disease resistant *Saccharum officinarum* var. Pindar subclones from tissue cultures. Proc Int Soc Sugarcane Technol 15:130-137
Kudirka DT. Schaeffer GW. Baenziger PS (1986) Wheat: Genetic variability through anther culture. In: Bajaj YPS (ed) Biotechnology in agriculture and forestry, vol 2. Crops I. Springer, Berlin Heidelberg New York Tokyo, pp 39-54
Kueh JSH. Bright SWJ (1981) Proline accumulation in barley mutant resistant to trans-4-hydroxy-L-proline. Planta 153:166-171
Kueh JSH. Bright SWJ (1982) Biochemical and genetical analysis of three proline-accumulating barley mutants. Plant Sci Lett 27:233-241
Kumashiro T. Kubo T (1984) Production of cytoplasmic male sterile tobacco by protoplast fusion. I. *Nicotiana debneyi* + *N. tabacum* cv. cosolation 402. Jpn J Breed (Suppl) 34:54-55
Landsmann J. Uhrig H (1985) Somaclonal variation in *Solanum tuberosum* detected at the molecular level. Theor Appl Genet 71:500-505
Larkin PJ. Scowcroft WR (1981) Somaclonal variation – a novel source of variability from cell culture for plant improvement. Theor Appl Genet 60:197-214
Larkin PJ. Brettell RIS. Scowcroft WR (1984) Heritable somaclonal variation in wheat. Theor Appl Genet 67:443-455
Latunde-Dada AO. Lucas JA (1983) Somaclonal variation and reaction to *Verticillium* wilt in *Medicago sativa* L. plants regenerated from protoplasts. Plant Sci Lett 32:205-211
Lazar MD. Chen THH. Gusta LV. Kartha KK (1988) Somaclonal variation for freeze tolerance in a population derived from norstar winter wheat. Theor Appl Genet 75:480-484
Lebrun L. Rajasekaran K. Mullins G (1985) Selection in vitro for NaCl-tolerance in *Vitis rupestris* Scheele. Ann Bot 56:733-739
Leshem B (1986) Carnation plantlets from vitrified plants as a source of somaclonal variation. Hort Sci 21:320-321
Linacero R. Vazquez AM (1986) Somatic embryogenesis and somaclonal variation in rye. Int Conf Biotechnology and agriculture in the Mediterranean Basin, Athens June 26-28, CEC
Ling DH (1987) A quintuple reciprocal translocation produced by somaclonal variation in rice. Cereal Res Commun 15:5-12
Ling DH. Ma ZR. Chen Wy. Chen MF (1987) Male sterile mutant from somatic cell culture of rice. Theor Appl Genet 75:127-131
Liu M. Chen W (1976) Tissue and cell culture as aids to sugar cane breeding. I. Creation of genetic variation through callus culture. Euphytica 25:393-403
Liu Ming-Chin. Yeh Huei-Sin (1982) Selection of NaCl tolerant line through stepwise salinized sugarcane cell cultures. In: Fujiwara A (ed) Plant tissue culture 1982. Maruzen, Tokyo, pp 477-478
Loo SW. Xu Zhi-Hong (1986) Rice: Anther culture for rice improvement in China. In: Bajaj YPS (ed) Biotechnology in agriculture and forestry, vol 2. Crops I. Springer, Berlin Heidelberg New York Tokyo, pp 139-156
Lourens AG. Martin FA (1987) Evaluation of in vitro propagated sugarcane hybrids for somaclonal variation. Crop Sci 27:793-796
Maddock SE. Semple JT (1986) Field assessment of somaclonal variation in wheat. J Exp Bot 37:1065-1078
Maddock SE. Risiott R. Parmar S. Jones MGK. Shewry PR (1985) Somaclonal variation in the gliadin patterns of grains of regenerated wheat plants. J Exp Bot 37:1065-1078
Maddox AD. Gonsalves F. Shields R (1983) Successful preservation of suspension cultures of three *Nicotiana* species at the temperature of liquid nitrogen. Plant Sci Lett 28:157-162

Matern U. Strobel GA (1987) In vitro production of potatoes bearing resistance to fungal diseases. In: Bajaj YPS (ed) Biotechnology in agriculture and forestry. vol 3. Potato. Springer. Berlin Heidelberg New York Tokyo. pp 298–317

Matern U. Strobel G. Shepard J (1978) Reaction to phytotoxins in a potato population derived from mesophyll protoplasts. Proc Natl Acad Sci USA 75:4935–4939

Mathews BF. Shye SCH. Widholm JM (1980) Mechanism of resistance of a selected carrot cell suspension culture to S(2-aminoethyl)-L-cysteine. Z Pflanzenphysiol 96:453–463

May CE (1983) Triticale x wheat hybrids and the introduction of speckled leaf blotch resistance to wheat. In: Proc 6th Int Wheat Genet Symp Kyoto Univ. Kyoto. pp 175–179

McCoy TJ. Phillips RL. Rines HW (1982) Cytogenetic analysis of plants regenerated from oat (*Avena sativa*) tissue culture: frequency of partial chromosome loss. Can J Genet Cytol 24:37–50

McHughen A. Swartz (1984) A tissue-culture derived salt-tolerant line of flax (*Linum usitatissimum*). J Plant Physiol 117:109–117

Mc Nay JW. Chourey PS. Pring DR (1984) Molecular analysis of genomic stability of mitochondrial DNA in tissue cultured cells of maize. Theor Appl Genet 67:433–437

Meins F (1983) Heritable variation in plant cell culture. Annu Rev Plant Physiol 34:327–346

Melchers G. Sacristan MD. Holder AA (1978) Somatic hybrid plants of potato and tomato regenerated from fused protoplasts. Carlsberg Res Commun 43:203–218

Meredith CP (1978) Selection and characterization of aluminium resistant variants from tomato cell cultures. Plant Sci Lett 12:25–34

Metakovsky EV. Novoselskaya A Yu. Sozinov AA (1987) Problems of interpreting results obtained in studies of somaclonal variation in gliadin proteins in wheat. Theor Appl Genet 73:764–766

Meulemans M. Duchene D. Fouarge G (1987) Selection of variants by dual culture of potato and *Physophthora infestans*. In: Bajaj YPS (ed) Biotechnology in agriculture and forestry. vol 3. Potato. Springer. Berlin Heidelberg New York Tokyo. pp 318–331

Miao S. Duncan DR. Widholm JM (1988) Selection of regenerable maize callus cultures resistant to 5-methyl-DL-tryptophan. S-2-aminoethyl-L-cysteine and high levels of lysine plus L-threonine. Plant Cell Tissue Culture (in press)

Miller OK. Hughes KW (1980) Selection of paraquat-resistant variants of tobacco from cell culture. In Vitro 16:1085–1091

Mix G. Wilson HM. Foroughi-Wehr B (1978) The cytological status of plants of *Hordeum vulgare* L. regenerated from microspore callus. Z Pflanzenzücht 80:89–99

Mok DWS. Mok MC. Rabakoarihanta A (1978) Interspecific hybridization of *Phaseolus vulgaris* with *P. lunatus* and *P. acutifolius*. Theor Appl Genet 52:209–215

Mohapatra D. Bajaj YPS (1984) In vitro hybridization in an incompatible cross – *Brassica juncea* × *Brassica hirta*. Curr Sci 53:489–490

Mohapatra D. Bajaj YPS (1987) Interspecific hybridization in *Brassica juncea* × *Brassica hirta* using embryo rescue. Euphytica 36:321–326

Morinaga T. Fukushima E (1935) Cytogenetical studies on *Oryza sativa* L. II Spontaneous autotriploid mutants in *Oryza sativa* L. Jpn J Bot 7:207–225

Mroginski LA. Fernandes A (1979) In vitro anther culture of species of *Arachis* (Leguminosae). Oleagineux 34:243–248

Mujeeb-Kazi A. Roldan S. Miranda JL (1984) Intergeneric hybrids of *Triticum aestivum* with *Agropyron* and *Elymus* species. Cereal Res Commun 12:75–79

Murakishi HH. Harris RR (1987) Testing somaclonal variants of potato for resistance to virus disease. In: Bajaj YPS (ed) Biotechnology in agriculture and forestry. vol 3. Potato. Springer. Berlin Heidelberg New York Tokyo. pp 332–345

Murakishi HH. Lesney MS. Carlson PS (1984) Protoplasts and plant viruses. In: Maramorosch K (ed) Advances in cell culture. vol 3. Academic Press. New York. pp 1–55

Murashige T. Nakano R (1966) Tissue culture as a potential tool in obtaining polyploid plants. J Hered 57:115–118

Murata M. Orton TJ (1984) Chromosome fusion in cultured cells of celery. Can J Genet Cytol 26:395–400

Nabors MW (1982) Plant tissue culture can help plant breeders produce stress-tolerant plants. Tissue Culture Crops Newslett 1:1–2

Nabors MW. Gibbs GE. Bernstein CS. Meis ME (1980) NaCl-tolerant tobacco plants from cultured cells. Z Pflanzenphysiol 97:13–17

Nabors MW. Kroskey CS. McHugh DM (1982) Green spots are predictors of high callus growth rates and shoot formation in normal and in salt-stressed tissue cultures of oat (*Avena sativa* L.). Z Pflanzenphysiol 105:341–349

Nag KK. Street HE (1975) Freeze preservation of cultured plant cells II. The freezing and thawing phases. Physiol Plant 34:261–265

Nagl W (1972) Evidence of DNA amplification in the orchid *Cymbidium* in vitro. Cytol 5:145–154

Negruk VI. Eisner GI. Redichkina TD. Dumanskaya NN. Cherny DI. Alexandrov AA. Shemyakin MF. Butenko RG (1986) Diversity of *Vicia faba* circular mt DNA in whole plants and suspension cultures. Theor Appl Genet 72:541–547

Negrutiu I. Cattoir-Reynaerts A. Verbruggen I. Jacobs M (1984) Lysine overproducer mutants with an altered dihydrodipicolinate synthase from protoplast culture of *Nicotiana sylvestris* (Spegazzini and Comes). Theor Appl Genet 68:11–20

Newell CA. Rhoads ML. Bidney DL (1984) Cytogenetic analysis of plants regenerated from tissue explants and mesophyll protoplasts of winter rape. *Brassica napus* L. Can J Genet Cytol 26:752–761

Niemirowicz-Szczytt K. Ciupka B. Malepszy S (1984) Polyploids from *Fragaria vesca* L. meristems. induced by colchicine in the in vitro culture. Bull Polish Acad Sci 32:58–63

Nyman LP. Gonzales CJ. Arditti J (1983) In vitro selection for salt tolerance of taro (*Colocasia esculenta* var. antiquorum). Ann Bot 51:229–236

O'Connell MA. Hosticka LP. Hanson MR (1986) Examination of genome stability in cultured *Lycopersicon*. Plant Cell Rep 5:276–279

Ogihara Y (1982) Characterization of chromosomal changes of calluses induced by callus cloning and para-fluorophenylalanine (PFP) treatment and regenerates from them. In: Fujiwara A (ed) Plant tissue culture 1982. Maruzen. Tokyo. pp 439–440

Ogino T. Hiraoka N. Tabata M (1978) Selection of high nicotine producing cell lines of tobacco callus by single cell cloning. Phytochemistry 22:2447–2450

Ogura H. Kyozuka J. Hayashi Y. Koba T. Shimamoto K (1987) Field performance and cytology of protoplast-derived rice (*Oryza sativa*): high yield and low degree of variation of four japonica cultivars. Theor Appl Genet 74:670–676

Ogura H. Kyozuka J. Hayashi Y. Shimamoto K (1988) Yielding ability and phenotypic uniformity in the selfed progeny of protoplast-derived rice plants. Jpn J Breed 39:47–56

Oksman-Caldentey KM. Strauss A (1986) Somaclonal variation of scopolamine content in protoplast-derived cell culture clones of *Hyoscyamus muticus*. Planta Medica 1:6–12

Ooms G. Karp A. Burrell MM. Twell D. Roberts J (1985) Genetic modification of potato development using R1 T-DNA. Theor Appl Genet 70:440–446

Oono K (1985) Putative homozygous mutations in regenerated plants of rice. Mol Gen Genet 198:377–384

Orton TJ (1980a) Comparison of salt tolerance between *Hordeum vulgare* and *H. jubatum* in whole plants and callus cultures. Z Pflanzenphysiol 98:105–118

Orton TJ (1980b) Chromosomal variability in tissue cultures and regenerated plants of *Hordeum*. Theor Appl Genet 56:101–112

Orton TJ (1983) Experimental approaches to the study of somaclonal variation. Plant Molec Biol Rep 1:67–76

Partanen CR (1963) Plant tissue culture in relation to developmental cytology. Int Rev Cytol 15:215–243

Pelcher LE. Kao KN. Gamborg OL. Yodder OC. Gracen VE (1975) Can J Bot 53:427

Pelletier C. Primand C. Vedel F. Chetrie P. Remy R. Rousselle P. Renard M (1983) Intergeneric cytoplasmic hybridization in Cruciferae by protoplast fusion. Mol Gen Genet 191:244–250

Peters BM. Cribbs DH. Stelzig DA (1978) Science 201:364–367

Pezzotti M. Arcioni S. Damiani F. Mariotti D (1985) Time-related behaviour of phenotypic variation in *Lotus corniculatus* regenerants under field conditions. Euphytica 34:619–623

Phillips GC. Collins GB. Taylor NL (1982) Interspecific hybridization of red clover (*Trifolium pratense* L.) with *T. sarosiense* Hazsl. using in vitro embryo rescue. Theor Appl Genet 62:17–24

Powell W. Hayter AM. Wood W. Dunwell JM. Huang B (1984) Variation in the agronomic characters of microspore-derived plants of *Hordeum vulgare* cv. Sabarlis. Heridity 52:19–23

Powell W. Borrino EM. Allison MJ. Griffiths DW. Asher MJC. Dunwell JM (1986) Genetical analysis of microspore derived plants of barley (*Hordeum vulgare*). Theor Appl Genet 72:619–626

Prat D (1983) Genetic variability induced in *Nicotiana sylvestris* by protoplast culture. Theor Appl Genet 64:223–230

Pua E-C, Thorpe TA (1986) Differential response of nonselected and Na_2SO_4-selected callus cultures of *Beta vulgaris* L. to salt stress. J Plant Physiol 123:241–248

Quak F (1977) Meristem culture and virus-free plants. In: Reinert J, Bajaj YPS (eds) Applied and fundamental aspects of plant cell, tissue, and organ culture. Springer, Berlin Heidelberg, New York, pp 598–615

Radin DN, Carlson PS (1978) Herbicide tolerant tobacco mutant selected in situ and recovered via regeneration from cell culture. Genet Res 32:85–89

Reich TJ, Iyer VN, Miki BL (1986) Efficient transformation of alfalfa protoplasts by the intranuclear microinjection of Ti plasmids. Bio/Technology 4:1001–1003

Reisch B, Bingham ET (1981) Plants from ethionine-resistant alfalfa tissue culture: Variation in growth and morphological characteristics. Crop Sci 21:783–788

Reisch B, Duke HS, Bingham ET (1981) Selection and characterization of ethionine-resistant alfalfa (*Medicago sativa* L.) cell lines. Theor Appl Genet 59:89–94

Renfroe MH, Berlyn GP (1985) Variation in nuclear DNA content in *Pinus taeda* L. tissue culture of diploid origin. J Plant Physiol 121:131–139

Rhodes CA, Pierce DA, Mettler IJ, Mascarenhas D, Detmer JA (1988) Genetically transformed maize plants from protoplasts. Science 240:204–207

Riccardi G, Cella R, Cumerino G, Cifferri O (1983) Resistance to azelidine-2-carboxylic acid and sodium chloride tolerance in carrot cell cultures and *Spirulina platensis*. Plant Cell Physiol 24:1073–1078

Rietveld RC, Hasegawa PM, Bressan RA (1987) Genetic variability in tuber disc-derived potato plants. In: Bajaj YPS (ed) Biotechnology in agriculture and forestry, vol 3. Potato. Springer, Berlin Heidelberg New York Tokyo, pp 392–407

Rines HW (1986) Origin of tissue culture selected resistance to *Helminthosporium victoriae* toxin in heterozygous susceptible oats. Int Congr plant tissue and cell culture. Abstr Univ Minnesota, Minneapolis, p 212

Robertson D, Palmer JD, Earle ED, Mutschler MA (1987) Analysis of organelle genomes in a somatic hybrid derived from cytoplasmic male-sterile *Brassica oleracea* and atrazine-resistant *B. campestris*. Theor Appl Genet 74:303–309

Rode A, Hartmann C, Benslimane A, Picard E, Quetier F (1987) Gametoclonal variation detected in the nuclear ribosomal DNA from double haploid lines of a spring wheat (*Triticum aestivum* L. cv. Cesar). Theor Appl Genet 74:31–37

Ruiz ML, Vazquez AM (1982) Chromosome number evolution in stem derived calluses of *Hordeum vulgare* L. cultured in vitro. Protoplasma 111:83–86

Rush DW, Epstein E (1976) Genotypic responses to salinity. Plant Physiol 57:162–166

Ryan SA, Larkin PJ, Ellison FW (1987) Somaclonal variation in some agronomic and quality characters in wheat. Theor Appl Genet 74:77–82

Sasse F, Buchholz M, Berlin J (1983) Selection of cell lines of *Catharanthus roseus* with increased tryptophan decarboxylase activity. Z Naturforsch 38c:916–922

Sato F, Yamada Y (1984) High berberine-producing cultures of *Coptis japonica* cells. Phytochemistry 23:281–285

Saunders JW, Doley WP (1986) One step shoot regeneration from callus of whole plant leaf explants of sugarbeet lines and a somaclonal variant for in vitro behavior. J Plant Physiol 124:473–479

Schaeffer GW, Sharpe FT (1981) Lysine in seed protein from S-amino ethyl-L-cysteine resistant anther-derived tissue cultures of rice. In Vitro 17:545–552

Schaeffer GW, Sharpe ET (1987) Increased lysine and seed protein in rice plants recovered from calli selected with inhibitory levels of lysine plus threonine and S-(2-aminoethyl)cysteine. Plant Physiol 84:509–515

Schaeffer RP, Yodder OC (1972) Phytotoxins in plant diseases. Academic Press, London, p 251

Schaeffer GW, Sharpe FT, Cregan PB (1984) Variability for improved protein and yield from rice anther culture. Theor Appl Genet 67:383–389

Scheller HV, Huang B, Goldsbrough PB (1986) Selection and characterization of cadmium-tolerant cell lines in *Lycopersicon esculentum* – tomato. Int Congr plant tissue and cell culture. Univ Minnesota, Minneapolis, p 77

Schenck HR, Röbbelen G (1982) Somatic hybrids by fusion of protoplasts from *Brassica oleracea* and *B. campestris* Z Pflanzenzücht 89:278–288

Schwartz D, Dennis E (1986) Transposase activity of the Ac controlling element in maize is regulated by its degree of methylation. Mol Gen Genet 205:476–482

Seitz U, Alfermann W, Reinhard E (1983) Stability of biotransformation capacity in *Digitalis lanata* cell cultures after cryogenic storage. Plant Cell Rep 2:273–276

Semal J (ed) (1986) Somaclonal variations and crop improvement. Nijhoff, Dordrecht

Shah DM, Horsch RB, Klee HJ, Kishore GM, Winter JA, Tumer NE, Hironaka CM, Sanders PR, Gasser CS, Aykent S, Siegel NR, Rogers SG, Fraley RT (1986) Engineering herbicide tolerance in transgenic plants. Science 233:478–481

Shahin EA, Spivey R (1986) A single dominant gene for *Fusarium* wilt resistance in protoplast-derived tomato plants. Theor Appl Genet 73:164–169

Shepard JF (1975) Regeneration of plants from protoplasts of potato virus X-infected tobacco leaves. Virology 66:492–501

Shepard JF, Bidney D, Shahin E (1980) Potato protoplasts in crop improvement. Science 208:17–24

Shepherd SLK, Sohndal MR (1986) Selection for early blight disease resistance in tomato: Use of tissue culture with *Alternaria solani* culture filtrate. 6th Int Congr plant tissue and cell culture. Abstr Univ Minnesota, Minneapolis, p 211

Sibi M (1976) La notion de programme génétique chez les végétaux supérieurs. II. Aspect expérimental: obtention de variants par culture de tissus in vitro sur *Lactuca sativa* L., apparition de vigueur chez les croisements. Ann Amélior Pl 26(4):523–547

Sibi M (1978) Multiplication conforme, non conforme. Sélectionneur Fr 26:9–18

Sibi M (1980) Variants épigéniques et cultures in vitro chez *Lycopersicon esculentum* L. In: Application de la culture in vitro à l'amélioration des Plantes Potagères. CR EUCARPIA CNRA 1980, Versailles, pp 179–185

Sibi M, Biglary M, Demarly Y (1984) Increase in the rate of recombinants in tomato (*Lycopersicon esculentum* L.) after in vitro regeneration. Theor Appl Genet 68:317–321

Sjodin C, Glimelius K (1986) Toxic metabolites produced by *Phoma lingam* and their effects on plant cells. 6th Int Congr plant tissue and cell culture. Abstr Univ Minnesota, Minneapolis, p 230

Skirvin RM, Janick J (1976) Tissue culture-induced variation in scented *Pelargonium* spp. J Am Soc Hort Sci 101(3):281–290

Smith MK, McComb JA (1983) Selection for NaCl tolerance in cell cultures of *Medicago sativa* and recovery of plants from a NaCl-tolerant cell line. Plant Cell Rep 2:126–128

Smith RH, Bhaskaran S, Schertz K (1983) Sorghum plant regeneration from aluminium selection media. Plant Cell Rep 2:129–132

Snappe JW, Sitch LA, Simpson E, Parker BB (1988) Tests for the presence of gametoclonal variation in barley and wheat doubled haploids produced using the *Hordeum bulbosum* system. Theor Appl Genet 75:509–513

Sopory SK, Bajaj YPS (1987) Anther culture and haploid production in potato. In: Bajaj YPS (ed) Biotechnology in agriculture and forestry, vol 3. Potato. Springer, Berlin Heidelberg New York Tokyo, pp 89–105

Spangenberg G, Schweiger H-G (1986) Controlled electrofusion of different types of protoplasts and subprotoplasts including cell reconstitution in *Brassica napus* L. Eur J Cell Biol 41:51–56

Spiegel-Roy P, Ben-Hayyim G (1985) Selection and breeding for salinity tolerance in vitro. Plant Soil 89:243–252

Sree Ramulu K, Dijkhuis P, Roest S (1983) Phenotypic variation and ploidy level of plants regenerated from protoplasts of tetraploid potato (*Solanum tuberosum* L. cv. Bintje). Theor Appl Genet 65:329–338

Sree Ramulu K, Dijkhuis P, Roest S, Bokelmann GS, de Groot B (1984) Early occurrence of genetic instability in protoplast cultures of potato. Plant Sci Lett 36:79–86

Sree Ramulu K, Dijkhuis P, Roest S, Bokelmann GS, De Groot B (1985) Patterns of DNA and chromosome variation during in vitro growth in various genotypes of potato. Plant Sci 41:69–78

Steen P, Keimer B, D'Halluin K, Pedersen HC (1986) Variability in plants of sugar beet (*Beta vulgaris* L.) regenerated from callus, cell-suspension and protoplasts. In: Horn W et al. (eds) Genetic manipulation in plant breeding. De Gruyter, Berlin, pp 633–634

Stelly DM, Altman DW, Kohel RJ, Rangan TS, Commiskey E (1989) Cytogenetic abnormalities of cotton somaclones from callus cultures. Genome (in press)

Sun CS, Wu SC, Wang CC, Chu CC (1979) The deficiency of soluble proteins and plastids robosomal RNA in the albino pollen plantlets of rice. Theor Appl Genet 55:193–197

Sun L-H, She J-M, Lu X-F (1986) In vitro selection of *Xanthomonas oryzae*-resistant mutants in rice. I. Induction of resistant callus and screening regenerated plants. Acta Genet Sin 13:188–193

Sun M (1985) Plants can be patented now. Science 230:303
Sun Z. Zhao C. Zheng K. Qi X. Fu Y (1983) Somaclonal genetics of rice. *Oryza sativa* L. Theor Appl Genet 67:67–73
Sundberg E. Glimelius K (1986) A method for production of interspecific hybrids within Brassicaceae via somatic hybridization. using resynthesis of *Brassica napus* as a model. Plant Sci 43:155–162
Sutter E. Langhans EW (1981) Abnormalities in chrysanthemum regenerated from long term cultures. Ann Bot 48:559–568
Tal M. Katz A (1980) Salt tolerance in the wild relatives of the cultivated tomato. The effect of proline on growth of callus tissue of *Lycopersicon esculentum* and *L. peruvianum* under salt and water stresses. Z Pflanzenphysiol 98:283–288
Tan BH. Stern WR (1985) Induction of somaclonal variation in subterranean clover via in vitro shoot culture. Abstract of results presented to the Pasture Production and Utilization Conference. Leura. N.S.W.. Australia. December 15–19. 1985
Tanda AS. Atwal AS. Bajaj YPS (1980) Reproduction and maintenance of root-knot nematode (*Meloidogyne incognita*) in excised roots and callus cultures of Okra (*Abelmoschus esculentus*). Indian J Exp Biol 18:1340–1342
Terada R. Kyozuka J. Nishibayashi S. Shimamoto S (1987a) Plantlet regeneration from somatic hybrids of rice (*Oryza sativa* L.) and barnyard grass (*Echinochloa oryzicola* Vasing). Mol Gen Genet 210:39–43
Terada R. Yamashita Y. Nishibayashi S. Shimamoto K (1987b) Somatic hybrids between *Brassica oleracea* and *B. campestris*: selection by the use of iodoacetamide inactivation and regeneration ability. Theor Appl Genet 73:379–384
Thanutong P. Furusawa I. Yamamoto M (1983) Resistant tobacco plants from protoplast-derived calluses selected for their resistance to *Pseudomonas* and *Alternatia* toxins. Theor Appl Genet 66:209–215
Tokumasu S. Kato M (1979) Variation of chromosome numbers and essential oil components of plants derived from anther culture of the diploid and the tetraploid in *Pelargonium roseum*. Euphytica 28:329–338
Torello WA. Symington AG. Rufer R (1984) Callus initiation. plant regeneration and evidence of somatic embryogenesis in red fescue. Crop Sci 24:1037–1040
Toriyama K. Hinata K. Kameya T (1987) Production of somatic hybrid plants. "Brassicomoricandia". through protoplast fusion between *Moricandia arvensis* and *Brassica oleracea*. Plant Sci 48:123–128
Toriyama K. Arimoto Y. Uchimiya H. Hinata K (1988) Transgenic rice plants after direct gene transfer into protoplasts. Biotechnology 6:1072–1074
Toyoda H. Hayashi H. Yamamoto K. Hirai T (1984a) Selection of tomato calli resistant to fusaric acid. Ann Phytopathol Soc Jpn 50:538–540
Toyoda H. Tanaka N. Hirai T (1984b) Effects of the culture filtrate of *Fusarium oxysporum* f. sp. lycopersici on tomato callus growth and selection of callus cells resistant to the filtrate. Ann Phytopathol Soc Jpn 50:53–62
Tuberosa R. Phillips RL (1986) Isolation of methotrexate-tolerant cell lines of corn. Maydica 31:215–225
Uhrig H. Wenzel G (1981) *Solanum gourlayi* Hawkes as a source resistance against the white potato cyst nematode *Globodera pallida* Stone. Z Pflanzenzücht 86:148–157
Uhrig H. Wenzel G (1987) Breeding for virus and nematode resistance in potato through microspore culture. In: Bajaj YPS (ed) Biotechnology in agriculture and forestry. vol 3. Potato. Springer. Berlin Heidelberg New York Tokyo. pp 346–357
van Swaaij AC. Jacobsen E (1986) Frost tolerant plants obtained from proline accumulating cell lines. In: Horn W. Jensen CJ. Odenbach W. Schieder O (eds) Proc Int Symp Genetic manipulation in plant breeding. Eucarpia 8–13 Sept 1985. W-Berlin. De Gruvter: Berlin. pp 355–357
van Swaaij AC. Wijbrandi J. Huitema H. Timmerije W (1984) Hydroxyproline-resistant cell lines of dihaploid potato: isolation and partial characterization. Acta Bot Neerl 33:357
van Swaaij AC. Jacobsen E. Feenstra WJ (1985) Effect of cold hardening. wilting and exogenously applied proline on leaf proline content and frost tolerance of several genotypes of *Solanum*. Physiol Plant 64:230–236
van Swaaij AC. Jacobsen E. Kiel JAKW. Feenstra WJ (1986) Selection. characterization and regeneration of hydroxyproline-resistant cell lines of *Solanum tuberosum:* Tolerance to NaCl and freezing stress. Physiol Plant 68:359–366
Waisel Y (1972) Biology of halophytes. Academic Press. New York

Wakasa K, Widholm JM (1982) Regeneration of resistant cells of tobacco and rice to amino acids and amino acid analogs. In: Fujiwara A (ed) Plant tissue culture 1982. Maruzen, Tokyo, pp 455–456
Wakasa K, Widholm JM (1987) A 5-methyltryptophan-resistant rice mutant MTRI, selected in tissue culture. Theor Appl Genet 74:49–54
Walbot V, Cullis CA (1985) Rapid genomic change in higher plants. Annu Rev Plant Physiol 36:367–396
Watanabe K, Midsuda H, Yamada Y (1983) Retention of metabolic and differentiation potential of green *Lavandula vera* callus after freeze preservation. Plant Cell Physiol 24:119–122
Webb KJ (1986) Transformation of forage legumes using *Agrobacterium tumefaciens* Theor Appl Genet 72:53–58
Weber G, Roth EJ, Schweiger HG (1983) Storage of cell suspensions and protoplasts of *Glycine max* (L.) Merr, *Brassica napus* (L.), *Datura innoxia* (Mill.), and *Daucus carota* (L.) by freezing. Z Pflanzenphysiol 109:213–220
Weller SC, Masiunas JB, Gressel J (1987) Biotechnologies of obtaining herbicide tolerance in potato. In: Bajaj YPS (ed) Biotechnology in agriculture and forestry, vol 3. Potato. Springer, Berlin New York Tokyo, pp 280–297
Wenzel G (1985) Strategies in unconventional breeding for disease resistance. Annu Rev Phytopathol 23:149–172
Wenzel G, Uhrig H (1981) Breeding for nematode and virus resistance in potato via anther culture. Theor Appl Genet 59:333–340
Wenzel G, Hoffmann F, Thomas E (1977) Anther culture as a breeding tool in rape. I. Ploidy level and phenotype of androgenetic plants. Z Pflanzenzücht 78:149–155
Widholm JM (1974) Cultured carrot cell mutants: 5-methyltryptophan-resistance trait carried from cell to plant and back. Plant Sci Lett 3:323–330
Widholm JM (1976) Selection and characterization of cultured carrot and tobacco cells resistant to lysine, methionine and proline analogs. Can J Bot 54:1523–1529
Widholm JM (1978) Selection and characterization of a *Daucus carota* L. cell line resistant to four amino acid analogues. J Exp Bot 29:1111–1116
William L, Collin HA (1976) Growth and cytology of celery plants derived from tissue cultures. Ann Bot 40:333–338
Williams EG, Collins GB, Myers JR (1990) Clovers (*Trifolium* species). In: Bajaj YPS (ed) Biotechnology in agriculture and forestry, vol 10. Legumes and oilseed crops I. Springer, Berlin Heidelberg New York Tokyo, pp 242–287
Wong C-K, Woo S-C, Ko S-W (1986) Production of rice plantlets on NaCl-stressed medium and evaluation of their progenies. Bot Bull Acad Sin 27:11–23
Yamabe M, Yamada T (1973) Studies on differentiation in cultured cells. 2. Chromosomes of *Haworthia* callus and of the plants grown from the callus. La Kromosoma 94:2923–2931
Zehr BE, Williams ME, Duncan DR, Widholm JM (1987) Somaclonal variation in the progeny of plants regenerated from callus of seven inbred lines of maize. Can J Bot 68:491–499
Zenk MH (1974) Haploids in physiological and biochemical research. In: Kasha KJ (ed) Haploids in higher plants – advances and potential. Univ Guelph, pp 332–353
Zhang HM, Yang H, Rech EL, Golds TJ, Davis AS, Mulligan BJ, Cocking EC, Davey MR (1988) Transgenic rice plants produced by electroporation-mediated plasmid uptake into protoplasts. Plant Cell Rep 7:379–384
Zheng G-Z, He J-B, Wang S-L (1982) An *Anisodus acutangulus* callus strain of high and stable growth rate selected from the callus after irradiation with Coγ ray. In: Fujiwara A (ed) Plant tissue culture 1982. Maruzen, Tokyo, pp 339–340
Zheng G-Z, He J-B, Wang S-L (1983) Cryopreservation of calli and their suspension culture cells of *Anisodus acutangulus*. Acta Bot Sin 25:512–517

I.2 Chromosome Variation in Plant Tissue Culture

H. OGURA[1]

1 Introduction

When plant tissues and cells are explanted on phytohormone-containing media, they may be induced to proliferate and form an unorganized tissue mass, the callus. Differing from organized intact tissues or organs, however, the callus is known to be genetically variable or unstable. The genetic variability or instability of callus cells is well characterized by the variation of chromosome number. Not only in callus cells, but in callus-derived regenerated plants, it has been shown that many showed chromosome number variation. Some studies suggest that the predominant appearance of eudiploid cells in callus and others showed predominantly eudiploid constitution of the regenerates. Here, the chromosomal constitution of cultured tissues and regenerates, anther-derived regenerates, some protoplast-derived regenerates, somatic hybrids, and some genetically engineered plants is presented. Although there are a considerable number of reports estimating ploidy level by DNA amount per cell by cytophotometry, here reports of usual cytological analyses of chromosome constitution are mainly cited and discussed. For the mechanism of chromosome variation of cultured tissues and regenerates, refer to the reviews by Bayliss (1980), Krikorian et al. (1983) and D'Amato (1985).

2 Chromosome Constitution of Callus and Suspension from Somatic Tissues

In regard to chromosomal constitution of cultured tissues, there have been two different types of results; one type of studies shows the majority of callus or suspension cells as eudiploids or polyploids, such as in carrot (Mitra et al. 1960), tobacco (Cooper et al. 1964), alfalfa (Clement 1964), *Haplopappus* (Kao et al. 1970; Singh et al. 1972), a relative of broad bean, *Vicia hajastana* (Singh et al. 1972), *Haworthia* sp. (Yamabe and Yamada 1973), and lily (Sheridan 1974, 1975). Among these studies, only eudiploid cells were observed, in alfalfa (Clement 1964), *Haplopappus* suspension (Kao et al. 1970), *Haworthia aristata* (Yamabe and Yamada 1973), *Lolium* sp. (Ahloowalia 1975), and *Brachycome lineariloba* (Gould 1982).

[1] Department of Agricultural Sciences, Ishikawa College of Agriculture, Nonoichi, Ishikawa 921, Japan
Present Address: Laboratory of Biology, Faculty of Education, Okayama University, 3-1-1, Tsushima-naka, Okayama City, Okayama 700, Japan

Biotechnology in Agriculture and Forestry, Vol. 11
Somaclonal Variation in Crop Improvement I (ed. by Y.P.S. Bajaj)

Another type emphasizes the appearance of aneuploid cells in maize (Straus 1954), tobacco (Fox 1963; Shimada and Tabata 1967), rye-grass (Norstog et al. 1969), onion (Yamane 1975), *Arabidopsis* (Negrutiu et al. 1975), broad bean (Ono 1979; Ogura 1982), luzula (Inomata 1982), celery (Murata and Orton 1982), and cotton (Bajaj and Gill 1985). The latter type of report seems to be increasing and to have become general in recent years. The chromosomal constitution of callus or suspension of various species is listed in Table 1. Bennici and D'Amato (1978) assumed the cause of chromosome number variation, especially of chromosome number reduction to be due to nuclear fragmentation at the time of callus induction; however, several factors are considered to play an important role in the chromosomal constitution of cultured tissues. They are as follows: (1) nuclear condition of the original explants; (2) composition of the medium, especially kinds and/or concentrations of the plant growth regulators; (3) age of culture; (4) variation due to plant species; and (5) karyotypic changes. The influences of these factors are briefly discussed here.

2.1 Nuclear Conditions of Original Explants

When a plant tissue, an organ, or a portion of an organ is explanted in vitro, the nuclear conditions in the culture partly reflect the nuclear conditions in vivo (in the original explant) and partly result from nuclear changes which may occur at the time of callus induction and/or during further growth of callus in vitro. Murashige and Nakano (1967) reported that in *Nicotiana tabacum* cv. Wisconsin 38, pith tissues of an older region of the stem (15.5–22.5 cm below the apex) contained predominantly tetraploid cells (about 70% of cells examined), and a small portion of diploid (9%), octoploid (16%), and aneuploid cells, while pith sampled from the more apical position (3.5–10.5 cm below the apex) showed no aneuploid dividing cells, but nearly equal percentages of diploids and tetraploids. Shimada and Tabata (1967) cultured pith parenchyma of *N. tabacum* cv. Bright Yellow, and observed nuclear conditions of dedifferentiation processes. The first mitotic division occurred after 5–7 days of culture, when chromosome number already varied from 40 to 215. 2n and 4n cells comprised 12.5% and 17.5% of the cells examined, respectively, the majority of cells being aneuploids of various ploidy levels (2n–10n). They assumed that the wide range of variation in chromosome number indicated pre-existing variation in situ in the original pith tissue, namely that the tobacco pith tissue in vivo consisted of cells with diploid, aneuploid, and polyploid chromosome number. By DNA cytophotometric analyses, Marchetti et al. (1975) investigated the nuclear conditions of tobacco pith segments and observed certain percentages of "aneuploid" DNA content cells. Heinz et al. (1969) cytologically investigated leaf tissue of five clones of sugarcane and found that one clone, H50–7209, showed variation in chromosome number over a range of 108–128, whereas four other clones exhibited constant chromosome numbers. Cytological observation of H50–7209 callus derived from stem parenchyma revealed the widest range of variation in chromosome number (2n = 51 to more than 300), whilst calli of other clones showed a comparatively narrow range. Wheeler et al. (1985) observed nine aneuploids out of 63 plantlets regenerated not from callus, but from cultured excised

Table 1. Chromosomal constitution of callus and suspension derived from somatic tissues

Species, chromosome no., mode of culture	Basal medium	Chromosome constitution of cultured tissues	Organ regeneration	Reference
Allium cepa var. *proliferum* (2n = 16), callus	B5	16	Yes	Fridborg (1971)
Arabidopsis thaliana (2n = 10), callus	Modified B5 (callus induction and subculture) Modified Gresshof and Doy (organ regeneration)	10–60 Polyploid, aneuploid (mode at 10)	Yes 10, 20	Negrutiu et al. (1975)
Crepis capillaris (2n = 6), callus	MS	3, 4, 5, 6, 7	Yes	Sacristán and Wendt-Gallitelli (1973)
Datura innoxia (2n = 12), callus	Modified B5	12 → 48	Yes, 16–48 (mode at 24)	Furner et al. (1978)
Daucus carrota (2n = 18) Suspension	Modified White	9, 18, 36, 72	Yes	Mitra et al. (1960)
Haplopappus gracilis (2n = 4), suspension from hypocotyl segments	B5	4 → 16 (mode at 4)	Not reported	Singh and Harvey (1975)
Haworthia aristata (2n = 14), callus	MS	14	Yes, 14	Yamabe and Yamada (1973)
H. setata (2n = 14), callus	MS	2x, 3x, 4x, 6x, 4x–1 = 27	Yes 2x, 4x–1, 4x	Yamabe and Yamada (1973)
Hordeum vulgare (2n = 2x = 14)				
Callus	B5	26–51		
Suspension		12–44	Not reported	Orton (1980)
H. jubatum (2n = 4x = 28)				
Callus	B5	20–30		
Suspension		19–83	Not reported	Orton (1980)
H. vulgare × *H. jubatum* (2n = 3x = 21)				

Table 1. *(Continued)*

Species, chromosome no., mode of culture	Basal medium	Chromosome constitution of cultured tissues	Organ regeneration	Reference
Callus (6 months)	B5	ca. 13–21	Yes, 13–22	Orton (1980)
Callus (15 months)		7–23		Orton and Steidl (1980)
Suspension (11 months)		12–ca. 130		
Lilium longiflorum (2n = 24), callus	RM-1964	24, 24 + isochromosome 24, including 1 isochr.	Yes, 24, 48 24 + 1 isochromosome	Sheridan (1974, 1975)
Lolium perenne (2n = 6), callus	White	18–50 (mode at 21 and 22)	Not reported	Norstog et al. (1969)
Luzula elegans (2n = 6), callus	RM-1964	CH medium 5–13 (mode at 6) YE medium 3–12 (mode at 6)	Not reported	Inomata (1982)
Medicago sativa (2n = 32), callus	Modified Heller	32	Not reported	Clement (1964)
Nicotiana tabacum (2n = 48)				
cv. Wisconsin 38, callus				
KX – 1 (auxin, kinetin-heterotrophic)	Fox	40 → 221	Not reported	Fox (1963)
X – 1 (auxin-heterotrophic)		122–275		
O – 1 (auxin, kinetin-autotrophic)		40–190		
cv. White Burley, callus				
Strain M223	Modified White	2n, 4n, 8n (mode at 2n)	Not reported	Cooper et al. (1964)
Strain M471		2n, 4n, 8n (mode at 2n)		
cv. Bright Yellow				
Pith-derived callus (1.5 year)	RM-1964	38–153	Not reported	Shimada (1971)
Root-tip derived callus (3–7 months)		45–96 (mode at 48)		
Root-tip derived callus (2 years)		46–198		
Oryza sativa (2n = 24)				
Callus	RM-1964	Wide variation	Yes, 24	Nishi et al. (1968)

Pelargonium zonale, haploid (n = x = 9), callus	RM-1964	9, 18, 36	Yes, 9, 18	Bennici (1974)
Prunus amygdalus (2n = 16) Callus	Half-strength modified MS	9, 15, 16, 24, 28, 32	Yes, 16	Mehra and Mehra (1974)
Prunus pomila (3n = 51) Callus from endosperm	MS	17–206 (mode at 37–46)	Not reported	Mu and Liu (1978)
Triticum aestivum (2n = 42) Callus	White or Risser and White	26–84 (mode at 42)	Yes, 42	Shimada et al. (1969)
cv. Norin 60, callus	White or MS	Wide variation	Yes, 42	Adachi and Katayama (1969)
cv. Maris Huntsman, Kleiber Sicco, Siete cerros	Half-strength MS	Regenerative callus 42 Nonregenerative callus 35–46	Yes, 42	Ahloowalia (1982)
Vicia faba (2n = 12) Suspension	Bonner and Devirian	Aneuploid, diploid Tetraploid-octoploid	Not reported	Venketeswaran (1963)
cv. Sanuki-Nagasaya Callus	Modified RM-1964	12–48 (mode at 18)	Not reported	Yamane (1975)
cv. Wase-soramame Callus	RM-1964	10–48 (mode at 12)	Not reported	Ogura (1982)
Vicia hajastana (2n = 10) Suspension	B5	Hypodiploid-hypertetraploid (mode at 10)	Not reported	Singh et al. (1972)

stem of potato, and suggested that this may be due to explant source or medium composition. These studies suggest that chromosome number variation in cultured tissues partly reflects the nuclear conditions of the original explants.

2.2 Effects of Medium Composition and/or Plant Growth Regulators on Chromosomal Constitution

Matthysse and Torrey (1967) reported that in root-tip meristem callus of *Pisum sativum* cv. Alaska, the division frequency of 4n cells was three times greater than of 2n cells on the addition of kinetin (1 mg/l) to the S-2 medium. Niizeki (1974) obtained haploid tobacco (cv. Wisconsin 38) plants derived from anther culture, and after eight transfers of subculture, placed the callus in four different kinds of media, as shown in Table 2. In medium A, which contains no growth regulators, only n and 2n cells are observed, whereas in media B, C, D, 4n and aneuploid cells can to a greater or lesser extent be recognized. Auxins or cytokinins, which are known to promote cell division, also caused chromosomal abnormalities. Some studies suggest that auxins and/or cytokinins induced chromosomal abnormalities, probably leading to chromosome number variation in cultured tissues (Bayliss 1973; Ogura 1982). Three representative types of chromosomal abnormalities observed in anaphase cells of *Vicia faba* callus and root tips, probably involved in chromosome number variation, are seen in Fig. 1.

On the other hand, Doležel and Novák (1984) tested the mutagenicity of the plant tissue culture medium and its hormonal composition, using the *Tradescantia* mutagen-sensitive system heterozygous for stamen-hair color (blue/pink, blue being dominant). Their results suggested that three kinds of the most frequently used growth regulators, IAA, kin, and 2,4-D, had no mutagenic effect and so had probably no direct effect on the induction of mutagens in plant tissue cultures, despite several studies showing that auxins and cytokinins caused chromosome

Table 2. Chromosome constitution changes of "haploid" tobacco callus by four different kinds of media. (Niizeki 1974)

Medium composition	Days in culture	No. cells examined	Chromosome no. (%)			
			24 (n)	48 (2n)	96 (4n)	Aneuploid
A. Miller basal medium	10	25	20 (80.0)	5 (20.0)		
	30	25	16 (64.0)	9 (36.0)		
B. Basal medium +4 mg/l kinetin	10	29	25 (86.2)	4 (13.8)		
	30	29	11 (39.9)	9 (31.0)	4 (13.8)	5 (17.2)
C. Basal medium +2 mg/l IAA	10	30	18 (60.0)	9 (30.0)	0	3 (10.0)
	30	28	6 (21.4)	7 (25.0)	5 (17.9)	10 (35.8)
D. Basal medium +2 mg/l kinetin	10	27	15 (55.6)	10 (37.0)	0	2 (7.4)
	30	30	3 (10.0)	11 (36.7)	6 (20.0)	10 (33.3)

+2 mg/l IAA

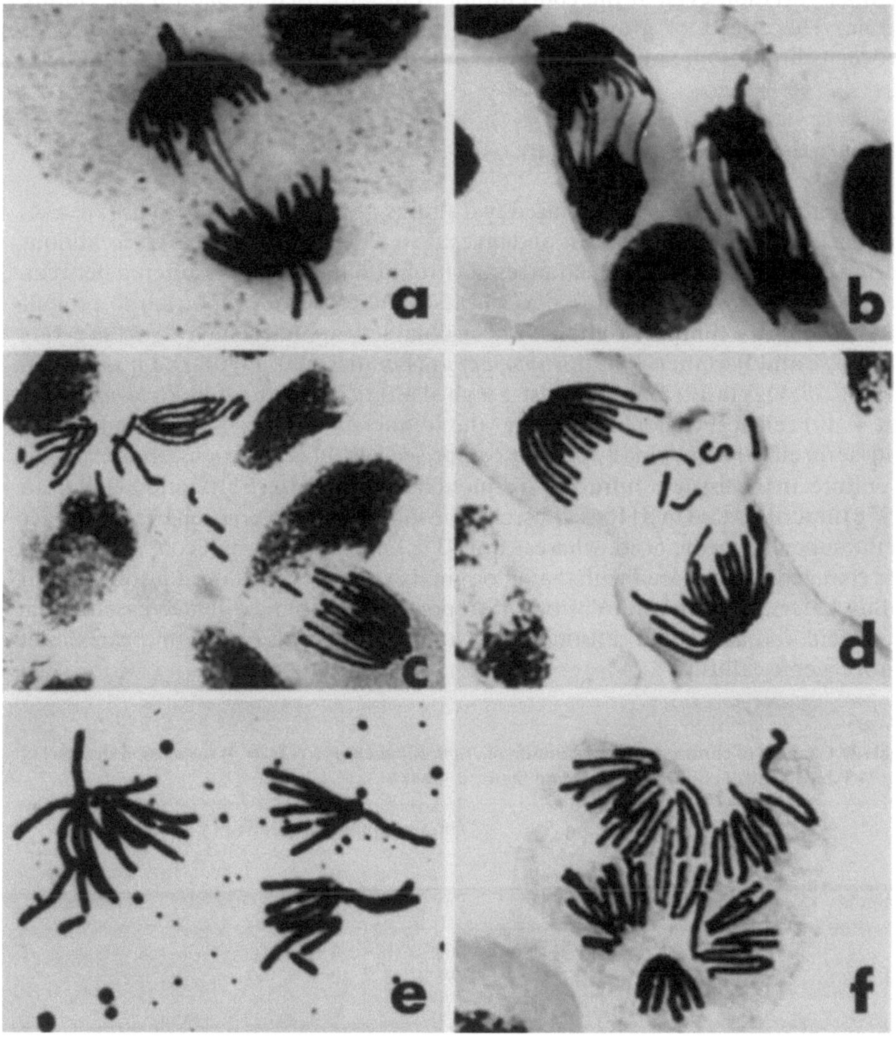

Fig. 1a-f. Three representative types of chromosomal abnormalities observed in anaphase cells of callus and growth regulator-treated root tips of *Vicia faba*: (**a, b**) chromosome bridges. (**c, d**) lagging chromosomes or laggards. (**e, f**) multipolar spindles. (Ogura 1982)

number variation even at the concentrations usually used in plant tissue culture media. The effects of growth regulators on cultured tissues and relationships between chromosome number variation and mutation should be further studied.

2.3 Variation Due to Age of Culture

Blakely and Steward (1964) cultured two strains of *Haplopappus gracilis* (2n = 4) suspension for more than 2 years and investigated the change in nuclear conditions during culture (Table 3). Appearances of chromosomal variation differed between two strains; however, chromosomal variation increased in proportion to prolongation of culture. Similar results have been obtained in studies, such as in tree peony (Demoise and Partanen 1969), *Haplopappus* (Bennici et al. 1970), and broad bean (Ono 1979; Ogura 1982). In contrast, Singh et al. (1972) reported in *Vicia hajastana* (2n = 10) cell suspension culture that the frequency of 2n cells increased during long-term culture, and finally 2n cells comprised 92% of cell suspension at 295 days of culture. In the tissue culture of two diploid strains of maize, S1104 and S-3, Balzan (1978) described that in S1104 callus, chromosome number variation became larger as the subculture advanced, whereas in H-3 callus, chromosomal constitution was less changed as compared with that of original explant, even at the 8th subcultured callus. Later, Swedlund and Vasil (1985) reported that diploid cells decreased, while tetraploid and aneuploid cells increased after 6 months of culture in pearl millet embryogenic callus.

Table 3. Changes of chromosome constitution of *Haplopappus gracilis* (2n = 4) suspension strains, DS and WS due to age of culture. (Blakely and Steward 1964)

Strain	Months in culture	No. of cells observed	Chromosome constitution (%)				
			2n	4n	8n	aneuploid	%Abnormal anaphases
	8	46	100	0	0	0	0
DS	12	64	98	2	0	0	0
	25	52	77	19	2	2	5
	10	113	59	41	0	0	0
WS	12	140	80	18	2	0	0
	25	47	0	0	0	100	35

2.4 Variation Due to Plant Species

Cells of many plant species become aneuploid or polyploid during culture. The degree of aneuploidy or polyploidy seems to depend on the species cultured. Shimada (1971) induced calli of three plant species, to investigate the relations between chromosomal variation and species. Of these, tobacco callus exhibited the most variable chromosome number. Two wheat species, *Triticum dicoccum*

(2n = 28) and *T. aestivum* (2n = 42), calli were comparatively stable in chromosome constitution. Zosimovich and Kunakh (1975) studied chromosomal aberrations of tissue cultures of *Nicotiana tabacum, Haplopappus gracilis,* and *Crepis capillaris* and reported that a high frequency of aberrant anaphase was found in tobacco tissue cultures, whereas aberrant anaphases of *H. gracilis* and *C. capillaris* were comparatively low. In the callus of cotton interspecific hybrids, Bajaj and Gill (1985) observed a wide range of chromosome number variation, ranging from the haploid to hexaploid level.

In calli of carrot (Mitra et al. 1960), pea (Torrey 1967), tree peony (Demoise and Partanen 1969), *Haworthia aristata* (Yamabe and Yamada 1973), and lily (Sheridan 1974, 1975), the majority of cells were eudiploid or polyploid; in contrast, a wide range of chromosomal variation, including various aneuploid cells, was encountered in calli of tobacco (Fox 1963; Nishiyama and Taira 1966; Shimada and Tabata 1967; Murashige and Nakano 1967; Sacristán and Melchers 1969), maize (Straus 1954), rice (Nishi et al. 1968), sugarcane (Heinz et al. 1969), rye-grass (Norstog et al. 1969), apple (Mu and Liu 1978), *Datura innoxia* (Furner et al. 1978), barley (Orton 1980), broad bean (Ono 1979; Ogura 1982), celery (Browers and Orton 1982), and cotton (Bajaj and Gill 1985).

2.5 Karyotypic Changes in Cultured Tissues

Chromosome structural changes or karyotypic changes of different types (deficiencies, translocations, isochromosomes, dicentric chromosomes) are not rare in cultured tissues; e.g., in carrot (Mitra et al. 1960), *Crepis capillaris* (Sacristán 1971, 1975; Sacristán and Wendt-Gallitelli 1973) and *Vicia faba* (Papeš et al. 1978). By using Giemsa banding techniques in suspension culture of *Brachycome dichromosomatica* (2n = 4), Gould (1982) reported that gross rearrangement of the karyotype was found in cultured cells, although chromosome number in culture was the same as the original explant tissue, and suggested that chromosome number appears to be a poor indicator of chromosomal stability in plant tissue culture. Ogihara (1982) showed that the most frequent karyotype of *Haworthia setata* callus was 7L + 1M + 6S (2n = 14), as compared with control diploid karyotype, 8L + 6S. Murata and Orton (1982, 1983, 1984) observed chromosome fusions in celery suspension culture. They reported that chromosome fusion seems to be a primary source of chromosomal aberrations such as translocation, duplication, and deficiency, which are probably involved in chromosome number variation of celery suspension culture. Doležel and Novák (1986) recently showed that sister chromatid exchanges (SCE's) in garlic (*Allium sativum* L.) callus cells was higher than those in meristem root-tip cells. High SCE's in garlic callus cells may correlate with the genetic instability of garlic callus.

By examining the karyotype of a diploid cell (2n = 14) of 6-month-old embryogenic callus of pearl millet, *Pennisetum americanum*, Swedlund and Vasil (1985) found that the karyotype was not different from that of a diploid cell in a control plant. Banerjee-Chattopadhyay et al. (1985) observed altered karyotypes in *Lycopersicon peruvianum* root-tip cells transformed by *Agrobacterium rhizogenes*. The cytology of genetically engineered plants is presented in Section 6.

Regarding karyotypic changes of regenerated plants from callus, some studies have shown karyotypic changes without alteration of chromosome number, such as *Hemerocallis* sp. (Krikorian et al. 1981), celery (Orton 1983), and *Triticale* (Armstrong et al. 1983; Lapitan et al. 1984).

3 Chromosomal Constitution of Plants or Shoots Regenerated from Callus or Cell Suspension

The chromosomal constitution of regenerates seems to be as highly dependent on species, kind of original explant, or conditions of regeneration from callus, etc., as is that of cultured tissues. Table 4 shows some examples of chromosomal constitution of regenerates, except those showing chromosome number chimerism or mixoploidy. The different types of regenerates may be classified into three groups: diploid, polyploid and aneuploid, mixoploid or chromosome number chimera.

3.1 Selective Advantage of Diploid Cells

In some species, e.g., carrot, rice, or wheat, despite the occurrence in culture of different ploidy levels and/or aneuploidy, only or mostly eudiploid plants have been regenerated (see D'Amato 1977, 1985). As studies of plant regeneration from callus by somatic embryogenesis have increased in recent years, examples of eudiploid regenerates have also increased. Vasil (1983) reported that regenerates by somatic embryogenesis were eudiploid and chromosomally stable and had no noticeable phenotypic changes. Despite a number of plant regeneration studies by somatic embryogenesis, few studies on cytological analyses of the regenerates have been made. Among such studies, eudiploid regenerates were reported in rice (Wernicke et al. 1981), proso millet (Rangan and Vasil 1983), sorghum (Dunstan et al. 1978, 1979), corn (Lu et al. 1982), wheat (Ozias-Akins and Vasil 1982), pearl millet (Vasil and Vasil 1981), hybrid triploid napiergrass (Rajasekaran et al. 1986), etc.

Karp and Maddock (1984), however, described how the plants regenerated through somatic embryogenesis from cultured immature embryos of four wheat cultivars (2n = 42) exhibited chromosome number variation. In total, 29% of the 192 plants examined were aneuploid, with a range in chromosome numbers of 38–45. Vasil (1985) explained the appearance of chromosomally variable regenerates by somatic embryogenesis in that the original explants retained chromosome number variation before the culture starts or plant regeneration occurs via both embryogenesis and organogenesis in the same culture. More detailed cytological and histological studies seem to be necessary on this topic.

Ogura et al. (1987) recently examined 126 regenerates derived from embryo protoplasts of four rice cultivars, regenerated partly through somatic embryogenesis. Twelve chromosome variants (9.5% of the total regenerates) were found, of which ten were tetraploids, one triploid, and one aneuploid (trisomic, 2n = 25). The remaining 114 plants were diploid, but 12 distinct somaclonal variants

Table 4. Examples of chromosome constitution of regenerated plants or shoots, excluding those exhibited chromosome number chimerism

Species, chromosome no. and mode of culture	Basal medium	Explant source	Chromosomal constitution of regenerates	Reference
Apium graveolens (2n = 18, 22) cv. Lathom Blanching, callus	MS	Inner petiole	18, 22, 42, 44	Williams and Collin (1976)
" F_1 hybrid callus	MS	Petiole segment	19–42 (mode at 21, 22)	Orton (1983)
Asclepias curassavica (2n = 22) callus	B5, MS	Excised node	18, 20, 22, 23	Pramanik and Datta (1986)
Asparagus officinalis (2n = 20) callus, suspension	RM-1964	Hypocotyl	20, 40	Wilmar and Hellendoorn (1968)
Avena sativa (2n = 42) cv. Lodi and Tippecanoe, callus	B5	Immature embryo	41, 42, 43 (mode at 42)	McCoy et al. (1982)
Hordeum vulgare (2n = 14) cv. Himalaya and Mari, callus	Modified MS and Cheng (1972)	Apical meristem	14	Cheng and Smith (1975)
Lolium multiflorum (2n = 14) × *Festuca arundinacea* (2n = 42) F_1 (2n = 28), callus	Kasperbauer and Collins (1972)	Internodal stem, peduncle	28, 56	Kasperbauer et al. (1979)
Medicago media cv. Heinrichs and Beaver, callus	Blaydes, SH	Root, hypocotyl	Aneuploids 64% (Heinrichs) " 21% (Beaver)	Natarajan and Walton (1987)
Nicotiana tabacum (2n = 48) cv. Bright Yellow, callus	RM-1964	Stem pith	48, 96, 97	Tabata et al. (1968)
" Su/Su mutant, suspension	1/2MS	Leaf	48	Evans and Gamborg (1982)
Nicotiana alata (2n = 18), callus	RM-1964	Stem pith	18, 19	Nishiyama and Taira (1966)
Nicotiana glauca (2n = 24) Suspension	1/2MS	Leaf	24, 48	Evans and Gamborg (1982)
Nicotiana otophora (2n = 24) *N. sylvestris* (2n = 24)	MS	Leaf protoplast	24 24	Banks and Evans (1976)

Table 4. *(Continued)*

Species, chromosome no. and mode of culture	Basal medium	Explant source	Chromosomal constitution of regenerates	Reference
N. tabacum (2n = 48), callus			48	
Oryza sativa (2n = 24) cv. Kyoto Asahi, callus	RM-1964	Ovary Embryo or shoot nodule Anther	2n, 4n (mostly 4n) 2n n, 2n, 3n, 4n, 5n	Nishi and Mitsuoka (1969)
" cv. Hua Han Zao and F_1 hybrids, callus	N6	Anther	Euploids 89.8% Aneuploids 10.2%	Chu et al. (1985)
Panicum maximum (2n = 32), callus	MS	Immature inflorescence	32	Hanna et al. (1984)
Panicum miliaceum (2n = 14) cv. Abarr, callus	RM-1964	Root, mesocotyl, leaf, stem and immature embryo	14	Heyser and Nabors (1982)
Pennisetum americanum (2n = 14), callus	MS	Immature inflorescence	13–46 (mode at 14)	Swedlund and Vasil (1985)
Solanum melongena (2n = 24), strain LVR, callus	RM-1964	Young leaf	22.5 ± 5.1	Alicchio et al. (1984)
Triticale (2n = 6x = 42) cv. Beagle, Rosner and Welsh, callus	B5, MS, Kao (1977)	Immature embryo	41, 42, 43	Nakamura and Keller (1982)
Triticum aestivum (2n = 42), callus	MS	Young inflorescence, Immature embryo	42	Ozias-Akins and Vasil (1982)
" Four cultivars, callus	MS	"	38–45 (mode at 42)	Karp and Maddock (1984)
Zea mays (2n = 20), callus	MS	Immature embryo	19–21 (mode at 20)	McCoy and Phillips (1982)

were found among the 114 diploid plants. These results can be interpreted that (1) plant regeneration occurred by both somatic embryogenesis and organogenesis in the same culture or (2) plant regeneration occurred only by somatic embryogenesis. but chromosomal and somaclonal variation arose.

3.2 Polyploidy and Aneuploidy

Polyploid and aneuploid regenerates are commonly known in a number of species. Aneuploid regenerates seem to be comparatively frequent in solanaceous species. Murashige and Nakano (1967) reported that 32 leafy shoots were regenerated from 45 callus strains derived from single isolated cells of tobacco. of which only one was diploid and the others tetraploid or aneuploid. Aneuploid leafy shoots were weak in vigor and did develop to plantlets. whereas tetraploid leafy shoots developed to be tetraploid plantlets. Nishiyama and Taira (1966) obtained three plants regenerated from pith-derived calli of *Nicotiana alata* ($2n = 18$). of which one plant was found to be trisomic. $2n + 1$. Chen and Lin (1976) observed 23 tetraploids among 165 regenerated plants of five *Japonica* varieties of rice. Bajaj and Bidani (1980. 1986) found polyploids and aneuploids in the regenerates derived from rice embryo callus. Among the regenerates derived from immature embryos of oat plants. McCoy et al. (1982) reported a number of cytogenetically abnormal plants by detailed meiotic analysis. The most common cytogenetic alteration was chromosome breakage. which resulted in partial chromosome loss. Johnson et al. (1984) compared two lines of protoplast-derived alfalfa plants with their parent clones. and found that only 30 and 45% of protoclones. respectively. had a parental chromosome number of $2n = 32$. Even in the regenerates derived not from callus. but from cultured excised stem. rachis. or leaf pieces of 13 potato cultivars. Wheeler et al. (1985) obtained nine aneuploids out of the total 63 regenerates. They reported that this may be due to explant source or medium composition. Recently. Natarajan and Walton (1987) reported that 64% of the regenerates from callus of *Medicago media* cv. Heinrichs were aneuploids.

There are other instances reporting polyploid or aneuploid regenerates obtained in tobacco (Sacristán and Melchers 1969; Tabata et al. 1968; Novák and Vyskot 1975). *Haworthia setata* (Yamabe and Yamada 1973; Ogihara 1981). *Arabidopsis thaliana* (Negrutiu et al. 1975; Kearthley and Scholl 1983). Guinea grass (Bajaj et al. 1981a). common millet (Bajaj et al. 1981a). peanuts (Bajaj et al. 1981b). rice (Bajaj et al. 1980). triticale (Nakamura and Keller 1982). celery (Williams and Collin 1976; Orton 1983). maize (Edallo et al. 1981; McCoy and Phillips 1982). winter rape (Newell et al. 1984). a forage legume. *Lotus corniculatus* (Damiani et al. 1985). a medicinal milkweed. *Asclepias curassavica* (Pramanik and Datta 1986). etc.

Chromosomal variations observed in the regenerates constitute an important part of somaclonal variation. For the subject of somaclonal variation. refer to J. Semal (1986). and see Chap. I.1 this Vol.

3.3 Chromosome Number Chimerism

Chromosome number chimerism or mixoploidy in regenerates from tissue cultures is known in monocot, sugar cane (Heinz and Mee 1971; Liu and Chen 1976), barley (Mix et al. 1978), durum wheat (Bennici and D'Amato 1978; Lupi et al. 1981), lily (Bennici 1979), etc. and in dicot, tobacco (Sacristán and Melchers 1969; Ogura 1975, 1976), marrow stem kale (Horák 1972), *Lycopersicon peruvianum* (Ancora and Sree Ramulu 1981), protoplast-derived *Datura innoxia* (Furner et al. 1978), and *Haworthia* (Ogihara 1981). Examples of chromosome number chimerism observed in regenerated shoots or plants are listed in Table 5. Sacristán and Melchers (1969) observed an extensive chromosome number chimerism in a few of the hundreds of tobacco plants regenerated from long-established callus. Heinz and Mee (1971) found a wide range of chromosomal chimerism in *Saccharum* species hybrids, although in this case, original explants of H50–7209 clone showed a wide range of chromosomal chimerism. A considerable number of mixoploid regenerates were also found in callus of Formosa sugarcane varieties (Liu and Chen 1976). Ancora and Sree Ramulu (1981) reported diplo-tetraploid chimerism in regenerates from stem-internode-derived callus of *Lycopersicon peruvianum.* Diplo-tetraploid chimerism has been observed in 2 out of 17 regenerated plants from *Brassica oleracea* callus (Horák 1972). Ogura (1976) observed extensive chromosomal chimerism in callus-derived regenerates of *Nicotiana tabacum* cv. Wisconsin 38. He also reported that chromosomal chimerism was transmitted to first and second selfed progeny plants. The inheritance pattern of chromosomal chimerism is discussed in the following section. In plants regenerated from anther-derived callus of *Hordeum vulgare,* Mix et al. (1978) reported extensive chromosomal chimerism in root-tip cells. Orton (1980) also described chromosome number chimerism in barley interspecific hybrids. Among 102 regenerated plants from seedling explants of four lines of winter rape, Newell et al. (1984) observed that 23 were chromosomally chimeric in root-tip mitosis; however, these plants were identified as not being chimeric by meiotic analysis. In this case, observation of mitosis and meiosis was not parallel. Doležel et al. (1986) analyzed 264 regenerated plants from garlic callus, of which 31% were mixoploid. Bennici and D'Amato (1978) investigated not only root-tip cells but shoot-tip cells of regenerates from calli of five *Triticum durum* cultivars, and found that most of the regenerates were chromosomally chimeric. Their following studies (Lupi et al. 1981) revealed that the majority of spikes of regenerated durum wheat plants are aneusomatic, being composed of eudiploid and aneuploid cells in different proportions. This suggests that chromosomal chimerism was maintained during plant development and was still present in young spikes. However, despite the very broad variation in levels of chromosome number variation among spikes, there is a tendency in advanced stages of spike development toward a clear reduction in, and even the disappearance of the aneuploid mitoses. This would imply that a mechanism operates which favors the selective advantage of diploid over aneuploid cells (diplontic selection), in contrast to the case of tobacco (Ogura 1976).

The following four possibilities can be considered as the cause of chromosomal chimerism: (1) regeneration occurred from single cultured cell (Backs-Hüseman and Reinert 1970). This process should give rise to only chromosomally uniform

Table 5. Examples of chromosome number chimerism of plants or shoots regenerated from callus or suspension cultures.

Species, chromosome number and type of culture	Chromosome number in culture	Chromosome constitution of regenerates	No. chimeras / All regenerates	(%)	Reference
Allium sativum (2n = 16), callus	11–56	2n, 4n, aneuploids, mixoploids		(31.0)	Doležel et al. (1986)
Brassica oleracea (2n = 18), stem-pith callus	Not reported	2n, 4n	2/17	(11.8)	Horák (1972)
Cichorium intybus (2n = 18), callus	Not reported	9, 10–17, 18, 19–35	Majority		Caffaro et al. (1982)
Datura innoxia (2n = 12), protoplast-derived suspension and callus	12–48	16, 24, 48, ca. 60	1/15	(6.7)	Furner et al. (1978)
Hordeum vulgare (2n = 14), anther-derived callus	Not reported	2n, 4n	11/12	(91.7)	Mix et al. (1978)
(*Hordeum vulgare* × *H. jubatum*)F_1 (2n = 3x = 21), callus	7–24 (mode at 19, 20) 8–24	13–20 (mode at 17, 20) 13–22	Not reported "		Orton (1980) Orton and Steidl (1980)
Lilium longiflorum (2n = 24), callus	Not reported	7–51	17/20	(85.0)	Bennici (1979)
Lycopersicon esculentum (2n = 24), callus	Not reported	20–26	Majority		Zagorska et al. (1986)
Lycopersicon peruvianum (2n = 36), callus	Not reported	2n, 4n	1/44	(2.3)	Ancora and Sree Ramulu (1981)
"	Not reported	36–48	Majority		Zagorska et al. (1986)
Nicotiana tabacum (2n = 48) cv. White Burley Tc: auxin autotrophic callus	35–37, 47–53, 68–70, 71–76, 80–82 (mode at 72)	60–70 (mode at 63)	4/138	(2.9)	Sacristán and Melchers (1969)
Nicotiana tabacum (2n = 48) cv. Trapezond, callus strain I	n–4n	14–60	Not reported		Zagorska et al. (1974)

Table 5. *(Continued)*

Species, chromosome number and type of culture	Chromosome number in culture	Chromosome constitution of regenerates	No. chimeras / All regenerates	(%)	Reference
cv. Wisconsin 38					
pith-derived callus	Not reported	9–ca. 120	10/10	(100)	Ogura (1976)
1st progeny plants					
No. 1	"	19–92	4/5	(80)	Ogura (1976)
No. 2	"	16–92	2/2	(100)	
2nd progeny plants					
No. 1	"	48 (stabilized)	0/5	(0)	
No. 2	"	24–76	4/4	(100)	
Saccharum species hybrid (2n = 108–128), callus	71–90, 101–110, 154–160, more than 300	17–118, 94–120	36/37	(97.3)	Heinz and Mee (1971)
Saccharum Formosa cultivars					
F 146 (2n = 110)		104–180	6/7	(85.7)	
F 156 (2n = 114)	Not reported	86–126	6/6	(100)	Liu and Chen (1976)
F 164 (2n = 108)		88–108	5/8	(62.5)	
Triticum durum (2n = 28), callus	Not reported	6 → 56	22/23	(95.7)	Bennici and D'Amato (1978)
T. durum, callus	Not reported	15–56	12/14	(85.7)	Lupi et al. (1981)

plants, unless cell contains gene(s) for chromosomal chimerism (Ogura 1978a), (2) regeneration arose from more than one cell – multicellular origin (Mix et al. 1978; Bennici 1979, 1986), (3) new chromosomal variability is generated after organ regeneration from cultured tissues, producing mixoploid tissues in conjunction with cases (1) and (2), which finally resulted in chromosomal chimerism (Orton 1980), and (4) the complex genetic nature of the primary explant itself, where asynchronous division of multinucleate cells frequently occurs (Heinz and Mee 1971). In this case, chromosome number chimerism is commonly observed in original explants, cultured tissues, and culture-derived regenerates. Even in the progeny plants of primary regenerates, chromosome number chimerism should be encountered, although no information on progeny was available in their studies. Among these, case 4 seems to be restricted in special species and cases 1 and 3 appear to be less common than case 2. By comparing the extent of chromosomal variability and of karyotypic alterations between callus and callus-derived regenerates in *Haworthia setata*, Ogihara (1981) considered case 2 to be most probable. Based on cytological and histological observations of both root-tip cells and shoot apex cells of regenerated durum wheat plants, Bennici (1986) stressed the multicellular origin of the root apex and chromosomal chimerism of the pericycle tissue. However, all or some of these cases may overlap as the cause of chromosomal variability in regenerates. Ogura (1978a) reported a case implying that the chromosome number variation of regenerates is under genetic control. This is discussed in the following section.

3.4 Genetic Control of Chromosomal Chimerism

As mentioned earlier, there are a number of reports on the appearance of chromosomally chimeric regenerates from cultured tissues, and this type of studies is still increasing. However, this trait seems to be undesirable for the plant itself, in general. In tobacco, after selfing callus-derived regenerates which exhibited an extensive chromosomal chimerism, most of the offspring stabilized to have a certain chromosome number (Ogura 1977, 1978b). Nevertheless, a plant regenerated from tobacco callus exhibited variation in its chromosome number, even in the second selfed generation. Since the chromosomal chimerism was transmitted to progeny plants, the chimerism was considered to be genetic (Ogura 1976). To investigate the mode of inheritance of the chimerism, a progeny plant of chromosomally chimeric lineage (designated No. 2) was reciprocally crossed with a normal (not through culture) Wisconsin 38 tobacco plant. These F_1's were found to be chromosomally chimeric. They were further selfed or backcrossed with each of their parents. Segregation data of chromosomally chimeric vs. stable plants in the F_2 and B_1 generations are presented in Table 6. Segregation data of F_2s and $(Ch \times S)F_1 \times S$ fits for a 1:1 segregation ratio, respectively as expected. However, that of $(S \times Ch)F_1 \times S$ does not fit for a 1:1 segregation ratio. The segregation data imply that the chromosomal chimerism is controlled by a single Mendelian gene (and some modifiers), probably induced by mutation in tissue culture processes. Supposing that a single (dominant) gene, *Ch* controls the chimeric character, all the B_1 plants backcrossed with the plants having *Ch* should be chimeric, although actually some stable segregants were found in two B_1 combinations. Since unequal separation of

Table 6. Segregation of chromosomally chimeric vs. stable plants in the B_1 and F_2 generations. (Ogura 1978a)

Combination	No. of progenies Chimeric	Stable	Total	Ratio (Ch : S)	x^2-Value
(S×Ch)F_1×S	65	41	106	1.59 : 1	5.43[a] (1 : 1)
(Ch×S) F_1×S	56	38	94	1.49 : 1	3.44 (1 : 1)
Subtotal	121	79	200	1.53 : 1	8.82[b] (1 : 1)
(S×Ch)F_1×Ch	40	7	47	5.71 : 1	
(Ch×S) F_1×Ch	10	5	15	2.00 : 1	
Subtotal	50	12	62	4.17 : 1	
(S×Ch)F_2	31	10	41	3.10 : 1	0.01 (3 : 1)
(Ch×S) F_2	20	6	26	3.33 : 1	0.05 (3 : 1)
Subtotal	51	16	67	3.20 : 1	0.04 (3 : 1)

[a] and [b]: Significant at the 5% and 1% levels, respectively. Ch : chromosomally chimeric plants. S : chromosomally stable plants.

chromosomes does occur in the first and second anaphase of meiosis, probably due to mispairing or other causes occurred in meiosis, whether or not the *Ch*-bearing chromosome (if a single *Ch* is responsible for the chimerism) has entered a given cell in both micro- and megasporogenesis, there is the chance that some gametic combinations which do not possess the *Ch*-bearing chromosome are produced. On this assumption, some stable plants are expected in the offspring of No. 2 lineage. The observation of anaphase bridges and laggards in the first anaphase of meiosis, dyad bridges, bridges in tetrad stage, and polyad formation in meiosis substantiate this assumption. Therefore, the appearance of some stable segregants in the B_1 combinations backcrossed with *Ch* may be explained. These processes are hypothetically summarized in Fig. 2 (Ogura 1978a).

There are a number of instances in the literature regarding chromosome number chimerism in intact (not by tissue culture) higher plants. Among them, some assumed the existence of gene(s) responsible for chromosome numerical chimerism in *Ribes nigrum* (Vaarama 1949), in a variety of common wheat (Watanabe 1962) and in triploid pearl millet (Pantulu and Narasimha Rao 1977). Therefore, it is likely during tissue culture processes that induction of genetic factor(s) or gene(s) involved in chromosome number variation occurs.

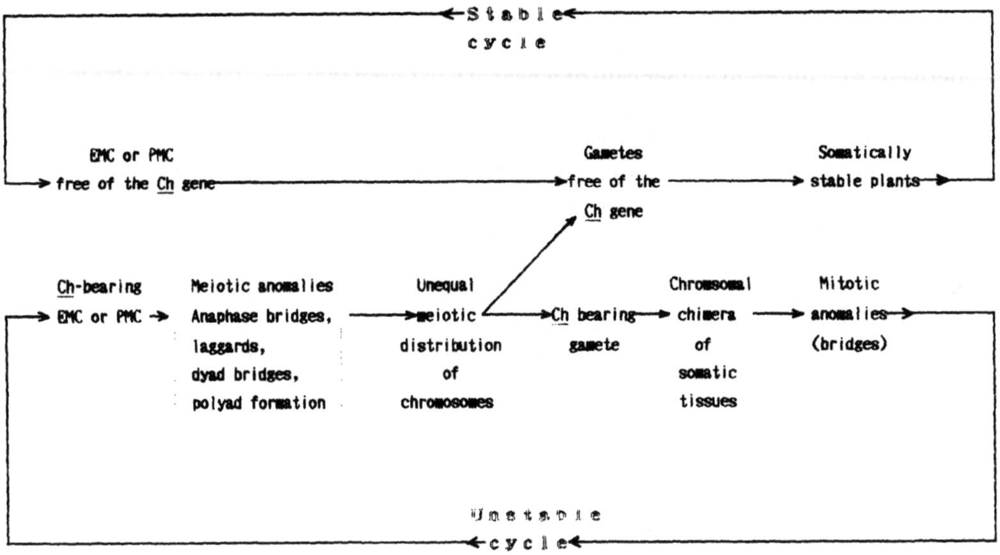

Fig. 2. Diagramatic presentation of a hypothetical consequence of chromosomal instability caused by a gene. *Ch*. (Ogura 1978a)

3.5 Chromosome Constitution of Regenerated Plants from Anther and Pollen Culture

Anther and pollen culture has made rapid progress (see Reinert and Bajaj 1977; Collins and Genovesi 1982; Bajaj 1983; Dunwell 1985). In the case of solanaceous species, haploid plantlet formation occurs mostly via embryogenesis, sometimes via both embryogenesis and organogenesis. Probably due to this type of development, most of the resulting plants seem to be comparatively stable, whereas when haploid plantlet formation occurs mostly via organogenesis, namely regeneration occurs through the nonembryogenic callus stage, the resulting plants sometimes became polyploids or aneuploids, such as in barley (Clapham 1973; Mix et al. 1978), corn (Ku et al. 1978; Ting 1985), rice (Academia Sinica 1974; Oono 1981; Chu et al. 1985; Mercy and Zapata 1986), wheat (Hu et al. 1979; De Bruyser and Henry 1980), rye (Wenzel and Thomas 1974), *Agropyron repens* (Zenkteller et al. 1975), *Solanum nigrum* (Harn 1972), digitalis (Corduan and Spix 1975), *Petunia violacea* (Gupta 1983), and others.

Chu et al. (1985) obtained a mean frequency of 10.2% of aneuploids from 1715 plants regenerated by anther culture from 1983 to 1985. Among the aneuploids obtained, the frequency of primary trisomics ranged from 5.4 to 6.7%, tetrasomics from 1.1 to 1.7%, monosomics from 0.9 to 1.3%, nullisomics from 0.5 to 1%, and double trisomics from 0.5 to 0.7%. About 90% of the regenerates were euploid and these plants can be used for breeding. Hu (1983) analyzed 444 pollen-derived lines of wheat obtained between 1972 and 1978, and found that 90% were completely homozygous, while 10% segregated for certain characteristics.

Regarding the appearance of aneuploid cells from wheat anther culture. Kudirka et al. (1983. 1986) studied chromosome numbers in root tips of plants derived from different anther calli of wheat. They found that populations of plants regenerated from two different anther calli both had cells with 20 and 21 chromosomes. However. in both cases. the frequency of cells with 20 chromosomes was greater than cells with 21 chromosomes. clearly indicating that eudiploid cells did not necessarily have a proliferative advantage over aneuploid cells. In respect to chromosomal stability and variability of anther or pollen-derived plants. the situation is similar to those of regenerates derived from calli of somatic tissues: however. regenerates from anther or pollen appear to be somewhat more chromosomally stable than callus culture regenerates of somatic tissues. The chromosomal constitution of androgenetic plants was reviewed by D'Amato (1985).

4 Nuclear Conditions of Protoplast-Derived Plants

Callus derived from single protoplast can be recognized to be of single cell origin. Therefore. plants regenerated from callus originating from a single protoplast are theoretically considered to have an identical genetic background. unless variation occurs during culture. Some studies suggest the stability or uniformity of protoplast-derived plants in several *Nicotiana* species (Banks and Evans 1976: Evans 1979). *Crotalaria juncea* (Ramanuja Rao et al. 1982). rice (Toriyama et al. 1986). and Trovita orange. *Citrus sinensis* (Kobayashi 1987). whereas other studies suggest the variability in chromosome number of protoplast-derived plants in *Atropa* (Bajaj et al. 1978). potato (Shepard et al. 1980: Thomas et al. 1982: Karp et al. 1982: Sree Ramulu et al. 1983: Gill et al. 1985. 1986. 1987: Sree Ramulu et al. 1986). alfalfa (Johnson et al. 1984). and *japonica* cultivars of rice (Ogura et al. 1987).

Gill et al. (1986) cytologically analyzed Russet Burbank potato protoclones that showed extensive chromosomal variability. and considered nuclear fragmentation. chromosome nondisjunction. and endomitosis as the cause of chromosome number variation in protoplast-derived potato plants. Johnson et al. (1984) compared two lines of alfalfa plants with their parent clones. and found that only 30 and 45% of protoclones. respectively. had the parental chromosome number of $2n = 32$. Karp et al. (1987) observed extensive numerical and structural chromosomal variation in dividing protoplasts of bread wheat. They isolated protoplasts from cell suspension cultures derived from immature embryo callus. If they directly use immature embryo of wheat as the material for protoplast isolation and subsequent protoplast division was possible. they may find out whether or not the material cell itself retains chromosomal variability.

Ogura et al. (1987) reported some chromosomal variants in protoplast-derived rice (Table 7). Examples of somatic and meiotic chromosomes of protoplast-derived rice plants are seen in Fig. 3. Ogura et al. (unpubl.) also examined chromosomal and phenotypic variations among protoclones (regenerates from a callus of single protoplast origin) of two *japonica* cultivars. Iwaimochi and Norin 14. and found that the coefficient of variation of several characters among protoclones was

Table 7. Somaclonal variation observed in protoplast-derived plants of four rice cultivars. (Ogura et al. 1987)

Cultivar	No. of plants	No. of normal plants (%)	Somaclonal variants: Chromosomal variants (%)	Somaclonal variants: Low seed fertility variants (%)	Somaclonal variants: Undetermined variants (%)
Nipponbare	73	61 (83.6)	4 (5.5)	8 (11.0)	
Fujisaka 5	11	9 (81.8)	1 (9.1)	1 (9.1)	
Norin 14	13	11 (84.6)	2 (15.4)		
Iwaimochi	29	21 (72.4)	5 (17.2)	2 (6.9)	1 (3.4)
Total	126	102 (81.0)	12 (9.5)	11 (8.7)	1 (0.8)

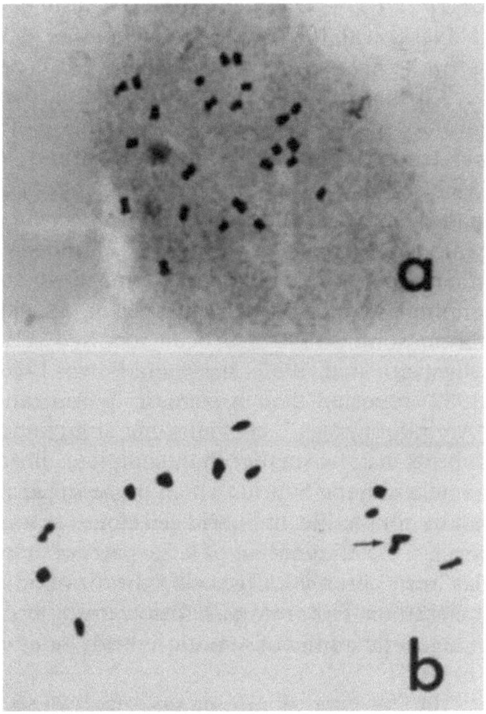

Fig. 3. a Metaphase chromosomes of a root-tip cell of protoplast-derived plant of a rice cultivar Nipponbare, showing normal 2n = 24 chromosome number. **b** Metaphase chromosome configuration of a PMC of a protoplast-derived trisomic plant of cv. Nipponbare, showing 1‴ + 11″. An *arrow* indicates a trivalent chromosome. (Ogura et al. 1987)

not necessarily smaller than in nonprotoclonal plants. This suggests that genetic and chromosomal variations occur during protoplast and/or callus culture of rice.

5 Karyological Analyses of Somatic Hybrids

Cytological instability apparently persists during morphogenesis in a number of somatic hybrid systems and has resulted in regenerates with variable chromosome numbers, e.g., in somatic hybrids of *Nicotiana glauca* + *N. langsdorffii* (Smith et al. 1976), *Nicotiana tabacum* mutant S + *N. tabacum* mutant V (Melchers and Labib 1974), *N. tabacum* + *N. rustica* (Nagao 1978; Douglas et al. 1981), *Petunia parodii* + *P. hybrida* (Power et al. 1976), *Lycopersicon esculentum* + *Solanum lycopersicoides* (Handley et al. 1986), *Brassica oleracea* + *B. campestris* (Terada et al. 1987), *Oryza sativa* + *O. brachyantha* (Hayashi et al. 1987), *Oryza sativa* + *O. officinalis* (Hayashi et al. 1987), *Solanum tuberosum* + *S. phureja* (Pijnacker et al. 1987), *Oryza sativa* + *Echinochloa oryzicola* (Terada et al. 1987b), and others (Table 8).

Of the 21 somatic hybrids between two haploid tobacco strains, 10 had 48 chromosomes, 7 had in the neighborhood of 70, one was tetraploid (4n = 96), one was triploid, and the remaining two had 46 and 90 chromosomes, respectively (Melchers and Labib 1974). A wide range of chromosomal variability was observed in the somatic hybrids of *Nicotiana tabacum* + *N. knightiana*, where all the hybrids exhibited intraplant chromosome number variation (Maliga et al. 1978). On the other hand, somatic hybrid systems in which all the regenerates had the expected amphiploid chromosome number or occasional loss or gain of only one or a few chromosomes have also been reported (Carlson et al. 1972; Evans et al. 1980; Ohgawara et al. 1985; Pirrie and Power 1986; Toriyama et al. 1987). Adachi et al. (1982) reported that in somatic hybridization between two remote species like "Arabidobrassica", chromosome complements or genetic material units of the hybrids may be smaller than complete chromosome complements of two species, namely somatic hybrids which possess true amphiploid chromosome number may not be obtainable. In hybrid cell clones or somatic hybrids of distantly related plant species, the elimination of large parts of or even of all chromosomes of one parent has been observed after cell hybridization, this phenomenon being called asymmetrization. For protoplast fusion and hybridization of distantly related species and cytogenetic studies of somatic hybrids, refer to the monograph by Gleba and Sytnik (1984).

In the case of intrageneric hybridization, Pijnacker et al. (1987) reported elimination of nucleolar chromosomes (satellited chromosomes) in *Solanum tuberosum* + *S. phureja* somatic hybrids. Terada et al. (1987) also found considerable variation in chromosome number in *Brassica* intrageneric somatic hybrids, probably due to the occurrence of multiple fusion and chromosome loss during the culture. Chromosomes of *Brassica* somatic hybrids are seen in Fig. 4. Hamill et al. (1985) produced somatic hybrids of *Nicotiana rustica* and *N. tabacum*. From a total of 17 somatic hybrids, three were partially self-fertile, while the others did not set seeds. Somatic chromosome constitution of these partially self-fertile hybrids was

Table 8. Examples of chromosomal constitution of somatic hybrids

Species and chromosome no.	Protoplast source	Chromosomal constitution of the hybrids	Reference
Arabidipsis thaliana (2n = 8x = 40) + *Brassica campestris* (2n = 20) (Arabidobrassica)	Cultured cell (AT) Leaf (BC)	30–45	Gleba and Hoffmann (1979, 1980) Hoffmann and Adachi (1981)
Brassica oleracea (2n = 18) + *Brassica campestris* (2n = 20)	Leaf	18–54	Schenck (1982)
"	Hypocotyl (BO) Leaf (BC)	33, 36, 38 49, 56, 57	Terada et al. (1987a)
Brassica oleracea (2n = 18) + *Moricandia arvensis* (2n = 28)	Cultured cell (MA) Leaf (BO)	46	Toriyama et al. (1987)
Citrus sinensis (2n = 18) + *Poncirus trifoliata* (2n = 18)	Cultured cell (CS) Leaf (PT)	36	Ohgawara et al. (1985)
Datura innoxia diploid (2n = 24), tetraploid (4n = 48) + *Atropa belladonna* (2n = 72)	Leaf	ca. 100 – ca. 175	Krumbiegel and Schieder (1979)
Datura innoxia (2n = 24) + *Datura discolor* (2n = 24)	Leaf	46, 48	Schieder (1978)
Datura innoxia (2n = 24) + *Datura innoxia* (2n = 24)	Leaf	34–108	Schieder (1977)
Lycopersicon esculentum (2n = 24) + *Solanum rickii* (2n = 24)	Leaf (LE) Cultured cell (SR)	44, 48, 49, 51, 61, 63 64, 67, 71, 130	O'Connell and Hanson (1986)
Lycopersicon esculentum (2n = 24) + *Solanum lycopersicoides* (2n = 24)	Leaf	48, 53, 54, 55 60, 64, 68	Handley et al. (1986)
Nicotiana glauca (2n = 24) + *N. langsdorffii* (2n = 18)	Leaf	42	Carlson et al. (1972)
"	"	56–64 (mode at 60)	Smith et al. (1976)
"	"	42 ± 6, 55 ± 2, 59, 62 63, 70, 78, 80	Chupeau et al. (1978)

Table 8. *(Continued)*

Species and chromosome no.	Protoplast source	Chromosomal constitution of the hybrids	Reference
Nicotiana glauca (2n = 24) + *N. tabacum* (2n = 48)	Leaf (NG) Cultured cell (NT)	72	Evans et al. (1980)
Nicotiana glutinosa (n = 12) + *N. tabacum* (2n = 48)	Anther (NG) Leaf (NT)	60	Pirrie and Power (1986)
Nicotiana tabacum (2n = 48) + *N. knightiana* (2n = 24)	Cultured cell (NT) Leaf (NK)	44–131	Maliga et al. (1978)
Nicotiana tabacum (2n = 48) + *N. rustica* (2n = 48)	Leaf	60–91 79, 84, 88	Nagao (1978)
"	Leaf	68–96	Douglas et al. (1981)
"	Cultured cell (NR) Leaf (NT)	63–67, 72–74, 73–76 74–77, 75–79, 82–87 etc.	Hamill et al. (1985)
Nicotiana tabacum (2n = 48) + *N. alata* (2n = 18)	Leaf	66, 71	Nagao (1979)
Nicotiana tabacum albino mutant (24 = 48) + *N. tabacum* NR mutant (n = 24)	Albino leaf (albino) Cultured cell (NR)	Petiolated 40–53 Sessile 46–87	Glimelius and Bonnett (1981)
Petunia hybrida (2n = 14) + *P. parodii* (2n = 14)	Leaf	24–28	Power et al. (1976)
Solanum tuberosum (n = 24) + *Lycopersicon esculentum* (2n = 24)	Cultured cell (ST) Leaf (LE)	50–72	Melchers et al. (1978)
Solanum tuberosum (2n = 24) + *S. phureja* (2n = 24)	Leaf and stem	45–47, 65–66, 69, 71, 72, 90–96	Puite et al. (1986)

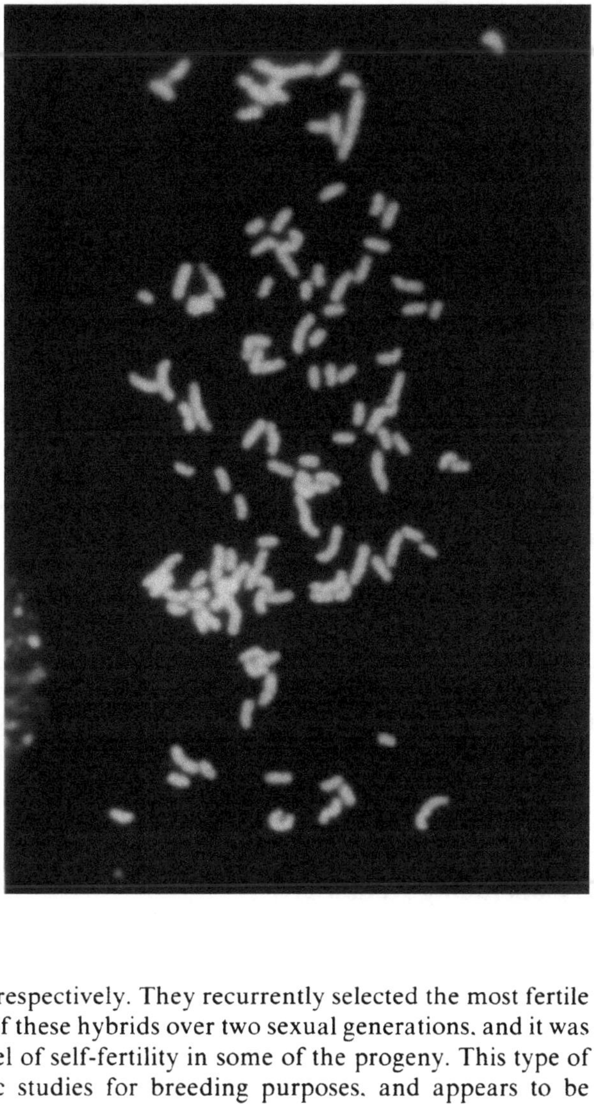

Fig. 4. Somatic chromosomes of rice and barnyard grass somatic hybrids, depicting 120 chromosomes, probably resulting from the fusion between chromosome doubled cells of both rice (2n = 24) and barnyard grass, *Echinochloa oryzicola* (2n = 36), (courtesy of Dr. Shimamoto)

2n = 74–77, 76–79, 82–87, respectively. They recurrently selected the most fertile among the progeny plants of these hybrids over two sexual generations, and it was possible to increase the level of self-fertility in some of the progeny. This type of study is necessary as basic studies for breeding purposes, and appears to be increasing gradually.

6 Karyology of Genetically Transformed Plants

Recent progress in plant biotechnology has made it possible to produce genetically transformed plants (see Bajaj 1989). However, most of transformed regenerates have been obtained by Ti or Ri plasmid method so far. Plant species for which Ti plasmid T-DNA transformed plants or shoots have been regenerated include tobacco (Turgeon et al. 1976; Wullems et al. 1981), *Arabidopsis thaliana* (Pavingerova et al. 1983), *Bidens alba* (Norton and Towers 1984), alfalfa (Marriotti et al. 1984; Webb 1986), *Medicago varia* (Deak et al. 1986), oilseed rape (Ooms et al. 1985b), tomato (McCormick et al. 1986), and cotton (Umbeck et al. 1987). Regenerated plants transformed with Ri plasmid T-DNA include carrot (Chilton et al. 1982), convolvulus (Tepfer 1984), tobacco (Ackermann 1977; Tepfer 1984), potato (Ooms et al. 1985; Hänisch ten Cate et al. 1987), and *Lycopersicon peruvianum* (Banerjee-Chattopadhyay et al. 1985).

Nevertheless, cytological analyses of these genetically manipulated plants are very few. Banerjee-Chattopadhyay et al. (1985) reported karyotypic changes in *Agrobacterium rhizogenes*-transformed roots of *Lycopersicon peruvianum*. Ooms et al. (1985) cytologically examined 42 Desiree potato plants regenerated from a hairy-root line after infection of *Agrobacterium rhizogenes*. Ten of the 42 transformed plants had 47 or 49 chromosomes instead of the normal 48 chromosomes. In two of these aneuploids, structural changes (probably due to the result of translocation or deletion) were observed. Ooms et al. (1987) also transformed two potato cultivars with disarmed T-DNA from *Agrobacterium tumefaciens* strains. The transformed Maris Bard plants were morphologically abnormal and highly aneuploid: among four transformed plants, one was $2n = 93$, another $2n = 92$, the third $2n = 73$ and the last was undetermined, whereas two transformed Desiree plants were morphologically normal, and one had euploid 48 chromosome number and another 47 chromosomes. Hänisch ten Cate et al. (1987) recently reported cytological analyses of *Agrobacterium rhizogenes* transformed hairy roots of the tetraploid potato cv. Bintje. They observed that 21 of the 27 primary hairy roots and all 16 subcultured hairy roots were tetraploid. This suggests that transformed roots are genetically stable and *Agrobacterium rhizogenes* transformation is a suitable system for genetically stable plant regeneration from transformed cells.

7 Concluding Remarks

The mode of shoot or plantlet regeneration from cultured tissues may be of two different types, namely organogenesis and embryogenesis. Plants regenerated via organogenesis are usually associated with genetic and chromosomal variability, but those via embryogenesis are related to genetic uniformity. In the case of embryogenesis, the majority of the regenerates seems to be chromosomally stable and only some chromosomally variable. In other words, the rate of occurrence of chromosomally variable regenerates is relatively low by regeneration via embryogenesis as compared with that via organogenesis.

Clonally propagated plants by meristem or shoot-tip culture have traditionally been known to show no variability. Whether minute karyotypic changes or structural rearrangements occur or not in clonally propagated plants will become gradually clearer by the chromosome banding technique.

Haploid production by anther, pollen, or ovary culture is now an accepted breeding technique of several crops, as discriminating and selecting mutants is easier in the haploid state than in the diploid, and because of the relatively short period and ease of obtaining homozygous diploids. The chromosomal stability and variability of pollen-derived plants are considered to be basically similar to those of regenerates originated from somatic tissues.

The uniformity and variability of protoplast-derived regenerates seem to be similar to those of regenerates from cultured tissues. Regenerates from a colony or a callus of single protoplast origin. namely protoclones. are considered to be in the identical genetic situation: however. chromosomal and genetic variations are found among the protoclones. This indicates that somaclonal variation does occur during the processes of culture. In the case of protoplast-derived rice. about 15 somaclonal variants with favorable phenotypes. and showing no chromosomal anomalies were obtained from 130 protoplast-derived regenerates. These favorable variants were selfed. and most progeny plants retained the favorable phenotypes. These progeny plants are being cultivated in the field for breeding purposes.

It is obvious that cytogenetic instability is commonly involved in somatic hybrid plants produced by fusion of protoplasts. Ignoring such an instability. and even if simply selecting useful hybrids showing favorable phenotypes such as high yield. nitrogen-fixing ability. or resistance to fungi or cold. problems of cytogenetic instability still seem to exist. For breeding purposes. somatic hybrid production by asymmetric fusion may become more useful.

Although detailed studies are still lacking. some have reported chromosomal variability in *Agrobacterium*-transformed plants. If cytogenetic instability is generally associated with *Agrobacterium*-transformed plants. genetic manipulation by the Ti or Ri plasmid is also considered disadvantageous. No data are yet available concerning cytogenetic analyses of genetically manipulated plants by the microinjection or electroporation method: however. a genetic manipulation system which causes no or little genetic variability to the manipulated plants will be necessary in the near future.

There are. however. a number of instances which show no or little chromosomal or genetic variations in the plants regenerated via embryogenesis from cultured tissues and several cases of plants regenerated even via organogenesis. Therefore. it does not seem impossible to make a culture system that enables one to obtain the desired results. along with cytogenetic studies on cultured tissues and the regenerates. Once we can realize such a cytogenetically stable system. the plant tissue culture method will be of direct benefit for plant breeding. By that time. the cause of chromosome number variation in cultured tissues and the regenerates may have been be adequately elucidated.

Acknowledgments. I wish to express deep thanks to Prof. Emeritus Dr. S. Akai. Director of the Institute for Agricultural Resources. Ishikawa College of Agriculture. Ishikawa Pref.. for his continuous encouragement. I also thank Dr. S. Tsuji. Laboratory of Pharmaceutical Sciences. Sumitomo Chemicals Co. Ltd.. Hyogo Pref.. for his technical assistances. and Prof. Y.P.S. Bajaj for his useful comments. Figure 4 was kindly provided by courtesy of Dr. K. Shimamoto. Plantech Research Laboratory. Kanagawa Pref.

References

Academia Sinica Second Division, 3rd Laboratory, Institute of Genetics (ed) (1974) Investigations on the induction and genetic expression of rice pollen plants. Sci Sin 17:209–226

Ackermann C (1977) Pflanzen aus *Agrobacterium rhizogenes* Tumoren an *Nicotiana tabacum*. Plant Sci Lett 8:23–30

Adachi T, Katayama Y (1969) Callus formation and shoot differentiation in wheat tissue culture. Bull Fac Agric Univ Miyazaki 16:77–82

Adachi T, Hoffmann F, Yokoh H (1982) Abnormality of chromosomal behavior and restriction of cell division in intergeneric and interspecific protoplast fusion products. In: Fujiwara A (ed) Plant tissue culture 1982. Maruzen, Tokyo, pp 441–442

Ahloowalia BS (1975) Regeneration of ryegrass plants in tissue culture. Crop Sci 15:449–452

Ahloowalia BS (1982) Plant regeneration from callus culture in wheat. Crop Sci 22:405–410

Alicchio R, Antonioli C, Palenzona D (1984) Karyotypic variability in plants of *Solanum melongena* regenerated from callus grown in presence of culture filtrate of *Verticillium dahliae*. Theor Appl Genet 67:267–271

Ancora G, Sree Ramulu K (1981) Plant regeneration from in vitro cultures of stem internodes in self-compatible triploid *Lycopersicon peruvianum* Mill. and cytogenetic analysis of regenerated plants. Plant Sci Lett 22:197–204

Armstrong CL, Phillips RL (1986) Genetic and cytogenetic stability of embryogenic and organogenic maize tissue cultures. In: Somers DA, Gengenbach BG, Biesboer DD, Hackett WP, Green CE (eds) Abstr 6th Int Congr Plant tissue cell culture. Univ Minnesota Press, pp 284

Armstrong KC, Nakamura C, Keller WA (1983) Karyotype instability in tissue culture regenerants of Triticale (x *Triticosecale* Whittmack) cv. Welsh from 6-month-old callus cultures. Z Pflanzenzücht 91:233–245

Backs-Hüseman D, Reinert J (1970) Embryobildung durch isolierte Einzellen aus Gewebekulturen von *Daucus carota*. Protoplasma 70:49–60

Bajaj YPS (1983) In vitro production of haploids. In: Evans DA, Sharp WR, Ammirato PV, Yamada Y (eds) Handbook of plant cell culture, vol 1. Techniques for propagation and breeding. MacMillan, New York, pp 228–287

Bajaj YPS (ed) (1989) Biotechnology in agriculture and forestry, vol 9. Plant protoplasts and genetic engineering II. Springer, Berlin Heidelberg New York Tokyo

Bajaj YPS, Bidani M (1980) Differentiation of genetically variable plants from embryo-derived callus cultures of rice. Phytomorphology 30:290–294

Bajaj YPS, Bidani M (1986) In vitro induction of genetic variability in rice (*Oryza sativa* L.). Int Symp New genetical approaches to crop improvement. PIDC, Karachi, pp 63–74

Bajaj YPS, Gill MS (1985) In vitro induction of genetic variability in cotton (*Gossypium* spp.). Theor Appl Genet 70:363–368

Bajaj YPS, Gosch G, Ottma M, Weber A, Grobler A (1978) Production of polyploid and aneuploid plants from anthers and mesophyll protoplasts of *Atropa belladonna* and *Nicotiana tabacum*. Indian J Exp Biol 16:947–953

Bajaj YPS, Saini SS, Bidani M (1980) Production of triploid plants from the immature and mature endosperm cultures of rice. Theor Appl Genet 58:17–18

Bajaj YPS, Sidhu BS, Dubey VK (1981a) Regeneration of genetically diverse plants from tissue cultures of forage grass – *Panicum* sps. Euphytica 30:135–140

Bajaj YPS, Ram AK, Labana KS, Singh H (1981b) Regeneration of genetically variable plants from the anther-derived callus of *Arachis hypogaea* and *Arachis villosa*. Plant Sci Lett 23:35–39

Balzan R (1978) Karyotype instability in tissue cultures derived from the mesophyll of *Zea mays* seedlings. Caryologia 31:75–88

Banerjee-Chattopadhyay S, Schwemmin AM, Schwemmin DJ (1985) A study of karyotypes and their alterations in cultured and *Agrobacterium*-transformed roots of *Lycopersicon peruvianum* Mill. Theor Appl Genet 71:258–262

Banks MS, Evans PK (1976) A comparison of the isolation and culture of mesophyll protoplasts from several *Nicotiana* species and their hybrids. Plant Sci Lett 7:409–416

Bayliss MW (1973) Origin of chromosome number variation in cultured plant cells. Nature (London) 246:529–530

Bayliss MW (1980) Chromosomal variation in plant tissues in culture. Int Rev Cytol Suppl 11A:113–144
Bennici A (1974) Cytological analysis of roots, shoots and plants regenerated from suspension and solid in vitro cultures of haploid *Pelargonium*. Z Pflanzenzücht 72:199–295
Bennici A (1979) A cytological chimera in plants regenerated from *Lilium longiflorum* tissues grown in vitro. Z Pflanzenzücht 82:349–353
Bennici A (1986) Durum wheat (*Triticum durum* Desf.). In: Bajaj YPS (ed) Biotechnology in agriculture and forestry, vol 2. Crops I. Springer, Berlin Heidelberg New York Tokyo, pp 89–)&4
Bennici A, D'Amato (1978) In vitro regeneration of durum wheat plants. I. Chromosome numbers of regenerated plants. Z Pflanzenzücht 81:305–311
Bennici A, Buiatti M, D'Amato F, Pagliai M (1970) Nuclear behaviour in *Haplopappus gracilis* cells grown in vitro on different culture media. Coll Int CNRS 193:245–250
Blakely LM, Steward FC (1964) Growth and organized development of cultured cells. VII. Cellular variation. Am J Bot 26:661–665
Bonner J, Devirian PS (1939) Growth factor requirements of four species of isolated roots. Am J Bot 26:661–665
Browers MA, Orton TJ (1982) A factorial study of chromosomal variability in callus cultures of celery (*Apium graveolens*). Plant Sci Lett 26:65–73
Caffaro L, Dameri M, Profumo P, Bennici A (1982) Callus induction and plantlet regeneration in *Cichorium intybus* L.: 1. A cytological study. Protoplasma 111:107–112
Carlson PS, Smith HH, Dearing RD (1972) Parasexual interspecific plant hybridization. Proc Natl Acad Sci USA 69:2292–2294
Chen CC, Lin MH (1976) Induction of rice plantlets from anther culture. Bot Bull Acad Sinica 17:18–24
Cheng TY (1972) Induction of indoleacetic acid synthetases in tobacco pith explants. Plant Physiol 50:723–727
Cheng TY, Smith HH (1975) Organogenesis from callus culture of *Hordeum vulgare*. Planta 123:307–308
Chilton MD, Tepfer DA, Petit A, David C, Casse-Delbart F, Tempe J (1982) *Agrobacterium rhizogenes* inserts T-DNA into the genomes of the host plant root cells. Nature (London) 295:432–434
Chu Q, Zhang Z, Gao Y (1985) Cytogenetical analysis on aneuploids from protoclones of rice (*Oryza sativa* L.). Theor Appl Genet 71:506–512
Chupeau Y, Missonier C, Hommel MC, Goujaud J (1978) Somatic hybrids of plants by fusion of protoplasts. Observations on the model system *Nicotiana glauca-Nicotiana langsdorffii*. Mol Gen Genet 165:239–245
Clapham D (1973) Haploid *Hordeum* plants from anthers in vitro. Z Pflanzenzücht 69:142–155
Clement WH (1964) Stability of chromosome numbers in tissue cultures of alfalfa, *Medicago sativa*. Am J Bot 51:670
Collins GB, Genovesi AD (1982) Anther culture and its application to crop improvement. In: Tomes DT, Ellis BE, Harney PM, Kasha KJ, Peterson RL (eds) Application of plant tissue culture to agriculture and industry. Univ Press, Guelph, pp 1–24
Cooper LS, Cooper CC, Hildebrandt AC, Riker AJ (1964) Chromosome numbers in single cell clones of tobacco tissue. Am J Bot 51:284–290
Corduan G, Spix C (1975) Haploid callus and regeneration of plants from anthers of *Digitalis purpurea*. L. Planta 124:1–11
D'Amato F (1977) Cytogenetics of differentiation in tissue and cell cultures. In: Reinert J, Bajaj YPS (eds) Applied and fundamental aspects of plant cell, tissue, and organ culture. Springer, Berlin Heidelberg New York, pp 343–357
D'Amato F (1985) Cytogenetics of plant cell and tissue cultures and their regenerates. CRC Crit Rev Plant Sci 3:73–112
Damiani F, Mariotti D, Pezzotti M, Arcioni S (1985) Variation among plants regenerated from tissue cultures of *Lotus corniculatus* L.Z Pflanzenzücht 94:332–339
Deak M, Kiss GB, Koncz C, Dudits D (1986) Transformation of *Medicago* by *Agrobacterium*-mediated gene transfer. Plant Cell Rep 5:97–100
De Bruyser J, Henry Y (1980) Induction of haploid and diploid plants through anther culture of haploid wheat (n = 3x = 21). Theor Appl Genet 57:57–58
Demoise CF, Partanen CR (1969) Effects of subculturing and physical condition of medium on the nuclear behavior of a plant tissue culture. Am J Bot 56:147–152
Doležel J, Novák FJ (1984) Effects of plant tissue culture media on the frequency of somatic mutations in *Tradescantia* stamen hairs. Z Pflanzenphysiol 114:51–58

Doležel J, Novák FJ (1986) Sister chromatic exchanges in garlic (*Allium sativum* L.) callus cells. Plant Cell Rep 5:280–283
Doležel J, Novák FJ, Havel L (1986) Cytogenetics of garlic (*Allium sativum* L.) in vitro culture. In: Int Symp Nuclear techniques and in vitro culture for crop improvement. IAEA, Vienna, pp 11–19
Douglas GC, Wetter LR, Nakamura C, Keller WA, Setterfield G (1981) Somatic hybridization between *Nicotiana rustica* and *N. tabacum*. III. Biochemical, morphological and cytological analysis of somatic hybrids. Can J Bot 59:228–237
Dunstan DI, Short KC, Thomas E (1978) The anatomy of secondary morphogenesis in cultured scutellum tissues of *Sorghum bicolor*. Protoplasma 97:251–260
Dunstan DI, Short KC, Dhaliwal H, Thomas E (1979) Further studies on plantlet production from cultured tissues of *Sorghum bicolor*. Protoplasma 101:355–361
Dunwell JM (1985) Anther and ovary culture. In: Bright SWJ, Jones MGK (eds) Cereal tissue and cell culture. Nijhoff/Junk, Dordrecht Boston Lancaster, pp 1–44
Edallo S, Zucchinali C, Parenzin M, Salamini F (1981) Chromosomal variation and frequency of spontaneous mutations associated with in vitro culture and plant regeneration in maize. Maydica 26:39–56
Evans DA (1979) Chromosome stability of plants regenerated from mesophyll protoplasts of *Nicotiana* species. Z Pflanzenphysiol 95:459–463
Evans DA, Gamborg OL (1982) Chromosome stability of cell suspension cultures of *Nicotiana* spp. Plant Cell Rep 1:104–107
Evans DA, Wetter LR, Gamborg OL (1980) Somatic hybrid plants of *Nicotiana glauca* and *Nicotiana tabacum* obtained by protoplast fusion. Physiol Plant 48:225–230
Fox JE (1963) Growth factor requirements and chromosome number in tabacco tissue cultures. Physiol Plant 16:793–803
Fridborg G (1971) Growth and organogenesis in tissue cultures of *Allium cepa* var. *proliferum*. Physiol Plant 25:436–440
Furner IJ, King J, Gamborg OL (1978) Plant regeneration from protoplasts isolated from a predominantly haploid suspension culture of *Datura innoxia* (Mill). Plant Sci Lett 11:169–176
Gamborg OL, Miller RA, Ojima K (1968) Nutrient requirements of suspension cultures of soybean root cells. Exp Cell Res 50:151–158
Gamborg OL, Shyluk JP, Fowke LC, Wetter LR, Evans DA (1979) Plant regeneration from protoplasts and cell cultures of *N. tabacum* sulfur mutant (Su/Su). Z Pflanzenphysiol 95:255–264
Gill BS, Kam-Morgan LNW, Shepard JF (1985) An apparent meiotic mutation in a mesophyll cell protoclone of the Russet Burbank potato. J Hered 76:17–20
Gill BS, Kam-Morgan LNW, Shepard JF (1986) Origin of chromosomal and phenotypic variation in potato protoclones. J Hered 77:13–16
Gill BS, Kam-Morgan LNW, Shepard JF (1987) Cytogenetic and phenotypic variation in mesophyll cell-derived tetraploid potatoes. J Hered 78:15–20
Gleba YY, Hoffmann F (1979) "Arabidobrassica": plant-genome engineering by protoplast fusion. Naturwissenschaften 66:547–554
Gleba YY, Hoffmann F (1980) "Arabidobrassica": a novel plant obtained by protoplast fusion. Planta 149:112–117
Gleba YY, Sytnik KM (1984) Protoplast fusion, genetic engineering in higher plants. In: Shoeman R (ed) Monographs on theoretical and applied genetics, vol 8. Springer, Berlin Heidelberg New York, pp 1–220
Glimelius K, Bonnett HT (1981) Somatic hybridization in *Nicotiana*: restoration of photoautotrophy to an albino mutant with defective plastids. Planta 153:497–503
Gould AR (1982) Chromosome stability in plant tissue cultures studied with banding techniques. In: Fujiwara A (ed) Plant tissue culture 1982. Maruzen, Tokyo, pp 431–432
Gresshoff PM, Doy CH (1972) Haploid *Arabidopsis thaliana* callus and plants from anther culture. Aust J Biol Sci 25:259–264
Gupta PP (1983) Microspore-derived haploid, diploid and triploid plants in *Petunia violacea* Lindl. Plant Cell Rep 2:255–256
Hamill JD, Pental D, Cocking EC (1985) Analysis of fertility in somatic hybrids of *Nicotiana rustica* and *N. tabacum* and progeny over two sexual generations. Theor Appl Genet 71:486–490
Handley LW, Nickels RL, Cameron MW, Moore PP, Sink KC (1986) Somatic hybrid plants between *Lycopersicon esculentum* and *Solanum lycopersicoides*. Theor Appl Genet 71:691–697

Hänisch ten Cate CH, Sree Ramulu K, Dijkhuis P, Groot B de (1987) Genetic stability of cultured hairy roots induced by *Agrobacterium rhizogenes* on tuber discs of potato cv. Bintje. Plant Sci 49:217–222

Hanna WW, Lu C, Vasil IK (1984) Uniformity of plants regenerated from somatic embryos of *Panicum maximum* Jacq. (Guinea grass). Theor Appl Genet 67:155–159

Harn C (1972) Production of plants from anthers of *Solanum nigrum* cultured in vitro. Caryologia 25:429–437

Hayashi Y, Kyozuka J, Shimamoto K (1987) Cell fusion between cultivated rice and four species of wild rice. Proc 10th Plant tissue culture Symp Jpn. Tohoku Univ Press, Sendai, p 185

Heinz DJ, Mee GWP (1971) Morphogenetic, cytogenetic and enzymatic variation in *Saccharum* species hybrid clones derived from callus tissue. Am J Bot 58:257–262

Heinz DJ, Mee GWP, Nickell LG (1969) Chromosome numbers of some *Saccharum* species hybrids and their cell suspension cultures. Am J Bot 56:450–456

Heller R (1953) Recherche sur la nutrition minérale des tissus végétaux cultivés in vitro. Ann Sci Nat (Bot) 14:1–223

Heyser JW, Nabors MW (1982) Regeneration of proso millet from embryogenic calli derived from various plant parts. Crop Sci 22:1070–1074

Hoffmann F, Adachi T (1981) "Arabidobrassica": chromosomal recombination in asymmetric intergeneric hybrid cells. Planta 153:586–593

Horák J (1972) Ploidy chimeras in plants regenerated from the tissue cultures of *Brassica oleracea* L. Biol Plant 14:423–426

Hu H (1983) Genetic stability and variability of pollen-derived plants. In: Sen SK, Giles KL (eds) Plant cell culture in crop improvement. Plenum, New York London, pp 145–157

Hu H, Xi Z, Zhuang J, Ouyang J, Zeng J, Jia S, Jia X, Jing J, Zhou S (1979) Genetic investigation on pollen-derived plants in wheat (*Triticum aestivum*). Acta Genet Sin 6:322–330

Inomata N (1982) Chromosome variation in callus cells of *Luzula elegans* Lowe with nonlocalized kinetochores. Jpn J Genet 57:29–64

Johnson LB, Stuteville DL, Schlarbaum SE, Skinner DZ (1984) Variation in phenotype and chromosome number in alfalfa protoclones regenerated from non-mutagenized calli. Crop Sci 24:948–951

Kao KN (1977) Chromosome behaviour in somatic hybrids of soybean-*Nicotiana glauca*. Mol Gen Genet 150:225–230

Kao KN, Miller RA, Gamborg OL, Harvey BL (1970) Variations in chromosome number and structure in plant cells grown in suspension culture. Can J Genet Cytol 12:297–301

Karp A, Maddock SE (1984) Chromosome variation in wheat plants regenerated from cultured immature embryos. Theor Appl Genet 67:249–255

Karp A, Nelson RS, Thomas E, Bright SWJ (1982) Chromosome variation in protoplast-derived potato plants. Theor Appl Genet 63:265–272

Karp A, Wu QS, Steele SH, Jones MGK (1987) Chromosome variation in dividing protoplasts and cell suspensions of wheat. Theor Appl Genet 74:140–146

Kasperbauer MJ, Collins GB (1972) Reconstitution of diploids from anther-derived haploids in tobacco. Crop Sci 12:98–101

Kasperbauer MJ, Buckner RC, Bush LP (1979) Tissue culture of annual ryegrass × tall fescue F_1 hybrids: Callus establishment and plant regeneration. Crop Sci 19:457–460

Kearthley DE, Scholl RL (1983) Chromosomal heterogeneity of *Arabidopsis thaliana* anther callus, regenerated shoots and plants. Z Pflanzenphysiol 112:247–255

Kobayashi S (1987) Uniformity of plants regenerated from orange (*Citrus sinensis* Osb.) protoplasts. Theor Appl Genet 74:10–14

Krikorian AD, Staicu SA, Kann RP (1981) Karyotype analysis of a daylily clone reared from aseptically cultured tissues. Ann Bot (London) 47:121–131

Krikorian AD, O'Conner SA, Fitter MS (1983) Chromosome number variation and karyotype stability in cultures and culture-derived plants. In: Evans DA, Sharp WR, Ammirato PV, Yamada Y (eds) Handbook of plant cell culture, vol 1. Techniques for propagation and breeding. MacMillan, New York, pp 541–581

Krumbiegel G, Schieder O (1979) Selection of somatic hybrids after fusion of protoplasts from *Datura innoxia* Mill. and *Atropa belladonna* L. Planta 145:371–375

Ku MK, Cheng WC, Kuo LC, Kuan YL, An HP, Huang CH (1978) Induction factors and morphocytological characteristics of pollen-derived plants in maize (*Zea mays*). In: Beijing Symp Plant tissue culture. Pitman, Boston London Melbourne, pp 35–42

Kudirka DT, Schaeffer GW, Baenziger PS (1983) Cytogenetic characteristics of wheat plants regenerated from anther calli of Centurk. Can J Genet Cytol 25:513–517
Kudirka DT, Schaeffer GW, Baenziger PS (1986) Wheat: genetic variability through anther culture. In: Bajaj YPS (ed) Biotechnology in agriculture and forestry, vol 2. Crops I. Springer, Berlin Heidelberg New York Tokyo, pp 39–54
Lapitan NLV, Sears RG, Gill BS (1984) Translocations and other karyotypic structural changes in wheat x rye hybrids regenerated from tissue culture. Theor Appl Genet 68:547–554
Linsmaier EM, Skoog F (1965) Organic growth factor requirements of tobacco tissue cultures. Physiol Plant 18:100–127
Liu MC, Chen WH (1976) Tissue and cell culture as aids to sugarcane breeding. I. Creation of genetic variation through callus culture. Euphytica 25:393–403
Lu C, Vasil IK, Ozias-Akins P (1982) Somatic embryogenesis in *Zea mays* L. Theor Appl Genet 62:109–112
Lupi MC, Bennici A, Baroncelli S, Gennai D, D'Amato F (1981) In vitro regeneration of durum wheat plants II. Diplontic selection in aneusomatic plants. Z Pflanzenzücht 87:167–171
Maliga P, Kiss ZR, Nagy AG, Lazar G (1978) Genetic instability in somatic hybrids of *Nicotiana tabacum* and *Nicotiana knightiana.* Mol Gen Genet 163:145–151
Marchetti S, Ancora G, Brunori A (1975) Time course of polyploidization in calli derived from stem and pith explants of *Nicotiana tabacum* studied on isolated nuclei. Z Pflanzenphysiol 78:307–313
Marriotti D, Davey MR, Draper J, Freeman JP, Cocking EC (1984) Crown gall tumorigenesis in the forage legume *Medicago sativa.* Plant Cell Physiol 25:474–482
Matthysse AG, Torrey JG (1967) Nutritional requirements for polyploids mitosis in cultured pea root segments. Physiol Plant 20:661–672
McCormick S, Niedermeyer J, Fry J, Barnson A, Horsch R, Fraley R (1986) Leaf disc transformation of cultivated tomato (*L. esculentum*) using *Agrobacterium tumefaciens.* Plant Cell Rep 5:81–84
McCoy TJ, Phillips RL (1982) Chromosome stability in maize (*Zea mays*) tissue cultures and sectoring in some regenerated plants. Can J Genet Cytol 24:559–565
McCoy TJ, Phillips RL, Rines HW (1982) Cytogenetic analysis of plants regenerated from oat (*Avena sativa*) tissue cultures; high frequency of partial chromosome loss. Can J Genet Cytol 24:37–50
Mehra A, Mehra PN (1974) Organogenesis and plantlet formation in vitro in almond. Bot Gaz 135:61–73
Melchers G, Labib G (1974) Somatic hybridization of plants by fusion of protoplasts. I. Selection of light resistant hybrids of "haploid" light sensitive varieties of tobacco. Mol Gen Genet 135:277–294
Melchers G, Sacristán MD, Holder AA (1978) Somatic hybrid plants of potato and tomato regenerated from fused protoplasts. Carlsberg Res Commun 43:203–218
Mercy ST, Zapata FJ (1986) Chromosomal behavior of anther culture derived plants of rice. Plant Cell Rep 3:215–218
Miller CO (1963) Kinetin and kinetin-like compounds. In: Linskens HF, Tracey MV (eds) Moderne Methoden der Pflanzenanalyse. Springer, Berlin Heidelberg New York, pp 192–202
Mitra J, Steward FC (1961) Growth induction in cultures of *Haplopappus gracilis.* II. The behavior of the nucleus. Am J Bot 48:358–368
Mitra J, Mapes MO, Steward FC (1960) Growth and organized development of cultured cells. IV. The behavior of the nucleus. Am J Bot 47:357–368
Mix G, Wilson HM, Foroughi-Wehr B (1978) The cytological status of plants of *Hordeum vulgare* L. regenerated from microspore callus. Z Pflanzenzücht 80:89–99
Mu SK, Liu SC (1978) Cytological observations on calluses derived from apple endosperm cultured in vitro. In: Beijing Symp Plant tissue culture. Pitman, Boston London Melbourne, pp 507–510
Murashige T, Nakano R (1967) Chromosome complement as a determinant of the morphogenetic potential of tobacco cells. Am J Bot 54:963–970
Murashige T, Skoog F (1962) A revised medium for rapid growth and bioassays with tobacco tissue culture. Physiol Plant 15:473–497
Murata M, Orton TJ (1982) Analysis of karyotypic changes in suspension culture of celery. In: Fujiwara A (ed) Plant tissue culture 1982. Maruzen, Tokyo, pp 435–436
Murata M, Orton TJ (1983) Chromosome structural changes in cultured celery cells. In Vitro 19:83–89
Murata M, Orton TJ (1984) Chromosome fusions in cultured cells of celery. Can J Genet Cytol 26:395–400
Nagao T (1978) Somatic hybridization by fusion of protoplasts. I. The combination of *Nicotiana tabacum* and *N. rustica.* Jpn J Crop Sci 47:491–498

Nagao T (1979) Somatic hybridization by fusion of protoplasts. II. The combinations of *Nicotiana tabacum* and *N. glutinosa* and of *N. tabacum* and *N. alata*. Jpn J Crop Sci 48:385–392
Nakamura C, Keller WA (1982) Callus proliferation and plant regeneration from immature embryos of hexaploid triticale. Z Pflanzenzücht 88:137–160
Natarajan P, Walton PD (1987) A comparison of somatic chromosomal instability in tissue culture regenerants from *Medicago media* Pers. Plant Cell Rep 6:109–113
Negrutiu I, Beeftink F, Jacobs M (1975) *Arabidopsis thaliana* as a model system in somatic cell genetics. I. Cell and tissue culture. Plant Sci Lett 5:293–304
Newell CA, Rhoads ML, Bidney DL (1984) Cytogenetic analysis of plants regenerated from tissue explants and mesophyll protoplasts of winter rape, *Brassica napus* L. Can J Genet Cytol 26:752–761
Niizeki M (1974) Studies on plant cell and tissue culture. V. Effects of different kinds of media on the variation of chromosome number in tobacco callus and regenerated plant. J Fac Agric Hokkaido Univ 57:357–367
Nishi T, Mitsuoka S (1969) Occurrence of various ploidy plants from anther and ovary culture of rice plant. Jpn J Genet 44:341–346
Nishi T, Yamada Y, Takahashi E (1968) Organ redifferentiation and plant restoration in rice callus. Nature (London) 219:508–509
Nishiyama I, Taira T (1966) The effects of kinetin and indoleacetic acid on callus growth and organ formation in two species of *Nicotiana*. Jpn J Genet 41:357–365
Norstog K, Wall WE, Howland GP (1969) Cytological characteristics of ten-year-old ryegrass endosperm tissue cultures. Bot Gaz 130:83–86
Norton RA, Towers GHN (1984) Transmission of nopaline crown gall tumor markers through meiosis in regenerated whole plants of *Bidens alba*. Can J Bot 62:408–413
Novák FJ, Vyskot B (1975) Karyology of callus cultures derived from *Nicotiana tabacum* L. haploids, and ploidy of regenerants. Z Pflanzenzücht 75:62–70
O'Connell, Hanson MR (1986) Regeneration of somatic hybrid plants formed between *Lycopersicon esculentum* and *Solanum rickii*. Theor Appl Genet 72:59–65
Ogihara Y (1981) Tissue culture in *Haworthia* IV. Genetic characterization of plants regenerated from callus. Theor Appl Genet 60:353–363
Ogihara Y (1982) Tissue culture in *Haworthia* V. Characterization of chromosomal changes in cultured callus cells. Jpn J Genet 57:499–511
Ogura H (1975) The effects of a morphactin, chlorflurenol, on organ redifferentiation from tobacco calluses cultured in vitro. Bot Mag (Tokyo) 88:1–8
Ogura H (1976) The cytological chimeras in original regenerates from tobacco tissue cultures and in their offsprings. Jpn J Genet 51:161–174
Ogura H (1977) Genetical studies on the cultured tissues and the regenerates in the genus *Nicotiana*. D Sc Diss, Kyoto Univ, pp 1–134
Ogura H (1978a) Genetic control of chromosomal chimerism found in a regenerate from tobacco callus. Jpn J Genet 53:77–90
Ogura H (1978b) Effects of exogenously supplied growth regulators on chromosomal constitution of tobacco regenerates. Bull Ishikawa Agric Coll 8:9–15
Ogura H (1982) Studies on the genetic instability of cultured tissues and the regenerated plants – effects of auxins and cytokinins on mitosis of *Vicia faba* cells. In: Fujiwara A (ed) Plant tissue culture 1982. Maruzen, Tokyo, pp 433–434
Ogura H, Kyozuka J, Hayashi Y, Koba T, Shimamoto K (1987) Field performance and cytology of protoplast-derived rice (*Oryza sativa*): high yield and low degree of variation of four japonica cultivars. Theor Appl Genet 74:670–676
Ohgawara T, Kobayashi S, Ohgawara E, Uchimiya H, Ishii S (1985) Somatic hybrid plants obtained by protoplast fusion between *Citrus sinensis* and *Poncirus trifoliata*. Theor Appl Genet 71:1–4
Ono H (1979) Chromosomal variation. In: Takeuchi M, Nakajima T, Furuya T (eds) Plant tissue culture, new edn. Asakura Syoten, Tokyo, pp 80–88 (in Japanese)
Ooms G, Karp A, Burrell M, Twell D, Reberts J (1985a) Genetic modification of potato development using Ri T-DNA. Theor Appl Genet 70:440–446
Ooms G, Bains A, Burrell M, Karp A, Twell D, Wilcox E (1985b) Genetic manipulation in cultivars of oil seed rape (*Brassica napus*) using *Agrobacterium*. Theor Appl Genet 71:325–329
Ooms G, Burrell MM, Karp A, Bevan M, Hille J (1987) Genetic transformation in two potato cultivars with T-DNA from disarmed *Agrobacterium*. Theor Appl Genet 73:744–750

Oono K (1975) Production of haploid plants of rice (*Oryza sativa*) by anther culture and their use for breeding. Bull Natl Inst Agric Sci D26:139–222
Oono K (1981) In vitro method applied to rice. In: Thorpe A (ed) Plant tissue culture methods and applications in agriculture. Academic Press, New York London, pp 273–298
Orton TJ (1980) Chromosomal variability in tissue cultures and regenerated plants of *Hordeum*. Theor Appl Genet 56:101–112
Orton TJ (1983) Spontaneous electrophoretic and chromosomal variability in callus cultures and regenerated plants of celery. Theor Appl Genet 67:17–24
Orton TJ, Steidl RP (1980) Cytogenetic analysis of plants regenerated from colchicine-treated callus cultures of an interspecific *Hordeum* hybrids. Theor Appl Genet 57:89–95
Ozias-Akins P, Vasil IK (1982) Plant regeneration from cultured immature embryos and inflorescences of *Triticum aestivum* L. (wheat): evidence for somatic embryogenesis. Protoplasma 110:95–105
Pantulu JV, Narasimha Rao GJ (1977) Genetically controlled chromosome numerical mosaicism in pearl millet. Proc Indian Acad Sci 86B:15–22
Papeš D, Jelaska S, Tomaseo M, Devide Z (1978) Triploidy in callus cultures of *Vicia faba* L. investigated by the Giemsa C-banding technique. Experientia 34:1016–1017
Pavingerova D, Ondrej M, Matousek J (1983) Analysis of progeny of *Arabidopsis thaliana* plants regenerated from crown gall tumors. Z Pflanzenphysiol 112:427–433
Pijnacker LP, Ferwerda MA, Puite KJ, Roest S (1987) Elimination of *Solanum phureja* nucleolar chromosomes in *S. tuberosum* + *S. phureja* somatic hybrids. Theor Appl Genet 73:878–882
Pirrie A, Power JB (1986) The production of fertile, triploid somatic hybrid plants [*Nicotiana glutinosa* (n) + *N. tabacum* (2n)] via gametic: somatic protoplast fusion. Theor Appl Genet 72:48–52
Power JB, Frearson EM, Hayward C, George D, Evans PK, Berry SF, Cooking EC (1976) Somatic hybridization of *Petunia hybrida* and *P. parodii*. Nature (London) 263:500–502
Power JB, Berry SF, Chapman JV, Cocking EC (1980) Somatic hybridization of sexually incompatible Petunias: *Petunia parodii, Petunia parviflora*. Theor Appl Genet 57:1–4
Pramanik TK, Datta SK (1986) Plant regeneration and ploidy variation in culture-derived plants of *Asclepias curassavica* L. Plant Cell Rep 3:219–222
Puite KJ, Roest S, Pijnacker (1986) Somatic hybrid potato plants after electrofusion of diploid *Solanum tuberosum* and *Solanum phureja*. Plant Cell Rep 5:262–265
Rajasekaran K, Schank SC, Vasil IK (1986) Characterization of biomass production cytology and phenotypes of plants regenerated from embryogenic callus cultures of *Pennisetum americanum* × *P. purpureum* (hybrid triploid napiergrass). Theor Appl Genet 73:4–10
Ramanuja Rao IV, Mehta U, Mohan Ram HY (1982) Whole plant regeneration from cotyledonary protoplasts of *Crotalaria juncea*. In: Fujiwara A (ed) Plant tissue culture 1982. Maruzen, Tokyo, pp 595–596
Rangan TS, Vasil IK (1983) Somatic embryogenesis and plant regeneration in tissue cultures of *Panicum miliaceum* L. and *Panicum miliare* Lamk. Z Pflanzenphysiol 109:49–53
Reinert J, Bajaj YPS (eds) (1977) Anther culture: haploid production and its significance. In: Applied and fundamental aspects of plant cell, tissue, and organ culture. Springer, Berlin Heidelberg New York, pp 251–267
Risser PG, White PR (1964) Nutritional requirements of spruce tumor cells in vitro. Physiol Plant 17:620–635
Sacristán MD (1971) Karyotypic changes in callus cultures from haploid plants of *Crepis capillaris* (L.) Wallr. Chromosoma 57:1–4
Sacristán MD (1975) Clonal development in tumorous cultures of *Crepis capillaris* Naturwissenschaften 62:139–140
Sacristán MD, Melchers G (1969) The caryological analysis of plants regenerated from tumorous and other callus cultures of tobacco. Mol Gen Genet 105:317–333
Sacristán MD, Wendt-Gallitelli MF (1971) Transformation to auxin-autotrophy and its reversibility in a mutant line of *Crepis capillaris* callus culture. Mol Gen Genet 110:355–360
Sacristán MD, Wendt-Gallitelli MF (1973) Tumorous cultures of *Crepis capillaris:* Chromosomes and growth. Chromosoma 43:279–288
Sangwan RS, Gorenflot R (1975) In vitro culture of *Phragmites* tissues. Callus formation, organ differentiation and cell suspension culture. Z Pflanzenphysiol 75:256–269
Schenck HR (1982) *Brassica napus* – successful resynthesis by protoplast fusion between *B. oleracea* and *B. campestris*. In: Fujiwara A (ed) Plant Tissue culture 1982. Maruzen, Tokyo, pp 639–640

Schieder O (1977) Hybridization experiments with protoplasts from chlorophyll-deficient mutants of some solanaceous species. Planta 137:253–257

Schieder O (1978) Somatic hybrids of *Datura innoxia* Mill. + *Datura discolor* Bernh. and of *Datura innoxia* Mill. + *Datura stramonium* L. var. *tatula* L. I. Selection and characterization. Mol Gen Genet 162:113–119

Semal J (ed) (1986) Somaclonal variations and crop improvement. Nijhoff, Dordrecht

Shepard JF, Bidney D, Shahin E (1980) Potato protoplasts in crop improvement. Science 208:17–24

Sheridan WF (1974) Long-term callus cultures of *Lilium:* relative stability of karyotype. J Cell Biol 63:313a

Sheridan WF (1975) Plant regeneration and chromosome stability in tissue cultures. In: Ledoux L (ed) Genetic manipulations with plant materials. Plenum, London, pp 263–295

Shimada T (1971) Chromosome constitution of tobacco and wheat callus cells. Jpn J Genet 46:235–241

Shimada T (1981) Haploid plants regenerated from the pollen callus of wheat (*Triticum aestivum* L.). Jpn J Genet 56:581–588

Shimada T, Tabata M (1967) Chromosome numbers in cultured pith tissue of tobacco Jpn J Genet 42:195–201

Shimada T, Sasakuma T, Tsunewaki K (1969) In vitro culture of wheat tissues. I. Callus formation, organ redifferentiation and single cell culture. Can J Genet Cytol 11:294–304

Singh BD, Harvey BL (1975) Selection for diploid cells in suspension cultures of *Haplopappus gracilis*. Nature (London) 253:453

Singh BD, Harvey BL, Kao KN, Miller RA (1972) Selection pressure in cell populations of *Vicia hajastana* cultured in vitro. Can J Genet Cytol 14:65–70

Smith HH, Kao KN, Combatti NC (1976) Interspecific hybridization by protoplast fusion in *Nicotiana*: Confirmation and development. J Hered 67:123–128

Sree Ramulu K, Dijkhuis P, Roest S (1983) Phenotypic variation and ploidy level of plants regenerated from protoplasts of tetraploid potato (*Solanum tuberosum* L. cv. Bintje). Theor Appl Genet 65:329–338

Sree Ramulu K, Dijkhuis P, Roest S, Bokelmann GS, Groot B de (1986) Variation in phenotype and chromosome number of plants regenerated from protoplasts of dihaploid and tetraploid potato. J Plant Breed 97:119–128

Straus J (1954) Maize endosperm tissue grown in vitro. II. Morphology and cytology. Am J Bot 41:833–839

Swedlund B, Vasil IK (1985) Cytogenetic characterization of embryogenic callus and regenerated plants of *Pennisetum americanum* (L.) K. Schum. Theor Appl Genet 69:575–581

Tabata M, Yamamoto H, Hiraoka N (1968) Chromosome constitution of mature plants derived from cultured pith of tobacco. Jpn J Genet 43:319–332

Tepfer D (1984) Transformation of several species of higher plants by *Agrobacterium rhizogenes* – sexual transmission of the transformed genotype and phenotype. Cell 37:959–967

Terada R, Yamashita Y, Nishibayashi S, Shimamoto K (1987a) Somatic hybrids between *Brassica oleracea* and *B. campestris*. Theor Appl Genet 73:379–384

Terada R, Kyozuka J, Nishibayashi S, Shimamoto K (1987b) Plantlet regeneration from somatic hybrids of rice (*Oryza sativa* L.) and barnyard grass (*Echinochloa oryzicola* Vasing). Mol Gen Genet 210:39–43

Thomas E, Bright SWJ, Franklin J, Lancaster VA, Miflin BJ (1982) Variation amongst protoplast-derived potato plants (*Solanum tuberosum* cv. Maris Bard). Theor Appl Genet 62:65–68

Ting TC (1985) Meiosis and fertility of anther culture-derived maize plants. Maydica 30:161–169

Toriyama K, Hinata K, Sasaki T (1986) Haploid and diploid plant regeneration from protoplasts of anther callus in rice. Theor Appl Genet 73:16–19

Toriyama K, Hinata K, Kameya T (1987) Production of somatic hybrid plants, Brassicomoricandia through protoplast fusion between *Moricandia arvensis* and *Brassica oleracea*. Plant Sci 48:123–128

Torrey JG (1967) Morphogenesis in relation to chromosome constitution in long term tissue cultures. Physiol Plant 20:265–275

Turgeon R, Wood HN, Braun AC (1976) Studies on the recovery of crown gall tumor cells. Proc Natl Acad Sci USA 73:3562–3564

Umbeck P, Johnson G, Barton K, Swain W (1987) Genetically transformed cotton (*Gossypium hirsutum* L.) plants. Biotechnology 5:263–266

Vaarama A (1949) Spindle abnormalities and variation in chromosome number in *Ribes nigrum*. Hereditas 35:194–206
Vasil IK (1983) Regeneration of plants from single cells of cereals and grasses. In: Lurquin PF, Kleinhofs A (eds) Genetic engineering in eukaryotes. Plenum, London, pp 233–252
Vasil IK (1985) Somatic embryogenesis and its consequences in the Gramineae. In: Henke RR, Hughes KW, Constantin MP, Hollaender A (eds) Tissue culture in forestry and agriculture. Plenum, New York, pp 31–47
Vasil IK, Hildebrandt AC (1966) Variations of morphogenetic behavior in plant tissue cultures. I *Cichorium endiva*. Am J Bot 53:860–869
Vasil IK, Vasil V (1981) Somatic embryogenesis and plant regeneration from tissue cultures of *Pennisetum americanum* and *P. americanum* × *P. purpureum*. Am J Bot 68:864–872
Venketeswaran S (1963) Tissue culture studies on *Vicia faba*. II. Cytology. Caryologia 16:91–100
Watanabe Y (1962) Chromosome mosaics observed in a variety of wheat, Shirahada Jpn J Genet 37:194–206
Webb KJ (1986) Transformation of forage legumes using *Agrobacterium tumefaciens*. Theor Appl Genet 72:53–58
Wei ZM, Kamada H, Harada H (1986) Transformation of *Solanum nigrum* L protoplasts by *Agrobacterium rhizogenes*. Plant Cell Rep 5:93–96
Wenzel G, Thomas E (1974) Observations on the growth in culture of anthers of *Secale cereale*. Z Pflanzenzücht 72:89–94
Wenzel G, Hoffmann F, Potrykus I, Thomas E (1975) The separation of viable rye microspores from mixed populations and their development in culture. Mol Gen Genet 138:293–297
Wernicke W, Brettell R, Wakizuka T, Potrykus I (1981) Adventitious embryoid and root formation from rice leaves. Z Pflanzenphysiol 103:361–365
Wheeler VA, Evans NE, Foulger D, Webb KJ, Karp A, Franklin J, Bright SWJ (1985) Shoot formation from explant cultures of fourteen potato cultivars and studies of the cytology and morphology of regenerated plants. Ann Bot (London) 55:309–320
White PR (1963) The cultivation of animal and plant cells. Ronald, New York, pp 1–228
Williams L, Collin HA (1976) Growth and cytology of celery plants derived from tissue culture. Ann Bot (London) 40:333–338
Wilmar C, Hellendoorn (1968) Growth and morphogenesis of asparagus cells cultured in vitro. Nature (London) 217:369–370
Wullems GJ, Molendijk L, Ooms G, Schilperoort RA (1981) Retention of tumor markers in F_1 progeny plants from in vitro-induced octopine and nopaline tumor tissues. Cell 27:719–727
Yamabe M, Yamada T (1973) Studies on differentiation in cultured cells. II. Chromosomes of *Haworthia* callus and of the plants grown from the callus. Kromosomo 94:2923–2931
Yamane Y (1975) Chromosomal variation in calluses induced in *Vicia faba* and *Allium cepa*. Jpn J Genet 50:353–355
Zagorska NA, Shamina ZB, Butenko RG (1974) The relationship of morphogenetic potency of tobacco tissue culture and its cytogenetic features. Biol Plant 16:262–274
Zagorska N, Abadjieva M, Chalukova M, Achkova Z, Nikova V (1986) Somaclonal variation in tobacco and tomato plants regenerated from tissue cultures. In: IAEA (ed) Nuclear techniques and in vitro culture for plant improvement. IAEA, Vienna, pp 349–356
Zenkteller M, Misiura E, Ponitka A (1975) Induction of androgenetic embryoids in the in vitro cultured anthers of several species. Experientia 31:289–291
Zosimovich VP, Kunakh VA (1975) Levels, types and origin of chromosome aberrations in cultures of isolated plant tissues. Sov Genet 11:37–46

I.3 Genetic Mosaics and Chimeras: Implications in Biotechnology

M. MARCOTRIGIANO[1]

1 General Account

1.1 Genetic Mosaics

Genetic mosaics are plants which are composed of tissues of two or more genotypes. They should not be confused with plant hybrids, which possess only one genotype; a genotype which is the product of recombination following fertilization. Mosaics can arise spontaneously or can be induced with chemical or physical mutagens. In mutagenized plants most of the mutant sectors arise outside the shoot apex (i.e., "extra-apical mosaicism", described by Bergann 1967).

The discovery of transposable elements has increased the awareness of genetic mosaicism. The action of mobile genetic elements is not confined to the shoot apex, and the distribution of mutant and nonmutant cells greatly resembles that found in plants which have been subjected to mutagens, in that the mutant cell clones can appear in isolated regions of the plant and may not be traceable to an event occurring in the shoot apex (see, e.g., Doodeman et al. 1984a,b). On occasion, transposable elements can result in stable "intra-apical mosaics" (see Sect. 1.1.2 below and Doodeman and Bianchi 1985).

"Extra-apical mosaics" can be important because it is possible to regenerate adventitious shoots from single cells or clusters of somatic cells which are no longer a part of the shoot apex. In some plants this is accomplished in vivo, as with shoot formation from leaf or root cuttings. In addition, in many species, plant tissue culture technology has allowed in vitro plant regeneration from virtually any somatic cell (see Bajaj 1986).

While many genetic mosaics are phenotypically obvious, any genetic change can result in mosaicism. Variegation is often confused with genetic mosaicism. Variegation can be a common expression of genetic mosaicism, but variegated plants need not be genetic mosaics (Kirk and Tilney-Bassett 1978; Marcotrigiano and Stewart 1984). In many species, the localization of pigments is either tissue- or organ-specific (e.g., occurs in petals but not in leaves) or position-dependent (e.g., occurs along the midrib of the leaf but not at the leaf edge) and is under the control of a single nuclear genome. For example, many cultivars of *Coleus* × *hybridus* Benth. have foliage with patterned variegation, yet the red, green, and yellow cells have the same genotype.

[1] Department of Plant and Soil Sciences, University of Massachusetts, French Hall, Amherst, MA 01003, USA

Biotechnology in Agriculture and Forestry, Vol. 11
Somaclonal Variation in Crop Improvement I (ed. by Y.P.S. Bajaj)

The mutations that give rise to mosaics need not be of nuclear origin. Mutations in chloroplasts or mitochondria can cause mosaicism. Chloroplast mutations which affect chlorophyll production can result in foliar variegation. In this case, all nuclei can be genetically identical, but cells can contain mixtures of normal and mutated chloroplasts (i.e., they are heteroplastidic). As the chloroplasts and cells divide, the chloroplasts will sort out, and eventually cells will be present which contain either all mutant or all normal chloroplasts (i.e., they become homoplastidic).

Nuclear genes can cause variegation patterns by genetically modifying some but not all plastids. For example, the recessive *iojap* mutant in corn, when homozygous recessive, results in variegated plants which possess both normal plastids and plastids with permanently defective ribosomes (Walbot and Coe 1979). Heterozygous *iojap* plants possess a normal phenotype, but when self-pollinated, yield ratios of 3:1 (green:variegated). If the variegated (i.e., the homozygous recessive) plants are used as females in crosses with green males, the female behaves like a typical plastid mutant which has not completely "sorted out". Offspring will contain green, albino, and variegated plants in no specific ratio and regardless of nuclear background. This cause of mosaicism is not common, but should not be ignored.

Specific variegation patterns can indicate a mosaic condition, but phenotypic evaluation alone is an unreliable technique for determining whether or not a plant is mosaic. Definitive proof can only be obtained by performing a genetic and morphogenetic analysis of the plant.

The cells in the shoot apex of a plant normally remain diploid. As differentiation proceeds, however, there is a tendency for endopolyploidy to occur (D'Amato 1964). All plants eventually become ploidy mosaics. Ploidy differences are generally tissue- or cell-specific (e.g., high ploidy levels for xylem and haploid numbers for gametes). The consequences of this fact will be discussed in Section 4. Infrequently (e.g., following exposure to mutagens), plants will consist of a randomly arranged array of cells of varying ploidy level. Such plants are generally referred to as mixoploid.

1.2 Chimeras

According to Poethig (1987), the term "chimera" has historically been used to describe plants with mutations in apical cell lineages (i.e., the "intra-apical mosaicism", described by Bergann 1967). The relative stability of a chimeral shoot is dependent on the original position of the mutation, the competition between mutant and nonmutant cells, the phyllotaxy of the species in question, and the fact that most higher plants possess apices which are composed of discrete cell layers (i.e., histogens). The existence of a layered apex is caused by the tendency for cells in the outer layers of the apex to divide anticlinally and therefore remain independent from the cell layers above or below them. In higher plants, there appear to be several apical initial cells within each layer of the apex. A mutational event in one of them normally leads to a chimeral plant possessing a region of genetically distinct tissue. Chimeras in which there is a mutated sector passing through all of the histogens are called sectorial chimeras (Fig. 1). Such mutated sectors can be lost if they occur in a region which does not form axillary shoots. Therefore, sectorial chimeras are extremely unstable. Sectorial chimeras can arise in early embryo ontogeny when a mutation in one cell contributes to many cell layers in the developing shoot apical meristem. Because the exact nature of the mutation is frequently unknown, or its phenotype is not clearly discernable in all cell layers, the term "sectorial" is frequently misused. More frequently, mericlinal chimeras are generated. A mericlinal chimera is a chimera which possesses a mutated sector in some, but not all layers of the meristem. In addition, the entire layer is not mutated (Fig. 1). This arrangement probably comes about when one apical initial in one cell layer mutates and generates a mutant cell lineage. Some axillary buds on mericlinal chimeras can possess a stable type of chimeral arrangement: the periclinal arrangement (Fig. 1). Periclinal chimeras are chimeras which have one or more (but not all) complete layers mutated. Mericlinal chimeras can revert to periclinal chimeras when random cell displacement occurs or if competition exists between the initials within a histogen. Periclinal chimeras are stable because, as mentioned previously, the strict planes of cell division (anticlinal) in the apex result in histogen independence. In addition, axillary buds will maintain the same histogenic arrangement as the terminal shoot which gave rise to them (Foster 1935; Reeve 1948). Therefore, most periclinal chimeras can be maintained by using vegetative propagation techniques

Fig. 1. Diagrammatic representation of a three-layered chimeral shoot apex with a complex mixture of normal green (*stippled*) and mutant white (blank) cell lines. The central shoot apex is in cross-section and the view is from above. Surrounding the apex are diagrammatic representations of five axillary shoots which have begun to grow following the decapitation of the central shoot apex. Longitudinal sections through the five axillary shoot apices are also shown. The five axillary shoots are (counter-clockwise beginning at the top) normal (G-G-G), sectorial chimera (G-G-G/W-W-W), fully mutant (W-W-W), mericlinal chimera (W-W-W/W-W-G), and periclinal chimera (W-W-G)

which utilize axillary buds (e.g., stem cuttings, grafting, or budding), but not adventitious buds (e.g., leaf cuttings or root cuttings).

When the histogens of the apex are different due to differences in ploidy level, the chimeras are called cytochimeras. Since increases in ploidy level generally result in increases in cell and nuclear size, longitudinal histological sections of shoot apices can indicate that such a condition exists (Satina et al. 1940). Another method to assess cytochimerism is to cytologically compare the chromosome number of pollen mother cells and root tips. Dissimilar chromosome counts indicate cytochimerism because the pollen mother cells do not originate from the same cell lineage as adventitious roots. A discussion of the derivatives of each apical cell layer is presented in Section 1.3.

The eventual sorting out of mutant and normal plastids can result in chimeras which possess layers of cells containing all mutant plastids and other cell layers containing cells with only normal plastids. This is the cause of most chimeras which possess foliar variegation (i.e., the "chloroplast chimeras").

Chimeras can also be caused by any nuclear mutation occurring in an apical cell layer. While this generally occurs in a single apical initial, following cell division and cell competition, a single mutation can result in an entire histogen of mutated cells.

The following discussions will deal with the anatomical reasons for the existence of plant chimeras, the causes for chimera formation and dissociation, and the implications chimeras have in plant anatomy, reproduction, and biotechnology. In many instances their occurrence becomes problematic because it is frequently the goal of biotechnology to create genetically uniform superior genotypes. In contrast, it must be mentioned that many horticulturally important species possess unique qualities because they are chimeral. Various fruit cultivars (Pratt 1983), foliage plants (Stewart and Dermen 1979) and floricultural crops (Sagawa and Mehlquist 1957; Stewart and Arisumi 1966; Stewart and Dermen 1970b; Pereau-Leroy 1974) are asexually propagated stable chimeras. In addition, useful information has been gained in developmental biology (Dermen 1960; Dulieu 1970; Stewart and Dermen 1970a, 1975; Stewart 1978; Poethig 1984, 1987; Marcotrigiano 1985), plant physiology (Heichel and Anagnostakis 1978) and entomology (Clayberg 1975) by using chimeral plants with phenotypically expressed genetic differences.

1.3 The Shoot Apex in Plants: Effects on the Persistence of a Chimeral State

The existence and perpetuation of a chimeral state in higher plants can be dependent on the relative persistence of shoot apical initials in shoot apical meristems. While considerable controversy exists regarding the number and permanence of such initials, a popular current opinion is that apical initials do not remain indefinitely fixed in the shoot apex, but instead are periodically replaced by their own descendents or by adjacent apical initials. Ultimately, it is the descendents of the shoot apical initials which make up the body of a plant. Therefore, barring new mutations, each shoot apical initial gives rise to a cell lineage of a specific genotype.

In many lower plants, such as in many pteridophytes, there is one terminal apical initial (Popham 1951), which is responsible for the entire body of the sporophytic generation. If mutation occurs in this initial, all subsequent daughter cells will be mutated. Therefore, plants with only one apical initial cannot maintain chimeral apices (Klekowski 1985).

Most angiosperms do, however, have several apical initial cells and these tend to exist in several layers in the apex. Dicots typically have three apical layers while monocots, depending on species, may have two or three (Tilney-Bassett 1963; Stewart and Dermen 1979). In addition, there is evidence that the number of layers may vary during shoot growth in a particular shoot or that certain species may possess more than the typical number of stable histogens (Tilney-Bassett 1963; Stewart et al. 1974; Stewart and Dermen 1975; Pohlheim 1982). This fluctuation is frequently ignored and can cause difficulties when attempting to interpret chimeras using phenotypic analysis.

The apices in gymnosperms are much more complex and variable (see Johnson 1951 for a review). Chimeras can be maintained in some families, such as the Cupressaceae, but they tend to behave as a two-layered apex where the outer apical layer frequently divides periclinally, invading the inner layer (see, e.g., Ruth et al. 1985). In other families there is little evidence of a stable outer layer. For more complete reviews on meristems see Foster (1939, 1941), Johnson (1951), Popham (1951), Clowes (1957), Romberger (1963), Gifford and Corson (1971), and Morel (1972).

1.4 The Derivatives of the Shoot Apical Initials

Plant chimeras with clearly discernable markers have been extensively used to trace cell lineages (Satina and Blakeslee 1941, 1943; Satina 1944, 1945; Dulieu 1970). By studying these chimeras, it has been possible to determine the fate of cells derived from the apical initials of the shoot apical meristem. The "tunica-corpus" concept, as introduced by Schmidt (1924), is generally accepted as the typical arrangement for cells in the shoot apex of the dicot. Most dicots possess a two-layered tunica covering the inner corpus. The outermost histogen of the tunica (the LI or TI) normally gives rise to the epidermis of the plant. The second apical layer (the LII or the TII) gives rise to the palisade parenchyma, the lower spongy parenchyma and all of the spongy parenchyma of the leaf margin. The LII is usually the layer responsible for the formation of male and female gametes. The third layer (LIII or corpus) gives rise to the upper and middle layers of the spongy parenchyma, as well as most of the central stem tissue such as pith. It generally makes no contribution to the leaf margin.

The contribution of apical layers in monocots is frequently similar to that in dicots. However, in most cases, the LI gives rise to all epidermal cells and the mesophyll cells along the edge of the leaf blade. In addition, there appear to be many monocots which possess only two apical layers, and therefore, the LII makes the major contribution to the body of the plant.

The relative contribution of each apical histogen to plant organs can vary depending on the species, and in chimeras depending on the severity of the mutation (Stewart et al. 1974; Tilney-Bassett 1986). In many cases, the ultimate structure of an organ is unchanged but the contribution that each layer makes is altered. In chimeras with extremely dissimilar tissue genotypes (e.g., interspecific and intergeneric chimeras) morphological abnormalities can occur (Jorgensen and Crane 1927; Marcotrigiano 1985).

1.5 Implications of Chimeras in Sexual Reproduction

Plant chimeras which originate following nuclear mutations cannot be perpetuated by sexual reproduction. Gametes are single cells which are ultimately derived from a select group of cells in the shoot apex. These cells are derived from one histogen (typically the LII). Therefore, offspring of a self-pollinated periclinal chimera will be nonchimeral and possess the genotype of the cell layer which gave rise to the pollen and eggs. For example, the White-White-Green (WWG) periclinal chimera in Fig. 1 would, upon self-fertilization, give rise to albino seedlings (unless replacement or displacement had occurred – see Sect. 3.1). Obviously, all embryos are derived from zygotes and, barring future mutation, should result in nonchimeral plants.

Mosaic plants can arise from seed if, at the time of egg formation, both mutant and normal proplastids exist in the egg cells (see, e.g., Burk et al. 1964). Such mosaic seedlings, following plastid sorting, can eventually become periclinal chimeras (see, e.g., Stewart and Burk 1970). If complete sorting occurs prior to egg formation, the seedling population of a self-fertilized plant would be nonchimeral because egg cells would contain only one type of proplastid. The inheritance of the plastid genotype is maternal in most plants but in some can be biparental. In plants with biparental inheritance the pollen contributes organelles and it is possible to recover mosaically variegated seedlings by performing sexual crosses between periclinal chimeras (e.g., crossing a GWG male to a WGG female). There is evidence in certain biparentally inherited species (Tilney-Bassett and Birky 1981) that the maternal plastids have a larger influence on the ultimate composition of the chloroplast genome in a developing seedling, and that wild-type plastids are generally more competitive than mutant types. For an extensive review of the chloroplast inheritance patterns in agriculturally important species see Kirk and Tilney-Bassett (1978).

2 Formation of Chimeras

Causes for the formation of chimeras are varied. Since chimerism may or may not be desirable, an understanding of the factors leading to their formation is critical.

2.1 Spontaneous and Induced Mutations

2.1.1 In Vivo

Mutations occurring in the shoot apices of higher plants generally result in the formation of chimeras. Mutations can occur spontaneously but infrequently, or they can be induced with physical or chemical mutagens to increase their frequency. The use of mutagens in mutation breeding programs has frequently resulted in chimeral plants. The technology of inducing mutations, pioneered by Stadler (1930), has been extensively discussed by Broetjes and Van Harten (1978).

2.1.2 *In Vitro*

Spontaneous mutations frequently occur in cultured plant cells. Thorough analysis of this subject has been performed, and a summary of work relevant to chimeras is presented in Section 4. Mutations can be induced in cell cultures with either chemical or physical mutagens (see King 1984). Sung (1976) concluded that the use of chemical mutagens to induce mutation in cultured cells resulted in a one-fold increase in the mutation frequency. Cell killing was found to be directly correlated with the mutation frequency. In some cases (see, e.g., Widholm 1974) it has been noted that spontaneous mutation rates often are high enough to eliminate the need for induced mutations. Nasim et al. (1981), studying the induction of mutation in cultured yeast cells, demonstrated that the recovery of either pure or mosaic cell colonies can be influenced by the type of mutagen. For example, agents like ultraviolet light produced mainly pure clones through a pre-replicative process involving an error-prone DNA repair process, while others such as ethyl methane sulfonate produced mainly mosaics due to the different nature of DNA lesions; lesions which may require a replication-dependent process for fixation of mutations.

2.2 Experimentally Synthesized Chimeras

Since many chimeral plants possess horticulturally desirable qualities, the synthesis of chimeras from known genotypes would be advantageous, since the components of the chimera could be genetically characterized prior to synthesis. This would allow for a degree of predictability regarding the characteristics of the chimeral plant.

2.2.1 *In Vivo*

The earliest known synthesized chimeras were those arising from the graft union of plants grafted for horticultural purposes. At first, it was thought that such peculiar shoots were somatic hybrids which resulted from cell fusions occurring at the graft union (see discussions in Neilson-Jones 1969 and Tilney-Bassett 1986). Later, Baur (1909) determined that the foliar variegation in certain cultivated geraniums was due to independent albino and green cell lines coexisting in a structured shoot apex. Additional studies with "graft hybrids" indicated that these plants were actually chimeras which arose when cells from both the understock and the scion gave rise to a single adventitious shoot of multicellular origin (see Neilson-Jones 1969). Previously known as "graft hybrids", these plants are now called "graft chimeras". Theoretically, any two plants that are graft-compatible and which can form adventitious shoots from the graft union, can form graft chimeras. This technique has been most successful in the Solanaceae (Jorgensen and Crane 1927; Krenke 1933; Clayberg 1975; Marcotrigiano and Gouin 1984b). Interspecific and intergeneric chimeras have arisen from the graft unions of woody plants commonly grafted for other reasons, and less frequently by experimental synthesis. Woody plant chimeras exist in the Rutaceae (Bizzaria, Tanaka 1927), Leguminosae (+ *Laburnocytissus adami*, Neilson-Jones 1969), Rosaceae (+ *Crataegomespilus* spp., Bergann and Bergann 1984), and Theaceae (*Camellia* + Daisy Eagleson, Stewart et al. 1972). For a thorough review of the occurrence and history of graft chimeras see Weiss (1930), Neilson-Jones (1969), and Tilney-Bassett (1986).

2.2.2 *In Vitro*

Attempts have been made at synthesizing chimeral plants by regenerating shoots from mixed cell cultures (Ball 1969; Carlson and Chaleff 1974; Koenigsberg and Langhans 1978; Marcotrigiano and Gouin 1984b). These experiments utilized heterogeneous callus cultures which were formed by mixing together genetically distinct cell lines and allowing the cells to "graft" in vitro. However, in all but one case, only nonchimeral plants of both genotypes were recovered following organogenesis. Carlson and Chaleff (1974) were able to isolate 28 chimeral plants from approximately 7000 callus-derived shoots. In their system, interspecific chimeral calli composed of *Nicotiana tabacum* and the amphiploid hybrid of *N. glauca* × *N. langsdorfii* were transferred to media which favored the regeneration of only one species. The regenerated shoots, which had morphological characteristics of the species unable to regenerate, were selected. When the same technology was attempted with *N. glauca*/*N. tabacum* cell mixtures, only nonchimeral shoots were recovered (Marcotrigiano and Gouin 1984b). It has been noted, however, that competition between genetically distinct cell lines can occur (Bayliss 1977). This phenomenon may reduce the probability of recovering chimeras from mixed callus cultures (Marcotrigiano and Gouin 1984a).

Recently, Binding and co-workers (1987) have recovered interspecific chimeras in *Solanum* following the co-culture of protoplasts. Periclinal, sectorial, and mericlinal chimeras were recovered, and while the frequency was not reported, this technique appears to hold promise when chimeral synthesis is desired. An important implication of these findings is the potential of densely plated protoplasts to form microcalli and eventually shoots derived from more than one cell. Therefore, even when fusion agents such as polyethylene glycol are used in an attempt to form somatic hybrids, a fraction of unfused but tightly packed protoplasts can form mosaic calli which regenerate into chimeral plants.

3 Separation of Chimeras

Plant chimeras can be dissociated into their component genotypes utilizing various techniques. This may or may not be desirable, depending on the qualities which are imposed by the state of chimerism.

3.1 In Vivo

Whether or not the chimeral condition persists can be dependent on the position of the mutant cell in the shoot apex and the severity of the mutation. Many chimeras are transient because the mutant cell is at a selective disadvantage and soon becomes eliminated from the apex. This phenomenon has been termed "diplontic selection" (Balkema 1971) when it is the competition between genetically dissimilar cells which causes the loss of chimerism, or "diplontic drift" when the chimeral state is lost due to normal developmental processes (Balkema 1972). Environmental factors which effect growth rate can influence the size and persistence of a mutant cell line (Buiatti et al. 1970). Klekowski and Kazarinova-Fukshansky have mathematically modeled the fate of "neutral" (1984a) and "disadvantaged" cell genotypes (1984b) in stochastic shoot apices. The results indicate that "neutral" mutations will persist for many nodes if selected randomly from a pool of apical initials. In contrast, a "disadvantaged" genotype can be quickly eliminated from the shoot apex. This elimination is dependent on the number of apical initials and on the number of cell divisions which occur before another set of apical initials is "chosen".

The stability of a chimeral state is also dependent on the arrangement of the genetically dissimilar tissue within the plant (e.g., periclinal vs. mericlinal chimera) and the number of shoot apical initials in a shoot apical meristem. Even the most stable arrangement, the periclinal chimera, can change when cells are displaced or replaced (Dermen 1960; Stewart and Dermen 1970a). The terms "perforation" and "reduplication" coined by Bergann and Bergann (1959) are synonymous for displacement and replacement, respectively. Displacement occurs when a cell from an inner layer takes over the position of a cell from the outer layer. Replacement occurs when there is a periclinal division of a cell in an outer layer followed by further anticlinal divisions resulting in an inner layer now having a sector or complete layer of cells of the genotype of the outer layer. These events most frequently occur during organ development and result in patches of tissue being ultimately derived from a different histogen than normal (Fig. 2). Less frequently, displacement or replacement occurs in the shoot apex. When it occurs, the arrangement of the chimera can be completely altered. Displacement or replacement occurring near the summit of the apex can result in a drastic change in histogenic composition, while those occurring near the base of the apex will result in a thin sector having an altered histogenic composition (Ruth et al. 1985). For example, a GWG periclinal shoot can become a GGW shoot if an LI apical initial divides periclinally and it and its descendents replace the initials in the layer below. The green LIII initials are then replaced by the white LII initials. Conversely, the GWG shoot can become a WGG if the LI initials are displaced by the LII initials below them. If this were to occur again, the white genotype could be completely eliminated and the apex would no longer be chimeral.

Displacement and replacement events periodically occur even when one genotype is not disadvantaged. In addition, it has been noted that environmental factors, which alter the growth rate of a plant, can affect cell division patterns in shoot apices (Popham 1951) and this can affect the persistence of a chimeral apex (Balkema 1972). Instabilities in the tunica layers of woody plant apices can be a seasonal occurrence (Reeve 1948; Pillai 1963).

Fig. 2. A An atypical bract from Annette Hegg poinsettia, a periclinal chimeral cultivar. The bracts of this cultivar are usually white with varying amounts of pink in the region of the midrib. The original periclinal shoot apex possesses a White-White-Red (W-W-R) histogenic arrangement. The unusual bract shown demonstrates the phenomenon of replacement. The red region (*R*) is R-R-R and is ultimately derived from the LIII. During bract formation the LIII derivatives have replaced the LII and LI derivatives. The dark pink region (*DP*) is W-R-R and originated when the red LIII replaced the white layer above it. The pale pink region (*P*) is W-W-R and is typical of the cultivar. The white region (*W*) is W-W and is composed of LI and LII derivatives and appears normal since this region of a bract or a leaf does not contain LIII derivatives. **B** Carnation flower taken from a W-R-R White Sim carnation plant. Flowers, unlike leaves or bracts, rarely have LIII derivatives in petal tissue. Therefore, the flower should be W-R. Since anthocyanins are only expressed in the epidermal cell layers in carnation, flowers of White Sim are generally white. This specimen, however, shows a region of red tissue (R) where, during the course of flower formation, the LI was replaced by LII derivatives causing a R-R sector to form. Therefore, this particular flower should be considered mericlinal (W-R/R-R). This flower is an example of a chimera which appears to be sectorial and emphasizes the difficulty in using phenotypic observations to determine the nature of the chimera. A true sectorial chimera (White-White/Red-Red) would look identical. The conclusion that this flower is mericlinal is made because axillary shoots on the "white side" of the plant occasionally give rise to flowers like this one, indicating that the red genotype is still present in the LII on the "white side" of the plant

The most common cause of chimeral dissociation is the production of adventitious shoots. The production of adventitious shoots can be encouraged by the surgical removal of all terminal and axillary shoot buds (i.e., disbudding). The adventitious shoots which eventually arise are usually nonchimeral. Such technology has been used to separate the components of cytochimeral apple trees (Dermen 1948), potato chimeras (Miedema 1973) and flower color chimeras in chrysanthemums (Stewart and Dermen 1970b). While adventitious shoots can be multicellular in origin, they frequently give rise to nonchimeral shoots. If the adventitious shoots are chimeral. the periclinal arrangement of the two genotypes in the shoot is frequently different than that of the original shoot (Stewart and Dermen 1970b; Bergann and Bergann 1982). Shoots which are ultimately derived from a single cell would be nonchimeral.

Since adventitious shoots can arise during vegetative propagation. such techniques can be useful to separate a chimeral plant into its constituent genotypes. Leaf cuttings (Bergann and Bergann 1959. 1982; Burk 1975) and root cuttings (Bateson 1916) have both been used to dissociate chimeras in vivo. The origin of shoots from leaf cuttings is variable and largely dependent on the species in question. For example. in leaf cuttings in *Saintpaulia*. most evidence indicates that shoots tend to originate from the epidermis (Pohlheim 1981) while in *Nicotiana* (Burk 1975) and *Sansevieria* (Marcotrigiano and Morgan 1988) derivatives of the LII and LIII tend to give rise to adventitious shoots. Adventitious roots typically arise from LIII derivatives and root cuttings will give rise to shoots with the genotype of the LIII of the original shoot (Tilney-Bassett 1963).

If the dissociation of a chimera is desirable. and if the species in question does not readily produce adventitious shoots. other techniques can be used. Foliar sprays of the cytokinin. 6-benzylaminopurine (BA). have been used to stimulate shoot production from *Cordyline terminalis* Celeste Queen. a chimeral foliage plant (Maene and Debergh 1982). The higher concentrations of BA led to the recovery of many nonchimeral shoots. Treatments such as radiation. which are capable of destroying cells in the outer most layers of the apex. have been used to cause high-frequency displacement in chrysanthemum (Bowen et al. 1962). potato (Howard 1964). and carnation (Dommergues and Gillot 1973). This technique has also been utilized to elucidate the chimeral nature of the White Sim and Pink Sim carnations (Sagawa and Mehlquist 1957).

3.2 In Vitro

While most of the factors which are responsible for the dissociation of chimeras mentioned in Section 3.1 also occur in tissue culture. in most situations they occur more frequently in vitro. With the exception of shoot tip cultures. plant tissue culture systems do not utilize axillary shoot formation. Systems which do not utilize axillary or terminal shoot apices generally cause chimeras to separate into their component genotypes. If chimeral shoots should arise. their histogenic composition is frequently altered. The phenomenon of chimeral separation associated with specific tissue culture systems will be discussed in Section 5.

4 Chimerism and Somaclonal Variation

Somaclonal variation has been defined by Larkin and Scowcroft (1981) as that variation which is the direct result of some form of cell culture (i.e.. tissue culture). Its causes are numerous and the subject has been extensively reviewed (Skirvin 1978. Binns 1981; Larkin and Scowcroft 1981; Meins 1983; Evans et al. 1984. see also Chap. I.1 and I.6. this Vol.).

One of the most overlooked causes or results of variation in tissue culture is chimerism. As mentioned previously, chimeras are easily dissociated in vitro and the recovered phenotypes could be considered somaclonal variants. Somatic mutations can be carried along in the shoot of a plant for extensive periods (Klekowski and Kazarinova-Fukshansky 1984a) and mutant cells can be incorporated into virtually any plant part. Somatic crossing-over (Evans and Paddock 1976) and endopolyploidy (D'Amato 1964) are common in plants, and there is a possibility that the explant chosen may be chimeral or cytochimeral. Since endopolyploidy occurs more frequently in mature tissues, choosing such explants could result in a higher incidence of chimera recovery.

A priori knowledge that a plant is chimeral should immediately eliminate many tissue culture techniques as methods for perpetuating chimeras. However, many genetic changes are not outwardly obvious, and occasionally, explants of a particular clone may be chimeral. There is evidence that genetic differences can preexist in young cotyledonary (Barbier and Dulieu 1983) and other tissues (Lorz and Scowcroft 1983). In some cases it is difficult to determine whether the genetic variability preexists in the explant or occurs after in vitro growth begins (see, e.g., Seeni and Gnanam 1981). Nevertheless, in some reports, the recovery of variants could result from the dissociation of a chimeral explant (especially "extra-apical mosaics").

While utilizing mosaic explants could be a reason for obtaining variants, the recovery of chimeras from nonchimeral explants is probably a more frequent occurrence. High-frequency genetic variation in tissue cultures can be caused by many factors such as somatic crossing over, deletions, inversions, translocations, and ploidy changes (Larkin and Scowcroft 1981). Variation is enhanced if cultures are maintained for long periods of time. If genetic changes occurred prior to organogenesis, the recovered plants would be chimeral only if they were ultimately derived from more than one cell and if, within the multicellular organizational event, the shoot arose from both normal and mutant cells. While there is still considerable controversy on the subject of single versus multicellular origin in the process of organogenesis (see, e.g., Broertjes and Keen 1980), Dulieu (1967) and Marcotrigiano (1986) have demonstrated that the multicellular origin of shoots can occur when internode or leaf explants are used. In both studies, it was also determined that shoots could arise from any cell layer in cultured leaf tissue and that cells derived from more than one layer (e.g., epidermal and mesophyll derivatives) could organize into a single shoot. When chimeras were recovered from chimeral explants they did not necessarily possess the same histogenic arrangement as the original explant.

There are numerous reports on the recovery of cytochimeras from tissue culture (e.g. Horak 1972; Ogura 1976; Bennici and D'Amato 1978; D'Amato 1978; Sree Ramulu et al. 1976; Mix et al. 1978; Bennici 1979; Hussey and Falavigna 1980; Keathley and Scholl 1983; Jagannathan and Marcotrigiano 1987). In contrast, Mehra and Mehra (1974) noted that while callus cultures of almond were extremely heterogenous in chromosome number, the regenerated shoots were all diploid suggesting that, in almond, chromosome aberrations may have a direct impact on the morphogenetic capability of the cells. D'Amato (1964) has noted that the composition of the media can favor diploid cell lines.

5 Direct Implications of Chimerism in Tissue Culture and Genetic Engineering

The development of plant tissue culture and genetic engineering technology has led to numerous tissue, cell, organ, and genetic manipulations which were previously impossible. Consequently, it is timely to discuss the ramifications of chimerism in biotechnology.

5.1 Chimeras and Meristem Culture

Meristem culture involves the excision of a shoot apical meristem less than or equal to 1 mm in length (Hartmann and Kester 1983). While it is known that the growth patterns of excised shoot apices in culture can vary from species to species (Morel 1972), little is known about the effect of meristem excision on the stability of the histogens of the shoot apical meristem. Marcotrigiano et al. (1987) excised strawberry runner tips from a Green-White-Green periclinal chimera of alpine strawberry, *Fragaria vesca*. Explants ranged in size from 0.6 to 5.6 mm in length. On medium which was not conducive to shoot multiplication, the integrity of the histogens remained unaltered in the developed plants. Cassells et al. (1980) obtained similar results with geranium chimeras. Therefore, if no damage is done to the apex, development appears to proceed normally. Skene and Barlass (1983) noted that fragmenting the shoot apices of a chimeral grape cultivar prior to culturing led to the breakdown of the chimeral structure. Johnson (1980) obtained similar results with chimeral carnations. In addition, it has been demonstrated that shoot apices which are subjected to exogenous hormone application possess unique morphology and cell distribution patterns (Smith 1968; Mauseth 1979; Pereira and Dale 1982). The impact of these findings on the maintenance of a chimeral apex has not been studied.

5.2 Chimeras and Shoot Cultures (Micropropagation)

The commercial micropropagation of most woody and herbaceous ornamentals, edible fruits, and foliage plants is based on the establishment and periodic subculture of shoot cultures, which are generally established from preexisting terminal or axillary buds. Therefore, in theory, the primary explant from a periclinal chimera should develop into a shoot which maintains its chimeral integrity. However, the micropropagation of periclinal chimeras generally leads to the dissociation of the chimeral state. There have been no detailed studies to determine whether this is caused by the production of adventitious shoots or whether the cell division planes in shoot apices are disturbed during rapid multiplication. An analysis of regenerants from shoot cultures of *Fragaria vesca* Albo-Marginata indicated that both processes were probably involved (Marcotrigiano et al. 1987). In some instances adventitious shoots develop in conjunction with axillary shoots (George and Sherrington 1984). It has also been proposed that the culture environment may cause enhanced competition between histogens, leading to the eventual elimination of the disadvantaged cell line (George and Sherrington 1984). Whatever the cause, it is evident that high-frequency multiplication of periclinal chimeras is difficult to achieve in vitro. While a certain percentage of shoots identical to the original explant are recovered, many economically valuable chimeral plants have been dissociated during the process of tissue culture (Table 1).

Chloroplast chimeras, which are not periclinal chimeras (i.e., the plastid types have not sorted out), can be propagated in vitro. However, the regenerants will be variable in appearance. Micropropagation has been successfully accomplished with variegated mutants of *Syngonium* and *Paulownia tomentosa* (Marcotrigiano unpubl.). In addition, this technique could be used to obtain large numbers of chloroplast chimeras from a single mutant seedling or shoot. These could be grown in vivo until the periclinal state was achieved. Conventional propagation could then be used to perpetuate the chimera.

5.3 Chimeras and Somatic Embryogenesis

Somatic embryos, being of single cell origin, should be nonchimeral if mutational events do not occur after the establishment of a structured shoot apical meristem. Fully mutant and fully normal plants should be recovered from chimeral explants. If, however, the mutation occurs after the embryo has begun to form, it is possible that some of the recovered plants would be chimeral. In many tissue culture systems it may be difficult to determine whether the regenerants are of embryogenic or organogenic origin. Indeed, it has been demonstrated that in some culture systems it is possible to regenerate plants via somatic embryogenesis and organogenesis (Lowe et al. 1985; Barwale et al. 1986). A criterion for determining whether or not shoots regenerate via organogenesis or embryogenesis, based on the origin of vascular tissue and the ontogeny of the developing plant, has been suggested by Haccius (1978). Determining which process has led to plant formation is critical since, as noted in Section 5.4, the recovery of chimeras following organogenesis may be a frequent occurrence. In addition, it is important to note that critical differences between the ontogeny of zygotic versus nonzygotic embryos can exist. Haccius and Bhandari (1975) have noted that more cell divisions occur before recognizable histogens appear in the apices of nonzygotic embryos, when compared to zygotic embryos. This phenomenon could influence the persistence and arrangement of genetically dissimilar cells if a mutation occurs shortly after the initiation of an embryo.

5.4 Chimeras and Organogenesis

Organogenesis differs from somatic embryogenesis in that the regeneration process results in the formation of an organ, not an embryo, and therefore, shoots or roots are obtained. In many cases, regenerated shoots quickly form adventitious roots and "whole" plants are recovered. Shoot organogenesis is common in vivo in species which can be propagated vegetatively from root or leaf cuttings. In vitro organogenesis is also common especially in certain families (e.g., the Solanaceae). While there may be instances where shoots are ultimately derived from one cell (Broertjes

Table 1. Species in which alterations in chimeral structure occurred following culture of explants derived from periclinal chimeras

Plant species and cultivars	Explant	Tissue culture system	Regenerants[a]	Reference
Aechmea fasciata Albo-marginata	Shoot tip	Shoot culture	Separation occurs. % not given	Jones and Murashige (1974)
Ananas comosus Variegatus				
Cryptanthus It				
Chrysanthemum morifolium Indianapolis	Shoot tip	Shoot organogenesis from callus	100% nonchimeral	Bush et al. (1976)
Dianthus caryophyllus S. Arthur Sim	Meristem tip	Shoot organogenesis from callus	7% identical chimeras. 93% nonchimeral	Johnson (1980)
Braun's Yellow Sim			29% identical chimeras. 47% nonchimeral. 24% new chimeral forms	
Dusty	Macerated shoot tip	Shoot cultures from shoot tips	85% identical chimeras. 10% non-chimeral. 5% new chimeral forms	
Pink Ice			80% identical chimera. 20% new chimeral forms	
Dianthus caryophyllus White Sim	Shoot apex	Not described	Separation occurs. % not given	Hackett and Anderson (1967)
Dianthus caryophyllus White Sim	Shoot tip	Shoot organogenesis from callus	Separation occurs. % not given	Dommergues and Gillot (1973)
Dracaena marginata Tricolor	Stem explants	Shoot organogenesis from callus	100% nonchimeral	Chua et al. (1981)
Episcia cupreata Pink Brocade	Leaf pieces	Shoot organogenesis	100% nonchimeral	Chin (1980)
Euphorbia pulcherrima Annemie	Shoot tips	Shoot organogenesis from cell suspensions	100% nonchimeral	Preil and Engelhardt (1982)

(*to be continued*)

Table 1. *(Continued)*

Plant species and cultivars	Explant	Tissue culture system	Regenerants[a]	Reference
Fragaria vesca Albo-marginata	Shoot tips	Shoot organogenesis and shoot cultures	8% identical chimeras, 82% nonchimeral, 10% new chimeral forms	Marcotrigiano et al. (1987)
Hosta sieboldiana Frances Williams	Flower scapes	Shoot organogenesis from callus[b]	45% identical chimeras, 55% nonchimeral	Meyer (1980)
Nicotiana tabacum Samsun GWG periclinal chimera	Internodal stem explant and leaf tissue	Shoot organogenesis	Nonchimeral and new chimeral forms, % not given	Dulieu (1967)
Interspecific *Nicotiana* periclinal chimeras	Leaf tissue	Shoot organogenesis	2% identical chimeras, 93% nonchimeral, 5% new chimeral forms	Marcotrigiano (1986)
Pelargonium × *zonale* Mme Salleron	Meristem petiole	Adventitious shoots[c] shoot organogenesis	100% identical chimeras 100% nonchimeral	Cassells et al. (1980)
Pelargonium graveolens Rober's Lemon Rose		Shoot organogenesis from callus	0.6% identical chimeras 99.4% nonchimeral	Skirvin and Janick (1976)
Pelargonium sp. chloroplast chimera	Leaf tissue	Shoot organogenesis from protoplasts	100% nonchimeral	Kameya (1975)
Rubus lacianthus Thornless Evergreen	Shoot tip	Shoot culture	˙60% identical chimeras 40% nonchimeral	McPheeters and Skirvin (1983)
Rubus sp. Thornless Loganberry L654	Shoot apical meristem	Not determined	Separation occurs percent not given	Hall et al. (1986)
Rubus sp. Thornless Logan	Shoot tips	Shoot culture (from terminal and axillary buds)	92.6% identical chimeras, 7.4% new chimeral forms	Rosati et al. (1986)

Saintpaulia ionatha Valencia	Variety of explants[d]	Shoot organogenesis	No identical chimeras, 66% nonchimeral, 34% new chimeral forms	Lineberger and Druckenbrod (1985)
	Inflorescence (incl. nodes)		46% identical chimeras, 36% nonchimeral, 18% new chimeral forms	
Daredevil	Variety of explants		1% identical chimeras, 87% nonchimeral, 12% new chimeral forms	
	Inflorescence (incl. nodes)		31% identical chimeras, 55% nonchimeral, 14% new chimeral forms	
Desert Dawn	Variety of explants		No identical chimeras, 100% nonchimeral	
	Inflorescence (incl. nodes)		20% identical chimeras, 80% nonchimeral	
Sansevieria trifaciata Bantel's Sensation Golden Hahnii Laurentii	Leaf pieces	Adventitious shoots (somatic embryos?)	100% nonchimeral	Marcotrigiano and Morgan (1988)
Yucca elephantipes (chimeral form)	Axillary shoot tips	Axillary branching	100% identical chimeras	Pierik and Steegmans (1983)
		Adventitious buds from shoot cultures	100% nonchimeral	

[a] Readers may consult original references, since in some cases results are open to interpretation. Identical chimeras refer to regenerants that possess chimeral apices identical to the explant, nonchimeral refers to the complete separation of chimeral explants, new chimeras refers to new chimeral arrangements of the original genotype. The table does not include chimeras which may have arisen de novo via somaclonal variation.
[b] No histological evidence was presented to eliminate the possibility of axillary origin.
[c] The authors state adventitious but the reader's interpretation is that the plants were derived from original meristems.
[d] Leaf, petiole, peduncle, petal, sepal, and subepidermal explants were employed.

and Keen 1980), there is also definitive proof that more than one cell can ultimately give rise to a shoot (Dulieu 1967; Marcotrigiano 1986). In addition, shoots may appear to have arisen from a single cell when it is possible that two genetically identical cells gave rise to the shoot. The recovery of nonchimeral mutant shoots following the induction of mutations in an explant may indicate that the visible plant is no longer chimeral, but does not prove that the shoot originated from a single cell. It could have arisen from a group of cells in which factors such as diplontic selection, diplontic drift, displacement, and replacement have eliminated one genotype from the apex. Therefore, whenever the pathway of organogenesis is followed, there is a possibility that chimeras will result. If this is undesirable it is possible to dissociate these plants into their component genotypes using procedures discussed in Section 3.

5.5 Chimeras and Androgenesis

Androgenesis is the production of plants from cultured anthers. In species where the pollen is the source of regenerants the regeneration pathway is generally embryogenic, in which case considerations in Section 5.3 should be noted. In some species, regenerants are recovered only after callus proliferation. In such cases, the regenerants could be derived via organogenesis and considerations in Section 5.4 should be noted. In addition, if the diploid tissue surrounding the pollen is included in the explant and plants are regenerated via organogenesis, it is possible to recover cytochimeras. Chromosome doubling can also occur in some cells of calli which have originated directly from pollen and, if an organogenic pathway is followed, cytochimeras could result.

Another possibility is that the tissue culture system itself can cause genetic variability. In some instances (see, e.g., Burk and Matzinger 1976), doubled haploids do not appear to be entirely homozygous even when the starting material is an inbred line. In addition, cytoplasmic mutations can be induced during the process of androgenesis (Matzinger and Burk 1984), and it is feasible that plastid or mitochondrial chimeras could be recovered.

5.6 Chimeras and Protoplast Fusion

Through protoplast fusion, the nuclei and organelles of both cells can be combined into a single cell. True hybrids form if the nuclear material of both partners becomes integrated into one nucleus. However, in some cases, the nuclei do not fuse and the cell becomes binucleate. Subsequent cell division can lead to chimeral cell clusters which, depending on the mode of regeneration (organogenesis vs. embryogenesis) and the amount of cell division before regeneration, could result in the formation of chimeral plants.

By definition, any successful cell fusion between cells with dissimilar mitochondrial and/or chloroplast genomes will result in a chimeral state. As mentioned previously, the chimeral state will persist until the sorting-out process is complete. While sorted cells contain only one genome, the cell clusters are frequently composed of cell of both types. Again, this phenomenon can result in the regeneration of chimeral plants if the organogenetic pathway is followed.

Plastid or mitochondrial chimeras may or may not result following plant regeneration from fused protoplasts. Although protoplast fusion causes cytoplasmic "mixing", it is not possible to predict which plastid or mitochondrial genome will eventually dominate. In some instances, it has been noted that chloroplasts sort out randomly and completely in cells which have hybrid nuclei (Scowcroft and Larkin 1981; Kumar and Cocking 1982). Plastid heterogeneity has occurred to a greater extent in some regenerants (Maliga et al. 1978, 1982), and to a much lesser extent in others (Chen et al. 1977). Obviously, the relative number, viability, and division rate of the plastids can greatly influence the sorting-out procedure. For example, fusing cell suspension protoplasts with leaf mesophyll protoplasts will result in a numerical advantage for the plastids of the leaf-derived protoplasts. Experiments by Fluhr et al. (1983) indicate that selection pressure against one plastid type can greatly accelerate the sorting out process, whereas in noncompetitive situations plastid heterogeneity can extend into the next sexual generation. Preliminary evidence (Iwai et al. 1981) indicates that pollen derived from heteroplastidic plants can yield heteroplastidic haploids following anther culture. For a more detailed discussion of the possible products of fusion see Gleba and Evans (1983).

Not all protoplasts fuse following fusion treatments. If plating densities are high, microcalli which are composed of fused and original cell types can form resulting in complex chimeras.

The above discussion clearly demonstrates that the products of protoplast fusion must be carefully studied to demonstrate if they are indeed hybrids. Since chimeras can possess unique morphology, can be infertile, or can possess intermediate phenotypes, they could easily be confused with somatic hybrids. In addition, since chimeral tissue contains cells of both genetic components, electrophoretic protein analysis of the tissue could indicate the presence of both genomes.

5.7 Chimeras and Secondary Products (Metabolites)

The extraction of biochemicals from cultured plant cells has important implications for the industrial production of medicinal alkaloids, food additives, dyes, perfumes, and insecticides. This subject has been covered in depth (Butcher 1977; Mantell et al. 1985; Staba 1980; also see Bajaj 1988). Plant tissue culture can be utilized to dissociate genetic or epigenetic mosaics. Therefore, if techniques are developed to identify and isolate cells which overproduce a desired compound, it will be possible to clone these superior cell lines. This has been demonstrated in carrot clones isolated from suspension-derived heterogenous callus cultures for their ability to accumulate anthocyanins (Dougall et al. 1980), and other medicinal plants (see Bajaj 1988).

5.8 Chimeras and Genetic Engineering

Using *Agrobacterium tumefaciens* as a vector, it is possible to transfer genes into plants (Chilton 1983). The first techniques which were developed relied on the infection of protoplasts with *Agrobacterium* and the subsequent regeneration of plants. Since many of the protoplasts are not transformed, it is necessary to use a selection system designed to eliminate nontransformed cells. This is accomplished by utilizing another gene in the system which will allow a selection pressure to be imposed. Genes for antibiotic resistance are most commonly used (e.g., kanamycin resistance).

Assuming that no spontaneous mutations occur following transformation, transformed protoplasts should yield nonchimeral transformed plants. However, the requirement of totipotency limits the use of protoplast transformation to improve many agriculturally important species. Horsch et al. (1985) have demonstrated that some of the cells in leaf discs exposed to genetically engineered *Agrobacterium tumefaciens* can be successfully transformed and transformed plants can be regenerated via shoot organogenesis. The system, like the protoplast system, is dependent on selection pressure to eliminate nontransformed cells. However, it is possible that the selection pressure is not as stringent in a multicellular system in which cross-feeding or physical exclusion from the selection agent is possible. A leaf disc composed of transformed and untransformed cells can be considered chimeral. Marcotrigiano (1986) has demonstrated that shoots arising from leaf discs can be multicellular in origin and that if the explant is chimeral it is possible to recover chimeral shoots possessing cells which should have been unable to regenerate under the imposed cultural conditions. Therefore, it is possible that the use of the leaf disc system may result in the recovery of plants which are chimeras (i.e., composed of both tranformed and untransformed cells). It is important that care be taken to note the position on the "transformed" plant where seed is collected because, if the plants were chimeral, Mendelian ratios could be altered by "pooling" seed from transformed and nontransformed sectors of the plant.

6 "Epigenetic Chimeras"?

It is well documented that some of the variation which is observed following tissue culture is caused by changes in gene expression rather than structural changes in the plant genome (Binns 1981; Meins 1983). It is also documented that more than one cell can be involved in shoot formation when an organogenic pathway to regeneration is followed (see, e.g., Marcotrigiano 1986). There should be no reason why a regenerated shoot could not contain a mixture of cells, some of which exhibit altered gene expression. In our laboratory we are currently regenerating *Coleus* × *hybridus* plants from shoot tip culture and have recovered several plants which appear to be chimeral because only certain sectors possess leaves with anthocyanins or leaves with albino midribs (Fig. 3). When a detailed genetic analysis is completed it will be possible to determine if the variants are altered because of true mutation or if the plant is an "epigenetic chimera" in which the shoot apical meristem possesses cells

A

Fig. 3A,B. "Chimeral" leaves from somaclonal variants of *Coleus* × *hybridus*. Leaf **A** is green and white on one side of the midrib and red on the other, while leaf **B** is green and white on one side of the midrib and all green on the other. Both leaves came from the same shoot culture. They originated from a single axillary bud of a cultivar which characteristically possesses green leaves with white centers and random blotches of red pigmentation. We are currently determining whether these variants are true mutants or differ only at the level of gene expression

exhibiting different degrees of gene expression. If the latter explanation is proved, to our knowledge it would be the first documented case of an "epigenetic chimera", and would add one more item to the list of causes of somaclonal variation.

7 Concluding Remarks

Genetic variation in tissue culture, whether considered a beneficial phenomenon with potential use in plant improvement, or detrimental, as in clonal propagation, will continue to be a phenomenon deserving serious investigation. An understanding of the development of mosaic or chimeral plants becomes essential if one is to fully appreciate the significance of genetic heterogeneity in a multicellular system. There is a strong possibility that a plant regenerated from tissue cultures will not be genetically homogeneous. This frequently ignored fact has undoubtedly clouded the interpretation of some investigators who are involved in the selection of mutant plants. In contrast, the spontaneous appearance or deliberate creation of genetically heterogeneous plants may result in novel plants with desirable characteristics, another fact frequently ignored.

It is important to realize that an accurate interpretation of genetic mosaicism and chimerism is dependent on a knowledge of the genetic and anatomical characteristics of the plant in question. False conclusions can easily result if a complete characterization of the plant is not performed prior to its use in genetic and morphological research. Frequently, genetic changes which occur in specific tissue layers are not expressed unless derivatives of these cell layers are displaced into other positions. It should not be assumed that a mature plant is genetically homogeneous. Somatic mutations and ploidy changes do occur and are generally tolerated. Because the body of higher plants is ultimately derived from more than one apical initial cell and more than one apical cell layer, genetic changes frequently result in chimeras. When mutations occur in nonapical cells, they are frequently of no consequence. However, the technology of tissue culture allows such mutations to be recovered, as plants can be regenerated from somatic cells. Tissue culture systems which result in plant formation from single cells can avoid the recovery of chimeral plants. However, systems which yield shoots of multicellular origin can yield chimeral regenerants. Understanding genetic mosaicism was previously important only to those involved in mutation breeding. Biotechnology has advanced so rapidly that a knowledge of the factors which lead to genetic mosaicism and chimerism is important for all plant researchers. This knowledge is deeply seated in classical anatomy, genetics, and developmental biology. Researchers are urged to gain such knowledge so as to avoid the misinterpretation of data which are being generated in unprecedented amounts.

Acknowledgments. The author greatly acknowledges the critical reviews of Drs. R. Scott Poethig, Robert M. Skirvin, and R. Daniel Lineberger, the proofreading and suggestions of Susan E. Bentz and Pamela A. Morgan, and the artistic talent of Dennis J. Harris who drew Fig. 1.

References

Bajaj YPS (ed) (1986) Biotechnology in agriculture and forestry, vol 2. Crops I. Springer, Berlin Heidelberg New York Tokyo

Bajaj YPS (ed) (1988) Biotechnology in agriculture and forestry, vol 4. Medicinal and aromatic plants I. Springer, Berlin Heidelberg New York Tokyo

Balkema GH (1971) Chimerism and diplontic selection. PhD Thesis, Landlouwhogeschool, Wageningen. Balkema, Rotterdam Cape Town, 173 pp

Balkema GH (1972) Diplontic drift in chimeric plants. Radiat Bot 12:51–55

Ball EA (1969) Histology of mixed callus cultures. Bull Torrey Bot Club 96:52–59

Barbier M, Dulieu H (1983) Early occurrence of genetic variants in protoplast cultures. Plant Sci Lett 29:201–206

Barwale VB, Kerns HR, Widholm JM (1986) Plant regeneration from callus cultures of several soybean genotypes via embryogenesis and organogenesis. Planta 167:473–481

Bateson W (1916) Root-cuttings, chimaeras and sports. J Genet 6:75–80

Baur E (1909) Das Wesen und die Erblichkeitsverhältnisse der "Varietates albomarginata hort" von *Pelargonium zonale*. Z Abstammungs Vererbungsl 1:330–351

Bayliss MW (1977) The causes of competition between two cell lines of *Daucus carota* in mixed cultures. Protoplasma 92:117–127

Ben-Jaacov J. Langhans RW (1972) Rapid multiplication of chrysanthemum plants by stem-tip proliferation. HortSci 7:289–290
Bennici A (1979) Cytological chimeras in plants regenerated from *Lilium longiflorum* tissues grown in vitro. Z Pflanzenzücht 82:349–353
Bennici A. D'Amato F (1978) In vitro regeneration of *Durum* wheat plants 1. Chromosome numbers of regenerated plantlets. Z Pflanzenzücht 81:305–311
Bergann F (1967) The relative instability of chimerical clones – the basis for further breeding. Abh Dtsch Akad Wiss Berlin 2:287–300
Bergann F. Bergann L (1959) Über experimentell ausgelöste vegetative Spaltungen und Umlagerungen an chimärischen Klonen, zugleich als Beispiele erfolgreicher Staudenauslese. 1. *Pelargonium zonale* Ait. 'Madame Salleron'. Züchter 29:361–374
Bergann F. Bergann L (1982) Zur Entwicklungsgeschichte des Angiospermenblattes 1. Über Periklinalchimären bei *Peperomia* und ihre experimentelle Entmischung und Umlagerung. Biol Zentralbl 101:485–502
Bergann F. Bergann L (1984) Gelungene experimentelle Synthese zweier neuer Pfropfchimären-die Rotdornmispeln von Potsdam: + *Crataegomespilus potsdamiensis* cv. Diekto. cv. Monekto. Biol Zentralbl 103:283–293
Binding H. Witt D. Monzer J. Mordhorst G. Kollmann R (1987) Plant cell graft chimeras obtained by co-culture of isolated protoplasts. Protoplasma 141:64–73
Binns AN (1981) Developmental variation in plant tissue culture. Environ Exp Bot 21:325–332
Bowen HJM. Cawse PA. Dick MJ (1962) The induction of sports of chrysanthemums by gamma radiation. Radiat Bot 1:297–303
Broertjes C. Keen A (1980) Adventitious shoots: do they develop from one cell? Euphytica 29:73–87
Broertjes C. van Harten AM (1978) Application of mutation breeding methods in the improvement of vegetatively propagated crops. Elsevier. New York
Buiatti M. Baroncelli R. Tesi R. Boscariol P (1970) The effect of environment on diplontic selection in irradiated gladiolus corms. Radiat Bot 10:531–538
Burk LG (1975) Clonal and selective propagation of tobacco from leaves. Plant Sci Lett 4:149–154
Burk LG. Matzinger DF (1976) Variation among anther-derived doubled haploids from an inbred line of tobacco. J Hered 67:381–4
Burk LG. Stewart RN. Dermen H (1964) Histogenesis and genetics of a plastid-controlled chlorophyll variegation in tobacco. Am J Bot 51:713–724
Bush SR. Earle ED. Langhans RW (1976) Plantlets from petal segments, petal epidermis, and shoot tips of the periclinal chimera. *Chrysanthemum morifolium* Indianapolis. Am J Bot 63:729–737
Butcher DM (1977) Secondary products in tissue cultures. In: Reinert J. Bajaj YPS (eds) Applied and fundamental aspects of plant cell, tissue, and organ culture. Springer. Berlin Heidelberg New York. pp 669–693
Carlson PS. Chaleff RS (1974) Heterogeneous associations of cells formed in vitro. In: Ledoux L (ed) Genetic manipulations with plant materials. Plenum. New York. pp 245–261
Cassells AC. Minas G. Long R (1980) Culture of *Pelargonium* hybrids from meristems and explants: chimeral and beneficially-infected varieties. In: Ingram DS. Helgeson JP (eds) Tissue culture methods for plant pathologists. Blackwell. Oxford. pp 125–130
Chen K. Wildman SG. Smith HH (1977) Chloroplast DNA distribution in parasexual hybrids as shown by polypeptide composition of fraction I protein. Proc Natl Acad Sci USA 74:5109–5112
Chilton MD (1983) A vector for introducing new genes into plants. Sci Am 248:50–59
Chin C-K (1980) Growth behavior of green and albino plants of *Episcia cupreata* Pink Brocade in vitro. In Vitro 16:847–850
Chua BU. Kunisaki JT. Sagawa Y (1981) In vitro propagation of *Dracaena marginata* Tricolor. HortSci 16(4):494
Clayberg CD (1975) Insect resistance in a graft-induced periclinal chimera of tomato. HortSci 10:13–15
Clowes FAL (1957) Chimeras and meristems. Heredity 11:141–148
D'Amato F (1964) Endopolyploidy as a factor in plant tissue development. Caryologia 17:41–52
D'Amato F (1978) Chromosome number variation in cultured cells and regenerated plants. In: Thorpe TA (ed) Frontiers of plant tissue culture. Int Assoc Plant tissue culture. Calgary. Can. pp 287–295
Dermen H (1948) Chimeral apple sports and their propagation through adventitious buds. J Hered 39:235–242

Dermen H (1960) The nature of plant sports. Am Hortic Mag 39:123-173
Dommergues P. Gillot J (1973) Obtention de clones génétiquement homogènes dans toutes leurs couchés ontogéniques à partir d'une chimère d'oeillet americain. Ann Amelior Plantes 23:83-93
Doodeman M. Bianchi F (1985) Genetic analysis of instability in *Petunia hybrida* 3. Periclinal chimeras resulting from frequent mutations of unstable alleles. Theor Appl Genet 69:297-304
Doodeman M. Boersma EA. Koomen W. Bianchi F (1984a) Genetic analysis of instability in *Petunia hybrida* 1. A highly unstable mutation induced by a transposable element inserted at the An 1 locus for flower colour. Theor Appl Genet 67:345-355
Doodeman M. Gerats AGM. Schram AW. de Vlaming P. Bianchi F (1984b) Genetic analysis of instability in *Petunia hybrida* 2. Unstable mutations at different loci as the result of transpositions of the genetic element inserted at the An 1 locus. Theor Appl Genet 67:357-366
Dougall DK. Johnson JM. Whitten GH (1980) A clonal analysis of anthocyanin accumulation by cell cultures of wild carrot. Planta 149:292-297
Dulieu H (1967) Étude de la stabilité et d'une déficience chlorophyllienne induite chez le tabac par traitement au méthane sulfonate d'éthyle. Ann Amelior Plantes 17:339-355
Dulieu H (1970) Emploi des chimères chlorophyliennes pour l'étude de l'ontogénie foliare. Bull Sci Bourgogne 25:1-72
Evans DA. Paddock EF (1976) Comparisons of somatic crossing over frequency in *Nicotiana tabacum* and three other crop species. Can J Genet Cytol 18:57-65
Evans DA. Sharp WR. Medina-Filho HP (1984) Somaclonal and gametoclonal variation. Am J Bot 71:759-774
Fluhr R. Aviv D. Edelman M. Galun E (1983) Cybrids containing mixed and sorted-out chloroplasts following interspecific somatic fusions in *Nicotiana*. Theor Appl Genet 65:289-294
Foster AS (1935) A histogenetic study of foliar determination in *Carya buckleyi* var. *arkansana*. Am J Bot 22:88-147
Foster AS (1939) Problems of structure. growth. and evolution in the shoot apex of seed plants. Bot Rev 5:454-470
Foster AS (1941) Comparative studies on the structure of the shoot apex in seed plants. Bull Torrey Bot Club 68:339-350
George EF. Sherrington PD (1984) Plant propagation by tissue culture – handbook and directory of commercial operations. Exergenics. Eversley Basingstoke
Gifford EM Jr. Corson GE Jr (1971) The shoot apex in seed plants. Bot Rev 37:143-229
Gleba YY. Evans DA (1983) Genetic analysis of somatic hybrid plants. In: Evans DA. Sharp WR. Ammirato PV. Yamada Y (eds) Handbook of plant cell culture. vol 1. Techniques for propagation and breeding. MacMillan. New York. pp 322-357
Haccius B (1978) Question of unicellular origin of non-zygotic embryos in callus cultures. Phytomorphology 28:74-81
Haccius B. Bhandari NN (1975) Delayed histogen differentiation as a common primitive character in all types of non-zygotic embryos. Phytomorphology 25:91-94
Hackett WP. Anderson JM (1967) Aseptic multiplication and maintenance of differentiated carnation shoot tissue derived from shoot apices. Proc Am Soc Hortic Sci 90:365-369
Hall HK. Quasi MH. Skirvin RM (1986) Isolation of a pure thornless loganberry by meristem tip culture. Euphytica 35:1039-1044
Hartmann HT. Kester DE (1983) Plant propagation-principles and practices. Prentice-Hall. Englewood Cliffs. NJ
Heichel GH. Anagnostakis L (1978) Stomatal response to light of *Solanum pennellii*, *Lycopersicon esculentum*, and a graft-induced chimera. Plant Physiol 62:387-390
Horak J (1972) Ploidy chimeras in plants regenerated from the tissue cultures of *Brassica oleracea* L. Biol Plant 14:423-426
Horsch RB. Fry JE. Hoffmann NL. Eichholtz D. Rogers SG. Fraley RT (1985) A simple and general method for transferring genes into plants. Science 227:1229-1231
Howard HW (1964) The use of X-rays in investigating potato chimeras. Radiat Bot 4:361-371
Hussey G. Falavigna A (1980) Origin and production of in vitro adventitious shoots in the onion, *Allium cepa* L. J Exp Bot 31:1675-1686
Iwai S. Nakata K. Nagao T. Kawashima N. Matsuyama S (1981) Detection of *Nicotiana rustica* chloroplast genome coding for the large subunit of fraction 1 protein in a somatic hybrid in which only the *N. tabacum* chloroplast genome appeared to have been expressed. Planta 152:478-480

Jagannathan L, Marcotrigiano M (1987) Phenotypic and ploidy status of *Paulownia tomentosa* trees regenerated from cultured hypocotyls. Plant Cell Tissue Org Cult 7:227–236
Johnson MA (1951) The shoot apex in gymnosperms. Phytomorphology 1:188–203
Johnson RT (1980) Gamma irradiation and in vitro induced separation of chimeral genotypes in carnation. HortSci 15:605–606
Jones JB, Murashige T (1974) Tissue culture propagation of *Aechmea fasciata* Baker and other bromeliads. Proc Int Plant Propagators' Soc 24:117–126
Jorgensen CA, Crane MB (1927) Formation and morphology of *Solanum* chimeras. J Genet 18:247–273
Kameya T (1975) Culture of protoplasts from chimeral plant tissue of nature. Jpn J Genet 50:417–420
Keathley DE, Scholl RL (1983) Chromosomal heterogeneity of *Arabidopsis thaliana* anther callus, regenerated shoots and plants. Z Pflanzenphysiol 112:247–255
King PJ (1984) Mutagenesis of cultured cells. In: Vasil IK (ed) Cell culture and somatic cell genetics of plants, vol 1. Laboratory procedures and their applications. Academic Press, New York London, pp 547–551
Kirk JTO, Tilney-Bassett RAE (1978) The plastids. Elsevier/North-Holland Biomedical Press, New York
Klekowski EJ Jr (1985) Mutations, apical cells, and vegetative reproduction. Proc R Soc Edinburgh 86B:67–73
Klekowski EJ Jr, Kazarinova-Fukshansky N (1984a) Shoot apical meristems and mutation: fixation of selectively neutral cell genotypes. Am J Bot 71:22–27
Klekowski EJ Jr, Kazarinova-Fukshansky N (1984b) Shoot apical meristems and mutation: selective loss of disadvantageous cell genotypes. Am J Bot 71:28–34
Koenigsberg S, Langhans RW (1978) Tissue culture studies with chrysanthemum chimeras. HortSci 13:349 (Abstr)
Krenke NP (1933) Wundkompensation, Transplantation und Chimären bei Pflanzen. Springer, Berlin
Kumar A, Cocking EC (1982) Restriction endonuclease analysis of chloroplast DNA in interspecies somatic hybrids of *Petunia*. Theor Appl Genet 62:377–383
Larkin PJ, Scowcroft WR (1981) Somaclonal variation – a novel source of variability from cell cultures for plant improvement. Theor Appl Genet 60:197–214
Lineberger RD, Druckenbrod M (1985) Chimeral nature of the pinwheel flowering african violets (*Saintpaulia*, Gesneriaceae). Am J Bot 78:1204–1212
Lorz H, Scowcroft WR (1983). Variability among plants and their progeny regenerated from protoplasts of Su/su heterozygotes of *Nicotiana tabacum*. Theor Appl Genet 66:67–75
Lowe K, Barnes D, Ryan P, Paterson KE (1985) Plant regeneration via organogenesis and embryogenesis in the maize inbred line 73. Plant Sci 41:125–132
Maene LJ, Debergh PC (1982) Stimulation of axillary shoot development of *Cordyline terminalis* Celeste Queen by foliar sprays of 6-benzylaminopurine. HortSci 17:344–345
Maliga P, Kiss ZR, Nagy AH, Lazar G (1978) Genetic instability in somatic hybrids of *Nicotiana tabacum* and *Nicotiana knightiana*. Mol Gen Genet 163:145–151
Maliga P, Lorz H, Lazar G, Nagy F (1982) Cytoplast-protoplast fusion for interspecific chloroplast transfer in *Nicotiana*. Mol Gen Genet 185:211–215
Mantell SH, Matthews JA, McKee RA (1985) Principles of plant biotechnology. Blackwell, Oxford
Marcotrigiano M (1985) Experimentally synthesized plant chimeras. 3. Qualitative and quantitative characteristics of the flowers of interspecific *Nicotiana* chimeras. Ann Bot (London) 57:435–442
Marcotrigiano M (1986) Origin of adventitious shoots regenerated from cultured tobacco leaf tissue. Am J Bot 73:1541–1547
Marcotrigiano M, Gouin FR, (1984a) Experimentally synthesized plant chimeras. 1. In vitro recovery of *Nicotiana tabacum* L. chimeras from mixed callus cultures. Ann Bot (London) 54:503–511.
Marcotrigiano M, Gouin FR (1984b) Experimentally synthesized plant chimeras. 2. A comparison of in vitro and in vivo techniques for the production of interspecific *Nicotiana* chimeras. Ann Bot (London) 54:513–521
Marcotrigiano M, Stewart RN (1984) All variegated plants are not chimeras. Science 223 (4635):505
Marcotrigiano M, Morgan PA, Swartz HJ, Ruth J (1987) Histogenic instability in tissue culture-proliferated strawberry plants. J Am Soc Hortic Sci 112:583–587
Marcotrigiano M, Morgan PA (1988) Chlorophyll-deficient cell lines which are genetically uncharacterized can be inappropriate for use as phenotypic markers in developmental studies. Am J Bot 75:985–989

Matzinger DF, Burk LG (1984) Cytoplasmic modification by anther culture in *Nicotiana tabacum* L. J Hered 75:167–170
Mauseth JD (1979) Cytokinin-elicited formation of the pith-rib meristem and other effects of growth regulators on the morphogenesis of *Echinocereus* (Cactaceae) seedling shoot apical meristems. Am J Bot 66:446–451
Mcpheeters K, Skirvin RM (1983) Histogenic layer manipulation in chimeral Thornless Evergreen trailing blackberry. Euphytica 32:351–360
Mehra A, Mehra PN (1974) Organogenesis and plantlet formation in vitro in almond. Bot Gaz 135:61–73
Meins FR Jr (1983) Heritable variation in plant cell culture. Annu Rev Plant Physiol 34:327–346
Meyer MM Jr (1980) In vitro propagation of *Hosta sieboldiana*. HortSci 15:737–738
Miedema P (1973) The use of adventitious buds to prevent chimerism in mutation breeding of potato. Euphytica 22:209–219
Mix G, Wilson HM, Foroughi-Wehr B (1978) The cytological status of plants of *Hordeum vulgare* L. regenerated from microspore callus. Z Pflanzenzücht 80:89–99
Morel GM (1972) Morphogenesis of stem apical meristem cultured in vitro: application to clonal propagation. Phytomorphology 22:265–277
Nasim A, Hannan MA, Nestmann ER (1981) Pure and mosaic clones – a reflection of differences in mechanisms of mutagenesis by different agents in *Saccharomyces cerevisiae*. Can J Genet Cytol 23:73–79
Neilson-Jones W (1969) Plant chimeras. Methuen, London
Ogura H (1976) The cytological chimeras in original regenerants from tobacco tissue cultures and their offspring. Jpn J Genet 51:161–174
Pereau-Leroy P (1974) Genetic interactions between the tissues of carnation petals as periclinal chimeras. Radiat Bot 14:109–116
Pereira MFA, Dale JE (1982) Effects of 2,4-dichlorophenoxyacetic acid on shoot apex development of *Phaseolus vulgaris*. Z Pflanzenphysiol 107:169–177
Pierik RLM, Steegmans HHM (1983) Vegetative propagation of a chimerical *Yucca elephantipes* Regel in vitro. Sci Hortic 21:267–272
Pillai K (1963) Structural and seasonal activity of the shoot apex of some *Cupressus* species. New Phytol 62:335–340
Poethig RS (1984) Cellular parameters of leaf morphogenesis in maize and tobacco. In: White RA, Dickison WC (eds) Contemporary problems in plant anatomy. Academic Press, New York London, pp 235–259
Poethig RS (1987) Clonal analysis of cell lineage patterns in plant development. Am J Bot 74:581–594
Pohlheim F (1981) Genetischer Nachweis einer NMH-induzierten Plastommutation bei *Saintpaulia ionantha* H. Wendl Biol Rundsch 19:47–50
Pohlheim F (1982) Klonvariabilität durch Chimärenumlagerung und Mutation bei *Dracaena dermensis* (N.E. BR.) ENGL. Arch Züchtungsforsch 12:399–409
Popham RA (1951) Principal types of vegetative shoot apex organization in vascular plants. Ohio J Sci 51:249–270
Pratt C (1983) Somatic selection and chimeras. In: Janick N, Moore J (eds) Methods in fruit breeding. Purdue Univ Press, West Lafayette, Indiana, pp 172–185
Preil W, Engelhardt M (1982) In vitro separation of chimeras by suspension cultures of *Euphorbia pulcherrima* wild. Gartenbauwissenschaft 47:241–244
Reeve RM (1948) The "tunica-corpus" concept and development of shoot apices in certain dicotyledons. Am J Bot 35:65–75
Romberger JA (1963) Meristems, growth and development in woody plants. Tech Bull 1293, US Dep Agric For Ser 214 pp
Rosati P, Gagglioni D, Giunchi L (1986) Genetic stability in micropropagated loganberry plants. J Hortic Sci 61:33–41
Ruth J, Klekowski EJ Jr, Stein O (1985) Impermanent initials of the shoot apex and diplontic selection in a juniper chimera. Am J Bot 72:1127–1135
Sagawa Y, Mehlquist GAL (1957) The mechanism responsible for some X-ray induced changes in flower color of the carnation, *Dianthus caryophyllus*. Am J Bot 44:397–403
Satina S (1944) Periclinal chimeras in *Datura* in relation to the development and structure (A) of the style and stigma (B) of calyx and corolla. Am J Bot 31:493–502

Satina S (1945) Periclinal chimeras in *Datura* in relation to the development and structure of the ovule. Am J Bot 32:72–81
Satina S, Blakeslee AF (1941) Periclinal chimeras in *Datura stramonium* in relation to development of leaf and flower. Am J Bot 28:862–871
Satina S, Blakeslee AF (1943) Periclinal chimeras in *Datura* in relation to the development of the carpel. Am J Bot 30:453–462
Satina S, Blakeslee AF, Avery AG (1940) Demonstration of the three germ layers in the shoot apex of *Datura* by means of induced polyploidy in periclinal chimeras. Am J Bot 27:895–905
Schmidt A (1924) Histologische Studien an phanerogamen Vegetationspukten. Bot Arch 8:345–404
Scowcroft WR, Larkin PJ (1981) Chloroplast DNA assorts randomly in interspecific somatic hybrids of *Nicotiana debneyi*. Theor Appl Genet 60:179–184
Seeni S, Gnanam G (1981) In vitro regeneration of chlorophyll chimeras in tomato (*Lycopersicon esculentum*). Can J Bot 59:1941–1943
Skene KGM, Barlass M (1983) Studies of the fragmented shoot apex of grapevine. IV. Separation of phenotypes in a periclinal chimera in vitro J Exp Bot 34:1271–1280
Skirvin RM (1978) Natural and induced variation in tissue culture. Euphytica 27:241–266
Skirvin RM, Janick J (1976) Tissue culture-induced variation in scented *Pelargonium* spp. J Am Soc Hortic Sci 101:281–290
Smith CW (1968) The effect of growth substances on growth of excised embryo shoot apices of wheat in vitro. Ann Bot (London) 32:593–600
Sree Ramulu K, Devreux M, Ancora G, Laneri U (1976) Chimerism in *Lycopersicum peruvianum* plants regenerated from in vitro cultures of anthers and stem internodes. Z Pflanzenzücht 76:299–319
Staba EJ (1980) Plant tissue culture as a source of biochemicals. CRC, Boca Raton, Fla
Stadler LJ (1930) Some genetic effects of x-rays in plants. J Hered 21:3–19
Stewart RN (1978) Ontogeny of the primary body in chimeral forms of higher plants. In: Subtelny S, Sussex IM (eds) The clonal basis of development. Academic Press, New York London, pp 131–160
Stewart RN, Arisumi T (1966) Genetic and histogenetic determination of pink bract color in poinsettia. J Hered 57:217–220
Stewart RN, Burk LG (1970) Independence of tissues derived from apical layers in ontogeny of the tobacco leaf and ovary. Am J Bot 57:1010–1061
Stewart RN, Dermen H (1970a) Determination of number and mitotic activity of shoot apical initial cells by analyses of mericlinal chimeras. Am J Bot 57:816–826
Stewart RN, Dermen H (1970b) Somatic genetic analysis of the apical layers of chimeral sports in *Chrysanthemum* by experimental production of adventitious shoots. Am J Bot 57:1061–1070
Stewart RN, Dermen H (1975) Flexibility in ontogeny as shown by the contribution of the shoot apical layers to leaves of periclinal chimeras. Am J Bot 62:935–947
Stewart RN, Dermen H (1979) Ontogeny in monocotyledons as revealed by studies of the developmental anatomy of periclinal chloroplast chimeras. Am J Bot 66:47–58
Stewart RN, Meyer FG, Dermen H (1972) Camellia + Daisy Eagleson, a graft chimera of *Camellia sasanqua* and *C. japonica*. Am J Bot 59:515–524
Stewart RN, Semeniuk P, Dermen H (1974) Competition and accommodation between apical layers and their derivatives in the ontogeny of chimeral shoots of *Pelargonium* × *hortorum*. Am J Bot 61:54–67
Sung ZR (1976) Mutagenesis of cultured plant cells. Genetics 84:51–57
Tanaka T (1927) Bizzaria – a clear case of periclinal chimera. J Gen 18:77–85
Tilney-Bassett RAE (1963) The structure of periclinal chimeras. Heredity 18:265–285
Tilney-Bassett RAE (1986) Plant chimeras. Arnold, London
Tilney-Bassett RAE, Birky CW Jr (1981) The mechanism of the mixed inheritance of chloroplast genes in *Pelargonium*. Theor Appl Genet 60:43–53
Walbot V, Coe EH Jr (1979) Nuclear gene iojap conditions a programmed change to ribosome-less plastids in *Zea mays*. Proc Natl Acad Sci USA 76:2760–2764
Weiss FE (1930) The problem of graft hybrids and chimeras. Bot Rev 5:231–271
Widholm JM (1974) Cultured carrot cell mutants: 5-methyltryptophan-resistance trait carried from cell to plant and back. Plant Sci Lett 3:323–330

I.4 Genetic Bases of Variation from in Vitro Tissue Culture

M. SIBI[1]

1 Variant Plants Resulting from Tissue Culture: History, Heredity, and Terminology

1.1 History of Variation Resulting from Tissue Culture

Many tissue culture studies initiated during the 1950's were summarized by Gautheret (1954, 1959, 1964). Since then, meristem and callus cultures of a large number of plant species have been successfully worked out, and phenotypic variation in callus regenerated plants has been reported (Bajaj 1986). In fact, variation resulting from meristem culture (Clare and Collin 1974) was only little reported, but it frequently occurred from callus regeneration or in strains. Many review articles are available on the subject, some of which attempt to explain the causes and find the origin of this variability (Sibi 1978; Brettell and Ingram 1979; Larkin and Scowcroft 1981; Reisch 1983; Orton 1984; Karp and Bright 1985; Gould 1986). This variation was termed differently according to authors. "Phenovariants" was first proposed by Sibi (1971) as describing the new regenerated phenotypes, while "somaclonal variation" was used by Larkin and Scowcroft (1981), and seems to be generally accepted for all types of variations coming from in vitro tissue culture.

The first *variant regenerated plants* from somatic tissue were reported before 1970, in tobacco (Lutz 1969; Mousseau 1970) and sugarcane (Heinz and Mee 1969), while *hereditary transmission after selfing* was first observed on lettuce (Sibi 1971, 1974, summarized in Sibi 1976) and tobacco (Binns and Meins 1973). Since then, this kind of analysis has been commonly used; for example, in tomato, many experiments presenting a large spectrum of results have been reported (Sibi 1979; Mikoko-Nsika 1982, Evans and Sharp 1983; Evans et al. 1984; Sibi et al. 1984; Buiatti et al. 1985). Whereas, *reciprocal crosses analyses* seem to have been scarcely achieved, they were first experimented in lettuce (Sibi 1974, 1976) and tomato (Sibi 1980, 1981, 1982, summarized in Sibi 1986).

Variant plants were also reported after gametophytic regeneration or haplodiploid plants from rice anther (Truong-André 1977; Odjo 1978; Kouadio 1979; Dossou-Yovo et al. 1982; Schaeffer 1982; Wakasa 1982; Schaeffer et al. 1984), wheat

[1]E.N.S.A.I.A. 2, Avenue de la Forêt de Haye, 54500 Vandoeuvre, France

Biotechnology in Agriculture and Forestry, Vol. 11
Somaclonal Variation in Crop Improvement I (ed. by Y.P.S. Bajaj)

(Picard 1984), tobacco (Matzinger and Burk 1984), and from either anthers or ovaries of barley (San-Noeum and Ahmadi 1982). Furthermore, as discussed in Section 4.1, the cells of a whole regenerated plant should not be regarded as a clone. Thus, "vitro variants" or "vitro variation" could be general and convenient terms to describe variability obtained by in vitro tissue culture.

1.2 Heredity and Terminology

The biological level implied in this vitro variation has been investigated by analysis of the "memory", stability, or transmissibility of the modified traits, and sometimes new terms have been created in order to describe what was observed.

From a biochemical point of view, animal and plant tissue strains showed evidence of changes that could not be explained as mutations. By comparing mutant and revertant frequencies in culture of human tissue, De Mars (1974) postulated a new biological state of the functioning of the material. This kind of modification was termed "perigenic" (De Mars 1974) to imply changes in a biological system located around the gene; while in situ variants of flax were described as "genotrophs" (Durrant 1962), and mechanisms implied in several similar modifications on maize were termed "paramutations" (Brink et al. 1968; discussed by Tartof 1973 and by Holliday 1987).

In accordance with this, Holliday (1987) proposed the term "epimutation" for DNA modifications arising in situ at very high frequency and, in so far as animal species are concerned, occurring in small, short-lived animals.

Meanwhile, in tobacco tissue culture, modifications, which could be either transient or stable throughout plant regeneration and the following sexual cycles, were described as "habituation" (Binns and Meins 1973; Meins and Binns 1977; Meins and Lutz 1980; Meins 1983) or "mutation-like adaptation" (Skokut and Filner 1980; Filner 1980) because these changes could not be resulting from conventional spontaneous mutations. Thus Filner (1980) emphasized the problem of terminology, because some of these modifications were classified as "epigenetic" (i.e., = nonpermanent) changes, although hereditarily transmissible through sexual cycles of the regenerated plants.

Since 1971, experiments on homozygous lettuce (Sibi 1974 to 1978) and tomato (Sibi 1979 to 1986) have shown that in vitro tissue culture might lead to an increase in mutants. More than 10% monogenic nuclear mutants were observed in lettuce, and 56% of regenerants showed either single-gene or multiple mutations in tomato. This is in accordance with the results of Evans et al. (1984) and Buiatti et al. (1985) on tomato, as they compiled more than 17% and 5.6% mutant regenerated plants respectively.

Furthermore, for both lettuce and tomato (Sibi 1974 to 1986), another category of variation could be observed, which *could not be accounted for either strictly Mendelian or cytoplasmic genetic material.* The hereditary stability and genetic fixity of this variation could be verified through several generations of selfing. The progenies from reciprocal crosses showed asymmetrical behavior, i.e., maternal or even paternal effects, and sometimes transgression (or vigor-like) effects with asymmetrical expression, these last observations being the most surprising, as the genuine genome of the basic material was homozygous and unmodified in every case.

A similar behavior of hereditary characteristics showing asymmetrical expression after reciprocal crosses was observed for haplodiploid rice (Kouadio 1979). and barley (San-Noeum and Ahmadi 1982).

According to all these facts the terms "epigenic variation" were proposed and discussed (Sibi 1974 to 1986; Demarly 1985) to describe genetically transmitted modifications associated with unusual genetic behavior.

2 Genetic Potential, in Vitro Phase Duration, and Variability

2.1 Genetic Background and in Vitro Tissue Culture Effect

The impact of genetic background on the success of in vitro culture has been investigated through quantitative analysis in red clover (Keyes et al. 1980). and petunia (Bergounioux et al. 1982). Evidence of a genetic structure effect has also been described for tomato species. Here. specific mutability has been stated to depend on the gene concerned and to interact with the rest of the genetic background of the material.

This has been demonstrated by analyzing gametophytic self-incompatibility mechanisms of *Lycopersicon peruvianum* (de Nettancourt and Devreux 1977). This species is allogamous. with a high level of heterozygosity. In this material. the authors reported that changes transforming one self-incompatibility allele into another (e.g.. $S_1 \rightarrow S_2$) could not be induced by the usual mutagenic treatments. whereas it became possible for inbred lines. as if related to a mutagenic or switching mechanism controlled by the level of homozygosity in the plant. Furthermore. somatic cells cultured in vitro generated spontaneous new S-alleles through the regenerated plants.

These analyses were continued (Sree Ramulu 1982) as *L. peruvianum* regenerated plants could be obtained from stem internodes or anthers in vitro culture (Sree Ramulu et al. 1976a). Plants originating from stem internodes and remaining heterozygous showed no modifications at the S-locus. whereas for plants regenerated from anthers. the authors observed new expressions of the S-allele system and sexually hereditary changes for new S-specificities. such as a switch from gametophytic to sporophytic action. or S-allele reversion. They confirm the stability of S-alleles under a relatively high degree of heterozygosity. the strong impact of inbreeding on the S-allele polygenic system. and they state that new S-specificities probably do not result from point mutations at the S-locus (Sree Ramulu 1980). Thus. a modification of genome regulations could be implied in these new S-specificities obtained in vitro.

2.2 Effect of in Vitro Phase Duration

The duration of the in vitro stage, and more precisely, of the nonmorphogenetic phase has long been supposed to be a factor influencing variability (Demarly 1974; Sibi 1978). Several experiments were conducted to demonstrate this relation. A high increase of mutants and wild-type expressions (Sibi 1974, 1976) have in fact been observed. The frequency of mutants seems to be related to the number of in vitro transfers (Skirvin and Janick 1976; Deshayes 1976; Brettell et al. 1980; McCoy and Phillips 1982; Mikoko-Nsika 1982; Lee 1984), as if the in vitro phase has had a mutagenic effect on the cultured cells, or the regulation of the DNA-repairing systems was modified (Sibi 1981).

For example, tomato plants obtained from apical shoot tip (Novàk and Maskovà 1979) or regenerated directly from leaf (Kartha et al. 1977) showed no variation, while the majority of authors observed to some extent variability of all types for plants regenerated from callus (Padmanabhan et al. 1972; de Langhe and de Bruijne 1976; Sibi 1979, 1981, 1986, Mikoko-Nsika 1982; Evans and Sharp 1983; Sibi et al. 1984; Buiatti et al. 1985; Evans 1986). The majority of tomato "vitro variants" were found to be nuclear genic mutants. As previously stated, the reported rates vary from 56% (Sibi 1979, 1981, 1986) to 5.6% (Evans et al. 1984), with the intermediate value of 17.04% (Buiatti et al. 1985).

The most surprising is the modification of recombinant rates observed after in vitro tissue culture (Sibi et al. 1984). The increase, recorded as reaching 25 to 30%, would imply either gene amplification between the marker-genes, as suggested by several reports on in vitro tissue culture (Nuti Ronchi 1971; Nagl 1972; Nagl et al. 1972; Buiatti 1977; Schimke et al. 1978), or (and) crossing-over control modifications (Holliday 1987).

While the emergence of mutations is shown to be very frequent, unusual new regulations of the cells seem to be expressed by several mechanisms like crossing-over control or mutability regulation.

The impact of the duration of the in vitro phase which can be observed on the rates of mutants does not seem to affect the frequency of "epigenic" modifications. In fact, they may appear by the first in vitro cycle, and their frequencies seem to be stabilized by the third or fourth transfer (Sibi 1976, 1981, 1986; Mikoko-Nsika 1982). In tomato, regenerated plants from this "early" phase have given rise to 17% fixed progenies, thus behaving as epigenic modified plants, and this value does not seem to increase in older strains.

Thus, callus formation and sometimes the number of in vitro cycles appears to be one of the main factors in inducing, or permitting preexistent variability, but the latter seems also to be frequently related to species or genotype plasticity (Sibi 1981), and to interact with genomic regulation.

2.3 The Kind and Uses of Modifications

Vitro variation obtained randomly has long been recorded at many levels, e.g. cytological, general aspect, etc. In vitro selective pressure experiments first succeeded on tobacco (Carlson 1973), but at that time it was difficult to understand the results. which were controverted. However. since then in vitro selections have been attempted more and more often for all types of specificities. and have been summarized by many authors. For example. resistance or tolerance to disease (Sacristàn 1986). herbicides (Crocomo and Ochoa-Alejo 1983; Hughes 1983; Chaleff 1985. 1986). salts (Rains 1986). cold (Chen and Gusta 1986). stress (Tal 1983). etc.. or the creation of specific qualities (Yamada and Sato 1983). All these experiments are very promising for plant improvement as long as the new advantageous potentialities remain in the regenerated plant and later in the vegetative (tuber etc.) or seed progenies.

3 Variant Plants Regenerated from In Vitro Cultured Cells

3.1 Cell Survival Functions, Explant Size, and Variability

In situ. a cell is included in differentiated organs constituting the whole plant. Its expression and regulation is very precisely programmed and correlated to the whole system.

The size of the explant to be used in vitro will imply modifications in structure or relationship between the cells. and play a main role in variability expression (Sibi 1978. 1981; Demarly 1986). Internode. preformed shoot or differentiated apex will already be "stratified" or programmed. and few modifications are to be expected from such explants. as reported in the literature. in so far as the composition of the medium does not lead to a strong stress or a callus phase; while excised tissues. dissociated cells. and protoplasts will give rise to wide variability. more especially as the callus phase will usually occur unless. maybe. direct somatic embryogenesis can be obtained (Table 1).

In fact. the cell may have many different ways of expression. as the potential chromosomal information can be used according to several modalities (Demarly 1976. 1985). In a nonmorphogenetic callus only survival functions are needed or used. and have to remain unmodified. Therefore changes in the biological structures and genetic bases of any other function are possible (and more likely to happen than in "normal" tissue). allowing a wide spectrum of variability (Sibi 1974). Thus. the impact. previously described. of the duration of the in vitro callus phase enhancing the possibility of modifications. seems to be logical.

3.2 Stability of the Changes and the Process of Organogenesis

When morphogenesis occurs. it may keep the already recorded changes. but will prevent further changes (Demarly 1972). The cell(s) will divide and evolve so as to form a bud. Consequently. the new potentialities of the involved cell(s) will be extended to the entire organism and be maintained as long as they remain in keeping with the regeneration process. Then the new genomic expressions can be visualized throughout a whole plant (Sibi 1974) and the genetic stability or transmissibility can be analyzed.

Table 1. In vitro tissue culture and variation

Explant	True to type	In vitro variation	
		Extrachromosomal modifications	Coding gene modifications
Somatic tissue			
Immature embryo		Naturally lethal interspecific crosses	
Somatic embryo	Propagation	Wide variability	
Bud	Propagation		
Meristem	Virus-free material and		
Apex	Propagation		
Tissue	Embryogenesis and	Creation of resistance to disease, cold, stress, etc. Epigenic variants	Nuclear coding gene modifications
Callus	Regeneration		
Cells			
	Biosynthesis of Substances Identical or Different Qualitatively or Quantitatively		
Protoplasts fusion etc.		No natural hybridization (usually) Introduction of new potentialities (cytoplasmic male sterility, herbicide resistance, etc.) using two different species or in the same species Introduction of organelles or DNA, etc.	
Gametophytes		Haplodiploidization	
Male	Direct genetic stability Choice of parents and	♂ Marked variability	Wild-type Genetic stabilization of interspecific crosses
Female	Creation of varieties	♀ Blurred out variability	Balanced genotypes from aneuploids

4 Plant Organization and Hereditary Transmissibility of the Original Cell Modifications

4.1 Plant Organization

A regenerated bud may originate from a single cell (Skirvin and Janick 1976; Skirvin 1977) but more frequently a pluricellular origin has been demonstrated. By regenerating petunia plants from albinos and green periclinal chimerae, three classes of phenotypes were obtained (Bergounioux 1974), pure green or white on one hand, and a medium shade on the other hand, showing that both albinos and green layers participated in the morphogenesis. This conclusion was confirmed on tobacco species (Carlson and Chaleff 1974), tomato (Sree Ramulu et al. 1976a,b; Buiatti et al. 1985), maize (Springer et al. 1979), and chimeric flower plants (Skirvin et al. 1982), and it was discussed by Benzion et al. (1986). *The cellular lines developed from the basic cells should thus not be considered as clones, but as a mosaic or a chimeric tissue* (Sibi 1974, 1976).

Furthermore, in many species, the plant cannot be taken as a whole, and it was demonstrated to be composed of at least three periclinal layers (Satina et al. 1940) termed L_1, L_2, and L_3 (Fig. 1), with the phenotype resulting from interactions between these layers (Cameron et al. 1964; Dermen and Stewart 1973; Péreau-Leroy 1974). The roots emerge from the deepest one L_3 (corpus), the embryos from L_2, and L_1 wraps the aerial parts. Both L_1 and L_2 layers, and sometimes L_3 are implied in seed or fruit parts (Sagawa and Melhquist 1957; Péreau-Leroy 1975). Thus, the germinal set, i.e., the progeny, does not originate from any solid constitutive tissue, but from L_2 specifically.

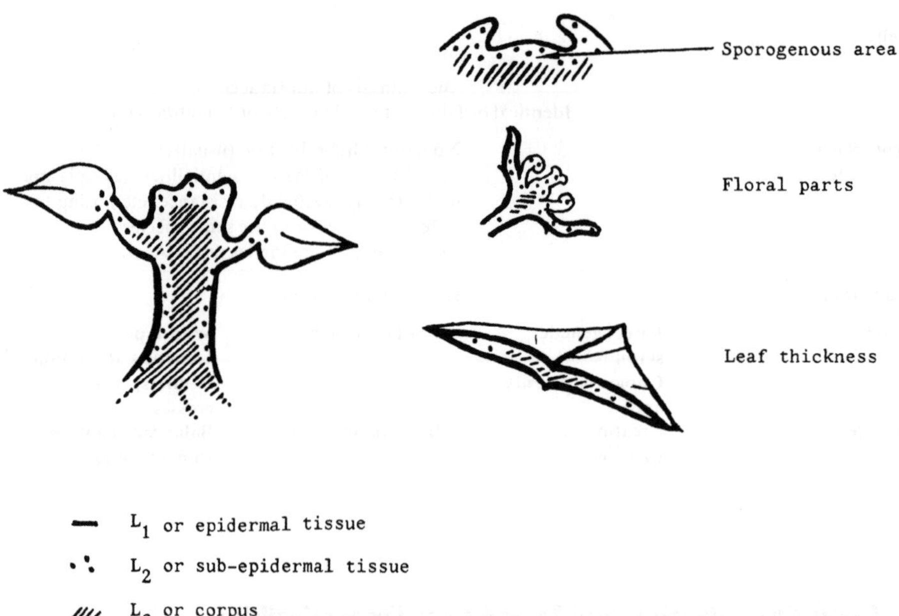

Fig. 1. Periclinal layers and plant constitution. Schematically, L_1 wraps aerial parts, and roots issue from L_3, while the sporogenous area or embryos are made of L_2

4.2 Hereditary Transmissibility of the Modifications

The behavior and stability of the modifications, whether observable or not, in the regenerated plant, will depend on at least two systems:

The kind of changes implied in the original cell

Unstable gene amplification (Buiatti 1977), reversible or evolving biological changes (Meins and Binns 1977; Reisch and Bingham 1979; Skokut and Filner 1980), new expression or silencing of genes (Sager and Kitchin 1975; Siminovitch 1976) through intensity, efficiency, and pattern of DNA methylation (Holliday and Pugh 1975; Bird et al. 1981; Holliday 1987) may behave as transient modifications, and will be able to disappear more or less quickly by some developmental stage of the regenerated plant, or after meiosis, i.e., in the progeny, or possibly later.

The layer(s) of the regenerated plant in relation to the modification(s)

Any kind of change will be lost at meiosis if not implicating each cell of the L_2 layer. Conversely, phenotypically normal regenerated plant may give rise to self-progenies showing new criteria, if the modifications occurred "silently" in L_2. A "late" effect can then be observed through expression of the cryptic potentialities of the plant.

According to these possibilities, several kind of tests can be made to identify the biological level of the modifications.

5 Experimental Approach of the Biological Systems Implied (Table 2)

The best experimental approach should include the following successive points:

- diploid species, i.e., genetic criteria following simple Mendelian laws
- homozygous genetic structure, i.e., genetically fixed material
- floral system allowing easy cross- or self-fertilization, i.e., not depending on self-incompatibility or male sterility
- chromosome countings easy to perform, i.e., low number of chromosomes
- material allowing in vitro callusing and regenerating conditions

N.B.
All the following progeny tests should comprise a reliable number of individuals for both controls and analyzed categories. Experimental designs such as repetitive randomized blocks made under good conditions (e.g., no interaction between each observed individual and its location in the plot), should supply reliable data.

5.1 Chromosome Countings of the Root Tips of Regenerated Plants

Karyotypic modifications such as aneuploidy or changes in ploidy level will be directly detected by chromosome countings in the cells of the strains or the root tip cells of the regenerated plants. This type of variability has been widely described in many review articles (D'Amato 1952, 1978, 1985; Reisch 1983; Karp and Bright 1985; Ahloowalia 1986). Here again, it should be emphasized that in "three-layer species" chromosome numbers have to be checked once again in the progeny issued from selfing, to be sure of the stability or karyotypic basis of the modified trait.

When the karyotype stability has been confirmed, although a modified phenotype is expressed, selfing analyses should be made.

Table 2. Biological "diagnosis" of regenerated plant modification (obtained from homozygous material)

Hypothetical basis of the modifications	Kind of test	Indication
Ploidy numbers, aneuploidy	Cytological karyotypic verification	Ploidy or chromosome modified numbers
Chromosomal rearrangements Recessive or dominant mutations of one chromosome of the pair	Selfing progeny	Segregating criteria
Double mutant: monogenic specific amplification or deletion conversion etc.	Diallel crosses of the control and "genetically homogeneous" parents	Identical behavior of the reciprocal crosses
	Progenies from selfing of the reciprocal cross	Segregating criteria of which parental types are represented
Gametic selection through male or female path	Diallel crosses of the control and "genetically homogeneous" parents	Asymmetrical behavior of the reciprocal crosses
	Progenies from selfing of the reciprocal cross	"Homogeneous-like" criteria (or segregating in case of switching-off mechanism) with vanishing of parental categories
Organelles modifications Genome expression modification by switching-on new regulations through extrachromosomal systems i.e., epigenic system modifications	Diallel crosses of the control and "genetically homogeneous" parents	Asymmetrical behavior of the reciprocal crosses, possible expression of maternal, paternal, or transgression effect, and sometime recovery of parental categories
	Progenies from selfing of the reciprocal cross	Remaining potentialities of the crosses, intraprogeny variance stability, and homogeneity

5.2 Selfing Analyses

Each regenerated plant is a genuine entity and the comparison between progenies from selfing of both homozygous control and regenerated plants will give rise to the following options:

5.2.1 Identical Selfing and Control Progenies

If the progeny from the regenerated plant shows no modification. it apparently demonstrates a true-to-type situation. Thus the new traits expressed by the regenerated plant can be stated as transient. i.e.. reversion or loss of the modifications. before differentiation of germinal cells. or not implicating the L_2 layer cells in the changes.

5.2.2 Segregation of Selfing

This can be performed through simple Mendelian segregation for qualitative criteria giving rise to several categories of phenotypes among which the control type is represented. or through a significant increase of intraprogeny variance compared to the control. for biometrical characteristics. each unit being directed by a set of genes following Mendelian laws.

Both situations are usually related to coding-gene modifications. on one chromosome of the pair for the locus involved. and account for most of the mutant cases noticed before. They give evidence of mutation events (contingently transposition effects: McClintock 1956). either recessive or dominant. possibly bound to quantitative traits. or of some chromosomal rearrangements (deletions. inversions. etc.) not easily detectable unless cytological banding studies are made.

Meanwhile it should be kept in mind that a wide spectrum of categories may also arise through segregating cytoplasmic elements (Ohta 1980; Birky 1983) which can occur simultaneously with (or independently of) the previous points.

5.2.3 Homogenous Variant Phenotype of the Whole Progeny

In this more unusual case. new aspects or new biometrical properties can be observed. but the control type never appears. although intraprogeny variance remains unmodified. Thus. wherever the changes occur in the cell. the two chromosomes of the pair are expressed in the same way. The vitro variant modifications may then be located strictly in the cytoplasmic compartment. but they may also affect nuclear structures concerning either the same locus pair. i.e.. double mutations. specific gene deletions or amplifications through particular conversions (Nagylaki and Petes 1982; Borst and Greaves 1987). or DNA-repairing mechanisms (Holliday and Pugh 1975). etc.. or new expressions of genes through extrachromosomal system modifications.

Furthermore. all the previous types of changes may sum up. and mutation events can occur simultaneously with other kinds of modifications. such as cytoplasmic ones. preventing or hiding the emergence of the control-type category. Thus. no clear hypothesis can be stated here. and reciprocal crosses will have to be carried out.

5.3 Comparison of the Reciprocal Cross Progenies

This kind of analysis is commonly used to provide evidence of asymmetrical transmission of parental genetic potentialities through the male or female part of an hermaphrodite plant. Thus, two parental entities will be confronted symmetrically, and the reciprocal crosses can be compared with one another and to the parents. Complete diallel design including control is often difficult to achieve, but it leads to more precise information.

Several mathematical models of analysis are proposed, but the Griffing method (1956) without parents seems to be one of the less restrictive. Additional information can be obtained from comparison of means by a "multiple range test" made on the parents (or selfings) and the crosses performances.

Value of the variances have to be checked here again to insure that the observed characteristics are reliable.

In diallel design, a set of reciprocal crosses originates from a single couple in which each individual is fertilized by the other. Several results can be obtained from data analyses, and each case will refer to a specific hypothesis, and can be summarized as follows:

5.3.1 Identical Behavior of the Reciprocal Crosses

The same behavior of reciprocal progenies for qualitative or quantitative traits will establish identical transmission of the genetic potentialities through the paternal or maternal path. The new parental phenotypes observed should be related to nuclear gene modifications. This can be checked by self-fertilization of the set of crosses, as presented in Fig. 2. All the symmetrical progenies from selfing will show the same categories of phenotypes (comprising the control-type), or identical variability, i.e., identical simple segregating traits for changes due to a few genes, and identical reciprocal mean and variance values for quantitative characteristics.

5.3.2 Different Behavior of the Reciprocal Crosses

The potential diversity and asymmetrical contribution of a set will be related to the biological bases (Preer 1971; Frankel et al. 1979; Gray 1982) and mechanisms leading to the different genetic content (Clauhs and Grun 1977; Lansman et al. 1983; Sager and Grabowy 1983) or expression, according to the male or female origin of each gamete. In fact, the phenotypes expressed through the progenies depend on modified biological genetic bases that are specific to each parent. This may occur during sexual differentiation (Tilney-Bassett and Abdel-Wahab 1982), before gametic cell maturation (Clauhs and Grun 1977), or during the stages (pollen tube migration, etc.) preceding fertilization, or in the zygote through material typing specified by sexual origin (Tilney-Bassett and Birky 1981; Tilney-Bassett and Abdel-Wahab 1982).

Furthermore, two kinds of effects must be distinguished: gametic selection, and extrachromosomal differential potentialities. For both effects reciprocal phenotypes for qualitative traits or mean values for quantitative ones, are different.

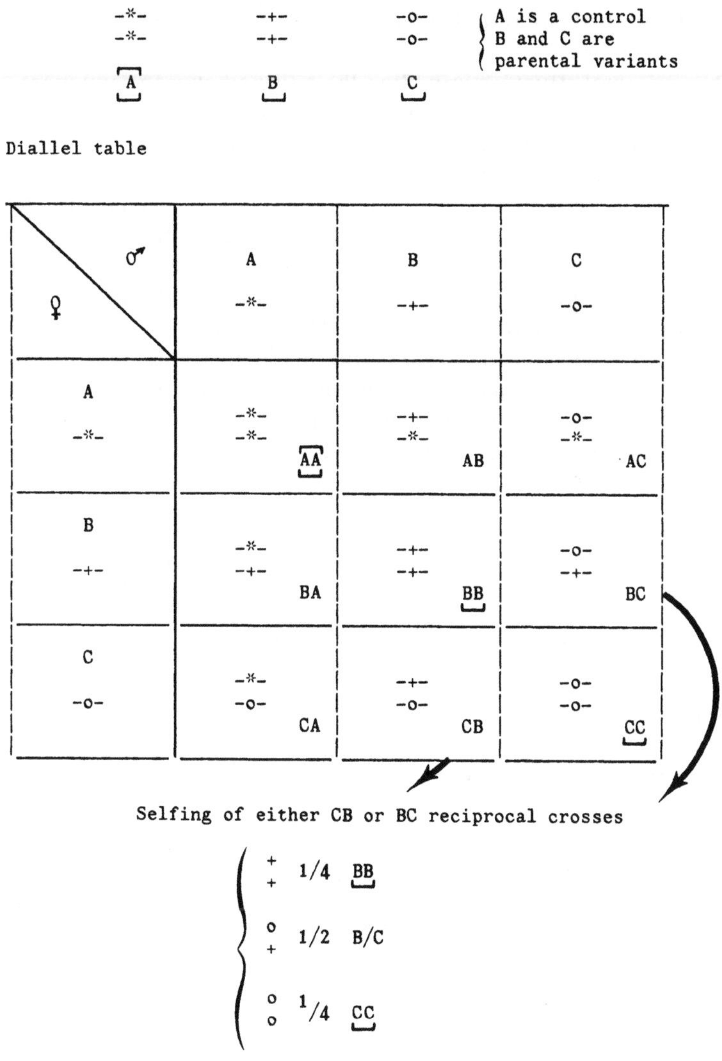

Fig. 2. Double mutant situation. The selfings of identical reciprocal crosses behave alike and give rise to the same categories. Parental variant or control categories can be recovered

Gametic Selection. The first hypothesis is related to differences in nuclear gene composition according to male or female gametes. Thus this case refers to the nuclear gene modifications previously described, but with additional specific regulations. This could imply direct selective effects during sexual differentiation through the survival of specific haploid genotypes according to their evolution in male or female gamete, or systematic specific corrections; this may happen through differential DNA-repairing systems (Holliday and Pugh 1975; Holliday 1987) sometimes going beyond crossing-over events (Nagylaki and Petes 1982; Sugawara and Szostak 1983) and give rise regularly to male and female gametes of specific different types. Indeed, crossing-over mechanisms could be restricted through chromosomal structure impeding the pairing between sequences, as described by Sugawara and Szostak (1983) or the events could be repaired systematically. This could explain the genetic stability of the progenies from selfing of heterozygote parents as presented in Fig. 3.

Furthermore, with this single and permanent mechanism, the direct parental characteristics could be maintained for variant-selfing through systematic gametic complementation. However, they could be definitely lost after crossing to another parent by giving intermediate parental forms through additional (or additive) inheritance, or producing new types through interactive effects.

Such a mechanism should maintain phenotypes of the crosses through selfing, and the former parental aspects or performances should never reappear, while if it is switched off, various segregating types should be observed respectively to each cross through the usual Mendelian laws and crossing-over events.

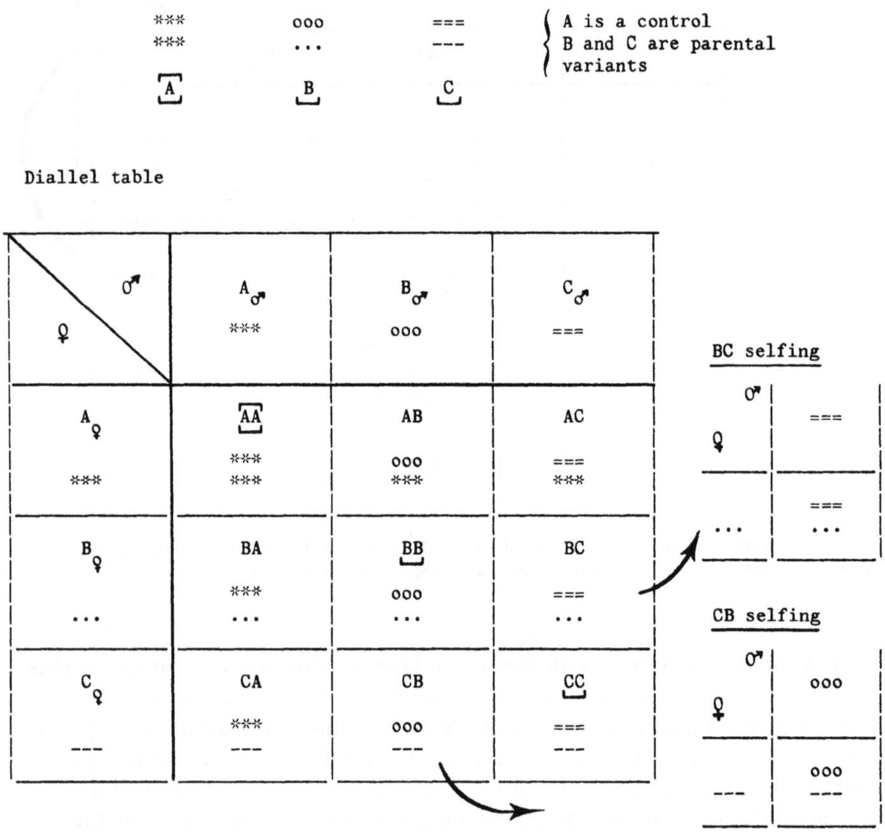

Fig. 3. Gametic selective and restrictive crossing-over effects. Through this mechanism the phenotypes of the reciprocal crosses are maintained and the parental or control type cannot be recovered by selfing

Extrachromosomal Differential Genetic Potentialities. The second hypothesis is based on extrachromosomal differential genetic potentialities. Here homogenous phenotypes or intraprogeny variance refer to nuclear genes stability, while significant differences between reciprocal crosses phenotypes or mean values lead to the conclusion that the same genome is expressed in different ways in respect to extrachromosomal structures. These may belong either to cytoplasmic systems or to nuclear parts different from Mendelian coding genes (Fig. 4).

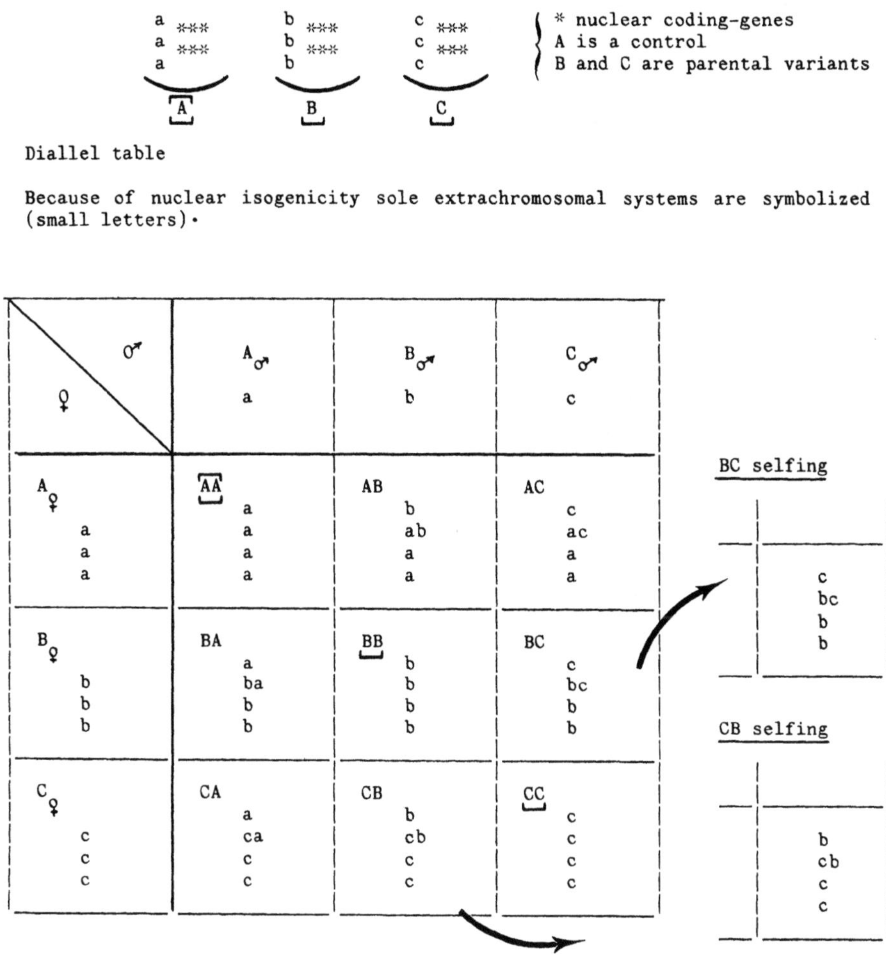

Fig. 4. Extrachromosomal or epigenic modifications. At fertilization, reciprocal crosses differ in quantitative and qualitative content. Different interactive effects are represented by pairs of letters with, e.g., cb≠bc. Complexity may be increased through complementary interactions, e.g., cbb. Each of these structures may evolve and stabilize at a specific equilibrium. The selfing of each cross may maintain the initial balance for the whole progeny. Direct reciprocal effects (maternal or paternal) are represented here, but in the case of specific evolution or interactions, parental or control types may be recovered and sometimes transgressive effects can be expressed

The specific potentialities of a parent will then be confronted with all the others in the diallel cross. For a given cross, every type of situation may be obtained according to the evolution, stabilitization, and relationship of the male and female gamete properties.

The same behavior for all the individuals from a cross should result in unmodified intraprogeny variance. Asymmetrical new behavior can be stratified at fertilization and expressed by the *maternal effect*. But in the case of the specific evolution of each of the parental contributions, amplifications of elements may occur through specific division kinetics. Furthermore synergetic effects may arise through the mixing of extrachromosomal potentials, and (or) new nucleo-cyto-plasmic interactions. Thus unusual behavior such as *paternal or transgression effects (or vigor-like) may be expressed*. Nevertheless, as the nuclear genetic bases are maintained, *parental variant or control expression can be recovered* in some cases.

The offspring from selfing of the reciprocal crosses will remain homogenous in so far as the extrachromosomal systems implied do not segregate (Birky 1983) after meiosis, and they may remain identical to the first cross-progenies if biological balance is recovered.

6 In Vitro Variants in Lettuce and Tomato

Experiments were made on two species: lettuce (*Lactuca sativa*, 2n = 18) and tomato (*Lycopersicon esculentum*, 2n = 24). This section is a synthesis of the author's work (Sibi 1971 to 1986), more especially Sibi (1976) on lettuce, and Sibi (1986) on tomato). In vitro tissue culture was achieved after genotypes were verified as behaving homozygously, i.e., the absence of segregating traits by selfing. It gave rise to a wide spectrum of variability through regenerated individuals.

6.1 Chromosome Numbers

Karyotypic analyses were made for both lettuce and tomato material, at each of the previously described stages (cf. Sect. 5.1). In vitro culture of lettuce never gave rise to altered karyotypes in either callus cells, vitro variants or self progeny root tips; while for tomato the variability in chromosome number was largely expressed in the callus and sometimes appeared in the regenerated plants. Of 70 individuals, 7 showed aneuploidy, and one was tetraploid. Regular karyotypes were observed in the rest, and remained stable through the next selfed progeny.

6.2 Progenies Analyses

The following successive statistical analyses were carried out from either qualitative or quantitative traits (color, shape, aspect of the leaves; leaf size, number of leaves, weight, height of plants, etc.) and refer to approximately 42,000 data for lettuce and 24,000 for tomato.

6.2.1 *Progenies from Selfing*

The next comparison included offspring from regular karyotype individuals specifically. It comprised selfed progenies of 15 variants and 7 controls for lettuce on the one hand, and on the other of 25 variants and 5 controls for tomato. A mutant situation was observed for 2 lettuce and 17 tomato progenies, while uniform new phenotypes appeared for the rest of the lettuce progenies and for 8 in tomato.

All the criteria of uniform progenies were followed through two successive generations and showed to be perfectly stable, i.e., homozygous. The easily recognizable qualitative traits of lettuce variant were followed in comparison to the controls and they remained stable throughout six successive generations. Such variation thus refers to the "homogenous variant phenotype" (cf. Sect. 5.2.3) described previously. Only well-characterized progenies and control were subjected to the following diallel crosses and analyses.

6.2.2 *Diallel Crosses*

The diallel designs included variants (three for lettuce and four for tomato) and a control. The progenies were compared, and data analyses gave the following results.

Lettuce specific variant traits were mostly maternally inherited, and so were some of the quantitative tomato traits. More surprising and characteristic was the frequent paternal inheritance specifically observed for tomato. While for both lettuce and tomato, the progenies from crosses between variants (or with the control), behaving like the control statistically, for a given trait, could exhibit marked transgressive effects on another trait. Furthermore, according to the partner or reciprocity, the same parent could give rise to either transgressive or unmodified homogenous progeny, the latter case referring to the integrity of parental potentialities. For any situation, intraprogeny variance remained unmodified.

6.2.3 *Selfing of the Crosses*

This kind of checking was only achieved on tomato diallel cross progenies. Selfing was carried out essentially on one "reciprocal row-column" set of the diallel design, reciprocal progenies affected by asymmetrical transgressive effects, and the parents. Data analyses from comparative trial showed the "memory" of asymmetrical transgressive expressions, and confirmed the variance stability of the progenies.

6.3 Implication of a Hypothesis

The results described are of great importance and do not support the hypothesis of either double mutations or gametic selection to explain the stability of the selfing of vitro variants. Since the crosses between variants or with the control never gave any segregating traits in the selfed progenies, the nuclear genome can be identified as homozygous and unmodified. Thus the transgressive expressions observed cannot be explained by residual heterozygosity or chromosome pair substitution in the parental vitro variant.

The asymmetrical transgressions and frequent paternal effects observed here favor extrachromosomal biological modifications with partially self-controlled inheritance. Nonetheless, the multiplicity of reactions for one parent in several different crosses suggests a multiplicity of mechanisms possibly acting simultaneously. These might imply genetic wavering in populations of cytoplasmic organelles or the dialogue between the DNA of nucleus and organelles (Whisson and Scott 1985) etc. Because they are genetically or sexually transmissible, such mechanisms have been grouped under the term "epigenic" modifications (Sibi 1976), that is to say, concerning biological structures or dynamic systems, not implying the coding genes directly, and not segregating in progeny.

7 Conclusions

Plant regeneration through callus in vitro culture has given rise to a wide spectrum of variability. The new phenotypes observed has long been among our main subject of interest and "vitro variants" or "vitro variation" have been suggested here to express this phenomenon.

The genetic heredity of the new traits is based on the kind of changes occurring in the cell destined to regenerate, and on the regenerated plant layer affected by the modifications. Only modifications committed to retain genetic traits and implying at least a sporogenous area through the L_2 layer of the regenerated plant will show hereditary transmission to the offspring.

In any case a systematic experimental approach is necessary to define the cellular compartment implied and the inheritability of the new phenotype. Chromosome countings of the regenerants and of their offsprings, or genetical analyses through selfings and reciprocal cross progenies can be complemented by the next selfings of the latter.

The conclusions arising from these types of analysis made on lettuce and tomato show that, beyond the classical hypotheses based on nuclear changes in chromosome numbers or frequent mutational events, extrachromosomal systems must be involved to explain the genetic behavior of the new traits.

Through successive selfings, genetic fixity has been expressed for either qualitative or quantitative criteria. The progenies of reciprocal crosses in some cases have shown asymmetrical transgressive expressions, remaining through selfings, and frequent paternal effects.

This means that either simple nuclear or strictly cytoplasmic modifications are insufficient hypotheses to account for the behavior expressed after in vitro tissue culture.

The term "epigenic" modifications has thus been created, i.e., genetic changes which follow neither classical Mendelian laws nor maternal cytoplasmic heredity.

More and more biochemical information and unexpected behavior of regenerants are being reported in the literature. The first steps before biochemical analyses investigating the molecular bases of such a phenomenon should be the detection of a cellular compartment implicated in the modifications. It can be presumed that several systems may be modified simultaneously or through inter-relations.

These results lead us to suppose that hereditary systems going beyond meiosis mechanisms and acting on regulations may be modified during tissue culture. These may be located not only in the cytoplasmic part, but also in the nuclear extra-chromosomal area or among noncoding nuclear genes, such as silent redundancy, as very little is yet known about their function and replication.

References

Ahloowalia BS (1986) Limitations of the use of somaclonal variation in crop improvement. In: Semal J (ed) CEC Symp Somaclonal variations and crop improvement. 1985, Gembloux, Belgium, pp 14–27

Bajaj YPS (ed) (1986) Biotechnology in agriculture and forestry, vol 2. Crops I. Springer, Berlin Heidelberg New York Tokyo

Benzion G, Phillips RL, Rines HW (1986) Case histories of genetic variability in vitro: oats and maize. In: Vasil IK (ed) Cell culture and somatic cell genetics of plants, vol 3. Plant regeneration and genetic variability. Academic Press, New York London, pp 435–448

Bergounioux C (1974) Déclenchement des phénomènes de régénération sur des pétales de *Petunia hybrida*. Ann Amelior Plantes 24:(1):55–62

Bergounioux C, Perennes C, Miloux B (1982) Analysis of the nucleoplasmic interactions in the phenomenon of heterosis. In: Earle ED, Demarly Y (eds) Variability in plants regenerated from tissue culture. Proc NSF-CNRS Congr 1980, Orsay. Praeger, New York, pp 331–342

Binns A, Meins F (1973) Evidence that habituation of tobacco pith cells for factors promoting cell division is heritable and potentially reversible. Proc Natl Acad Sci USA 70:2660–2662

Bird AP, Taggart MH, Gehring CA (1981) Methylated and unmethylated ribosome RNA genes in the mouse. J Mol Biol 152:1–17

Birky CW, Jr. (1983) Relaxed cellular controls and organelles heredity. Science 222:468–475

Borst P, Greaves DR (1987) Programmed gene rearrangements altering gene expression. Science 235:658–667

Brettell RIS, Ingram DS (1979) Tissue culture in the production of novel disease-resistant crop plants. Biol Rev 54:329–345

Brettell RIS, Thomas E, Ingram DS (1980) Reversion of Texas male-sterile cytoplasm maize in culture to give fertile, T-toxin resistant plants. Theor Appl Genet 58:55–58

Brink RA, Styles ED, Axtell JD (1968) Paramutation: directed genetic change. Science 159:159–161

Buiatti M (1977) DNA amplification and tissue culture. In: Reinert J, Bajaj YPS (eds) Applied and fundamental aspects of plant cell, tissue, and organ culture. Springer, Berlin Heidelberg New York, pp 358–374

Buiatti M, Marcheschi G, Tognoni F, Lipucci di Paola M, Collina Grenci F, Martini G (1985) Genetic variability induced by tissue culture in the tomato *Lycopersicon esculentum*. Z Pflanzenzücht 94:162–165

Cameron JW, Soost R, Olson F (1964) Chimeral basis for color pink and red grapefruit. J Hered 55(1):23–28

Carlson PS (1973) Methionine-sulfoximine resistant mutants of tobacco. Science 180:1366–1368

Carlson PS, Chaleff RS (1974) Heterogeneous associations of cells formed in vitro. In: Ledoux L (ed) Genetic manipulations with plant materials. Plenum, New York, pp 245–261

Chaleff RS (1985) Selection for herbicide-resistant mutants. In: Evans DA, Sharp WR, Ammirato PV (eds) Handbook of plant cell culture, vol 4. Techniques and applications. Macmillan, New York, pp 133–148

Chaleff RS (1986) Isolation and characterization of mutant cell lines and plants: herbicide-resistant mutants. In: Vasil IK (ed) Cell culture and somatic cell genetics of plants, vol 3. Plant regeneration and genetic variability. Academic Press, New York London, pp 499–512

Chen THH, Gusta LV (1986) Isolation and characterisation of mutant cell lines and plants: cold tolerance. In: Vasil IK (ed) Cell culture and somatic cell genetics of plants, vol 3. Plant regeneration and genetic variability. Academic Press, New York London, pp 527–547
Clare MV, Collin HA (1974) Meristem culture of brussels sprouts. Hortic Res 13(2/3):111–118
Clauhs RP, Grun P (1977) Changes in plastid and mitochondrion content during maturation of generative cells of *solanum* (solanaceae). Am J Bot 64(4):377–383
Cromoco OJ, Ochoa-Alejo N (1983) Herbicide tolerance in regenerated plants. In: Evans DA, Sharp WR, Ammirato PV, Yamada Y (eds) Handbook of plant cell culture, vol 1. Techniques for propagation and breeding. Macmillan, New York, pp 770–781
D'Amato F (1952) Polyploidy in the differentiation and function of tissues and cells in plants. A critical examination of literature. Caryologia (4):311–357
D'Amato F (1978) Chromosome number variation in cultured cells and regenerated plants. In: Thorpe TA (ed) Frontiers of plant tissue culture. IAPTC 1978, Univ Press, Calgary, pp 287–295
D'Amato F (1985) Cytogenetics of plant cell and tissue cultures and their regenerates. CRC Crit Rev Plant Sci 3:73–112
Demarly Y (1972) Régulation et hétérosis. Ann Amélior Plantes 22:143–166
Demarly Y (1974) Les effets cytoplasmiques et l'amélioration des plantes. Select Fr 19:61–88
Demarly Y (1976) La notion de programme génétique chez les végétaux supérieurs. I. Aspects théoriques. Ann Amelior Plantes 26:117–138
Demarly Y (1985) L'épigénique. Bull Soc Bot Fr 132 Actual Bot 3/4:79–94
Demarly Y (1986) Experimental and theoretical approach of in vitro variations. In: Semal J (ed) CEC Symp Somaclonal variations and crop improvement. Gembloux, Belgium, pp 84–99
De Mars R (1974) Resistance of cultured human fibroblasts and other cells, to purine and pyrimidine analogues in relation to mutagenesis detection. Mutat Res 14:335–364
Dermen H, Stewart RN (1973) Ontogenic study of floral organs of peach utilizing cytochimeral plants. Am J Bot 60(3):283–291
Deshayes A (1976) Effets de cycles successifs de néoformation de bourgeons in vitro sur l'aptitude à la variation somatique chez un mutant chlorophyllien de *Nicotiana tabacum* var. *Samsun*. Mutat Res 35:331–346
Dossou-Yovo S, Prioul JL, Demarly Y (1982) Croissance et photosynthèse de phénovariants de Riz. Effet de l'ombrage. Agronomie 2:493–502
Durrant A (1962) The environmental induction of heritable change in *Linum*. Heredity 17:27–61
Evans DA (1986) Case histories of genetic variability in vitro: tomato. In: Vasil IK (ed) Cell culture and somatic cell genetics of plants, vol 3. Plant regeneration and genetic variability. Academic Press, New York London, pp 419–434
Evans DA, Sharp WR (1983) Single gene mutations in tomato plants regenerated from tissue culture. Science 221:949–951
Evans DA, Sharp WR, Medina-Filho HP (1984) Somaclonal and gametoclonal variation. Am J Bot 71:759–774
Filner P (1980) Possible origins of heritable variations in plant cell cultures. In Vitro (Abstr) (16)3:232
Frankel R, Scowcroft WR, Whitfeld PR (1979) Chloroplast DNA variation in isonuclear male-sterile lines of *Nicotiana*. Mol Gen Genet 169:129–135
Gautheret RJ (1954) Catalogue des cultures de tissus végétaux. Rev Gen Bot 61:672–702
Gautheret RJ (1959) La culture des tissus végétaux. Masson, Paris, 863 pp
Gautheret RJ (1964) La culture des tissus végétaux: son histoire, ses tendances. Rev Cytol Biol Veg 27:99–220
Gould AR (1986) Factors controlling generation of variability in vitro. In: Vasil IK (ed) Cell culture and somatic cell genetics of plants, vol 3. Plant regeneration and genetic variability. Academic Press, New York London, pp 549–567
Gray MW (1982) Mitochondrial genome diversity and the evolution of mitochondrial DNA. Can J Biochem 60:157–171
Griffing B (1956) Concept of general and specific combining ability in relation to diallel crossing systems. Aust J Biol Sci 9:463–493
Heinz DJ, Mee GWP (1969) Creating variability through callus tissue cultures. Annu Rep Hawaiian Sugar Planter's Assoc
Heinz DJ, Mee GWP (1971) Morphologic, cytogenetic and enzymatic variation in *Saccharum* species hybrid clones derived from callus tissue. Am J Bot 58:257–262

Holliday R (1987) The inheritance of epigenetic defects. Science 238:163–170
Holliday R, Pugh JE (1975) DNA modification mechanisms and gene activity during development. Developmental clocks may depend on the enzymatic modification of specific bases in repeated DNA Sequences. Science 187:226–232
Hughes KW (1983) Selection for herbicide resistance. In: Evans DA, Sharp WR, Ammirato PV, Yamada Y (eds) Handbook of plant cell culture, vol 1. Techniques for propagation and breeding. Macmillan, New York, pp 442–460
Karp A, Bright SWJ (1985) On the causes and origins of somaclonal variation. Oxford Surv Plant Mol Cell Biol 2:199–234
Kartha KK, Champoux S, Gamborg OL, Pahl J (1977) In vitro propagation of tomato by shoot apical meristem culture. J Am Soc Hortic Sci 102(3):346–349
Keyes GJ, Collins GB, Taylor NL (1980) Genetic variation in tissue culture of red clover. Theor Appl Genet 58:265–271
Kouadio YS (1979) Analyse génétique de la variabilité observée dans une descendance de lignée pure de riz (*Oryza sativa* L. var Cigalon) traitée par androgenèse in vitro. Thèse Doc-Ing. Amélior Plantes, Univ Paris-Sud, Orsay, 161 pp
Langhe de E, de Bruijne E (1976) Continuous propagation of tomato plants by means of callus cultures. Sci Hortic 4:221–227
Lansman RA, Avise JC, Huettel MD (1983) Critical experimental test of the possibility of "paternal leakage" of mitochondrial DNA. Proc Natl Acad Sci USA 80:1969–1971
Larkin PJ, Scowcroft WR (1981) Somaclonal variation – a novel source of variability from cell cultures for plant improvement. Theor Appl Gent 60:197–214
Lee M (1984) Cytogenetic analysis and progeny evaluation of maize (*Zea mays* L.) plants regenerated from organogenic callus cultures. MS Thesis, Univ Minnesota, Minneapolis
Lutz A (1969) Etude des aptitudes morphogénétiques des cultures de tissus. Analyse par la méthode des clones d'origine unicellulaire. Rev Gen Bot 76:309–359
Matzinger DF, Burk LG (1984) Cytoplasmic modification by anther culture in *Nicotiana tabacum* L. Heredity 75:167–170
McClintock B (1956) Intranuclear systems controlling gene action and mutation. Brookhaven Symp Biol 8:58
McCoy TJ, Phillips RL (1982) Chromosome stability in maize (*Zea mays* L.) tissue cultures and sectoring in some regenerated plants. Can J Genet Cytol 24:559–565
Medina FJ, Risueño MC, Rodriguez-Garcia MI (1981) Evolution of the cytoplasmic organelles during female meiosis in *Pisum sativum* L. Planta 151:181–225
Meins F (1983) Heritable variation in plant cell culture. Annu Rev Plant Physiol 34:327–346
Meins F, Binns A (1977) Epigenetic variation of cultured somatic cells: evidence for gradual changes in the requirement for factors promoting cell division. Proc Natl Acad Sci USA 74:2928–2932
Meinz F, Lutz J (1980) The induction of cytokinin habituation in primary pith explants of tobacco. Planta 149:402–407
Mikoko-Nsika E (1982) Analyse de la variabilité dans la descendance des plantes obtenues par culture in vitro de tissus somatiques de tomate (*Lycopersicon esculentum* Mill). Thèse Spéc Amélior, Univ Paris-Sud, Orsay, 118 pp
Mousseau J (1970) Fluctuation induite par la néoformation de bourgeons in vitro. Coll Int CNRS Les cultures de tissus de plantes, 1969 Strasbourg, pp 235–239
Nagl W (1972) Evidence of DNA amplification in the orchid *Cymbidium* in vitro. Cytobiology 5:145–154
Nagl W, Hendon J, Rucker W (1972) DNA amplification in *Cymbidium* protocorms in vitro as it relates to cytodifferentiation and hormone treatment. Cell Differ 1:229–237
Nagylaki T, Petes TD (1982) Intrachromosomal gene conversion and the maintenance of sequence homogeneity among repeated genes. Genetics 100:315–337
Nettancourt de D, Devreux M (1977) Incompatibility and in vitro cultures. In: Reinert J, Bajaj YPS (eds) Applied and fundamental aspects of plant cell, tissue, and organ culture. Springer, Berlin Heidelberg New York, pp 426–464
Novàk FJ, Maskovà I (1979) Apical shoot tip culture of tomato. Sci Hortic 10:337–344
Nuti Ronchi V (1971) Amplificazione genica in cellules differenziate di cotiledoni di *Lactuca sativa* in vitro. Atti Assoc Genet Ital 16:47–50

Odjo JA (1978) Analyse biométrique de niveaux de variations observées chez des plantes de riz (*Oryza sativa* L. var. Cigalon) obtenues par androgenèse in vitro. Thèse Doc-Ing. Amélior Plantes. Univ Paris-Sud. Orsay. 123 pp

Ohta T (1980) Two-locus problems in transmission genetics of mitochondria and chloroplasts. Genetics 96:543–555

Orton TJ (1984) Somaclonal variation: theoretical and practical considerations. In: Gustafson JP (ed) Gene manipulation in plant improvement. Plenum. New York. pp 427–468

Padmanabhan V. Paddock EF. Sharp WR (1972) Plantlet formation from *Lycopersicon esculentum* leaf callus. Am J Bot 52:1429–1432

Péreau-Leroy P (1974) Genetic interaction between tissues of carnation petals as periclinal chimeras. Radiat Bot 14:109–116

Péreau-Leroy P (1975) Recherches radiobiologiques sur des chimères d'oeillet. *Dianthus caryophyllus* L. Thèse Docès-Sci. Univ Clermont-Ferrand. 169 pp

Picard E (1984) Contribution à l'étude de l'hérédité et de l'utilisation en sélection de l'haploidisation par androgenèse in vitro chez une céréale autogame: *Triticum aestivum* L. Thèse Docès-Sci. Univ Paris Sud. Orsay. 292 pp

Preer JR. Jr. (1971) Extrachromosomal inheritance: hereditary symbionts. mitochondria. chloroplasts. Annu Rev Genet 5:361–406

Rains DW. Croughan SS. Croughan TP (1986) Isolation and characterisation of mutant cell lines and plants: salt tolerance. In: Vasil IK (ed) Cell culture and somatic cell genetics of plants. vol 3. Plant regeneration and genetic variability. Academic Press. New York London. pp 537–547

Reisch B (1983) Genetic variability in regenerated plants. In: Evans DA. Sharp NR. Ammirato PV. Yamada Y (eds) Handbook of plant cell culture. vol 1. Techniques for propagation and breeding. Macmillan. New York. pp 748–769

Reisch B. Bingham ET (1979) In vitro selection of ethionine-resistant variants of diploid alfalfa. Agron Abstr:74

Reisch B. Bingham ET (1981) Plants from ethionine-resistant alfalfa tissue cultures: variations in growth and morphological characteristics. Crop Sci 21:783–788

Sacristàn MD (1986) Isolation and characterisation of mutant cell lines and plants: disease resistance. In: Vasil IK (ed) Cell culture and somatic cell genetics of plants. vol 3. Plant regeneration and genetic variability. Academic Press. New York London. pp 513–525

Sagawa Y. Melhquist GAL (1957) The mechanism responsible for some X-ray induced changes in flower color of the carnation. *Dianthus caryophyllus* Am J Bot. 44:397–403

Sager R. Grabowy C (1983) Differential methylation of chloroplast DNA regulates maternal inheritance in a methylated mutant of *Chlamydomonas*. Proc Natl Acad Sci USA 80:3025–3029

Sager R. Kitchin R (1975) Selective silencing of eukaryotic DNA. Science 189:426–433

San-Noeum LH. Ahmadi N (1982) Variability of doubled haploids from in vitro androgenesis and gynogenesis in *Hordeum vulgare*. In: Earle ED. Demarly Y (eds) Variability in plants regenerated from tissue culture. Proc NSF-CNRS Congr 1980. Orsay. Praeger. New York. pp 273–283

Satina S. Blakeslee AF. Avery AG (1940) Demonstration of the three germ layers in the shoot apex of *Datura* by means of induced polyploidy in periclinal chimeras. Am J Bot 27:895–905

Schaeffer GW (1982) Recovery of heritable variability in anther-derived doubled haploid rice. Crop Sci 22:1160–1164

Schaeffer GW. Sharpe FT Jr. Cregan PB (1984) Variation for improved protein and yield from rice anther culture. Theor Appl Genet 67:383–389

Schimke RT. Kaufman RJ. Alt FW. Fellems R (1978) Gene amplification and drug resistance in cultured murine cells. Science 202:1051–1055

Sibi M (1971) Création de variabilité par culture de tissus in vitro chez *Lactuca sativa*. DEA Amelior Plantes. Univ Paris-Sud. Orsay

Sibi M (1974) Création de variabilité par culture de tissus in vitro chez *Lactuca sativa*. Thèse Spéc Amélior Plantes. Univ Paris-Sud. Orsay. 142 pp

Sibi M (1976) La notion de programme génétique chez les végétaux supérieurs. II. Aspect expérimental: obtention de variants par culture de tissus in vitro sur *Lactuca sativa* L. apparition de vigueur chez les croisements. Ann Amelior Plantes 26(4):523–547

Sibi M (1978) Multiplication conforme. non conforme. Select Fr 26:9–18

Sibi M (1979) Expression of cryptic genetic factors in vivo and in vitro. In: Zeven AC. van Harten AM (eds) Proc Conf Broadening genetic base crops. 1978 Wageningen. Pudoc. Wageningen. pp 339–340

Sibi M (1980) Variants épigéniques et cultures in vitro chez *Lycopersicon esculentum* L. In: Application de la culture in vitro à l'amélioration des plantes potagères. CR EUCARPIA CNRA, Versailles, pp 179–185

Sibi M (1981) Hérédité de variants epigéniques obtenus par culture de tissus in vitro chez les végétaux supérieurs. Thèse Doctès-Sci, Univ Paris-Sud, Orsay, 280 pp

Sibi M (1982) Heritable epigenic variation from in vitro tissue culture of *Lycopersicon esculentum* (var. Monalbo). In: Earle ED, Demarly Y (eds) Variability in plants regenerated from tissue culture. Proc NSF-CNRS Congr. 1980 Orsay. Praeger, New York, pp 228–244

Sibi M (1986) Non-mendelian heredity. Genetic analysis of variant plants regenerated from in vitro culture: epigenetics and epigenics. In: Semal J (ed) CEC Symp Somaclonal variation and crop improvement, 1985 Gembloux, Belg, pp 53–83

Sibi M, Biglary M, Demarly Y (1984) Increase in the rate of recombinants in tomato (*Lycopersicon esculentum*) after in vitro regeneration. Theor Appl Genet 68:317–321

Siminovitch L (1976) On the nature of hereditable variation in cultured somatic cells. Cell 7:1–11

Skirvin RM (1977) Separation of phenotypes in a periclinal chimera. J College Sci Teach 7:33–35

Skirvin RM, Janick J (1976) Tissue culture-induced variation in scented *Pelargonium* spp. J Am Soc Hortic Sci 101:281–290

Skirvin RM, Carlson CL, Gorske S (1982) Natural and tissue cultured-induced variation in *Portulaca* hybrid. In: Earle ED, Demarly Y (eds) Variability in plants regenerated from tissue culture. Proc NSF-CNRS Congr 1980, Orsay. Praeger, New York, pp 245–267

Skokut TA, Filner P (1980) Slow adaptative changes in urease levels of tobacco cells cultured on urea and other nitrogen sources. Plant Physiol 65:995–1003

Springer WD, Green CE, Kohn KA (1979) A histological examination of tissue cultures initiated from immature embryos of maize. Protoplasma 101:269–281

Sree Ramulu K (1980) Failure of EMS to induce S-locus mutations in *Nicotiana alata* Link and Otto. Environ Exp Bot 20:149–155

Sree Ramulu K (1982) Genetic instability at the S-locus of *Lycopersicon peruvianum* plants regenerated from in vitro culture of anthers: generation of new S-specificities and S-allele reversions. Heredity 49:319–330

Sree Ramulu K, Devreux M, Ancora G, Laneri U (1976a) Chimerism in *Lycopersicum peruvianum* plants regenerated from in vitro cultures of anthers and stem internodes. Z Pflanzenzücht 76:299–319

Sree Ramulu K, Devreux M, De Martinis P (1976b) Origin and genetic analysis of plants regenerated in vitro from periclinal chimeras of *Lycopersicum peruvianum*. Z Pflenzenzücht 77:116–124

Sugawara N, Szostak JW (1983) Recombination between sequences in nonhomologous position. Proc Natl Acad Sci USA 80:5675–5679

Tal M (1983) Selection for stress tolerance. In: Evans DA, Sharp NR, Ammirato PV, Yamada Y (eds) Handbook of plant cell culture, vol 1. Techniques for propagation and breeding. Macmillan, New York, pp 461–488

Tartof KD (1973) Unequal mitotic sister chromatid exchange and disproportionate replication as mechanisms regulating ribosomal RNA-gene redundancy. Cold Spring Harbor Symp Quant Biol 38:491–500

Tilney-Bassett RAE, Abdel-Wahab OAL (1982) Irregular segregation at the *Pr* locus controlling plastid inheritance in *Pelargonium*: gametophytic lethal or incompatibility system? Theor Appl Genet 62:185–191

Tilney-Bassett RAE, Birky CW Jr (1981) The mechanism of the mixed inheritance of chloroplast genes in *Pelargonium*: evidence from gene frequency distributions among the progeny of crosses. Theor Appl Genet 60:43–53

Truong-André I (1977) Variabilité des plantes issues de l'androgenèse in vitro: tentatives d'application directe en sélection par cette variabilité et de la mutagenèse par voie haploide chez le riz, *Oryza sativa*. Thèse Spéc Amélior Plantes, Univ Paris-Sud, Orsay, 188 pp

Wakasa R (1982) Application of tissue culture to plant breeding. Method improvement and mutant production. Bull Natl Inst Agric Sci D 33:121–200

Whisson DL, Scott S (1985) Nuclear and mitochondrial DNA have sequence homology with a chloroplast gene. Plant Mol Biol 4:267–273

Yamada Y, Sato F (1983) Selection for photoautotrophic cells. In: Evans DA, Sharp WR, Ammirato PV, Yamada Y (eds) Handbook of plant cell culture, vol 1. Techniques for propagation and breeding. Macmillan, New York, pp 489–500

I.5 Molecular Basis of Somaclonal Variation

S.G. BALL[1]

1 Introduction

Somaclonal variation has been extensively reviewed (Larkin and Scowcroft 1983; Scowcroft 1985; see also Chap. I.1. this Vol.). Genotypic changes of all possible types involving qualitative or quantitative traits. from single Mendelian mutations to dramatic changes in the ploidy level. aneuploidy. chromosomal aberrations. and even mutations of cytoplasmic origin have been reported.

This review. will be focused on molecular hypotheses that have been put forward to explain somaclonal variation. as well as the existing evidence supporting them. Practical implications or descriptive accounts of cases of somaclonal variation can be found elsewhere (Larkin and Scowcroft 1983; Scowcroft 1985). It is our belief that a molecular investigation of this problem will some day give the clue to the origin of this intriguing and complex phenomenon. A clear understanding of the basis of somaclonal variation will probably warrant a better control of its manifestations. avoiding or optimizing them when necessary.

2 Is Somaclonal Variation Culture-Induced?

Before investigating the molecular origin of somaclonal variation. it is worth wondering where and when the changes originate. In other words. is the variation preexisting in somatic cells and revealed by enabling these to regenerate in differentiated plants by tissue culture. or is the variation induced by in vitro techniques per se?

One way to answer this problem would be to follow plants heterozygous for a set of easily scored genetic markers through one cycle of protoplast culture and subculturing of the protoclones. Plants regenerated from each subculture are scored phenotypically. Protoclones are defined as parental if they give rise only to parental plants regenerated from each subculture of the protoclone. Other protoclones will be called homogeneous or heterogeneous variants if they give rise only to heterozygotes or to a mixture of homozygotes and heterozygotes in the different subclones respectively. The presence of heterogeneous variants is indicative of

[1] Chaire de Génétique et d'Amélioration des Plantes. Faculté des Sciences Agronomiques de l'Etat. 5800 Gembloux. Belgium. Present address: Bldg. C9. Laboratoire de Chimie Biologique. 59655 Villneuve d'Aseq CEDEX. France

Biotechnology in Agriculture and Forestry. Vol. 11
Somaclonal Variation in Crop Improvement I (ed. by Y.P.S. Bajaj)

variation induced during in vitro culture, while homogeneous variants suggest its preexistence.

Using this system with two sets of chlorophyll markers, Barbier and Dulieu (1983) have scored up to 17% heterogeneous tobacco protoclones, while in a similar analysis of one semi-dominant pigmentation mutant Lörz and Scowcroft (1983) report a frequency ranging from 1.4 to 6% of heterogeneous protoclones, which always exceeded homogeneous protoclone frequencies, that ranged from 0.14 to 1.8%. It thus seems that variation occurs predominantly in the tissue culture phase. In a later study, Dulieu et al. (1986, pers. commun.) were able to further define the occurrence of the variations at the earliest in the differentiated leaf cell and at the latest, before the second division in tissue culture. This seems to somewhat contradict those previous results that clearly demonstrate the predominance of heterogeneous versus homogeneous protoclones. However, it should be pointed out that heterogeneous protoclones can arise by mitotic segregation of a mutation that preexisted as a single strand lesion in the parental somatic cell. In other studies, (Sree Ramulu et al. 1984) chromosome duplication, polyploidization, and aneuploidization were traced back by cytophotometric and cytological analysis to events occurring at the very beginning of the culture process, and continuing throughout the whole callus phase. However, in these experiments, the changes occurring in potato are mainly chromosomal and seem to be of very different nature than those reported by Dulieu et al. in tobacco.

Definite proof of the occurrence of variation during tissue culture comes from studies performed on microspores culture (Kudirka et al. 1986). In these experiments, the haploid microspore shows spontaneous chromosome doubling during the culture phase and thus generates dihaploid cells and plants. Any mutation that precedes this event will be recovered as homozygous. If a variation occurs after the doubling event, then it can be kept in a heterozygous state. Thus, the presence of heterozygous mutations in a dihaploid regenerated plant can be taken as proof that variations do occur during the culture phase. Segregation of heterozygous mutations in dihaploid plants was indeed observed in rape (Hoffman et al. 1982). *Nicotiana sylvestris* (De Paepe et al. 1981; Prat 1983) and *Nicotiana tabacum* (Brown et al. 1983). The variation observed in in vitro culture from gametophytic cells is sometimes called gametoclonal variation (Evans et al. 1984).

Somaclonal variation thus appears to cover both preexisting and culture-induced changes. In the latter, some experiments seem to pinpoint a very early event occurring in the culture; (Dulieu et al. pers. commun.), while some gametoclonal variations appear after doubling of the haploid chromosome line. This dual nature (preexisting and culture-induced) of the observed changes must be kept in mind when searching for the molecular origins of somaclonal variation.

3 Molecular Aspects: The Problem of Cause and Effect

As stated above, many different types of changes can be recovered in somaclonal variation experiments. Chromosomal aberrations and changes in ploidy levels are universally attested as being the most common changes in tissue culture (Bayliss

1980; D'Amato 1985). However, numerous reports have also been published where none or few of these changes has been observed, while many somaclonal variants seem to segregate in a simple Mendelian fashion (Scowcroft 1985). Thus, mitotic spindle instabilities and abnormal chromosome segregation are clearly insufficient to explain these complex phenomena.

Whenever a single biochemical technique was available to investigate somaclonal variation at the molecular level, it has been used, and justly so. Repetitive DNA elements are by nature extremely easy to clone. Copy number experiments can thus be performed and they have been shown to differ in somaclonal variants (Brettell et al. 1986 a, b; Cullis 1985; Breiman et al. 1987a; Landsmann and Uhrig 1985). In some cases, spacer DNA length between rDNA coding sequences can be easily scored by choosing a suitable restriction enzyme in Southern blotting experiments. rDNA spacer length variation has in fact been found (Rode et al. 1987). Organelle DNA can be purified with relative ease and has sequence complexity several orders of magnitude lower than the nuclear genome, making it an attractive subject for molecular investigation. Again, organelle DNA changes were found in somaclonal variants (Kemble and Shepard 1984). Some restriction enzymes can easily distinguish between internal and external cytosine methylation at the restriction site. Subsequently, grossly altered methylation patterns have been reported in somaclonal variants (Brown and Lorz 1986). Finally, while certain enzymatic patterns vary according to the organ under investigation or to the physiological state of the plant, making them often unsuitable markers for somaclonal variation studies, some isozyme or protein profiles are taken to be sufficiently reproducible to be used as criteria for cultivar indentification. Indeed, qualitative and quantitative changes in these proteins have been reported in plants regenerated from tissue culture (Maddock et al. 1985; Cooper et al. 1986).

It thus seems that whenever a biochemical technique is amenable to analysis of somaclonal variation, some report of change at the molecular level will appear. When faced with these facts, a number of legitimate questions arise. First, to what extent are the reported changes linked to somaclonal variation? Second, does this overwhelming diversity of molecular changes reflect a similar complexity of molecular origins of somaclonal variation?

The first question can be easily answered by assaying the stability of the sequences under investigation in a sufficiently large sample of suitable controls. Since somaclonal variation is defined by hereditary changes occurring in plants derived from in vitro culture, individuals from the same original cultivar propagated by conventional means through the seed or tuber should be suitable controls. Unfortunately, when reported, the number of controls assayed is often much smaller than the number of variants under study, making it sometimes difficult to identify the relative importance of the detected changes.

The second question is somewhat more difficult to answer. Once these changes are proven to be much more frequent in somaclonal variants than in sexually or vegetatively propagated controls, many of them can be taken as a potential origin of somaclonal variation. For instance, altered methylation patterns can lead to differences in gene control expression, that in turn can switch on or off other mechanisms, leading to permanent changes in the genome. Point mutations or other alterations (deletions or insertions) in the same genes can also trigger a similar cascade of events. Thus, it will not be an easy task to ascertain that a molecular event is the origin or a consequence of somaclonal variation. On the other hand, the dual nature (preexisting and tissue-culture-induced) and the diversity of the changes reported make it likely that more than one molecular explanation might be found for somaclonal variation

4 The Use of Zymograms to Investigate the Origin of Somaclonal Variation

Well-defined genetic markers were clearly helpful in settling to some extent, the preexisting versus culture-induced variation debate. From a molecular point of view, the selection of specific phenotypic changes from in vitro culture is particularly useful if the responsible gene has already been cloned. One can thus select the mutated DNA sequences in a genomic bank by DNA hybridization and sequence the altered gene. Easily scored phenotypes where the implicated structural genes have been cloned are not as frequently found in higher plants as in lower organisms such as yeast.

However, a first successful attempt has been recently reported in maize by Brettell et al. (1986b) by using ADH (alcohol dehydrogenase) zymograms together with the cloned ADH1 and ADH2 sequences to fish out the variations. In a total of 645 primary regenerant plants from embryo cultures of maize hybrids, only one stable mutant was detected at the ADH1 locus that was characterized by slower electrophoretic mobility in starch gels. The mutant had full enzymatic activity and segregated as an ordinary Mendelian marker in subsequent crosses. Southern hybridization showed no differences in the restriction pattern of this mutant gene. The latter was thus cloned with the help of the ADH1 probe and completely sequenced. The ADH1 sequences from the somaclonal variant were compared to those from the parental allele from which it was derived. The only detected change was a single base pair alteration that resulted in a change from glutamic acid to valine in the last codon of exon 6. This modification leads to the loss of a single negative charge that is sufficient to explain the slower mobility of the polypeptide.

It must be said that this is the first report to pin down somaclonal variation to the nucleotide level. Surprisingly, none of the favored molecular explanations seems to be at work in this somaclonal variant, that clearly highlights the importance of single point mutations in at least some of the variations generated by tissue culture. Interestingly, other plants that were regenerated from the same embryo culture did not display any alteration in their zymograms, suggesting thus that the mutation did not preexist in the immature embryo. It is yet too early to overemphasize the importance of mechanisms that generate point mutations as a major cause of somaclonal variation. In fact, the frequency of changes occurring in the roots of seed-propagated plants was not given in this report, making it difficult to define the change as a typical culture-induced one. However, we must acknowledge that the frequency of appearance of this variant is by at least two orders of magnitude higher than that to be expected for similar changes in classical genetics studies.

Even when clones of the corresponding structural genes are not available. zymograms have been used to give additional insight into the nature of tissue culture-induced changes. For instance. Ryan and Scowcroft (1987) recently reported a culture induced-change in the zymogram pattern of β-amylase in the developing wheat grain. Out of 149 plants regenerated from immature embryo tissue culture. 22 could be characterized by the loss of several isozyme bands that could be traced back to a chromosome loss event.

However. one plant showed a complex variant pattern consisting of several additional bands. together with the loss of a single isozyme band. All spikes of this regenerated plant showed the same pattern and segregated approximately according to Mendelian laws for heterozygous mutations. In this variant family. no chromosomal aberrations could be detected. either through mitosis or meiosis. As in the previous case. no other mutant plant could be regenerated from the embryo callus that gave rise to the somaclonal variant. Thus again. the change probably occurred during the tissue culture phase. In these experiments over 220 parental seeds that had not undergone any in vitro culture were used as controls and scored for changes in their zymograms patterns. No alterations could be found. thus nailing down the fact that the variant pattern is indeed a molecular manifestation of somaclonal variation. The great complexity of the enzymatic pattern prevented both determining the dominance or codominance of the mutation and deducing a possible single base pair change in a β-amylase structural gene as a molecular explanation for the observed changes (Ryan and Scowcroft 1987).

In contrast with the results obtained for ADH zymograms in maize. similar experiments conducted on triploid wheat endosperm tissue for ADH activity could not trace back the changes to single base pair changes (Davies et al. 1986). In fact. out of a total of 551 regenerants from immature embryo cultures. 17 showed altered zymograms. The majority of these (13) could be attributed to chromosome loss. while two were shown to be reciprocal translocations. resulting in one case in duplication. and in the other in deletion of a set of different markers. including ADH-A1. Another euploid mutant was shown to possess an isochromosome. In each case. no plant showing the same disturbed phenotypes could be regenerated from the immature embryo callus. suggesting once again that the variation is culture-induced (Davies et al. 1986).

Gliadin proteins have been extensively used as molecular markers for somaclonal variation in wheat. Both qualitative and quantitative changes have been detected in plants regenerated from immature embryo callus cultures (Maddock et al. 1985; Cooper et al. 1986). The relevance of the data has been severely questioned (Metakovsky et al. 1987). In the latter report. cross-pollination was often evoked as a source of variant electrophoretic spectrum of gliadins. Simultaneous changes in several gliadin loci are taken by these authors to be very unlikely. In any case. in contrast with the ADH or β-amylase studies. the genetic basis of the mutant pattern remains unclear.

In barley some variations have been detected in number and mobilities of hordeins in the progeny of plants regenerated from tissue culture. However. these changes did not yield altered patterns in Southern hybridization experiments using the appropriate hordein cDNA clone (Breiman et al. 1987b). Other soluble protein electrophoretic profiles were examined both in alfalfa and in potato somaclones

(Baertlein and McDaniel 1987; Ball et al. submitted). In alfalfa, statistical treatment of quantitative changes in the protein profile was used to distinguish two somaclonal variants from a sample of 22 in comparison of four repeats of the control plant (Baertlein and McDaniel 1987). However, if such changes are significant, they have not proven to be hereditary, and could very well be epigenetic.

In our experiments, we compared 58 somaclones to 58 parent plants for electrophoretic patterns of soluble tuber proteins. (Ball et al. submitted). We also detected subtle quantitative differences; the problem being that such small changes are also detected at about the same frequency in our control sample. It is our belief that subtle quantitative changes such as these are not workable phenotypes in comparison with the clear-cut differences observed in the zymogram studies.

5 Overall DNA Amounts and Repetitive DNA Copy Number Changes

Both overall alteration in the amount of DNA and quantitative changes in copy number of specific repetitive sequences, have been documented (see Nagl Chap. I.6, this Vol.). Differences in total amount of DNA per cell was reported for regenerated plants in *Nicotiana sylvestris* (De Paepe et al. 1982) and in *Nicotiana tabacum* (Berlyn 1982; Dhillon et al. 1984) and for cultured callus in *N. tabacum* (Berlyn 1982), *N. glauca* (Durante et al. 1983), and potato (Sree Ramulu et al. 1984, 1985; Carlberg et al. 1984). This amount was shown to increase in population and individual cells during the in vitro callus phase. The changes could in some cases be recovered to a somewhat lesser extent in regenerated plants.

In potato, the cytophotometric studies were often correlated to extensive cytological observations, that clearly show endoreplication as a main source of polyploidization, followed by chromosome lagging and nondisjunction as a subsequent source of aneuploidy (Sree Ramulu et al. 1985). Moreover, it was shown in one study (Sree Ramulu et al. 1985) that the rate of change in the callus phase clearly depended on the genotype of the parent plant and perhaps, more simply, on its initial ploidy level.

Other cytogenetic studies using chromosome counts and BrdC Giemsa labeling (Pijnacker et al. 1986; Gill et al. 1986) lead to the same conclusions. Thus, specific sequence amplification does not seem to be at work to explain the extent of DNA variation in potato.

However, both in *Nicotiana sylvestris* (De Paepe et al. 1982) and *N. glauca* (Durante et al. 1983), thermal denaturation profiles suggest selective amplification of repeated G-C- and A-T-rich sequences that could occur very early during the callus phase and be maintained in the regenerated plants.

In rice, it has been shown using cloned repeated DNA sequences that very dramatic events of specific amplification do occur during the culture phase (Zheng et al. 1987). Seventy-fivefold differences were indeed observed between leaves and in vitro-cultured cells, while two to threefold differences occurred between leaf cells from different cultivars. The chromosome number was shown to be rather stable with occasional aneuploidy, which is clearly insufficient to explain the dramatic amplification levels. This, however, might not hold true for the identified ribosomal DNA sequences, where copy number increase is somewhat less striking. The probes were mostly of the clustered tandem repeat type. However, one interspersed cloned sequence also showed culture-induced amplification (Zheng et al. 1987). It would be of great interest to perform such studies on seed cultures from which somaclonal variant plants have been regenerated (Sun ZX et al. 1983; Oono 1985). One would thus be able to test to what extent these differences are stably maintained in adult plants and their progenies.

DNA variation associated with in vitro culture was also reported in flax (Cullis 1985). However, in this case the most dramatic effect reported for a satellite DNA sequence is in the twofold range, and leads to sequence deamplification. Interestingly, the reduction in copy number, which was also observed for 5S rDNA sequences, was limited to a Taq1-resistant cluster of the genome. These experiments on calli clearly mimic the variation reported in the progeny of plants grown under stressful conditions (genotrophs). The overall effect, in this case, is a reduction by about 15% of the total DNA quantity (Cullis 1985).

In wheat, three reports (Bretell et al. 1986a; Breiman et al. 1987a; Rode et al. 1987) have described alterations in one or more of the three mapped Nor (nucleolus organizer) loci. These consist of tandem arrays of rDNA sequences that can be distinguished by their different spacer lengths.

In one study (Brettell et al. 1986a), on a total of 192 plants derived from 26 immature embryo cultures, one regenerated plant showed a marked reduction in rDNA copy number at one specific locus, as evidenced by hybridization of Taql-digested genomic DNA with rDNA probes. The beauty of this study is that numerous chromosome counts and cytological observations were performed to rule out the possibility of dealing with chromosome loss events. Reduced staining at the position of the Nor locus was clearly observed using the C banding technique. Other plants regenerated from the callus that gave rise to the variant were analyzed for their Nor loci rDNA content: none showed the same pattern of alterations, suggesting that the variation occurred during the tissue culture phase.

These authors were also able to assign the variant locus to the 1R chromosome using both biochemical and classical genetic techniques. One can only regret in such elegant studies that insufficient stability controls make it difficult to realize to what extent this rDNA copy number reduction is a manifestation of somaclonal variation.

In similar experiments, Breiman et al. (1987a) report quantitative and qualitative differences in rDNA spacer length that are more likely to be interpreted as modification in rDNA copy number in the various Nor loci. However, the alterations could also very well be due to chromosome loss events, since no cytological observations were performed. Aneuploidy causing modifications of Nor loci rDNA copy number were in fact reported by Brettell et al. (1986a) and cannot be ruled out.

In this case, control experiments were performed that suggest that the modifications can be accounted for by somaclonal variation. However, control versus variant numbers were not reported, making it still difficult to acknowledge that we are dealing here with somaclonal variation.

Dramatic changes in physical length of the spacer region of doubled haploid lines, has been recently reported in wheat cv. César (Rode et al. 1987). This gametoclonal variation produces, in all first-cycle double haploids, a completely altered pattern of restriction fragments when probed with a DNA clone covering the spacer region. These modifications, that remain stable throughout subsequent cycles of in vitro androgenesis, cannot be explained by chromosome loss, since old bands are replaced by new ones that were not previously detectable.

Moreover, summing up restriction fragments in the new pattern gives an increase of 150 base pairs for the entire spacer region! Whether this change is introduced during androgenesis or through in vitro culture remains unknown. These staggering alterations are thus introduced in only one cycle of culture, and lead to the complete replacement of hundreds of copies of the parental pattern by new ones in a highly reproducible fashion. It is hard to imagine a molecular mechanism that could generate both conversion on such a large scale and reproducibility of the converted pattern. Further experiments, including sequencing of the spacer regions and crosses between cv. Benoit, for instance, that was reported not to show this spacer mutability and the César parental cultivar, will be needed to confirm and explain these fascinating changes. On the other hand, these need not be at work in other examples of somaclonal variation, since, as pointed out by the authors, they might prove to be highly specific aspects of androgenesis (Rode et al. 1987).

Experiments performed in barley (Breiman et al. 1987b) have shown, as in wheat, that several rDNA spacer TaqI fragments disappear in some plants regenerated from tissue culture. In this particular case, the change was attributed to spacer length variation from results obtained using other restriction enzymes. With these enzymes, new bands could be detected in the variant pattern, thus chromosome loss does not seem to be involved. Since it had been pointed out in other experiments in barley that some morphogenic callus cells were observed that showed chromosome number alterations (Singh 1986), the authors have set about confirming their molecular analysis conclusions by direct cytological observations. In no cases could chromosomal variations in number or structure be detected (Breiman et al. 1987b).

In potato, 36 individual plants taken from the tuber progeny of 12 somaclones were analyzed in Southern hybridizations with 26 tester DNA clones representing 167 kbp of potato DNA (Landsmann and Uhrig 1985). Two of the 12 somaclones showed progenies with reduced copy number of sequences homologous to a probe that turned out to be 25S rDNA. The plants contained approximately 70% of the wild-type amount. No qualitative change was detected with this particular or other probes. Again, analysis of a sufficiently large sample of controls was not reported, and could thus not warrant that this reduction in copy number is a typical origin or consequence of somaclonal variation. On the other hand, no convincing arguments are presented to rule out chromosome loss as a possible explanation for this apparent deamplification. In *S. tuberosum*, polyploidization and aneuploidy in regenerated plants from tissue culture are known to be extremely frequent (Karp et al. 1982; Jacobsen et al. 1983; Sree Ramulu et al. 1983; Fish and Karp 1986). We have also observed copy number reduction of repetitive DNA sequence in two somaclones with a probe that was selected as a highly repetitive and transcribed sequence: probably rDNA (Ball and Seilleur 1986). In these experiments, the fivefold reductions were in the relative intensities of two hybridizations signals.

Of course. the same remarks concerning chromosomal instabilities also apply here.

It seems from all this body of work that a convincing case is being made for at least some specific sequence amplification or deamplification during in vitro tissue culture.

In the author's opinion. further work is necessary. first to clearly pin point some of these changes. and second to estimate somewhat more precisely to what extent they are correlated with or can even be a cause of somaclonal variation.

6 Plant Transposable Elements and Somaclonal Variation

Between all molecular hypotheses that have been put forward to explain somaclonal variation at the molecular level. transposon activity seems to be a particularly favored one. It must be said that today the evidence supporting it is particularly scarce. and no convincing molecular proof has been reported concerning transposition-induced somaclonal variation. However. the temptation remains great to seek a hallmark of transposition activity in the diversity and far-reaching consequences of culture-induced changes.

One can only recall the genetic puzzle that characterized hybrid dysgenesis-induced changes before it could be traced back to P element activity in *Drosophila*. Can all variations in DNA amount. specific sequence amplifications. single gene mutations. chromosome aberrations. altered methylation patterns. and organelle DNA modifications be a consequence of transpositional activity?

Some authors take the simultaneous change of several characters in a single individual as a symptom of transposon-induced changes (Scowcroft 1985). In fact. we believe that many other molecular mechanisms can lead to such a cascade of events. Chromosome aberrations or instability and single gene mutations can be easily explained by evoking transposon activity. Despite transposon insertion. which is more likely to inactivate a gene. minor changes such as an insertion of short duplicated sequences (transposon footprints) that are left over after transposable element excision in plants can lead to stable single gene mutations (Wessler et al. 1986; for a review see Doring and Starlinger 1986).

In the one case in maize where a genetic marker was used together with the corresponding DNA clone to analyze somaclonal variation. no transposable element or footprint was found placed in the middle of the altered gene (Brettell et al. 1986b). However. one can argue that in this case. a bias existed toward harvesting point mutations. since enzymatic activity had to be conserved to lead to altered mobility of bands in the ADH zymograms (see above).

Another reason for relating transposon to somaclonal variation is the known effect of environmental stresses in a burst of tranpositional activity. Wide intra- or interspecific crosses. together with some viral infections. stimulate activity of previously silent transposons. (Peterson 1985; Doring and Starlinger 1986).

In vitro culture is indeed an environmental stress and one can wonder to what extent it can mimic these stresses above. The high frequency of chromosomal aberrations reported in tissue culture also favors the transposition hypothesis (for a review see Bayliss 1980). Many authors think sufficiently of them to suggest transpositional element activity as an explanation for their observations (Ahloowalia and Sherington 1985).

Unstable expression is also a hallmark of transposon-induced changes and has been reported in a number of cases such as wheat (Ahloowalia and Sherington 1985), tobacco (Lörz and Scowcroft 1983), and alfalfa (Groose and Bingham 1984). More circumstantiated evidence of tissue culture as a possible trigger of transposition bursts comes from experiments performed in maize by Peschke et al., (1985). In this case, AC activity was investigated by crossing plants regenerated from immature embryo cultures with a Ds tester. Transposition was indeed recovered in 5% of the cases, suggesting that tissue culture can act as environmental stress to give rise to transposition bursts.

7 DNA Methylation and Plant Tissue Culture

Plant DNA is known to be highly methylated. Evidence has been accumulated that indicates the role of methylation in gene expression, active genes being predominantly undermethylated (for a review see Yisraeli and Szyf 1984; Vanyushin 1984).

Moreover, methylation has been recently shown to silence the transpositional activity of some ac elements in maize (Schwartz and Dennis 1986). Thus, differences in the methylation state of the genome can also generate transpositional bursts, leading in turn to more permanent changes. In addition, altered methylation can provide a very convenient explanation for epigenetic changes that are sometimes observed in tissue culture experiments.

For these and other reasons, a number of authors have studied the impact of methylation in the particular sequences under investigation during tissue culture and after regeneration of somaclonal variants (Grisvard 1985; Rode et al. 1987; Brown and Lörz 1986). All of the studies use comparative digestion of DNA with the HpaII and MspI isoschizomers.

Both enzymes cleave at the CCGG sequence. MspI does not cut when the external cytosine is methylated, while HpaII cannot cleave when the internal C is methylated. In general, cytosine is more likely to be modified when it is beside a guanine, thus MspI usually restricts more plant DNA than HpaII.

Comparative HpaII MspI digestions of melon satellite DNA has revealed differences in the extent of methylation between hypocotyl and callus tissues. The latter showed extensive digestion by both enzymes, and are thus clearly undermethylated (Grisvard 1985).

In another study, gross methylation patterns of total DNA were compared in maize plants regenerated from tissue culture and showing disturbed phenotypes (Brown and Lörz 1986). Some of these somaclonal variants were shown to be overmethylated with respect to control plants, while others were preferentially digested by HpaII, suggesting a surprising switch of methylation specificity from the internal to the external cytosine. These results were interpreted as gross alterations in methylation due to tissue culture. Again, control versus somaclonal variant numbers were not reported.

We have conducted similar experiments in potato by comparing 55 plants regenerated from protoplast culture to 57 vegetatively reproduced parental individuals. Results are summarized in Fig. 1. In fact, the same gross alterations in methylation patterns were detected. However, if one conducts a statistical analysis with the control population, it is clear that these show exactly the same type of variability. Thus, we conclude that in potato, at least, gross alterations of methylation patterns are not molecular manifestations of somaclonal variation. These observations do not rule out the possibility of methylations being involved in tissue culture-induced changes of specific genes. We have tested this possibility in only one case by hybridizing the same sample of controls and somaclonal variants digested with HpaII and MspI with our highly repeated DNA probe (Fig. 2).

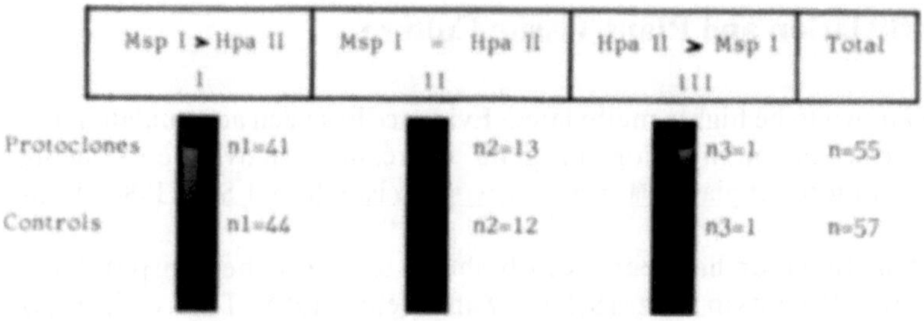

Fig. 1. Five μg of total genomic DNA were loaded on 1% agarose gels after digestion with HpaII or MspI. Fifty-five protoclones and 57 controls could be ordered in one of the three following classes.
Class *I* : samples showing a higher degree of methylation of the internal cytosine.
Class *II* : samples showing the same degree of methylation the two cytosines corresponding both to hypo- or hypermethylation of the DNA.
Class *III* : samples showing a higher degree of methylation for the external cytosine.
No significant differences could be detected between the control and protoclone populations

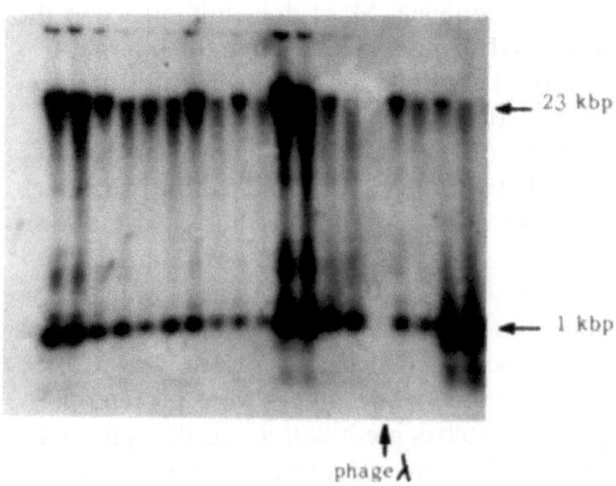

Fig. 2. Five μg of total genomic DNA from nine protoclones digested with HpaII or MspI were loaded on a 1% agarose slab gel and transferred to a Hybond-N filter through the Southern procedure. The samples were then probed with p47 a highly repetitive and transcribed DNA labeled at 5×10^7 cpm/μg. The filter was autoradiographed overnight using an X-Omat S film. Each pair of samples shows to the *left* the HpaII followed to the *right* by the corresponding MspI digest. The same large-scale variation in methylation patterns was observed with both protoclones and control populations

Again, a variable pattern of methylation was produced that could not distinguish statistically somaclonal variants and the control population. The probe most probably represents rDNA sequences, suggesting that their methylation is not involved in somaclonal variation.

Another similar type of comparison in ribosomal DNA methylation, was made between a flax tumor-derived tissue and the genotroph line from which it was derived. No differences were detected except that proportionally to their decreased copy number, a higher percentage of the tumor line sites were undermethylated (Blundy et al. 1987). However, the total number of these sites seem on average to be the same in the two lines. The authors suggest that these hypomethylated sites may represent the active rDNA genes.

Methylation analysis was also performed in the ribosomal DNA spacer region of first and second cycle androgenesis-derived gametoclonal variants. Once again, no significant differences were seen between these and the control plants (Rode et al. 1987).

One must not be led by these experiments to believe that methylation has no role to play in somaclonal variation. In fact, undermethylation, which seems to occur in callus tissues, could very well lead to activation of transposable elements that in turn will generate the vast array of changes that are commonly reported in somaclonal variants.

8 Organelle DNA Changes

Plants are characterized by the presence of two low complexity cytoplasmic genomes: the chondriosome and the plastome. The chloroplast genome typically consists of molecules of about 150 kbp containing in most cases a 20 to 25 kbp inverted repeat (for a review see Weil 1987). The organization of this genome seems to have been well conserved during evolution, the overall rate of variation of the plastome being somewhat lower than those reported for the chondriosome (Holwerda et al. 1986).

In most cases of somaclonal variation where chloroplast DNA sequences have been studied, stability seems to remain the rule. For instance, in potato, six different restriction enzymes were used to score 47 protoclones for cpDNA polymorphisms: no differences were found (Kemble and Shepard 1984).

However, chloroplast DNA changes have been reported in some cases of gametoclonal variation (Day and Ellis 1984, 1985). Anther culture in wheat is known to generate a high but variable percentage of albino plants. Leaf cells from control and albino plants were compared as to the size and organization of the plastome. Large deletions were shown to occur, leading to as much as 80% loss of sequences in one case (Day and Ellis 1984). Although it is not known to what extent these changes preexist or may be culture-induced, the authors hypothesize that the underlying mechanisms are the same as those responsible for generating maternal inheritance of chloroplast genes.

Similar studies were performed in albino barley plants derived from anther culture (Day and Ellis 1985). DNA molecules, in some cases possibly linear in structure, had, as in wheat, suffered a major deletion followed by duplication of some sequences to restore a unit genome size. Loss of the same fragment seems to be involved in both wheat and in barley (Day and Ellis 1985).

In other cases where green double haploid wheat plants were obtained from tissue culture, cpDNA was shown to be stable (Rode et al. 1985). However, in these experiments, the small size (four lines) of the analyzed sample makes it difficult to realize to what extent there is plastome stability before, during, or after anther culture.

Evidence for cytoplasmic inheritance of the albino character was also reported in rice (Sun et al. 1979; Sun Zong-xiu et al. 1983). In one case, albinism was derived from seed calli, while anther cultures were reported to generate rRNA-deficient plastids (Sun et al. 1979). The latter report is reminiscent of gametoclonal albinism observed in wheat. In fact, some nuclear encoded proteins in the stroma were also shown to be variant in this case. Thus, in rice, albinism seems to be a consequence of a dual nuclear and cytoplasmic control.

In contrast with the relative stability of the plastome, a number of reports have been published that suggest culture-induced changes in the chondriosome (Brettell et al. 1980; Gegenbach et al. 1981; Kemble et al. 1982; Umbech and Gegenbach 1983; McNay et al. 1984; Kemble and Shepard 1984; Negruk et al. 1986). This is, in fact, not too surprising, since the structure of plant mitochondrial DNA shows much more diversity between species and is known to evolve at a higher rate than the plastome (McClean and Hanson 1986; for a review see Pring and Lonsdale 1985; Walbot and Cullis 1985).

In maize, cytoplasmic male sterility is correlated with changes in mitochondrial DNA structure. In a particular male sterile cytoplasm (the T cytoplasm) male sterility co-segregates with sensitivity to *Helminthosporum maydis* race T pathotoxin (for a review see Laughnan and Gobay-Laughnan 1983). Because of the usefulness of male sterility in maize breeding programs and seed production, many attempts have been made to recombine the chondriosome in order to produce toxin-resistant male sterile T cytoplasms. Some of these attempts involved in vitro culture techniques together with T pathotoxin selection procedures. The results obtained from these studies strongly suggest that toxin sensitivity and male sterility are pleiotropic effects of the same cytogene in the T cytoplasm.

In fact, in the only case reported where male sterile T toxin-resistant recombinant plants were obtained, viable seed could not be produced (Gegenbach et al. 1977, 1981). Restriction analyses performed on mitochondrial DNA extracted from four culture-induced T toxin selected male fertile revertants revealed clearcut differences between them and the control plant (Gegenbach et al. 1981). It is difficult to assess the frequency of these culture-induced changes because of the bias caused by T toxin selection procedures.

In a subsequent study, Kemble et al. (1982) reported a revertant regenerated from tissue culture without toxin selection, that showed alterations in the XhoI digestion profile of its mitochondrial DNA. Umbech and Gegenbach (1983) have regenerated eight such revertants from a total of 169 plants from tissue culture without applying any selective pressure. In similar experiments, Brettell et al. (1980) have obtained 22 out of 23 such plants regenerated from tissue culture of T cytoplasm immature embryo cells! Umbech and Gegenbach (1983) have further characterized their revertants by restriction analysis. In five cases, an XhoI 6.6 kb fragment had disappeared, no alterations being detected with the sixth somaclonal variant.

Analysis of seed-propagated control plants had to await a further report (Oro et al. 1985) in which ten plants sampled from a third generation inbred line were subjected to restriction digestion with five different enzymes. In addition, Southern analysis was undertaken using the two maize linear mitochondrial plasmids as probes. In no cases could qualitative or quantitative changes be detected.

Before concluding that mitochondrial DNA rearrangements are a consequence or possible origin of somaclonal variation in maize, a number of observations have to be made. First, the number of control plants reported by Oro et al. is rather low, making it still difficult to evaluate chondriosome stability in seed-propagated plants. Second, the individuals that were analyzed at the DNA level are not a random sample of in vitro culture-regenerated plants. In fact, these were nearly always revertants that were suspected to have undergone mitochondrial DNA variation. In some cases, the high frequency of appearance of such revertants seems to argue against this (Brettell et al. 1980). However, there could also be a selective pressure in the culture phase to regenerate normal plants.

Mitochondrial DNA variation was also investigated in cultured cells of maize (Chourey and Kemble 1982; McNay et al. 1984). In a first study (Chourey and Kemble 1982) involving the S cytoplasm genotype, a unique correlation was found between the friable aspect of the callus and the loss of the S1 and S2 mitochondrial plasmids. However, when studies were performed in similar cultures of the Black Mexican sweet line, no mitochondrial plasmid loss or rearrangement could be detected (McNay et al. 1984). In fact, only minor quantitative changes seem to be involved. Thus, chondriosome stability in the culture phase might be strongly genotype-dependent.

Clearcut quantitative differences were also detected in restriction fragment patterns of *Vicia faba* mitochondrial DNA extracted from suspension cultures in comparison with whole plants. In this study, the differences could be traced back to a shift in size of subgenomic mitochondrial DNA circular molecules (Negruk et al. 1986).

In potato, 47 plants regenerated from protoplast culture (protoclones) were analyzed for alterations in their mitochondrial DNA (Kemble and Shepard 1984). Twenty-six of these protoclones had been selected for their vegetative vigor, while the remainder were a random sample of somaclonal variants. One vigor and one unselected clone were shown to contain an episome that was clearly absent from the parental Russet Burbank plant. In addition, 5 out of 47 protoclones showed restriction fragment polymorphisms, while none of these modifications appeared in the control population. Once again, the authors have omitted to report the size of their control sample, making it difficult for the reader to figure out how strongly these changes are correlated with somaclonal variation.

We have undertaken a similar study in potato using random protoclone and control populations of the same size from cultivar Désirée (Ball, this work). The presence of an episome of mitochondrial origin and restriction fragment polymorphisms was assayed by Southern hybridization of digested or undigested total genomic DNA with probes consisting of several fragments of cloned mitochondrial DNA or of the whole purified chondriosome labeled to a very high specific activity (10^3 cpm/μg). The size of our sample (58 in each population) would have enabled us to detect a change occurring at a frequency of 4% with a probability of 0.9. This frequency is below those that were reported for episome excision (4.3%) and well under the one reported for restriction fragment polymorphisms (10.6%). We found neither episome excision nor restriction fragment polymorphisms.

These contradictory results can be explained by differential chondriosome stability according to the genotype. However, all manifestations of somaclonal variations were detected phenotypically in both cases. We thus suggest that, in potato at least, mitochondrial DNA alterations are not necessarily a manifestation of somaclonal variation.

Other reports, both in wheat and barley, seem to indicate chondriosome stability in somaclonal variants. In barley, plants regenerated from tissue culture were assayed for restriction fragment polymorphisms (Breiman et al. 1987b), while in wheat four plants were investigated for both quantitative or qualitative changes in restriction patterns (Rode et al. 1985). In no case could differences be scored in the somaclonal variants.

It thus seems that more work will be necessary to have a clear idea of the impact of organelle DNA changes in somaclonal variation.

9 Conclusions

Chromosomal aberrations, point mutations, alterations of methylation patterns or DNA amounts, selective sequence amplification or deamplification, tissue culture-induced transposition activity, and modifications of organelle genomes have all been attested in plants regenerated from tissue culture.

It thus seems that considerable progress has been made in our perception of somaclonal variation at the molecular level. However, a really convincing case has been made for only a few of these changes. In fact, many alterations might very well not be linked at all to somaclonal variation, be it as a cause or a consequence.

Too much work, including our own, has been devoted to hunting for differences in whatever can be easily detected at the molecular level. This, in turn, leaves us with too many reported changes to have a clear idea of their relative importance in the phenotypic alterations that define somaclonal variation. In other words, what is the biological relevance of all the reported changes?

We now feel that the problem should be taken backwards. Looking at the whole or a small part of the genome for a particular type of change and reporting it does not always yield a clear picture of the mechanisms generating phenotypic diversity.

A better approach would be first to focus on a particular well-studied and mapped genetic locus and generate for this phenotype as many somaclonal variations as possible, for instance through protoplast culture.

Second, somaclonal variants should then be analyzed with the cloned gene in Southern experiments. If some information seems to be lost, cytological analysis and chromosome counts could be performed.

Third, when justified, the mutated gene will have to be cloned and sequenced. In comparison with suitably large control populations, one will then be able to produce statistics on the types of changes observed and clearly state their relevance in somaclonal variation. This approach has been successfully undertaken in a few studies. However, most of these concerned isozyme markers, where analysis at the DNA level could not be performed, probably because the cloned genes were not available. In only one case, in maize, was the analysis carried right down to the nucleotide level (Brettell et al. 1986b).

One can only regret, in higher plants, the scarcity of easily scored cloned genetic markers. Although this situation is steadily improving, one way to circumvent the problem is simply to introduce such a marker in the plant genome by genetic transformation. For instance, kanamycine resistance could be easily introduced by using a number of different available TDNA vectors. Kanamycine-sensitive revertants generated by somaclonal variation can then be analyzed by using the vector as a probe.

In a recent report concerning somaclonal variation in shooty crown gall tumor lines unambiguously demonstrated a somaclonal variation event that could be traced back to deletion and subsequent TDNA amplification (Peerbolte et al. 1987).

Could these events be a true reflection of somaclonal variation?

We believe that, in these experiments, abnormal states of hormonal balance generated by TDNA genes might introduce a particular bias, making this system an unsuitable model for studying conventional somaclonal variation.

However, disarmed TDNA vectors can now be used, avoiding these disagreements. These sequences behave exactly as Mendelian markers and are maintained stably in the plant genome (Budar et al. 1986; Chyi et al. 1986; Ooms et al. 1987). It is obvious that more experiments involving plant or foreign introduced genetic markers will be needed to clarify the now rather confusing picture that we have of somaclonal variation. It is true that in this particular field, hypotheses are generated more quickly than hard molecular data. For instance, the enhancement of somaclonal variation in protoclones regenerated from evacuolated protoplasts has recently led some researchers to hypothesize a direct role of the cytosqueletton in genome stability (Lawrence and Davies 1987). It is our belief that this multiplicity of hypotheses only reflects the dynamic aspects of this growing and fascinating field.

References

Ahloowalia BS, Sherington J (1985) Transmission of somaclonal variation in wheat. Euphytica 34:525–537

Baertlein DA, McDaniel RG (1987) Molecular divergence of alfalfa somaclones. Theor Appl Genet 73:575–580

Barbier M, Dulieu H (1983) Early occurrence of genetic variants in protoplast cultures. Plant Sci Lett 29:201–206

Ball SB (1987) Molecular genetics of potato. In: Bajaj YPS (ed) Biotechnology in agriculture and forestry, vol 3. Potato. Springer, Berlin Heidelberg New York Tokyo, pp 155–173

Ball SB, Seilleur P (1986) Characterization of somaclonal variations in potato: a biochemical approach. In: Semal F (ed) Somaclonal variations and crop improvement. Nijhoff, Dordrecht Lancaster, pp 229–235

Bayliss MW (1980) Chromosomal variation in plant tissue culture. Int Rev Cytol 11A:113–143
Berlyn MB (1982) Variation in nuclear DNA content of isonicotinic acid hydrazide-resistant cell lines and mutant plants of *Nicotiana tabacum*. Theor Appl Genet 63:57–63
Blundy KS, Cullis CA, Hepburn A (1987) Ribosomal DNA methylation in a flax genotroph and a crown gall tumour. Plant Mol Biol 8:217–225
Breiman A, Felsenburg T, Galun E (1987a) Nor loci analysis in progenies of plants regenerated from the scutellar callus of bread-wheat. Theor Appl Genet 73:827–831
Breiman A, Rotem-Abarbanell D, Karp A, Shaskin H (1987b) Heritable somaclonal variation in wild barley (*Hordeum spontaneum*). Theor Appl Genet 74:104–112
Brettell RIS, Thomas E, Ingram DS (1980) Reversion of Texas male-sterile cytoplasm maize in culture to give fertile, T-toxin resistant plants. Theor Appl Genet 58:55–58
Brettell RIS, Pallotta MA, Gustafson JP, Appels R (1986a) Variation at the Nor loci in triticale derived from tissue culture. Theor Appl Genet 71:637–643
Brettell RIS, Dennis ES, Scowcroft WR, Peacock JW (1986b) Molecular analysis of a somaclonal mutant of maize alcohol dehydrogenase. Mol Gen Genet 202:235–239
Brown JS, Wernsmann EA, Schnell RJ (1983) Effect of a second cycle of anther culture on flue-cured lines of tobacco. Crop Sci 23:729–733
Brown PTH, Lörz H (1986) Molecular changes and possible origins of somaclonal variation. In: Semal F (ed) Somaclonal variations and crop improvement. Nijhoff, Dordrecht Lancaster, pp 148–156
Budar F, Thia-Toong L, Van Montagu M, Hernalsteens J-P (1986) *Agrobacterium*-mediated gene transfer results mainly in transgenic plants transmitting T-DNA as a single mendelian factor. Genetics 114:303–313
Carlberg I, Glimelius K, Eriksson T (1984) Nuclear DNA-content during the initiation of callus formation from isolated protoplasts of *Solamum tuberosum* L. Plant Sci Lett 35:225–230
Chourey PS, Kemble RJ (1982) Transpositionment in tissue cultured cells of maize. In: Fujiwara A (ed) Plant tissue culture 1982. Maruzen, Tokyo, pp 425–426
Chyi YS, Jorgensen RA, Goldstein D, Tanksley SD, Loaiza-Figueroa F (1986) Locations and stability of *Agrobacterium*-mediated T-DNA insertions in the *Lycopersicon* genome. Mol Gen Genet 204:64–69
Cooper DB, Sears RG, Lookhart GL, Jones BL (1986) Heritable somaclonal variation in gliadin proteins of wheat plants derived form immature embryo callus culture. Theor Appl Genet 71:784–790
Cullis CA (1985) Sequence variation and stress. In: Holm B, Denis ES (eds) Genetic flux in plants. Springer, Berlin Heidelberg New York Tokyo, pp 158–168
D'Amato F (1985) Cytogenetics of plant cell and tissue cultures and their regenerates. CRC Crit Rev Plant Sci 3:73–112
Davies PA, Pallotta MA, Ryan SA, Scowcroft WR, Larkin PJ (1986) Somaclonal variation in wheat: genetic and cytogenetic characterization of alcohol dehydrogenase 1 mutants. Theor Appl Genet 72:644–653
Day A, Ellis THN (1984) Chloroplast DNA deletions associated with wheat plants regenerated from pollen: possible basis for maternal inheritance of chloroplasts. Cell 39:359–368
Day A, Ellis THN (1985) Deleted forms of plastid DNA in albino plants from cereal anther culture. Curr Genet 9:671–678
De Paepe R, Belton D, Gnangbe F (1981) Basis and extent of genetic variability among doubled haploid plants obtained by pollen culture in *Nicotiana sylvestris*. Theor Appl Genet 59:117–184
De Paepe R, Prat D, Huquet T (1982) Heritable nuclear DNA changes in doubled haploid plants obtained by pollen culture on *Nicotiana sylvestris*. Plant Sci Lett 28:11–28
Dhillon SS, Wernsman EA, Mischke JP, Henry L (1984) Relationship between nuclear changes and plant productivity in anther derived doubled haploids of tobacco. Plant Physiol Suppl 75:120
Doring HP, Starlinger P (1986) Molecular genetics of transposable elements in plants. Annu Rev Genet 20:175–200
Durante M, Geri C, Grisvard JA, Goodman HM (1984) Herbicide resistant alfalfa cells: an example of gene amplification in plants. J Mol Appl Genet 2:621–635
Evans DA, Sharp WR, Medina-Filho HP (1984) Somaclonal and gametoclonal variation. Am J Bot 71:759–774
Fish N, Karp A (1986) Improvements in regeneration from protoplasts of potato and studies on chromosome stability. Theor Appl Genet 72:405–412

Gegenbach BG, Green CE, Donovan CM (1977) Inheritance of selected pathotoxin resistance in maize plants regenerated from cell cultures. Proc Nat Acad Sci 74:5113–5117
Gegenbach BG, Connelly JC, Pring DR, Conde MF (1981) Mitochondrial DNA variation in maize plants regenerated during tissue culture selection. Theor Appl Genet 59:161–167
Gill BS, Kam-Morgan LNW, Shepard JF (1986) Origin of chromosomal and phenotypic variation in potato protoclones. J Hered 77:13–16
Grisvard J (1985) Different methylation pattern of melon satellite DNA sequences in hypocotyl and callus tissues. Plant Sci 39:189–193
Groose RW, Bingham ET (1984) Variation in plants regenerated from tissue cultures of tetraploid alfalfa heterozygous for several traits. Crop Sci 24:655–658
Hoffmann F, Thomas E, Wenzel G (1982) Anther culture as a breeding tool in rape. II. Progeny analyses of androgenetic lines and induced mutants from haploid cultures. Theor Appl Genet 61:225–232
Holwerda BC, Jana S, Crosby WL (1986) Chloroplast and mitochondrial DNA variation in *Hordeum vulgare* and *Hordeum spontaneum*. Genetics 114:1271–1291
Jacobsen E, Tempelaar MJ, Bijmolt EW (1983) Ploidy levels in leaf callus and regenerated plants of *Solanum tuberosum* determined by cytophotometric measurements of protoplasts. Theor Appl Genet 65:113–118
Karp A, Nelson RS, Thomas E, Bright SWJ (1982) Chromosome variation in protoplast-derived potato plants. Theor Appl Genet 63:265–272
Kemble RJ, Flavell RB, Brettell RIS (1982) Mitochondrial DNA analyses of fertile and sterile maize plants derived from tissue culture with the Texas male sterile cytoplasm. Theor Appl Genet 62:213–217
Kemble RJ, Shepard JF (1984) Cytoplasmic DNA variation in a potato protoclonal population. Theor Appl Genet 69:211–216
Kemble RJ, Carlson JE, Erickson LR, Sernyk JL, Thompson DJ (1986) The *Brassica* mitochondrial DNA plasmid and large RNAs are not exclusively associated with cytoplasmic male sterility. Mol Gen Genet 205:183–185
Kudirka DT, Schaeffer GW, Baenziger PS (1986) Wheat: genetic variability through anther culture. In: Bajaj YPS (ed) Biotechnology in agriculture and forestry, vol 2. Crops I. Springer, Berlin Heidelberg New York Tokyo, pp 39–54
Landsmann J, Uhrig H (1985) Somaclonal variation in *Solanum tuberosum* detected at the molecular level. Theor Appl Genet 71:500–505
Larkin PJ, Scowcroft WR (1983) Somaclonal variation and crop improvement. CSIRO, Div Plant Ind, Canberra, Australia
Laughnan JR, Gobay-Laughnan S (1983) Cytoplasmic male sterility in maize. Annu Rev Genet 17:27–48
Lawrence WA, Davies DR (1987) Variation in plants regenerated from vacuolate and evacuolate protoplasts. Plant Sci 50:125–132
Lörz H, Scowcroft WR (1983) Variability among plants and their progeny regenerated from protoplasts of Su/su heterozygotes of *Nicotiana tabacum*. Theor Appl Genet 66:67–75
Maddock SE, Risiott R, Parmar S, Jones MGK, Shewry PR (1985) Somaclonal variation in the gliadin patterns of grains of regenerated wheat plants. J Exp Bot 36:1976–1984
Maliga P (1984) Isolation and characterization of mutants in plant cell culture. Annu Rev Plant Physiol 35:519–542
McClean EP, Hanson M (1986) Mitochondrial DNA sequence divergence among *Lycopersicon* and related *Solanum* species. Genetics 112:649–667
McNay JW, Chourey PS, Pring DR (1984) Molecular analysis of genomic stability of mitochondrial DNA in tissue cultured cells of maize. Theor Appl Genet 67:433–437
Meins F (1983) Heritable variation in plant cell culture. Annu Rev Plant Physiol 34:327–46
Metakovsky EV, Novoselskaya AY, Sozinov AA (1987) Problems of interpreting results obtained in studies of somaclonal variation in gliadin proteins in wheat. Theor Appl Genet 73:764–766
Negruk VI, Eisner GI, Redichkina TD, Dumanskaya NN, Cherny DI, Alexandrov AA, Shemyakin MF, Butenko RG (1986) Diversity of *Vicia faba* circular mtDNA in whole plants and suspension cultures. Theor Appl Genet 72:541–547
Ooms G, Burrell MM, Karp A, Bevan M, Hille J (1987) Genetic transformation in two potato cultivars with T-DNA from disarmed *Agrobacterium*. Theor Appl Genet 73:744–750

Oono K (1985) Putative homozygous mutations in regenerated plants of rice. Mol Gen Genet 198:377–384
Oro AE, Newton KJ, Walbot V (1985) Molecular analysis of the inheritance and stability of the mitochondrial genome of an inbred line of maize. Theor Appl Genet 70:287–293
Peerbolte R, Ruigrok P, Wullems G, Schilperoort R (1987) T-DNA rearrangements due to tissue culture: somaclonal variation in crown gall tissues. Plant Mol Biol 9:51–57
Peschke VM, Phillips RL, Gegenbach BG (1985) Discovery of AC activity among progeny of regenerated maize plants. Maize Genet Coop Newslett 59:91
Peterson PA (1985) Virus-induced mutations in maize: on the nature of stress-induction of unstable loci. Genet Res Cambridge 46:207–217
Pijnacker LP, Walch K, Ferwerda MA (1986) Behaviour of chromosomes in potato leaf tissue cultured in vitro as studied by BrdC-Giemsa labelling. Theor Appl Genet 72:833–839
Prat D (1983) Genetic variability induced in *Nicotiana sylvestris* by protoplast culture. Theor Appl Genet 64:223–230
Pring DR, Lonsdale DM (1985) Molecular biology of higher plant mitochondrial DNA. Int Rev Cytol 97:1–47
Rode A, Hartmann C, Dron M, Picard E, Ouetier F (1985) Organelle genome stability in anther-derived doubled haploids of wheat (*Triticum aestivum* L., cv. "Moisson"). Theor Appl Genet 71:320–324
Rode A, Hartmann C, Benslimane A, Picard E, Quetier F (1987) Gametoclonal variation detected in the nuclear ribosomal DNA from double haploid lines of a spring wheat (*Triticum aestivum* L., cv. César). Theor Appl Genet 74:31–37
Ryan SA, Scowcroft WR (1987) A somaclonal variant of wheat with additional β-amylase isozymes. Theor Appl Genet 73:459–464
Schwartz D, Dennis E (1986) Transposase activity of the Ac controlling element in maize is regulated by its degree of methylation. Mol Gen Genet 205:476–482
Scowcroft WR (1985) Somaclonal variation: the myth of clonal uniformity. In: Hohn B, Dennis S (eds) Genetic flux in plants. Springer, Berlin Heidelberg New York Tokyo, pp 218–245
Singh RJ (1986) Chromosomal variation in immature embryo derived calluses of barley (*Hordeum vulgare* L.). Theor Appl Genet 72:710–716
Sree Ramulu K, Dijkhuis P, Roest S (1983) Phenotypic variation and ploidy level of plants regenerated from protoplasts of tetraploid potato (*Solanum tuberosum* L. cv. "Bintje"). Theor Appl Genet 65:329–338
Sree Ramulu K, Dijkhuis P, Roest S, Bokelmann GS, De Groot (1984) Early occurrence of genetic instability in protoplast culture of potato. Plant Sci Lett 36:79–86
Sree Ramulu K, Dijkhuis P, Hanisch Ten Cate ChH, De Groot B (1985) Patterns of DNA and chromosome variation during in vitro growth in various genotypes of potato. Plant Sci 41:69–78
Sun CS, Wu SC, Wang CC, Chu CC (1979) The deficiency of soluble proteins and plastid ribosomal RNA in the albino pollen plantlets of rice. Theor Appl Genet 55:193–197
Sun Zong-xiu, Cheng-zhang Z, Kang-le Z, Xiu-fang Q, Ya-ping F (1983) Somaclonal genetics of rice, *Oryza sativa* L. Theor Appl Genet 67:67–73
Umbech PF, Gegenbach BG (1983) Reversion of male-sterile T-cytoplasm maize to male fertility in tissue culture. Crop Sci 23:548–588
Vanyushin BF (1984) Replicative DNA methylation in animals and higher plants. Curr Top Microbiol Immunol 108
Vasil IK (1986) Developing cell and tissue culture systems for the improvement of cereal and grass crops. J Plant Physiol 128:193–218
Walbot V, Cullis CA (1985) Rapid genomic change in higher plants. Annu Rev Plant Physiol 36:367–396
Weil JH (1987) Organization and expression of the chloroplast genome. Plant Sci 489:149–157
Wessler SR, Baran G, Varagona M, Dellaporta L (1986) Excision of Ds produces *Waxy* proteins with a range of enzymatic activities. EMBO J 5:2427–2432
Yisraeli J, Szyf M (1984) Gene methylation patterns and expression. In: Razin A, Cedar H, Rigg AD (eds) DNA methylation: biochemistry and biological significance. Springer, Berlin Heidelberg New York, pp 353–376
Zheng KL, Castiglione S, Biasini MG, Biroli A, Morandi C, Sala F (1987) Nuclear DNA amplification in cultured cells of *Oryza sativa* L. Theor Appl Genet 74:65–70

I.6 Gene Amplification and Related Events

W. Nagl[1]

1 Introduction

The term *gene amplification* refers to the multiple extra replication of a DNA sequence including a gene. This sequence, or amplicon, is normally much larger than the coding gene itself, and a gene is not in every case involved at all. The term, although referring to an event, has the result in mind, rather than the mechanism, which is still subject to speculations (see Sect. 7).

Although gene amplification and differential DNA replication are a comparatively new field of research, several aspects of genome variation already represent part of applied biology and biotechnology. Especially important is that DNA amplification offers unique opportunities and new strategies for gene technology in programs of crop improvement. Nevertheless, our present knowledge is more connected with basic science, and these aspects will be the central theme of this chapter, although the evident relationships with applied gene technology will be indicated as often as possible. Emphasis will be given to results that have been obtained since the publication of earlier reviews on this topic (Buiatti 1977; Nagl 1978; Schimke 1982).

The rationale of this chapter is the following. After an overview of today's understanding of the eukaryotic genome and some technical comments, unstable (or transitory) events of DNA amplification will be discussed, as they relate to differentiation, dedifferentiation, maldifferentiation (carcinogenesis), and to increased cell function. Amplification related to resistance phenomena lead to semi-stable and stable events that create a new genome organization after integration of the extra copies of a sequence, hence representing a step in evolution. Finally, some special cases, aspects of the mechanisms involved and the regulation of gene amplification, will be discussed, and conclusions for the technical application of the findings will be drawn.

On reading this chapter, it must be kept in mind that any description of biological events cuts the natural continuity (i.e., the fluid genome) into pieces, and that the phenomena discussed here represent the variation of one theme (i.e., the independent replication of certain DNA sequences). However, the same instrument may be used for quite different kinds of work.

[1] Division of Cell Biology, Department of Biology and Biotechnology Program, The University of Kaiserslautern, 6750 Kaiserslautern, Federal Republic of Germany

Biotechnology in Agriculture and Forestry, Vol. 11
Somaclonal Variation in Crop Improvement I (ed. by Y.P.S. Bajaj)

2 The Plasticity of the Plant Genome

After the detection of DNA as the genetic material, it was assumed that this molecule must possess high stability in order to function as the basis of heredity. For somatic cells, the first cytophotometric measurements of nuclear DNA contents led to the establishment of the paradigm of DNA constancy (Swift 1950). The spirit of this dogma masked all deviating findings for a long time, putting them into line with methodical errors, irreproducible results, and expression of pathological conditions. But the detection of somatic polyploidy as the rule rather than the exception in many animals and plants (Geitler 1939; Tschermak-Woess 1971; D'Amato 1977; Nagl 1978; Brodsky and Uryvaeva 1985) and of transposable elements in maize (McClintock 1951, and others) was the origin of a new paradigm, that of genome variability and fluidity. More recently, differential DNA replication, i.e., under-replication and amplification (extra-replication) of certain DNA sequences, could be demonstrated to be equally common, and to be an important part of cell differentiation and function. Some of the evidence for quantitative and qualitative changes of the genome is rather old, but it lasted more than 40 years for these events to be incorporated into the mainstream of biological thinking (Walbot and Cullis 1985).

In contrast to higher animals with a rather limited range of epigenetic variation, plants are open to many more changes at the DNA and chromosome levels. This is particularly evident from cell and tissue cultures, and now generally ascribed to *somaclonal variation* (Larkin and Scowcroft 1981; see Sect. 6). This genetic variability represents, on the one hand, an important source for plant engineering, and, on the other hand, the way in which plants adapt by their plasticity as developmentally open systems: the changed DNA sequences and chromosomes can pass through meiosis and contribute to evolution, as demonstrated by plant regeneration from tissue cultures with changed genomes (Buiatti 1977).

The dichotomy of thinking in terms of either ontogeny and somatic differentiation on the one hand, or of phylogeny and evolutionary divergence on the other, thus broke down one after the other. Today it is clear that both ontogenetic and phylogenetic mechanisms of diversification do actually have much in common; they are the two sides of the same coin (Galau et al. 1976). In vitro studies contributed a good deal to this knowledge. Hence, plasticity of the genome is now widely accepted as reflecting an intrinsic and important property of the DNA molecule.

The recent theory of genome variation as it relates to ontogenetic and phylogenetic changes understands increase and decrease, diversification, and orientation of sequences as a general spectrum of possibilities, which (dependent on different factors) may lead to a periodic, transient story during cell differentiation and function, or to an instable, semi-permanent change for some cell generations, or to a stable, permanent and heritable change in evolution. In human and mammalian cells, instable amplification is often recognized by the appearance of small chromatin bodies called "double minutes" (DM), and stable amplification is visible in the form of chromosomes which are elongated by a "homogeneously staining region" (HSR) (see Cowell 1982 for a review, and Fig. 1). Figure 2 shows some aspects of evolutionary DNA increase. The amplified sequences (genes) may

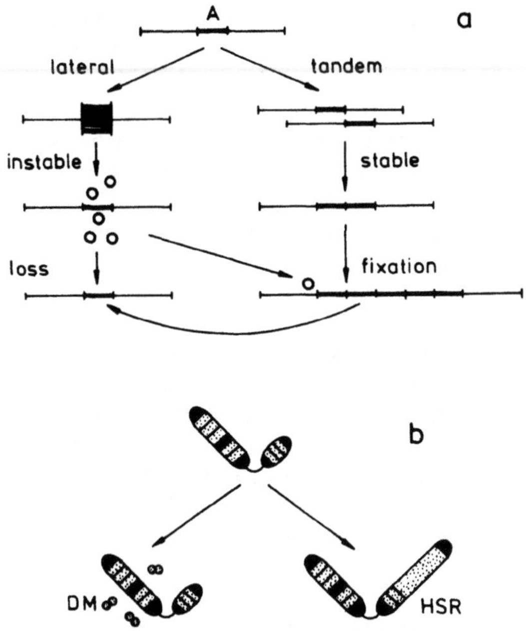

Fig. 1a,b. Diagrams to illustrate possible mechanisms and karyotypical expression of DNA amplification. **a** An amplicon (*A*) may be either laterally extra replicated (selective polyteny, cf. Fig. 14) leading to unstable extra copies, or it could be multiplicated by unequal crossing-over, leading to evolutionary stable DNA increase. But also circular extra copies might be inserted into the genome, or originally integrated copies might be lost by excision. **b** In several (mainly transformed) cell lines, instable amplification (comparable to that in oocytes and somatic cells) is expressed in the form of double minutes (*DM*), while stable amplification (which is inherited) leads to an elongation of the marker chromosome bearing the amplified sequence. The latter is visible, after Giemsa banding, as homogeneously staining region (*HSR*)

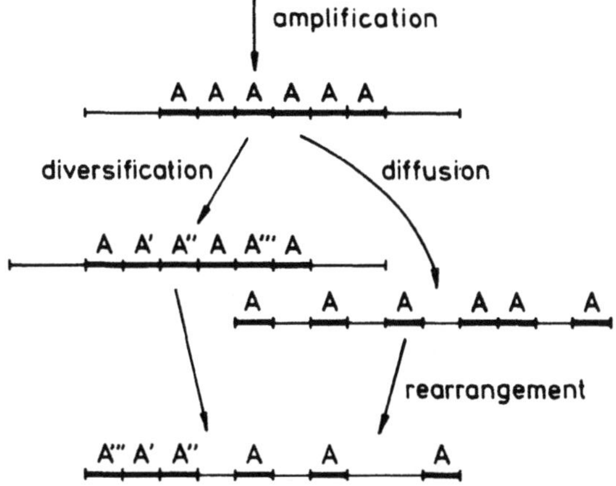

Fig. 2. Diagram to illustrate the plasticity of the genome. After amplification of an amplicon (*A*), the multiplied copies may undergo diversification by mutations, or diffusion through the genome by transposition, or both. Rearrangements, and new rounds of amplification, etc. can follow

undergo individual diversification by "traditional" mutations, or they may diffuse throughout the genome by changes in position (rearrangements), or both. In vitro, such changes occur much more rapidly than in vivo, hence allowing their study within the laboratory.

While certain genomic changes can be clearly related to stress situations, others are thought to be part of the normal developmental program, and others to occur "spontaneously". However, any event must have its cause, and so-called spontaneous and random events are, in my opinion, simply those whose causes we have not realized. In any case, differential DNA replication can be seen as part of a feed-back response of the genome to changes of the internal or external environment.

Presumably each of the independent replicons of a plant's chromosome set can act independently as the site of over- or under-replication, and hence contribute to the genomic flexibility. Some replicons, however, appear to be centers of instability.

The events which can contribute to quantitative and qualitative (positional) changes of the nuclear genome are listed in Table 1. Gene amplification in the genome of chloroplasts and mitochondria will not be discussed here (see, e.g., Smith et al. 1987; Ozias-Akins 1988; Shirzadegan et al. 1989).

Table 1. Ontogenetic and phylogenetic dynamics of the eukaryotic genome. (After Nagl 1985b)

Ontogenetic changes	Corresponding phylogenetic changes
Polyploidization	Polyploidization
Endopolyploidy	Generative or germ-line polyploidy
Polyteny	Auto-, allopolyploidy
Under-replication	Cryptopolyploidy
	Dysploidy
DNA elimination	DNA elimination (loss)
Chromosome elimination	
Chromatin elimination	
Chromatin diminution	
Differential DNA replication	DNA amplification
Under-replication	Gene duplication
Amplification	Internal gene enlargement
	Amplification of noncoding sequences
	(evolution of repetitive sequences,
	diversification and dispersion of them)
DNA sequence rearrangement	Translocation of mobile genetic elements
Mobile genetic elements	Transposons
Mutator genes	Immunoglobulin genes
Chromatin restructuring	Copia sequences etc.
	"Jumping genes"
Karyotype repatterning	Karyotype repatterning
Chromosome rearrangement	Robertsonian translocation etc.
Specific chromatin condensation	Chromosome mutation
Somatic gene mutation	Gene mutation
DNA conformation changes	Chromatin conformational changes
(e.g., B to Z transition)	(new regulatory domains)

3 Unstable Gene Amplification and Cell Function

Unstable, or transitory, gene amplification represents an effective mechanism to increase cell function in a highly specific way, in order to enable the cell to produce a specific product in huge amounts within a short time, or in order to undergo a step of differentiation rapidly.

3.1 Methods of Detection

Rather briefly, some technical and phenomenological comments will be given. The following methods indicate differential DNA replication; the arrangement given is in order to show their increasing accuracy and reliability.

- Structural analysis of nuclei and chromosomes (Figs. 1b, 4, 7).
- In vivo ^{3}H-thymidine incorporation and autoradiography.
- Cytophotometry of DNA content (Feulgen, fluorochromes), flow cytometry, cell (or chromosome) sorting.
- In situ hybridization (Fig. 5).
- CsCl density gradient centrifugation of DNA (Fig. 9).
- Thermal DNA denaturation (Fig. 8).
- DNA reassociation kinetics, (Fig. 10).
- Restriction analysis (Fig. 3, 15).
- Southern blotting (with varying dilution of the probe, and varying exposure times; Figs. 5, 6, 11).

Examples will be given and discussed in the text. It should be noted that, in any case, more than one technique has to be employed in order to avoid misinterpretations; at least one microscopic and one molecular method have to be used (see Nagl 1987a).

In addition, indirect evidence can be obtained by physiological investigations, e.g., growth of resistant cells, increased RNA synthesis, protein accumulation, etc. But it must be noted that these events can also be the result of different mechanisms, for instance gene mutation, degradation of a toxic agent, increased rate of transcription, reduced degradation of a gene product, etc.

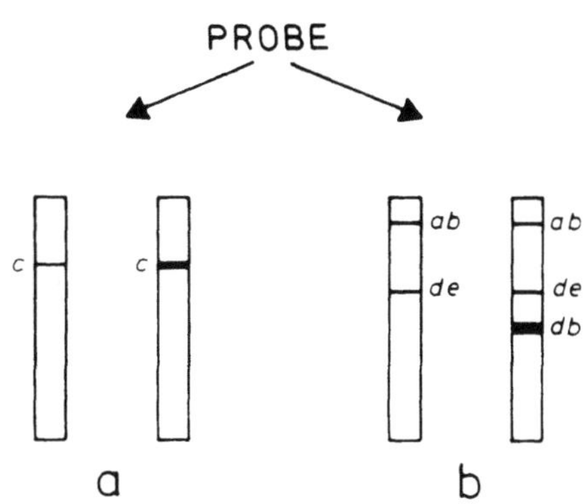

Fig. 3a,b. Analysis of DNA amplification by restriction and Southern hybridization. An in vitro-labeled cloned fragment of the amplicon is used as a probe. **a** The unchanged sequence is amplified. **b** New sequences (restriction fragments) were created by recombination events during amplification. (After Stark and Wahl 1984)

Some more technical details are given and discussed by Nagl (1979a) and Stark and Wahl (1984).

3.2 Gene Amplification and Cell Function

An extremely high productivity of a cell in a relatively short space of time evidently requires the existence of multiple copies of the genes encoding these products. Evidence is best established for amplification of the ribosomal cistrons. This results in a strong increase of the number of ribosomes in a given cell, and thus of the cell's capacity for protein synthesis in general. Especially the oocytes of many animals have been shown to need large amounts of preformed ribosomes for the very short post-fertilization cleavage division cycles, which do not allow de novo ribosome synthesis (Brown and Dawid 1968; Gall 1968). A similar function can be attributed to the trophocytes or nurse cells in insects with ovarioles of the meroistic type. In the latter, the trophocytes take over the supply of the oocytes with ribosomes (and informosomes) to a greater or lesser extent (Bier 1969; Nagl 1977). However, there are also somatic cells whose specific function is achieved by gene amplification, e.g., the salivary gland cells of the dipteran, *Rhynchosciara americana*: Here puffs containing amplified levels of DNA form at several polytene chromosome sites during the late larval stages (Glover et al. 1982). The appearance of the puffs coincides with the spinning of a communal cocoon within which the larvae pupate. At least some of these DNA puffs contain genes that encode cocoon proteins. Another example are the follicular cells just before and during choriogenesis in *Drosophila melanogaster*. The chorion or eggshell is deposited around each oocyte, towards the end of oogenesis, and it consists of about 20 proteins (six of them major), encoded by genes which are single-copy in the haploid genome. Choriogenesis is extremely rapid and exceeds by nearly two orders of magnitude the efficiency limited by the genome (although the follicle cells are polyploid). As shown by Spradling and Mahowald (1980), the deficit is made up by differential gene amplification in the follicular cells just before and during choriogenesis (see also Spradling 1981; Osheim and Miller 1983). Another example of somatic gene amplification was detected in ciliates. The rDNA is amplified during a specific developmental stage of the polyploid macronucleus (e.g., Pan and Blackburn 1981; Yao et al. 1985). The studied examples are summarized in Nagl (1978). Ribosomal gene amplification was also suggested to occur in plant cell cultures, although evidence for rDNA was somewhat indirect in some of the reports. The appearance of a satellite DNA has been noted at certain stages in the culture of pith cells from stems of *Nicotiana glauca* and has been interpreted in terms of rDNA amplification in relation to "dedifferentiation" of the cells before proliferation (Parenti et al. 1973; Martini and Nuti Ronchi 1974; Nuti Ronchi et al. 1974), coinciding with the time of fragmentation of the endopolyploid nuclei in these cells (Nuti Ronchi et al. 1973; see also Sect. 4). Ribosomal RNA gene amplification in human cells mediates a selective advantage with regard to rapid growth (Roberts et al. 1987). A variable content of a GC-rich satDNA has been reported to occur in tumorous and normal cultures of *Crepis capillaris* by Sacristán and Dobrigkeit (1973).

A good part of a role of DNA amplification for cell function comes from tissues involved in the nourishment of the embryo of plants and of germinating seeds, respectively (see Sect. 4).

3.3 DNA Amplification in Differentiation and Dedifferentiation

The role of DNA amplification (and under-replication) in differentiation was specifically reviewed by Nagl (1978). The bulk of evidence therein comes from the animal kingdom. New results from plants were, for instance, given by Natali et al. (1986). These authors induced cell dedifferentiation in roots of *Vicia faba* by removing the whole meristem. The resulting dedifferentiating tissue developed a heavy satellite DNA not occurring in the differentiated tissues. These results are in good agreement with other findings: in the differentiating root of *Vicia faba*, respectively heterochromatin and DNA under-replication was detected by cytochemical and biochemical techniques (Kononowicz et al. 1983; Bassi et al. 1984). Contrarily, Altamura et al. (1987) observed in cultured tissues of *Nicotiana tabacum* that loss of DNA sequences progressively impaired the morphogenetic capability, while amplification of repetitive DNA sequences was a factor of regeneration.

It is interesting to realize that in plants mainly noncoding DNA sequences are amplified, or in some other way differentially replicated, during differentiation steps. This suggests that the extremely large amount of noncoding DNA comprising the genome of higher plants generally plays some role in the control of development (and evolution). It was suggested that the noncoding, predominately repetitive DNA, is engaged as "conformational DNA" in the establishment of the 3-D architecture of the nucleus. The created amount and pattern of chromatin condensation determines the gross domains enclosing groups of genes to be silenced and liberated for transcription, respectively (Nagl 1982a, 1983b; Nagl et al. 1983). Also the typical gene amplification in mammals is characterized by co-amplification of an amount of DNA as large as up to 3000 kb (kilobase pairs), many times larger than a common gene. The large size of the amplified units ("amplicons") may be, therefore, generally better conveyed by the term "DNA" or "sequence" amplification than by "gene" amplification (Stark and Wahl 1984). The latter term seems to me to be rather an expression of some over-emphasis of structural genes versus that of noncoding sequences, probably because most geneticists are working with typical genes; but with increasing knowledge on the importance of all the other sequences in the control of genes, this view is changing. No wonder that the gene is in search of a new identity (Falk 1984; Nagl 1987b).

The amplification of noncoding sequences located in constitutive heterochromatin is dramatically exemplified in tissue cultures of the orchid *Cymbidium* (Fig. 4). Tissue pieces develop into organized protocorms, in that certain cells undergo transient heterochromatin amplification during differentiation, as could be shown by Feulgen-DNA measurements, fluorescence microscopy, autoradiography, and thermal denaturation profiles (Nagl 1972; Nagl and Rücker 1972, 1976; Nagl et al. 1972). The heterochromatin contains AT-rich DNA but no ribosomal sequences (Schweizer and Nagl 1976). After differentiation, the extra-DNA is released from the nuclei and degraded. A strong correlation between amplification and the

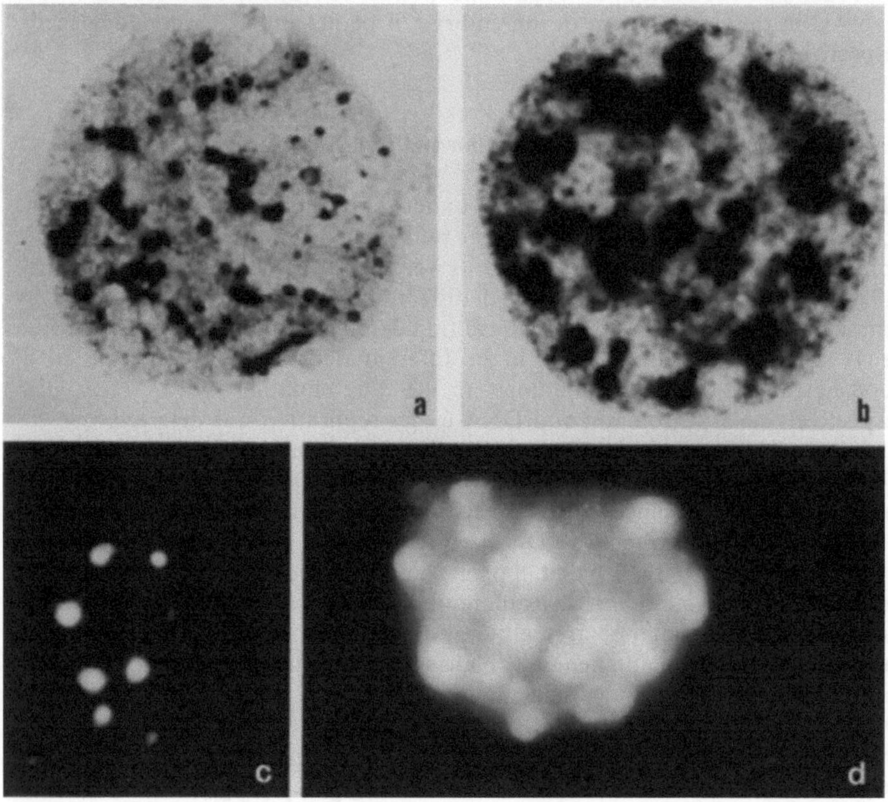

Fig. 4a-d. Evidence for heterochromatin amplification in differentiating protocorms of the orchid *Cymbidium* in vitro. **a** and **b** Giemsa-stained (C-banding technique) and **c**; **d** quinacrine-stained (AT-specific fluorochrome) polyploid nuclei. **a** and **c** Nuclei before (respectively after) DNA amplification, **b** and **d** nuclei during amplification (× 1000; from Nagl 1982c)

pattern of morphogenesis of the cultures could be ascertained in this system, supporting the concept of a regulatory role of this noncoding "conformational" DNA.

The view of DNA amplification as a rather frequent event during differentiation and/or dedifferentiation is now receiving permanent support. Kikuchi et al. (1987) cloned two types of variable copy number DNA sequences from the rice embryo genome. One of these sequences was amplified about 50-fold during callus formation and diminished in copy number to the embryonic level during regeneration. The other clone showed the reciprocal pattern. The copy number of both sequences changed even in the early developmental stages of the plants and were eliminated from nuclear DNA along with growth. In suspension cultures of rice, Zheng et al. (1987) detected amplification of highly repeated nuclear DNA sequences by dot hybridization. In hybrids of wheat and rye regenerated from tissue culture, Lapitan et al. (1988) observed enlarged C-bands in some of the rye chromosomes and evidenced by in situ hybridization, using telomeric repetitive DNA sequences from rye, stable amplification of these sequences in the hybrid cells and plants. Murry et al. (1987) described amplification of a repetitive DNA sequence in root hairs of *Hordeum vulgare*. The amplified DNA separated as a satellite band in CsCl density gradients, it exhibited a unique restriction pattern, and it did not hybridize with ribosomal, chloroplast, or mitochondrial DNAs, but in part it was evidently transcribed. The amplification, divergence, and rearrangement of series of discrete classes of reiterated sequences were also observed in cultivated cells of *Rauwolfia serpentina* (Solovyan et al. 1987).

3.4 Gene Amplification and Misdifferentiation

Although the relationship between oncogene amplification and carcinogenesis is restricted to the animal kingdom, it is of general interest to molecular and cell biologists and should, therefore, not be omitted in this chapter. The topic has been reviewed by several authors (e.g., Arrighi 1983; Graham 1984; Marx 1984; Der 1987). Table 2 and Fig. 5 show examples of oncogene amplification in human

Table 2. Oncogene amplification in tumors and transformed cell lines (selected examples)

Amplified gene	Tumor/cell type	Reference
c-*abl*	Leukemia	Collins and Groudine (1983)
c-*abl*	Leukemia	Selden et al. (1983)
c-*erb* B-1	Stomach cancer	Nomura et al. (1986)
HER-2/*neu*	Breast cancer	Slamon et al. (1987)
myc family	Lung cancer	Johnson et al. (1988)
c-*myc*	T-cell lymphoma	Ohno et al. (1988)
c-*myc*	Leukemia	Nowell et al. (1983), Misawa et al. (1987)
c-*myc*	Lung cancer	Little et al. (1983)
c-*myc*	Colon carcinoma	Alitalo et al. (1983), Cherif et al. (1988)
c-*myc*	Stomach cancer	Nomura et al. (1986)
N-*myc*	Rhabdomyosarcoma	Mitani et al. (1986)
N-*myc*	Neuroblastoma	Brodeur et al. (1984)
N-*myc*	Neuroblastoma	Schwab et al. (1984)
N-*myc*	Retinoblastoma	Lee et al. (1984)
N-*myc*	Retinoblastoma	Sakai et al. (1988)
N-*myc*	Lung cancer	Nau et al. (1984)
N-*myc*	Wilm's tumor	Norris et al. (1988)
HER-2/*neu*	Breast cancer	Slamon et al. (1987)
c-Ha-*ras*	Cervix carcinoma	Riou et al. (1984)
K-*ras*	Embryonal cancer	Wang et al. (1987)
K-*ras*	Osteosarcoma	Nardeux et al. (1987)
Other genes:		
gli	Glioma	Kinzler et al. (1987)
hst-1	Esophagal cancer	Tsuda et al. (1988)
Ornithine decarboxylase	Myeloma	Leinonen et al. (1987)

Oncogene amplification was reviewed by Marx (1984), Alitalo et al. (1987), Gebhart et al. (1987), and others.

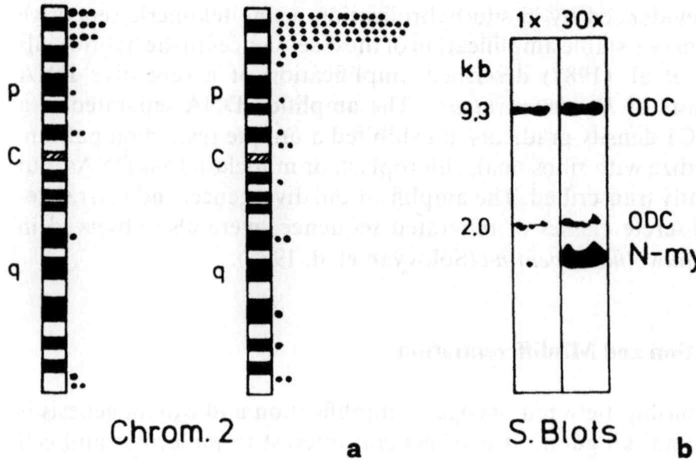

Fig. 5a,b. Evidence for amplification of the N-*myc* oncogene in carcinogenesis. **a** Results of in situ hybridization of a restriction fragment containing the N-*myc* sequence to chromosome no. 2 of normal lymphocytes and neuroblastoma cells, respectively (in this case, no HSR can be seen). The *dots* indicate the average number of silver grains. **b** Estimation of the copy number by Southern hybridization analysis of normal and lung carcinoma cell DNA (*ODC* gene for ornithine decarboxylase. (After Shiloh et al. 1985; and Alitalo et al. 1987)

cancers, but also repetitive DNA sequences may be involved (e.g., Baskin et al. 1987) In many cases, gene amplification is accompanied by chromosomal aberrations (Atkin et al. 1983; Schwab et al. 1983; Fougère-Deschatrette et al. 1984; Fearon et al. 1984). In general, the position of a gene appears to play an important role in the induction of amplification (e.g., Wahl et al. 1984). This is consistent with ideas on "the chromosome field" (Lima-de-Faria 1983; Lima-de-Faria and Mitelman 1986) and aspects of gene regulation in interphase nuclei depending on their location within certain chromatin domains (e.g., Nicolini 1979; Nagl 1983b).

The correlation of amplification of an oncogene with survival was suggested to be of greater prognostic value than most currently used prognostic factors (Slamon et al. 1987), although the influence of these findings is related to basic studies rather than to therapeutic application.

Besides oncogenes, the amplification of resistance genes is related to the tumor problem, as it represents a major cause of failure in the chemotherapeutic treatment of human cancers. This point will be discussed, together with other aspects of resistance, in the following Section.

In plant tumors, differential DNA replication also seems to play some role, both amplification and under-replication of certain DNA sequences apparently taking place and contributing to the observed changes of genome organization (Durante et al. 1986).

3.5 Gene Amplification and Resistance

Resistance phenomena are of increasing interest for biotechnology, including gene technology. The worldwide use of herbicides, antibiotics, and other bactericides, pesticides, and fungicides, and the use of different chemical compounds against nematodes etc. call for a better understanding of resistance phenomena, for the selection of resistant plants in vivo and in vitro, the attempts to transfer resistance genes, and the investigation of the ecological consequences of crop resistance to the environment. Although plants dispose of defence mechanisms against viruses and parasites (e.g., phytoalexins), the reduction of yield by parasites is one of the greatest problems in agriculture.

Resistance can result from at least the following mechanisms:

- an altered transport of the toxic/parasitic agent,
- reduced affinity,
- single gene mutation,
- overproduction of the gene product that is affected, by
  1. increased translation and decreased degradation,
  2. increased transcription rate,
  3. dose-related quantitative change in gene copy number (amplification).

Among these, gene amplification is a well-documented mechanism by which cells become able to meet the demand for increased quantities of specific gene products under toxic or metabolic stress (Stark 1986; Biedler et al. 1988; Lewis et al. 1988). The wide field of engineering herbicide resistance in plants by other mechanisms and its ecological problems cannot be discussed here in general.

In principle, gene amplification can contribute to resistance in two ways. On the one hand, the gene, whose product is affected (bound, or changed) by the toxic agent, can be amplified, so that the cell is able to overproduce the protein and hence to overcome the toxic affect (see below). On the other hand, the genes for polypeptides involved in membrane transport can be amplified. The *mdr* genes (mdr = multidrug resistance) of mammals show strong homology to bacterial transport protein genes, and Gros et al. (1986) suggested that an energy-dependent transport mechanism is responsible for the multidrug-resistant phenotype (see also Ames 1986; Chen et al. 1986).

Examples for amplification of genes, the product of which is disturbed by toxins in some way, are mainly known from mammalian and human cell cultures and represent a severe problem of cancer therapy (e.g., Warr and Atkinson 1988). Selective findings are given in Table 3. The most studied case is the amplification of the dihydrofolate reductase (*dhfr*) genes in methotrexate (MTX)-resistant cells (for reviews see Fox 1984; Schimke 1984). Dihydrofolate reductase catalyzes the reduction of dihydrofolic acid to tetrahydrofolic acid, and hence a metabolic step underlying nucleic acid synthesis. Inhibitors of the enzyme, such as trimethoprim, dihydrotriazine, aminopterin, and methotrexate are antifolate drugs, commonly used, e.g., as cytostatica in cancer therapy. MTX has a very high affinity for dhfr. Resistance of mammalian cell lines to MTX can be ascribed, besides reduced cell permeability and lower affinity of *dhfr* to the drug, to increased content of the target enzyme as a result of gene amplification (Alt et al. 1978). Methotrexate-resistant

Table 3. Gene amplification resulting in resistance (examples mainly from human and mammalian cell lines)

Amplified gene for	Resistance to	Reference
Alcohol dehydrogenase	Antimycin A	Walton et al. (1986)
Adenosine deaminase	Deoxycoformycin	Hunt and Hoffee (1983)
Antifreeze protein	Freezing	Scott et al. (1988)
ATPase (Na, K)	Quabain	Emanuel et al. (1986)
CAD[a]	N-(Phosphonacetyl)-L-asparatate	Wahl et al. (1979); Ardeshir et al. (1983); Meinkoth et al. (1987)
Copper chelatin	Copper	Karin et al. (1984); Aladjem et al. (1988)
Dihydrofolate reductase	Methotrexate	Brown et al. (1983); Srimatkandada et al. (1983) (review: Schimke 1984)
Glutamine synthetase	L-phosphinothricin	Donn et al. (1984)
Glutathione S-transferase	Nitrogen mustards	Lewis et al. (1988)
Glycoprotein-P[b]	Multidrug resistance	Van der Bliek et al. (1986)
Hydroxy-3-methyl-glutaryl CoA reductase	Compactin	Ryan et al. (1981); Chin et al. (1982); Luskey et al. (1983)
MDRA[b]	Multidrug resistance	Scotto et al. (1986); Martinsson and Levan (1987) Ueda et al. (1987); Stahl et al. (1987)
Metallothionine	Cadmium	Beach and Palmiter (1981)
Ornithine decarboxylase,	Difluoromethylornithine	Leinonen et al. (1987)
Ribonucleotide reductase	Hydroxyurea	Lewis and Srinivasan (1983)
Thymidylate synthetase	5-Fluorodeoxuridine	Rossana et al. (1982); Garvey and Santi (1986)
Thymidylate synthetase	Methotrexate	Coderre et al. (1983)
UMP synthetase	Pyrazofurin	Kanalas and Suttle (1984)
Unknown genes	Aminopterin	Watson et al. (1988)
	Actinomycin D	Jakobsson et al. (1987)
	Aresenite	Detke et al. (1989)
	Hydroxyurea	Tonin et al. (1987)

[a]CAD = multifunctional protein containing the enzymatic activities *c*arbamyl-P synthetase, *a*spartate transcarbamylase, and *d*ihydroorotase, which catalyze the first three reactions of de novo UMP biosynthesis;
[b]MDRA = *m*ulti*d*rug *r*esistance *a*ssociated genes, including the genes for the p-glycoprotein and other membrane glycoproteins.

cells could also be selected in plant cell cultures, e.g., from *Daucus carota* (Cella et al. 1984, 1987) and *Glycine max* (Weber et al. 1985), but the mechanism of resistance is not yet clearly established. In transgenic *Petunia* plants (transformed with an *Agrobacterium tumefaciens* strain carrying a chimeric mouse *dhfr* gene fused to the cauliflower mosaic virus 35S promoter and nopaline synthetase polyadenylation site), the resistance was evidently due to mutation of the *dhfr* gene (Eichholtz et al. 1987). Still unclear is also the mechanism which is responsible for resistance to p-fluorophenylalanine (PFP) in certain tobacco cell lines. PFP is a structural analog of phenylalanine and tyrosine that has proven toxic to a variety of prokaryotic and eukaryotic cells. Exposure of cultured plant cell populations to toxic levels of PFP has resulted in the recovery of cell lines which display resistance to previously lethal levels of the analog (Palmer and Widholm 1975; Quesnel and Ellis 1987). Suspension cultures selected for resistance exhibit an elevated synthesis of phenylalanine-derived secondary metabolites; in *Nicotiana tabacum* an overproduction of caffeoyl and feruoyl amides of putrescine has been found, and in parallel, an increased activity of phenylalanineammonia lyase (PAL) and related enzymes (Berlin et al. 1982). The resistant cultures exhibit an increased nuclear DNA content, increased chromosome numbers, and altered chromatin ultrastructure, but the amplification of genes (e.g., the *pal* gene) could so far not be substantiated (Nagl et al. 1984).

A clear example of resistance due to gene amplification has been found in alfalfa by Donn et al. 1984) and further studied by Deak et al. (1988). These authors selected *Medicago sativa* cells resistant to the nonselective herbicide L-phosphinotricin, an amino acid analog produced by some *Streptomyces viridochromogenes* strains which acts as a mixed competitive inhibitor of glutamine synthetase (GS). It was found that the resistance is a consequence of a four- to elevenfold amplification of one *gs* gene resulting in about an eightfold increase in mRNA levels, and an increased enzyme synthesis sufficient to overcome the toxic effect of the inhibitor (Fig. 6).

Another well-studied example in plants is resistance to glyphosate (N-[phosphonomethyl]glycine), a broad-spectrum herbicide strongly inhibiting EPSP synthase (5-enolpyruvylshikimic acid-3-phosphate synthase, or 3-phosphoshikimate-1-carboxyvinyltransferase). An adaptation of *Daucus* and *Petunia* cells to glyphosate is accompanied by large increases in the activity of this shikimic acid pathway enzyme (e.g., Nafziger et al. 1984; Steinrücken et al. 1986). The cells overproduce the enzyme and its messenger RNA as a result of a 20-fold amplification of the gene (Shah et al. 1986). A transient amplification of the gene for

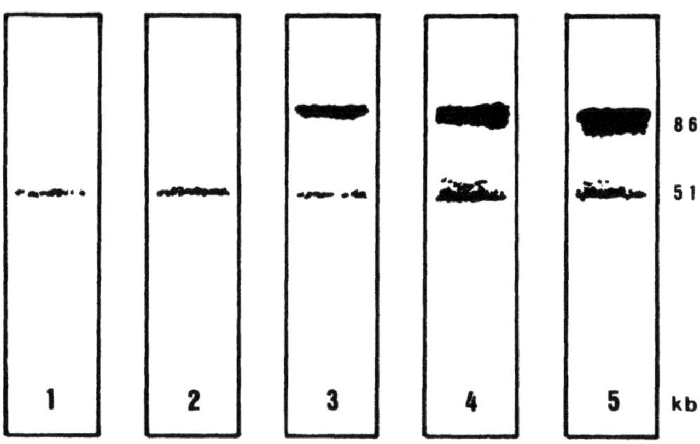

Fig. 6. Southern blot of *Hind*III-digested alfalfa DNA hybridized to the alfalfa gene for glutamine synthetase (*GS*). The DNA was purified from wild-type leaves (slot *1*), wild-type suspension cultures (slot *2*), and the herbicide-resistant sublines selected on B5 medium containing various concentrations of L-phosphinothricin (*L-PPT*): 0.6 mM (slot *3*), 2.0 mM (slot *4*), and 3.0 mM (slot *5*). (According to a blot by Donn et al. 1984)

acetolactate synthase was observed in chlorsulfuron-resistant cell cultures of *Datura innoxia* by Xiao et al. (1987). Similarly, amplification events in insects have been found to correlate with insecticide resistance (e.g. Mouchès et al. 1986; Field et al. 1988).

Several attempts show that resistance can be transferred from one plant to another by gene technology (see the *Petunia* example given below; Eichholtz et al. 1987). Transfer of resistance by somatic cell hybridization was undertaken by Fox (1984) and Schimke (1984) in animal cells, and chromosome-mediated gene transfer was successfully undertaken by Lewis and Srinivasan (1983; see also Bostock and Clark 1983). In carrot cell lines, resistance could be transferred by protoplast fusion (Cella et al. 1987). This method is particularly successful in the case of mutation-dependent resistance. However, Shah et al. (1986) were able to construct a chimeric EPSP synthase gene from an amplification cell line with the use of the cauliflower mosaic virus 35S promoter, and to obtain glyphosate-tolerant *Petunia* cells and regenerated transgenic plants.

Nevertheless, there are still questions left. Are all genes able to amplify? Do they do so at random and permanently, or are they induced by stress, particularly by inhibition of "normal" DNA replication? Evidently not all inhibitors select amplification. Moreover, severe problems still exist in regenerating plants from resistant cell lines in many species.

4 Differential DNA Replication

The amplification of a gene can be envisaged as simply a specific case of differential DNA replication. This term signifies an event in that one, or several, replicons are more or less often replicated than the remainder of the genome. Most frequently, noncoding DNA sequences are the subject of individual rounds of extra- or of under-replication in both ontogenetic and phylogenetic changes of the genome (the latter will be discussed in the following section). There is increasing evidence that, within the same cell, a certain DNA sequence can be amplified, while another may be under-replicated (Brady 1973). In addition, the degree of under-replication can vary among sequences (Hammond and Laird 1985). Events like these support the concept of a highly flexible, dynamic, and plastic genome.

Differential or selective DNA replication has been repeatedly reviewed during recent years (Nagl 1978, 1979a, 1982b, 1985b). The phenomenon could be detected in virtually all groups of organisms, ranging from algae and protists to angiosperms and mammals. DNA sequences involved are rRNA genes, mRNA genes, signal sequences within promoters, specific repetitive sequences (like the *Alu* sequences), and satellite DNA. At the microscopic level, heterochromatin is often affected. In the present chapter, only some of the newer findings and a few model examples will be shown.

4.1 Differential Replication in Plant Development

The under-replication of satellite DNA and heterochromatin, respectively, has been known from insects since the early studies of polytene chromosomes, and could be demonstrated repeatedly by modern techniques (Zacharias 1979, 1986). In angiosperms, DNA under-replication was described as occurring during the development of cotyledons, leaves, and fruits of *Cucumis* (Pearson et al. 1974), the embryo-suspensor of *Tropaeolum* (Nagl et al. 1976; Figs. 7, 8), the endosperm and cotyledons of germinating castor beans (Galli et al. 1986), and the raphide crystal idioblasts of *Vanilla* (Kausch and Horner 1984). However, DNA amplification was revealed in maize endosperm during kernel development (Kowles and Phillips 1985). The amplification of GC-rich nuclear satellite DNA was observed during *Vicia faba* root dedifferentiation (Natali et al. 1986), under-replication of highly repetitive sequences in differentiating root cells of *Vicia* (Bassi et al. 1984), rapid amplification of fixation of sequences in the ribosomal DNA repeat unit in derivatives of a cross between maize and *Tripsacum* (Lin et al. 1985), and amplification of extrachromosomal circular DNA in nuclei of intact bean leaves (Kunisada et al.

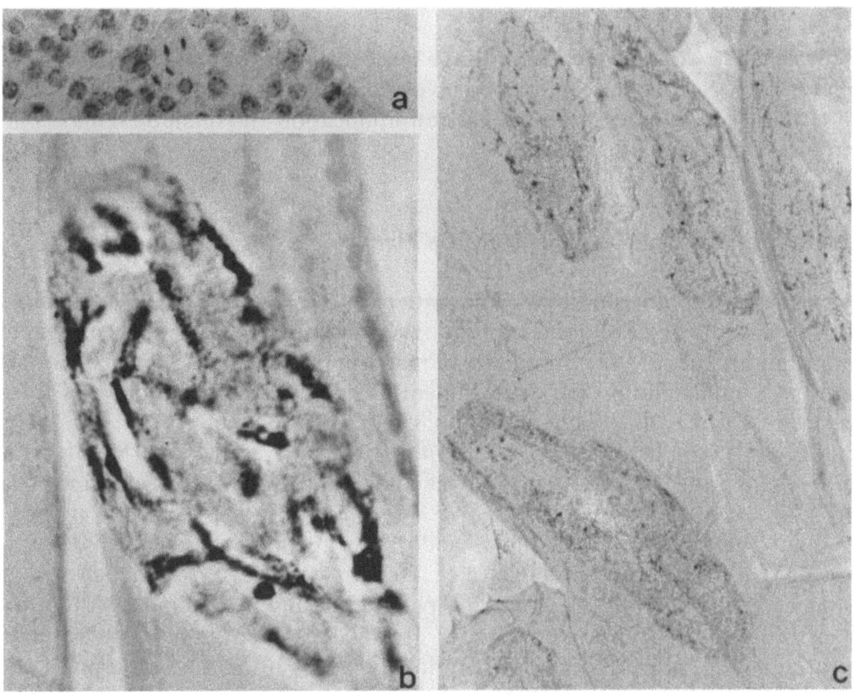

Fig. 7a-c. Heterochromatin under-replication in the embryo-suspensor of *Tropaeolum majus*. **a** Diploid tissue of the embryo, for comparison. **b** Highly endopolyploid nucleus, whose heterochromatin forms polytenic structures. **c** Polyploid nuclei whose heterochromatin is under-replicated (actually, nonreplicated during polyploidization; Feulgen, phase contrast, × 1200; Nagl et al. 1976)

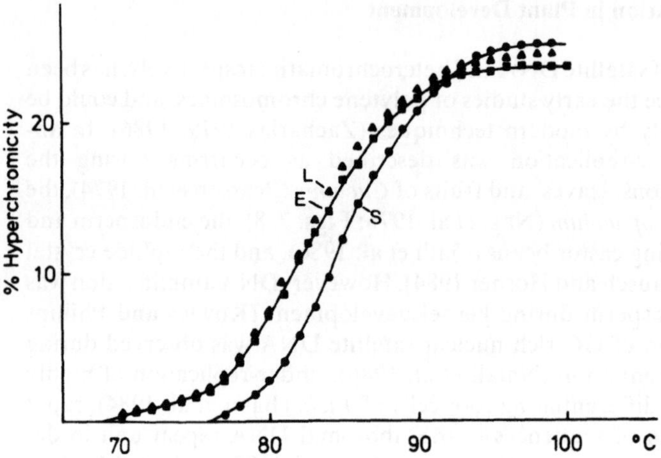

Fig. 8. Thermal denaturation profiles of DNA isolated from leaves (*L*), embryos (*E*), and suspensors (*S*) of *Tropaeolum majus*. The suspensor DNA exhibits a higher melting temperature than the DNA's from the other sources, because the AT-rich DNA (which is located in the heterochromatin) is under-represented in the suspensor cells, due to under-replication in many nuclei. (Nagl 1982b)

1986). These examples clearly indicate the extremely different events in which differential DNA replication may be involved.

Table 4 summarizes findings of gene amplification and related events in plants and plant cell and tissue cultures.

4.2 DNA Increase and the Yield of Crop Plants

Applied aspects of the studies on differential DNA replication are given in the case of crop plants. The developmental biology of maize (*Zea mays* L.) is particularly important because of the prominence of maize in plant genetics research and plant industry. Understanding the development of the endosperm becomes especially important because this tissue makes up 85–90% of the mature kernel dry weight. Endosperm tissue obviously plays a critical role in overall kernel development, embryogeny, and germination (Kowles and Phillips 1985; Sheridan and Clark 1987).

Endopolyploidization of maize endosperm nuclei, and even polyteny, was recorded already many years ago (e.g., Tschermak-Woess and Enzenberg-Kunz 1965). More recently, Kowles and Phillips (1985) found that increased DNA levels in centrally located endosperm nuclei are related to endosperm development. Mitotic activity sharply decreases in endosperm cells 10–12 days after pollination. At this time, nuclear size and DNA content per nucleus (where C = haploid content) sharply increase until peak levels are reached at about 14–18 days after pollination. Mean DNA content per endosperm nucleus in strain A 188 was shown by Feulgen cytophotometry to increase to about 90C by this peak stage, with the pattern being remarkably consistent over four consecutive growing seasons. Some individual

Table 4. Examples of plant organs and plant cell cultures, for which evidence or indications of DNA amplification and related events have been found (based upon Nagl 1978, and completed up to March 1989 for further examples see Table 1 in Buiatti 1977)

Species	Cell type	DNA species[a]	Event[b]	Reference
Lower plants				
Acetabularia cliftonii	Primary nucleus	rDNA	A	Scheer et al. (1976)
A. major	Primary nucleus	rDNA	A	Spring et al. (1974)
A. mediterranea	Primary nucleus	rDNA	A	Spring et al. (1974)
Chlamydomonas reinhardii	Cell culture	rDNA	A	Howell (1972)
Chlorella phyrenoidosa	Thermophile strain	?	A	Hopkins et al. (1972)
Euglena gracilis	Chloroplasts	?	A	Mielenz and Hershberger (1974)
Aspergillus nidulans	Hyphae	?	A	Parag and Roper (1975)
Physarum polycephalum	Plasmodium	rDNA	A	Guttes (1974)[3]
Equisetum arvense	Shoot apical cell	?	A	D'Amato (1975)
Blechnum brasiliense	Root apical cell	?	A	D'Amato (1975)
B. gibbum	Root apical cell	?	A	D'Amato (1975)
Ceratopteris thalictroides	Root apical cell	?	A	D'Amato (1975)
Marsilea drummondi	Root apical cell	?	A	Sossountzov (1969)
M. strigosa	Root apical cell	?	A	Avanzi and D'Amato (1970); D'Amato (1975)
Gymnosperms				
Pinus banksiana	Seed	Repet. DNA	A	Miksche and Hotta (1973)
P. glauca	Seed	Repet. DNA	A	Miksche and Hotta (1973)
P. resinosa	Seed	Repet. DNA	A	Miksche and Hotta (1973)
Angiosperms				
Allium cepa	Metaxylem (root)	rDNA	A	Innocenti and Avanzi (1971); Avanzi et al. (1973); Innocenti (1975)
Arachis hypogaea	Seed	?	A	Aldana et al. (1972)
Avena fatua	Seed (aleuron)	?	A	Maherchandani and Naylor (1971)
Calystegia soldanella	Quiescent center	?	A	Avanzi et al. (1974)
Crepis capillaris	Tissue culture	Sat. DNA	A	Sacristán and Dobrigkeit (1973)
Cucumis melo	Fruit	Heterochr.	U,A	Pearson et al. (1974)
		Sat. DNA	U,A	Ingle and Timmis (1975)
	Tissue culture	Sat. DNA		Grisvard and Tuffet-Anghileri (1980)
Cymbidium (hybrid)	Protocorms, roots	Heterochr.	A	Nagl (1972); Nagl and Rücker (1972); Nagl et al. (1972); Schweizer and Nagl (1976)
		AT-rich DNA	A	Nagl and Rücker (1976)
C. ceres	Protocorms	Heterochr.	A.	W. Nagl (unpublished)
Datura innoxia	Cell culture	ALS gene	A	Xiao et al. (1987)
Daucus carota	Cell culture	?	D	Schäfer and Neumann (1978); Hase et al. (1979)
Hedera helix	Phase change	Repet. DNA	A	Schäffner and Nagl (1979)
Helianthus annuus	Intraspecific	Various sequ.	D	Cavallini et al. (1986)
Hordeum vulgare	Root hairs	Repet. DNA		Murry et al. (1987)
Linum usitatissimum	Whole plant, Cell culture	rDNA, Repet. DNA	D	Cullis (1977, 1983, 1986); Cullis and Cleary (1986)
Lycopersicon esculentum	Internodi	?	A	Hurst and Graham (1975)
Medicago sativa	Cell culture	GS gene	A	Donn et al. (1984)
Nicotiana glauca	Tissue culture	Sat. (r)DNA	A	Parenti et al. (1973); Durante et al. (1974); Nuti Ronchi et al. (1974)
N. paniculata	Leaves	rDNA	A	Lighfoot (1972)
N. tabacum	Leaves	?	A	Siegel (1975)
	Floral buds	?	A	Wardell and Skoog (1973)
	Tumorous tissue	Sat. DNA	A	Guillé and Quetier (1970); Guillé and Grisvard (1971); Durante et al. (1986)
	Tissue culture	Repet. DNA	A,U	Altamura et al. (1987)

(to be continued)

Table 4. *(Continued)*

Species	Cell type	DNA species[a]	Event[b]	Reference
Oryza sativa	Cell Cult., Reg.	Repet. DNA	A,U	Kikuchi et al. (1987)
	Suspension	Repet. DNA	A	Zheng et al. (1987)
Petunia hybrida	Zygote	?	A	Essad et al. (1975)
	Cell culture	EPSPS gene	A	Shah et al. (1986)
Phaseolus coccineus	Embryo suspensor	Sat. DNA	A	Lima-de-Faria et al. (1975)
		Heterochr.	A	Avanzi et al. (1970, 1971); Cremonini and Cionini (1977)
		rDNA	U	Lima-de-Faria et al. (1975)
Phaseolus vulgaris	Leaves	Extrachr. DNA	A	Kunisada et al. (1986)
Pisum sativum	Seedling	?	A	Wildon et al. (1971); Bryant et al. (1974)
	Soma	rDNA	A?	Cullis and Davies (1975)
	Epicotyl	Repet. DNA	U	van Oostveldt and van Parijs (1972, 1976); Broekaert et al. (1979)
	Root tip	Extrachrom. DNA	A?	Van't Hof (1987), Krimer and Van't Hof (1983)
Rauwolfia serpentina	Cell culture	Repet. DNA	A	Solovyan et al. (1987)
Rhoeo discolor	Flower buds	?	A	Frölich and Nagl (1979)
Ricinus communis	Endosp., Cotyled.	?	U	Galli et al. (1986)
Scilla bifolia	Antipodals	rDNA	A	Nagl (1976a)
		?		Frisch and Nagl (1979)
Scorconera hispanica	Tumorous tissue	?	A	Guillé and Quetier (1970); Guillé and Grisvard (1971)
Secale cereale (× *Triticum*)	Cell culture, plants	Repet. DNA	A	Lapitan et al. (1988)
Sinapis alba	Hypocotyl	Heterochr.	A?	Bopp and Capesius (1971); Capesius and Stöhr (1974)
Solanum nigrum	Epidermis	?	A	Landré (1975, 1976)
Spinaceae oleracea	Internodia	?	A	Anker et al. (1971)
Triticum aestivum	Germinating embryo	rDNA	A	Chen and Osborne (1970)
Tropaeolum majus	Embryo suspensor	Heterochr.	U	Nagl (1976c); Nagl et al. (1976)
		AT-rich DNA	U	Deumling and Nagl (1978)
Vanda sanderina	Embryo	?	A	Alvarez (1968, 1969)
Vanilla planifolia	Crystal idioblasts	Heterochr.	U	Kausch and Horner (1984)
Vicia faba	Root initials	?	A	Macleod (1973); Jalouzot et al. (1975)
	Root (dediff.)	Sat. DNA	A	Natali et al. (1986)
	Root (diff.)	Repet. DNA	U	Kononowicz et al. (1983) Bassi et al. (1984)
Zea mays	Endosperm	Heterochr.	A	Kowles and Phillips (1985)
(cross)	Whole plant	rDNA	A	Lin et al. (1985)
	Cell culture, seedlings	Various sequ.	D	Nagl (1988)
	Suspension	Mitoch. DNA	A	Smith et al. (1987)

[a] ? = indicates that the sequences are not known. [b] A = amplification, D = differential DNA replication, U = underreplication.

nuclei achieved levels of > 200C. Most other strains compared during one growing season averaged even higher peak levels of DNA per nucleus than did A 188. Individual nuclei in those strains reached levels as high as 690C. A decrease in DNA level was observed in older endosperms with most strains. Cytological studies revealed much variation in chromatin strandedness. A form of DNA amplification, i.e., selective polytenization, appears therefore to be occurring during endosperm development.

Besides the described ontogenetic DNA changes, phylogenetic ones may also contribute to yield variation between cultivars. Different strains of maize exhibit different levels of DNA amounts in the endosperm (Tschermak-Woess and Enzenberg-Kunz 1965). According to the well-known relationship between nuclear DNA content and cell size and storage capability, the extent of DNA increase may determine the size and weight of the kernel. Such an association of DNA accumulation in the endosperm with grain weight was demonstrated, for instance, in wheat (Chojecki et al. 1986a,b; for related results in the cotyledons of the legume *Lupinus* see Le Gal et al. 1986). The development of the endosperm may be affected, on the other hand, by the degree of polyploidy of the antipodal cells, which are also apparently highly active cells, and might also undergo DNA amplification (e.g., Nagl 1976a). The different genetic regulatory ability of maize cultivars is also expressed by intraspecific qualitative and quantitative variation in histone in genome organization (Hake and Walbot 1980) and in histone H1 subfractions (e.g., Faulde and Nagl 1987). The former authors found that sequence divergence within *Zea mays* and between maize and near relatives is at least an order of magnitude greater than expected, reflecting rapid evolutionary changes in vivo. Similarly, in vitro studies indicate inbred line-specific somaclonal variation due to mutations and genomic rearrangements which even allow to construct phylogenetic trees from tissue culture data and regenerated plants (e.g., Lee and Phillips 1987; Zehr et al. 1987). Moreover, changes in the genome composition could be induced by phytohormones and auxin-related herbicides like 2,4-D (2,4-dichlorophenoxyacetic acid), 2,4,5-T (2,4,5-trichlorophenoxyacetic acid), MCPA (4-chloro-2-methylphenoxyacetic acid), and Picloram (or Tordon, 4-amino-3,5,6-trichloropicolinic acid) in vivo and in vitro (for plants and callus cultures of *Zea mays*, and cell suspension cultures of *Nicotiana tabacum*, see Nagl 1988 and Figs. 9–11). Effects of auxin on protein synthesis and phosphorylation have also been recorded (Pérez et al. 1987).

Several other aspects make maize an excellent candidate for biotechnical work on food improvement. The genes for the major storage protein of corn endosperm, zein, and of another storage protein, glutelin, are well characterized and cloned (Pintor-Toro et al. 1982; Boronat et al. 1986; Ottoboni and Steffensen 1987;

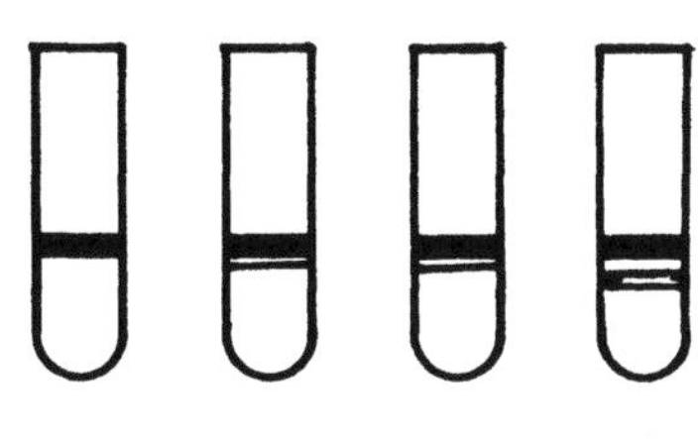

Fig. 9a-d. Banding of maize DNA after ultracentrifugation in CsCl/ethidium bromide, showing the appearance of a satellite fraction in certain tissues, and after certain treatments with herbicides, due to selective amplification of DNA. **a** DNA of untreated leaves; **b** DNA of untreated roots; **c** DNA of calli grown on a medium containing 2,4-D; **d** DNA of calli grown on a medium containing MCPA, 4-chloro-2-methyl-phenoxyacetic acid (0.2 mg/ml each). The buoyant densities were 1.701 g/ml for the main band, 1.702 for the satellites in *b* and *c*, and respectively 1.705 and 1.706 for the satellites in *d*

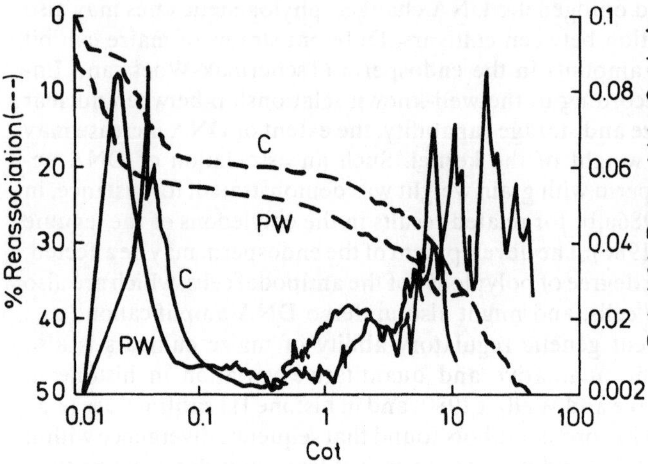

Fig. 10. DNA reassociation kinetics (cot curves, and their first derivatives, respectively) in maize without (*C* control) and after treatment with herbicides (Picloram, 0.2 mg/ml). After treatment with the auxin-related herbicide, the DNA is evidently differentially replicated

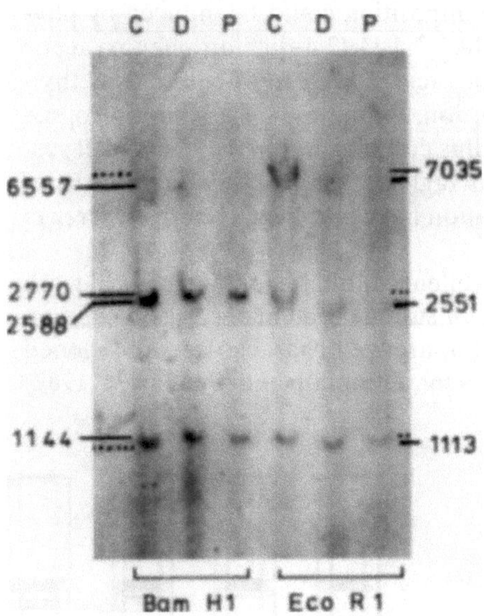

Fig. 11. Auxin and auxin-herbicide induced amplification and rearrangement of ribosomal RNA genes in tobacco TX4 cell suspension cultures as visualized by Southern blot hybridization. The DNA of control cultures (*C*), and of cultures treated with either 28.9×10^{-6} M 2,4-dichlorophenoxyacetic acid (*D*), or $0.8 \times 10°$ -6 M Picloram (*P*) was digested with the restriction enzymes *Bam*H 1 and *Eco*R 1, transferred to nylon filters, and hybridized with ^{32}P-labeled ribosomal DNA (cloned in pTA 250). There are differences in both location and quantity of ribosomal bands between the DNA samples of control and treated cultures; the fragment lengths in base pairs are given at the margins (B. Michel and W. Nagl, unpubl.)

reviewed by Heidecker and Messing (1986). The genomic responses of the maize genome to environmental stress are well documented (see above for DNA and histones; for heat-shock response see, e.g., Rees et al. 1986; Yacoob and Filion 1986). The transposable elements have been intensively studied and their effects analyzed. Callus and suspension cultures are well established, and even *Agrobacterium*-mediated transfer of DNA into maize plants was possible (Grimsley et al. 1987).

Among the *dicotyledonous* food plants, the genus *Phaseolus* was studied with regards to phylogenetic and ontogenetic DNA amplification, DNA variation in general and especially with respect to embryogeny. In contrast to maize, in the Leguminosae the embryo proper (i.e., its cotyledons) acts as storage system. For its own nourishment, the embryo develops a haustorium-like organ, the suspensor, which appears to play the key role in the synthesis and transfer of material from the endosperm cavity and the surrounding tissues to the developing embryo, and hence in the determination of the final size of the cotyledons and the seeds (Nagl 1974a, 1985b, Brady and Walthall 1985).

DNA amplification apparently represents a step in the evolution of the genus *Phaseolus*. Beridze (1972) found that the species of American origin possess satellite DNA, while this fraction cannot be found in species of Asian origin. More important, however, may be the DNA increase in the developing suspensor cells.

The basal suspensor cells contain highly endopolyploid nuclei exhibiting polytene chromosomes (Nagl 1969, 1974a, 1981). This circumstance allows a much better analysis of the genome at the cytological level than would be possible on the rather small mitotic chromosomes. In addition, the polytene chromosomes develop under periods of DNA under-replication and amplification (Fig. 12; Avanzi et al. 1970; Lima-de-Faria et al. 1975; Forino et al. 1976; Cremonini and Cionini 1977; Tagliasacchi et al. 1983, 1984; Frediani et al. 1986).

Recent success in the in vitro propagation of bean (e.g., Allavena and Rossetti 1986; Saunders et al. 1987), and progress in the understanding of the relevance of the suspensor for phytohormone synthesis in embryogenesis in vitro (e.g., Walthall and Brady 1986; Picciarelli and Alpi 1986), open new avenues for biotechnological breeding of *Phaseolus*. In addition, the nucleotide sequence of the genes and the mRNA's coding for specific phaseolin-type storage proteins has been determined (Slightom et al. 1983, 1985), as well as those for phytohemagglutinin genes (Voelker et al. 1986), and hemoglobin genes of legumes and other plants (Landsmann et al. 1986). The improvement of the nutritional quality of seed proteins represents one of the economically important functions of crop plant breeding. Genetic engineering promises to be a valuable new tool for this aim by directed changes of the seed protein composition resulting in a more balanced composition of the group of essential amino acids (Müntz 1987a,b). Figure 13 shows the strategy for sequence modification in the legumin-B gene. A better understanding of DNA amplification in the embryo-suspensor and its experimental control may help to increase storage protein synthesis in beans by conventional plant breeding and by in vitro regeneration of somaclonal variants, especially after the introduction of genes modified by gene technology; such a gene transfer using storage protein genes has already been successfully performed in some species (cf. Müntz 1987b). The additional introduction of an "amplificator" sequence (cf. Giulotto et al. 1987) or an amplifying episome (cf. Carroll et al. 1987) could lead to a new area in plant technology. Moreover experiments with the TI plasmid or viral vectors (Hayes et al. 1988) may be successful, the main problem still being to find a suitable selection system.

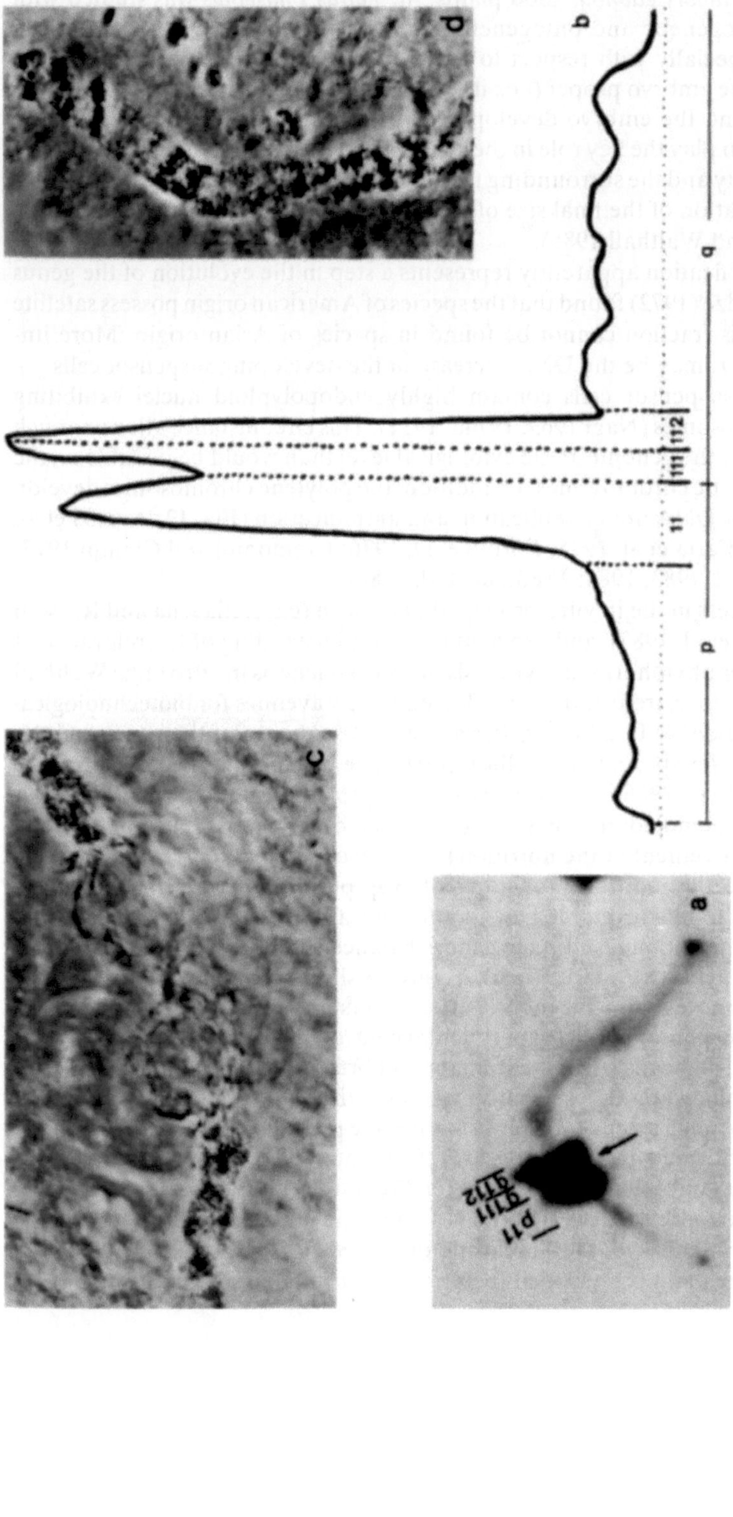

Fig. 12a-d. Differential DNA and RNA synthesis in polytene chromosomes of the embryo-suspensor of *Phaseolus coccineus*. **a** and **b** Feulgen-stained chromosome no. VII and densitogram thereof, respectively, showing a prominent DNA puff. (After Tagliasacchi et al. 1983). **c** Polytene chromosome showing two RNA puffs. **d** Nearly inactive polytene chromosome (×1200)

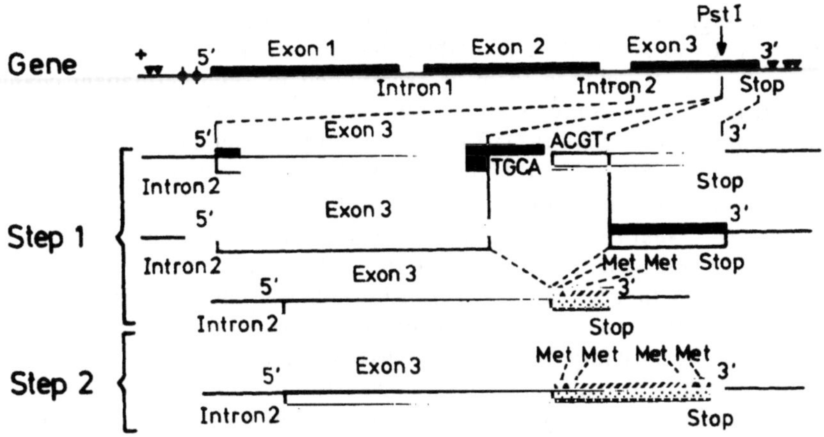

Fig. 13. Diagram to illustrate the genetic engineering of storage proteins. (After Müntz 1987b). The isolated legumin-B gene of *Vicia faba* was cut with the restriction enzyme *Pst* I within an evolutionary highly variable region of exon 3. Then, in step 1, the single-stranded ends were removed and the remaining parts of the exon joint. By this, the reading frame was changed so that the downstream part now contains two methionin codons and moved-up stop codon. In step 2, the latter was modified into a lysine codon by in vitro mutagenesis, so that two more methionin codons become part of the gene sequence. A similar increase in the essential amino acid was possible by the construction of chimeric genes

5 Stable (Permanent) DNA Amplification as a Factor of Evolution

DNA (gene) amplification is now recognized as one of the most important mechanisms in genome evolution, and hence in the evolution of new physiological or morphological phenotypes. This principle is valid for bacteria (McBeth and Shapiro 1984), plants (Oliver et al. 1983; Kushnir and Halloran 1983; Bonas et al. 1984; Dührssen et al. 1984; Karin et al. 1984), animals and man (Kominami et al. 1983; Taub et al. 1983; Brown 1984; Anagnou et al. 1984). Differential DNA replication, especially gene amplification, could be induced by environmental factors in vivo (reviewed by Cullis 1983) and in vitro (e.g., Cowell 1982; Lin et al. 1984).

Most of the phylogenetic changes in the genome are brought about by the same mechanisms as the ontogenetic changes, i.e., polyploidization, DNA amplification and elimination, mutation and rearrangement, and by additional fixation of the new genomic and karyotypic state (Table 1; Nagl and Capesius 1977; Nagl 1978, 1985b; compare also Keim and Lark 1987, and Sen et al. 1987). The sentence of Galau et al. (1976) is recalled, that "the study of genomic regulation and the study of evolution must be considered the two sides of the same coin". A point with which embryologists have been familiar for decades: Haeckel (1866) already suggested that early ontogenesis (embryogeny) recapitulates phylogenesis in its principle lines. This suggestion, which resulted from anatomical and morphological comparisons, was repeatedly confirmed at the molecular level (summarized by Lima-de-Faria 1983).

The suggestion that the same mechanisms are effective in ontogenetic and phylogenetic variation of nuclear DNA quantity and quality, and that a functional continuum exists between them, is exemplified by semi-permanent changes in nuclear DNA, such as the magnification of ribosomal DNA in *bobbed* mutants of *Drosophila* (Ritossa et al. 1971; Tartof 1971) and in other animals and plants (e.g., *Saccharomyces*: Kaback and Halvorson 1977), and by the origin of genotrophs in flax (Durrant and Jones 1971; Cullis 1977, 1983; for reviews of these events see Buiatti 1977; Nagl 1978). Such phenomena clearly indicate that it is only a matter of more or less stringent fixation of the differentially replicated DNA which determines whether we can find a transitory developmental increase in DNA, a semi-permanent, or a permanently fixed increase (see Figs. 1 and 2).

One of the most important distinctions to be made in analyzing mechanisms of evolution is that of anagenesis and cladogenesis. The confusion of these two events (and terms) has led to many misunderstandings of the whole process: the one relates to evolutionary trends, minor molecular and phenotypic, slow changes, and to adaptation; the other relates to evolutionary origins (such as speciation), to major and sudden changes, to cladistic or phyletic lineages. Nagl (1979b) tried to show that evolutionary diversification of bio-matter is accompanied (or even caused?) by increase and diversification of noncoding, repetitive DNA (this does not exclude that also loss of DNA can occur during speciation, e.g., in adaptation to a single, narrow niche). Such an amplification process may lead to regulatory changes, which allow speciation over short periods of time.

The evolution of repetitive DNA has been extensively studied in some Gramineae species (cereals) by reassociation kinetics in the following way. Heteroduplices were formed between the repeated sequence DNA fractions from each of the species, so that the velocity of renaturation (corresponding to the extent of double strand formation) could be taken as a measure of the degree of homology between the repetitive DNA's of these species (Smith and Flavell 1974; Flavell et al. 1977; Rimpau et al. 1978; Bedbrook et al. 1980). The number of DNA sequence groups, which are homologous in two of the four species (*Triticum aestivum, Secale cereale, Hordeum vulgare,* and *Avena sativa*), is the greater the more closely the species are related to each other. Based on the origin of new repetitive sequence groups, a scheme of evolution can be drawn for these plants, which agrees well with that outlined from taxonomic and genetic considerations (Fig. 14). Group I sequences, which are found in all of the four genomes, have evidently already evolved within a common ancestor species. The amplification of group II sequences has led to the separation of *Avena* from the ancestor lineage of the other three species, but before the separation of *Hordeum* from the ancestor of *Triticum* and *Secale*. Similarly, the origin of the other repetitive sequence groups, III–VII, can be explained. The systematic relationship can therefore be seen as the result of the position of saltatory replication (amplification) processes in evolutionary trees (interpretation by W. Nagl). The changes in the repetitive DNA sequences indicate, in contrast to gene mutations, splitting events rather than the age of a lineage. This can be well understood in terms of the Britten-Davidson model of gene regulation and evolution. The amplification and dispersion of repetitive sequences were recently studied by McIntyre et al. (1988) in Triticeae; for other plant taxa see Bonierbale et al. (1988) and Kuriyan and Narayan (1988).

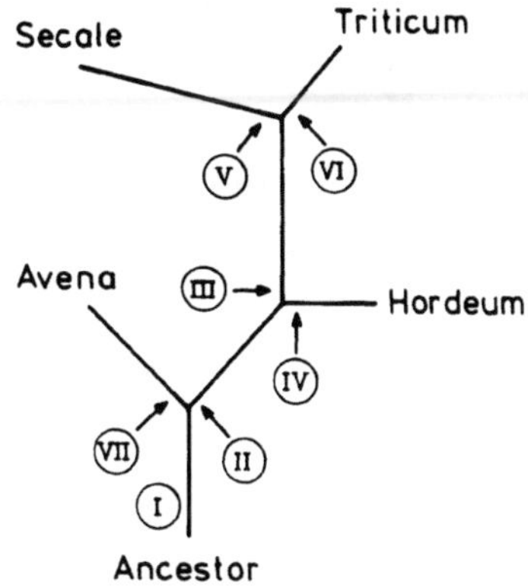

Fig. 14. Diagram to illustrate the evolution of four cereal species from a common ancestor, as estimated from the occurrence and cross-hybridization of repetitive DNA sequence groups (*Roman numerals*). (Data taken from Smith and Flavell 1974 and Rimpau et al. 1978)

A detailed analysis of quantitative and qualitative changes of the genome is facilitated, if part of the repetitive sequences form a satellite band upon ultracentrifugation and can, therefore, be easily isolated. Amplification and rearrangement of satellite sequences were reviewed by Stephan (1987). The satellite DNA of the nasturtium, *Tropaeolum majus*, was studied by restriction and sequencing by O. Bill, B. Knapp and W. Nagl (unpubl.), and a high intra-specific and intra-individual variation in quantitative and qualitative terms was established. Figure 15 shows, for instance, the result of a secondary restriction experiment. The smallest 3′ end-labeled *Hpa* II fragments of the satellite (20 to 170 bp in length) were eluted from the agarose gels and restricted with *Hae* III, separated by 8% polyacrylamide gel electrophoresis and autoradiographed. Each of the two parallel runs, representing DNA samples of different individuals, exhibits a different pattern.

Genomic changes in evolution of plants may actually occur rapidly (Walbot and Cullis 1985), a fact that might explain the discontinuities found in macroevolutionary processes. The molecular changes in DNA and chromosome organization are closely related to, and probably the cause of, phenotypic variation (Flavell 1981; Cullis 1986). There is now, however, stringent evidence for a dominant role of noncoding DNA in phenotypic evolution rather than for genes. In other words, the evolution of new species and taxa is the result of regulatory changes, in which structural gene mutations may play a secondary role (Ayala 1976; Hinegardner 1976; Wilson 1976; Nagl 1978, 1979b; Lima-de-Faria 1983; Nagl et al. 1983). Therefore, an important discrepancy between sequence evolution in genes and proteins on the one hand, and evolution at the organismic, phenotypic level on the other, could frequently be observed. However, in the light of noncoding DNA as conformational DNA (for chromatin structure, Nagl 1983a,b, 1985a) and hence regulatory DNA (for gene activity, Britten and Davidson 1969, 1971), it is not surprising that many coding genes are essentially the same in bacteria, plants, and man, while the noncoding DNA underwent dramatic alterations in qualitative and quantitative terms (MacGregor et al. 1976; this holds also for the promoter sequences of extremely conserved genes). These fundamentals should also be taken into consideration in future biotechnical approaches in crop improvement. In this connection it may also be remembered that, in higher organisms, genes represent

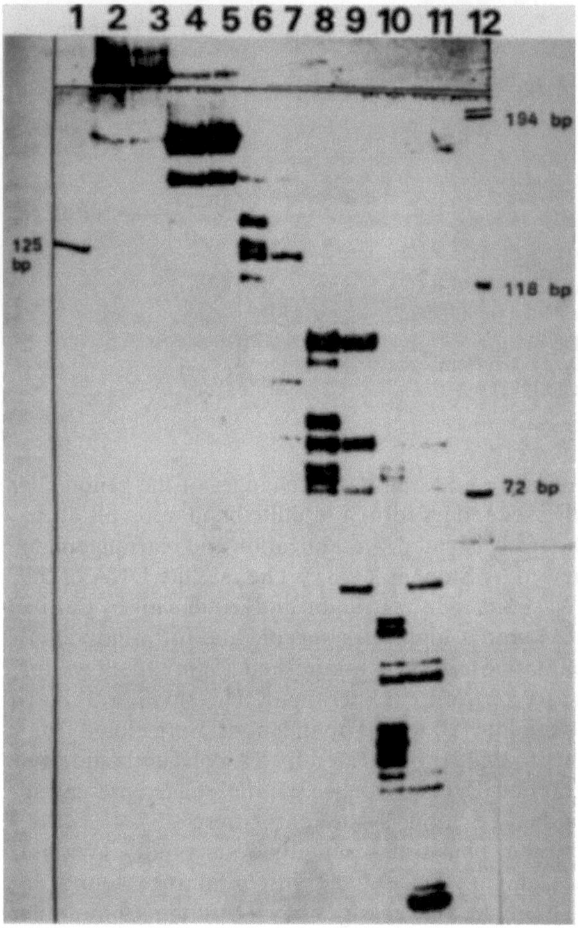

Fig. 15. Example for the intraspecific variation of the base sequence of the *Tropaeolum* satellite DNA as visualized by autoradiograms of secondary restriction fragments generated by *Hae* III digested. 3'end-labeled small fragments of primary *Hpa* II digested satellite DNA. Slot *1*: Marker (*Hind* III digested λ DNA); slots *2* and *3*: 197 bp *Hpa* II fragment; slots *4* and *5*: 131 bp fragment; slots *6* and *7*: 88 bp fragment; slots *8* and *9*: 76 bp fragment; slots *10* and *11*: 40 bp fragment; slot *12*: marker (*Hae* III digested ϕX174-RF DNA). The photo represents a very hard print, to show, for clearness, the heavily labeled bands only; actually, 60 bands were visible. Note that each of the two parallel runs (slots) of samples of different individuals exhibit different patterns of secondary restriction, indicating rapid diversification of the satellite DNA (O. Bill, B. Knapp, W. Nagl, unpubl.)

only 1% of the nuclear DNA, or even less (down to 0.1%; Hinegardner 1976; Nagl 1983a,b). It is, to my mind, nonsensical to describe all these sequences as "selfish", "parasitic", or "junk", as has been done by some workers (e.g., Dawkins 1976; Doolittle and Sapienza 1980; Orgel and Crick 1980). This can be emphasized from both a selectionistic (Darwinistic) and an energetic (physico-deterministic) point of view (Lima-de-Faria 1983; Nagl 1983a,b, 1985a, 1986).

Cell and tissue cultures can act as excellent models for evolution, because genomic changes occur very rapidly, their fixation (inheritance) can be followed up for many generations within short periods of time, and the phenotypic consequences can be studied immediately after regeneration of the plants. Cell and tissue cultures demonstrate, in addition, the close relationship between somatic and phylogenetic genome changes, as they can be seen in both ways, and the culture conditions, which might be in some way stressful for the cells in any case, reflect a changing environment in nature as well. Summarizing, the mechanisms that generate rapid genomic changes can be envisaged as a metabolic feed-back process which may induce adaptation (Walbot and Cullis 1985), or as the response to genetic stress (McClintock 1984). Experiments with phytohormones in vitro and with auxin-derived herbicides in vivo demonstrate this correspondingly (Nagl 1972, 1988).

6 Role of Gene Amplification in Somaclonal Variation in Plants

6.1 Chromosomal Variation

Although cloning of identical plants was originally the principal use for plant cell cultures, it has become increasingly clear that the great genetic variability of plant cells under appropriate cultural conditions can be the source of new and rapid breeding and improvement techniques. As protoplasts, callus tissue, and tissue explants show variability and can be regenerated via organogenesis or embryogenesis, new strategies in agriculture could be developed (see Bajaj 1986). The genetic variability of somatic cells in culture was termed *somaclonal variation* by Larkin and Scowcroft (1981) and reviewed by Semal (1986).

Somaclonal variation has received interesting applications in plant breeding for better quality and yield (Evans and Sharp 1986; Gavazzi et al. 1987; Tyagi and Naqvi 1987; Mizrahi 1988; Evans 1989). In addition, it can be used for screening of freezing tolerance (Lazar et al. 1988) and resistance to parasites (Heath-Pagliuso et al. 1988).

The genetic variation was first detected as altered chromosome numbers, and chromosome instability has been well characterized in many species (reviewed by D'Amato 1975, 1977; Sunderland 1977; Bayliss 1980; Chaleff 1981; Flick and Evans 1983). The most frequently observed variation has been polyploidy, attributed either to the selective growth of normally nondividing polyploid cells that preexisted in the original explant, or to the induction of polyploidization by the types of phytohormones used in the culture medium. Aneuploid changes, typically involving the gain or loss of a few chromosomes due to spindle misfunction, have also been frequently reported in plant cell cultures, but in many cases these lines are incapable of plant regeneration. In addition to numerical chromosome changes, the structure of plant chromosomes is modified in many cultures, and evidence of recovery of single gene mutations produced via somaclonal variation has also been represented (see reviews by D'Amato 1985; Vasil 1986). Somaclonal variation in crop plants demonstrates the potential application of this method in crop improvement. Although most variants reported to date have been chromosomal (e.g. Natali and Cavallini 1987; Zehr et al. 1987; Gaponenko et al. 1988; Pohler et al. 1988; Wersuhn and Sell 1988; and many others), a detailed genetic analysis has been completed for only a few somaclonal variants, and changes other than chromosomal may be more important for crop improvement of sexually propagated crops. At present, there is some evidence that differential DNA replication may also contribute to somaclonal variation.

Arnholdt-Schmitt et al. (1987) and K.-H. Neumann (pers. commun.) analyzed the genome of carrot lines and their tissues in vivo, and callus in vitro, by restriction enzyme analysis and observed a great variation between cell and plant lines and tissues (see also Ball, Chap. I.5, this Vol.). These results clearly indicate that DNA amplification and related events contribute to "phylogenetic" and "ontogenetic" diversity of plants and tissues in vitro and in vivo. While in the latter studies application of kinetin seemed to play a key role in induction of variability, earlier experiments on carrot (Schäfer and Neumann 1978; Schäfer et al. 1978), in orchids (Nagl and Rücker 1974), and recent ones in maize (Nagl 1988) indicate that also other phytohormones affect the genome composition (due to the induction of differential DNA replication) in vitro, and herbicides and other environmental stress in vivo (see also the work done in flax, e.g., Cullis and Cleary 1986).

Experimental molecular approaches to the study of somaclonal variation are, at present, in rapid development and promise a great deal of new information on DNA sequence amplification and rearrangement related to this specific genetic aspect of cell and tissue cultures (e.g., Dhillon et al. 1983; Orton 1983; Peerbolte et al. 1987; an overview is given by Vasil 1986). Selection strategies have also been introduced for amplified and hence resistant varieties and somatic hybrids (Brunold et al. 1987; Ye et al. 1987).

Somaclonal variation finds its expression also at the level of gene products, for instance in isozyme variation (Allicchio et al. 1987). Single cell clones derived from tobacco callus exhibit some significant differences in esterase isozymes and the production of the secondary metabolite, cinnamoyl putrescine (U. Arbeiter and W. Nagl, unpubl.; Fig. 16). Other consequences of somaclonal variation are differences in regeneration ability (Singh 1986; Bebeli et al. 1988) and morphology (Furze et al. 1987; Qu et al. 1988; Osifo et al. 1989; Sree Ramulu et al. 1989).

The problems with new plant varieties produced by either somaclonal variation or gene technology was briefly discussed by Yang (1988).

6.2 Restriction Fragment Length Polymorphism (RFLP)

Genetic variability can now also be analyzed at the molecular level. Cloned DNA sequences can be used to probe specific regions of eukaryotic genomes for the presence of polymorphism at the DNA sequence level. This polymorphism is detected as variation of the restriction fragments isolated from genomic DNA and hybridized with a DNA probe. Such variation was first used for genetic analysis of adenovirus mutants (Grodzicker et al. 1974) and termed "restriction fragment length polymorphism" (RFLP).

Although the majority of the methods, probes and applications of RFLP have been developed for human genetics and diagnosis of diseases (originally for sickle-cell anemia by Kan and Dozy 1978), genetic markers have also been developed for several plant species (Burr et al. 1983; Landry and Michelmore 1987). RFLP analyses are powerful tools in gene mapping and the construction of linkage maps (Tanksley et al. 1989; for maize see Burr et al. 1988 and Helentjaris 1987, for lettuce Landry et al. 1987, for tomato Bernatsky and Tanksley 1986, for *Arabidopsis* Chang et al. 1988; for soybean Apuya et al. 1988). RFLPs are also used to obtain genetic markers for resistance genes (Landry et al. 1987; Nienhuis et al. 1987).

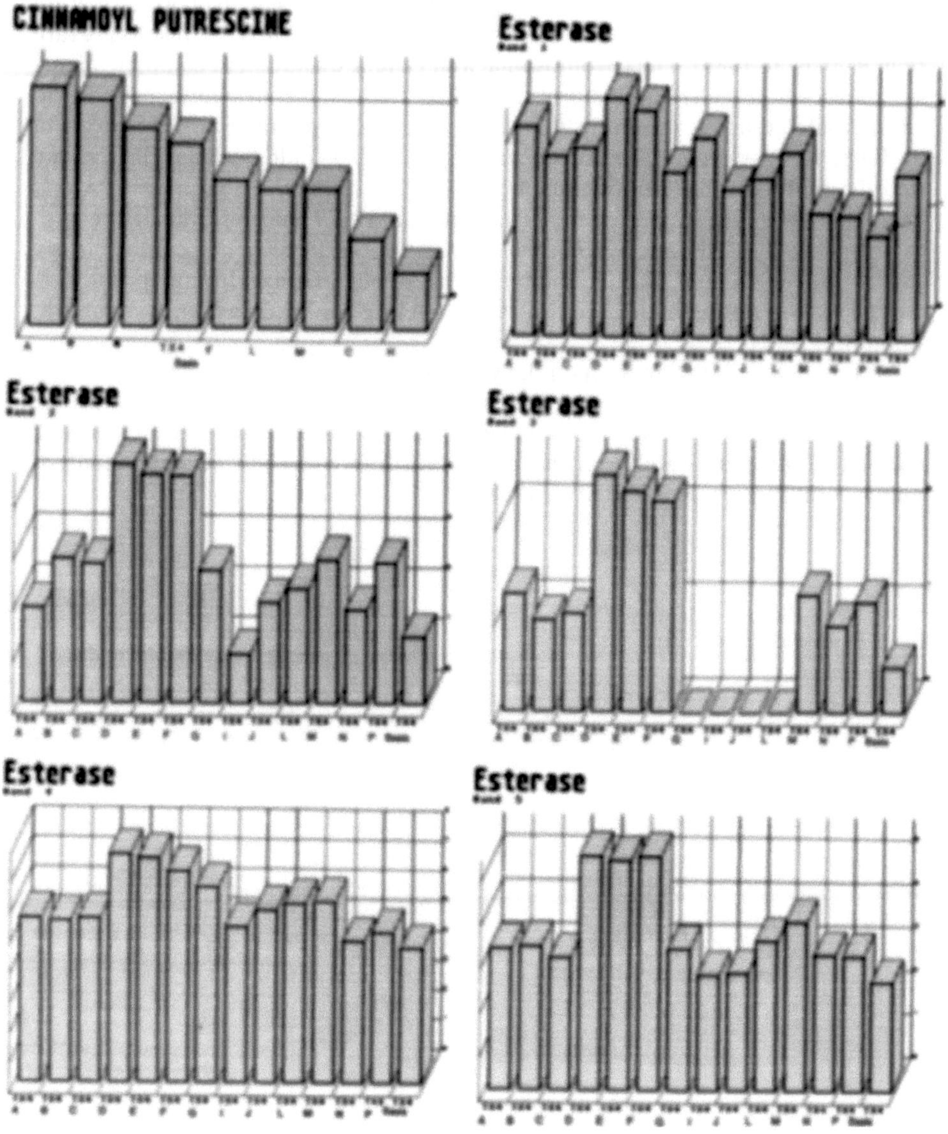

Fig. 16. Evidence for somaclonal variation at the level of primary gene products and a secondary natural product: Variation of the amount of the secondary metabolite, cinnamoyl putrescine, and of the five main esterase isozyme bands (as visualized by densitogram data of polyacrylamide gels) among single cell clones of *Nicotiana tabacum*, line TX4. The stem line culture is marked as "*basis*", the individual clones by *letters* (U. Arbeiter and W. Nagl, unpubl.)

RFLP markers can, of course, also be used in conjunction with plant cell culture technology. Studies at this level are in progress worldwide, but so far results are rare (e.g. Roth et al. 1989). The chromosomal and genetic evolution in complete plants was, however, already successfully studied by RFLPs, e.g. in potato and tomato (Bonierbale et al. 1988), and in the genus *Brassica* (Figdore et al. 1988; Song et al. 1988). In *Gossypium*, variation of gene copy number could be determined by gel and dot blot hybridization of cloned cDNAs (Galau et al. 1988), and in *Petunia* McLean et al. (1988) analyzed the actin gene families. The bean cultivars belonging respectively to the species *Phaseolus coccineus* and *P. vulgaris* can be clearly distinguished by their RFLP obtained with a phytohemagglutinin probe (i.e. a gene for one of the major seed storage proteins; W. Nagl and M. Redenbach, unpubl.; Fig. 17).

More detailed information on the variation and evolution of DNA sequences even between individuals can be obtained in humans by a mini-satellite probe, resulting in highly specific "DNA fingerprints" (Jeffreys 1986). Recently, it was shown that such mini-satellite probes also reveal differences between individuals of angiosperm species (Rogstad et al. 1988).

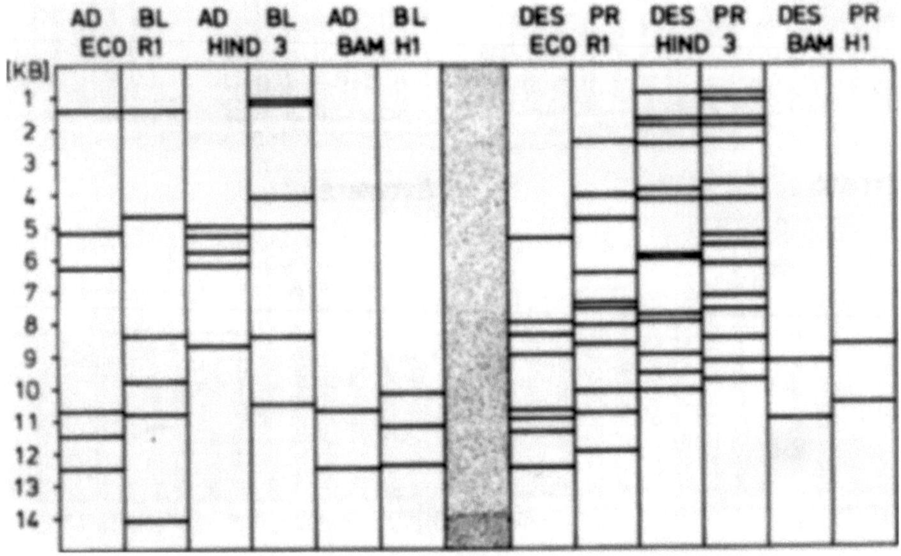

Fig. 17. RFLP (restriction fragment length polymorphism) between cultivars of *Phaseolus vulgaris* and *P. coccineus* as probed with a phytohemagglutinin (PHA-L) clone (pdlec2; Voelker et al. 1986). Genomic DNA was digested with various restriction enzymes, alkali-transferred to Hybond-N, and hybridized with the ^{32}P-labeled probe. The results were evaluated by a computer program for the determination of restriction fragment lengths. *AD* = *Ph. vulgaris*, cv. Admires; *BL* = *Ph. vulgaris*, cv. Blauhilde; *DES* = *Ph. coccineus*, cv. Desiree; *PR* = *Ph. coccineus*, cv. Preisgewinner (W. Nagl and M. Redenbach, unpubl.)

7 Mechanisms of Amplification

In spite of the manifold phenomenological data on DNA amplification, recent approaches of molecular dissection of amplified sequences, and genetic attempts to understand the induction of drug-resistant cell populations, only little is known of the underlying mechanisms. It appears that major differences in the modes of amplification between that of rDNA, structural genes and heterochromatin, between various cell types and organisms, and between the response to general genetic stress and that to agents or treatments which affect a single gene (gene product) have to be assumed. One common feature is that the extra replicated unit, the "amplicon", is much larger than a structural gene, and may even be free of a coding sequence. Amplification of the *dhfr* domain in CHO cells evidently initiates at two distinct sites, one of which is a repetitive sequence element (Anachkova and Hamlin 1989; see also Ma et al. 1988 and Leu and Hamlin 1989).

Moreover, the amplicon may be composed of a nested set of sequences, whose dosage forms a gradient, so that centrally located sequences are amplified more than those distal to the "origin" of DNA amplification (Shiloh et al. 1985). While the highly amplified core sequences, which are approximately between 200 and 2000 base pairs in length, may be relatively similar in sequence content, due to recombination events "novel joints", and thus new restriction fragments (cf. Fig. 3), lead to a much more complex structure of amplified DNA than a linear repeat of identical units.

DNA endoreduplication and differential DNA replication, including under-replication and amplification, can mechanistically be seen as progressive steps of "short cuts" of the mitotic cell cycle (Nagl 1974b, 1976b), i.e., the successive omission of mitotic stages and phases of the S period, resulting respectively in total and local polyploidy (or polyteny), whereby the replicated portion became more and more reduced down to a single gene (amplicon). A similar model was put forward by Lima-de-Faria (1974), who defined amplification as a local polytenizaton. Quite recently, over-replication was again discussed in relation to the cell cycle and the position a gene holds in the temporal pattern of the S period (Schimke et al. 1986). This idea can be combined with various modes of recombination (Holden et al. 1987; Hyrien et al. 1988) and rearrangement (Radinsky et al. 1988; cf. Figs. 1 and 18C, and Nagl 1988). The independent activation of replication origins in the eukaryotic chromosome, a typical multireplicon structure, is clearly consistent with, and expressed by, the replication pattern of chromosomes, and is consistent with re-replication or extra-replication of a certain sequence, as well as with the nonreplicaton or under-replication of another sequence (see also Roberts et al. 1983; Hamlin et al. 1984).

A frequently shown model representing a modification of the local polytenization model was proposed by Bullock and Botchan (1982): the generation of an "onion skin" structure, which is then resolved into amplified units by recombination (Fig. 18B).

Although the mechanism of DNA amplification is still somewhat obscure and possibly is not identical in all cases, its stimulation by stress is now generally accepted. It seems that perturbations of DNA replication, e.g., induction by inhibitors, represent a general step (Schimke 1988; see also the discussion by Nagl

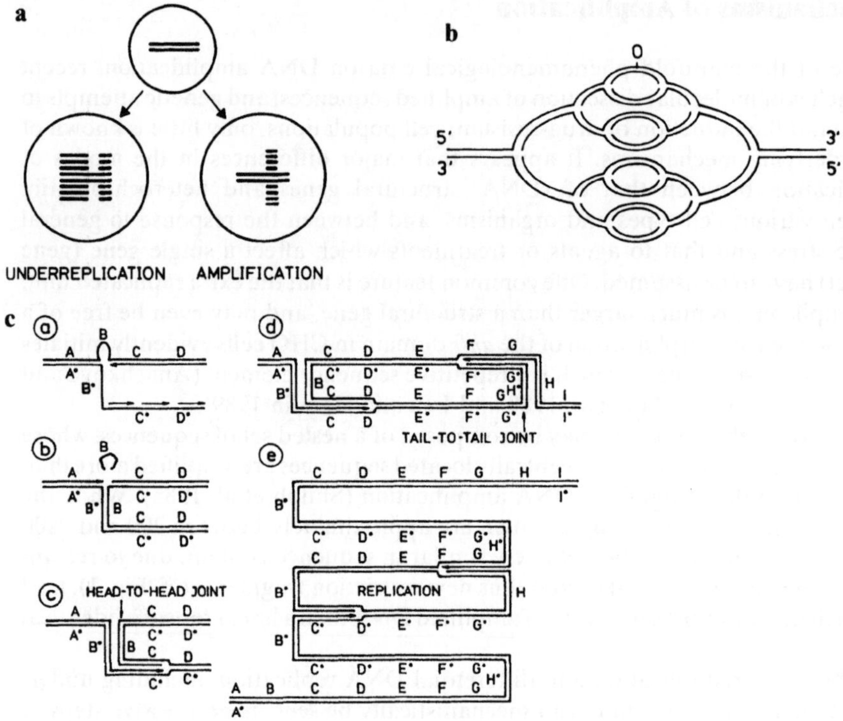

Fig. 18a-c. Basic models of differential DNA replication. **a** The principles of DNA under-replication and DNA amplification. (After Nagl 1979a). **b** A more detailed scheme of DNA amplification. (After Stark and Wahl 1984, and the references therein), as it can be seen as the basis of all recent hypotheses (local polytenization, or unscheduled DNA synthesis, or re-replication). This process can generate, by recombination, linear intrachromosomal amplified arrays, extrachromosomal circles, or extrachromosomal linear duplexes (double minutes), as shown in Fig. 1. Bidirectional replication at an origin (O) generates bubbles that can undergo further rounds of extra-replication, resulting in a nest of partially replicated duplexes (i.e. noncontiguous chromatids). **c** Model for DNA amplification by inverted duplication: *a-c* Formation of a head-to-tail joint by copy-choice recombination. *d* Structure obtained if a head-to-head joint and a tail-to-tail joint are formed by the same events occurring at both ends of the replication bubble. *e* The resulting inverted duplication that may be amplified by intra-chromosomal double rolling circle replication (only part of the model, according to Hyrien et al. 1988; from Nagl 1989)

1989 with regard to extrachromosomal DNA in plant cells after blockage of DNA replication). Normally, all sequences of each chromosomal DNA molecule are replicated once and only once during a single cell cycle in the S-phase. That control is apparently lost after perturbation. Chinese hamster cells blocked with hydroxyurea re-replicate certain sequences within a single cell cycle upon release from the block, indicating that perturbation of DNA synthesis results in misfiring of replicon initiation, and subsequent amplification of these genes (Hoy et al. 1987). The specificity of sequences responding to perturbation by over-replication may be established by "amplification control elements", i.e. some enhancing regions (Delidakis and Kafatos 1989). In cases of gene amplification during normal development, certain replication origins must be, however, under specific regulation stimuli. These ideas are consistent with findings that replication origins are fairly homologous with transcription enhancers (briefly reviewed by Nagl 1989).

Two main questions are still under discussion: the one is that as to the relationship between DNA amplification and karyotype repatterning (or chromosome aberrations). Schimke et al. (1986) suggested that chromosomal alterations are the consequence of recombination events involving the strands of overreplicated DNA, and not that the additional DNA occurs after the generation of chromosomal aberrations. Thus, the same initial processes of over-replication and recombination can result in a number of chromosomal alterations, which have major consequences for surviving somatic and germ cells. However, in the case of oncogene amplification, it was suggested that the chromosomal structural alterations may translocate an oncogene to an effective promoter of transcription and replication, so that amplification could be the result of recombination (Morgan et al. 1986; see also Nagl 1987a). It should be remembered that many human tumors can be characterized by specific chromosomal anomalies (numerical and structural), so that they exhibit a tumor-specific karyotype evolution (reviewed by Forman and Rowley 1982; German 1983; Mitelman 1983). Recently, evidence was accumulated that oncogenes are involved in those chromosome aberrations (e.g., Varmus 1984; Kuhn et al. 1985; Bigner et al. 1987; Haluska et al. 1987; Heim and Mitelman 1987; Ibson et al. 1987). Hence, our view of the interrelationships between gene amplification and chromosome aberration becomes somewhat rounded off, although still many details remain to be elucidated.

In this connection it is worthwhile to mention that some progress has also been achieved in the general understanding of the relationship between changes at the genome (DNA) level and the karyotype (chromosome) level in plants (e.g., Pijnacker et al. 1986; Nagl 1987a; Nagl and Pfeifer 1988; see also Nuti Ronchi et al. 1986).

The other open question relates to a fundamental problem of evolution. Is the amplification of specific sequences, e.g., in drug-resistant cells, the result of a *selection* process (Scotto et al. 1986), or an induced *response* by the cells (Quesnel and Ellis 1987)? On the one hand, there is no close quantitative relationship between DNA amplification and resistance, on the other hand, DNA changes and resistance can be routinely induced by stress or drugs. For an answer to this question we evidently do not yet know enough on the regulation of amplification events, but an objective scientist should be open-minded and see new findings in an unbiased way.

8 Concluding Remarks

The amplification, and more generally, the differential replication, of certain DNA sequences is now well recognized in a great number of plant and animal species and cell types in vivo and in vitro. From the body of data obtained so far, the following conclusions can be drawn.

1. The genome can be considered to be in a state of permanent plasticity, due to its composition of independent replicative units and their ability for selective replication.

2. A continuous spectrum of events must be seen between transitory, somatic changes, and stable, evolutionary changes, both being the result of sequence amplification, diversification, diffusion (transposition), and loss. Genome and karyotype changes are closely interdependent.
3. Genomic changes by differential DNA replication occur as steps of normal differentiation, but also of tumorigenesis. Gene amplification enables a cell a peak output of a specific product, and it serves as dynamic genomic adaptation to stressful conditions (natural or experimental ones), resulting, for instance, in resistance.
4. A better knowledge of the mechanisms of DNA amplification may open new strategies for its biotechnological utilization, e.g., in the improvement of the quality and the yield of food seeds, and the induction of resistance. Cell and tissue cultures appear to be an ideal tool for the study of DNA changes, and for their application in plant breeding. The selection of an "amplificator" phenotype, or the transfection of protoplasts with an amplifying episome, combined with gene engineering, could be a promising step in this direction. A suitable selection system should be possible to be constructed for instance by the aid of an universal hybridizer.
5. Somaclonal variation and RFLPs clearly show the manifold, and often interconnected genetic mechanisms involved in the adaptation and evolution of new cell lines and plant varieties, from the DNA sequence level to the ploidy level. New plant engineering and breeding programs should consider all of them as a useful source of improvement.

References

Aladjem MI, Koltin Y, Lavi S (1988) Enhancement of copper resistance and CupI amplification in carcinogen-treated yeast cells. Mol Gen Genet 211:88–94

Aldana AB, Fites RC, Pattee HE (1972) Changes in nucleic acids, protein and ribonuclease activity during maturation of peanut seeds. Plant Cell Physiol 13:515–521

Alhonen-Hongisto L, Leinonen P, Sinervirta R, Laine R, Winqvist R, Alitalo K, Jaenne OA, Jaenne J (1987) Mouse and human ornithine decarboxylase genes. Methylation polymorphism and amplification. Biochem J 242:205–210

Alitalo K, Schwab M, Lin CC, Varmus HE, Bishop JM (1983) Homogeneously staining chromosomal regions contain amplified copies of an abundantly expressed cellular oncogene (c-*myc*) in malignant neuroendocrine cells from a human colon carcinoma. Proc Natl Acad Sci USA 80:1707–1711

Alitalo K, Koskinen P, Mäkelä TP, Saksela K, Sistonen L, Winqvist R (1987) *Myc* oncogenes: activation and amplification. Biochim Biophys Acta 907:1–32

Allavena A, Rossetti L (1986) Micropropagation of bean (*Phaseolus vulgaris* L.): effect of genetic, epigenetic and environmental factors. Sci Hortic 30:37–46

Allichio R, Antonioli C, Graziani L, Roncarati R, Vannini C (1987) Isozyme variation in leaf-callus regenerated plants of *Solanum tuberosum*. Plant Sci 53:81–86

Alt FW, Kellems RE, Bertino JR, Schimke RT (1978) Selective multiplication of dihydrofolate reductase genes in methotrexate-resistant variants of cultured murine cells. J Biol Chem 253:1357–1370

Altamura MM, Bassi P, Cavallini A, Cionini G, Cremonini R, Monacelli B, Pasqua G, Sassoli O, Tran Than Van K, Cionini PG (1987) Nuclear DNA changes during plant development and the morphogenetic response in vitro of *Nicotiana tabacum* tissues. Plant Sci 53:73–79

Alvarez MR (1968) Quantitative changes in nuclear DNA accompanying postgermination embryonic development in *Vanda* (Orchidaceae). Am J Bot 55:1036–1041

Alvarez MR (1969) Cytophotometric study of proteins and nucleic acids in parenchymatous tissues of the orchid embryo. Exp Cell Res 57:179–184

Ames GFL (1986) The basis of multidrug resistance in mammalian cells: Homology with bacterial transport. Cell 47:323–324

Anachkova B, Hamlin JL (1989) Replication in the amplified dihydrofolate reductase domain in CHO cells may initiate at two distinct sites, one of which is a repetitive sequence element. Mol Cell Biol 9:532–540

Anagnou NP, O'Brien SJ, Shimada T, Nash WG, Chen M-J, Nienhuis AW (1984) Chromosomal organization of the human dihydrofolate reductase genes: dispersion, selective amplification, as a novel form of polymorphism. Proc Natl Acad Sci USA 81:5170–5174

Anker P, Stroun M, Greppin H, Fredj M (1971) Metabolic DNA in spinach stems in connexion with ageing. Nature New Biol 234:184–186

Apuya NR, Frazier BL, Keim P, Jill Roth E, Lark KG (1988) Restriction fragment length polymorphism as genetic markers in soybean, *Glycine max* (L.) merrill. Theor Appl Genet 75:889–901

Ardeshir F, Giulotto E, Zieg J, Brison O, Liao WSL, Stark GR (1983) Structure of amplified DNA in different Syrian hamster cell lines resistant to N-(phosphonacetyl)-L-aspartate. Mol Cell Biol 3:2076–2088

Arnholdt-Schmitt B, Sander S, Dührssen E, Neumann K-H (1987) The influence of hormonal treatments on DNA amplification of *Daucus carota*. Abstr 3-09-3, 14th Int Botanical Congr, Berlin

Arrighi FE (1983) Gene amplification in human tumor cells. In: Mirand EA, Hutchinson WB, Michich E (eds) 13th Int Cancer Congr, Seattle, WA, USA; Progr Clin Biol Res 1321 Alan R Liss, New York, pp 259–268

Atkin NB, Baker MC, Ferti-Passantonopoulou A (1983) Chromosome changes in early gynecologic malignancies. Acta Cytol 27:450–453

Avanzi S, D'Amato F (1970) Cytochemical and autoradiographic analyses on root primordia and root apices of *Marsilea strigosa*. Caryologia 23:335–345

Avanzi S, Cionini PG, D'Amato F (1970) Cytochemical and autoradiographic analyses on the embryo suspensor cells of *Phaseolus coccineus*. Caryologia 23:605–638

Avanzi S, Buongiorno-Nardelli M, Cionini PG, D'Amato F (1971) Cytological localization of molecular hybrids between rRNA and DNA in the embryo suspensor cells of *Phaseolus coccineus*. Atti Accad Naz Lincei, Sci Fis Mat Nat, Ser VIII 50:357–361

Avanzi S, Durante M, Cionini PG, D'Amato F (1972) Cytological localization of ribosomal cistrons in polytene chromosomes of *Phaseolus coccineus*. Chromosoma 39:191–203

Avanzi S, Maggini F, Innocenti AM (1973) Amplification of ribosomal cistrons during the maturation of metaxylem in the root of *Allium cepa*. Protoplasma 76:197–210

Avanzi S, Bruni A, Tagliasacchi AM (1974) Synthesis of DNA and RNA in the quiescent center of the primary root of *Calystegia soldanella*. Protoplasma 80:393–400

Ayala FJ (ed) (1976) Molecular evolution. Sinauer, Sunderland, MA

Bajaj YPS (ed) (1986) Biotechnology in agriculture and forestry, vol 2. Crops I. Springer, Berlin Heidelberg New York Tokyo

Baskin F, Grossman A, Bhagat SG, Burns D, Davis RM, Warmoth LA, Rosenberg RN (1987) Frequent alterations of specific reiterated DNA sequences abundances in human cancer. Cancer Genet Cytogenet 28:163–172

Bassi P, Cionini PG, Cremonini R, Seghizzi P (1984) Under-replication of nuclear DNA sequences in differentiating root cells of *Vicia faba*. Protoplasma 123:70–77

Bayliss MW (1980) Chromosomal variation in plant tissues in culture. Int Rev Cytol Suppl 11A:113–143

Beach LR, Palmiter RD (1981) Amplification of the methallothionein-1 gene in cadmium-resistant mouse cells. Proc Natl Acad Sci USA 78:2110–2114

Bebeli P, Karp A, Kaltsikes PJ (1988) Plant regeneration and somaclonal variation from cultured immature embryos of sister lines of rye and triticale differing in their content of heterochromatin. Theor Appl Genet 75:929–936

Bedbrook JR, O'Dell M, Flavell RB (1980) Amplification of rearranged repeated DNA sequences in cereal plants. Nature (London) 288:133–137

Beridze T (1972) DNA nuclear satellites of the genus *Phaseolus*. Biochim Biophys Acta 262:393–396

Berlin J, Knobloch K-H, Höfle G, Witte L (1982) Biochemical characterization of two tobacco cell lines with different levels of cinnamoyl putrescines. J Nat Prod 45:83–87

Bernatzky R. Tanksley SD (1986) Towards a saturated linkage map in tomato based on isozymes and random cDNA sequences. Genetics 112:887–898

Biedler JL. Chang TD. Scotto KW. Melera PW. Spengler BA (1988) Chromosomal organization of amplified genes in multidrug-resistant Chines hamster cells. Cancer Res 48:3179–3187

Bier K (1969) Oogenesetypen bei Insekten und Vertebraten. ihre Bedeutung für die Embryogenese und Phylogenese. Zool Anz Suppl 33:7–29

Bigner SH. Wong AJ. Mark J. Muhlbaier LH. Kinzler KW. Vogelstein B. Bigner DD (1987) Relationship between gene amplification and chromosomal deviations in malignant human gliomas. Cancer Genet Cytogenet 29:165–170

Bird AP (1978) A study of early events in ribosomal gene amplification. Cold Spring Harbor Symp Quant Biol 42:1179–1183

Bonas U. Sommer H. Saedler H (1984) The 17-kb *Tam* 1 element of *Antirrhinum majus* induces a 3-bp duplication upon integration into the chalcone synthase gene. EMBO J 3:1015–1019

Bonierbale MW. Plaisted RL. Tanksley SD (1988) RFLP maps based on a common set of clones reveal modes of chromosomal evolution in potato and tomato. Genetics 120:1095–1103

Bopp M. Capesius I (1971) Markierungsmuster nach ^{3}H-Thymidineinbau in Kernen der Hypokotylzellen von *Sinapis alba*. Chromosoma 33:386–395

Boronat A. Martinez MC. Reina M. Puigdomenech P. Palau J (1986) Isolation and sequencing of a 28 kD glutelin-2 gene from maize. Common elements in the 5′flanking regions among zein and glutelin. Plant Sci 47:95–102

Bostock CJ. Clark EM (1983) Gene amplification in methotrexate-resistant mouse cells. V. Intact amplified units can be transferred to and amplified in methotrexate-sensitive mouse L. cells. Chromosoma 88:31–41

Brady T (1973) Feulgen cytophotometric determination of the DNA content of the embryo proper and suspensor cells of *Phaseolus coccineus*. Cell Differ 2:65–75

Brady T. Walthall E (1985) The effect of the suspensor and gibberellic acid on *Phaseolus vulgaris* embryo protein content. Dev Biol 107:531–536

Britten RJ. Davidson EH (1969) Gene regulation for higher cells: a theory. Science 165:349–357

Britten RJ. Davidson EH (1971) Repetitive and non-repetitive DNA sequences and a speculation on the origins of evolutionary novelty. Q Rev Biol 46:111–138

Brodeur GM. Seeger RC. Schwab M. Varmus HE. Bishop JM (1984) Amplification of N-*myc* in untreated human neuroblastomas correlates with advanced disease stage. Science 224:1121–1124

Brodsky V. Ya. Uryvaeva IV (1985) Genome multiplication in growth and development. Univ Press. Cambridge

Broekaert D. van Oostveldt P. van Parijs R (1979) Differential DNA replication in *Pisum sativum* L. seedlings at the onset of germination. Biochem Physiol Pflanzen 174:629–640

Brown AL (1984) On the origin of the *Alu* family of repeated sequences. Nature (London) 312:106

Brown DD. Dawid I (1968) Specific gene amplification in oocytes. Science 160:272–280

Brown PC. Tlsty TD. Schimke RT (1983) Enhancement of methotrexate resistance and dihydrofolate reductase gene amplification by treatment of mouse 3T6 cells with hydroxyurea. Mol Cell Biol 3:1097–1107

Brunold C. Krüger-Lebus S. Saul MW. Wegmüller S. Potrykus I (1987) Combination of kanamycin resistance and nitrate reductase deficiency as selectable markers in one nuclear genome provides a universal somatic hybridizer in plants. Mol Gen Genet 208:469–473

Bryant JA. Wildon DC. Wong D (1974) Metabolic labile DNA in aseptically grown seedlings of *Pisum sativum* L. Planta 118:17–24

Buiatti M (1977) DNA amplification and tissue cultures. In: Reinert J. Bajaj YPS (eds) Applied and fundamental aspects of plant cell. tissue. and organ culture. Springer. Berlin Heidelberg New York. pp 358–374

Bullock P. Botchan M (1982) Molecular events in the excision of SV40 DNA from the chromosomes of cultured mammalian cells. In: Schimke RT (ed) Gene amplification. Cold Spring Harbor Lab. New York. pp 215–224

Burr B. Evola SV. Burr FA. Beckmann JS (1983) The application of restriction fragment length polymorphism to plant genetic engineering. In: Setlow JK. Hollaender A (eds) Genetic engineering. Vol 5. Plenum. New York. pp 45–59

Burr B. Burr FA. Thompson KH. Albertson MC. Stuber CW (1988) Gene mapping with recombinant inbreds in maize. Genetics 118:519–526

Capesius I, Stöhr M (1974) Endopolyploidisierung während des Streckungswachstumes der Hypokotyle von *Sinapis alba*. Protoplasma 82:147–153
Carroll SM, Gaudray P, De Rose ML, Emery JF, Meinkoth JL, Nakkim E, Subler M, von Hoff DD, Wahl GM (1987) Characterization of an episome produced in hamster cells that amplify a transfected CAD gene at high frequency: functional evidence for a mammalian replication origin. Mol Cell Biol 7:1740–1750
Cavallini A, Zolfino C, Cionini G, Cremonini R, Natali L, Sassoli O, Cionini PG (1986) Nuclear DNA changes within *Helianthus annuus* L.: cytophotometric, karyological and biochemical analyses. Theor Appl Genet 73:20–26
Cella R, Albani D, Biasini MG, Carbonera D, Parisi B (1984) Isolation and characterization of a carrot cell line resistant to methotrexate. J Exp Bot 35:1390–1397
Cella R, Albani D, Carbonera D, Etteri L, Maestri E, Parisi B (1987) Selection of methotrexate-resistant cell lines in *Daucus carota*: Biochemical analysis and genetic characterization by protoplast fusion. J Plant Physiol 127:135–146
Chaleff RS (1981) Genetics of higher plants – applications of cell culture. Univ Press, Cambridge
Chang C, Bowman JL, DeJohn AW, Lander ES, Meyerowitz EM (1988) Restriction fragment length polymorphism linkage map for *Arabidopsis thaliana*. Proc Natl Acad Sci USA 85:6856–6860
Chen D, Osborne DJ (1970) Ribosomal genes and DNA replication in germinating wheat embryos. Nature (London) 225:336–340
Chen C-j, Chin JE, Ueda K, Clark DP, Pastan I, Gottesman MM, Roninson IB (1986) Internal duplication and homology with bacterial transport proteins in the *mdr*1 (P-glycoprotein) gene from multidrug-resistant human cells. Cell 47:381–389
Cherif D, Le Coniat M, Suarez HG, Bernheim A, Berger R (1988) Chromosomal localization of amplified *c-myc* in a human colon adenocarcinoma cell line with a biotinylated probe. Cancer Genet Cytogenet 33:245–249
Chin DJ, Luskey KL, Anderson RGW, Faust JR, Boldstein JK, Brown MC (1982) Appearance of crystalloid endoplasmic reticulum in compactin-resistant Chinese hamster cells with a 500-fold increase in 3-hydroxy-3-methylglutaryl-coenzyme A reductase. Proc Natl Acad Sci USA 79:1185–1189
Chojecki AJS, Bayliss MW, Gale MD (1986a) Cell production and DNA accumulation in the wheat endosperm, and their association with grain weight. Ann Bot 58:809–817
Chojecki AJS, Gale MD, Bayliss MW (1986b) The number and sizes of starch granules in the wheat endosperm, and their association with grain weight. Ann Bot (London) 58:819–831
Coderre A, Beverley M, Schimke T, Santi DV (1983) Overproduction of a bifunctional thymidylate synthetase-dihydrofolate reductase and DNA amplification in methotrexate-resistant *Leishmania tropica*. Proc Natl Acad Sci USA 80:2132–2136
Collins S, Groudine M (1982) Amplification of endogenous *myc*-related DNA sequences in a human myeloid leukaemia cell line. Nature (London) 298:679–681
Cowell JK (1982) Double minutes and homogeneously staining regions: Gene amplification in mammalian cells. Annu Rev Genet 16:21–59
Cremonini R, Cionini PG (1977) Extra DNA synthesis in embryo suspensor cells of *Phaseolus coccineus*. Protoplasma 99:303–313
Cullis CA (1977) Molecular aspects of the environmental induction of heritable changes in flax. Heredity 38:129–154
Cullis CA (1983) Environmentally induced DNA changes in plants. CRC Crit Rev Plant Sci 1:117–131
Cullis CA (1986) Phenotypic consequences of environmentally induced changes in plant DNA. Trends Genet 2:307–309
Cullis CA, Cleary W (1986) DNA variation in flax tissue culture. Can J Genet Cytol 28:247–251
Cullis CA, Davies DR (1975) Ribosomal DNA amounts in *Pisum sativum*. Genetics 81:485–492
D'Amato F (1975) The problem of genetic stability in plant tissue and cell cultures. In: Frankel O, Hawkes JG (eds) Crop resources for today and tomorrow. Univ Press, Cambridge, pp 333–348
D'Amato F (1977) Nuclear cytology in relation to development. Univ Press, Cambridge
D'Amato F (1985) Cytogenetics of plant cell and tissue cultures and their regenerates. CRC Crit Rev Plant Sci 3:73–112
Davidson AD, Manners JM, Simpson RS, Scott KJ (1987) cDNA cloning of mRNAs induced in resistant barley during infection by *Erysiphe graminis* f.sp. *Hordei*. Plant Mol Biol 8:77–85
Dawkins R (1976) The selfish gene. Univ Press, Oxford

Deak M. Donn G. Feher A. Dudits D (1988) Dominant expression of a gene amplification-related herbicide resistance in *Medicago* cell hybrids. Plant Cell Rep 7:158–161
Delidakis C. Kafatos FC (1989) Amplification enhancers and replication origins in the autosomal chorion gene cluster of *Drosophila*. EMBO J 8:891–901
Der CJ (1987) Cellular oncogenes and human cancer. Clin Chem 33:641–646
Detke S. Katakura K. Chang K-P (1989) DNA amplification in arsenite-resistant *Leishmania*. Exptl Cell Res 180:161–170
Deumling B. Nagl W (1978) DNA characterization. satellite DNA localization. and nuclear organization in *Tropaeolum majus*. Cytobiologie 16:412–420
Dhillon SS. Wernsman EA. Miksche JP (1983) Evaluation of nuclear DNA content and heterochromatin changes in anther-derived dihaploids of tobacco (*Nicotiana tabacum*) cv. Coker 1139. Can J Genet Cytol 25:169–1
Donn G. Tischer E. Smith JA. Goodman M (1984) Herbicide-resistant alfalfa cells: an example of gene amplification in plants. J Mol Appl Genet 2:621–635
Doolittle WF. Sapienza C (1980) Selfish genes. the phenotype paradigma and genome evolution. Nature (London) 284:601–603
Dührssen E. Schäfer A. Neumann K-H (1979) Qualitative differences in the DNA of some higher plants. and aspects of selective DNA replication during differentiation. Plant Syst Evol Suppl 2:95–103
Dührssen E. Lanzendorfer M. Neumann K-H (1984) Comparative investigations on DNA organization of some varieties of *Daucus carota* L. Z Pflanzenphysiol 113:223–229
Durante M. Giorgi L. Parenti R (1974) Preparative electrophoresis of plant tissue DNA. Anal Biochem 60:626–630
Durante M. Geri C. Buiatti M. Ciomei M. Cecchini E. Martini G. Parenti R. Giorgi L (1986) A comparison of genome modifications leading to genetic and epigenetic transformation in *Nicotiana* spp. tissue cultures. Dev Genet 7:51–64
Durrant A. Jones TWA (1971) Reversion of induced changes in amount of nuclear DNA in *Linum*. Heredity 27:431–439
Eichholtz A. Rogers G. Horsch RB. Klee HJ. Harford M. Hoffmann NL. Braford SB. Fink C. Flick J. O'Connell KM. Fraley RT (1987) Expression of mouse dihydrofolate reductase gene confers methotrexate resistance in transgenic *Petunia* plants. Somat Cell Mol Genet 13:67–76
Emanuel JR. Garetz S. Schneider J. Ash JF. Benz EJ Jr. Levenson R (1986) Amplification of DNA sequences coding for the Na. K-ATPase α-subunit in ouabain-resistant C^+ cells. Mol Cell Biol 6:2476–2481
Essad S. Vallade J. Cornu A (1975) Variations interphasiques de la teneur en DNA et du volume nucléaires du zygote de *Petunia hybrida* – conséquences métaboliques. Caryologia 28:207–224
Evans DA (1989) Somaclonal variation – genetic basis and breeding applications. Trends Genet 5:46–50
Evans DA. Sharp WR (1986) Applications of somaclonal variation. Bio/Technol 4:528–532
Falk R (1984) The gene in search of an identity. Human Genet 68:195–204
Faulde M. Nagl W (1987) Cultivar. organ. and herbicide treatment-specific variation of histone H1 in maize (*Zea mays* L.). J Plant Physiol 129:415–424
Fearon ER. Vogelstein B. Feinberg AP (1984) Somatic deletion and duplication of genes on chromosome 11 in Wilm's tumours. Nature (London) 309:176–178
Field LM. Devonshire AL. Forde BG (1988) Molecular evidence that insecticide resistance in peach-potato aphids (*Myzus persicae* Sulz.) results from amplification of an esterase gene. Biochem J 251:309–312
Figdore SS. Kennard WC. Song KM. Slocum MK. Osborn TC (1988) Assessment of the degree of restriction fragment length polymorphism in *Brassica*. Theor Appl Genet 75:833–840
Flavell RB (1981) Molecular changes in chromosomal DNA organisation and origins of phenotypic variation. Chrom Today 7:42–54
Flavell RB. Rimpau J. Smith DB (1977) Repeated sequence DNA relationships in four cereal genomes. Chromosoma 63:205–222
Flick CE. Evans DA (1983) Genetic variation derived from plant cell cultures. In: Randall DD (ed) Current topics in plant biochemistry and physiology. vol 2. Univ Missouri Press. Columbia. Ohio. pp 200–210
Forino LMC. Cremonini R. Tagliasacchi AM (1976) Frequenza di cellule con "DNA puffs" nel suspensore di *Phaseolus coccineus* in momenti diversi dello sviluppo embrionale. G Bot It 110:337–346
Forman D. Rowley J (1982) Chromosomes and cancer. Nature (London) 300:403–404

Fougere-Deschatrette C, Schimke RT, Weil D, Weiss MC (1984) A study of chromosomal changes associated with amplified dihydrofolate reductase genes in rat hepatoma cells and their dedifferentiated variants. J Cell Biol 99:497–502

Fox M (1984) Gene amplification and drug resistance. Nature (London) 307:212–213

Frediani M, Forino LMC, Tagliasacchi AM, Cionini PG, Durante M, Avanzi S (1986) Functional heterogeneity, during early embryogenesis, of *Phaseolus coccineus* ribosomal cistrons in polytene chromosomes of embryo suspensor. Protoplasma 132:51–57

Frisch B, Nagl W (1979) Patterns of endopolyploidy and 2C nuclear DNA content (Feulgen) in *Scilla* (Liliaceae). Plant Syst Evol 131:261–276

Frölich E, Nagl W (1979) Transitory increase in chromosomal DNA (Feulgen) during floral differentiation in *Rhoeo discolor*. Cell Differ 8:11–18

Furze JM, Hamill JD, Parr AJ, Robins RJ, Rhodes MJC (1987) Variations in morphology and nicotine alkaloid accumulation in protoplast-derived hairy root cultures of *Nicotiana rustica*. J Plant Physiol 131:237–246

Galau GA, Chamberlin ME, Hough BR, Britten RJ, Davidson EH (1976) Evolution of repetitive and non-repetitive DNA. In: Ayala JF (ed) Molecular evolution. Sinauer, Sunderland, MA, pp 200–224

Galau GA, Bass HW, Hughes DW (1988) Restriction fragment length polymorphisms in diploid and allotetraploid *Gossypium*: assigning the late embryogenesis-abundant (*Lea*) alloalleles in *G. hirsutum*. Mol Gen Genet 211:305–314

Gall JG (1968) Differential synthesis of the genes for ribosomal RNA during amphibian oogenesis. Proc Natl Acad Sci USA 60:553–560

Galli MG, Balzaretti R, Sgorbati S (1986) Autoradiographic and cytofluorometric analysis of DNA synthesis in endosperm and cotyledons of germinating castor bean. J Exp Bot 37:1716–1724

Gaponenko AK, Petrova TF, Isakov AR, Sozinov AA (1988) Cytogenetics of in vitro cultured somatic cells and regenerated plants of barley (*Hordeum vulgare* L.). Theor Appl Genet 75:905–911

Garvey EP, Santi DV (1986) Stable amplified DNA in drug-resistant *Leishmania* exists as extrachromosomal circles. Science 233:535–540

Gavazzi G, Tonelli C, Todesco G, Arreghini E, Raffaldi F, Vecchio F, Barbuzzi G, Biasini AG, Sala F (1987) Somaclonal variation versus chemically induced mutagenesis in tomato (*Lycopersicon esculentum* L.). Theor Appl Genet 74:733–738

Gebhart E, Augustus M, Brüderlein S (1987) Homogeneously staining regions (HSR) on chromosomes of human solid carcinomas. Biol Zentralbl 106:191–200

Geitler L (1939) Die Entstehung der polyploiden Somakerne der Heteropteren durch Chromosomenteilung ohne Kernteilung. Chromosoma 1:1–22

German J (ed) (1983) Chromosome mutation and neoplasia. Liss, New York

Giulotto E, Knights E, Stark GR (1987) Hamster cells with increased rates of DNA amplification, a new phenotype. Cell 48:837–845

Glover DM, Zaha A, Stocker AJ, Santelli RV, Pueyo MT, De Toledo SM, Lara FJS (1982) Gene amplification in *Rhynchosciara* salivary gland chromosomes. Proc Natl Acad Sci USA 79:2947–2951

Graham J (1984) Cancer and evolution: amplification. J Theor Biol 107:341–343

Grimsley N, Hohn T, Davies JW, Hohn B (1987) *Agrobacterium*-mediated delivery of infectious maize streak virus into maize plants. Nature (London) 325:177–179

Grisvard J, Tuffet-Anghileri A (1980) Variations in the satellite DNA content of *Cucumis melo* in relation to differentiation. Nucl Acids Res 8:2843–2858

Grodzicker T, Williams J, Sharp P, Sambrook J (1974) Physical mapping of temperature-sensitive mutations of adenoviruses. Cold Spring Harb Symp Quant Biol 39:439–446

Gros P, Croop J, Housman D (1986) Mammalian multidrug resistance gene: complete cDNA sequence indicates strong homology to bacterial transport proteins. Cell 47:371–380

Guillé E, Grisvard J (1971) Modifications of genetic information in crown-gall tissue cultures. Biochem Biophys Res Commun 44:1402–1409

Guillé E, Quetier F (1970) Le crown-gall: modèle expérimental pour l'application du mécanisme de régulation quantitative de l'information génétique à l'évènement néoplastique. Bull Cancer 57:217–238

Guttes E (1974) Continuous nucleolar DNA synthesis in late-interphase nuclei of *Physarum polycephalum* after transplantation into post-mitotic plasmodia. J Cell Sci 15:131–143

Haeckel E (1866) Generelle Morphologie der Organismen. Allgemeine Grundzüge der organischen Formwissenschaft, mechanisch begründet durch die von Ch. Darwin reformierte Deszendenztheorie. Reimer, Berlin

Hake S, Walbot V (1980) The genome of *Zea mays*, its organization and homology to related grasses. Chromosoma 79:251–270

Haluska FG, Tsujimoto Y, Croce CM (1987) Mechanisms of chromosome translocation in B- and T-cell neoplasia. Trends Genet 3:11–15

Hamlin JL, Milbrandt JD, Heintz NH, Azizkhan JC (1984) DNA sequence amplification in mammalian cells. Int Rev Cytol 90:31–82

Hammond MP, Laird CD (1985) Chromosome structure and DNA replication in nurse and follicle cells of *Drosophila melanogaster*. Chromosoma 91:267–286

Hase Y, Yakura K, Tanifuji S (1979) Differential replication of satellite and main band DNA during early stages of callus formation in carrot root tissue. Plant Cell Physiol 20:1461–1469

Hastie N (1987) More to transposable elements than meets the eye. Trends Genet 3:85–86

Hayes RJ, Petty ITD, Coutts RHA, Buck KW (1988) Gene amplification and expression in plants by a replicating geminivirus vector. Nature 334:179–182

Heath-Pagliuso S, Pullman J, Rappaport L (1988) Somaclonal variation in celery: screening for resistance to *Fusarium oxysporum* f.sp. *apii*. Theor Appl Genet 75:446–451

Heidecker G, Messing J (1986) Structural analysis of plant genes. Annu Rev Plant Physiol 37:439–466

Heim S, Mitelman F (1987) Nineteen of 26 cellular oncogenes precisely localized in the human genome map to one of the 83 bands involved in primary cancer-specific rearrangements. Human Genet 75:70–72

Helentjaris T (1987) A genetic linkage map for maize based on RFLPs. Trends Genet 3:217–221

Hinegardner R (1976) Evolution of genome size. In: Ayala JF (ed) Molecular evolution. Sinauer, Sunderland, MA, pp 179–199

Holden JJA, Hough MR, Reimer DL, White BN (1987) Evidence for unequal crossing-over as the mechanism for amplification of some homogeneously staining regions. Cancer Genet Cytogenet 29:139–149

Hopkins HA, Flora JB, Schmidt RR (1972) Periodic DNA accumulation during the cell cycle of a thermophilic strain of *Chlorella pyrenoidosa*. Arch Biochem Biophys 153:845–849

Howell SH (1972) The differential synthesis and degradation of ribosomal DNA during the vegetative cell cycle in *Chlamydomonas reinhardii*. Nature New Biol 240:264–267

Hoy CA, Rice GC, Kovacs M, Schimke RT (1987) Over-replication of DNA in S phase Chinese hamster ovary cells after DNA synthesis inhibition. J Biol Chem 262:11927–11934

Hunt SW III, Hoffee PA (1983) Amplification of adenosine deaminase gene sequences in deoxycoformycin-resistant rat hepatoma cells. J Biol Chem 258:13185–13192

Hurst PR, Graham PB (1975) Turnover of DNA in ageing tissues of *Lycopersicon esculentum*. Ann Bot (London) 39:71–76

Hyrien O, Debatisse M, Buttin G, Robert De Saint Vincent B (1988) The multicopy appearance of a large inverted duplication and the sequence at the inversion joint suggest a new model for gene amplification. EMBO J 7:407–417

Ibson JM, Waters JJ, Twentyman PR, Bleehen NM, Rabbitts PH (1987) Oncogene amplification and chromosomal abnormalities in small cell lung cancer. J Cell Biochem 33:267–288

Ingle J, Timmis JN (1975) A role for differential replication of DNA in development. In: Markham R, Davies DR, Hopwood DA (eds) Modification of the information content of plant cells. Elsevier/North-Holland Biomedical Press, Amsterdam, pp 37–52

Innocenti AM (1975) Cyclic changes of histone/DNA ratio in differentiating nuclei of metaxylem cell line in *Allium cepa* root tip. Caryologia 28:225–228

Innocenti AM, Avanzi S (1971) Some cytological aspects of the differentiation of metaxylem in the root of *Allium cepa*. Caryologia 24:283–292

Jakobsson A-H, Arnason U, Levan A, Martinsson T, Hanson C, Levan G (1987) Novel cytogenetic expression of gene amplification in actinomycin D-resistant somatic cell hybrids: transfer of resistance by centric chromatin bodies. Chromosoma 95:408–418

Jalouzot R, Lechenault H, Gontachareff M (1975) Autoradiographic and cytophotometric study of the earliest stage of cellular activation during initiation of adventitious roots. CR Acad Sci Paris Ser D 280:1733–1736

Jeffreys AJ (1986) Highly variable minisatellites and DNA fingerprints. Biochem Soc Transact 15:309–317

Johnson BE, Makuch RW, Simmons AD, Gazdar AF, Burch D, Cashell AW (1988) Myc family DNA amplification in small cell lung cancer patients' tumors and corresponding cell lines. Cancer Res 48:5163–5166

Kaback DB, Halvorson HO (1977) Magnification of genes for rRNA in *Saccharomyces cerevisiae*. Proc Natl Acad Sci USA 74:1177–1180

Kafatos FC, Orr W, Delidakis C (1985) Developmentally regulated gene amplification. Trends Genet 1:301–306

Kan YW, Dozy AM (1978) Antenatal diagnosis of sickle-cell anaemia by DNA analysis of amniotic fluid cells. Lancet 2:910–912

Kanalas JJ, Suttle DP (1984) Amplification of the UMP synthase gene and enzyme overproduction in pyrazofurin-resistant rat hepatoma cells. Molecular cloning of a cDNA for UMP synthase. J Biol Chem 259:1848–1853

Karin M, Najarian R, Haslinger A, Valenzuela P, Welch J, Fogel S (1984) Primary structure and transcription of an amplified genetic locus: the CUP 1 locus of yeast. Proc Natl Acad Sci USA 81:337–341

Kausch AP, Horner HT (1984) Increased nuclear DNA content in raphide crystal idioblasts during development in *Vanilla planifolia* L. (Orchidaceae). Eur J Cell Biol 306:385–387

Keim P, Lark KG (1987) The sequence of the 3.3-kilobase repetitive element from *Dipodomys ordii* suggests a mechanism for its amplification and interspersion. J Mol Evol 25:65–73

Kikuchi S, Takaiwa F, Oono F (1987) Variable copy number DNA sequences in rice. Mol Gen Genet 210:373–380

Kinzler KW, Bigner SH, Bigner DD, Trent JM, Law ML, O'Brien SJ, Wong AJ, Vogelstein B (1987) Identification of an amplified, highly expressed gene in a human glioma. Science 236:70–73

Kominami R, Urano Y, Mishima Y, Muramatsu M, Moriwaki K, Yoshikura M (1983) Novel repetitive sequence families showing size and frequency polymorphism in the genome of mice. J Mol Biol 165:209–228

Kononowicz AK, Olszewska MJ, Waldoch E (1983) Changes in heterochromatin content during differentiation of some root tissues in *Vicia faba* L., subsp. *major* and subsp. *minor*. Biol Zentralbl 102:675–685

Kowles RV, Phillips RL (1985) DNA amplification patterns in maize endosperm nuclei during kernel development. Proc Natl Acad Sci USA 82:7010–7014

Krimer DB, Van't Hof J (1983) Extrachromosomal DNA of pea (*Pisum sativum*) root-tip cells replicates by strand displacement. Proc Natl Acad Sci USA 80:1933–1937

Kuhn EM, Therman E, Denniston C (1985) Mitotic chiasmata, gene density, and oncogenes. Human Genet 70:1–5

Kunisada T, Yamagishi H, Kinoshita I, Tsuji H (1986) Amplification of extrachromosomal circular DNA in intact bean leaves treated with benzyladenine. Plant Cell Physiol 27:355–361

Kuriyan PN, Narayan RKJ (1988) The distribution and divergence during evolution of families of repetitive DNA sequences in *Lathyrus* species. J Mol Evol 27:303–310

Kushnir U, Halloran GM (1983) Evidence on the saltatory origin of the G genome in wheat: the description of a *Triticum timopheevi*-like mutant. Ann Bot (London) 51:561–569

Landré P (1975) DNA content of the different types of cells of the epidermal layer. Abstr 12th Int Botanical Congr, vol 1, Leningrad, p 221

Landré P (1976) Teneurs en DNA nucléaire de quelques types cellulaires de l'épiderme de la morelle noire (*Solanum nigrum* L.) au cours du développement de la feuille. Ann Sci Nat Bot Biol Veg Ser 12 17:5–104

Landry BS, Michelmore RW (1987) Methods and applications of restriction fragment length polymorphism analysis to plants. In: Bruening G, Harada J, Kosuga T, Hollaender A (eds) Tailoring genes for crop improvement. Plenum, New York, pp 25–44

Landry BS, Kesseli RV, Farrara B, Michelmore RW (1987) A genetic map of lettuce (*Lactuca sativa* L.) with restriction fragment length polymorphism, isozyme, disease resistance and morphological markers. Genetics 116:331–337

Landsmann J, Dennis ES, Higgins TJV, Appleby CA, Kortt AA, Peacock WJ (1986) Common evolutionary origin of legume and non-legume plant haemoglobins. Nature (London) 324:166–168

Lapitan NLV. Sears RG. Gill. BS (1988) Amplification of repeated DNA sequences in wheat x rye hybrids regenerated from tissue culture. Theor Appl Genet 75:381–388

Larkin PJ. Scowcroft WR (1981) Somaclonal variation: a novel source of genetic variability from cell cultures for plant improvement. Theor Appl Genet 60:197–214

Lazar MD. Chen THH. Gusta LV. Kartha KK (1988) Somaclonal variation for freezing tolerance in a population derived from Norstar winter wheat. Theor Appl Genet 75:480–484

Lee M. Phillips RL (1987) Genomic rearrangements in maize induced by tissue culture. Genome 29:122–128

Lee M. Phillips RL (1988) The chromosomal basis of somaclonal variation. Ann Rev Plant Physiol Plant Mol Biol 39:413–437

Lee W-H. Murphree AL. Benedict WF (1984) Expression and amplification of the N-*myc* gene in primary retinoblastoma. Nature (London) 309:458–460

Le Gal MF. Lecocq FM. Hallet JN (1986) The reserve proteins in the cells of mature cotyledons of *Lupinus albus* var. Lucky. II. Relationship with the nuclear DNA content. Protoplasma 130: 128–137

Leinonen P. Alhonen-Hongisto L. Laine R. Jänne OA. Jänne J (1987) Human myeloma cells acquire resistance to difluoromethylornithine by amplification of ornithine decarboxylase gene. Biochem J 242:199–203

Leu T-H. Hamlin JL (1989) High-resolution mapping of replication fork movement through the amplified dihydrofolate reductase domain in CHO cells by in-gel renaturation analysis. Mol Cell Biol 9:523–531

Lewis WH. Srinivasan PR (1983) Chromosome-mediated gene transfer of hydroxyurea resistance and amplification of ribonucleotide reductase activity Mol Cell Biol 3:1053–1061

Lewis AD. Hickson ID. Robson CN. Harris AL. Hayes JD. Griffiths SA. Manson MM. Hall AE. Moss JE. Wolf CR (1988) Amplification and increased expression of alpha class glutathione S-transferase-encoding genes associated with resistance to nitrogen mustards. Proc Natl Acad Sci USA 85:8511–8515

Lightfoot D (1972) Quantitation of ribosomal genes in developing Chinese cabbage and tobacco plants. PhD Thesis. Univ Arizona

Lima-de-Faria A (1974) The molecular organization of the chromomeres of *Acheta* involved in ribosomal DNA amplification. Cold Spring Harbor Symp Quant Biol 38:559–571

Lima-de-Faria A (1983) Molecular evolution and organization of the eukaryotic chromosome. Elsevier/North-Holland Biomedical Press. Amsterdam

Lima-de-Faria A. Mitelman F (1986) The chromosome territory of human oncogenes. Biosci Rep 6:349–354

Lima-de-Faria A. Pero R. Avanzi S. Durante M. Ståhle U. D'Amato F. Granström H (1975) Relation between ribosomal RNA genes and the DNA satellites of *Phaseolus coccineus*. Hereditas 79:5–20

Lin CR. Cheu WS. Kruiger W. Stolarsky LS. Weber W. Evans RM. Verma JM. Gill GN. Rosenfeld MG (1984) Expression cloning of human EGF receptor complementary mNA: gene amplification and three related mRNA products in A431 cells. Science 224:843–848

Lin L-S. Ho T-hD. Harlan JR (1985) Rapid amplification and fixation of new restriction sites in the ribosomal DNA repeats in the derivatives of a cross between maize and *Tripsacum dactyloides*. Dev Genet 6:101–112

Little CD. Nau MM. Carney DN. Gazdar AF. Minna JP (1983) Amplification and expression of the c-*myc* oncogene in human lung cancer cell lines. Nature (London) 306:194–196

Luskey KL. Faust JR. Chin DJ. Brown MS. Goldstein JL (1983) Amplification of the gene for 3-hydroxy-3-methyl-glutarylcoenzyme A reductase. but not for the 53-kDA protein. in UT-1 cells. J Biol Chem 258:8462–8469

Ma C. Looney JE. Leu T-H. Hamlin JL (1988) Organization and genesis of dihydrofolate reductase amplicons in the genome of a methotrexate-resistant Chinese hamster ovary cell line. Mol Cell Biol 8:2316–2327

Macgregor HC. Mizuno S. Vlad M (1976) Chromosomes and DNA sequences in salamanders. Chrom Today 5:331–339

Macleod RD (1973) ^{3}H-Thymidine labelled DNA in the primary root of *Vicia faba* L. Z Pflanzenphysiol 68:379–381

Maherchandani NJ. Naylor JM (1971) Variability in DNA content and nuclear morphology of the aleuron cells of *Avena fatua* (wild oats). Can J Genet Cytol 13:578–584

Martini G, Nuti Ronchi V (1974) Microdensitometric and autoradiographic analysis of cell proliferation in primary culture of *Nicotiana glauca* pith tissue. Cell Differ 3:239–247
Martinsson T, Levan G (1987) Localization of multidrug resistance-associated 170 kDa P-glycoprotein gene to mouse chromosome 5 and to homogeneously staining regions in multidrug-resistant mouse cells by in situ hybridization. Cytogenet Cell Genet 45:99–101
Marx JL (1984) Oncogenes amplified in cancer cells. Science 223:40–41
McBeth DL, Shapiro JA (1984) Reversal by DNA amplification of an unusual mutation blocking alkane and alcohol utilization in *Pseudomonas putida*. Mol Gen Genet 197:384–391
McClintock B (1951) Chromosome organization and gene expression. Cold Spring Harbor Symp Quant Biol 16:13–47
McClintock B (1984) The significance of responses of the genome to challenge. Science 226:792–801
McIntyre CL, Clarke BC, Appels R (1988) Amplification and dispersion of repeated DNA sequences in the Triticeae. Plant Syst Evol 160:39–59
McLean M, Baird WV, Gerats AGM, Meagher RB (1988) Determination of copy number and linkage relationships among five actin gene subfamilies in *Petunia hybrida*. Plant Molec Biol 11:663–672
Meinkoth J, Killary AM, Fournier REK, Wahl GM (1987) Unstable and stable CAD gene amplification: importance of flanking sequences and nuclear environment in gene amplification. Mol Cell Biol 7:1415–1424
Mielenz JR, Hershberger CL (1974) Are segments of chloroplast DNA differentially amplified? Biochem Biophys Res Commun 58:769–777
Miksche JP, Hotta Y (1973) DNA base composition and repetitious DNA in several conifers. Chromosoma 41:29–36
Misawa S, Staal SP, Testa JR (1987) Amplification of the c-*myc* oncogene is associated with an abnormally banded region on chromosome 8 or double minute chromosomes in two HL-60 human leukemia sublines. Cancer Genet Cytogenet 28:127–135
Mitani K, Kurosawa H, Suzuki A, Hayashi Y, Hanada R, Yamamoto K, Komatsu A, Kobayashi N, Nakagome Y, Yamada M (1986) Amplification of N-*myc* in a Rhabdomyosarcoma. Jpn J Cancer Res (Gann) 77:1062–1065
Mitelman F (1983) Catalogue of chromosome aberrations in cancer. Cytogenet Cell Genet 36:1–515
Mizrahi A (ed) (1988) Biotechnology in agriculture. Alan R Liss, New York
Morgan WF, Bodycote J, Fero ML, Hahn PJ, Kapp LN, Pantelias GE (1986) A cytogenetic investigation on DNA replication after HU treatment: implications for gene amplification. Chromosoma 93:191–196
Mouchès C, Pasteur N, Berge JB, Hyrien O, Raymond M, Robert de Saint Vincent B, de Silvestri M, Georghiou GP (1986) Amplification of an esterase gene is responsible for insecticide resistance in California *Culex* mosquito. Science 233:778–780
Müntz K (1987a) Developmental control of storage protein formation and its modulation by some internal and external factors during embryogenesis in plant seeds. Biochem Physiol Pflanzen 182:93–116
Müntz K (1987b) "Engineering" pflanzlicher Speicherproteine. BioEngineering 2:36–43
Murry LE, Christianson ML, Alfinito SH, Garger SJ (1987) Characterization of the nuclear DNA of *Hordeum vulgare* root hairs: amplification disappears under salt stress. Amer J Bot 74:1779–1786
Nafziger ED, Widholm JM, Steinrücken HC, Killmer JL (1984) Selection and characterization of a carrot cell line tolerant to glyphosate. Plant Physiol 76:571–574
Nagl W (1969) Banded polytene chromosomes in the legume *Phaseolus vulgaris*. Nature (London) 221:70–71
Nagl W (1972) Evidence of DNA amplification in the orchid *Cymbidium* in vitro. Cytobios 5:145–154
Nagl W (1974a) The *Phaseolus* suspensor and its polytene chromosomes. Z Pflanzenphysiol 73:1–44
Nagl W (1974b) DNA synthesis in tissue and cell cultures. In: Street HE (ed) Tissue culture and plant science 1974. Academic Press, New York London, pp 19–42
Nagl W (1976a) The polytenic antipodal cells in *Scilla bifolia*: DNA replication pattern and possibility of nucleolar DNA amplification. Cytobiologie 14:165–170
Nagl W (1976b) Nuclear organization. Annu Rev Plant Physiol 27:39–69
Nagl W (1976c) Early embryogenesis in *Tropaeolum majus* L.: evolution of DNA content and polyteny in the suspensor. Plant Sci Lett 7:1–8
Nagl W (1977) The evolution of chromosomal DNA redundancy: ontogenetic lateral versus phylogenetic tandem changes. Nucleus 20:10–27

Nagl W (1978) Endopolyploidy and polyteny in differentiation and evolution. Elsevier/North Holland Biomedical Press, Amsterdam

Nagl W (1979a) Differential DNA replication: a critical review. Z Pflanzenphysiol 95:283–314

Nagl W (1979b) Search for the molecular basis of diversification in phylogenesis and ontogenesis. Plant Syst Evol Suppl 2:3–25

Nagl W (1981) Polytene chromosomes of plants. Int Rev Cytol 73:21–53

Nagl W (1982a) Condensed chromatin: species-specificity, tissue-specificity, and cell cycle-specificity, as monitored by scanning cytometry. In: Nicolini C (ed) Cell growth. Plenum, New York, pp 171–218

Nagl W (1982b) Cell growth and nuclear DNA increase by endoreduplication and differential DNA replication. In: Nicolini C (ed) Cell growth. Plenum, New York, pp 619–665

Nagl W (1982c) DNA endoreduplication and differential DNA replication. In: Encyclopedia of plant physiology, vol 14b. Springer, Berlin Heidelberg New York, pp 111–124

Nagl W (1983a) Evolution: theoretical and physical considerations. Biol Zentralbl 102:257–269

Nagl W (1983b) Physical aspects of gene regulation. Kew Chrom Conf 2:55–61

Nagl W (1985a) Chromatin organization and the control of gene activity. Int Rev Cytol 94:21–56

Nagl W (1985b) Chromosomes in differentiation. In: Sharma AK, Sharma A (eds) Advances in chromosome and cell genetics. Oxford & IBH, New Delhi, pp 135–172

Nagl W (1986) Molecular phylogeny. In: Raup DM, Jablonski D (eds) Patterns and process in the history of life. Dahlem-Konferenzen, Springer, Berlin Heidelberg New York Tokyo, pp 223–232

Nagl W (1987a) Replication. In: Progress in botany, vol 49. Springer, Berlin Heidelberg New York Tokyo, pp 181–191

Nagl W (1987b) Gentechnologie und die Grenzen der Biologie. Wissenschaftliche Buchgesellschaft, Darmstadt

Nagl W (1988) Genome changes induced by auxin-herbicides in seedlings and calli of *Zea mays* L. Environm Exptl Bot 28:197–206

Nagl W (1989) Replication. Progr Bot 51: 173–180

Nagl W, Capesius I (1977) Repetitive DNA and heterochromatin as factors of karyotype evolution in phylogeny and ontogeny of orchids. Chrom Today 6:141–152

Nagl W, Pfeifer M (1988) Karyological stability and microevolution of a cell suspension culture of *Brachycome dichromosomatica.* Plant Syst Evol 158:133–139

Nagl W, Rücker W (1972) Beziehungen zwischen Morphogenese und nuklearem DNS-Gehalt bei aseptischen Kulturen von *Cymbidium* nach Wuchsstoffbehandlung. Z Pflanzenphysiol 67:120–134

Nagl W, Rücker W (1974) Shift of DNA replication from diploid to polyploid cells in cytokinin-controlled differentiation. Cytobios 10:137–144

Nagl W, Rücker W (1976) Effects of phytohormones on thermal denaturation profiles of *Cymbidium* DNA: indication of differential DNA replication. Nucleic Acids Res 3:2033–2039

Nagl W, Hendon J, Rücker W (1972) DNA amplification in *Cymbidium* protocorms, as it relates to cytodifferentiation and hormone treatment. Cell Differ 1:229–237

Nagl W, Peschke C, van Gyseghem R (1976) Heterochromatin under-replication in *Tropaeloum* embryogenesis. Naturwissenschaften 63:198–199

Nagl W, Jeanjour M, Kling H, Kühner S, Michels I, Müller T, Stein B (1983) Genome and chromatin organization in higher plants. Biol Zentralbl 102:129–148

Nagl W, Ribicki R, Mertler H-O, Hezel U, Jacobi R, Bachmann E (1984) The fluorophenylalanine sensitive and resistant tobacco cell lines, TX1 and TX4. 1. DNA contents, chromosome numbers, nuclear ultrastructures, and effects of spermidine. Protoplasma 122:138–144

Nardeux PC, Daya-Grosjean L, Landin RM, Andeol Y, Suarez HG (1987) A c-Ki-ras oncogene is activated, amplified, and overexpressed in a human osteosarcoma cell line. Biochem Biophys Res Comm 146:395–402

Natali L, Cavallini A (1987) Nuclear cytology of callus and plantlets regenerated from pea (*Pisum sativum* L.) meristems. Protoplasma 141:121–125

Natali L, Cavallini A, Cremonini R, Bassi P, Cionini PG (1986) Amplification of nuclear DNA sequences during induced plant cell dedifferentiation. Cell Differ 18:157–161

Nau MM, Carney DN, Batley J, Johnson B, Little C, Gazdar A, Minna JD (1984) Amplification, expression, and rearrangement of c-*myc* and N-*myc* oncogenes in human lung cancer. In: Potter M, Melchers F, Weigert M (eds) Oncogenes in B-cell neoplasia. Springer, Berlin Heidelberg New York, pp 172–177

Nicolini C (1979) Chromatin structure, from Ångstrom to micron levels, and its relationship to mammalian cell proliferation. In: Nicolini C (ed) Cell growth. Plenum, New York, pp 613–666

Nienhuis J, Helentjaris T, Slocum M, Ruggero B, Schaefer A (1987) Restriction fragment length polymorphism analysis of loci associated with insect resistance in tomato. Crop Sci 27:797–803

Nomura N, Yamamoto T, Toyoshima K, Ohami H, Akimaru K, Sasaki S, Nakagami Y, Kanauchi H, Shoji T, Hiraoka Y, Matsui M, Ishizaki R (1986) DNA amplification of the c-*myc* and c-*erb B*-1 genes in a human stomach cancer. Jpn J Cancer Res (Gann) 77:1188–1192

Norris MD, Brian MJ, Vowels MR, Stewart BW (1988) N-myc amplification in Wilms' tumor. Cancer Genet Cytogenet 30:187–189

Nowell P, Finan J, Dalla Favera R, Gallo RC, Ar-Rushdi A, Romanczuk H, Selden JR, Emanuel BS, Rovera G, Croce CM (1983) Association of amplified oncogene c-*myc* with an abnormally banded chromosome 8 in a human leukaemia cell line. Nature (London) 306:494–497

Nuti Ronchi V (1971) Amplificazione genica in cellule sdifferenziate di cotiledoni di *Lactuca sativa* coltivati in vitro. Atti Assoc Genet It 16:47–50

Nuti Ronchi V, Bennici A, Martini G (1973) Nuclear fragmentation in dedifferentiating cells of *Nicotiana glauca* pith tissue grown in vitro. Cell Differ 2:77–85

Nuti Ronchi V, Martini G, Parenti R, Geri C, Giorgi L, Grisvard J (1974) Early cytological and biochemical events in plant tissue dedifferentiation. 3rd Int Congr Plant tissue cell culture, Leicester, Abstr, p 134

Nuti Ronchi V, Bonatti S, Durante M, Turchi G (1986) Preferential localization of chemically induced breaks in heterochromatic regions of *Vicia faba* and *Allium cepa* chromosomes. II. 4-epoxyethyl-1,2-epoxy-cyclohexane interacts specifically with highly repetitive sequences of DNA in *Allium cepa*. Environ Exp Bot 26:127–135

O'Hare K (1987) Chromosome plasticity and transposable elements in *Drosophila*. Trends Genet 3:87–88

Ohno H, Fukuhara S, Arita Y, Doi S, Takahashi R, Fujii H, Honjo T, Sugiyama T, Uchino H (1988) Establishment of a peripheral T-cell lymphoma cell line showing amplification of the c-*myc* oncogene. Cancer Res 48:4959–4963

Oliver JL, Martinez-Zapater JM, Pascual L, Enriquez AM, Ruiz-Rejon C, Ruiz-Rejon M (1983) Different genome amplification of multilocus isoenzymes in hexaploid wheat. In: Rattazzi MC, Scandalios JG, Whitt GS (eds) Isoenzymes. Liss, New York, pp 341–363

Orgel LE, Crick FCH (1980) Selfish DNA: the ultimate parasite. Nature (London) 284:604–607

Orton TJ (1983) Experimental approaches to the study of somaclonal variation. Plant Mol Biol Rep 1:67–76

Osheim YN, Miller OL Jr (1983) Novel amplification and transcriptional activity of chorion genes in *Drosophila melanogaster* follicle cells. Cell 33:543–553

Osifo EO, Webb JK, Henshaw GG (1989) Variation amongst callus-derived potato plants, *Solanum brevidens*. J Plant Physiol 134:1–4

Ottoboni LMM, Steffensen DM (1987) Localization of zein genes in maize. Biochem Genet 25:123–142

Ozias-Akins P, Tabaeizadeh Z, Pring DR, Vasil IK (1988) Preferential amplification of mitochondrial DNA fragments in somatic hybrids of the Gramineae. Curr Genet 13:241–245

Palmer JF, Widholm J (1975) Characterization of carrot and tobacco cell cultures resistant to p-fluoro-phenylalanine. Plant Physiol 56:233–238

Pan W-C, Blackburn EH (1981) Single extrachromosomal rRNA gene copies are synthesized during amplification of the rDNA in *Tetrahymena*. Cell 23:459–466

Parag Y, Roper JA (1975) Genetic control of chromosome instability in *Aspergillus nidulans* as a mean for gene amplification in eukaryotic microorganisms. Mol Gen Genet 140:275–287

Parenti R, Guille E, Grisvard J, Durante M, Giorgi L, Buiatti M (1973) Transient DNA satellite in dedifferentiating pith tissue. Nature New Biol 246:237–239

Pearson GG, Timmis JN, Ingle J (1974) The differential replication of DNA during plant development. Chromosoma 45:281–294

Peerbolte R, Ruigrok P, Wullems G, Schilperoort R (1987) T-DNA rearrangements due to tissue culture: somaclonal variation in crown gall tissues. Plant Mol Biol 9:51–58

Pérez L, Aguilar R, Sánchez-de-Jiménez E (1987) Effect of an exogenous auxin on maize tissues. Alteration of protein synthesis and phosphorylation. Physiol Plant 69:517–522

Picciarelli P, Alpi A (1986) Gibberellins in suspensors of *Phaseolus coccineus* L. seeds. Plant Physiol 82:298–300

Pijnacker LP, Hermelink JHM, Ferwerda MA (1986) Variability of DNA content and karyotype in cell cultures of an interdihaploid *Solanum tuberosum*. Plant Cell Rep 5:43–46
Pintor-Toro JA, Langridge P, Feix G (1982) Isolation and characterization of maize genes coding for zein protein of the 21000 dalton size class. Nucleic Acids Res 10:3845–3860
Pohler W, Schumann G, Sulze M, Kummer I-M (1988) Cytological investigation of callus tissue and regenerated plants from triticale anther culture. Biol Zbl 107:643–652
Quesnel AA, Ellis BE (1987) Population responses in tobacco cell cultures during selection for resistance to parafluorophenylalanine. Plant Sci 49:223–229
Radinsky R, Kraemer PM, Proffitt MR, Culp LA (1988) Clonal diversity of the Kirsten-ras oncogene during tumor progression in athymic nude mice: mechanisms of amplification and rearrangement. Cancer Res 48:4941–4953
Rees CAB, Hogan NC, Walden DB, Atkinson BG (1986) Identification of mRNAs encoding low molecular mass heatshock proteins in maize (*Zea mays* L.) Can J Genet Cytol 28:1106–1114
Rimpau J, Smith DB, Flavell RB (1978) Sequence organization analysis of the wheat and rye genomes by interspecies DNA/DNA hybridization. J Mol Biol 123:327–359
Riou G, Barrois M, Tordjman I, Dutronguay V, Orth G (1984) Detection of papillomavirus genomes and evidence for amplification of the oncogenes c-*myc* and c-Ha-*ras* in invasive squamous cell carcinoma of the uterine cervix. CR Acad Sci Paris Ser III 299:575–580
Ritossa F, Malva C, Boncinelli E, Graziani F, Polito L (1971) The first steps of magnification of DNA complementary to rRNA in *Drosophila melanogaster*. Proc Natl Acad Sci USA 68:1580–1584
Roberts JM, Buck LB, Axel R (1983) A structure for amplified DNA. Cell 33:53–63
Roberts C, Brasch J, Tattersall MHN (1987) Ribosomal RNA gene amplification: a selective advantage in tissue culture. Cancer Genet Cytogenet 29:119–127
Rochaix J-D, Bird A, Bakken A (1974) Ribosomal RNA gene amplification by rolling circles. J Mol Biol 87:473–487
Rogstad SH, Patton JC, Schaal BA (1988) A human minisatellite probe reveals RFLPs among individuals of two angiosperms. Nucleic Acids Res 16:11378
Rossana E, Rao LG, Johnson LF (1982) Thymidylate synthetase overproduction in 5-fluorodeoxyuridine-resistant mouse fibroblasts. Mol Cell Biol 2:1118–1125
Roth EJ, Frazier BL, Apuya NR, Lark KG (1989) Genetic variation in an inbred plant: variation in tissue cultures of soybean (*Glycine max* [L.] Merrill). Genetics 121:359–368
Ryan J, Hardeman EC, Endo A, Simoni RD (1981) Isolation and characterization of cells resistant to ML236B (compactin) with increased levels of 3-hydroxy-3-methylglutaryl coenzyme A reductase. J Biol Chem 256:6762–6768
Sacristán MD, Dobrigkeit I (1973) Variable content of a (GC)rich satellite DNA in tumorous and normal cultures of *Crepis capillaris*. Z Naturforsch 28c:564–567
Sakai K, Tanooka H, Sasaki MS, Ejima Y, Kaneko A (1988) Increase in copy number of N-*myc* in retinoblastoma in comparison with chromosome abnormality. Cancer Genet Cytogenet 30:119–126
Saunders JW, Hosfield GL, Levi A (1987) Morphogenetic effects of 2,4-dichlorophenoxyacetic acid on pinto bean (*Phaseolus vulgaris* L.) leaf explants in vitro. Plant Cell Rep 6:46–49
Schäfer A, Neumann K-H (1978) The influence of gibberellic acid on reassociation kinetics of DNA of *Daucus carota* L. Planta 143:1–4
Schäfer A, Blaschke JR, Neumann K-H (1978) On DNA metabolism of carrot tissue cultures. Planta 139:97–101
Schäffner K-H, Nagl W (1979) Differential DNA replication involved in transition from juvenile to adult phase in *Hedera helix*. Plant Syst Evol Suppl 2:105–110
Scheer U, Franke WW, Trendelenburg HF, Spring H (1976) Classification of loops of lampbrush chromosomes according to the arrangement of transcriptional complexes. J Cell Sci 22:503–519
Schimke RT (ed) (1982) Gene amplification. Cold Spring Harbor Lab, New York
Schimke RT (1984) Gene amplification, drug resistance, and cancer. Cancer Res 44:1735–1742
Schimke RT (1988) Gene amplification in cultured cells. J Biol Chem 263:5989–5992
Schimke RT, Sherwood SW, Hill AB, Johnston RN (1986) Overreplication and recombination of DNA in higher eukaryotes: Potential consequences and biological implications. Proc Natl Acad Sci USA 83:2157–2161
Schwab M, Alitalo K, Varmus HE, Bishop JM, George D (1983) A cellular oncogene (c-Ki-*ras*) is amplified, overexpressed, and located within karyotypic abnormalities in mouse adrenocortical tumor cells. Nature (London) 303:497–501

Schwab M, Ellison J, Busch M, Rosenau W, Varmus HE, Bishop JM (1984) Enhanced expression of the human gene N-*myc* consequent to amplification of DNA may contribute to malignant progression of neuroblastoma. Proc Natl Acad Sci USA 81:4940–4944

Schwarz-Sommer Z, Shepherd N, Tacke E, Gierl A, Rohde W, Leclercq L, Mattes M, Berndtgen R, Peterson PA, Saedler H (1987) Influence of transposable elements on the structure and function of the A1 gene of *Zea mays*. EMBO J 6:287–294

Schweizer D, Nagl W (1976) Heterochromatin diversity in *Cymbidium*, and its relationship to differential DNA replication. Exp Cell Res 98:411–423

Scott GK, Davies PL, Kao MH, Fletcher GL (1988) Differential amplification of antifreeze protein genes in the Pleuronectinae. J Mol Evol 27:29–35

Scotto KW, Biedler JL, Melera PW (1986) Amplification and expression of genes associated with multidrug resistance in mammalian cells. Science 232:751–755

Selden JR, Emanuel BS, Wang E, Cannizzaro L, Palumbo A, Erikson J, Nowell PC, Rovera G, Croce CM (1983) Amplified C_λ and c-*abl* genes are on the same marker chromosome in K562 leukemia cells. Proc Natl Acad Sci USA 80:7289–7292

Semal J (ed) (1986) Somaclonal variations and crop improvement. Nijhoff, Dordrecht

Sen S, Teeter LD, Kuo T (1987) Specific gene amplification associated with consistent chromosomal abnormality in independently established multidrug-resistant Chinese hamster ovary cells. Chromosoma 95:117–125

Shah DM, Horsch RB, Klee HJ, Kishore GM, Winter JA, Tumer NE, Hironaka CM, Sanders PR, Gasser CS, Aykent S, Siegel NR, Rogers SG, Fraley RT (1986) Engineering herbicide tolerance in transgenic plants. Science 233:478–481

Sheridan WF, Clark JK (1987) Maize embryogeny: a promising experimental system. Trends Genet 3:3–6

Shiloh Y, Shipley J, Brodeur GM, Bruns G, Korf B, Donlon T, Schreck RR, Seeger R, Sakai K, Latt SA (1985) Differential amplification, assembly, and relocation of multiple DNA sequences in human neuroblastomas and neuroblastoma cell lines. Proc Natl Acad Sci USA 82:3761–3765

Shirzadegan M, Christey M, Earle ED, Palmer JD (1989) Rearrangement, amplification, and assortment of mitochondrial DNA molecules in cultured cells of *Brassica campestris*. Theor Appl Genet 77:17–25

Siegel A (1975) Gene amplification in plants. In: Markham R (ed) Modification of the information content of plant cells. Elsevier North-Holland Biomedical Press, Amsterdam, pp 15–26

Singh RJ (1986) Chromosomal variation in immature embryo derived calluses of barley (*Hordeum vulgare* L.). Theor Appl Genet 72:710–716

Slamon DJ, Clark GM, Wong SG, Levin WJ, Ullrich A, McGuire WL (1987) Human breast cancer: correlation of relapse and survival with amplification of the HER-2/*neu* oncogene. Science 235:177–182

Slightom JL, Sun SM, Hall TC (1983) Complete nucleotide sequence of a French bean storage protein gene: phaseolin. Proc Natl Acad Sci USA 80:1897–1901

Slightom JL, Drong RF, Klassy RC, Hoffman LM (1985) Nucleotide sequences from phaseolin cDNA clones: the major storage proteins from *Phaseolus vulgaris* are encoded by two unique gene families. Nucleic Acids Res 13:6483–6498

Smith DB, Flavell RB (1974) The relatedness and evolution of repeated nucleotide sequences in the genomes of some Gramineae species. Biochem Genet 12:243–256

Smith AG, Chourey PS, Pring DR (1987) Replication and amplification of the small mitochondrial DNAs in a cell suspension of black Mexican sweet maize. Plant Mol Biol 10:83–90

Solovyan VT, Kostenyuk IA, Kunakh VA (1987) Genome changes in *Rauwolfia serpentina* Benth. cells cultivated in vitro. Genetika (USSR) 23:1200–1208

Song KM, Osborn TC, Williams PH (1988) *Brassica* taxonomy based on nuclear restriction fragment length polymorphisms (RFLPs). 1. Genome evolution of diploid and amphidiploid species. Theor Appl Genet 75:784–788

Sossountzov L (1969) Incorporation de précurseurs tritiés des acides nucléiques dans les méristèmes apicaux de la Fongère aquatique *Marsilea drummondii* A. Br Rev Genet Bot 76:109–156

Spradling AC (1981) The organization and amplification of two chromosomal domains containing *Drosophila* chorion genes. Cell 27:193–201

Spradling AC, Mahowald AP (1980) Amplification of genes for chorion proteins during oogenesis in *Drosophila melanogaster*. Proc Natl Acad Sci USA 77:1096–1100

Spradling AC, De Cicco DV, Wakimoto BT, Levine JF, Kalfayan LJ, Cooley L (1987) Amplification of the X-linked *Drosophila* chorion gene cluster requires a region upstream from the s38 chorion gene. EMBO J 6:1045–1053

Spring H, Trendelenburg MF, Scheer U, Franke WW, Herth W (1974) Structural and biochemical studies of the primary nucleus of two green algal species, *Acetabularia mediterranea* and *Acetabularia major*. Cytobiologie 10:1–65

Sree Ramulu K, Dijkhuis P, Roest S (1989) Patterns of phenotypic and chromosome variation in plants derived from protoplast cultures of monohaploid, dihaploid and diploid genotypes and in somatic hybrids of potato. Plant Sci 60:101–110

Srimatkandada S, Medina WD, Cashmore AR, Whyte M, Engel D, Moroson BA, Franco CT, Dube SK, Bertino JR (1983) Amplification and organization of dihydrofolate reductase genes in a human leukemic cell line, K-562, resistant to methotrexate. Biochemistry 22:5774–5781

Stahl F, Sandberg P, Martinsson T, Skoog J, Dahllof B, Wittergren Y, Bjursell G, Levan G (1987) Isolation of selectively amplified DNA sequences from multidrug-resistant SEWA cells. Hereditas 106:97–105

Stark GR (1986) DNA amplification in drug resistant cells and in tumours. Cancer Surv 5:1–24

Stark GR, Wahl GM (1984) Gene amplification. Annu Rev Biochem 53:447–491

Steinrücken HC, Schulz A, Amrhein N, Porter CA, Fraley RT (1986) Overproduction of 5-enolpyruvylshikimate-3-phosphate synthase in a glyphosate-tolerant *Petunia hybrida* cell line. Arch Biochem Biophys 244:169–178

Stephan W (1987) Quantitative variation and chromosomal location of satellite DNAs. Genet Res Camb 50:41–52

Sunderland N (1977) Nuclear cytology. In: Street HE (ed) Plant cell and tissue culture, vol 2. Univ Cal Press, Berkeley, pp 177–206

Swift H (1950) The constancy of DNA in plant nuclei. Proc Natl Acad Sci USA 36:643–654

Tagliasacchi AM, Forino LMC, Frediani M, Avanzi S (1983) Different structure of polytene chromosomes of *Phaseolus coccineus* suspensors during early embryogenesis. 2. Chromosome pair VII. Protoplasma 115:95–103

Tagliasacchi AM, Forino LMC, Cionini PG, Cavallini A, Durante M, Cremonini R, Avanzi S (1984) Different structure of polytene chromosomes of *Phaseolus coccineus* suspensors during early embryogenesis. 3. Chromosome pair VI. Protoplasma 122:98–107

Tanksley SD, Young ND, Paterson AH, Bonierbale MW, (1989) RFLP mapping in plant breeding – new tools for an old science. Bio/Technol 7:257–266

Tartof KD (1971) Increasing the multiplicity of ribosomal RNA genes in *Drosophila melanogaster*. Science 171:294–297

Taub RA, Hollis GF, Hieter PA, Korsmeyer S, Waldmann TA, Leder P (1983) Variable amplification of immunoglobulin λ light-chain genes in human populations. Nature (London) 304:172–174

Tocchini-Valentini GP, Mahdavi V, Brown R, Crippa M (1974) The synthesis of amplified ribosomal DNA. Cold Spring Harbor Symp Quant Biol 38:551–558

Tonin PN, Stallings RL, Carman MD, Bertino JR, Wright JA, Srinivasan PR, Lewis WH (1987) Chromosomal assignment of amplified genes in hydroxyurea-resistant hamster cells. Cytogenet Cell Genet 45:102–108

Tschermak-Woess E (1971) Endomitose. In: Altmann H-W (ed) Handbuch der allgemeinen Pathologie 2/2/1. Springer, Berlin Heidelberg New York, pp 569–625

Tschermak-Woess E, Enzenberg-Kunz U (1965) Die Struktur der hochendopolyploiden Kerne im Endosperm von *Zea mays*, das auffallende Verhalten ihrer Nukleolen und ihr Endopolyploidiegrad. Planta 64:149–169

Tsuda T, Nakatani H, Matsumura T, Yoshida K, Tahara E, Nishihira T, Sakamoto T, Yoshida M, Terada M, Sugimura T (1988) Amplification of the hst-1 gene in human esophageal carcinomas. Jap J Cancer Res (GANN) 79:584–588

Tyagi BR, Naqvi AA (1987) Relevance of chromosome number variation to yield and quality of essential oil in *Mentha arvensis* L. Cytologia 52:377–385

Ueda K, Clark DP, Chen C-j, Roninson IB, Gottesman MM, Pastan I (1987) The human multidrug resistance (*mdr* 1) gene. cDNA cloning and transcription initiation. J Biol Chem 262:505–508

Van der Bliek AM, van der Velde-Koerts T, Ling V, Borst P (1986) Overexpression and amplification of five genes in a multidrug-resistant Chinese hamster ovary cell line. Mol Cell Biol 6:1671–1678

Van Oostveldt P, van Parijs R (1972) Reiterated nucleotide sequences in DNA of dividing and elongating plant cells. Arch Int Physiol Biochem 80:416–418

Van Oostveldt P, van Parijs R (1976) Underreplication of repetitive DNA in polyploid cells of *Pisum sativum*. Exp Cell Res 98:210–221

Van't Hof J (1987) Extrachromosomal DNA and cell differentiation in cultured pea roots (*Pisum sativum*). In: Green CE, Somers DA, Hackett WP, Biesboer WP (eds) Plant tissue and cell culture. Alan R Liss, New York, pp 33–44

Varmus HE (1984) The molecular genetics of cellular oncogenes. Annu Rev Genet 18:553–612

Vasil IK (ed) (1986) Cell culture and somatic cell genetics of plants, vol 3. Plant regeneration and genetic variability. Academic Press, New York London

Voelker TA, Staswick P, Chrispeels MJ (1986) Molecular analysis of two phytohemagglutinin genes and their expression in *Phaseolus vulgaris* cv. Pinto, a lectin-deficient cultivar of the bean. EMBO J 5:3075–3082

Wahl GM, Padgett RA, Stark GR (1979) Gene amplification causes overproduction of the first three enzymes of UMP synthesis in N-(phosphonacetyl)-L-aspartate-resistant hamster cells. J Biol Chem 254:8679–8689

Wahl GM, de St Vincent BR, De Rose ML (1984) Effect of chromosomal position on amplification of transfected genes in animal cells. Nature (London) 307:516–520

Walbot V, Cullis CA (1985) Rapid genomic change in higher plants. Annu Rev Plant Physiol 36:367–396

Walthall ED, Brady T (1986) The effect of the suspensor and gibberellic acid on *Phaseolus vulgaris* embryo protein synthesis. Cell Differ 18:37–44

Walton JD, Paquin CE, Kaneko K, Williamson VM (1986) Resistance to antimycin A in yeast by amplification of ADH 4 on a linear, 42 kb palindromic plasmid. Cell 46:857–863

Wang L-C, Vass W, Gao C, Chang KSS (1987) Amplification and enhanced expression of the c-Ki-ras-2 protooncogene in human embryonal carcinomas. Cancer Res 47:4192–4198

Wardell WI, Skoog F (1973) Flower formation in excised tobacco stem segments. III. DNA content in stem tissue of vegetative and flowering tobacco plants. Plant Physiol 52:215–220

Warr JR, Atkinson GF (1988) Genetic aspects of resistance to anticancer drugs. Physiol Rev 68:1–26

Watson JM, McKay LM, Graves JAM (1988) Genome instability in interspecific cell hybrids II. Aminopterin resistance and gene amplification in lines arising from fusions of cells from divergent mammalian species. J Genet (India) 67:75–86

Weber G, de Groot E, Schweiger HG (1985) Methotrexate-resistant somatic cells of *Glycine max* (L.) Merr.:selection and characterization. J Plant Physiol 117:339–353

Wersuhn G, Sell B (1988) Numerical chromosomal aberrations in plants regenerated from pollen, protoplasts and cell cultures of *Nicotiana tabacum* L. cv. "Samsun". Biol Zbl 107:439–445

Wildon DC, Wong D, Bryant JA (1971) Turnover of DNA in seedlings of *Pisum sativum*. Biochem J 124:12P

Wilson AC (1976) Gene regulation in evolution. In: Molecular evolution. Ayala FJ (ed) Sinauer, Sunderland, MA, pp 225–234

Wilson AC, Carlson SS, White TJ (1977) Biochemical evolution. Annu Rev Biochem 46:573–639

Xiao W, Saxena PK, King J, Rank GH (1987) A transient duplication of the acetolactate synthase gene in a cell culture of *Datura innoxia*. Theor Appl Genet 74:417–422

Yacoob RK, Filion WG (1986) Temperature-stress response in maize: a comparison of several cultivars. Can J Genet Cytol 28:1125–1131

Yang N-S (1988) Genotypic and phenotypic changes in new plant varieties. Tree 3:S21–S22

Yao M-C, Zhu S-G, Yao C-H (1985) Gene amplification in *Tetrahymena thermophila*: formation of extrachromosomal palindromic genes coding for rRNA. Mol Cell Biol 5:1260–1267

Ye J, Hauptmann RM, Smith AG, Widholm JM (1987) Selection of a *Nicotiana plumbaginifolia* universal hybridizer and its use in somatic hybrid formation. Mol Gen Genet 208:474–480

Zacharias H (1979) Underreplication of a polytene chromosome arm in the chironomid *Prodiamesa olivaceae*. Chromosoma 72:23–51

Zacharias H (1986) Tissue-specific schedule of selective replication in *Drosophila nasutoides*. Roux's Arch Dev Biol 195:378–388

Zehr BE, Williams ME, Duncan DR, Widholm JM (1987) Somaclonal variation in the progeny of plants regenerated from callus cultures of seven inbred lines of maize. Can J Bot 65:491–499

Zheng KL, Castiglione S, Biasini MG, Biroli A, Morandi C, Sala F (1987) Nuclear DNA amplification in cultured cells of *Oryza sativa* L. Theor Appl Genet 74:65–70

I.7 Optical Techniques to Measure Genetic Instability in Cell and Tissue Cultures

G.P. Berlyn, A.O. Anoruo, and R.C. Beck[1]

Introduction

Biotechnology has great potential in forestry, particularly in plantation forestry, where rapid reforestation is financially and ecologically imperative, and especially in situations where the desirable species grows well in an exotic environment but does not reproduce there. The potential benefits of biotechnology include: rapid multiplication of selected genotypes; cloning; somatic hybridization; increased disease, insect, and herbicide resistance; germplasm preservation; triploid production; rapid haploid and pure line production; somatic embryogenesis; and somaclonal variation (see Berlyn et al. 1986a; Bajaj 1986). However, there are potential problems which include: low yield of somatic embryos, especially in conifers; inability to regenerate almost all forest trees from protoplasts, which prevents application of somatic hybridization; delayed expression of undesirable traits; morphological incompatibility of plant organs in genetically engineered plants; out-of-phase development, that alters timing mechanisms such as bud flushing, dormancy induction, initiation of earlywood and latewood formation and reproductive development; and genetic instability of regenerated plantlets and consequent inability to regenerate true to type. As has been documented, high levels of auxins employed in plant tissue culture for root initiation and organogenesis can cause DNA content polymorphisms (Renfroe and Berlyn 1985; Berlyn et al. 1987). These effects are probably deleterious, especially in conifers, but the long-term significance of these changes is not well understood. It is possible that in the course of time normal cells are able to supplant the abnormal ones. However, at this time we simply do not know what the consequences of elevated DNA levels are. The genetic variation that is created in tissue culture plants may be benign (a "normal" consequence of development, dedifferentiation, rejuvenation, or redifferentiation), may be transient, may be persistent but harmless, may be responsible for mortality during development and early establishment, or may fail at some stage of normal development later in the life of the stand. Whatever the case, it is important to find out and to genetically screen tissue culture-produced plants.

This chapter focuses on optical techniques to measure genetic variation (somaclonal variation). These techniques can resolve DNA changes of less than 18 genes and can measure DNA of intact organelles, as well as whole nuclei. In combination with immunocytochemistry and biotinylated or radioactive probes, it

[1]Yale School of Forestry and Environmental Studies, New Haven, CT 06511, USA

Biotechnology in Agriculture and Forestry, Vol. 11
Somaclonal Variation in Crop Improvement I (ed. by Y.P.S. Bajaj)

gives the potential to identify individual genes on specific chromosomes. Fluorescent probes specific for individual bases may also have potential in this system of analysis. Somaclonal variation has potential in biotechnology for isolation of somaclones and mutants of various types, as well as in the amplification of available genetic variation in such processes as somatic hybridization and ploidy increase. However, once a desirable somaclone or somatic hybrid is isolated, it is important to be able to perpetuate this genetic line and to have a screening process to monitor any changes. Since optical techniques have sensitivity to as little as 18 genes and permit selection of individual cells, they are a fast and efficient method of genetic screening. The specific techniques for the optical measurement of nuclear and organellar DNA amount to be discussed in this chapter are absorption microspectrophotometry and microspectrofluorometry. Several different types of instruments used to make these DNA measurements will be discussed and analyzed.

Karyotyping and chromosome banding are also techniques of use in analyzing genetic changes in culture. However, chromosome counts can only be made on dividing cells, and thus the sample is limited and potentially biased. Furthermore, chromosome counts do not measure changes in chromosome volume and DNA content, and therefore there may be significant changes in DNA amount and quality that chromosome counts do not reveal (see review by Karp and Bright 1985). Nevertheless, they are necessary to distinguish polyploidy and aneuploidy from elevated DNA levels. Microspectrofluorophotometric techniques measure cells in all phases of the cell cycle, and thus a more representative sample is available with this technique. Banding and other in situ localizations and probes for plant chromosomes have not been fully developed and are in need of additional investigation. Flow cytometry is probably the most rapid method, but most of these instruments do not permit individual cell selection, do not retain a permanent record of the cell measured as is the case in the other optical techniques, and are limited to fluorescence dyes. The flow cytometers are also much more expensive than the other optical techniques and are not as versatile. They are, however, quite suitable for large, well-funded single-purpose usage.

2 Basic Information

The amount of DNA in the sperm or egg cell is termed the C-value. A typical somatic cell has the 2N chromosome (diploid) number and the corresponding amount of DNA in the nucleus would appear to be 2C. However, in the normal cycle, the C-value doubles before each cell division, so that a normal cell varies from 2C to 4C during the cell cycle. Thus any somatic cell whose DNA content is greater than 4C has "elevated" DNA content.

The C-value is not constant for all the cellular phenotypes of an individual in many plant and animal species (Miksche 1968; D'Amato 1977; Doerschug et al. 1978; Dhillon et al. 1978; Nagl 1982). Causes can be increased chromosome volume or number (polyploidy or hyperaneuploidy) or extrachromosomal DNA. Some antipodal plant cells have a C-value of 1000 or more (Nagl 1982). In *Scilla decidua*, 71% of the somatic cells have been reported to have DNA contents greater than 4C

(Frisch and Nagl 1979). Clonal fidelity from such a plant would be questionable, and since this type of polysomaty is quite common in angiosperms there is indeed reason for concern about clonal fidelity (D'Amato 1977; Nagl 1978; Berlyn et al. 1986a).

Unfortunately, tissue culture increases nuclear DNA (nDNA) content polymorphism (Renfroe and Berlyn 1985; Berlyn et al. 1987, see also Nagl Chap. I.6, this Vol.), a situation that has been largely neglected in some forest trees. Its effect on organellar DNA has not been investigated as far as we know. In addition to plantlet production, more attention to long-term genetic consequences is needed (see Bajaj 1986). Genetic instability is especially critical in forest trees of long rotation age because of the possibility of delayed expression of genetic defects. Loss of a standing crop in the middle of a long rotation is financially devastating.

3 Variation in Culture

Three major types of variation observed in cell cultures are: (1) genetic, (2) developmental (epigenetic), and (3) physiological. Genetic variation is generally considered to be: not readily reversible, stable through meiosis, stochastic, infrequent (low mutation rate) but it can be increased by chemical mutagens or the tissue culture environment, and is clonally propagated. Developmental variation is: not readily reversible except in those organisms and/or organs capable of cellular dedifferentiation and redifferentiation, not sexually transmitted but it may be stable through mitosis, nonrandom, can be environmentally dependent, relatively frequent, and stable through clonal propagation. The term "physiological variation" refers to changes in culture induced by the culture environment. Physiological variation is not as well characterized, and may interact with developmental/genetic programs and become habituated or even heritable in some cases (see Berlyn et al. 1986a).

It is a common practice to add mutagens to plant cell cultures to induce mutants, but presently it is known that this is often unnecessary because of the high frequency of spontaneous mutations in plant cell and tissue cultures (Chaleff 1981, 1983; Chaleff and Keil 1981). Over the past decade it has become clear that factors promoting genetic instability in culture are: prolonged culture time, high levels of growth substances and/or plant growth regulators, and the callus form of growth.

4 Microspectrophotometry

A microspectrophotometer is essentially a microscope inside a spectrophotometer. The specific application of absorption spectrophotometry principles to microspectrophotometry has been reviewed by Berlyn and Cecich (1976) and Berlyn and Miksche (1976) among many others (see volumes edited by Wied 1966; Wied and Bahr 1970), and will not be dealt with here. Although microspectrophotometry can be used in a wide variety of quantitative analyses at the microscopic level ranging from photosynthesis to enzymatic activity (Berlyn 1969; Berlyn and Miksche 1976; Berlyn and Beck 1989), this review will consider only the measurement of DNA content. In this technique the Feulgen technique employing the Schiff reagent is used to provide a visible light chromophore on the DNA. Its accuracy and precision have been adequately proven over the past six decades since it was first proposed by Feulgen and Rossenbeck in 1924. It is accurate to at least the picogram level (10^{-12} g) and, according to Bartels (1966), is capable of 5×10^{-14} g in the limiting case. Nevertheless, alterations in stoichiometry have been reported in cases where the DNA is especially compacted, condensed, or dispersed (see Miksche and Dhillon 1984). We have found the egg nucleus of *Pinus* to be Schiff-negative even though it is stainable with acridine orange and other dyes. This is probably owing to dispersion due to the high volume of the egg nucleus (Berlyn and Passof 1965; Berlyn 1967).

4.1 Tissue and Cell Preparation

Fixation for microspectrophotometry is usually accomplished by incubation of the tissue or cells in acidified alcohol. The agent used in our laboratory is Carnoy's No. 2 (1:3:6, acetic acid:chloroform:ethyl alcohol). Probably the most commonly used fixative is Farmer's fluid (1:3, acetic acid:ethyl alcohol). These fixatives should be prepared fresh just before fixing to avoid ketone formation in the solution. Fixation time can vary with the material of interest, but fixation times of 1–2 h are common. For conifers we generally fix at 0° C, as this provides good cellular detail and good quantitation. After fixation the tissue is hydrated to water at 4° C, using steps of 95, 70, 50 and 30% ethyl alcohol. At this point the tissue is placed in 0.01 M sodium citrate at pH 5.0 and brought to room temperature. Cell separation can be accomplished by enzymatic or mechanical means. Fixatives containing aldehydes should be avoided or exhaustively washed out of the tissue because they can crosslink DNA (Greenwood and Berlyn 1968).

DNA can also be measured in sectioned material but a section thickness that provides the maximum number of whole nuclei must be determined and used. DNA values below 2C in somatic cells should be rejected. Great care must be exercised to insure that the measuring beam does not pass through any nuclear material that is out of the plane of focus. Despite the difficulties, sections provide information on tissue organization and morphogenesis that cannot be obtained from isolated cells.

In our laboratory we generally use enzymatic maceration for cell separation to minimize mechanical damage to nuclei (Berlyn et al. 1979). The sodium citrate prevents DNase activity if present as a contaminant in the crude pectinase, while not inhibiting the pectinase activity. A 0.001 M solution of sodium EDTA may be used in place of the sodium citrate. For conifer tissue we find 1–2% pectinase plus 0.5–1.0% cellulase or macerase in the buffer (pH 5.0) sufficient for most tissue. Variability in pectinase effectiveness from different manufacturers is a problem. The addition of liginases may facilitate cell separation in lignified tissue. The tissue is generally incubated overnight at 50° C. Mechanical cell separation can be accomplished by tapping with a glass rod or applying pressure on a coverslip (see Miksche and Dhillon 1984). The latter measure requires coverslip removal generally by freezing in dry ice or liquid CO_2 (Conger and Fairchild 1953; Bowen 1956). A Ten Broek tissue grinder can also be used (very carefully) to separate the cells, but again the possibility of mechanical damage and subsequent loss of DNA from the nuclei is enhanced. Only cells with stain confined to the nucleus may be used for measurement!

Following maceration, cell suspensions are sedimented in a clinical centrifuge and the enzyme solution is removed by repeated washing and sedimenting in the centrifuge. The cells are then suspended in a small volume of water and small drops are placed on subbed microscope slides that already have adhesive and a small smear of chicken red blood cells (fixed in Carnoy's No. 2 solution) on them as an internal standard (Dhillon et al. 1977). Internal standards of plant material have been used, but they are more variable than the chicken erythrocytes. (Miksche et al. 1979a, Berlyn et al. 1986b).

A convenient way to prepare the internal standard is to mix fresh chicken blood 2:1 with an aqueous solution of 2% sodium EDTA. Microscope slides are dipped in the solution so that only a small area at one end of the slide is immersed. The slides are then kept vertical, while the bottom of the slide is blotted to drain off excess blood. Lipshaw (Detroit, Michigan, USA) makes plastic holders that permit five slides to be processed at a time (see Berlyn and Miksche 1976, p. 66). The slides (in their holders) are then dried, preferentially in a forced air drying oven at ca. 40–50° C. They are then placed in Carnoy's No. 2 fixative for 1 h, washed in absolute or 95% ethyl alcohol and air-dried. These slides with the fixed chicken blood can be stored in sealed, desiccated refrigeration storage for several months. They are then ready for the application of adhesive, which may also be done before application of the red blood cells.

The adhesive used is gelatin-chromium potassium sulfate (chrome alum) as described in Berlyn and Miksche (1976). Slides are dipped (subbed) in the adhesive solution (2.5 g gelatin dissolved in 500 ml of water at 30–35° C to which solution is added 0.25 g chrome alum) and dried in the upright position, preferentially before any cells are added on the slide. This adhesive, unlike formalin-containing formulas such as Haupt's adhesive, does not crosslink the DNA (Greenwood and Berlyn 1968). After the cells are placed on the slides, they must be dried overnight on a warming pan or in a forced-air drying oven at 40–50° C.

4.2 Stain Preparation

The Schiff reagent used in this Feulgen technique is best prepared according to the procedures described in Berlyn and Miksche (1976). Unlike a number of other procedures, this method, adapted from Lillie (1951), does not fade rapidly and is very consistent, batch after batch, in terms of absorption spectra with a given tissue (Miksche and Dhillon 1984). The dye solution also remains good in sealed refrigeration storage for months. To prepare Schiff reagent add 2 g of certified (CI 42510, 42500) basic fuchsin (pararosanilin) plus 3.8 g of sodium or potassium metabisulfite to 200 ml of 0.15 M hydrochloric acid (HCl) in a flask. Wrap the flask in aluminum foil or shake in the dark for 2 h. This can conveniently be done on a ceramic-surfaced magnetic stirrer that does not heat up during operation (e.g.,

Corning models). Decolorize by adding 1 g of activated charcoal and stir for an additional 5 min. Vacuum filter through Whatman No. 1 filter paper. Repeat the decolorizing filtration at least once. The resulting solution should be completely clear or slightly straw-colored.

A bleaching solution (made up just before use) is prepared by adding 30 ml of 1 M HCl and 30 ml of 10% sodium or potassium metabisulfite to 540 ml of distilled water. This makes enough for three bleaching baths of 200-ml in large staining dishes.

The staining procedure is as follows:

1. Hydrolyze the material on slides in 5 M HCl at room temperature (25°C). The hydrolysis time that yields maximum stain intensity should be used, and this must be determined for each material, but 30 min is usually on the broad peak (Fig. 1) that results with this procedure. Hydrolysis times are more critical in the older procedures that used 1 M HCl at 60°C, where a sharp peak in intensity obtained.
2. Place slides immediately in distilled water at 4°C to stop the hydrolysis and wash in two changes of distilled water.
3. Stain for 2 h in the dark in the 200 ml Schiff reagent in a large coplin jar at room temperature or for 3–4 min on medium low setting in a microwave oven (temperature $\leq$ 60°C).
4. Place immediately in the bleach and move through the three bleaching jars at 10-min intervals.
5. Wash in running tap water for 5–10 min.
6. Rinse in distilled water and dehydrate through a standard alcohol-xylene series and mount in resin (e.g., Per mount, Preservaslide) using No. 1 1/2 coverslips. Alternatively the slides can be air-dried (no heat as the dye is heat-labile) and then mounted in immersion oil (index of refraction of 1.515). A summary of the fixing and staining procedure is shown in Fig. 2.

The chemistry of the Feulgen reaction is reviewed in Berlyn and Cecich (1976). The acid hydrolysis selectively removes the purines in the DNA, leaving aldehyde groups in their place. When the cells are incubated in the Schiff reagent, two of these aldehyde groups condense with each molecule of the Schiff reagent, forming a stable, magenta-colored, dye complex.

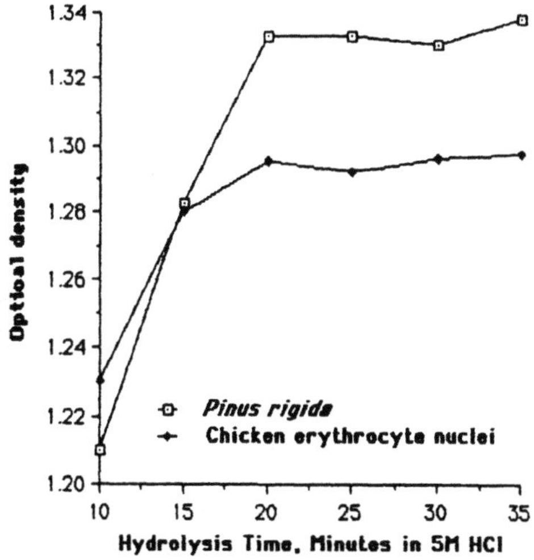

Fig. 1. Optical density vs. hydrolysis time in 5M HCl at 25°C for *Pinus caribaea* and chicken red blood cells

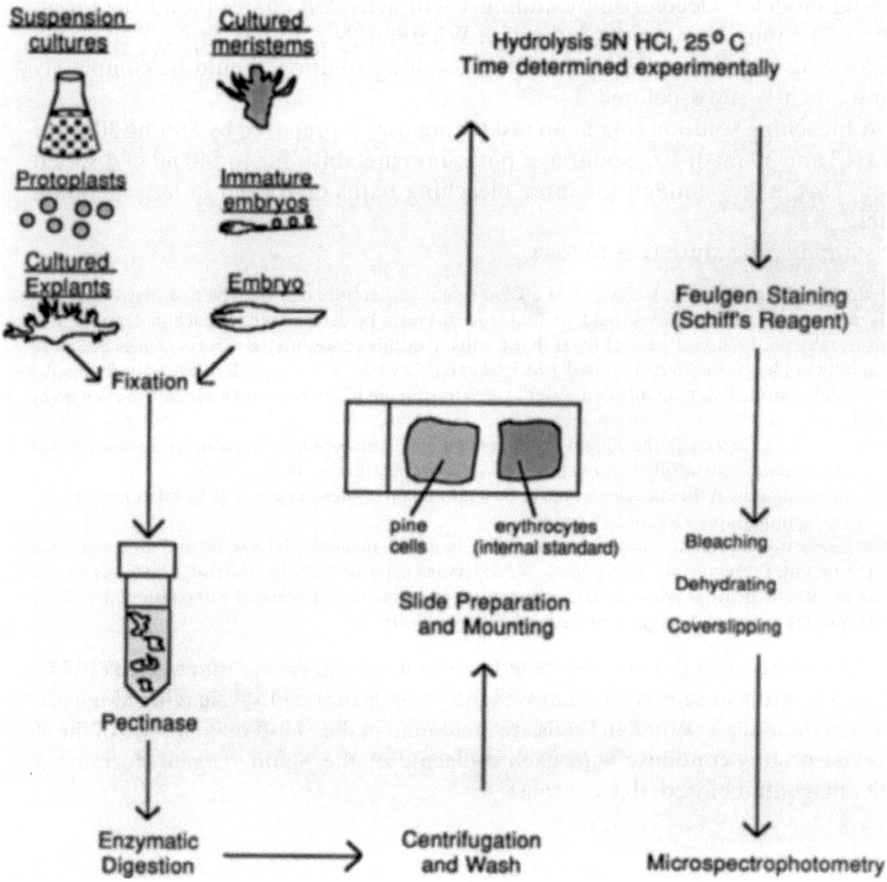

Fig. 2. Tissue fixation and staining procedure for microspectrophotometry. (Renfroe 1984)

4.3 Instrumentation

For DNA microspectrophotometry there are basically two types of units: the single beam and the scanning type (Fig. 3). The scanning microspectrophotometer (also called a microdensitometer) is either mechanical or optical in design. The mechanical scanning unit has a motorized stage that moves the specimen across the light beam while in the optical scanners the object remains stationary, while the light beam scans the specimen at a rate of ca. 12,000 spot measurements per second. The main advantage of the optical scanners is speed and ease of use. In a test of several hundred nuclei we found no difference in the accuracy between a Vickers M86 optical scanner and a single beam Leitz MPV-1 microspectrophotometer. However, the Vickers was much faster. The mechanical scanners are even slower than the single beam units, and surpass the single beam instruments mainly in ease of use, and price. Since the scanners measure almost every point in the nucleus, there is little

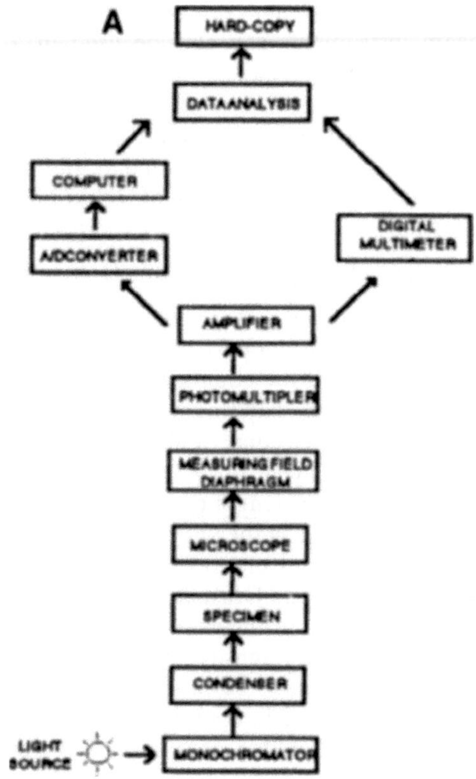

Fig. 3. A Diagram of single beam type microspectrophotometer. **B** An optical scanning microspectrophotometer

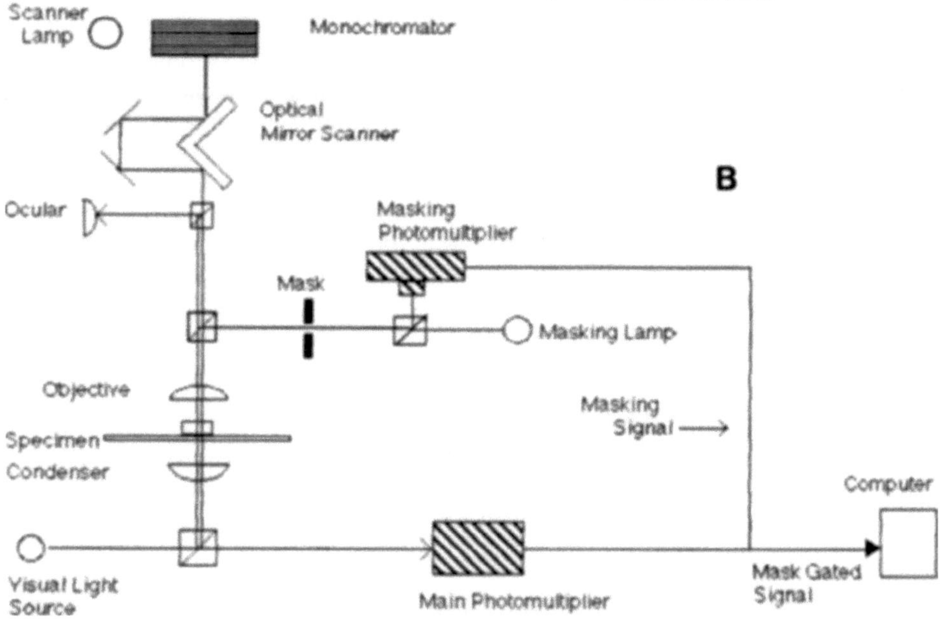

error due to variation in distribution of the DNA in the nucleus and a single measuring wavelength can be used, usually 550–580 nm (at or near peak). Single-beam instruments can also be used in a manner that eliminates distribution error, but this requires measurement of the same nucleus at two different wavelengths.

Use of an internal standard permits the instrument output, relative Feulgen units (fu), to be changed to actual pg amounts of DNA. The chicken red blood cells (rbc) have a constant amount of DNA (2.8–3.2 pg in different batches of blood with a mean of 3.0 pg) and therefore can used in the following equation:

$$\text{plant-DNA}_{pg} = 3.0_{pg} \times \text{plant-DNA}_{fu}(\text{rbc-DNA}_{fu})^{-1}$$

It should be noted that for the two-wavelength method this equation assumes that measuring diaphragms for rbc and plant nuclei are equal. If this is not the case, then the right side of the equation must be multiplied by the ratio of the area of the plant-measuring diaphragm to the rbc-measuring diaphragm. In the scanning system this is not a concern, since similar scan masks are generally used for both the plant and rbc nuclei.

The two-wavelength method permits the measurement of the relative mass of an object of any shape with greatly reduced distribution error and without any assumptions about, or models of, specimen shape. It was developed independently by Patau (1952) and Ornstein (1952) and, although not widely appreciated, was one of the more significant advances in science. It does assume that the absorption laws hold, i.e., optical density must increase linearly with concentration, and this assumption needs to be checked for each system to which the technique is to be applied. However, the conformity of DNA in plant nuclei to the absorption laws has been substantiated for many species by numerous workers.

$$\text{Optical density } (E) = \log 1/T = kcd = \text{kmass}/A, \qquad \text{(Lambert-Beer Law)}$$

where T = transmittance/I/I_0 and I is light transmitted through the specimen and I_0 is light transmitted through a clear spot (background) near the specimen; k = extinction coefficient, c = chromophore concentration, d = specimen thickness, A = measuring area, kmass = relative mass.

Figure 4 shows a nucleus of irregular shape (A_2) that is to be measured. The total measuring area is A and the clear area is A_1. Thus $A = (A_1 + A_2)$. To minimize scattering error (Schwarzschild-Villiger effect, where a small dark area is surrounded by a light area, resulting in some of the light from the light area being scattered into the dark area), A should be only slightly larger than A_1. Therefore, we can define $F_1 = A_1/A$ and $F_2 = A_2/A = (1 - F_1)$. Consider two wavelengths λ_a and λ_b.

$$T_b = T_{1b} F_1 + (1 - F_1) T_{2b} = F_1 + (1 - F_1) T_{2b},$$

where T_b = transmittance at wavelength b through the whole measuring area, A, that includes both A_1 and A_2; T_{2b} = transmittance at wavelength b through A_2; T_{1b} = transmittance at wavelength b through A_1 (clear space) = 1.0.

Solving for T_{2b} and similarly for T_{2a} we find

$T_{2b} = T_b - F_1/(1 - F_1)$ and $T_{2a} = T_a - F_1/(1 - F_1)$.

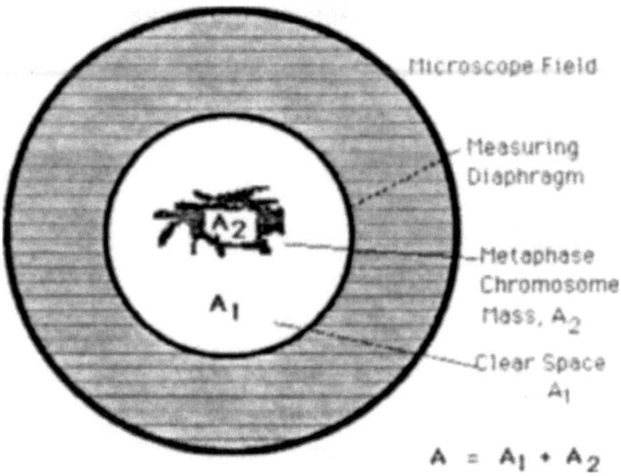

Fig. 4. Figure of nucleus of irregular shape

If we set $T_{2b} = T_{2a}^2$, we can solve for $F_1 = T_b - T_a^2/1 + T_b - T_a$.

Therefore, $(1 - F_1) = (1 - T_a)^2/1 + T_b - 2T_a$, and

$T_b - F_1 = (T_a - T_b)^2/1 + T_b - 2T_a$.

Thus, $kmass_b = A_2 \log 1/T_{2b} = A(1 - F_1) \log [(1 - F_1)/(T_b - F_1)]$.

If we substitute in the parenthetical quantities we obtain a final equation of

$kmass_b = A \times [(1 - T_a)^2/(1 + T_b - 2T_a)] \times 2 \times \log [(1 - T_a)/(T_a - T_b)]$.

Note that in the final equation we are able to determine the mass of the DNA (in this case) by simply measuring the transmission of light through the whole measuring area (A) at wavelengths a and b. As previously mentioned, A is set slightly larger than the nucleus to be measured. The only restriction is that:

$$\log 1/T_{2b} = \log 1/T_{2a}^2 = 2 \log 1/T_{2a},$$

i.e., the optical density at wavelength b must be twice that at wavelength a. Thus, to determine the wavelengths it is only necessary to prepare an absorption curve (optical density vs. wavelength, Fig. 5) and select wavelength b at or near the peak, divide its optical density by 2, and read down the curve to locate wavelength a at this half optical density ($0.5\ E_{2b}$). Step-by-step measuring procedures can be found in Berlyn and Miksche (1976) and Berlyn and Cecich (1976). It should be noted that the derivation given here is similar to the Ornstein derivation and uses wavelength b, while the tables of Mendelsohn (1958) are based on wavelength a as in the Patau derivation, and thus the relative mass (kmass) values will differ from each other by the constant (the extinction coefficient varies with wavelength). However, since an internal standard should always be used, this difference cancels out, since the standard and unknown are determined on the same basis. The tables are not really necessary today since almost everyone has access to computers and the equations are easily programmed. However, it is important to note that the two wavelengths used are generally different for the internal standard and the plant DNA. Thus separate absorption curves must be made for the standard and plant species of interest in order to determine the two wavelengths for each type.

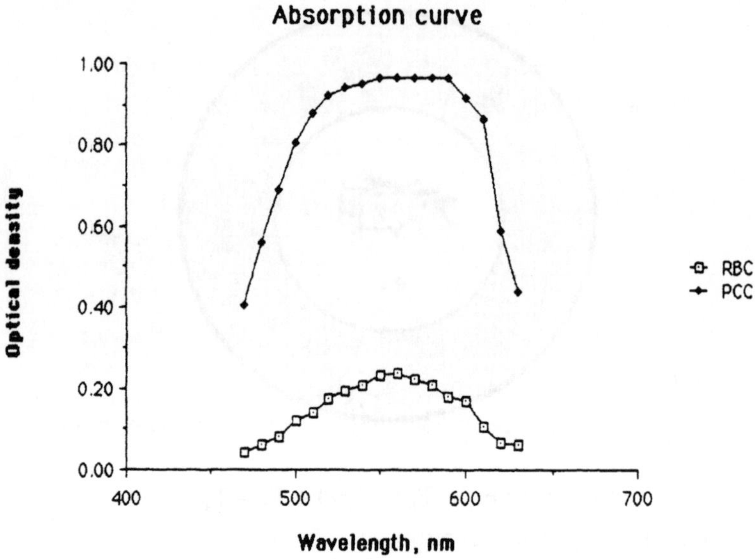

Fig. 5. Optical density vs. wavelength for *Pinus caribaea* var. *caribaea* (*PCC*) and chicken red blood cells (*RBC*)

4.4 Scanning Microspectrophotometry

The mechanical scanners operate just like the single beam units diagramed in Fig. 3 with the addition of a motorized stage and software and computer to run the stage in a scan mode. They are very slow, and mechanical and electronic stability become factors that limit accuracy. The scanning mode is also subject to error from glare, and while this can be corrected for by scanning a clear area near the specimen, it is slow and tedious to do it. It is seldom that the electrical supply is stable over the many minutes required to make a single measurement. In fact, many laboratories use electrical line conditioners or do their readings at night when the electrical power tends to be more stable. The readings have to be slow in order to minimize distribution error, i.e., a small aperture is required in order to insure that the DNA is homogeneous in the spot area. The larger the spot size, the faster the scan, but the larger the probability of distribution error.

The light path of an optical scanner of the Vickers M 85 type is shown in Fig. 6. The scanning unit contains a pair of independently vibrating mirrors that deflect the light coming from the monochromator to produce a raster pattern in the specimen plane. The light is collected by the mirror below the substage condenser and directed to the main photomultiplier, where the light flux is proportionally converted to an electrical signal. A masking photomultiplier is used as a photoelectric gating system that monitors the scan and permits only the light signals coming from the mask area to be fed into the electronic integrator that calculates the optical density. To permit selection of the nucleus to be measured, an image of the mask and the scanning spot are superimposed on the specimen image in the microscope. The Vickers, with its on-board computer, has the capacity to measure the amount of optical density lying above or below specified density levels, i.e., in a given nucleus it can separate densities between 0.1 to 0.2, 0.2 to 0.3, 0.3 to 0.4, etc. This can be useful in separating changes in euchromatin to heterochromatin during development. Histograms of density distribution within individual nuclei can be plotted. The Vickers can also be used to make spot measurements such as on portions of individual chromosomes.

Recently, the Leitz MPV3 scanning microspectrophotometer became available, and this instrument is capable of both mechanical and optical scanning. It also uses a pair of vibrating mirrors for the optical scan and can operate at more than 600 measuring points per second. It has the flexibility to operate with incident as well as transmitted light and can also function as a microspectrophotofluorometer.

The errors inherent in the scanning systems have been reviewed in a number of papers (e.g., Goldstein 1970, 1971, 1975; Miksche et al. 1979; Dhillon et al. 1983) and will merely be summarized here. With attention to detail and proper procedures, minimized repeatable accuracy within a picogram is rather easily attainable. The principal errors are due to: improper alignment and resulting uneven or incorrect wavelength illumination, glare, conical light, bandwidth, spot size and focus, and the specimen itself. Glare is a result of stray light entering the system, along the ray path, and from multiple reflections from glass to air surfaces in the microscope objective (objective flare). Glare is minimized by reducing the field and measuring diaphragms until they are just slightly larger than the specimen. Some instruments permit electronic compensation for glare, but it requires additional time and care. The two-wavelength method intrinsically compensates for glare by the selection of the wavelengths and the four measurements required for each nucleus (I_{0a}, I_a, I_{0b}, I_b).

Conical light is inherent in the Koehler system of illumination, but can be minimized by lower numerical aperture of the condenser (e.g., 0.4 in the H20x objective-condenser supplied with the Leitz MPV1) relative to the oil immersion objective usually used in Feulgen cytophotometry. Thus only the central, mostly parallel rays will be used in the light measurement. Some oil-immersion photometry objectives have an internal iris that can reduce the objective aperture to exclude all but the most paraxial rays.

Bandwidth errors can occur with a chromophore that has steep changes in optical density with wavelength. The obvious solution is to reduce bandwidth, but there is a limit to the reduction below which other errors are induced by diffraction effects. Another possible correction is to select a measuring wavelength on a plateau or in a more gently sloping part of the absorption curve – if such exists at or near the peak. With Feulgen-DNA measurements, the blue side of the curve is much more gently sloping than the red side, and thus the wavelengths chosen should be on the blue side.

According to Goldstein, the spot size in the scanners should not be greater than half the limit of resolution. This means that the ideal spot size is about 0.15 μm in diameter for green light and thus some error due to spot size occurs in most systems where the smallest practical spot diameter is 0.3 to 0.5 μm. Increased spot size leads to an optical density value that is less than the actual value. The reduction tends to be linear and the error can be corrected by using a series of spot sizes and extrapolating back to 0.1 to 0.2 μm spot size to determine a correction term that can be applied to the data. Deviation from perfect focus of the spot in relation to the specimen focus effectively increases spot size, so that the absorbance readings are lower than they should be. Careful focusing of the spot is thus essential.

Specimen errors are perhaps the least appreciated and most important. First the absorption laws must hold in the range of chromophore concentrations encountered in the specimens (see Berlyn 1969 for methods of testing absorption law conformation with the Schiff reagent). One of the advantages of microspectrophotometry is that one can visually select the nuclei to be measured, and this needs to be done astutely to avoid broken or crushed nuclei and to select nuclei of the cell type you are studying. For example, if you are studying spherical meristematic nuclei, do not include the elongated nuclei of developing xylem cells, since in some species these have been reported to have amplified nDNA values. Other specimen errors include the presence of interfering substances on the slide, lack of precision in selecting clear areas for background settings, stain variation in specimen or in the internal standard, and nonspecific staining. There must be an ample supply of measurable cells on the slide so that the operator is not tempted to select marginal or inappropriate cells. The first consideration is to start with excellent preparations. If the preparations are not excellent, the best choice is to start over again and continue to do so until they are excellent. No amount of care in measurement can compensate for a poor or marginal specimen.

5 DNA Cytofluorometry

This is the technique that has seen the most development in recent years. Its increased sensitivity permits femtogram (10^{-15} g) quantities of DNA to be measured and thus chloroplast and mitochondrial DNA as well as minute changes in the nuclear genome may be determined. Another reason for the resurgence of this technique, in addition to its ability to achieve and quantify the femtogram accuracy through the use of internal standards, is the development of new and more quantitative dyes, namely, DAPI (4,6-Diamidino-2-phenylindole), Hoechst-33258 (Bisbenzimide, 2′-[4-Hydroxyphenyl]-5-[4-methyl-1-piperazinyl]-2,5′-bi-1H-benzimidazole), Hoechst-33342 (Bisbenzimide, 2′-[4-Ethoxyphenyl]-5-[4-methyl-1-piperazinyl]-2,5′-bi-1H-benzimidazole) and mithramycin. The techniques to use these dyes are still under development and improvements are needed. Fluorescent probes, especially those involving the specificity of antibodies, have also contributed to the greater use of cytofluorometry. There are a number of recent applications of microspectrofluorometry to DNA measurement (e.g., Laloue et al. 1980; Coleman et al. 1981; Hull et al. 1982; Kawano et al. 1983; Miyakawa et al. 1984; Jackson 1985; Lawrence and Possingham 1986; Miyamura et al. 1986).

As far as genetic instability in tissue culture is concerned, there is little known about effects of culture conditions on plastid DNA content or its significance. Recent papers have shown that plastid nucleoids can increase in DNA content during development, and that cell types may differ in amount of DNA per chloroplast. Also DNA synthesis in nucleoids may not be closely linked to mitosis, possibly leading to different amounts of total plastid DNA between cells.

The basic instrumentation has been reviewed by a number of authors (e.g., Thaer 1966; Ruch 1970; von Sengbusch and Thaer 1973; Berlyn and Miksche 1976). Recent developments have been the use of scanning stages and microsecond shutters. The equipment is far from optimal, and better instruments could be produced with the technology that has already been developed for macro-level fluorometers. These instruments feature cooled photomultiplier tubes and built-in facilities for fluorescence polarization measurements. The basic instrument is diagramed in Fig. 6 A and B. Figure 6A illustrates transmitted light systems and Fig. 6B the incident light configuration. Figure 7 represents some of the additional features that are now available on more advanced instruments.

Fluorescence refers to the almost immediate re-emission of absorbed light (in less than 10^{-4} s). At room temperature, most molecules exist in the ground electronic

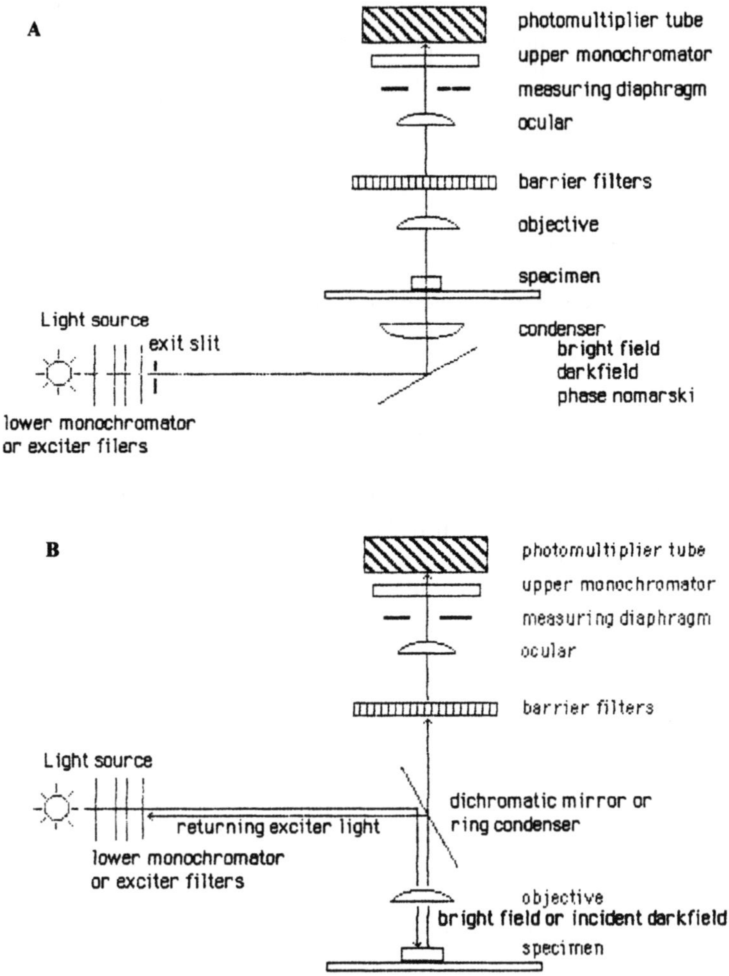

Fig. 6A,B. Microspectrofluorometry systems. **A** Transmitted light **B** Incident light

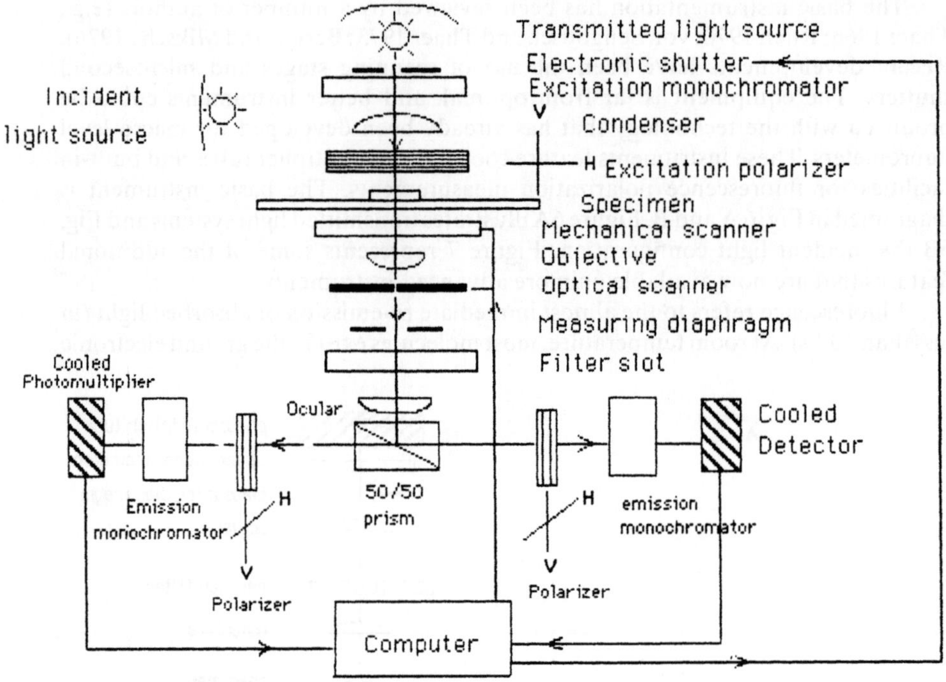

Fig. 7. Microspectrophotofluorometer equipped to measure flash, static and dynamic fluorescence, and fluorescence polarization

state, but there may be a number of excited vibrational states within the ground state. This is a consequence of the fact that the energy of molecules is displayed through four categories which can be listed in descending order as electronic, vibrational, rotation, and translational. The specificity of absorption of light by a substance is a consequence of its molecular configuration. Light energy is absorbed in discrete (whole) quanta and only those absorbing molecules transformed into an excited state above the ground state are capable of fluorescing. A molecule usually returns to the ground state in several jumps, and fluorescence occurs only in the final jump from the lowest "allowed" elevated energy level to the ground level (Fig. 8). The ground state is normally singlet (paired electron spins, + & –) and fluorescence is normally associated with singlet-singlet transitions.

Quantum efficiency or quantum yield, $\emptyset_f$, is defined as the number of quanta fluoresced divided by the number of quanta absorbed. From the Lambert-Beer law given above (all symbols defined above will not be redefined), it follows that:

$$I = I_o e^{-kcd}$$

and the amount of light absorbed is

$$(I_o - I) = (1 - T) = I_o (1 - e^{-kcd}).$$

Thus, the amount of light fluoresced, **F**, is given by:

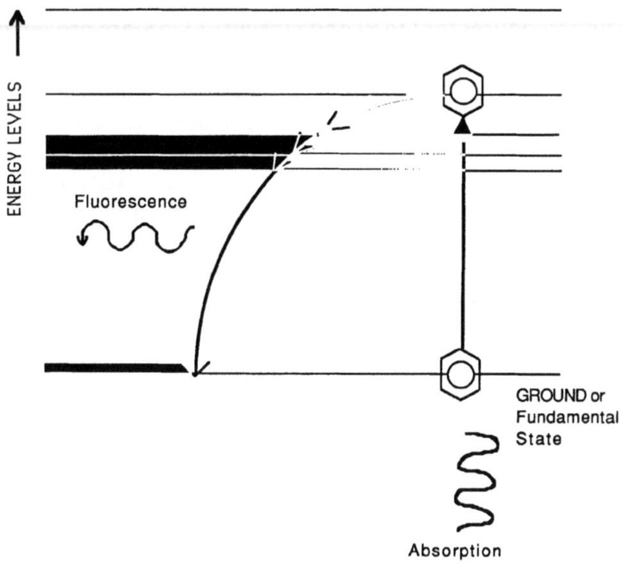

Fig. 8. Hypothetical absorption, excitation, and fluorescence of a model substance. (Berlyn and Miksche 1976)

$$F = \emptyset_f I_o (1 - e^{-kcd}).$$

However, when the amount of light absorbed is small (low optical density), as is the case with most of the DNA fluorochromes, the first term of the series expansion of e^{-kcd} may be substituted in the equation to give:

$$F = \emptyset_f I_o(kcd) = \mathbf{K}c.$$

In this case, the fluorescence is directly proportional to concentration. That is, for a given substance of sufficiently low absorptivity, at a given quantum efficiency, excitation light intensity, and thickness (path length), the fluorescence is linearly related to concentration with slope **K**. It is important that the internal standard and the plant nuclei be measured at their respective fluorescence maxima. It is also best that the same measuring diaphragm area be used for both the standard and the specimen, otherwise a correction may be necessary. Intensity of fluorescence varies greatly with wavelength, but quantum efficiency tends to be invariant with wavelength. However, quantum efficiency varies with temperature, pH, ion concentration and types of ions, dye concentration, chemical composition, and viscosity and properties of mounting media. Quantum efficiencies can vary from 0.95 to less than 0.01. Generally low temperature and high mountant viscosity increase the quantum efficiency by minimizing energy loss through heat transfer to solvent or neighboring molecules. Certain functional groups in the milieu or mountant increase quantum efficiency while others lower it (Bartels 1961). Hydrogen ion concentration is especially important, and may also affect stain specificity. Hydrogen ion concentration can be controlled with the use of buffers, but these can also cause fluorescence quenching so they must be tested for this effect before use.

5.1 Stain Preparation

The procedures for DAPI and Hoechst 33258, are provided here. Both of these dyes have proven to be quantitatively accurate in our procedures and with the species we are currently working with, viz., *Pinus caribaea* (three varieties), *Pinus coulteri, Pinus rigida, Pinus taeda, Populus deltoides,* and *Populus* hybrids. In our procedures, Hoechst shows much greater fluorescence maximum at a given excitation energy, and thus permits measurement at lower excitation energy, which decreases fluorescence fading. Unlike ethidium bromide, Hoechst 33258 does not stain primary cell walls to any extent and thus does not require the use of protoplasts. The dye in our system gives zero background and thus diaphragm size changes should not induce significant error.

5.2 Hoechst 33258 Procedure

Buffer is phosphate buffered saline (10 mM dibasic sodium phosphate, 150 mM NaCl). To this is added 1% mercaptoethanol and 0.015–0.5 μg/ml dye. The pH is adjusted to 4.0–7.0, depending on species. For *Pinus caribaea* we have found pH 5.0 to give lower background fluorescence than pH 4.0. Mercaptoethanol has been shown to retard fading and thus is a functional component of the technique. We stain the tissue for 1 h in the dark, but minimal staining time has not been determined. After staining, the tissue may be mounted in a drop of stain solution, coverslip sealed with nail varnish, and read immediately. Alcohol-xylene dehydration and mounting in fluormount (Edward Gurr Ltd., London) results in higher fluorescence intensity in some species and has the added advantage that the slides remain useful for at least several days. Spurr's nonfluorescing aqueous mountant may also be useful (see Berlyn and Miksche 1976, p. 191) (distilled water 40:cadmium iodide 34:polyvinyl alcohol 18:fructose 8). Adding 1% mercaptoethanol to this solution should help to retard fading. A low dye concentration increases sensitivity while high dye concentration causes fluorescence quenching.

5.3 DAPI Procedure

The dye is dissolved in McIlvaines citrate-phosphate buffer for pH 4.0 (Colowick and Kaplan 1955 p. 141) at a working concentration of 0.015 to 0.5 μg/ml with an addition of 1% mercaptoethanol (Coleman et al. 1981). The pH is readjusted to 4.0 after this addition. The tissue can be immersed directly in the dye solution or it can be applied in drops directly on the microscope slide (Lawrence and Possingham 1986). Staining time is generally 30 to 60 min in the dark at room temperature. The excess stain can be blotted off and coverslips sealed with nail varnish (Lawrence and Possingham 1986) or the tissue can be mounted in one of the ways described above for the Hoechst 33258 Bisbenzimide.

Miyamura et al. (1986) use 0.3 μg/ml DAPI and 0.3% glutaraldehyde in NS buffer (1 mM $CaCl2$, 1 mM $MgCl2$, 0.1 mM $ZnS04$, 0.25 M sucrose, 0.8 mM phenylmethyl sulfonylfluoride, 6 mM 2-mercaptoethanol, 1 mM EDTA, 20 mM

Tris-HCl, with pH adjusted to 7.6. A caution with procedure is that in other work glutaraldehyde has been shown to cross-link DNA, making it unavailable to acid hydrolysis (Greenwood and Berlyn 1968), and high pH was less specific than pH 4.0 in the work of Coleman et al. (1981). Miyamura et al. (1986) and Lawrence and Possingham (1986) found that DAPI faintly stained plant cell walls, and so they removed the walls before staining.

5.4 Internal Standard

The selection of the internal standard should theoretically be based on the amount of DNA to be measured. In our experience, chicken erythrocytes are the best standard for nuclear DNA measurements and this includes fluorescence as well as absorption measurements (Dhillon et al. 1977, 1978; Miksche et al. 1979; Berlyn et al. 1986b). Mature chicken erythrocytes do not replicate their DNA, and thus each nucleus contains a rather constant amount (mean value from interferometric studies is ca. 3.0 pg). Plant standards are widely used but are much more variable (Miksche et al. 1979; Berlyn et al. 1986b). Whether chicken erythrocytes are useful for femtogram amounts of DNA has not been tested, but theoretically it would be better to use a standard that was also in the femtogram range. Thus, Lawrence and Possingham (1986), in a study of chloroplast DNA, selected a bacterium, *Pediococcus damnosus (cerevisiae)* as their internal standard. They calculated that the absolute amount of DNA in this standard is 7.7×10^{-15} g (7.7 fg). *Pediococcus* cells vary in DNA according to cell type (4.51–18.7 fg) and thus it was necessary to select a particular cell type. They chose a "type 2", ovoid-shaped cell, from stationary phase cultures (5–7-day-old). The authors state that about a third of the cells in the culture were of the selected type. *Pediococcus* was selected over several other microorganisms tested because it has linearity between DNA content and fluorescence, and because it has a DNA mass and base composition similar to that of higher plant chloroplasts.

Miyamura et al. (1986) used T-phage DNA as an internal standard and calculated plastid DNA as the number of copies of the T genome. They also used a very sensitive photon counting microscope (sensitivity of 10 to 10^9 photons/mm^2/s) and minimized fading by grabbing the image in an image analyzer within 2s after exposure.

5.5 Measuring Procedure

The main problems in making accurate and precise fluorescence measurements are: (1) fluorescence fading, (2) obtaining sufficient fluorescence for required accuracy, (3) background fluorescence, (4) keeping excitation energy equivalent for specimen and internal standard, and (5) inclusion of scattered (diffracted) fluorescent light from fluorescent objects near or inside the measuring area.

Fluorescence fading is partly controlled by the presence of antioxidants in the stain and/or mountant. It can be further reduced by locating and positioning the cell for measurement under visible light of low intensity. This can be facilitated by using phase or interference contrast for cell selection and switching to the excitation light only just before measurement. The lowest possible intensity of excitation light should be used that will still provide enough fluorescence intensity so that the photomultiplier receives sufficient energy to render an accurate reading, i.e., if the multimeter reads from 0 to 1900 relative fluorescence units, the specimen fluorescence should read at least several hundred units above background, especially if the readings are noisy. Lowering the fluorescence intensity with neutral density filters on the exciting light (not on the fluoresced light!) can lower background fluorescence and increase precision; however, the precision may be at the expense of accuracy, and it may mask some of the true variation that exists. A compromise has to be reached whereby the background is reduced to the lowest value commensurate with the required accuracy. Obviously, these problems become extremely critical at the femtogram level. Some workers use an electronic shutter so that the tissue is exposed to the excitation light for only a fraction of a second – just long enough for the photomultiplier to make a reading. On the other hand, some specimens give an initial peak and then yield a constant intensity for some time. In this case some workers open the shutters on the excitation light and photomultiplier tube and make a reading after a set period of time, e.g., 1–10 s. It is quite possible to make accurate readings with minimal equipment, but it is, unfortunately, also possible to obtain inaccurate data with a highly sophisticated and expensive instruments.

6 Summary and Conclusions

Somaclonal variation can be used to provide the genetic variation necessary to fully utilize the potential of biotechnology in forestry and agriculture. However, once the desired somaclones are isolated and identified, it is necessary to be able to replicate them true to type. This requires methods to monitor and screen clones for any genetic changes subsequently induced in the somaclones. Optical techniques are a relatively rapid and inexpensive method of doing this and they are sensitive to DNA changes of ca. 1 fg (18 genes). They can also be used as a rapid first step in isolating and classifying somaclones before the much more time-consuming task of establishing genomic libraries.

The optical techniques discussed in detail are microspectrophotometry and microspectrofluorometry. Both scanning and two-wavelength Feulgen-microspectrophotometry are considered. The fluorescence methods discussed include incident and transmitted light instruments as well as different fluorochromes such as DAPI and Hoechst. Procedures are discussed from fixation through staining and measurement. The theory, advantages, disadvantages, and potential problems with each technique are analyzed and discussed. Picogram changes in DNA amount (10^{-12} g) are best measured by microspectrophotometry because the technique is well developed, faster, and relatively simple to use. The technique requires only a standard transmitted light microscope that is incorporated into a spectrophotometer. The microspectrofluorometric techniques are appropriate for femtogram changes in DNA (10^{-15} g). This approach is facilitated by incident light illumination with a Ploem type of illuminator. The technique is as yet not fully developed, but has great potential, and may possibly be useful for individual base determination as more specific dyes and light sensing devices become available.

References

Bajaj YPS (ed) (1986) Biotechnology in agriculture and forestry, vol 1. Trees I. Springer, Berlin Heidelberg New York Tokyo

Bartels PH (1961) Fluorescence microscopy. Leitz, New York

Bartels PH (1966) Sensitivity and evaluation of microspectrophotometric and microinterferometric measurements. In: Wied GL (ed) Introduction to quantitative cytochemistry, vol 1. Academic Press, New York London, pp 93–105

Berlyn GP (1967) The structure of germination in *Pinus lambertiana*. Bull 71, Yale Univ School For Environ Stud, New Haven, Conn

Berlyn GP (1969) Microspectrophotometric investigations of free space in plant cell walls. Am J Bot 56:498–506

Berlyn GP, Beck RC (1989) Cytophysical and cytochemical analysis of cell wall structure in relation to microbial and enzymatic degradation. In: Iqbal M (ed) KA Choudhury Memorial Volume, Aligarh Muslim University, Aligarh India (in press)

Berlyn GP, Cecich RA (1976) Optical techniques for measuring DNA quantity. In: Miksche JP (ed) Modern methods in forest genetics. Springer, Berlin Heidelberg New York, pp 1–18

Berlyn GP, Miksche JP (1976) Botanical microtechnique and cytochemistry. Iowa State Univ Press, Ames

Berlyn GP, Passof PC (1965) Cytoplasmic fibrils in proembryo formation in *Pinus*. Can J Bot 43:175–177

Berlyn GP, Dhillon SS, Miksche JP (1979) Feulgen cytophotometry of pine nuclei II. Effect of pectinase used in cell separation. Stain Technol 54:201–204

Berlyn GP, Beck RC, Renfoe MH (1986a) Tissue culture and the genetic improvement of conifers: problems and possibilities. Tree Physiol 1:224–240

Berlyn GP, Berlyn MK, Beck RC (1986b) A comparison of internal standards for plant cytophotometry. Stain Technol 61:297–302

Berlyn GP, Anoruo AO, Beck RC, Cheng J (1987) DNA content polymorphism and tissue culture regeneration in Caribbean pine. Can J Bot 65:954–961

Bowen CC (1956) Freezing by liquid carbon dioxide in making slides permanent. Stain Technol 31:87–90

Chaleff RS (1981) Genetics of higher plants: applications of cell culture. Cambridge Univ Press, New York, 184 pp

Chaleff RS (1983) The isolation of agronomically useful mutants from plant cell cultures. Science 219:676–682

Chaleff RS, Keil RL (1981) Genetic and physiological variability among cultured cells and regenerated plants of *Nicotiana tabacum*. Mol Gen Genet 181:254–258

Coleman AW, Maguire MJ, Coleman JR (1981) Mithramycin and 4′-6-damidino-2-phenylindole (DAPI) – staining for fluorescence microspectrophotometric measurements of DNA in nuclei, plastids, and virus particles. J Histochem Cytochem 29:959–968

Colowick SP, Kaplan NO (1955) Methods in enzymology, vol 1. Academic Press, New York London

Conger AD, Fairchild LM (1953) A quick-freeze method for making smear slide permanent. Stain Technol 28:281–283

D'Amato F (1977) Cytogenetics of differentiation in tissue and cell culture. In: Reinert J, Bajaj YPS (eds) Applied and fundamental aspects of plant cell, tissue, and organ culture. Springer, Berlin Heidelberg New York, pp 343–357

Dhillon SS, Berlyn GP, Miksche JP (1977) Requirement of an internal standard for microspectrophotometric measurements of DNA. Am J Bot 64:117–121

Dhillon SS, Berlyn GP, Miksche JP (1978) Nuclear DNA concentrations in populations of *Pinus rigida*. Am J Bot 65:192–196

Dhillon SS, Miksche JP, Cecich RA (1983) Microspectrophotometric applications in plant science research. In: Bechtel DB (ed) New frontiers in food microstructure. Am Assoc Cereal Chem, St. Paul. Mn. pp 27–69

Doerschug EB, Miksche JP, Palmer R (1978) DNA content and number of ribosomal RNA genes as related to protein content in cultivars and plant introductions of *Glycine max*. Can J Genet Cytol 20:531–538

Feulgen R, Rossenbeck H (1924) Mikroscopisch-chemischer Nachweis einer Nucleinsäure vom Typus Thymonucleinsäure und die darauf beruhende Elektiv-Färbung von Zellkernen in mikroscopischen Preparaten. Z Physiol Chem 135:203–248

Frisch B, Nagl W (1979) Patterns of endopolyploidy and 2C nuclear DNA content (Feulgen) in *Scilla* (Liliaceae). Plant Syst Evol 131:261–276

Goldstein DJ (1970) Aspects of scanning microdensitometry I. Stray light (glare). J Microsc 92:1–6

Goldstein DJ (1971) Aspects of scanning microdensitometry II. Spot size, focus, and resolution. J Microsc 93:15–42

Goldstein DJ (1975) Aspects of scanning microdensitometry III. The monochromator system. J Microsc 105:33–56

Greenwood MS, Berlyn GP (1968) Feulgen cytophotometry of pine nuclei: Effects of fixation role of formalin. Stain Technol 111–117

Hull HM, Howshaw RW, Wang J-C (1982) Cytofluorometric determination of nuclear DNA in living and preserved algae. Stain Technol 57:273–282

Jackson JF (1985) The nucleus-cytological methods and isolation for biochemical studies. In: Linskens HF, Jackson JF (eds) Modern methods of plant analysis, NS, vol 1. Cell components. Springer, Berlin Heidelberg New York Tokyo, pp 353–374

Karp A, Bright SWJ (1985) On the causes and origins of somaclonal variation. Oxford Surv Plant Mol Cell Biol 2:199–234

Kawano S, Nishibayashi, Shiraishi N, Miyehara M, Kuroiwa T (1983) Variance of ploidy in mitochondrial nucleus during spherulation in *Physarum polycephalum*. Exp Cell Res 149:359–363?

Laloue M, Courtois D, Manigault P (1980) Convenient and rapid fluorescent staining of plant cell nuclei with "33258" Hoechst. Plant Sci Lett 17:175–179

Lawrence ME, Possingham JV (1986) Direct measurement of femtogram amounts of DNA in cells and chloroplasts by quantitative microspectrofluorometry. J Histochem Cytochem 34:761–768

Lillie RD (1951) Simplification of the manufacture of Schiff reagent for use in histochemical procedures. Stain Technol 26:163–165

Mendelsohn ML (1958) The two-wavelength method of microspectrophotometry 2 A set of tables to facilitate calculations. J Biophys Biochem Cytol 4:415–424

Miksche JP (1968) Quantitative study of intraspecific variation of DNA per cell in *Picea glauca* and *Pinus banksiana*. Can J Genet Cytol 10:590–600

Miksche JP, Dhillon SS (1984) Microspectrophotometric analysis. In: Vasil IK (ed) Cell culture and somatic cell genetics of plants, vol 1. Academic Press, New York London, pp 744–752

Miksche JP, Dhillon SS, Berlyn GP, Landauer KJ (1979) Nonspecific light loss and intrinsic DNA variation problems associated with Feulgen DNA cytophotometry. J Histochem Cytochem 26:1377–1379

Miyakawa I, Aoi H, Sands N, Kuroiwa T (1984) Fluorescence microscopic studies of mitochondrial nucleoids during meiosis and sporulation in yeast, *Saccharomyces cerevisiae*. J Cell Sci 66:21–38

Miyamura S, Nagata T, Kuroiwa T (1986) Quantitative fluorescence microscopy on dynamic changes of plastid nucleoids during wheat development. Protoplasma 133:66–72

Nagl W (1978) Endopolyploidy and polyteny in differentiation and evolution. Elsevier North-Holland Biomedical Press, Amsterdam, 283 pp

Nagl W (1982) Cell growth and nuclear DNA increase by endoreduplication and differential DNA replication. In: Nicolini C (ed) Cell growth. Plenum, New York, pp 619–651

Ornstein L (1952) The distribution error in microspectrophotometry. Lab Invest 1:250–262

Patau K (1952) Absorption microphotometry of irregular-shaped objects. Chromosoma 5:241–262

Renfroe MH (1984) Organogenesis and nuclear DNA content in pine tissue cultures. Doctoral dissertation, Yale University Library, New Haven, Connecticut, USA

Renfroe MH, Berlyn GP (1985) Variation in nuclear DNA content in *Pinus taeda* L. tissue culture of diploid origin. J Plant Physiol 121:131–139

Ruch F (1970) Principles and some applications of cytofluorometry. In: Wied GL, Bahr GF (eds) Introduction to quantitative cytochemistry, vol 2. Academic Press, New York London, pp 431–450

Sengbusch G von, Thaer AA (1973) Some aspects of instrumentation and methods as applied to fluorometry at the microscale. In: Thaer AA, Sernetz M (eds) Fluorescence techniques in cell biology. Springer, Berlin Heidelberg New York, pp 31–39

Thaer AA (1966) Instrumentation for microfluorometry. In: Wied GL (ed) Introduction to quantitative cytochemistry, vol 1. Academic Press, New York London, pp 409–426

Wied GL (ed) (1966) Introduction to quantitative cytochemistry, vol 1. Academic Press, New York London

Wied GL, Bahr GF (eds) (1970) Introduction to quantitative cytochemistry, vol 2. Academic Press, New York London

I.8 Environmentally Induced Variation in Plant DNA and Associated Phenotypic Consequences

C.A. CULLIS[1]

1 Introduction

The organization of the life cycle of plants affects the interpretation of their genetic behavior. The higher plant cannot be considered to be a single organism as it is, in reality, an assemblage of competing units, the meristems, each of which is capable of contributing to the next generation. The orderly growth of the plant is achieved by the interaction of the competing meristems, usually through apical dominance. However, damage to the dominant meristem can result in the release of alternative units which can subsequently contribute to the next generation. Thus, in higher plants, there is no clear separation of the germ line and the soma. The consequence of this is that genetic variation arising from a mutation in any somatic cell has the potential to be transmitted to the next generation. This form of life strategy and its consequences for variation in higher plants has been reviewed recently (Walbot and Cullis 1985; Walbot 1986).

The evidence which has been accumulated over the past few years has indicated that the plant genome is in a state of flux, and a range of phenomena can be responsible for the rapid changes observed. These phenomena include the activity of transposable elements, and frequent amplification and deletion events. However, the events which result in a restructuring of the genome do not always occur at a constant rate. Their frequency can be influenced by a number of factors. Conditions which appear to be particularly conducive to increasing the rate of variation include the disruption of chromosomes (Dellaporta and Chomet 1985), the formation of particular F_1 hybrids (Gerstel and Burns 1966), the external environment in which the growth takes place (Durrant 1962), and the passage of cells through rounds of tissue culture and regeneration (Scowcroft and Larkin 1985). An insight into the mechanism by which the genome varies can be gained from the study of the instances where rapid genomic changes have occurred. Three such instances of conditions under which rapid genomic changes occur are considered here.

2 The Plant Genome

The DNA content per diploid nucleus varies greatly among eukaryotic organisms (Bennett and Smith 1976) with variation in excess of two- to threefold common

[1]Department of Biology, Case Western Reserve University, Cleveland, OH 44106, USA

Biotechnology in Agriculture and Forestry, Vol. 11
Somaclonal Variation in Crop Improvement I (ed. by Y.P.S. Bajaj)

among congeneric species. Within the angiosperms there is nearly a thousandfold range of variation, and there does not appear to be any correlation between genome size and organismal complexity. Thus the vast majority of the genome in most higher plants cannot be responsible for direct coding sequences. The bulk of the noncoding DNA is made up of repetitive sequences which can be arranged either in tandem arrays of a simple, or complex, repeating unit, or as dispersed repeating sequences (Flavell 1980). However, even those sequences which are present as tandem arrays can also be thought of as dispersed sequences, as separate blocks of such arrays may reside at various positions on the chromosomes (Cullis and Creissen 1987).

How has this wide range of variation in DNA content arisen? While a large genome size and heterogeneity seem to be indicators of evolutionary flexibility and progressivity, the decline in total DNA content apparently accompanies evolutionary specialization and adaptation to certain ecological niches (Nagl et al. 1983). Genome evolution is predominantly a molecular process governed by sequence amplification, divergence, dispersal, and loss. Thus the genome which is characterized is a result of the relative frequencies with which these events have occurred over evolutionary time. The comparison of related species, especially the cereals (Flavell 1982), has indicated the extent to which these processes have occurred over evolutionary time leading to the restructuring of the genome. However, the existence of changes occurring over a much shorter time-scale has been observed in a number of systems which are described below. The characterization of these rapidly varying systems facilitates experimental investigation into the precise mechanisms by which these genomic alterations can occur.

3 Environmentally Induced Heritable Changes in Flax

One of the most extensively studied systems in which rapid DNA changes can be generated in response to the environment is that observed in the flax variety Stormont Cirrus (Durrant 1962; Cullis 1977, 1983). The changes can occur in the genome when plants are exposed to various environmental stimuli with the generation of stable, genetically different lines. These lines have been termed genotrophs (Durrant 1962). The genotrophs differed from each other and the original line from which they were derived in a number of characters. These included plant weight, height (Durrant 1962) (Fig. 1), total nuclear DNA content as determined by Feulgen staining (Evans et al. 1966), and the isozyme band patterns for peroxidase and acid phosphatase (Fieldes and Tyson 1973). The alteration of the peroxidase isozyme band pattern was controlled at a single locus, with a dominant and a recessive allele. This gene showed the expected dominance in the F_1 and a 3:1 segregation of the dominant:recessive pattern in the F_2 in crosses between genotrophs.

3.1 Nuclear DNA Variation

The changes in the nuclear DNA associated with the environmentally induced heritable changes have been extensively characterized. It was shown by renatura-

Fig. 1. Fourth generation of two extreme flax phenotypes induced by fertilizer treatments. (Durrant 1962)

tion analysis that the majority of the variation occurred in the highly repeated fraction of the genome (Cullis 1983). Subsequently, representatives from all the highly repeated sequence families have been cloned (Cullis and Cleary 1986a). This set of cloned probes has been used to further characterize the differences between the genomes of the genotrophs. It has been shown that all but one of the highly repetitive sequence families varied, and that the variation in any one family was independent of the variation of any other family (Table 1).

The genes coding for the 5S ribosomal RNA (5S DNA) are one of the variable, highly repetitive sequences. The 5S DNA of flax is arranged in tandem arrays with a repeating unit of 350 base pairs. There can be more than 100,000 copies of this sequence in the genome, which amounts to more than 3% of the total nuclear DNA. This sequence can vary in copy number by more than twofold, from 117,000 copies in Pl (the original line from which the genotrophs were generated) to 49,600 in LH (Goldsbrough et al. 1981). In situ hybridization has shown that the 5S RNA genes are dispersed through the genome at many chromosomal sites (Cullis and Creissen 1987). In addition, the tandem arrays in which they are organized are of varying length, flanked by non-5S DNA. This latter conclusion is based on the characterization of a dozen 5S DNA containing recombinants isolated from a flax library. The use of a series of restriction enzymes, notably TaqI and BamHI, has identified

Table 1. A comparison of the proportion of the nuclear DNA (percent) from leaf and callus tissue of given lines homologous to particular cloned probes. (Cullis and Cleary 1986b)

Source of DNA	Probe								
	rDNA	5SDNA	pCL21	pCL13	pCL53	pCL2	pCL8	pBG87	pDC7
P1	1.5 ± 0.05	3.0 ± 0.3	5.6 ± 0.5	2.2 ± 0.14	2.3 ± 0.17	4.2 ± 0.20	2.6 ± 0.16	2.5 ± 0.18	3.2 ± 0.18
P1 callus	0.9 ± 0.04[a]	2.6 ± 0.24	5.9 ± 0.4	1.0 ± 0.1[a]	1.3 ± 0.15[a]	4.0 ± 0.20	2.4 ± 0.2	2.5 ± 0.15	2.5 ± 0.2[a]
L1	1.4 ± 0.06	2.5 ± 0.26	5.0 ± 0.4	1.6 ± 0.12	1.8 ± 0.15	4.2 ± 0.22	1.6 ± 0.2	2.4 ± 0.17	2.2 ± 0.17
L1 callus	1.4 ± 0.05	2.6 ± 0.22	5.7 ± 0.5	1.5 ± 0.11	1.6 ± 0.11	4.2 ± 0.21	1.7 ± 0.2	2.3 ± 0.20	2.3 ± 0.15
L3	1.5 ± 0.05	2.8 ± 0.19	4.9 ± 0.4	1.6 ± 0.13	2.0 ± 0.12	4.8 ± 0.24	2.6 ± 0.12	2.3 ± 0.23	2.6 ± 0.11
L3 callus	1.2 ± 0.06[a]	2.3 ± 0.19[a]	5.0 ± 0.4	1.7 ± 0.17	1.9 ± 0.12	3.6 ± 0.23[a]	2.1 ± 0.15[a]	2.3 ± 0.21	2.4 ± 0.17
L6	1.0 ± 0.03	1.4 ± 0.21	4.8 ± 0.5	1.5 ± 0.10	1.9 ± 0.15	3.6 ± 0.27	2.1 ± 0.17	2.3 ± 0.19	2.1 ± 0.13
L6 callus	1.1 ± 0.04	1.5 ± 0.16	4.6 ± 0.2	1.9 ± 0.11[a]	2.1 ± 0.17	4.0 ± 0.19	2.7 ± 0.16[a]	1.5 ± 0.14[a]	2.6 ± 0.15[a]
LH	1.6 ± 0.07	3.2 ± 0.17	5.8 ± 0.3	1.5 ± 0.17	1.1 ± 0.11	5.0 ± 0.3	3.3 ± 0.21	2.3 ± 0.22	2.8 ± 0.2
LH callus	1.5 ± 0.06	3.1 ± 0.22	5.7 ± 0.5	1.5 ± 0.12	1.5 ± 0.14[a]	8.0 ± 0.4[a]	3.3 ± 0.2	2.5 ± 0.21	2.4 ± 0.21
S3	1.2 ± 0.05	2.7 ± 0.21	4.5 ± 0.3	1.5 ± 0.11	1.8 ± 0.13	2.8 ± 0.19	2.5 ± 0.2	2.3 ± 0.13	1.6 ± 0.2
S3 callus	0.9 ± 0.04[a]	2.7 ± 0.19	4.4 ± 0.4	1.5 ± 0.15	1.8 ± 0.14	2.7 ± 0.22	2.2 ± 0.2	1.7 ± 0.15[a]	2.2 ± 0.17[a]
S6	0.9 ± 0.07	2.6 ± 0.16	4.9 ± 0.4	2.0 ± 0.11	2.2 ± 0.12	3.4 ± 0.12	2.3 ± 0.14	2.0 ± 0.14	2.7 ± 0.15
S6 callus	0.8 ± 0.06	2.6 ± 0.22	4.9 ± 0.5	2.0 ± 0.14	2.3 ± 0.17	3.4 ± 0.21	2.4 ± 0.10	2.0 ± 0.17	3.0 ± 0.08
C1	1.5 ± 0.04	2.0 ± 0.16	5.6 ± 0.3	1.3 ± 0.12	1.5 ± 0.16	3.8 ± 0.3	2.5 ± 0.25	2.4 ± 0.11	2.5 ± 0.13
C1 callus	1.4 ± 0.04	1.9 ± 0.13	5.6 ± 0.3	1.3 ± 0.13	1.5 ± 0.11	4.3 ± 0.24	2.8 ± 0.18	1.7 ± 0.20[a]	2.6 ± 0.15
S1	1.1 ± 0.06	3.0 ± 0.30	5.2 ± 0.4	1.9 ± 0.11	2.4 ± 0.14	3.9 ± 0.22	2.9 ± 0.21	2.3 ± 0.21	2.8 ± 0.2
FT37/1[b]	0.3 ± 0.03[a]	2.7 ± 0.22	5.0 ± 0.5	1.6 ± 0.14	1.3 ± 0.11[a]	2.1 ± 0.19[a]	1.8 ± 0.16[a]	1.3 ± 0.17[a]	1.1 ± 0.08[a]

[a]Values which differ significantly ($P < 0.01$) between leaf and callus DNA's from the same line.
[b]Agrobacterium-transformed flax callus from Dr. A.G. Hepburn.

two subsets of the 5S DNA's, one of which appears to vary rapidly, while the other remains more constant. Thus, within a repetitive sequence family it appears that there are subsets which can be differentially modulated. However, the mechanism by which these subsets are recognized within the genome is not known.

The overall impression from the data at present supports the contention that within the flax genome there are stable and labile subsets. All the rapid variations appear to take place in the labile subset. This conclusion is further strengthened by the results obtained from material which has been through a passage of tissue culture and regeneration. Once again, changes in the nuclear DNA were observed and were found to be localized in the same labile subset of genome as that identified for the environmentally induced changes.

3.2 Induction of Changes

Many of the phenotypic and DNA characteristics have been compared between genotrophs, which are stable lines resulting from environmentally induced heritable changes. These types of data indicate the end result of the induction process, but neither how nor when the changes actually occur during growth under inducing conditions. This information can only be obtained by the characterization of plants actually growing under inducing conditions. Since the environmental perturbations themselves affect the phenotypic characters, the nuclear DNA, or some component thereof, is the most appropriate marker for following changes occurring during growth under inducing conditions.

The induction of the total nuclear DNA changes and the variation in the number of ribosomal RNA genes have been followed in plants grown under inducing conditions (Evans et al. 1966; Cullis and Charlton 1981). It is clear from these experiments that the changes occur during the vegetative growth of these plants under inducing conditions. By the time the plants flower, the changes which will be observed in the next generation are already apparent in the apical cells. Thus, these changes do not appear to be generated during meiosis, although events at this stage of the life cycle may have been important in the stabilization and subsequent heritability of the induced changes (Durrant and Jones 1971).

Observations on plants grown under inducing conditions suggest that there may be some form of selection in the generation of the variation. The plants grow suboptimally under the most effective inducing conditions, although all the individuals survive and contribute to the next generation. However, after some time of growth under the inducing conditions, the growth rate of some part of the plant can increase dramatically. This portion of the plant grows as if it were under more favorable conditions, and generally contributes most to the next generation (Cullis 1986, 1987). One explanation for this observation is that there has been a change in the composition of the cells in either the apical and/or adventitious meristems (the meristem which is the basis for the increased growth which can be any of those present). This alteration is such that the cells in that meristem are now more fitted to grow under the prevailing environmental conditions. Since all the progeny of a single individual in which heritable changes have been induced are similar, all the cells in the meristem must contain the same changes.

4 Stability of the Genome in Inbreds and Hybrids

Extensive quantitative variation exists between species, but within a species large differences are rare. Is the limited variation within a species due to a lack of variation in the repetitive sequence families, or is it due to some mechanism which constrains the absolute amount of DNA while allowing the individual sequence families to float freely? The study of inbred lines of maize (Rivin et al. 1986) has allowed a partial answer to this question. This study compared the quantitative variation in ten inbred lines for nine repetitive sequences. These families represented about 50% of the most highly repetitive sequence families of the maize genome, but only a very small proportion of the mid-repetitive sequence families. Comparisons between inbred lines demonstrated variation for most of the sequence families tested, with the ribosomal RNA genes being the most variable. In contrast to the variation observed between inbred lines, the individual members of an inbred line were not detectably different. Each genotype had its own pattern of relative amounts of each of the sequence families. This pattern was characteristic for all the individuals of the family.

If each of the inbred lines has a stable set of repeated sequence families, then how are the differences between lines generated? One mechanism would be the destabilization of sequence families on outcrossing the inbred lines. The altered sets could then be reset to new, different values on the repeated inbreeding required to establish new inbred lines. Such a mechanism has been noted in maize, *Microseris* spp. and flax. Wide crosses have been shown to activate normally quiescent transposable elements in maize (McClintock 1950). The 2C nuclear DNA content of intra- and interspecific hybrids in *Microseris* spp. are not always at the level expected from the values known for the parents (Price et al. 1983). Crosses between flax genotrophs which differed in the number of ribosomal RNA cistrons showed F_1 values which were not at the expected mid-parent value (Cullis 1979). Thus it would appear that intraspecific crossing between inbred lines is a factor which can precipitate genomic variation. It is not yet clear whether or not this form of genomic instability is restricted to a particular subset of the genome. If it transpires that there is a labile set of sequences involved in the genomic variability, then this hybrid effect may be generated by a mechanism analogous to that responsible for the environmentally induced changes in flax.

5 Somaclonal Variation

An understanding of the genetic and/or molecular basis for the generation of somaclonal variation has practical value in addition to satisfying scientific curiosity. On occasions it would be desirable to optimize the production of specific genotypes where variability was desired, while for the plant propagator seeking uniformity, an avoidance of such variation would be vital. One of the impacts of molecular genetic analysis has been the demonstration that the eukaryotic genome exists in a state of genomic flux. Many of the mechanisms by which the genome is reorganized have been observed in cells in tissue culture or in plants regenerated from such cultures.

A detailed description and understanding of the mechanisms by which variation can occur may allow the identification of the determinants which control the production of variation. This will, in turn, allow the plant breeders to select material which best suits their purpose, in terms of the stability of that material through tissue culture.

The genomic changes that have been observed to occur in tissue culture include aneuploidy, chromosomal rearrangements such as translocations, inversions, deletions, gene amplification and deamplification, activation of transposable elements, point mutations, cytoplasmic genome rearrangements, and changes in ploidy level (Larkin and Scowcroft 1983; Orton 1983, 1984; Evans et al. 1984). There is also accumulating evidence that specific DNA sequences can be altered in copy number.

Flax cultures showed both increase and decrease in highly repetitive sequences when compared with leaf DNA from the same line (Cullis and Cleary 1986b). The nine cloned probes representing all the highly repeated sequence families in the flax genome were used to hybridize to DNA from leaf and callus tissues and the results shown in Table 1. In two of the lines, L1 and S6, there were no differences between the leaf and callus DNA's. However, for the other seven pairs there was a difference for at least one probe. The only highly repetitive sequence not to vary was the light satellite DNA, which has been shown to be the most stable highly repetitive sequence in the flax genome (Cullis and Cleary 1986a).

In one comparison, that between L3 leaf and callus, the number of 5S RNA genes varied. A comparison of BamHI and TaqI restriction enzyme digests of these two DNA's hybridized with a cloned 5S DNA is shown in Fig. 2. It is clear that the reduction in the 5S DNA occurred in that subfraction which is resistant to TaqI digestion. From a quantitation of this experiment it could be shown that all the reduction in the 5S DNA in the callus line occurred in this TaqI resistant fraction. This result is consistent with those observed in comparisons between the genotrophs. In each case where the 5S RNA gene number has been shown to decrease,

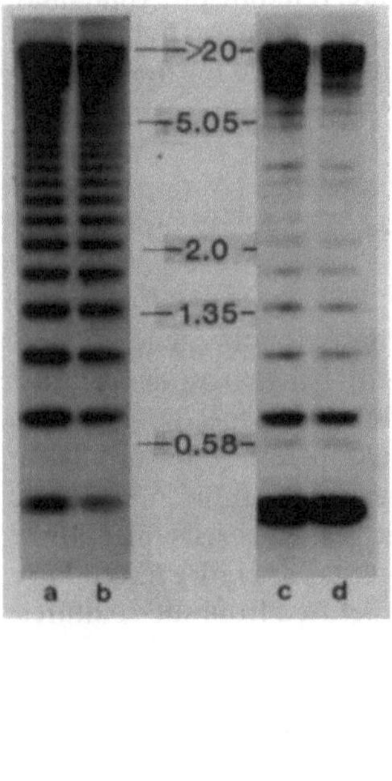

Fig. 2a-d. One microgram of DNA from leaves (**a, c**) and callus (**b, d**) from L3 digested to completion with the restriction endonucleases BamHI (**a, b**) and TaqI (**c, d**), separated on a 1.5% agarose gel, blotted onto nitrocellulose, and hybridized with nick translated plasmid pBG13. The sizes are given in kilobase pairs the decrease has been observed in the TaqI-resistant fraction (Goldsbrough et al. 1981; Cullis and Cleary 1986a).

DNA was prepared from the leaves of individual plants of the selfed progeny of four plants regenerated from P1 (Stormont Cirrus, the progenitor of the genotrophs). The proportions of the genomes of these plants that were complementary to eight of the genomic probes are given in Table 2. It is clear that the leaf DNA's from the regenerated plants could differ from that of the original line or from the callus DNA, or from both. It is also clear that there were a large number of differences between the DNA's from the individual progeny from a single regenerated plant. Thus, either the regenerated plants were heterozygous for copy number polymorphisms which were segregating, or they were still undergoing genomic rearrangement after regeneration.

Sequences other than those in the highly repetitive can also vary. The dramatic amplification of an intermediately repetitive sequence in Stormant Cirrus is shown in Fig. 3. The arrowed bands are present in the regenerated plants but absent from the original line. In addition, none of the arrowed bands cross-hybridize with any of the highly repeated sequences in the flax genome. Thus they appear to represent a significant amplification of a sequence originally present in intermediate or low copy number.

In a crown gall callus culture of flax, the number of ribosomal RNA genes decreased (to 300 copies compared to 800 in the original plant from which the culture was initiated) (Blundy et al. 1987). In rice suspension cultures several highly repeated sequences were amplified up to seventy-fivefold (Zheng et al. 1987). The methylation status of specific sequences can also alter in culture. The methylation patterns of 5S RNA genes in soybean callus and suspension cultures showed less methylation in newly initiated cultures than in whole plants, and this pattern could change in some cultures with time. In contrast to this observation, the methylation pattern of the rDNA in the flax crown gall callus was quantitatively unchanged in spite of the massive reduction in the number of these sequences (Blundy et al. 1987).

Table 2. A comparison of the DNA's from individual progeny (designated a-c) of four plants (numbered 1-4) regenerated from P1 callus. (Cullis and Cleary 1986b)

Source of DNA	Probe							
	rDNA	5SDNA	pCL13	pCL53	pCL2	pCL8	pBG87	pCL21
P1 leaf	1.5 ± 0.05	3.0 ± 0.3	2.2 ± 0.14	2.3 ± 0.17	4.2 ± 0.2	2.6 ± 0.16	2.5 ± 0.2	5.6 ± 0.5
P1 callus	0.9 ± 0.04	2.6 ± 0.24	1.0 ± 0.1	1.3 ± 0.15	4.0 ± 0.2	2.4 ± 0.2	2.5 ± 0.15	5.9 ± 0.4
1a	1.4[a]	2.3[a]	2.6[b]	3.3[ab]	3.3[a]	2.9[b]	2.7	5.7
2a	2.0[ab]	3.2[b]	2.6[b]	2.8[ab]	4.4	3.5[ab]	2.7	5.4
2b	1.4[b]	2.6	2.2[b]	1.6[a]	4.2	2.7	1.9[ab]	5.6
2c	1.5[b]	2.5	2.2[b]	2.3[b]	3.7	2.5	2.4	5.6
3a	1.5[b]	3.2[b]	2.5[b]	1.6[a]	2.7[ab]	1.9[ab]	2.9	5.7
3b	1.4[b]	2.4[a]	2.3[b]	1.2[a]	2.6[ab]	2.0[a]	2.7	5.2
3c	1.9[ab]	3.3[b]	3.4[ab]	1.5[a]	3.8	2.3	3.1[ab]	5.4
4a	1.8[ab]	2.5	–	3.1[ab]	2.2[ab]	–	–	5.3
4b	1.7[b]	–	–	3.3[ab]	2.2[ab]	–	–	5.5

[a]Values significantly different ($P < 0.01$) from P1.
[b]Values significantly different ($P < 0.01$) from P1 callus.

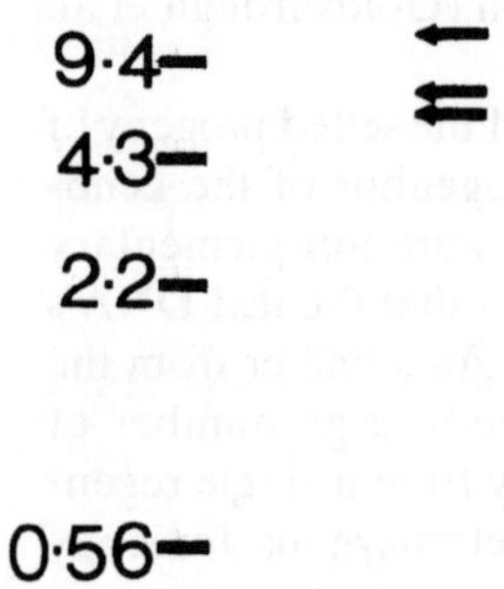

Fig. 3. Ethidium bromide stained gel of HaeIII digested DNA's from (Stormont Cirrus (lane *a*) and a plant regenerated from callus derived from Stormont Cirrus (lane *b*) showing the appearance of three bands in the regenerated plant DNA which are absent in the original line under these conditions.

In addition to the changes in the repetitive sequences, changes in the copy number of specific genes can also be selected in tissue culture. Cells with an amplified number of genes have been selected using resistance to the herbicides phosphinothricin (Donn et al. 1984) and glyphosate (Shah et al. 1986; Ye et al. 1987). These results show that gene amplifications can, indeed, be selected for in tissue culture systems. However, it has not been determined whether or not such amplifications are restricted to tissue culture systems.

Although specific instances of gene amplification have been described, most of the DNA alterations are found in the highly repetitive sequences. In studies with oats, Johnson et al. (1987) showed that the heterochromatic regions of root-tip chromosomes were late replicating. If chromatid separation occurred during anaphase, before replication was complete, then these heterochromatic regions were the sites of chromosome breakage. Analysis of the meiotic chromosomes of one regenerated plant did show a break, occurring in the heterochromatin. The consequence of this break was the formation of a heteromorphic chromosome pair.

In maize cultures, meiotic analysis showed that no chromosomal abnormalities were present in plants regenerated after 3 to 4 months in culture, while about half those regenerated after 8 to 9 months were cytologically abnormal (Lee and Phillips 1987). The breakpoints of these abnormalities were primarily on chromosome arms containing large blocks of heterochromatin such as knobs. Thus, in this case, the heterochromatic blocks appeared to be prime sites for the generation of chromosome abnormalities, and the number formed was dependent on the length of time the cells had been in culture.

Studies on regenerated wheat-rye hybrids showed a number of chromosomal abnormalities (Lapitan et al. 1984). Twelve of the 13 breakpoints involved in the deletions and interchanges characterized in this study were located in heterochromatin.

From results such as those described above, Johnson et al. (1987) and Lee and Phillips (1987) have proposed that one of the mechanisms by which chromosome breaks may be induced in culture is by the late replication of heterochromatin. Thus, if the replication of heterochromatin extended into anaphase, chromosome breaks could be induced. Such late replication of heterochromatin has been demonstrated in whole plants, but not in tissue cultures to date.

One of the consequences of a chromosome breakage in maize crosses is the activation of transposable elements (McClintock 1950). Since one of the proposed possible causes of somaclonal variation is the activation of transposable elements, studies to confirm this have been undertaken. Apparent activation of transposable elements has been found in studies with maize (Pesche et al. 1987) and alfalfa. With these results in mind, Lee and Phillips (1987) have proposed a model, in maize, whereby the incomplete replication of maize knob heterochromatin can lead to the formation of chromosome bridges, leading to chromosome breakage. The end result of this breakage cycle would be the activation of transposable elements, chromosome rearrangements, duplications and deletions, all of which could lead to somaclonal variation. It is still not clear whether the activation of transposable elements is the major cause of somaclonal variation in systems other than maize or even in maize itself.

6 Conclusions

The plant genome can be considered to be in a state of flux, with many changes occurring during both the mitotic and meiotic cycles. Some of the variation which can be observed is related to development, such as the endoreduplication in endosperm or cotyledon cells, and appears to affect the whole of the genome equally. Clearly, this type of variation does not change the genotype of the individual concerned. However, the variation associated with the environmental induction of heritable changes in flax and with somaclonal variation does alter the genotype of the plants. The rapid changes in the genome appear to be mediated by a variety of processes including transpositions, amplifications, deletions, and translocations.

The genomic alterations which have been described all take place under conditions which may be considered to be stressful to the individual. McClintock (1984) has proposed that the genomic response to stress, with the production of new genotypes, is related to an ability to generate diversity under stress conditions. If the genomic changes are produced in response to stress and result in an altered phenotype, then how are the phenotypic changes controlled by the genomic changes? It has been shown that the DNA changes occur mainly in the highly repetitive fraction of the genome, much of which is not transcribed (the exception being the genes coding for the ribosomal RNAs). Under these circumstances, the altered sequences cannot mediate their effect through products encoded by their sequence, but must generate diversity in some other way. One possible mechanism would be through some form of position effect variegation, where change in the DNA at a specific site alters the expression of sequences adjacent to that site. Examples of such effects are in the pleiotropic effects of the distal X heterochromatin in *Drosophila* (Hilliker and Appels 1982) and the effects of telomeric heterochromatin on seed development in triticale (Bennett and Gustafson 1982).

In conclusion, it is clear that the plant genome is variable, and the rate at which variation is generated can be increased by certain stressful perturbations. The alterations generated in response to these perturbations are limited to a subset of the genome. However, how this variation is generated, its direct phenotypic consequences and the way in which those phenotypic alterations are mediated by the changes in DNA are currently not understood.

References

Bennett MD, Gustafson JP (1982) The effect of telomeric heterochromatin from *Secale cereale* on triticale (× *Triticosecale* Wittmack). II. The presence or absence of blocks of heterochromatin in isogenic backgrounds. Can J Genet Cytol 24:93–100

Bennett MD, Smith JB (1976) Nuclear DNA amounts in angiosperms. Philos Trans R Soc London Ser B 274:227–240

Blundy KS, Cullis CA, Hepburn AG (1987) Ribosomal DNA methylation in a flax genotroph and a crown gall tumor. Plant Mol Biol 8:217–225

Cullis CA (1977) Molecular aspects of the environmental induction of heritable changes in flax. Heredity 38:129–154

Cullis CA (1979) Quantitative variation of ribosomal RNA genes in flax genotrophs. Heredity 42:237–246

Cullis CA (1983) Environmentally induced DNA changes in plants. CRC Crit Rev Plant Sci 1:117–129

Cullis CA (1986) Phenotypic consequences of environmentally induced changes in plant DNA. Trends Genet 2:307–309

Cullis CA (1987) The generation of somatic and heritable variation in response to stress. Am Nat 130:S62–S73

Cullis CA, Charlton LM (1981) The induction of ribosomal DNA changes in flax. Plant Sci Lett 20:213–217

Cullis CA, Cleary W (1986a) Rapidly varying DNA sequences in flax. Can J Genet Cytol 28:252–259

Cullis CA, Cleary W (1986b) DNA variation in flax tissue culture. Can J Genet Cytol 28:247–251

Cullis CA, Creissen GP (1987) Genomic variation in plants. Ann Bot 60 (Suppl 4):103–113

Dellaporta SL, Chomet PS (1985) The activation of maize controlling elements. Plant Gene Res 2:169–216

Donn G, Tischer E, Smith JA, Goodman HM (1984) Herbicide resistant alfalfa cells: an example of gene amplification in plants. J Mol Appl Genet 2:621–635

Durrant A (1962) The environmental induction of heritable changes in *Linum*. Heredity 17:27–61

Durrant A, Jones TWA (1971) Reversion of the induced changes in the amount of DNA in *Linum*. Heredity 27:431–439

Evans GM, Durrant A, Rees H (1966) Associated nuclear changes in the induction of flax genotrophs. Nature (London) 212:697–699

Evans DA, Sharp WR, Medina-Filho HP (1984) Somaclonal and gametoclonal variation. Am J Bot 71:759–774

Fieldes MA, Tyson H (1973) Activity and relative mobility of peroxidase and esterase isozymes of flax (*Linum usitatissimum*) genotrophs. I. Developing main stems. Can J Genet Cytol 15:731–744

Flavell RB (1980) The molecular characterization and organization of plant chromosomal DNA. Annu Rev Plant Physiol 31:596–596

Flavell RB (1982) Amplification, deletion and rearrangement: major sources of variation during species divergence. In: Dover GA, Flavell RB (eds) Genome evolution. Academic Press, New York London, pp 301–324

Gerstel DU, Burns JA (1966) Chromosomes of unusual length in hybrids between two species of *Nicotiana*. Chromos Today 1:41–56

Goldsbrough PB, Ellis THN, Cullis CA (1981) Organization of the 5S RNA genes in flax. Nucleic Acids Res 9:5895–5904

Hilliker AJ, Appels R (1982) Pleitropic effects associated with the deletion of heterochromatin surrounding rDNA on the X chromosome of *Drosophila*. Chromosoma 86:469–490

Johnson SS, Phillips RL, Rines HW (1987) Possible role of heterochromatin in chromosome breakage induced by tissue culture in oats (*Avena sativa* L.). Genome 29:439–446

Lapitan NLV, Sears RG, Gill BS (1984) Translocations and other karyotypic structural changes in wheat × rye hybrids regenerated from tissue culture. Theor Appl Genet 68:547–554

Larkin PJ, Scowcroft WR (1981) Somaclonal variation – a novel source of variability from cell culture for plant improvement. Theor Appl Genet 60:197–214

Larkin PJ, Scowcroft WR (1983) Somaclonal variation and crop improvement. In: Kosuge T, Meredith CP, Hollaender A (eds) Genetic engineering of plants. Plenum, New York, pp 289–314

Lee M, Phillips RL (1987) Genomic rearrangements in maize induced by tissue culture. Genome 29:122–128

McClintock B (1950) The origin and behaviour of mutable loci in maize. Proc Natl Acad Sci USA 36:344–355

McClintock B (1984) The significance of responses of the genome to challenge. Science 26:792–801

Nagl W, Jeanjour M, Kling H, Kuhner S, Michels I, Muller T, Stein B (1983) Genome and chromatin organizaton in higher plants. Biol Zentralbl 102:129–148

Orton TJ (1983) Experimental approaches to the study of somaclonal variation. Plant Mol Biol Rep 1:67–76

Orton TJ (1984) Somaclonal variation: theoretical and practical considerations. In: Gustafson JP (ed) Genetic manipulation and plant improvement. Plenum, New York, 427 pp

Pesche VM, Phillips RL, Gegenbach BG (1987) Discovery of transposable element activity among progeny of tissue culture-derived maize plants. Science 238:804–807

Price HJ, Chambers KL, Bachmann K, Riggs J (1983) Inheritance of nuclear 2C DNA content in intraspecific and interspecific hybrids of *Microseris* (Asteraceae). Am J Bot 70:1133–1138

Rivin CJ, Cullis CA, Walbot V (1986) Evaluating quantitative variation in the genome of *Zea mays*. Genetics 113:1009–1019

Scowcroft WR, Larkin PJ (1985) Somaclonal variation, cell selection and genotype improvement. In: Moo-Young M (ed) Comprehensive biotechnology. Pergamon, Oxford New York, pp 153–168

Shah DM, Horsch RB, Klee HJ, Kishore GM, Winter JA, Tumer NE, Hironaka CM, Sanders PR, Gasser CS, Aykent S, Siegel NR, Rogers SG, Fraley RT (1986) Engineering herbicide tolerance in transgenic plants. Science 233:478–481

Walbot V (1986) On the life strategies of plants and animals. Trends Genet 1:165–169

Walbot V, Cullis CA (1985) Rapid genomic change in higher plants. Annu Rev Plant Physiol 36:367–396

Ye J, Hauptman RM, Smith AG, Widholm JM (1987) Selection of a *Nicotiana plumbaginifolia* universal hybridizer and its use in intergenic somatic hybrid formation. Mol Gen Genet 208:474–480

Zheng KL, Castiglione S, Biasini MG, Biroli A, Morandi C, Sala F (1987) Nuclear DNA amplification on cultured cells of *Oryza sativa* L. Theor Appl Genet 74:65–70

I.9 Somaclonal Variation for Salt Resistance

M. TAL[1]

1 Introduction

Excess salt in the soil or in the water resources is an ever-increasing problem in the world today. In recent years attempts have been made to supplement conventional breeding directed toward the production of salt-resistant plants (Epstein et al. 1980; Norlyn 1980; Ramage 1980; Shannon 1985) with variability existing in tissue or cell culture (Rains 1982; Stavarek and Rains 1984). Such a supplement is especially important in species in which the gene pool is poor or completely lacks such variability.

The subject of variability existing in plant tissue and cell culture has received much attention in recent years. Attempts are therefore, being made to define the different aspects of this phenomenon. According to Chaleff (1981, 1983), tissue culture variability may result from: (a) genetic changes, such as those effecting nucleotide sequence of DNA and chromosome structure and number, which are transmitted through sexual crosses; (b) epigenetic changes, which are considered as changes in gene expression. They are relatively stable throughout mitotic divisions and may, therefore, persist after the removal of the inducing conditions. (c) physiological adaptation may also result from a change in gene expression which is maintained throughout mitotic divisions but disappears with the cessation of the inducing conditions. Unlike genetic changes, both epigenetic and physiological adaptation changes are usually not expressed in regenerated plants (referred to as R_0 generation) or their selfed progeny (referred to as R_1, R_2, etc). Thus, phenotypic expression in R_0, R_1 and R_2 plants is the criterion for distinguishing between the first class of changes and the other two classes, while stability at the cell level distinguishes between the second and the third classes. According to Chaleff (1983) these definitions should be considered as operational only, and therefore are recommended not to be applied too rigidly.

Handa et al. (1982b) distinguish between two classes of gene alterations: (1) genetic or mutational; (2) epigenetic or expressional. The latter are (a) environmentally induced; (b) relatively frequent and more easily reversible; (c) inherited only through mitosis and not through meiosis and may be stable. The term "adaptation" will be used throughout this chapter to define the latter class of gene alterations.

The term "variant" has been proposed to be used when the nature of the change – genetic, epigenetic, or only adaptive – is not known, whereas for the cases in which a trait is transmitted meiotically according to well established laws of inheritance the term "mutant" has been suggested (Chaleff 1981; Meins 1983). Widholm (1974) and Henke (1981) proposed four criteria for determining a mutant: (1) transmittance through sexual crosses and linkage; (2) low frequency of appearance; (3) stability through consecutive cell generations away from the selection agent; (4) biochemical mechanism connected to an altered gene product.

Scowcroft (1985) distinguished arbitrarily between tissue culture instability and somaclonal variation. Culture instability may appear on different levels (karyotypic, morphological, biochemical, or molecular) in callus or suspension cultures. Somaclonal variation (defined by Larkin and Scowcroft 1981) refers to variation observed among plants regenerated from tissue culture. According to Scowcroft (1985), there is evidence indicating that most of the variation found in regenerated plants occurs during the stage of tissue culture and did not originate in the cells of the donor explant.

Several authors have reviewed the subject of tissue culture instability and somaclonal variation and its potential for plant improvement in recent years. Further details can be found in these reviews (Maliga 1980; Chaleff 1981, 1983; Larkin and Scowcroft 1981, 1983; Meins 1983; Evans et al. 1984; Meredith 1984; Orton 1984; Scowcroft 1985; Semal 1986). The potential of and the problems associated with the application of cell culture methodology for the selection of stress-resistant cell lines and for studying the cellular mechanisms for stress resistance were discussed earlier (Nabors 1976; Croughan et al. 1981; Tal 1983; Stavarek and Rains 1984; Chandler and Thorpe 1986).

Here the main accomplishments in the search for salt-resistant somaclones, the difficulties and problems involved, and the perspectives for the future are presented.

2 Salt-Resistant Somaclones

Cell lines with enhanced resistance to salt have been isolated from many plant species (Table 1). Stability of salt resistance during mitosis has been determined in many experiments, and in some of them either embryos, plantlets, or whole plants were regenerated. It is important to remember that a basic assumption underlying such a study is that at least some of the mechanisms of salt resistance operate in both cultured cells and cells of the whole plant.

The main conclusions that can be drawn from Table 1 are:

1. The cell culture approach has been proved effective in obtaining salt-resistant cell lines in many di- and monocotyledoneous annual and some perennial glycophytic plant species from various families.
2. The selective agent (salt) was applied in most cases to previously prepared callus tissue or cells in suspension. In some cases the selection system included various explants from which callus (Nabors et al. 1982; Yano et al. 1982; Lebrun et al. 1985; Li et al. 1987), embryos (Lebrun et al. 1985), or plantlets (Li et al. 1987; Mathur et al. 1980) were induced to form directly in saline media.

3. Most of the cells exposed to salt were diploid or polyploid, and only very few of them were haploid (Zenk 1974; Dix and Street 1975; Croughan et al. 1981; Tyagi et al. 1981). In diploid and polyploid cells only dominant or co-dominant nuclear mutations, or mutations in organelles can be expressed phenotypically, whereas in haploid cells recessive mutations can also be expressed. However, in one case, where comparisons were performed, contrary to expectation, no striking differences were found between haploid and diploid cells in the rate of occurrence of resistant lines (Croughan et al. 1981).

Adaptation can, possibly, explain the following facts: (1) the extremely great success in obtaining salt-resistant cell lines in diploid and polyploid tissue and cell cultures of many plant species (Table 1); (2) lack of striking differences between diploid and haploid cells of alfalfa in the acquisition of salt resistance.

4. Almost all cases in which salt-resistant cell lines were obtained have relied on spontaneously occurring variation. When mutagens were employed (Nabors et al. 1975; Kochba et al. 1982; Gosal and Bajaj 1984; van Swaaij et al. 1986), no increase in the recovery rate of resistant lines above the spontaneous rate was detected except in the third case. Adaptation can possibly explain this finding also. It is interesting to mention in this connection the finding of McCoy (1987b), who showed that NaCl may enhance somaclonal variation.

5. The salt used as the selection agent in most of the studies was NaCl. Other salts used for selection included Na_2SO_4 (Pua and Thorpe 1986), synthetic seawater (Nyman et al. 1983), seawater (Yano et al. 1982), or a mixture of sulfate salts (McHughen and Swartz 1984). McHughen and Swartz (1984) included in their selection medium salt mixture of the same composition as that of high salt soil in Saskatchewan, where the main salts are sodium and magnesium sulfates. In some of the experiments, cross-resistance was tested by exposing NaCl-resistant cells to other salts. While NaCl-resistant cells of citrus were sensitive to KCl but grew well on various sulfate salts (Kochba et al. 1982). NaCl-resistant alfalfa cells were sensitive to these salts (Stavarek and Rains 1984). In most of the experiments the selection media contained hormones; however, in one case (McHughen and Swartz 1984), the selection media contained no hormones. McHughen and Swartz (1984) suggest that hormones can influence the response of the cells to salt and hence may lead to misinterpretation of the results. A possible effect of the medium constituents on the response of the cells to salt stress was suggested by Flowers et al. (1985) as an explanation for the unexpected higher salt resistance of cultured rice cells compared with that of leaf cells.

Another approach to the acquisition of salt-resistant cells may be based on the selection of cell lines which overproduce proline. Cultured cells of potato selected for resistance to proline analogs were found to be more resistant to freezing and in some cases also to salt (van Swaaij et al. 1986). Callus pieces of *Brassica napus* which were selected for proline overproduction survived much better on saline (Na_2SO_4) medium than unselected callus (Chandler and Thorpe 1987).

6. In some of the experiments, adaptation of the cells to osmotic stress preceded their exposure to salt stress. Since osmotic effect is part of the effect of salt, osmotically adapted cells can be exposed to conditions of high ionic stress without being osmotically shocked (Harms and Oertli 1985). Osmotically adapted cell cultures of carrot (Harms and Oertli 1985), tomato (Bressan et al. 1981), and potato

Table 1. Variation for salt resistance and regeneration of plants from cultured tissues and cells of various plant species

Species	Explant[a]	Salt(s) and concentration (mM)[b]	Exposure to salt OS = one step G = gradual	Stability in salt-free medium T = no. of transfers G = no. of generations	Regeneration (R_0)[c]	Resistance of R_0	Sexual transmittance	References
1. *Avena sativa*	Callus	170	G		+(plants)			Nabors et al. (1982)
2. *Brassica napus*	Callus	176(Na_2SO_4)	OS and G		+(roots)			Chandler et al. (1986)
3. *B. napus*	Callus	140(Na_2SO_4)		+(13T)				Chandler and Thorpe (1987)
4. *Capsicum annuum*	Callus and suspension and callus from suspension	171 and 342	OS	+(3T)				Dix and Street (1975)
5. *Cicer arietinum* *Pisum sativum* *Vigna radiata*	Callus and suspension	427(M)	G and OS		+(roots)			Gosal and Bajaj (1984)
6. *Cicer arietinum*	Callus	100(H)	OS	+(3T)				Pandey and Ganapathy (1984)
7. *Citrus sinensis*	Callus and suspension	200(M)	G	+(3T)	+(embryos and plantlets)			Kochba et al. (1982); Spiegel-Roy and Ben-Hayyim (1985)
8. *C. aurantium*	Callus and suspension	150(M)	G	+(3T)	+(embryos)			Kochba et al. (1982); Spiegel-Roy and Ben-Hayyim (1985)
9. *Colocasia esculenta*	Callus and suspension	Synthetic seawater	G		+(plantlets)			Nyman et al. (1983)
10. *Datura innoxia*	Callus (haploid)	171	OS	+(1 month)	+(plantlets)	+(callus)		Tyagi et al. (1981)
11. *Daucus carota*	Callus				+(plants)	+		Komizerko and Khretonova (1973)
12. *D. carota*	Callus	Seawater or various salts						Goldner et al. (1977)

(to be continued)

Table 1. *(Continued)*

Species	Explant[a]	salt(s) and concentration (mM)[b]	Exposure to salt OS = one step G = gradual	Stability in salt-free medium T = no. of transfers G = no. of generations	Regeneration (R_0)[c]	Resistance of R_0	Sexual transmittance	References
13. *D. carota*	Suspension	(DR → SR)	G					Harms and Oertli (1985)
14. *Glycine max*	(haploid)	64($NaNO_3$)						Jia-Ping et al. (1981)
15. *Ipomoea batatas*	Suspension	171	OS	+(3T)				Salgadogarciglia et al. (1985)
16. *Kickxia ramosissima*	Various explants	240		+(3T)	+(plantlets)			Mathur et al. (1980)
17. *Linum usitatissimum*	Callus	Sulfate salts	OS		+(plants)	+	+	McHughen and Swartz (1984)
18. *Lycopersicon esculentum*	Suspension	(DR → SR)						Bressan et al. (1981)
19. *L. esculentum*	Cotyledons	128	OS		+(shoots)			Li et al. (1987)
20. *Medicago sativa*	Callus from suspension	171(H)	OS		+(plants)			Croughan et al. (1978); Stavarek and Rains (1984)
21. *M. sativa*	Suspension				+(plants)	–		Smith and McComb (1983)
22. *M. sativa*	Callus	85 and 171	G	+(4T)	+(plants)			McCoy (1987b)
23. *Nicotiana sylvestris*	Suspension (haploid)	171	G					Zenk (1974)
24. *N. sylvestris*	Callus and suspension and callus from suspension (haploid and diploid)	171 and 342	OS	+(3T)				Dix and Street (1975)

25. *N. tabacum*	Suspension	150(M)	G		+(plants)	+	+	Nabors et al. (1975, 1980)
26. *N. tabacum*	Suspension and callus	130(SR → DR)	G					Heyser and Nabors (1981b)
27. *N. tabacum*	Suspension	171	G	−(1T)				Hasegawa et al. (1980)
28. *N. tabacum*	Suspension	600 and 770	G					Binzel et al. (1985); Hasegawa et al. (1986)
29. *N. tabacum*	Suspension	428	G	+(100G)	+(plants)	+	+	Bressan et al. (1985, 1987)
30. *N. tabacum*	Suspension	200 and 500	G	+(24G and 56G)				Watad et al. (1983, 1985)
31. *N. tabacum*	Callus	70(Na_2SO_4)	OS		+(plants)	+	+	Pua et al. (1985); Pua and Thorpe (1986)
32. *N. tabacum*	Suspension	Selenocystine and selenomethionine	OS					Flashman and Filner (1978)
33. *Oryza sativa*	Callus and suspension (haploid and diploid)	256 and 342(H)	OS					Croughan et al. (1981); Stavarek and Rains (1984)
34. *O. sativa*	Callus	Seawater	OS		+	+		Yano et al. (1982)
35. *Pennisetum americanum*	Embryogenic suspension	200	G	+(7T)	+(embryos)			Rangan and Vasil (1983)
36. *P. purpureum*	Embryogenic callus	214 and 342(H)	OS and G	−	+(plants)	−		Chandler and Vasil (1984)
37. *P. purpureum*	Embryogenic callus				+(plants)	+		Ashton (1986); Hildreth (1986)
38. *P. purpureum*	Callus	342	OS	+	+(plants)	+		Bajaj and Gupta (1986, 1987)
39. *Saccharum*	Suspension and callus	257(H)	G					Liu and Yeh (1982)
40. *Solanum melongena*	Callus	171(H)	G	+(4 months)				Jain et al. (1987)
41. *Solanum tuberosum*	Suspension (diploid)	(M)						van Swaaij et al. (1986)
42. S. tuberosum	Callus and suspension	(DR → SR)	G and OS	+(5T)				Sabbah and Tal (1987)

(to be continued)

Table 1. *(Continued)*

Species	Explant[a]	Salt(s) and concentration (mM)[b]	Exposure to salt OS = one step G = gradual	Stability in salt-free medium T = no. of transfers G = no. of generations	Regeneration (R_o)[c]	Resistance of R_o	Sexual transmittance	References
43. *Sorghum bicolor*	Callus	86	OS		+ (plants)		+?	Bhaskaran et al. (1983, 1986)
44. *Triticum aestivum*	Embryogenic callus	103	G		+ (plants)			Faizi and Ferguson (1986)
45. *Vitis rupensis*	Callus, suspension and embryos	150	OS and G	+ (3T)	+ (embryos)			Lebrun et al. (1985)
46. Colt cherry	Callus and protoplasts	200(NaCl, KCl, (Na_2SO_4)	OS		+ (plants)		+	Ochatt and Power (1988)
47. *O. sativa*	Callus	342(M)	OS		+ (plants)		+	Woo et al. (1985)

[a]Species 1 and 34 – callus was formed in saline media; 3 – cell lines which overproduce proline were more tolerant to Na_2SO_4; 16 – plantlets differentiation was induced in semisolid saline medium; 19 – callus proliferation and plantlets differentiation were made in saline media; 41 – hydroxyproline-resistant cell lines showed increased NaCl and frost resistance; 45 – callus was proliferated, suspension cells were formed and embryos were induced in saline media.

[b]The selective agent is NaCl if another salt is not mentioned. The concentration given is either the only one used or the upper limit found in a series of concentrations; M-mutagenic treatment; SR→ DR and DR→ SR – salt-adapted cells were exposed to drought stress and vice versá, respectively; H – halophytic response of cells.

[c]Species 20 – stunted and slow growing weak plants; 22 – abberant phenotypes, unbalanced chromosome sets and sterility; 26 – early senescence and death; 28 – differential tolerance to NaCl of the embryo tissues; 29 – several of altered characteristics; 34 – low fertility, unstable resistance; 43 – many plantlets died during acclimation, some were albinos, the green ones had low fertility.

(Sabbah and Tal 1987) showed an increased resistance to salt stress as compared with nonadapted cultures. Selection for salt resistance in tobacco cells conferred increased tolerance to osmotic stress produced by nonpenetrating osmotica (Heyser and Nabors 1981b). According to Harms and Oertli (1985), the use of osmotically adapted cell cultures is advantageous for the study of the physiological mechanisms of salt resistance, and for the selection of salt-resistant cell lines. Hence, in a two-stage selection procedure, the cells are first exposed to nonpenetrating osmoticum for the selection of osmotically adapted cells, and then to salt for the selection of cells resistant to ionic stress. The compounds used as first-step osmoticum in such experiments should be nonpenetrating and nontoxic. The finding of Thompson et al. (1986) that mannitol is taken up and metabolized in carrot and tobacco cells makes its applicability doubtful. Polyethylene glycol (Handa et al. 1982a) and melibiose (Dracup et al. 1986) were found suitable for this purpose.

7. The concentration of salt used for selection varied considerably. The highest reported NaCl concentration to which cells were able to adapt was 770 mM (Hasegawa et al. 1986), which is much higher than the NaCl concentration in seawater. Bressan et al. (1985) suggested that in the tobacco cells they studied, resistance to a moderate level of NaCl (171 mM) results from a totally reversible adaptation, while resistance to higher levels (428 mM) of NaCl results from both a reversible adaptation and a selective enrichment of the cell population with pre-existing, stable, nonreversible resistant cells. The question of whether a threshold concentration (above which enrichment of cells of the latter type occurs) also exists in other species should be clarified. The possibility that manipulation of other factors besides salt concentration may help to distinguish between true mutants and adapted cells should not be ruled out. For example, Henke (1981) suggests selecting mutants in a single step, and epigenetic variants stepwise.

8. The calli or cells were exposed to salt in one step or gradually. Harms and Oertli (1985) suggest that the latter procedure is preferred, since the former primarily tests the ability of cells to tolerate osmotic shock and to recover from it rather than their ability to withstand a certain level of salt or osmotic stress. According to them, the sudden shift to high stress conditions does not leave enough time for the cells to express their inherent capacities for stress response and adaptation, which can be remarkable. McHughen and Swartz (1984) claim, on the contrary, that the gradual method is ineffective for the selection of salt-resistant mutants, since nontolerant cells with a labile metabolism will have enough time to adapt to the stress and will, therefore, be positively selected. A similar opinion was expressed by Chandler and Vasil (1984). Bowman (1987) suggests that genetic differentiation with respect to growth and survival is expressed better under a rapidly induced short-term salinity treatment rather than gradually imposed salinity.

9. Salt resistance has generally been determined by measuring the increment of fresh or dry weight in callus and cell suspensions or by measuring packed cell volume or cell number in cell suspension. Visual appearance is an additional aspect for the description of callus growth. Some precautionary measures were suggested (Watad et al. 1983; Binzel et al. 1985) while measuring growth in tissue or cell culture. These include careful washing of ions present in the wall before their content within the cells is determined, and estimation of the volume occupied by water present outside the cell when fresh weight is determined. Flowers et al. (1985)

criticized the techniques used for determining growth in culture on the ground that there may be changes in cell size due to increase in water or ion contents which may affect fresh and dry weight quite independently of their effects on growth per se. They suggested, therefore, using respiration change in rice, which might reflect salt damage to this plant, as a much more direct indicator of cell state than the fresh or dry weight or cell number.

In most of the cases, calli or cells selected for salt resistance grow at a lower rate (in absolute terms) in the saline media than do nonselected cells in nonsaline medium. However, in some of the experiments, the growth of the former was not lower than that of the control. Such cases were the salt-resistant callus of *Datura* (Tyagi et al. 1981), rice (Croughan et al. 1981), and chickpea (Pandey and Ganapathy 1984), and cells of tobacco (Binzel et al. 1985; Watad et al. 1985) and potato (Sabbah and Tal 1987). NaCl-selected *Sorghum* callus grew equally well in the presence and absence of NaCl for up to 8 weeks (Bhaskaran et al. 1983).

10. Stability of salt resistance in culture is determined by transferring the resistant cells to a medium without salt for a number of generations and then back again to a saline medium. Stability was tested in a great number of experiments and in almost all of them the salt resistance of the cells was found to be stable during mitotic divisions in the salt-free media. Stability has been considered by several investigators as a means to clarify the type of change the resistant cells have undergone (Ben-Hayyim and Kochba 1982; Bressan et al. 1985, 1987; Watad et al. 1985). Ben-Hayyim and Kochba (1982) suggested stability as an indicator for the selection of a true genetic variant. A similar opinion was expressed by Dix and Street (1975). According to Watad et al. (1985), lack of stability indicates that physiological adaptation effected the change, while sustained resistance is compatible with an epigenetic change. On the basis of clonal analysis, Bressan et al. (1985) suggested that the stability observed in their material is the result of enrichment of the cell population with pre-existing cells having a stable resistance to high salt levels. The exact meaning of stability will probably be clarified only when the mechanisms responsible for its expression are known. At the present stage, the critical and most reliable test for the genetic basis of salt resistance in cultured cells remains the regeneration of salt-resistant somaclones and the demonstration of sexual inheritance of this characteristic.

11. The regeneration of salt-resistant plants from resistant cell lines and the demonstration of the sexual inheritance of this resistance are considered as the critical test and ultimate proof for the isolation of a salt-resistant genetic variant or a mutant. Unfortunately, although salt-resistant lines are easily selected for, and whole plants (or plants still not completely differentiated) were regenerated in about half of the experiments, there have been only a few cases where salt resistance of the regenerants was determined or its sexual inheritance was verified. The main reason for the lack of information is the fact that many of regenerants are defective in their development and/or fertility. Thus, the statement of Nabors (1976) that "the principal difficulty in adapting tissue culture breeding to food crop plants has heretofore been in obtaining reliable plant regeneration techniques", is still correct to a large extent at the present time, especially if the ending of that sentence is changed into . . . "in obtaining reliable techniques for the regeneration of normal and fertile plants". In many plants (*Brassica*, alfalfa, and rice, for example, refs.

[1–3] in Kavi Kishor 1987) the capacity for regeneration diminishes rapidly with time in culture. Loss of ability to regenerate or to undergo embryogenesis of tissues which have been exposed to high concentration of NaCl has been found in carrot (Komizerko and Khretonova 1973), alfalfa (Smith and McComb 1983), *Sorghum* (Bhaskaran et al. 1983), and *Pennisetum* (Rangan and Vasil 1983; Chandler and Vasil 1984; Bajaj and Gupta 1987). In some species chromosomal abberations were found to accompany the loss of regenerability (McCoy 1987b). According to Croughan et al. (1981), the loss of regeneration capability seems to be primarily a technical problem, which can be overcome by development of an appropriate media sequence. The main approaches commonly used to overcome the problems encountered in regeneration include the manipulation of hormones (Stavarek et al. 1980; Tatchell and Binns 1986), carbohydrates (Lupotto 1983; Kavi Kishor 1987), which can be used as energy source or as osmotica, and Na_2SO_4 (Pua et al. 1985). The latter two may exert their effect indirectly by influencing the endogeneous hormonal balance. Sometimes NaCl even enhances embryogenesis (Lebrun et al. 1985; Spiegel-Roy and Ben-Hayyim 1985). Additional possible approaches to overcome the regeneration problem are: (a) decreasing the time of exposure of the cells to salt by a short-term selection (McHughen and Swartz 1984); (b) producing calli or suspension cells directly in saline media (Nabors et al. 1982; Yano et al. 1982; Lebrun et al. 1985); (c) differentiating embryos or plantlets from various explants directly in saline media (Mathur et al. 1980; Lebrun et al. 1985; Li et al. 1987); (d) using cells which have a high inherent capacity for regeneration (Rangan and Vasil 1983; Smith and McComb 1983; Chandler and Vasil 1984).

3 Physiology and Genetics of Salt Resistance in Cell Culture

3.1 Physiology

Physiological investigation of cultured cells may help in elucidating the mechanisms involved in salt resistance of cultured cells (Hasegawa et al. 1986; Binzel et al. 1987a). A better understanding of these mechanisms may help to improve the techniques used to develop stress-resistant crop plants, in general, and salt-resistant in particular from cultured cells (Handa et al. 1982b). Some examples are listed:

a) Chandler and Vasil (1984) suggest that investigation of the uptake, transport, and distribution of the selective salt could contribute to better understanding of the increased sensitivity to salt of *Pennisetum* plants regenerated from NaCl-adapted callus.
b) Pua et al. (1985) state that knowledge of the balance of endogenous phytohormones is important, because they might be involved in the response of the cells to salt during the process of selection and in the success of the regeneration.
c) According to Ben-Hayyim (1987), the performance of cells under water stress may give useful information on the role of osmotic adaptation occurring during selection for NaCl resistance, since water stress is one aspect of salt stress. Knowledge of the response of cells to salt in general and the lethal concentration in particular is also important while planning the strategy of selection.

In addition, preliminary knowledge of the role of proline in salt resistance of the cells may affect the decision of what strategy to adopt, i.e., using salt or proline analogs as the selective agent.

While using cultured cells for studying the cellular mechanisms of salt resistance (as well as for the selection of resistant cell lines) the factors which might influence the response of such cells to salt should be considered. These factors include, besides the medium constituents, the stage of growth cycle, the inoculum density and the time under stress (Binzel et al. 1985; Bressan et al. 1985).

The main approach to study the cellular mechanisms of salt resistance includes the use of glycophytic cells which acquire salt resistance through in vitro adaptation. Some progress has recently been made in this area, mainly using tobacco cell culture. Since most of the findings have recently been reviewed (Hasegawa et al. 1986), only a short summary of those subjects which have been investigated most intensively in tissue or cell culture in relation to the physiology of salt resistance is presented here (Table 2). The most detailed study on the physiological mechanisms of salt resistance in cultured glycophytic cells was performed by the Purdue group in tobacco (Handa et al. 1983; Binzel et al. 1985; Handa et al. 1986; Binzel et al. 1987a,b; Bressan et al. 1987). According to this group, its results indicate that cultured cells derived from glycophytes have a genetic potential for the physiological and biochemical processes required for adaptation to high salinity. The adaptation of the tobacco cells involved considerable osmotic adjustment with an altered relationship between turgor and cell expansion, i.e., increase of turgor with

Table 2. Physiological research on salt resistance in cultured plant tissues and cells

Subject	Plant	Reference
Cell size and ultra-structure	Tobacco	Binzel et al. (1985, 1987a); Bressan et al. (1987)
	Citrus	Ben-Hayyim and Kochba (1983)
	Taro	Nyman et al. (1987)
Membranal mechanisms	Tobacco	Pesci et al. (1986)
Osmotic adjustment	Tobacco	Binzel et al. (1987a); Hasegawa et al. (1986); Heyser and Nabors (1981a); Watad et al. (1983)
	Citrus	Ben-Hayyim (1987)
Ion regulation	Tobacco	Binzel et al. (1987a); Hasegawa et al. (1986)
	Citrus	Ben-Hayyim et al. (1987)
	Alfalfa	Stavarek and Rains (1984)
Cellular compartmentation	Tobacco	Binzel et al. (1987b)
Water relations	Tobacco	Binzel et al. (1985); Hasegawa et al. (1986)
Hormonal effects	Tobacco	LaRosa et al. (1985); Singh et al. (1986)
Proline	Tobacco	Binzel et al. (1987a); Dix and Pearce (1981); Watad et al. (1983)
	Potato	van Swaaij et al. (1986)
	Tomato	Rosen and Tal (1981); Tal and Katz (1980)
	Brassica	Chandler and Thorpe (1987)
	Distichlis	Dains and Gould (1985)
Energy balance	Alfalfa	Croughan et al. (1981); Stavarek and Rains (1984)

reduction of cell expansion. The latter characteristic was found to be stable in the absence of NaCl and was transmitted to the regenerated plants.

Dains and Gould (1985) suggest developing tissue cultures from halophytes in which salt resistance has a significant cellular basis as an alternative to the "adapted glycophytic cell" approach. According to them, the use of cell cultures which express salt resistance without preceding long-term adaptation to high salt conditions may help identify more clearly some of the cellular strategies involved in this important trait. Tissue or cell cultures which express salt resistance spontaneously were derived from various glycophytic and halophytic salt-resistant plants (Table 3). Table 3 presents all possible correlations between the response to salt of the whole plant and the response of tissue or cell culture developed from that plant. A positive correlation, where both the whole plant and the cultured cells are resistant to salt, is interpreted as an indication for the operation of cellular mechanisms of salt resistance on the whole plant level. A negative correlation, where the whole plant is salt-resistant but the isolated cells are sensitive, was regarded as an indication for the operation of mechanisms depending on the organization of the cells in tissue and/or organs in the whole plant. One of the unsolved dilemmas in this respect is the different response of cultured cells of different halophytes to salt in culture media; some are resistant and some are sensitive. There seems, however, a correlation between this response and the type of medium used: sensitive cells were found on solid medium, resistant in liquid medium. Whether or not this correlation is causal is still an open question. Another negative correlation, where the plant is sensitive but the isolated cells are resistant, was found in rice (Flowers et al. 1985) and beans (Gale and Boll 1978). Both cases are used as examples for the possible effects of the components of the medium on the cell response to salt and as a warning against potential mistakes while interpreting the results of experiments performed in tissue or cell culture, and especially when trying to extrapolate from cultured cells to whole plants.

3.2 Genetics

Some of the questions raised by breeders for salt resistance as well as by those trying to obtain salt-resistant somaclones concern: (1) the extent of genetic variability as regards salt resistance; (2) the means to increase this variability; (3) the nature of this variability. The existence of genetic variability for salt resistance in cell culture is considered to be indicated by the clonal analysis as performed in tobacco cells by Bressan et al. (1985), and the production of salt-resistant somaclones which transmit this characteristic to their progenies (Nabors et al. 1980; McHughen and Swartz 1984). The amount and nature of potential spontaneous genetic variability for salt resistance in cell culture is not known. This lack of knowledge results from the occurrence of physiological adaptation, which hampers the identification and selection of the pre-existing stable mutated salt-resistant cells, and most of all, from the difficulties in obtaining normal and fertile regenerants, which can be selfed or crossed, as required for genetic analysis.

Salt resistance is considered as a complex character (Ramage 1980) which is controlled by nuclear and occasionally also by organellar genes. Action of the latter

Table 3. Correlations between the response of the whole plant and of cultured tissue or cells derived from it, to salt stress

Plant species	Explant	*Positive correlation* Plant resistant Cells resistant	*Positive correlation* Plant sensitive Cells sensitive	*Negative correlation* Plant resistant Cells sensitive	*Negative correlation* Plant sensitive Cells resistant	Reference
Lycopersicon esculentum	Callus, protoplasts and shoot apices		+			
L. pennellii	Callus and shoot apices	+				Rosen and Tal (1981); Tal et al. (1978); Taleisnik-Gertel et al. (1983)
L. peruvianum	Callus, protoplasts and shoot apices	+				
L. pennellii	Callus			+		Hanson (1984)
Phaseolus vulgaris	Callus		+			Smith and McComb (1981a)
Beta vulgaris	Callus	+				Smith and McComb (1981a)
Medicago sativa line W75RS	Callus	+				Smith and McComb (1981b)
Trifolium repens	Callus		+			Smith and McComb (1981b)
T. fraqiferum	Callus		+			Smith and McComb (1981b)
Brassica napus	Callus	+				Chandler et al. (1986)
B. campestris	Callus		+			Chandler et al. (1986)
Hordeum vulgare	Callus		+			Orton (1981)
H. jubatum	Callus	+				Orton (1981)
Suaeda maritima	Suspension	+				von Hedenström and Breckle (1974)
Salicornia europaea	Suspension	+				von Hedenström and Breckle (1974)
Distichlis spicata	Suspension	+				Warren and Gould (1982)
Spartina pectinata	Suspension	+				Warren et al. (1985)
Salicornia herbacea	Callus			+		Strogonov (1973)
Atriplex undulata	Callus			+		Smith and McComb (1981a)
Suaeda australis	Callus			+		Smith and McComb (1981a)
Medicago marina	Callus			+		McCoy (1987a)
Phaseolus vulgaris	Suspension				+	Gale and Boll (1978)
Oryza sativa	Callus				+	Flowers et al. (1985)

was demonstrated in sunflower (Shevyakova 1982). According to Bressan et al. (1987), the segregation of genetically altered organelles in regenerated plants may produce a large array of combinations of wild-type and altered organelles with a consequent polygenic-like expression. Tal (1984 and 1985) discussed the possibility that, as with other complex characters, salt resistance may mainly be controlled by a few major genes. The existence of genes which seem to have a major effect on salt resistance was suggested in the fern *Ceratopteris richardii* (Hickok and Warne 1987) and in pepper (Tal 1985). Hickok and Warne (1987) selected two mutants resistant to NaCl in the gametophytic generation of the fern, for which a control by a single nuclear gene was indicated. The decrease of Na^+ exclusion in the pepper mutant *Scabrous diminutive* as compared with that in the wild type seems to be the cause for its higher sensitivity to NaCl. It is interesting to mention in this connection the report of Tonelli (1987) on the selection of putative water-stress resistant mutants of tomato in the second generation of mutagenized plants or in plants regenerated from cultured cells. Binzel et al. (1987a) suggest that the multiple phenotypical and metabolic changes which characterize plant response to salt stress can be induced by a simple primary response, the genetic basis of which might be rather simple. The limitations or lack of genetic variability for salt resistance in crop species may result from the fact that most cultivated species were evolved in nonsaline environments where salt-resistance traits could have been lost from their gene pools (Maas and Nieman 1978). Such changes from salt resistance to salt sensitivity may occurred by one or few mutations in major genes involved in the inheritance of central components of the mechanism of salt resistance. The possible occurrence of such mutational events during the evolution of cultivated plants means that the reverse can occur at present.

The possible involvement of genes with a major effect on salt resistance is also indicated by molecular investigation of the products (protein) of genes expressed under condition of salt stress. Bressan et al. (1987), for example, found that one of the noticeable biochemical changes that ocurr during the adaptation of tobacco cells to salt is the accumulation of 26 kDa protein (referred to as osmotin-I), which results from a permanent change in the expression of a specific gene. This accumulation was found to be stable in the absence of NaCl, and was transmitted to R_0 plants. It is not yet known which of these gene products are involved in salt resistance and what are their precise functions (Hasegawa et al. 1986). Such studies, which may help identify the genes involved in salt resistance, are only at their very beginning.

4 Discussion and Perspectives

There is an increasing awareness of the potential and limitations of the technique of tissue and cell culture for the production of new genotypes with valuable characteristics for agricultural use, particularly in relation to salt resistance (Nabors et al. 1980; Smith and McComb 1981a; Handa et al. 1982b; Nyman et al. 1983; Chandler and Vasil 1984; McHughen and Swartz 1984; Lebrun et al. 1985; Pua and Thorpe 1986; Bajaj and Gupta 1987; McCoy 1987b). At the present stage, with the appreciable amount of experience and knowledge that has been accumulated, an analysis of the potential of this approach for the development of salt-resistant plants can be made.

Attempts to select salt-resistant cells in culture are based on the assumption that cultured cells represent physiologically, at least in part, the cells of the whole plant, i.e., that cellular mechanisms which are responsible for salt resistance in cultured cells also operate in the whole plant. The conclusions on the level – cellular or whole plant – of operation of the mechanisms of salt resistance, which is studied by comparing salt resistance in the whole plant and in cultured cells derived from it, are also based on this conclusion (Table 3). Thus, some of the most important conclusions and decisions made depend on the validity of this assumption. However, due to the limited success in obtaining salt-resistant somaclones, its validity is frequently questioned. Such doubts can be supported by the facts that, nutritionally and physically, cultured cells are to a large extent in a different environment than the cells of the organized multicellular plant; moreover they may be in a different state of differentiation. The phenomenon of adaptation which uniquely characterizes cells in culture also supports such doubts.

The main possible causes for the limited success can be classified as technical or biological.

Technical causes: (a) Lack of knowledge of the phenomenon of adaptation to salt stress in culture and how to control its occurrence. Such knowledge may enable a distinction between mutant and adapted cells and thus may help to improve the efficiency of selection of salt-resistant mutant cells. Watad et al. (1985) suggested that adaptation, a step-wise phenomenon, might result from gene amplification. This possibility was also discussed by Cullis (1985). (b) In many plant species, techniques for the regeneration of normal and fertile plants from salt-resistant cell lines are still unavailable or are not developed enough. Additional study is required to overcome the technical obstacles.

Biological causes (IK Vasil, pers. commun.): (a) Lack of correlation between the mechanism of resistance operating in cultured cells and those of the whole plant (a possible example for this was demonstrated by Lebrun et al. 1985 in *Vitis*). Consequently, successful selection at the cellular level does not necessarily assure recovery of a true genetically inherited salt-resistant plant. (b) Multigenicity of salt resistance, which makes selection for this trait in culture impossible because of the lack of appropriate strategies. However, Nabors (1976) suggests that alterations involving several genes could be obtained in tissue or cell culture by sequential selection. The subject of the genetics of traits which are considered multigenic, continuous or quantitative and the arguments supporting the idea that such traits may mainly be controlled by a few major genes were discussed in short in the previous section and in more detail by Tal (1984, 1985).

It is very important to verify the validity of these biological causes, since it may determine whether the efforts made to develop salt-resistant plants using tissue or cell culture techniques are worthwhile. It seems that this verification may be achieved only by an interdisciplinary research which will include:

1. Physiological study aimed at clarifying whether and to what extent cultured cells represent the whole plant (or specific cells in the whole plant) in their response to salt. Such a study may clarify what functions are shared by both and may explain

why adaptation (as defined in this chapter) is expressed only in cultured cells and not in cells of the whole plant. It may also allow control of this phenomenon and thereby improvement of selection efficiency.
2. Establishment of selection procedures which will prevent the development of adapted cells and thus will enable the enrichment of the cell population with mutated salt-resistant cells.
3. Solving of the problem of regeneration, which is considered one of the central obstacles for developing normal and fertile salt-resistant regenerants – this is a prerequisite for a genetic analysis of salt resistance evolved in cultured cells.
4. Molecular study to characterize, isolate, and manipulate genes expressed in salt-resistant cell lines under salt stress – these genes may prove to be those which control the expression and inheritance of salt resistance.

5 Protocols

At the present state of knowledge and accomplishments, it is impossible to recommend a definite protocol for the selection of salt-resistant cells and for the development of salt-resistant somaclones. Because of the large variety of approaches, which differ in details or in principle, four protocols which have resulted in the production of salt-resistant plants, and two additional are presented here. These protocols include both short-and long-term, one-step, and gradual procedures. Most procedures include the exposure of cells to the selective agent, directly to the highest concentration which allow their survival and growth. This treatment may prevent the development of adaptive cells. The short-term procedures may ensure that the ability of cells to regenerate will be maintained.

1. *Short-term*, following, with some modifications, McHughen and Swartz 1984 and Woo et al. 1985 who, respectively, reported the regeneration of salt-resistant plants from cultures of flax and rice.
   a) Production and establishment of either haploid, diploid, or polyploid callus or cells. The use of embryogenic cells or cells with high regeneration ability is recommended.
   b) Exposure of callus or cells to a range of salt concentrations for the determination of the "highest concentration" allowing cell growth. The salt used is that which best characterizes field conditions.
   c) Exposing (once) freshly prepared callus or cells, with or without previous mutagenic treatment, to the "highest concentration" of salt.
   d) Regeneration of plants from selected cells in an appropriate medium, with or without salt.
   e) Determination of salt resistance in R_0 plants and also in cultured cells derived from them.
   f) Production of R_1 and R_2 plants and determination of their resistance to salt under greenhouse and, if possible, under field conditions after their hardening.

2. *Use of salt-adapted feeder cells*, suggested by H.R. Lerner from the Hebrew University of Jerusalem, pers. commun.
   a) Production and establishment of either haploid, diploid, or polyploid callus or cells.
   b) Preparation of cells adapted to high salt level ("feeder cells") by gradual long-term exposure to salt or gradual exposure to mannitol and then to salt.
   c) Inoculation of solid medium containing high salt level with feeder cells and overlaying them with freshly prepared cells, treated or untreated with a mutagen. The two cell layers are separated by a filter paper.

   Other operations follow steps (d-f) of Protocol No. 1.

3. *Direct recurrent selection*, after Ochatt and Power 1988, who reported the regeneration of salt-resistant plants from culture of colt cherry.
   a) As step (a) in Protocol No. 1.
   b) Establishment of the callus by subculturing it, at least six times, on saline media without changing the salt concentration.

c) Exposure of the callus to at least three cycles, each cycle consists of alternate two-subculture passages (3 weeks each) on a medium with and then without the salt.

Other operations follow steps (d-f) Protocol No. 1.

4. *Long-term procedure* (after Bressan et al. 1985, who reported the generation of salt-resistant plants from a tobacco culture)

a) As step (a) in Protocol No. 1.

b) Transfer the cells gradually to media containing increasing concentration of NaCl (g/l):0–10–14–20–25 (S-25 cells). Exposure of the S-25 cells to NaCl for over 50 generations with at least 25 generations in media containing 25 g/l NaCl.

Other operations follow steps (d-f) of Protocol No. 1.

5. *Long-term or long-term, step-wise selection*, following Nabors 1983 and 1987 (in Ketchum et al. 1987), who reported the regeneration of salt-resistant plants from cultures of tobacco, oats, and rice.

a) As step (a) in Protocol No. 1.

b) Repeated subculture of the callus on medium containing NaCl concentration which reduces growth rate by 90%. Salt concentration can be increased every second to third passage or only when the culture doubles its fresh weight during the previous passage.

c) Regeneration is attempted after six passages, or earlier if high concentrations are being reached.

d) As steps (e) and (f) in Protocol No. 1.

6. *Use of proline analogs*, after Riccandi et al. (1983) and van Swaaij et al. (1986).

a) As step (a) in Protocol No. 1.

b) Exposure of mutagenized or nonmutagenized cells to different concentrations of the proline analogues hydroxyproline and azetidine-2-carboxilic acid. Subculturing of resistant colonies at the original concentrations of the analogs.

c) Transfer of the resistant cells to regeneration medium.

d) Exposure of regenerants to the analogs for the selection of the analog-resistant plants.

e) Testing the latter plants for salt resistance (and proline production) by exposure to elevated salt concentrations.

f) As step (f) in Protocol No. 1.

Acknowledgments. The author gratefully acknowledges the helpful criticism of Drs. HR Lerner and G.A.M. Van Marrewijk, and Ms. Dorot Imber.

References

Ashton B (1986) Pearl millet regeneration of somatic embryogenesis. Int Plant Biotech Network Newslett 6:4

Bajaj YPS, Gupta RK (1986) Different tolerance of callus cultures of *Pennisetum americanum* and *Pennisetum purpureum* Schum. to sodium chloride. J Plant Physiol 125:491–495

Bajaj YPS, Gupta RK (1987) Plants from salt tolerant cell lines of napier grass *Pennisetum purpureum* Schum. Ind J Exp Biol 25:58–60

Ben-Hayyim G (1987) Relationship between salt tolerance and resistance to polyethylene glycol-induced water stress in cultured citrus cells. Plant Physiol 85:430–433

Ben-Hayyim G, Kochba J (1982) Growth characteristics and stability of tolerance of citrus callus cells subjected to NaCl stress. Plant Sci Lett 27:87–94

Ben-Hayyim G, Kochba J (1983) Aspects of salt tolerance in NaCl selected stable line of *Citrus sinensis*. Plant Physiol 72:685–690

Ben-Hayyim G, Kafkafi U, Ganmore-Neumann R (1987) Role of internal potassium in maintaining growth of cultured *Citrus* cells on increasing NaCl and $CaCl_2$ concentrations. Plant Physiol 85:434–439

Bhaskaran S, Smith RH, Schertz K (1983) Sodium chloride tolerant callus of *Sorghum bicolor* (L.) Moench. J Plant Physiol 112:459–463

Bhaskaran S. Smith RH. Schertz KF (1986) Progeny screening of *Sorghum* plants regenerated from sodium chloride-selected callus for salt tolerance. J Plant Physiol 122:205–210

Binzel ML. Hasegawa PM. Handa AK. Bressan RA (1985) Adaptation of tobacco cells to NaCl. Plant Physiol 79:118–125

Binzel ML. Hasegawa PM. Rhodes D. Handa S. Handa AK. Bressan RA (1987a) Solute accumulation in tobacco cells adapted to NaCl. Plant Physiol 84:1408–1415

Binzel ML. Hess FD. Bressan RA. Hasegawa PM (1987b) Intercellular compartmentation of Na^+ and Cl^- in salt-adapted tobacco cells. 14th Int Botanical Congr. Berlin (West). p 65 (Abstr)

Bowman WD (1987) Physiological differentiation to salt stress in the C_4 nonhalophytic *Andropogon glomeratus*. 14th Int Botanical Congr. Berlin (West). p 74 (Abstr)

Bressan RA. Hasegawa PM. Handa AK (1981) Resistance of cultured higher plant cells to polyethylene glycol-induced water stress. Plant Sci Lett 21:23–30

Bressan RA. Singh NK. Handa AK. Kononowicz A. Hasegawa PM (1985) Stable and unstable tolerance to NaCl in cultured tobacco cells. In: Freeling M (ed) Plant genetics: proceedings of the third Annual ARCO plant cell research institute-UCLA symposium on plant biology. April 13–19. 1985. UCLA symposia on molecular and cellular biology: New Series A.R. Liss. New York. v. 35. pp 755–769

Bressan RA. Singh NK. Handa AK. Mount R. Clithero J. Hasegawa PM (1987) Stability of altered genetic expression in cultured plant cells adapted to salt. In: Monti L. Porceddu E (eds) Drought resistance in plants. Physiological and genetic aspects. Commiss Eur Communities. Brussels Luxembourg. pp 41–57

Chaleff RS (1981) Genetics of higher plants. Applications of cell culture. Univ Press. Cambridge

Chaleff RS (1983) Isolation of agronomically useful mutants from plant cell cultures. Science 219:676–682

Chandler SF. Thorpe TA (1986) Variation from plant tissue cultures: Biotechnological applications to improving salinity tolerance. Biotech Adv 4:117–135

Chandler SF. Thorpe TA (1987) Proline accumulation and sodium sulphate tolerance in cellular cultures of *Brassica napus L.* cv. Westar. Plant Cell Rep 6:176–179

Chandler SF. Vasil IK (1984) Selection and characterization of NaCl tolerant cells from embryonic cultures of *Pennisetum purpureum* schum. (Napier Grass). Plant Sci Lett 37:157–164

Chandler SF. Manda BB. Thorpe TA (1986) Effect of sodium sulphate on tissue cultures of *Brassica napus* cv. Westar and *Brassica campestris* L. cv Tobin. J Plant Physiol 126:105–117

Croughan TP. Stavarek SJ. Rains DW (1978) Selection of NaCl tolerant line of cultured alfalfa cells. Crop Sci 18:959–963

Croughan TP. Stavarek SJ. Rains DW (1981) In vitro development of salt resistant plants. Environ Exp Bot 21:317–324

Cullis CA (1985) Environmentally induced DNA changes. In: Pollard JW (ed). Evolutionary theory: paths into the future. A Wiley-Interscience. pp 203–216

Dains RJ. Gould AR (1985) The cellular basis of salt tolerance studied with tissue cultures of the halophytic grass *Distichlis spicata*. J Plant Physiol 119:269–280

Dix PG. Pearce RS (1981) Proline accumulation in NaCl-resistant and sensitive cell lines of *Nicotiana sylvestris*. Z Planzenphysiol 102:243–248

Dix PJ. Street HE (1975) Sodium chloride-resistant cultured cell lines from *Nicotiana sylvestris* and *Capsicum annuum*. Plant Sci Lett 5:231–237

Dracup M. Gibbs J. Greenway H (1986) Melibiose. a suitable. non-permeating osmoticum for suspension-cultured tobacco cells. J Exp Bot 37:1079–1089

Epstein E. Norlyn JD. Rush DW. Kingsbury RW. Kelley DB. Cunningham GA. Wronn AF (1980) Saline culture of crops: a genetic approach. Science 210:399–404

Evans DA. Sharp WR. Medina-Filho HP (1984) Somaclonal and gametoclonal variation. Am J Bot 71:759–774

Faizi F. Ferguson S (1986) Selection for salt tolerance in wheat. Int Plant Biotech Network Newslett 6:3

Flashman SM. Filner P (1978) Selection of tobacco cell lines resistant to selenoamino acids. Plant Sci Lett 13:219–229

Flowers TJ. Lachno DR. Flowers SA. Yeo AR (1985) Some effects of sodium chloride on cells of rice cultured in vitro. Plant Sci Lett 39:205–211

Gale J. Boll WG (1978) Growth of bean cells in suspension culture in the presence of NaCl and protein-stabilizing factors. Can J Bot 57:777–782

Goldner R, Umiel N, Chen Y (1977) The growth of carrot callus cultures of various concentrations and compositions of saline water. Z Pflanzenphysiol 85:307–317

Gosal SS, Bajaj YPS (1984) Isolation of sodium-chloride resistant cell lines in some grain-legumes. Ind J Exp Biol 22:209–214

Handa AK, Bressan RA, Handa S, Hasegawa PM (1982a) Characteristics of cultured tomato cells after prolonged exposure to medium containing polyethylene glycol. Plant Physiol 69:514–521

Handa AK, Bressan RA, Handa S, Hasegawa PM (1982b) Tolerance to water and salt stress in cultured cells. In: Fujiwara A (ed) Plant tissue culture 1982. Maruzen, Tokyo, pp 471–474

Handa S, Bressan RA, Handa AK, Carpita NC, Hasegawa PM (1983) Solutes contributing to osmotic adjustment. Plant Physiol 73:834–843

Handa S, Handa AK, Hasegawa PM, Bressan RA (1986) Proline accumulation and adaption of cultured plant cells to water stress. Plant Physiol 80:938–945

Hanson MR (1984) Cell culture and recombinant DNA methods for understanding and improving salt resistance of plants. In: Staples RC, Toenniessen GH (eds) Salinity tolerance in plants. Strategies for crop improvement. Wiley & Sons, New York, pp 335–359

Harms CT, Oertli JJ (1985) The use of osmotically adapted cell cultures to study salt tolerance in vitro. J Plant Physiol 120:29–38

Hasegawa PM, Bressan RA, Handa AK (1980) Growth characteristics of NaCl-selected and nonselected cells of *Nicotiana tabacum* L. Plant Cell Physiol 21:1347–1355

Hasegawa PM, Bressan RA, Handa AK (1986) Cellular mechanisms of salinity tolerance. HortSci 21:1317–1324

Henke RR (1981) Selection of biochemical mutants in plant cell cultures: some considerations. Environ Exp Bot 21:247–357

Heyser JW, Nabors MW (1981a) Osmotic adjustment of cultured tobacco cells (*Nicotiana tabacum* var. Samsum) grown on sodium chloride. Plant Physiol 67:720–727

Heyser JW, Nabors MW (1981b) Growth, water content and solute accumulation of two cell lines cultured on sodium chloride, dextran and polyethylene glycol. Plant Physiol 68:1454–1459

Hickok LG, Warne TR (1987) Selection characterization of sodium chloride tolerant mutants in the fern *Ceratopteris richardii.* 14th Int Botanical Congr, Berlin (West), p 75 (Abstr)

Hildreth G (1986) Salt-tolerant napier grass *Pennisetum purpureum.* Int Plant Biotech Network Newslett 6:4

Jain RK, Dhawan RS, Sharma DR, Chowdhury JB (1987) Salt tolerance and proline content: a comparative study in NaCl-treated and wild type cultured cells of eggplant. 14th Int Botany Congr, Berlin (West), p 75 (Abstr)

Jia-Ping Z, Roth EJ, Terzaghi W, Lark KG (1981) Isolation of sodium dependent variants from haploid soybean cell culture. Plant Cell Rep 1:48–51

Kavi Kishor PB (1987) Energy and osmotic requirements for high frequency regeneration of rice plants from long-term cultures. Plant Sci 48:189–191

Ketchum JLF, Gamborg OL, Hanning GE, Nabors MW (1987) Tissue culture for crops project progress report 1987. Colorado State University, Fort Collins, Colorado, pp 81

Kochba J, Ben-Hayyim G, Spiegel-Roy P, Saad S, Neumann H (1982) Selection of stable salt-tolerant callus cell lines and embryos in *Citrus sinensis* and *C. aurantium.* Z Pflanzenphysiol 106:111–118

Komizerko EI, Khretonova TI (1973) Effect of NaCl on the process of somatic embryogenesis and plant regeneration in carrot tissue culture. Fiziol Rast 20:268–276

Larkin PJ, Scowcroft WR (1981) Somaclonal variation – a novel source of variability from cell cultures for plant improvement. Theor Appl Genet 60:197–214

Larkin PJ, Scowcroft WR (1983) Somaclonal variation and crop improvement. In: Kosuge T, Meredith CP, Hollaender A (eds) Genetic engineering of plants. An agricultural perspective. Plenum, New York, pp 289–314

LaRosa PC, Handa AK, Hasegawa PM, Bressan RA (1985) Abscisic acid accelerates adaptation of cultured tobacco cells to salt. Plant Physiol 79:138–142

Lebrun L, Rajasekaran K, Mullins MG (1985) Selection in vitro for NaCl-tolerance in *Vitis rupestris* Scheele. Ann Bot (London) 56:733–739

Li NJ, Filippone E, Cardi T, Monti LM (1987) In vitro and in vivo evaluation of salinity tolerance in 9 tomato genotypes. 10th Eucarpia Meet Tomato, 2–6 Sept 1987, Pontecagnano, It, p 127

Liu M-C, Yeh H-S (1982) Selection of a NaCl tolerant line through stepwise salinized sugarcane cell cultures. In: Fujiwara A (ed) Plant tissue culture 1982. Maruzen, Tokyo, pp 477–478

Lupotto E (1983) Propagation of an embryogenic culture of *Medicago sativa* L. Z Pflanzenphysiol 111:95–104
Maas EV, Nieman RH (1978) Physiology of plant tolerance to salinity. In: Jung GA (ed) Crop tolerance to suboptimal land conditions. Am Soc Agron Publ 32, pp 277–299
Maliga P (1980) Isolation, characterization and utilization of mutant cell lines in higher plants. Int Rev Cytol Suppl 11A:225–250
Mathur AK, Ganapathy PS, Johri BM (1980) Isolation of sodium chloride-tolerant plantlets of *Kichxia ramosissima* under in vitro conditions. Z Pflanzenphysiol 99:287–294
McCoy TJ (1987a) Tissue culture evaluation of NaCl tolerance in *Medicago* species: Cellular versus whole plant response. Plant Cell Rep 6:31–34
McCoy TJ (1987b) Characterization of alfalfa (*Medicago sativa* L.) plants regenerated from selected NaCl tolerant cell lines. Tissue Cult Assoc Meet, Washington DC, pp 92–96
McHughen A, Swartz (1984) A tissue-culture derived salt-tolerant line of flax (*Linum usitatissimum*). J Plant Physiol 117:109–117
Meins F Jr (1983) Heritable variation in plants cell culture. Annu Rev Plant Physiol 34:327–346
Meredith CP (1984) Selecting better crops from cultured cells. In: Gustafson JP (ed) Gene manipulation in plant improvement. 16th Stadler Genetics Symp, Univ Missouri, Agric Exp Stn, Columbia, Miss, pp 503–528
Nabors MW (1976) Using spontaneously occurring and induced mutations to obtain agriculturally useful plants. BioSci 26:761–768
Nabors MW (1983) Increasing the salt and drought tolerance of crop plants. In: Randall DD, Blevins DG, Larson RL and Rapp BJ (eds) Current topics in plant biochemistry and physiology vol 2, University of Missouri-Columbia, pp 165–184
Nabors MW, Daniels A, Nadolny L, Brown C (1975) Sodium chloride tolerant lines of tobacco cells. Plant Sci Lett 4:155–159
Nabors MW, Gibbs GE, Bernstein CS, Meis ME (1980) NaCl-tolerant tobacco plants from cultured cells. Z Pflanzenphysiol 97:13–17
Nabors MW, Kroskey CS, McHugh DM (1982) Green spots are predictors of high callus growth rates and shoot formation in normal and in salt stressed tissue cultures of oat (*Avena sativa* L.). Z Pflanzenphysiol 105:341–349
Norlyn JD (1980) Breeding salt-tolerant crop plants. In: Rains DW, Valentine RC, Hollaender A (eds) Genetic engineering of osmoregulation. Impact on plant productivity for food, chemicals and energy. Plenum, New York, pp 293–309
Nyman LP, Gonzales CJ, Arditti J (1983) In vitro selection for salt tolerance of Taro (*Colocasia esculenta* var. antiquorum). Ann Bot (London) 51:229–236
Nyman LP, Walter RJ, Donovan RD, Berns MW, Arditti J (1987) Effects of artificial seawater on the ultrastructure and morphometry of taro (*Colocasia esculenta*, Araceae) cells in vitro. Environ Exp Bot 27:245–252
Ochatt SJ, Power JB (1988) Selection of salt/drought tolerant colt cherry plants (*Prunus avium* × *pseudocerasus*) from protoplast and explant-derived tissue cultures. Tree Physiol (in press)
Orton TJ (1981) Comparison of salt tolerance between *Hordeum vulgare* and *H. jubatum* in whole plants and callus cultures. Z Pflanzenphysiol 98:105–118
Orton TJ (1984) Somaclonal variation: theoretical and practical considerations. In: Gustafson JP (ed) Gene manipulation in plant improvement. 16th Stadler Genetics Symp, Univ Missouri, Agric Exp Stn, Columbia, pp 427–468
Pandey R, Ganapathy PS (1984) Isolation of sodium chloride-tolerant callus line of *Cicer arietinum* L. cv. BG-203. Plant Cell Rep 3:45–47
Pesci P-A, Reinhold L, Lerner HR (1986) Proton fluxes as a response to external salinity in wild type and NaCl-adapted *Nicotiana* cell lines. Plant Physiol 81:454–459
Pua EC, Thorpe TA (1986) Differential Na_2SO_4 tolerance in tobacco plants regenerated from Na_2SO_4-grown callus. Plant Cell Environ 9:9–16
Pua EC, Ragolsky E, Chandler SF, Thorpe TA (1985) Effect of sodium sulfate on in vitro organogenesis of tobacco callus. Plant Cell Tissue Org Cult 5:55–62
Rains DW (1982) Developing salt tolerance. Cal Agric 36:30–31
Ramage RT (1980) Genetic methods to breed salt tolerance in plants. In: Rains DW, Valentine RC, Hollaender A (eds) Genetic engineering of osmoregulation. Impact on plant productivity for food, chemicals and energy. Plenum, New York, pp 311–318

Rangan TS, Vasil IK (1983) Sodium chloride tolerant embryogenic cell lines of *Pennisetum americanum* (L.) K. Schum. Ann Bot (London) 52:59–64
Riccardi G, Cella R, Camerino G, Ciferri O (1983) Resistance to azetidine-carboxylic acid and sodium chloride tolerance in carrot cell cultures and *Spirulina platensis*. Plant Cell Physiol. 24:1073–1078
Rosen A, Tal M (1981) Salt tolerance in the wild relatives of the cultivated tomato: responses of protoplasts isolated from leaves of *Lycopersicon esculentum* and *L. peruvianum* plants to NaCl and proline. Z Pflanzenphysiol 102:91–94
Sabbah S, Tal M (1987) Responses of tissue culture of *Solanum tuberosum* to high NaCl and mannitol concentrations. 14th Int Bot Congr, Berlin (West), p 120 (Abstr)
Salgado-garciglia R, Lopez-Gutierrez F, Ochoa-alejo N (1985) NaCl-resistant variant cells isolated from sweet potato cells suspensions. Plant Cell Tissue Org Cult 5:3–12
Scowcroft WR (1985) Somaclonal variation: the myth of clonal uniformity. In: Hohn B, Denis ES (eds) Plant gene research. Genetic-fluxes in plants. Springer, Berlin Heidelberg New York Tokyo, pp 217–245
Semal J (ed) (1986) Somaclonal variation and crop improvement. Nijhoff, Dordrecht
Shannon MC (1985) Principles and strategies in breeding for high salt tolerance. Plant Soil 89:227–241
Shevyakova NI (1982) Salt tolerance of plastome chlorophyl mutants of sunflower. Fiziol Rast 29:317–324
Singh NK, LaRosa PC, Handa AK, Hasegawa PM, Bressan RA (1987) Hormonal regulation of protein synthesis associated with salt tolerance in plant cells. Proc Natl Acad Sci USA 84:739–743
Smith MK, McComb JA (1981a) Effect of NaCl on the growth of whole plants and their corresponding callus cultures. Aust J Plant Physiol 8:267–275
Smith MK, McComb JA (1981b) Use of callus cultures to detect NaCl tolerance in cultivars of three species of pasture legumes. Aust J Plant Physiol 8:437–442
Smith MK, McComb JA (1983) Selection for NaCl tolerance in cell cultures of *Medicago sativa* and recovery of plants from a NaCl-tolerant cell line. Plant Cell Rep 2:126–128
Spiegel-Roy P, Ben-Hayyim G (1985) Selection and breeding for salinity tolerance in vitro. Plant Soil 89:243–252
Stavarek SJ, Rains DW (1984) Cell culture techniques: selection and physiological studies of salt tolerance. In: Staples RC, Toenniessen GH (eds) Salinity tolerance in plants. Strategies for crop improvement. Wiley & Sons, New York, pp 321–334
Stavarek SJ, Croughan TP, Rains DW (1980) Regeneration of plants from long-term cultures of alfalfa cells. Plant Sci Lett 19:253–261
Strogonov BP (1973) Structure and function of plant cells in saline habitats. Isr Program Sci Transl. Wiley & Sons, New York
Tal M (1983) Selection for stress tolerance. In: Evans DA, Sharp WR, Ammirato PV, Yamada Y (eds) Handbook of plant cell culture, vol 1. Techniques for propagation and breeding. Macmillan, New York, pp 461–488
Tal M (1984) Physiological genetics of salt resistance in higher plants: studies on the level of the whole plant and isolated organs, tissues and cells. In: Staples RC, Toenniessen GH (eds) Salinity tolerance in plants. Strategies for crop improvement. Wiley & Sons, New York, pp 301–320
Tal M (1985) Genetics of salt tolerance in higher plants: theoretical and practical considerations. Plant Soil 89:199–226
Tal M, Katz A (1980) Salt tolerance in the wild relatives of the cultivated tomato: the effect of proline on growth of callus tissue of *Lycopersicon esculentum* and *L. peruviantum* under salt and water stresses. Z Pflanzenphysiol 98:283–288
Tal M, Heikin H, Dehan K (1978) Salt tolerance in the wild relatives of the cultivated tomato: Responses of callus tissues of *Lycopersicon esculentum, L. peruvianum* and *Solanum pennellii* to high salinity. Z Pflanzenphysiol 86:231–240
Taleisnik-Gertel E, Tal M, Shannon MC (1983) The response to NaCl of excised fully differentiated and differentiating tissues of the cultivated tomato, *Lycopersicon esculentum*, and its wild relatives, *L. peruviantum* and *Solanum pennellii*. Physiol Plant 59:659–663
Tatchell S, Binns AN (1986) A modified MS media for regeneration of direct explants and long term callus cultures of tomato. Tomato Genet Coop Rep 36:35–36
Thompson MR, Douglas TJ, Obata-Sasamoto H, Thorpe TA (1986) Mannitol metabolism in cultured plant cells. Physiol Plant 67:365–369

Tonelli C (1987) Genetic improvement of tomato: selection strategy for detecting water stress tolerant mutants. 10th Meet Eucarpia Tomato Work Group, 2–6 Sept 1987, Pontecagnano, It, pp 51–55
Tyagi AK, Rashid A, Maheshwari SC (1981) Sodium chloride resistant cell line from haploid *Datura innoxia*. A resistant trait carried from cell to plantlet and vice versa in vitro. Protoplasma 105:327–332
van Swaaij AC, Jacobsen E, Kiel JAKW, Feenstra WJ (1986) Selection characterization and regeneration of hydroxyproline-resistant cell lines of *Solanum tuberosum*: Tolerance to NaCl and freezing stress. Physiol Plant 68:359–366
von Hedenström H, Breckle SW (1974) Obligate halophytes ? A test with tissue culture methods. Z Pflanzenphysiol 74:183–185
Warren RS, Gould AR (1982) Salt tolerance expressed as a cellular trait in suspension cultures developed from the halophytic grass *Distichlis spicata*. Z Pflanzenphysiol 107:347–356
Warren RS, Baird LM, Thompson AK (1985) Salt tolerance in cultured cells of *Spartina pectinata*. Plant Cell Rep 4:84–87
Watad AA, Reinhold L, Lerner HR (1983) Comparison between a stable NaCl-selected *Nicotiana* cell line and the wild type K^+, Na^+ and proline pools as a function of salinity. Plant Physiol 73:624–629
Watad AA, Lerner HR, Reinhold L (1985) Stability of the salt-resistance character in *Nicotiana* cell lines adapted to grow in high NaCl concentrations. Physiol Veg 23:887–894
Widholm JM (1974) Selection and characterization of biochemical mutants of cultured plant cells. In: Street HE (ed) Tissue culture and plant science. Academic Press, New York London, pp 287–299
Woo S-C, Ko S-W, Wong CK (1985) In vitro improvement of salt tolerance in a rice cultivar. Bot Bull Acad Sin 26:97–104
Yano S, Ogawa M, Yamada Y (1982) Plant formation from selected rice cells resistant to salt. In: Fujiwara A (ed) Plant tissue culture 1982. Maruzen, Tokyo, pp 495–496
Zenk MH (1974) Haploids in physiological and biochemical research. In: Kasha KJ (ed) Haploids in higher plants – advances and potential. Univ Press, Guelph, pp 339–353

I.10 Somaclonal Variation for Nematode Resistance

G. Fassuliotis[1]

1 Introduction

The development of disease-resistant cultivars through plant breeding depends on the extent of variability in a plant population with resistance-bearing traits that can be selected for transfer into new cultivars. Breeders have been relatively successful in transferring nematode resistance into some crop species, through selection of variants occurring in genotypes and through introgression. For some crop species, the gene base for nematode resistance is extremely wide, and resistance can be easily transferred when it is expressed by a single dominant gene; in others, the base is so narrow that nematode resistance is nonexistent (Fassuliotis 1987). To increase the natural variation in plant populations, mutation techniques have produced some lines with increased nematode resistance (Fassuliotis 1987). The use of wild species as parental sources for resistance has been extremely successful in some crops. The tomato (*Lycopersicon esculentum*), to which root-knot nematode (*Meloidogyne* spp.) resistance was transferred by introgression of genes from the wild species *Lycopersicon peruvianum*, is an excellent example. F_1 hybrids were brought to maturity through embryo culture (Smith 1944). Eggplant, *Solanum melongena* lacks resistance to root-knot nematodes, and attempts to cross it with the resistant wild species, *S. sisymbriifolium*, have been unsuccessful (Fassuliotis 1975). In potato breeding *Solanum vernei* has served as an important source of resistance against the potato cyst nematode, *Globodera pallida*.

In vitro cell and tissue culture has been recognized as a potentially important source of variability for some traits not found in mutation breeding (Wenzel 1985). Regenerated plants from callus (somaclones) and from protoplasts (protoclones) often exhibit considerable phenotypic variability, which has allowed the recovery of increased levels or new sources of resistance to various viral and fungal diseases (Wenzel 1985). On the other hand, some in vitro work with nematode-resistant plant species has resulted in increased susceptibility (Fassuliotis and Bhatt 1981; Fassuliotis 1984).

We report here work using in vitro technology to transfer root-knot nematode resistance into eggplant and cyst nematode resistance into potato.

[1] US Department of Agriculture, Agriculture Research Service, US Vegetable Laboratory, Charleston, SC, USA

Biotechnology in Agriculture and Forestry, Vol. 11
Somaclonal Variation in Crop Improvement I (ed. by Y.P.S. Bajaj)

2 Eggplant (*Solanum melongena*)

2.1 Pathogenicity of Root-Knot Nematode

The usual phenotypic response of root-knot nematode infection in eggplant is the formation of root galls soon after penetration by the infective stage juveniles. Upon entering, the juveniles feed on transformed xylem cells called giant cells, from which they derive nourishment for growth and development. Females deposit eggs in a gelatinous sac outside the body.

The resistance expressed in the wild species *S. sisymbriifolium* is post-infectional. Although root swellings are produced after invasion by the infective stage juvelines, development of the nematode is severely retarded (Fassuliotis and Dukes 1972), due to a reduction in the number and size of giant cells.

2.2 Regenerants of *Solanum sisymbriifolium* from callus, and Their Characteristics

S. sisymbriifolium callus derived from stem pith parenchyma possesses good regenerative capacity on culture media that contain indole-3-acetic acid (IAA) and 6-[(3,3-dimethylallyl)amino]-purine (2iP). Organogenesis is induced over a wide range of hormone and auxin concentrations, the best being 73.8 μM 2iP and 0.03–1.7 μM IAA (Fassuliotis 1975).

Analysis of plants regenerated from *S. sisymbriifolium* callus suggested that they were variants for a number of morphological traits. The leaves of the regenerated plants were broader and less deeply lobed, fleshier, and generally darker green (Fig. 1). Stomata were larger than those of the donor plant and contained twice as many chloroplasts within each pair of guard cells. Flowers were slightly larger, anthers were longer and broader, the style sturdier and the stigma more massive and dark green. Pollen grains were larger and had high stain viability. Histologically, the upper and lower epidermal leaf cells of the callus plants were larger and had approximately half the number of cells per lineal unit than the donor plants. The palisade cells were about 30% longer and contained an increased number of chloroplasts which contributed to the dark green color of the leaves. Increased thickness of the leaf was due to an increase in the number of layers of spongy mesophyll cells interspersed with larger and more numerous intercellular spaces. Cytologically, microspores had a chromosome complement of $n = 24$, indicating the tetraploid nature of the regenerated plants (Fassuliotis 1977).

2.3 Regenerants of *Solanum melongena* from Callus and Their Characteristics

Unlike the wild species, Florida Market eggplant lacked the regenerative capacity to differentiate into whole plants. Callus produced only compact green buds in similar combinations of 2-iP and IAA that induced leaflets from *S. sisymbriifolium* callus (Fassuliotis 1975). Subsequently, Fassuliotis et al. (1981) developed a mul-

Fig. 1A,B. *Solanum sisymbriifolium* plant and variant plant after regeneration from callus derived from stem pith

tistep media sequence through which shoot morphogenesis of the eggplant cultivar was obtained (Fig. 2). A 3-year-old callus culture was passed through a liquid culture phase on Murashige and Skoog (MS) basal medium (1962) containing 13.5 μM 2,4-D. Large compact pale green clumps of callus formed. Upon transfer to an agar medium supplemented with IAA and 2-iP, the resulting calli produced masses of intense green nodules that failed to regenerate shoots.

Under the supposition that the endogenous auxin level in the green nodules suppressed shoot development, Fassuliotis et al. (1981) included the anti-auxins p-chlorophenoxy isobutyric acid (PCIB) or filter-sterilized ascorbic acid with 2-iP in the medium. Shoot differentiation was not induced. However, in one batch of medium in which the ascorbic acid additive was accidently autoclaved with the rest

Cell culture on MS medium

↓

Modified MS medium (0.298 mM $MgSO_4 \cdot 7H_2O$ and
1.98 mM KNO_3) with 13.5 μM 2,4-D
(FRIABLE CELL SUSPENSION)

↓

MS medium with 13.5 μM 2,4-D
(LARGE AGGREGATES)

↓

MS agar medium with various combinations
and concentrations of IAA and 2iP
(GREEN NODULES)

↓

Modified MS medium (MS salts, thiamine, ascorbic acid + 2iP)
(BUDS DEVELOP INTO ORGANIZED SHOOTS)

↓

MS basal medium with salts at 1/3 concentration
(PLANTS WITH ROOTS AND LEAVES)

Fig. 2. Flow chart of the procedure for organogenesis of eggplant from suspension cultures (Fassuliotis et al. 1981)

of the medium, shoots differentiated from the callus, resulting in the recovery of whole plants. Many repeat experiments with old callus cultures and freshly initiated pith callus cultured on MS medium containing filter-sterilized or autoclaved ascorbic acid confirmed that the latter activated the green nodules of Florida Market eggplant callus to regenerate into shoots. The degradation products of the autoclaved ascorbic acid apparently acted as an auxin that promoted shoot development.

Regenerated Florida Market eggplant (Fassuliotis et al. 1981) also showed similar morphological variations from the donor plant. The leaf morphology differed phenotypically from that of plants derived from seed. Stomates were larger and the leaves were thicker and deeper green in color. The microspores contained 24 chromosomes as compared to 12 chromosomes in seed-derived plants. Spontaneous doubling of chromosomes through endomitosis is quite common in tissue culture, possibly as an additive effect from the internal phytohormone concentration and the auxin in the medium.

2.4 Pathogenicity of Root-Knot Nematodes on Regenerants of *S. sisymbriifolium* and Eggplant

The typical phenotypic root galling responses common to *Solanum sisymbriifolium* and eggplant were expressed when selfed progeny of regenerants (R_1) derived from stem pith parenchyma callus, and plants derived from the parental seed were analyzed for their reaction to root-knot nematode infection. Based on a 1–5 scale with 5 equalling abundant galling or reproduction, regenerants, and parental seed plants of *S. sisymbriifolium* showed a moderate amount of root swellings, but no reproduction of *Meloidogyne incognita*. With *M. javanica* infection the reproduction on R_1 plants increased 20-fold over the seed parent. No differences in the gall or reproduction index to *M. incognita* (Table 1) or to *M. javanica* were noted in regenerated eggplant (Fassuliotis and Bhatt 1982).

Table 1. Response of eggplant (*Solanum sisymbriifolium* and *S. melongena*), propagated from seed and tissue culture, to *Meloidogyne incognita* and *M. javanica*. (Fassuliotis and Bhatt 1982)

		M. incognita[a]			*M. javanica*[a]		
Host	Source	Gall index	Repr. index	Eggs/g root	Gall index	Repr. index	Eggs/g root
S. sisymbriifolium	Seed	2.7	1.0	0.0	3.0	1.0	484
	R.P.[b]	2.7	1.0	0.0	3.0	4.5	8,715
S. melongena	Seed	4.0	5.0	112,800	4.5	5.0	.....
	R.P.[b]	4.0	5.0	44,400	4.0	4.5	.....

[a] Gall index and reproduction index based on a scale of 1–5:1 = no galling and/or reproduction, 5 = abundant galling and/or reproduction.
[b] R.P. = regenerated plants, R_1 progeny.

2.5 Protoplast Culture and Somatic Hybridization of Eggplant and *S. sisymbriifolium*

As an alternative to sexual hybridization to transfer root-knot nematode resistance into eggplant, Fassuliotis (1975) proposed protoplast fusion to bypass the complex sexual barriers between these species. Methodology was developed for isolation and culture of protoplasts from eggplant leaves (Bhatt and Fassuliotis 1981) and from cell suspensions of *S. sisymbriifolium*. The latter being colorless were used in fusion experiments with the green-colored leaf mesophyll protoplasts (Fig. 3).

There have been several reports on the physiological conditions required for culture of eggplant protoplasts. Plant and leaf age, photoperiodicity, and light intensity affected the efficacy of the extraction method, viability, and culture of mesophyll protoplasts (Bhatt and Fassuliotis 1981; Jia and Potrykus 1981; Saxena et al. 1981, 1987). Protoplasts were sensitive to the osmoticum source in the medium. Mannitol was detrimental to protoplasts. Sucrose encouraged cell wall regeneration, but mitotic divisions were inhibited. A combination of glucose and sucrose was

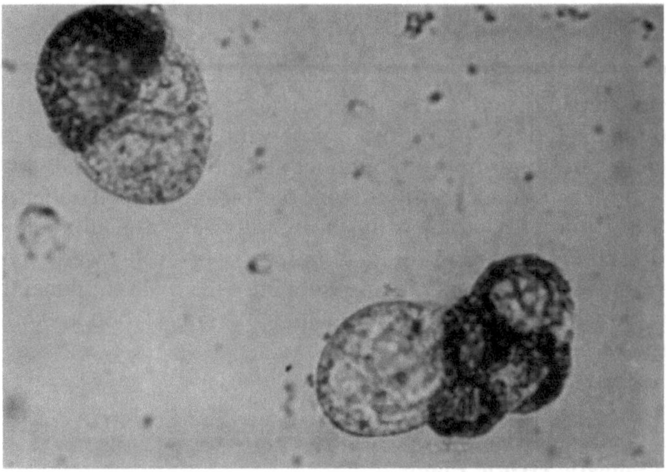

Fig. 3. Fused protoplasts of eggplant derived from leaf mesophyll cells (*dark*) and protoplasts from cell suspension cultures of *S. sisymbriifolium*

Fig. 4. Regenerated plants of *Solanum melongena* (*left*), somatic hybrid plants (*center*), and *S. sisymbriifolium* (*right*). (Gleddie et al. 1986)

necessary to maintain mitotic divisions of the cell colonies (Bhatt and Fassuliotis 1981). Gleddie et al. (1986) successfully fused *S. sisymbriifolium* and Imperial Black Beauty eggplant protoplasts and regenerated them to produce hybrid plants (Fig. 4). Using eggplant cell suspensions capable of growth in the presence of 1 mM 6 azauracil (AU) and leaf explant cultures of *S. sisymbriifolium* forming anthocyanin on a zeatin-supplemented MS medium, cell clones were selected from which 26 hybrid plants were regenerated from calli.

2.6 Characteristics of Somatic Hybrid Plants

Somatic hybrid plants regenerated by Gleddie et al. (1986) had morphological characteristics that were intermediate between those of eggplant and *S. sisymbriifolium,* but resembling more closely those of eggplant. The flowers of the hybrid plants were abnormal and contained separated petals which varied in number. Very few flowers produced anthers, none of which contained pollen. Other abnormalities were also noted. Cytologically, the hybrids were often mixoploids composed of cells with different chromosome numbers. Most cells were hypo-aneuploids with chromosome numbers ranging from 2n = 38 to 48. In pot culture, the roots of the hybrid plants developed poorly, with fewer lateral roots and branches than either of the parents (Gleddie et al. 1985).

2.7 Pathogenicity of Root-Knot Nematodes on Somatic Hybrid Plants

Gleddie et al. (1985) tested the response of somatic hybrid plants derived from the fusion of *S. sisymbriifolium* and eggplant protoplasts to infection by *M. incognita.* The hybrid plants inoculated with eggs of *M. incognita* did not support nematode reproduction and displayed a similar type of reaction as the resistant parent. Several galls found on the hybrid plants contained vacuolated, sexually undeveloped juveniles, which suggests that the resistance was transferred from the wild species to the hybrid via protoplast fusion (Table 2).

3 Potato (*Solanum tuberosum*)

A major nematode parasite of the potato (4x = 48) is the potato cyst nematode, *Globodera pallida.* Several wild species carry resistance to this nematode, which can provide plant breeders with a broad genetic base for potato cyst nematode resistance (Chavez et al. 1988). *Solanum vernei,* a diploid has served as one of the more important sources of resistance against this nematode in potato breeding programs, but transfer of resistance is difficult because inheritance of resistance is polygenic and quantitative with no clearcut segregation between resistant and susceptible plants. Highly variable results necessitate the need for large seedling populations in testing for resistance in order to obtain significant and reproducible results. Transfer of resistance is hampered also by the low fertility of *S. vernei* and the low seed set in crosses between *S. vernei* and *S. tuberosum* (Uhrig and Wenzel 1981).

3.1 Pathogenicity of the Potato Cyst Nematode

The potato cyst nematode, *Globodera pallida*, does not induce gall formation on potato roots. During its migration through the root, the juvenile leaves a path of destroyed cortical cells and feeds on tissues adjacent to the stele, which are

Table 2. Response of *Solanum melongena*, *S. sisymbriifolium*, and their somatic hybrids to inoculation with *Meloidogyne incognita*. Values are the means of five replications for the parents and one replicate for each hybrid plant. (Gleddie et al. 1985)

Plant	Mean no. of galls	Mean no. of egg masses	Gall index	Reproduction index
Eggplant cv.				
Imperial Black Beauty	52.5	40.5	3.75	3.25
S. sisymbriifolium	4.0	0	1.1	0
Somatic hybrid				
line 7	2.0	1	1	0
line 12	0	0	0	0
line 14	0	0	0	0
line 15	0	0	0	0
line 16	3	0	2	0

transformed into multinucleate cells analagous to the giant cells of root knot. The female also becomes saccate, but retains her eggs inside the swollen body.

3.2 Anther Culture of *Solanum tuberosum* and Nematode Response

In order to facilitate hybridization of *S. vernei* with potato, Wenzel et al. (1979) proposed a multistep breeding scheme which combines parthenogenetic, androgenetic, and combination breeding methods. After pollination with *S. phureja*, the tetraploid level of potato was reduced to 2x through parthenogenesis, and subsequently through anther culture to the monohaploid condition.

Most androgenetic clones derived from microspores doubled spontaneously to produce a homozygous population of fertile dihaploid ($2n = 24$) clones. Wenzel and Uhrig (1981) demonstrated that doubled monohaploids can be produced which possess resistance to the potato cyst nematode in the homozygous condition. From a highly *G. pallida*-resistant anther donor clone, other clones derived from it showed a good level of resistance, indicating that quantitatively inherited resistance to *G. pallida*, extracted from *S. vernei*, was retained during the reduction of the ploidy level from 2x to 1x. Resistance was transferred to the next generation with a very high efficiency.

4 Discussion and Conclusions

At present, we have no evidence that plants derived through organogenesis will provide somaclonal variants with increased resistance to plant-parasitic nematodes. On the contrary, there is some evidence that in vitro culture may shift the reaction to root-knot nematodes toward susceptibility. When leaf discs of Patriot tomato, which contains the Mi gene for resistance to root-knot nematodes, were regenerated on MS medium containing 0.1 μM NAA and 10 μM BA, the resulting plants exhibited the same response to galling and reproduction as plants regenerated from a susceptible cultivar (Fassuliotis and Bhatt 1981). Similarly, plants regenerated from leaf discs of NC-95 tobacco on MS medium containing three levels of BA showed that the level of resistance decreased as the concentration of BA in the medium was increased. Plants regenerated on MS plus 1 μM BA remained resistant to *M. incognita*, while those regenerated on 3 and 5 μM BA showed a significant shift toward the susceptible reaction (Fassuliotis 1984). Huettel and Hammerschlag (1986) found that the level of BA in the culture medium influenced the initial in vitro response of resistant Nemaguard peach plantlets to *M. incognita* infection at a concentration of 8.8 μM BA. The roots exhibited a galling response to infection by the nematode, but the nematode failed to complete its life cycle. This initial reaction was not observed at the 0.9 μM BA level.

While there is some evidence that the phytohormones may affect the host response to nematode infection, Ammati et al. (1984) found that *Lycopersicon* sp. plants regenerated from cotyledons remained resistant to *M. incognita*.

There appears to be a low correlation between expression of root-knot nematode resistance and the level of phytohormones in the plant tissues. The level of endogenous cytokinin in root-knot nematode-resistant peach and tomato roots was found to be lower than in susceptible roots (Kochba and Samish 1972; Van Staden and Dimalla 1977). Upon infection with the nematode, the level of cytokinin increased in both resistant and susceptible tomato (Van Staden and Dimalla 1977).

Exogenously applied phytohormones reversed resistance to root-knot nematodes. Kochba and Samish (1972) demonstrated that 20 mg/l NAA stimulated root galling and egg mass production on root-knot nematode-resistant peach types. Similarly, application of cytokinin to root-knot nematode-resistant tomatoes increased their susceptibility to nematode infection (Dropkin et al. 1969).

Unlike some other disease organisms, in which it is possible to screen large populations of cells for resistance by adding toxins isolated from the pathogens, this methodology is not appropriate for nematodes, since no such toxins have been identified or isolated. Methods for screening plants for resistance to either root-knot or cyst nematodes require an in toto root system to determine whether a plant expresses resistance or not. Reliable results have been obtained with excised root cultures for evaluating tomato (Orion and Pilowsky 1984) and soybean (Lauritis et al. 1982). The tomato seedlings from which the primary root is excised can be re-rooted and planted into soil for those that are resistant. The disadvantage for the plant breeder of the in vitro method with soybean is the inability to regenerate whole plants from the root explants. However, remnant seed can be used if the germplasm is homozygous for resistance.

Protoplast fusion is a viable method for transferring nematode resistance from a resistant plant to a susceptible one. It is clear that the value of somatic hybridization will depend on whether fertile hybrid plants can be obtained (Gleddie et al. 1986).

The haploid method has been used to incorporate disease resistance into a number of crops (Wenzel 1985). A stepwise reduction in the ploidy level of potato offers the advantages of a simpler inheritance and a better chance to combine qualitatively inherited characters in this plant. It is encouraging that, after reduction of the ploidy level, polygenically inherited resistance to *Globodera pallida* is still maintained at all ploidy levels (Wenzel 1985).

As more workers include nematode resistance in biotechnological research, improved in vitro culture methods may provide for more rapid and accurate methods for screening germplasm and for transferring nematode resistance from incompatible wild species to crop species.

References

Ammati M, Murashige T, Thomason IJ (1984) Retention of resistance to the root-knot nematode, *Meloidogyne incognita*, by *Lycopersicon* plants reproduced through tissue culture. Plant Sci Lett 35:247–250

Bhatt DP, Fassuliotis G (1981) Plant regeneration from leaf mesophyll protoplasts of eggplant. Z Pflanzenphysiol 104:81–89

Chavez R, Jackson MT, Schmiediche PE, Franco J (1988) The importance of wild potato species resistant to the potato cyst nematode, *Globodera pallida*, pathotypes P_4A and P_5A, in potato breeding. I. Resistance studies. Euphytica 37:9–14

Dropkin VH, Helgeson JP, Upper CD (1969) The hypersensitivity reaction of tomato resistant to *Meloidogyne incognita*: Reversal by cytokinins. J Nematol 1:55–56

Fassuliotis G (1975) Regeneration of whole plants from isolated stem parenchyma cells of *Solanum sisymbriifolium*. J Am Soc Hortic Sci 100:636–638

Fassuliotis G (1977) Tetraploid plants regenerated from callus of *Solanum sisymbriifolium*. In Vitro 13:146–147

Fassuliotis G (1984) Reversal of resistance to *Meloidogyne incognita* of NC 95 tobacco plants regenerated from leaf discs. Proc 1st Int Cong Nematology, Abstr 78, Guelph, Ont

Fassuliotis G (1987) Genetic basis of plant resistance to nematodes. In: Veech JA, Dickson DW (eds) Vistas on nematology. Painter De Leon Springs, Fl, pp 364–370

Fassuliotis G, Bhatt DP (1981) The effect of cloning on the resistance of Patriot tomato to *Meloidogyne incognita* race 1. Nematropica 10:66

Fassuliotis G, Bhatt DP (1982) Potential of tissue culture for breeding root-knot nematode resistance into vegetables. J Nematol 14:10–14

Fassuliotis G, Dukes PD (1972) Disease resistance of *Solanum melongena* and *S. sisymbriifolium* to *Meloidogyne incognita* and *Verticillium alboatrum*. J Nematol 4:222

Fassuliotis G, Nelson BV, Bhatt DP (1981) Organogenesis in tissue culture of *Solanum melongena* cv. Florida Market. Plant Sci Lett 22:119–125

Gleddie S, Fassuliotis G, Keller WA, Setterfield G (1985) Somatic hybridization as a potential method of transferring nematode and mite resistance into eggplant. Z Pflanzenzücht 94:348–351

Gleddie S, Keller WA, Setterfield G (1986) Production and characterization of somatic hybrids between *Solanum melongena* L. and *S. sisymbriifolium* Lam. Theor Appl Genet 71:613–621

Huettel RN, Hammerschlag FA (1986) Influence of cytokinin on in vitro screening of peaches for resistance to nematodes. Plant Disease 70:1041–1044

Jia J, Potrykus I (1981) Mesophyll protoplasts from *Solanum melongena* var. *depressum* Bailey regenerate to fertile plants. Plant Cell Rep 3:247–249

Kochba J, Samish RM (1971) Effect of kinetin and 1-naphtylacetic acid on root-knot nematodes in resistant and susceptible peach rootstocks. J Am Soc Hortic Sci 96:458–461

Kochba J, Samish RM (1972) Level of endogenous cytokinins and auxin in roots of nematode resistant and susceptible peach rootstocks. J Am Soc Hortic Sci 97:115–119

Lauritis JA, Rebois RV, Graney LS (1982) Screening soybean for resistance to *Heterodera glycines* Ichinohe using monoxenic cultures. J Nematol 14:593–594

Murashige T, Skoog F (1962) A revised medium for rapid growth and bioassays with tobacco tissue cultures. Physiol Plant 15:473–497

Orion D, Pilowsky M (1984) Excised tomato root culture as a tool for testing root-knot nematode resistance. Phytoparasitica 12:71–73

Saxena PK, Gill R, Rashid A, Maheshwari SC (1981) Plantlet formation from isolated protoplasts of *Solanum melongena* L. Protoplasma 106:355–361

Saxena PK, Gill R, Rashid A (1987) Optimal conditions for plant regeneration from mesophyll protoplasts of eggplant (*Solanum melongena* L.). Sci Hortic 31:185–194

Smith PG (1944) Embryo culture of a tomato species hybrid. Proc Amer Soc Hort Sci 44:413–416

Uhrig H, Wenzel G (1981) *Solanum gourlayi* Hawkes as a source of resistance against the white potato cyst nematode *Globodera pallida* Stone. Z Pflanzenzücht 86:148–157

Van Staden J, Dimalla GG (1977) A comparison of the endogenous cytokinins in the roots and xylem exudate of nematode-resistant and susceptible tomato cultivars. J Exp Bot 28:1351–1352

Wenzel G (1985) Strategies in unconventional breeding for disease resistance. Annu Rev Phytopathol 23:149–172

Wenzel G, Uhrig H (1981) Breeding for nematode and virus resistance in potato via anther culture. Theor Appl Genet 59:333–340

Wenzel G, Scheider O, Przewozny T, Sopory SK, Melchers G (1979) Comparison of single cell culture derived *Solanum tuberosum* L. plants and a model for their application in breeding programs. Theor Appl Genet 55:49–55

Section II Cereals
(For Wheat and Triticale see Volume 13)

II.1 Somaclonal Variation in Cereals

C. TONELLI[1]

1 Introduction

Somaclonal variation has been extensively documented (Larkin and Scowcroft 1981; Larkin et al. 1985, Sala and Biasini 1985; Tonelli 1985; Brown and Lorz 1986; Ahloowalia 1986). This variation seems not to be species- or organ-specific, but ubiquitous among regenerated plants, and appears to cover a wide spectrum of morphological, physiological, and biochemical characters, some of which may be of importance in plant improvement as long as the variation is stable and sexually transmitted.

This chapter will focus on somaclonal variation reported in cereals, and describe the various classes of mutants obtained and the different phenomena, including molecular changes, occurring during culture and possibly associated with somaclonal variation. The implications of these events for plant biology and plant breeding will be illustrated.

2 Terminology

Although the term "somaclones" to indicate variability displayed in plants regenerated from culture and somaclonal variation, as proposed by Larkin and Scowcroft (1981), has been widely accepted, a certain confusion exists on the symbols adopted to indicate the plants regenerated from tissue culture and their progenies obtained by selfing. The symbols adopted in some papers are shown in Table 1. In this review the R_1, R_2 and R_3 terminology will be followed, since it will reflect ideally the F_1, F_2, F_3 and M_1, M_2 and M_3 generations used respectively in Mendelian and mutagenesis experiments.

3 Examples of Somaclonal Variation in Cereals

Tissue cultures with the potential to regenerate plants have been established in a wide range of cereal species and from several different types of explant (Gamborg

[1]Dipartimento di Genetica e di Biologia dei Microorganismi, Università degli Studi di Milano, Via Celoria 26, 20133 Milano, Italy

Biotechnology in Agriculture and Forestry, Vol. 11
Somaclonal Variation in Crop Improvement I (ed. by Y.P.S. Bajaj)

Table 1. Terminology used by different authors to indicate regenerated plants and the successive generations

Regenerated plant	Plants from first self-pollination[a]	Plants from second self-pollination[b]	Reference
D_1	D_1	D_3	Oono (1978, 1985)
T_1	T_2	T_3	Sun et al. (1983)
SC_1	SC_2	SC_3	Larkin et al. (1984)
R_1	R_1S_1	R_1S_2	Everett and Flashman (1984)
R	R_1	R_2	Chaleff (1981)
R_0	R_1	R_2	Edallo et al. (1981); Evans and Sharp (1983)
R_1	R_2	R_3	Yurkova et al. (1982); Gavazzi et al. (1987)
F_1	F_2	F_3	Standard genetic symbols
M_1	M_2	M_3	Symbols used in mutagenesis

[a] Progeny obtained by self-fertilizing the regenerated plant.
[b] Progeny by self-fertilizing the plants obtained in [a].

et al. 1977; Gosch-Wackerle et al. 1979; Rines and McCoy 1981; Nabors 1982; Hu and Zeng 1984; Kaur-Sawhney and Galston 1984; King and Shimamoto 1984; Schaeffer et al. 1984; Yamada and Loh 1984; Hanzel et al. 1985; Kyozuka et al. 1987, see also the volume Crops I of this Series).

Cultures which retain morphogenetic or embryogenetic capacity have been derived from immature and mature embryos, immature inflorescences, very young leaves, and seeds (see Table 2).

Phenotypic variations have been observed among regenerated plants (R_1), and in some cases genetic analysis has been undertaken to test the heritability of the phenotypic variation as well as the pattern of transmission of the mutant character.

In Table 2, examples of somaclonal variants reported in cereals are listed. In most cases the variants obtained were similar to those either already present in natural germplasm and gene pools or selected after physical and chemical mutagenesis. Moreover, in some cases, the variants obtained have been novel: i.e., supernumerary head variants in wheat and triple head variant in barley.

3.1 Rice

In rice, Bajaj and Bidani (1980) established callus tissue cultures from roots, shoots, and mesocotyls. However, plant regeneration was better obtained from mesocotyl and embryo-derived calli and shown to be genotypic-dependent. A wide range of genetic variability was observed among regenerated plants. Extensive analysis of the somaclones has been performed by Oono (1978, 1981) on 1121 regenerated plants among which 7.4% were albino. Genetic analysis performed in the R_2 and R_3 generation showed variation in a wide number of traits including plant height, panicle shape, heading date, fertility. Similar results were obtained by Sun et al. (1983), who studied the inheritance of variant traits in more than 2000 somaclones derived from 18 varieties of rice. The analysis of the R_2 and R_3 generations suggested

Table 2. Somaclonal variants reported in cereal plants

Species	Explant	Phenotype of the variant	Reference
Rice	Mature embryo	Dwarf, grain weight, heading date, panicle, albino, no tillers	Sun et al. (1983)
	Seeds	early heading, albino, short culm sterility	Fukui (1983)
	Seeds	plant height, albino, heading date sterility	Oono (1978, 1981, 1985)
	Protoplasts from embryo or seed calli	panicles, spikelets, grain yield	Ogura et al. (1987)
Oats	Immature embryo	Seed set, sterility, height, heading date, albino, yellow striped, developmental variants	Cummings et al. (1976)
Sorghum	Immature embryo	Albino, waxy midrib, poor seed set, dwarfs, tallness	Hungtu Ma et al. (1987)
Maize	Immature embryo	Abphylsyndrome, reduced fertility, albino, defective kernel, *Helminthosporium maydis* resistance, cytoplasmic male sterility/fertility	Green (1977) Edallo et al. (1981) Gengenbach et al. (1977, 1981) Dixon et al. (1982) Brettel et al. (1980,1986a) Earle et al. (1987)
Triticale	Immature embryo	Spike length reduced, seed proteins	Jordan and Larter (1985)
Barley	Immature embryo	Albino plant height, tillering and fertility, seed protein, resistance to *Helminthosporium sativus*	Dale and Deambrogio (1979) Deambrogio and Dale (1980) Breiman et al. (1987b) Chawla and Wenzel (1987)
Wheat	Immature embryo	Height, heading date, reduced fertility, supernumerary spikes, awns, number of tillers, grain and glume color, waxiness, gliadin proteins, amylase regulation kernel weight, total dry weight	Ahloowalia (1982) Larkin et al. (1984) Maddock et al. (1983, 1985) Ahloowalia and Sherington (1985) Cooper et al. (1986); Bajaj (1986) Ryan and Scowcroft (1987) Ryan et al. (1987)
Millets	Embryo, shoot tip, young inflorescens	Size, leaf shape tillering salt tolerance	Bajaj et al. (1981) Bajaj and Gupta (1987)

mutations of nuclear and cytoplasmic genes as well as in qualitative and quantitative traits. The percent of variants was high (72%), as in the study by Oono (75.6%). An interesting work is that by Fukui (1983), where he was able to demonstrate the sequential induction of mutations during the growth of a seed-derived callus. Twelve plants were regenerated from a single callus, and four independent mutations due to unlinked recessive genes were observed in the progeny. The segregation patterns showed that each mutation was independently induced at different times during the stage of callus growth until the beginning of plant regeneration. Each new mutation accumulated over the preexisting one, as is shown by the fact that some plants were normal while others carried one or more mutant characters.

By analysis of their specific combination, it was concluded that the four mutations appeared in the callus in this sequence: early heading, albino, short culm, and sterility.

A lower frequency of variants has been reported in protoplast-derived plants (Ogura et al. 1987) compared to the previous works on plants regenerated by seed or embryo-derived calli (Oono 1978, 1981, Sun et al. 1983).

The results indicate that protoplast-derived rice plants were generally "normal" but produced a higher grain yield. A total of 19% of the regenerated plants showed variation like: fewer spikelets per panicle, low seed fertility, and increased panicle number.

3.2 Oats

Callus capable of plant regeneration is usually initiated by immature embryos, but excised apical meristem tissues and germinating mature embryos are also useful alternatives to the use of immature embryos. Cummings et al. (1976) regenerated 133 plants of 16 different genotypes. Although the growth of most of the regenerated plants appeared normal, some phenotypic and genotypic variability was observed. Variation for seed set, height, and heading date occurred within progenies of some regenerated plants.

Occasional regenerated plants exhibited various developmental abnormalities. One plant had two fused identical culms with the third leaf of each sharing a common midrib. Other abnormalities included two leaves or two panicles growing from a single node. These plants were observed in the progeny of normal-appearing regenerated plants. Albino and yellow stripe plants were also observed, but no data on genetic transmissions of these traits are available.

Plant regeneration from embryo-derived calli has been obtained by Rines and McCoy (1981) in *Avena sativa* and also in two wild oat species. Cytogenetic analysis (McCoy et al. 1982) of the regenerants showed a high frequency of abnormalities.

3.3 Sorghum

Sorghum somaclonal variants have been recently reported by Hungtu Ma et al. (1987). Immature embryos of 20 sorghum genotypes were cultured. The results showed that the ability of calli to differentiate varied among genotypes, and acts as a dominant heritable trait that seems to be controlled by at least two complementary genes.

The analysis was done on 251 R_1 plants and their progenies. Most of the R_1 plants were normal in appearance and fertile. However, some morphological variations have been observed among the R_1 plants, including albinos, onion leaf, satellited panicle branches, chimera, slow growth, poor seed set, tall plant, mixoploid, and waxy midrib. Only mutants for tallness and waxy midrib were observed in R_2 plants, thus confirming them to be dominant heritable traits. The appearance of male and female sterile plants and dwarfs in the R_2 (but not in R_1) generation suggested the presence of recessive mutations in R_1 plants. Minor variations in peroxidase banding patterns were found among R_1 plants.

3.4 Maize

In maize, plant regeneration is routinely obtained from immature embryo-derived calli, although it is genotype-dependent, as observed in other species. Green (1977) regenerated 85 plants and about 10% of them showed gross morphological changes like the abphyl syndrome (decussate leaf and ear arrangement giving twice the number of leaves on a normal number of nodes), twin stalks from a single node and reduced pollen fertility.

A very high frequency of variants in R_1 generations is reported also by Edallo et al. (1981), where on the average, each R_1 plants produced 0.8–1.2 mutations. Genetic analysis was carried out on R_3 progenies and only mutants segregating with simple Mendelian ratio were considered. Endosperm mutants such as defective endosperm, opaque, etched, white cap, germless, and seedling mutants such as lethal plant, abnormal growth, yellow green, pale green, albino, virescent, dwarf, and viviparous were recovered and resembled those isolated by conventional breeding (Coe and Neuffer 1977).

Somaclonal variation has also been observed to affect mitochondrial genome in maize. *Helminthosporium maydis* "race" T produces a toxin to which maize plants carrying the Texas male sterile cytoplasm (cms-T) are susceptible, while those carrying nonsterile cytoplasm are resistant. Callus cultures initiated from immature embryos of cms-T genotype were selected in the presence of progressively higher concentrations of toxin in the medium (Gengenbach et al. 1977). After five selection cycles, the plants regenerated from resistant cell lines showed toxin resistance, but most of them were male fertile. The remaining "male sterile" resistant plants showed abnormal morphology with deformed floral organs, short stature, and abnormal leaf development.

Resistance was cotransmitted with male fertility and resistance to *Phyllosticta maydis* pathotoxin. Similar cases of apparent pleiotropy were reported in *N. tabacum* (Chaleff 1981) during selection for resistance to Picloram, but in neither case has the basis of the association been elucitated.

Brettel et al. (1980), repeating these experiments, confirmed these results, but added the important finding that numbers of male fertile, T-toxin resistant plants were also obtained from the unselected (control) culture, grown in the absence of race T-toxin.

The restored male fertility and toxin resistance were shown to be cytoplasmically inherited. Recent experiments conducted by Earle et al. (1987) indicate that fertile revertants can readily be recovered from tissue culture of several sources of cmsS. All cultures examined gave revertants.

3.5 Triticale

Triticale (× *Triticosecale* Wittmack) is a man-made crop species produed by the hybridization of wheat (*Triticum* sp. L.) and rye (*Secale cereale* L.). Plant regeneration is achieved from callus initiated from 15-day-old embryos and a wide range of variability is observed (Jordan and Larter 1985). Short stature, increased number of fertile tillers, spike length, and lower fertility were the traits observed in R_1. As far as the transmission of these traits, the analysis of R_2 and R_3 generations showed that, while plant heights and number of fertile tillers did not vary significantly either within or between generations, significant differences occurred for spike length and fertility, thus indicating a genetic transmission of these last two traits. Variation was observed also in seed proteins, in fact prolamin analysis revealed a significant increase in percent in the seeds produced by two regenerated plants and in their progenies.

3.6 Barley

Plant regeneration from immature embryo-derived calli was reported in several barley species: *Hordeum vulgare*, *H. spontaneum*, *H. bulbosum* and *H. jubatum* (Orton 1979; Vazquez and Ruiz 1986; Breiman 1985); however, the morphogenic ability was found to be genotype-dependent in *H. vulgare* (Dale and Deambrogio 1979; Hanzel et al. 1985).

Information on heritable somaclonal variation in *H. vulgare* comes from the work of Deambrogio and Dale (1980). They reported variability for plant height, numbers of tillers, and fertility in plants derived from a 4-week-old callus culture. Karp et al. (1983) assayed the progeny of regenerated plants for isoenzymes, ribosomal DNA spacer length polymorphism and hordein patterns, and found relative genetic stability. More recently, Breiman et al. (1987b) observed differences in the number and mobilities of the B and C hordeins in the progenies of regenerated plants of *H. spontaneum*.

Using a selection procedure at cell level, Chawla and Wenzel (1987) screened calli derived from immature embryos of *H. vulgare* for the filtrate produced by the fungus *Helminthosporium sativum* P.K. and B. The majority of plants regenerated from callus lines surviving the toxin treatment were revealed after in vivo testing against pathogen to be less sensitive.

3.7 Wheat

In wheat plant, regeneration has been obtained by initiation of callus from different tissues (for a review see Bajaj and Gosal 1986). However, immature embryo-derived calli proved to be the most efficient for regeneration purposes (Larkin et al. 1984). Therefore this method has been adopted as a standard procedure to regenerate wheat from tissue culture. Compared to other cereals, a wealth of studies have been reported on somaclonal variation in wheat. Different degrees of variation have been documented at the cytological, morphological, biochemical, and molecular level, and some were of agronomic interest.

Variation affected the following traits: height, heading date, fertility, stem thickness, supernumerary spikes and spikelets, awns, tiller number, grain color, glume color, waxiness, gliadin, protein, and β-amylase regulation, (Ahloowalia 1982; Maddock et al. 1983; Larkin et al. 1984; Ahloowalia and Sherington 1985; Bajaj 1986; Ryan and Scowcroft 1987).

An extensive analysis was conducted by Larkin et al. (1984). Seventy-one genotypes were assessed for their morphogenic ability, and 2864 plants were regenerated from the ten best accessions. The progenies of 142 plants regenerated from tissue cultures of a Mexican line were analyzed to study the heritability of the variant traits. The results demonstrate that both morphological and biochemical traits were heritable through two seed generations, and include traits under simple and quantitative genetic control. Segregation data suggest that mutations both from dominance to recessiveness (awns, grain color) and from recessiveness to dominance or co-dominance (glume color and gliadins) occurred. Most mutations in the primary regenerants (R_1) were in the heterozygous state, but some were true-breeding and presumed to be homozygous. The mutations affected characters, the major genes for which are known to be located on all seven homoeologous chromosome groups.

A sequential occurrence of mutations, as in the case described by Fukui (1983) in rice, was suggested following the observation that within a group of plants regenerated from the same embryo some changes were particularly widespread and others were relatively infrequent.

A paper of Ryan et al. (1987) reported that significant somaclonal variation can be generated for a number of agronomic and quality characters. The analysis was carried out on 256 selected lines in R_3 or R_4 generations that were evaluated for field performance characteristics in replicated hill plot experiments. Significant variations were found in all the characters tested: height, grain number per spike, kernel weight, yield, total dry weight, and harvest index. All lines selected for these characters (except for high yield and harvest index) maintained significant improvements over their parental controls also in the following generations, thus representing potentially useful improvements.

3.8 Millet

Bajaj et al. (1981) induced callus cultures of three species of *Panicum* using different types of explant such as segment of young inflorescences, shoot tips, and excised embryos. Plants were then regenerated. The response in vitro was genotypically oriented, and clear differences in the growth and differentiation pattern were observed among species and cultivars, i.e., *P. miliaceum* showed a better response than *P. maximum*. Regenerated plants showed a wide range of variation in morphological characters such as size, leaf shape, and tillering. Recently, Bajaj and Gupta (1987) selected salt-tolerant cell lines in *Pennisetum purpureum* and regenerated plants from them.

4 Possible Origin of Somaclonal Variation

An understanding of the genetic changes that occur in somaclones is important both with a view to enhancing the level of variation if genetic variability is the desired objective, and to controlling and suppressing it if clonal propagation of generally identical plants is desired. When does this variation arise? Does it occur during the culturing period or is it pre-existing? What are the genetic bases of these variations? These are the issues of practical as well as basic significance that are still awaiting elucidation.

The origin of the genetic changes could be either a pre-existing variation present in vivo in the explant tissues and then propagated in vitro and/or induced during the in vitro culture. Evidence for both origins is substantially reviewed by D'Amato (1986). As far as the factors that are possibly responsible for genetic variability in vitro there are no data available. The composition of culture media in terms of hormones (2,4-D), deficiency in essential elements (e.g., calcium), anoxia, automutagenesis, and in vitro stress might all be potential mutagenic factors.

5 Nature of Genetic Changes in Somaclones

Following numerous reports of extensive phenotypic variation observed among regenerated plants, a number of studies have indicated that much of this variation could be due to a whole spectrum of genetic changes, ranging from ploidy level, aneuploidy, chromosome breakage and rearrangement, to gene amplification and loss, mobilization of transposable elements, and point mutation, that can act either separately or simultaneously. Some of these different processes will be briefly discussed here. Further details are provided by Larkin et al. (1985), D'Amato (1985) and Brown and Lorz (1986).

5.1 Polyploidy and Aneuploidy

As seen from Table 3, polyploids and aneuploids are often present among regenerated plants of different species. Polyploidy, although it may pre-exist in the culture in the form of in vivo endoreduplicated cells, is, however, frequently induced in vitro through two different mechanisms, chromosome endoreduplication and spindle failure leading to restitutional mitoses.

Aneuploidy is generally induced in vitro through either nuclear fragmentation (amitosis) followed by mitosis or defective chromosome behavior during mitosis. More variability in the different types of aneuploids is found in the regenerated

Table 3. Cytogenetic changes in somaclonal plants

Species	Chromosomal variants	References
Rice	Polyploidy, aneuploidy	Bajaj and Bidani (1980)
	polyploidy	Sun et al. (1983)
		Ogura et al. (1987)
Oat	Tripolar spindle, deletions,	Cummings et al. (1976)
	aneuploidy, chromosome	Rines and McCoy (1981)
	breakage, deletions	McCoy et al. (1982)
Triticale	Deletions, fusions,	Armstrong et al. (1983)
	translocation, deletion,	Lapitan et al. (1984)
	heterochromatin amplifications	Jordan and Larter (1985)
	deletions	
Barley	Polyploidy	Dale and Deambrogio (1979)
	polyploidy, aneuploidy	Orton (1980)
	chromosomal rearrangements	Vazquez and Ruiz (1986)
Panicum	Polyploidy, aneuploidy	Bajaj et al. (1981)
Wheat	Polyploidy, aneuploidy	Bennici and D'Amato (1978)
	haploids, inversions	Ahloowalia (1982)
	aneuploidy, rearrangements	Karp and Maddock (1984)
	deletions, isochromosome	
Maize	Polyploidy, aneuploidy, deletions	Green (1977)
		Edallo et al. (1981)
	Deficiency	McCoy et al. (1982)
	interchanges, deficiency	Lee and Phillips (1987)
	heteromorphic pairs	Lee and Phillips (1987)

plants of polyploid species than in diploids. In cereal, as expected, more variation is observed in hexaploid wheat, oats, and triticale than in maize.

5.2 Chromosome Structural Changes

This could be a major mechanism to generate somaclonal variation. The vast majority of these changes originate in vitro. The examination of meiosis in pollen mother cells and the karyotype analyses on mitotic cells in plants regenerated from cell cultures of different species indicates the presence of reciprocal translocations, deletions, inversions, centric and acentric fragments, isochromosomes, telocentric, and dicentric chromosomes (Table 3, and for reviews D'Amato 1985 and Larkin et al. 1985).

Each of these changes could have a genetic consequence affecting an individual phenotype. Such rearrangements may result in loss of genes and in activation of others that were previously silent. For example, a rearrangement may delete or otherwise switch off a dominant allele, allowing the expression of the recessive one. A chromosome breakage, besides affecting the gene at the break site, may alter the expression of neighboring genes, particularly those for which transcription may be coordinately regulated.

Translocation, inversion, duplication, and deletion could also induce a novel phenotype by placing the gene in a new position. This indirect effect, called "position effect", can alter gene expression in absolute terms, as well as changing the timing and tissue specificity of gene action.

Dicentric chromosomes, because of their mitotic propagation through the breakage-fusion-bridge cycle, are active producer of deficiency-duplication events. These changes could result also from asymmetric sister chromatid exchange.

5.3 Point Mutations

Many somaclonal mutants have been reported which appear to involve single genes due to their Mendelian segregations in selfed and test crosses (see Sect. 3). Analysis of the selfed progenies reveals that regenerates are generally heterozygous for mutation, as expected if the mutations originated in a diploid cells. Some cases, however, show homozygous mutants, these will be discussed in Section 5.8.

Although chromosomal rearrangement seems to be the major source for somaclonal variation, more cryptic changes, like point mutation, are also expected to occur.

The best approach to rule out if this type of mutation does occur in culture is to analyze variants at a defined locus. In maize, Brettel et al. (1986a), out of 645 regenerated plants tested, isolated a stable mutant which produces an electrophoretic variant of the ADH1 enzyme. A detailed analysis of this mutant has revealed that the change in electrophoretic mobility is due to a change in a single base in the exon 6 of the ADH1 gene. This change results in the substitution of one amino acid, glutamic acid, by another, valine, thus resulting in the loss of a negative charge and the production of a protein with slower electrophoretic mobility.

Since four other regenerants from the same cultured embryo were normal, it is possible to conclude that the mutation occurred during culture, and that single base changes are one of the events associated with the phenomenon of somaclonal variation.

5.4 Gene Amplification and Loss

Preferential replication of a single genetic locus results in gene amplification. Brown (1981) suggested that any gene which cannot modulate its expression is a good candidate for gene amplification if the right selective pressure is applied.

In plants there is evidence that such processes can occur. For example, Cullis (1979) observed quantitative changes in the ribosomal DNA of a flax variety in response to different environmental conditions. If such amplification or loss of DNA sequence copies also occurs in plant cell culture, it may account at least in part for somaclonal variation. Several data have been reported which are consistent with this hypothesis (De Paepe et al. 1982; Durante et al. 1983). Gene amplification in response to selection has been demonstrated in alfalfa cell cultures screened with increasing levels of the herbicide L-phosphinothricin (Donn et al. 1984). In cereals, amplification and reduction have been shown in the ribosomal DNA of regenerated plants of barley, wheat, and triticale (Brettel et al. 1986b; Breiman et al. 1987a,b;

May and Appels 1987). The analyses were conducted on genomic DNA digested with the restriction endonuclease TaqI and probed with specific rDNA fragments by Southern blots.

5.5 Transposable Elements Activation

Transposable elements, first recognized in maize by McClintock (1950), have been found in many organisms for which detailed genetic analysis is possible (Calos and Miller 1980; Federoff 1983). These genetic elements can move in the genome from one locus to another, and by their excision or insertion can directly affect the expression of neighboring genes. For example, the insertion of transposable elements can elicit phenotypic changes by blocking the function of the genes into which they are intercalated and may exert polar effects on neighboring genes. Their excision can restore gene activity completely or partially, or can determine a stable loss of function.

These unstable mutations can be sexually transmitted to the progeny. The key feature of these elements is that they provide a regulated disruption of the chromosome in the cells in which they are active, causing duplication and complex rearrangements. Thus, transposable elements not only create an initial mutation upon insertion, but remain the focus for continued chromosomal instability, resulting from chromosome rearrangement (Walbot and Cullis 1985).

Movement of these elements may be associated with genomic shock (McClintock 1950, 1978). In culture, plant cells are subject to different genomic stresses, ranging from the new environment (i.e., presence of mutagenic compound such as 2, 4-D) to the high frequency of chromosome rearrangements that may activate previously silent elements. Furthermore, considering that the activity of these elements has been recently correlated to the degree of their DNA methylation (Chandler and Walbot 1986; Chomet et al. 1987), and that change in gene methylation has been observed as a result of tissue culture (Brown and Lorz 1986), it is conceivable that cryptic elements could be activated in culture also by a mechanism of DNA modification. Unstable mutations have been shown in some cases resulting from tissue culture as the semi-dominant aurea mutant of sulfur in tobacco (Lorz and Scowcroft 1983) and flower color in alfalfa (Groose and Bunham 1984).

In maize, Peschke et al. (1987) investigated the possibility that the element *Ac* was released or activated by the process of tissue culture and plant regeneration. They screened for the presence of an active *Ac* the progeny of 301 regenerated plants originally derived from the crosses of three lines and a *Ds*-containing tester stock. Ten regenerated plants from two independent embryo cell lines contained an active *Ac* transposable element. No active *Ac* elements were present in the explant sources.

5.6 Altered Expression of Multigene Families

Since some agronomically important genes (e.g., gliadins, zeins, glutenins, β-amylase, hordeins) are coded on multigene families, it is important from a breeding point of view to investigate if variants can be generated through somaclonal variation.

In wheat, inheritable variation in the electrophoretic pattern of gliadins (Larkin et al. 1984; Maddock et al. 1985; Cooper et al. 1986) and β-amylase (Ryan and Scowcroft 1987) was observed in regenerants. No changes were found in control grains from the donor plants. The changes concern both deleted protein bands and proteins in new positions. Therefore it was postulated (Larkin et al. 1985) that rearrangements induced in culture could alter the expression of the family, leading to repression of previously expressed genes and to the expression of previously silent ones.

5.7 Cytoplasmic Genome Rearrangements

A rearrangement in the chloroplast and mitochondrial DNA has been observed in regenerated plants. The plastid genome (130 kb) of higher plants shows a very high degree of conservation in terms of DNA sequences and gene order among highly divergent species. There are very few examples of change in chloroplast DNA (ctDNA) as result of tissue culture stress. Sun et al. (1983) found in rice a number of albino plantlets, whose frequency was decreasing with the advance of generations, thus suggesting the involvement of cytogene mutations. Day and Ellis (1984) regenerated wheat plants from pollen by anther culture, and compared the restriction patterns of ctDNA from green and albino plants. The analysis showed that large regions of the ctDNA were deleted. The mitochondrial DNA (mtDNA) is more variable (from 218 to 2500 kb) and complex than the plastid genome, and contains one or more small linear or circular plasmids. The presence of two linear plasmid S_1 and S_2 in the S cytoplasm of maize has been correlated with male sterility. Tissue culture appears to increase the variation in mtDNA or to allow selection for unusual type. The most notable example is the reversion of maize with the T-cms to fertility (see Sect. 3.4 and Brown and Lorz 1986 for a review). The mutation to male fertility and toxin insensitivity has been shown to be a frameshift mutation in the mt DNA (Wise et al. 1987).

5.8 Putative Homozygous Mutations

Completely unexpected is the frequent recovery of homozygous "solid" mutants among regenerated plants. Evidence is lacking on their origin, and the phenomenon affects different fixed traits such as jointless pedicel and leaf shape in tomato (Evans and Sharp 1983; Gavazzi et al. 1987) awns, glume color, and grain color in wheat (Larkin et al. 1984), and plant height and glaborous seeds in rice (Oono 1981, 1985). These mutations not only bred true on selfing, but also behaved as a single recessive gene in test crosses, as in the case of tomato jointless pedicel. As far as the dwarf mutants described by Oono (1985), genetic studies showed that the character is transmitted in a stable manner up to the eighth generation by selfing; however, the character was not transmitted in reciprocal crosses with control plants or normal regenerated plants. This distorted segregation pattern is indeed difficult to explain in terms of classical mutation theory. Oono thus proposed that the mechanism responsible for this phenomenom is controlled by a process of a heritable gene inactivation induced by the in vitro culture. Whatever the basis of these events, the possibility of isolating homozygous mutants in regenerated plants represents a rather attractive tool for breeders to accelerate the selection scheme.

6 Somaclonal Variation Versus Chemical Mutagenesis

The main sources of variability for the breeder have been mutation and crosses with unrelated or wild species followed by selection for the desired character. Recently, somaclonal variation appears appealing to the breeder because it often occurs at a higher frequency than chemically induced mutagenesis. Therefore it is of particular interest to establish whether these two sources of variability are also qualitatively different, since differences in the genetic target may ultimately produce different classes of mutants and thus contribute differently to plant breeding. In a recent work, Gavazzi et al. (1987) compared the genetic variability recovered in the progeny of tomato plants regenerated after one passage in vitro to that induced by chemical mutagenesis with ethyl methane sulfonate. The comparison included type and frequency of mutational events as well as segregation patterns. The analysis was done on 120 R_2 and 1000 M_2 families. Several mutants were recovered in the progeny of regenerated and mutagenized plants of the two cultivars of tomato. They can be grouped into the following categories: seedling lethality, male sterility, resistance to *Verticillium*, short stature, and change in number of lateral shoots or in leaf shape. The results show that regeneration from in vitro culture leads to a higher number of mutations than the application of the chemical mutagen to either seeds or pollen or both. The spectrum of mutants is also different, with some mutant classes (e.g., potato leaf) arising exclusively in the case of somaclonal variation. The third difference was the frequent recovery of homozygous "solid" mutants only in the R_2 generation.

Thus, these results indicate that somaclonal variation and chemical mutagenesis differ in their effects, changing the spectrum and frequency of mutants as well as, at least in some cases, the pattern of transmission of the mutant character.

7 Concluding Remarks

All the data collected show unequivocally that the variant traits expressed by plants regenerated from tissue cultures, can in most cases be stably inherited by the progeny through successive generations. Thus, tissue culture appears to be an unexpectedly powerful mutagenic treatment and represents a rich and novel source of genetic variability. Since the techniques are relatively simple with respect to those of recombinant DNA, and do not imply a cloned gene for the desired character, it is expected that, in future, somaclonal variation will be widely used for crop improvement. A possible strategy is to introduce into cell culture the best available variety, to regenerate plants and then to select in R_2 generation for new variants. Since simple genetic changes have been produced by somaclonal variation, the new variants may retain all the favorable qualities of the parental variety, while adding one additional trait (Gavazzi et al. 1987).

Furthermore, if the desirable character can be selected in vitro or at early stages of seedling development rather than in the field, somaclonal variation could be more efficient and less time-consuming. This approach appears quite promising for the selection of resistance to pathotoxins, herbicides, and physiological stress (Gengenbach et al. 1977; Chaleff 1981; Tonelli 1987), where a correlation between cell and whole plant response seems feasible. To this extent, somaclonal variation provides a powerful option for crop improvement.

References

Ahloowalia BS (1982) Plant regeneration from callus in wheat. Crop Sci 22:405–410

Ahloowalia BS (1986) Limitation to the use of somaclonal variation in crop improvement. In: Semal J (ed) Somaclonal variations and crop improvement. Nijhoff, Dordrecht, pp 14–27

Ahloowalia BS, Sherington J (1985) Transmission of somaclonal variation in wheat. Euphytica 34:525–537

Armstrong KC, Nakamura C, Keller WA (1983) Karyotype instability in tissue culture regenerants of triticale (X *Triticosecale* Wittmack) cv. "Welsh" from 6-month-old callus cultures. Z Pflanzenzücht 91:233–245

Bajaj YPS (1986) In vitro regeneration of diverse plants and the cryopreservation of germplasm of wheat (*Triticum aestivum* L.). Cereal Res Commun 14:305–311

Bajaj YPS Bidani M (1980) Differentiation of genetically variable plants from embryo derived callus cultures of rice. Phyromorphology 30:290–294

Bajaj YPS, Gosal SS (1986) Biotechnology of wheat improvement. In: Bajaj YPS (ed) Biotechnology in agriculture and forestry, vol 2. Crops I. Springer, Berlin Heidelberg New York Tokyo, pp 3–38

Bajaj YPS, Gupta RK (1987) Plants from salt-tolerant cell lines of napier grass (*Pennisetum purpureum* Schum.). Indian J Exp Biol 25:58–60

Bajaj YPS, Sidhu BS, Dubey VK (1981) Regeneration of genetically diverse plants from tissue cultures of forage grass-*Panicum* sps. Euphytica 30:135–140

Bennici A, D'Amato F (1978) In vitro regeneration of *durum* wheat plants. 1. Chromosome numbers of regenerated plantlets. Z Pflanzenzücht 81:305–311

Breiman A (1985) Plant regeneration from *Hordeum spontaneum* and *H. bulbosum* immature embryo derived calli. Plant Cell Rep 4:70–73

Breiman A, Felsenburg T, Galun E (1987a) *Nor* loci analysis in progenies of plant regenerated from the scutellar callus of bread-wheat. Theor Appl Genet 73:827–831

Breiman A, Rotem-Abarbanell D, Karp A, Shaskin H (1987b) Heritable somaclonal variation in wild barley (*Hodeum spontaneum*). Theor Appl Genet 74:104–112

Brettel RIS, Thomas E, Ingram DS (1980) Reversion of Texas male-sterile cytoplasm maize in culture to give Fertile T-toxin resistant plants. Theor Appl Genet 58:55–58

Brettel RIS, Dennis ES, Scowcroft WR, Peacock WJ (1986a) Molecular analysis of a somaclonal mutant of maize alcohol dehydrogenase. Mol Gen Genet 202:235–239

Brettel RIS, Pallotta MA, Gustafson JP, Appels R (1986b) Variation at the *Nor* loci in triticale derived from tissue culture. Theor Appl Genet 71:637–643

Brown DD (1981) Gene expression in eukaryotes. Science 211:667–674

Brown PTH, Lorz H (1986) Molecular changes and possible origins of somaclonal variation. In: Semal J (ed) Somaclonal variations and crop improvement. Nijhoff, Dordrecht, pp 148–159

Calos MP, Miller JH (1980) Transposable elements. Cell 20:579–595

Chaleff RS (1981) Genetics of higher plants: applications of cell culture. Univ Press, Cambridge

Chandler VL, Walbot V (1986) DNA modification of a maize transposable element correlated with loss of activity. Proc Natl Acad Sci USA 83:1767–1771

Chawla MS. Wenzel G (1987) In vitro selection of barley and wheat for resistance against *Helminthosporium sativum*. Theor Appl Genet 74:841–845

Chomet PS. Wessler S. Dellaporta SL (1987) Inactivation of the maize transposable element activator is associated with its DNA modification. EMBO J 6:295–302

Coe EH. Neuffer MG (1977) The genetics of corn. In: Sprague GF (ed) Corn and crop improvement. Am Soc Agron. Madison. Wisc. pp 111–124

Cooper DB. Sears RG. Lookhart GL. Yones BL (1986) Heritable somaclonal variation in gliadin proteins of wheat plants derived from immature embryo callus culture. Theor Appl Genet 71:784–790

Cullis CA (1979) Quantitative variation of ribosomal RNA genes in flax genotrophes. Hereditary 42(2):237–246

Cummings DP. Green CE. Stuthman DD (1976) Callus induction and plant regeneration in Oats. Crop Sci 16:465–470

Dale PJ. Deambrogio E (1979) A comparison of callus induction and plant regeneration from different explants of *Hordeum vulgare*. Z Pflanzenphysiol 94:65–77

D'Amato F (1985) Cytogenetics of plant cell and tissue cultures and their regenerates. CRC Crit Rev Plant Sci 3:73–112

D'Amato F (1986) Spontaneous mutations and somaclonal variation. Proc Int Symp Nuclear techniques and in vitro culture for plant improvement. IAEA. Vienna. pp 3–10

Davies PA. Pallotta MA. Ryan SA. Scowcroft WR. Peacock WJ (1986) Somaclonal variation in wheat: genetic and cytogenetic characterisation of alcohol dehydrogenase 1 mutants. Theor Appl Genet 72:644–653

Day A. Ellis THN (1984) Chloroplast DNA deletions associated with wheat plants regenerated from pollen: possible basis for maternal inheritance of chloroplasts. Cell 39:359–368

Deambrogio E. Dale PJ (1980) Effect of 2.4-D on the frequency of regenerated plants in barley and on genetic variability between them. Cereal Res Commun 8:417–423

De Paepe R. Prat D. Huget T (1982) Heritable nuclear DNA. changes in double haploid (DH) plants obtained by pollen culture of *Nicotiana sylvestris*. Plant Sci Lett 28:11–28

Dixon LK. Leaver CJ. Brettel RIS. Gengenbach BG (1982) Mitochondrial sensitivity to *Drechslera maydis* T-toxin and the synthesis of a variant mitochondrial polypeptide in plants derived from maize tissue cultures with Texas male-sterile cytoplasm. Theor Appl Genet 63:75–80

Donn G. Tischer E. Smith JA. Goodman HM (1984) Herbicide resistant alfalfa cells: an example of gene amplification in plants. J Mol Appl Genet 2:621–635

Durante M. Geri C. Grisvard J. Guille E. Parenti R. Buiatti M (1983) Variation in DNA complexity in *Nicotiana glauca* tissue cultures. 1. Pith de-differentiation in vitro. Protoplasma 114:114–118

Earle ED. Gracen VE. Best VM. Batts LA. Smith ME (1987) Fertile revertants from S-type male-sterile maize grown in vitro. Theor Appl Genet 74:601–609

Edallo S. Zucchinali C. Perenzin M. Salamini F (1981) Chromosomal variation and frequency of spontaneous mutations associated with in vitro culture and plant regeneration in maize. Maydica 26:39–56

Evans DA. Sharp WR (1983) Single gene mutations in tomato plants regenerated from tissue culture. Science 221:949–951

Everett NP. Flashman SM (1984) Plants from culture. Nature (London) 307:115

Federoff NV (1983) Controlling elements in maize. In: Shapiro J (ed) Mobile genetic elements. Academic Press. New York. London. pp 1–63

Fukui K (1983) Sequential occurrence of mutations in a growing rice callus. Theor Appl Genet 65:225–230

Gamborg OL. Shyluk JP. Brar DS. Constable F (1977) Morphogenesis and plant regeneration from callus of immature embryos of sorghum. Plant Sci Lett 10:67–74

Gavazzi G. Tonelli C. Todesco G. Raffaldi F. Vecchio F. Barbuzzi G. Biasini MG. Sala F (1987) Somaclonal variation *versus* chemically induced mutagenesis in tomato (*Lycopersicon esculentum* L.). Theor Appl Genet 74:733–738

Gengenbach BG. Green CE. Donovan CM (1977) Inheritance of selected pathotoxin resistance in maize plants regenerated from cell culture. Proc Natl Acad Sci USA 74:5113–5117

Gengenbach BG. Connelly JA. Pring DR. Conde MF (1981) Mitochondrial DNA variation in maize plants regenerated during tissue culture selection. Theor Appl Genet 59:161–167

Gosch-Wackerle G. Avivi L. Galun E (1979) Induction. culture. and differentiation of callus from immature rachises. seeds. and embryos of *Triticum*. Z Pflanzenphysiol 91:267–278

Green CE (1977) Prospects for crop improvement in the field of cell culture. HortSci 12:131–134
Groose RW, Bingham ET (1984) Variation in plants regenerated from tissue cultures of tetraploid alfalfa heterozygous for several traits. Crop Sci 24:655–658
Hanzel JJ, Miller JP, Brinkman MA, Fendons E (1985) Genotype and media effects on callus formation and regeneration in barley. Crop Sci 25:27–31
Hu H, Zeng JE (1984) Development of new varieties via anther culture. In: Sharp WR, Evans DA, Ammirato PV, Yamada Y (eds) Handbook of plant cell culture, vol 3. McMillan, New York, pp 65–90
Hungtu Ma, Minghong Gu, Liang GM (1987) Plant regeneration from cultured immature embryos of *Sorghum bicolor* (L) Moench. Theor Appl Genet 73:389–394
Jordan MC, Larter EN (1985) Somaclonal variation in triticale (X *Triticosecale* Wittmack) cv. Carman. Can J Genet Cytol 27:151–157
Karp A, Maddock SE (1984) Chromosome variation in wheat plants regenerated from cultured immature embryos. Theor Appl Genet 67:249–255
Karp A, Steele SM, Parmar S, Yones GK, Shewry PR (1983) Relative stability among barley plants regenerated from cultured immature embryos. Genome 29:405–412
Kaur-Sawhney R, Galston AW (1984) Oats. In: Sharp WR, Evans DA, Ammirato PV, Yamada Y (eds) Handbook of plant cell culture, vol 2. McMillan, New York, pp 92–107
King PJ, Shimamoto K (1984) Maize. In: Sharp WR, Evans DA, Ammirato PV, Yamada Y (eds) Handbook of plant cell culture, vol 2. McMillan, New York, pp 69–91
Kyozuka J, Hayashi Y, Shimamoto K (1987) High frequency plant regenerations from rice protoplasts by novel nurse culture methods. Mol Gen Genet 206:408–413
Lapitan NLV, Sears RG, Gill BS (1984) Translocation and other karyotype structural changes in wheat × rye hybrids regenerated from tissue culture. Theor Appl Genet 68:547–554
Larkin PJ, Scowcroft WR (1981) Somaclonal variation – a novel source of variability from cell cultures for plant improvement. Theor Appl Genet 60:197–214
Larkin PJ, Ryan SA, Brettel RIS, Scowcroft WR (1984) Heritable somaclonal variation in wheat. Theor Appl Genet 67:443–455
Larkin PJ, Brettel RIS, Ryan SA, Davies PA, Pallotta MA, Scowcroft WR (1985) Somaclonal variation: Impact on plant biology and breeding strategies. In: Zaitlin M, Day P, Hollaender A (eds) Biotechnology in plant science. Academic Press, New York London, pp 83–100
Lee M, Phillips RL (1987) Genetic variants in progeny of regenerated maize plants. Genome 29:834–838
Lorz H, Scowcroft WR (1983) Variability among plants and their progeny regenerated from protoplasts of Su/su heterozygotes of *Nicotiana tabacum*. Theor Appl Genet 66:67–75
Maddock SE, Lancaster VA, Risiott R, Franklin J (1983) Plant regeneration from cultured immature embryos and inflorescens of 25 cultivars of Wheat (*Triticum aestivum*). J Exp Bot 34:915–926
Maddock SE, Risiott R, Parmar S, Yones MGK, Shewry PR (1985) Somaclonal variation in the gliadin pattern of grains of regenerated wheat plants. J Exp Bot 36:1976–1984
May CE, Appels R (1987) Variability and genetics of spacer DNA sequences between the ribosomal-RNA genes of hexaploid wheat (*Triticutum aestivum*). Theor Appl Genet 74:617–624
McClintock B (1950) The origin and behaviour of mutable loci in maize. Proc Natl Acad Sci USA 36:344–355
McClintock B (1978) Mechanisms that rapidly reorganize the genome. Stadler Genet Symp 10:25–48
McCoy TJ, Phillips RL, Rines HW (1982) Cytogenetic analysis of plants regenerated from oat (*Avena sativa*) tissue cultures: high frequency of partial chromosomal loss. Can J Genet Cytol 24:37–50
Nabors MW (1982) Cereal tissue culture at Colorado State University. In: Rice tissue culture planning conf. IRRI, Los Baños, pp 91–93
Ogura H, Kyozuka J, Mayashi Y, Koba T, Shimamoto K (1987) Field performance and cytology of protoplasts-derived rice (*Oryza sativa*): high yield and low degree of variation of four japonica cultivars. Theor Appl Genet 74:670–676
Oono K (1978) Test tube breeding of rice by tissue culture. Trop Agric Res Ser, Minist Agric For Jpn 11:109–124
Oono K (1981) In vitro methods applied to rice. In: Thorpe TA (ed) Plant tissue culture. Academic Press, New York London, pp 273–298
Oono K (1985) Putative homozygous mutations in regenerated plants of rice. Mol Gen Genet 198:377–384
Orton TJ (1979) A quantitative analysis of growth and regeneration from tissue cultures of *Hordeum vulgare*, *H. jubatum* and their interspecific hybrid. Environ Exp Bot 19:319–335

Orton TJ (1980) Chromosomal variability in tissue cultures and regenerated plants of *Hordeum*. Theor Appl Genet 56:101–112
Orton TJ (1984) Genetic variation in somatic tissues: Method or madness. Adv Plant Pachol 2:153–185
Peschke VM, Phillips RL, Gengenbach BG (1987) Discovery of transposable elements activity among progeny of tissue culture-derived maize plants. Science 238:804–807
Rines HW, McCoy TJ (1981) Tissue culture initiation and plant regeneration in hexaploid species of oats. Crop Sci 21:837–842
Ryan SA, Scowcroft WR (1987) A somaclonal variant of wheat with additional β-amylase isozymes. Theor Appl Genet 73:459–464
Ryan SA, Larkin PJ, Ellison FW (1987) Somaclonal variation in some agronomic and quality characters in wheat. Theor Appl Genet 74:77–82
Sala F, Biasini MG (1985) Heritable variability induced by the in vitro culture and genetic improvement of cultivated plants. G Bot It 120:43–54
Schaeffer GW, Lazar MD, Baenziger PS (1984) Wheat. In: Sharp WR, Evans DA, Ammirato PV, Yamada Y (eds) Handbook of plant cell culture, vol 2. McMillan, New York, pp 108–136
Sears RG, Deckart E (1982) Tissue culture variability in wheat: callus induction and plant regeneration. Crop Sci 22:546–550
Sun Z-X, Cheng-zhang Z, Kang-le Z, Xiu-fang Q, Ya-ping F (1983) Somaclonal genetics of rice, *Oryza sativa* L. Theor Appl Genet 67:67–73
Tonelli C (1985) In vitro selection and characterization of cereal mutants. Genet Agric 39:445–470
Tonelli C (1987) Genetic Improvement of tomato: selection strategy for detecting water stress tolerant mutants. In: Modern trends in tomato genetics and breeding. Eucarpia Tomato Working Group, Pontecagnano, It, pp 51–55
Vazquez AM, Ruiz ML (1986) Barley: induction of genetic variability through callus cultures. In: Bajaj YPS (ed) Biotechnology in agriculture and forestry, vol 2. Crops I. Springer, Berlin Heidelberg New York Tokyo, pp 204–219
Walbot V, Cullis C (1985) Rapid genomic change in higher plants. Annu Rev Plant Physiol 36:367–396
Wise RP, Pring DR, Gegenbach BG (1987) Mutation of male fertility and toxin insensitivity in Texas (T) cytoplasm maize is associated in a frameshift in a mitochondrial oper reading frame. Proc Natl Acad Sci 84:2858–2862
Yamada Y, Loh WH (1984) Rice. In: Ammirato PV, Evans DA, Sharp WR, Yamada Y (eds) Handbook of plant cell culture, vol 3. McMillan, New York, pp 151–170
Yurkova GN, Levenko BA, Novozhlov PV (1982) Plant regeneration in wheat tissue culture. Biochem Physiol Pflanzen 177:337–344

II.2 Somaclonal Variation in Rice

Z-X SUN[1] and K-L ZHENG[2]

1 Introduction

1.1 Importance and Distribution of Rice

Rice is the staple food of more than half of the world's population, and its present cultivation spans from latitude 53° N to 40° S (Lu and Chang 1980), from −23 to +7000 feet above sea level. It is cultivated under various conditions, including upland, deep water, and soils with pH 3.5 to 9.5 (Khush 1984). There are 22 species in the genus *Oryza*, most of which are diploid (2n = 24), and seven species are tetraploid (4n = 48) (Chang 1985). *O. sativa* L. and *O. glaberrima* Steud are the two cultivated species. *O. sativa* is usually divided into three eco-species: *indica*, *japonica* (*sinica*), and *javanica*.

Rice was one of the earliest domesticated crops in the world. Carbonated rice grains dated at 5000–4500 B.C., which were found at Ho-Mu-Do, Zhejiang, China, showed that rice has been cultivated for about 7000 years (Anonymous 1978a).

Rice provides 20% of calories and 13% of protein for human consumption on a worldwide basis, and for most consumers in Asia, who depend on this food for basic diet, it provides 40 to 80% calories and more than 40% proteins. Due to its importance in human life, much attention has been paid to its improvement.

1.2 Objectives in Rice Improvement

Today, rice production in the world is steadily increasing. World production of rough rice was 473 million t in 1986, harvested from about 145 million ha (FAO 1988). From 1975 to 1985, the average annual growth rate of total rice production in the world increased by 3%, that of unit yield increased by about 3%, and that of total land devoted to rice production increased by 0.2% (FAO 1986).

[1]Plant Genetics and Breeding Department, China National Rice Research Institute, Hangzhou 310006, China

[2]Biotechnology Department, China National Rice Research Institute, Hangzhou 310006, China

Abbreviations: Plants regenerated from somatic tissues or cells are referred to as SC_1, the selfed progeny of SC_1 is referred to SC_2 plants, subsequent selfed generations are termed SC_3, SC_4, etc. The plants regenerated from androgenesis are referred to as PC_1, the selfed progenies of PC_1 are termed PC_2, PC_3, PC_4, etc.

Biotechnology in Agriculture and Forestry, Vol. 11
Somaclonal Variation in Crop Improvement I (ed. by Y.P.S. Bajaj)

World population has, however, also been increasing especially in the rice-growing regions. The need for food is the most urgent and potentially explosive problem in more than half of the nations in the world today. Although 4–6 t/ha are produced in some areas where modern varieties are grown and improved agronomic practices are followed, rice yields range only between 1.5–2.0 t/ha or even lower in the tropics and subtropics.

The most important yield-limiting factors are:

1. Climatic problems, such as low or high temperature, flooding, drought, and strong wind.
2. Adverse soil conditions, such as salt or aluminium toxicity and zinc, iron, or phosphorus deficiency.
3. Pests and diseases, such as brown plant hopper, *Nilaparvata lugens* Stal, and rice gall midge, *Orseolia oryzae* Wood-Mason, rice blast and bacterial leaf blight caused by *Pyricularia oryzae* and *Xanthomonas oryzae* (Uyeda and Ishiyama) Dowson, respectively.

Besides the yield, the quality, nutritional value and hygienic deterioration are also affected by pests and diseases.

Two ways to increase rice productivity are considered possible: either modifying the environment or improving the rice itself. The former is usually much more expensive, or in some cases impossible, the latter, however, is the most effective way.

Major objectives in improvement suggested by plant breeders (Khush and Virmani 1985; Ling et al. 1986) are as follows:

1. High yield,
2. Early maturity,
3. Good grain quality,
4. Incorporation of genetic resistance to biological stresses, e.g., diseases, pests, nematodes and viruses,
5. Incorporation of genetic tolerance to environmental stresses, e.g., low or high temperature, soil toxicities, moisture stress, or nutritional imbalances,
6. Incorporation of novel traits from alien species, and
7. Streamlining of current breeding methods and selection procedures.

It is obvious that breeders in different rice-growing areas face different kinds of problems, so that they modify the objectives to meet the local requirements.

1.3 Available Genetic Variability and Need to Induce Somaclonal Variation

Genetic variability is the basis of crop improvement. Both spontaneous and induced heritable variations are widely used by breeders. The ancient farmers selected and domesticated plants for cultivation from natural plant populations, thus narrowing the gene pool and increasing the yield of crops. Lately, experienced breeders have been able to select some useful mutants and cross them in the attempt to combine the characteristics of both parents in the progeny.

The most significant examples are that the semi-dwarf gene of rice was discovered, evaluated, and utilized in China (Huang et al. 1960; Anonymous 1966) and at the International Rice Research Institute (Chang et al. 1965; IRRI 1967), resulting in the Green Revolution, and that a mutant plant with aborted pollen from a population of *O. sativa* f. *spontanea* was found in 1970 (Yuan 1972), and that the genetic tools (viz. cytoplasmic male sterile, maintainer, and restorer lines) essential to develop F_1 rice hybrids were successfully developed in 1973. Hybrid rice has caused the rice yield to increase dramatically in China in the past few years (Yuan and Virmani 1986).

Due to the very low frequency of spontaneous mutation, mutagens such as ionizing radiation and chemical mutagens were introduced to the breeding program, resulting in mutation breeding. Up to 1985, 336 mutant varieties of crops, belonging to 50 species, had been released for cultivation in 33 countries (Sharma 1986). Today, Yuan-Feng-Zao (Zao 1980), Calrose-76 (Rutger and Peterson 1976), and many other rice cultivars developed by mutation breeding are widely grown in the field. Additionally, interspecies and intraspecies crosses are also widely used to introduce certain gene(s). In brief, conventional plant breeding has an outstanding record of increasing productivity.

Although 50% of the crop yield increase during the past 50 years has been attributed to genetic improvement (Evans and Sharp 1983), the efficiency of conventional breeding is low because of the long term of selection and backcrosses (generally seven generations), the large scale of field trials (usually 4000–5000 F_2 plants or F_3 lines) and the labor intensity. Thus the cost for each new cultivar released can be up to 1 million dollars (Oram 1982).

Biotechnology, on the other hand, offers great potential for crop improvement (see Bajaj 1986a). Somaclonal variation, as one aspect of biotechnology, has been applied to rice breeding. It is already known that somaclonal variation includes genetic, epigenetic, and other kinds of variation. Genetic variation, both Mendelian and non-Mendelian, results from preexisting variation in the donor explant and/or from in vitro culture. Although the genetic mechanism is not yet completely understood, it is a novel source of variability for plant improvement and has been used in breeding sugarcane (Heinz et al. 1977), potato (Shepard et al. 1980), tomato (Evans and Sharp 1983), maize (Earle and Gracen 1985), and rice (Zhao et al. 1984).

Comparing this technique with the normal backcross program, Evans et al. (1984) concluded that the potential benefit of somaclonal and gametoclonal variations included the broadening of variation and the acceleration of the breeding process. The considerable differences between sexual self-pollinated and androgenetic lines (Dattee et al. 1984), between F_2 and SC_1 population (Nagai and Schnell 1986), between chemically induced mutation and somaclonal variation (Arreghini et al. 1986) have been found in both mutation spectra and frequencies, respectively.

The advantages of somaclonal variation in breeding programs can be considered as:

1. Wider spectrum and higher frequency.
2. Rapid stabilization. Stable lines could be obtained in one or two generations.
3. Small population. Promising lines would be selected from 100–150 somaclones.
4. High efficiency. Selection of certain characters could be done in vitro.

2 In Vitro Culture and Somaclonal Variation in Rice

2.1 Brief Review of the in Vitro Studies on Rice

Rice tissue culture work began in 1956 (Ameniya et al. 1956). Twelve years later, several Japanese groups reported plant regeneration from different explant-derived callus (Maeda 1968; Tamura 1968; Kawata and Ishihara 1968; Niizeki and Oono 1968; Nishi et al. 1968). So far, great success has been achieved in rice by biotechnology, and complete plants can be regenerated from different parts of the plant by different forms of cultures, including single cells and protoplasts (Table 1).

It is not easy to make distant crosses because of incompatibility and/or embryo abortion. Test-tube fertilization and embryo rescue could resolve the problems. Recently, foreign genes have been successfully transferred to *O. sativa* by embryo culture combined with backcross.

Anther culture of *japonica* rice has been widely used in breeding programs and, in this way, many new cultivars have been released (Chen 1986a,b; Zhang and Yin 1985).

Research on somaclonal variation and selection in vitro is opening a new pathway to rice improvement. Cryopreservation of organ, tissue, or cell (Bajaj 1986b), and storage of shoots in vitro in minimal growth conditions would be effective auxiliary paths for germplasm preservation of rice.

Regeneration of rice plants from protoplasts is considered as one of the most important advances in rice biotechnology. Whole plantlets regenerated not only from protoplasts of *japonica* and *indica* rice, but also from the fused- protoplasts between cultivated and wild rice, rice and *Echinochloa oryzicola*. Recently transgenic rice plants have been regenerated from foreign DNA inserted protoplasts (Table 1). Besides the most important cultivated species, *O. sativa* and *O. glaberrima*, tissue culture work on wild rice has also been carried out.

2.2 In Vitro Regeneration of Plants

2.2.1 Embryos

The rescue of immature embryos from distant crosses was the first application of tissue culture techniques in rice (Table 1). In early research, the cultured embryos were usually older than 10 days. Since many embryos of wide hybrids aborted in an earlier stage of embryo development, Davoyan and Smetanin (1979) and Lai (1982) improved the culture technique so that they could rescue 4–6-day-old embryos after pollination. Recently, Jena and Khush (1986) obtained some fully fertile alien addition lines through the combination of embryo culture and backcross of the embryo culture-derived distant hybrid of *O. sativa* × *O. officinalis* with *O. sativa* for several cycles, thus overcoming the chromosome inpairing in early generations of the hybrid. The resulting plants had *O. sativa* appearance and brown plant hopper resistance of *O. officinalis*.

Table 1. Organ, tissue, cell, and protoplast culture studies on rice

Species	Initiated material	Response in culture	Reference
Oryza sativa	Immature embryo	Plants	Ameniya et al. (1956)
O. sativa × *O. minuta*	Hybrid embryo	"	Nakajima and Morishima (1958)
O. paraguaiensis × *O. brachyantha*	"	"	H. Li et al. (1961)
O. sativa × *O. schweinfurthiana*	"	"	Bounharmont (1961)
O. sativa × *O. officinalis*	"	"	Iyer and Govila (1964)
O. sativa × *O. australiensis*	"	"	Jena and Khush (1984)
O. sativa × *O. latifolia*	"	"	Nowick (1986)
O. glumaepatula × *O. latifolia*	"	"	Nowick (1986)
O. sativa × *O. punctata*	"	"	Kostylev and Yatsyna (1986)
O. eichingeri	"	"	
O. officinalis	"	"	
O. schweinfurthiana	"	"	
O. malampuzhaensis	"	"	
O. latifolia	"	"	
O. alta	"	"	
O. australiensis	"	"	
O. sativa	Embryo	Callus and plants	Bajaj and Bidani (1980, 1986)
O. sativa L. f. *spontanea*	Proembryo	Plants	Lai (1982)
O. sativa	Proembryo	Plants	Lai (1982)
O. sativa	Immature embryo	Plants	Lai and Liu (1982)
O. sativa	Root	Continuous growth	Fujiwara and Ojima (1955)
O. sativa subsp. *jap.*	Stem node	Callus	Furuhashi and Yatazawa (1964)
O. sativa subsp. *jap.*	Seed	Callus and plants	Maeda (1968); Tamura (1968)
O. sativa	Root	"	Nishi et al. (1968)
O. sativa	Seminal root	"	Kawata and Ishihara (1968)
O. sativa	Seed	"	Furuhashi and Yatazawa (1970)
O. sativa subsp. *ind*	Leaf sheath	"	Yan and Zhao (1979)
O. sativa subsp. *jap*	Leaf sheath	"	Bhattacharya and Sen (1980)
O. sativa subsp. *ind*	Leaf blade	"	Bhattacharya and Sen (1980)
O. sativa	Leaf	Embryoid, plants	Wernicke et al. (1981)
O. sativa	Shoot apex	Callus and plants	Tsai (1978)
O. sativa	Immature inflorescence	"	Tang and Sun (1979)

O. sativa (n)	"	"	L. Sun and She (1982)
O. sativa (4n)	"	"	Qin et al. (1984)
O. sativa × *O. latifolia*	"	Embryogenic cluster and plants	Ling et al. (1984)
O. perennis	"	E callus and plants	M. Wang et al. (1987)
O. sativa (n)	Internode	"	Reddy (1983)
O. sativa (n)	Stem and internode	"	L. Sun and She (1982)
O. sativa (n)	Immature spikelet	Callus and plants	Heszky and Li (1984)
O. sativa	Immature endosperm	Callus and plants	Nakano et al. (1975)
O. sativa	Mature endosperm	Callus and plants	Bajaj et al. (1980)
O. austtraliensis	Seed	Callus	Mori (1986)
O. brachyantha	"	"	
O. breviligulata	"	"	
O. glaberrima	"	"	
O. gradiglumis	"	"	
O. latifolia	"	"	
O. minuta	"	"	
O. nivara	"	"	
O. officinalis	"	"	
O. perennis	"	"	
O. punctata	"	"	
O. alta	stem node	Callus and plants	Chu et al. (1986)
O. eichingeri	"	"	
O. grandiglumis	"	"	
O. latifolia	"	"	
O. minuta	"	"	
O. nivara	"	"	
O. officinalis	"	"	
O. paraguagensis	"	"	
O. perennis	"	"	
O. punctata	"	"	
O. ridleyi	"	"	
O. rufipogon	"	"	
O. minuta × *O. sativa*	"	"	
O. sativa × *O. rufipogon*	"	"	
O. sativa subsp. *jap.*	Anther	"	Niizeki and Oono (1968)
O. sativa subsp. *ind*	Anther	"	Guha et al. (1970)

Table 1. *(Continued)*

Species	Initiated material	Response in culture	Reference
O. sativa × *O. perennis*	Hybrid anther	"	Woo et al. (1978)
O. perennis	Anther	"	Wakasa and Watanabe (1979)
O. glaberrima	Anther	"	Woo and Huang (1980)
O. sativa × *O. glaberrima*	Hybrid anther	"	Woo and Huang (1980)
O. berviligulat	Anther	"	Wakasa (1982)
O. sativa subsp *jap.* (4n)	Anther	"	Takagi et al. (1982)
O. sativa	Pollen	Embryoids	Guha-Mukherjee (1972)
O. sativa	Pollen	Callus	Kuo et al. (1977)
O. sativa	Pollen	Callus, albino plants	R. Wang et al. (1979)
O. sativa	Pollen	Callus, albino and Green plants	Y. Chen et al. (1980)
O. sativa	Unpollinated ovary	Callus and plants	Nishi and Mitsuoka (1969)
O. sativa	Unpollinated Embryo sac	Callus and plants	Asselin de Beauville (1980); Zhou and Yang (1981)
O. sativa	Unpollinated ovary	Callus and plants	Zhou and Yang (1980)
O. sativa	Test-tube Fertilization	Plants	Zhang (1983); Yue et al.(1986)
O. sativa	Frozen suspension cell	Survived	Sala et al. (1979)
O. sativa	Frozen anther	Callus and plants	Bajaj (1980)
O. sativa	Frozen embryo and Endosperm	Callus, plants	Bajaj (1981)
O. sativa	Frozen callus	Survived	Finklė et al. (1982)
O. sativa	Frozen pollen embryo and zygotic embryo	Survived Plants	Bajaj (1984)
O. sativa	Shoots	Stored and survived	Z. Sun et al. (1988)
O. sativa f. *spontanea*	Shoots	Stored and survived	
O. sativa	Seedling	Single cells	Maeda (1965)
O. sativa	Suspension	Callus and plants	Ye (1984)
O. sativa subsp. *jap.*	Suspension	Callus and plants	Zimny and Lorz (1986)
O. sativa subsp. *ind.*	Suspension	Callus and plants	Abe and Futsuhara (1986b)
O. sativa × *O. latifolia*	Suspension	Callus and plants	Zou et al. (1986)
O. sativa subsp. *jap.*	Seminal root	Protoplasts	Maeda (1971)
O. sativa subsp. *jap.*	Protoplast	Callus	Deka and Sen (1976)

O. sativa subsp. *jap.*	Protoplast	Callus and plants	Fujimura et al. (1985); Toriyama and Hinatak (1986); Shimamoto et al. (1986); Yamada et al. (1986); Coulibaly and Demarly (1986); Cocking et al. (1986); Lei et al. (1986); G. Wang and Hsia (1986)
O. sativa + *Glycine max*	Protoplast	Fused protoplasts and callus	Niizeki and Kita (1981); Niizeki et al. (1986)
O. sativa subsp. *indica*	Protoplast	Plants	Kyozuka et al. (1988)
O. sativa + *Pisum sativum*	Protoplast	Fused protoplasts survived at −196 °C	Bajaj (1983)
O. sativa	Protoplast	Asymmetric fusion	Yang et al. (1988)
O. sativa + *O. officinalis*, *O. sativa* + *O. eichingeri*, *O. sativa* + *O. brachyantha*, *O. sativa* + *O. perrieri*,	Fused protoplast	Plants	Shimamoto (1988)
O. sativa + *Echinochloa oryzicola*	Fused Protoplast	Plants	Terada et al. (1987)
O. sativa	Protoplast	Expression of foreign DNA in rice cells	Zimny et al. (1986); Uchimiya et al. (1986); Ou-Lee et al. (1986); Baba et al. (1986)
O. sativa	Protoplast	Transgenic plants	Zhang et al. (1988); Zhang and Wu (1988)
O. sativa	Intact cell	Transgenic cell	Wang et al. (1988)

In addition, embryo culture has been used as a means to:

1. Enhance the germination ability of seeds, which were usually weak and hard to germinate in the field after long storage (Kucherenko 1986).
2. Accelerate the breeding cycle. Using embryo culture, Kucherenko and Loce (1982) obtained three generations of rice within one year when the normal growth period of the materials was less than 140 days.
3. Induce callus because of high regeneration frequency (Lai and Liu 1982).
4. Select in vitro. Embryos were treated with irradiation or mutagens and subsequently cultured in nutrient medium with or without selection stress, thus the influence of endosperm or maternal tissues was eliminated (Shome and Bhaduri 1982).

2.2.2 Callus and Suspension Culture

Callus can be induced from almost all parts of the rice plant. However, the capability of callus induction and plant regeneration depends considerably on the genotype (Zhao et al. 1981; Abe and Futsuhara 1986a) and also on the original parts of a plant which were used as explants (Davoyan and Smetanin 1979; Li and Heszky 1986). Generally, the explants most in use were immature embryos, immature inflorescences, and mature seeds. The immature inflorescences and embryos provided a higher frequency of callus induction and plant regeneration, but they were not available all year round. On the other hand, seeds could be routinely used although the capability for culture in vitro is not so good. There are two types of morphogenesis from tissue culture, i.e., organogenesis and somatic embryogenesis. Both possibilities were found in rice tissue culture (Kawata and Ishihara 1968; Heyser et al. 1983; Abe and Futsuhara 1985). That plants derived from somatic embryogenesis in the Gramineae were essentially free from genotypic variation was very important to keep somaclones from variation.

The successful suspension culture of rice was considered as one of the critical prerequisites for protoplast culture, selection in vitro, and in the future, for the production of artificial seeds and even for secondary metabolism products, because of the possibility of obtaining highly dispersed and homogeneous cell populations. Recently, plantlets were regenerated from cell suspension cultures of rice (Table 1). However, rice suspension culture, like anther culture and protoplast culture, significantly depended on the genotypes. The green plant regeneration frequency was, moreover low. It is necessary to improve the suspension culture in order to obtain more green plants, especially from suspension culture of *indica* rice.

Recently, several excellent reviews dealing with factors influencing the capability of callus initiation and plant regeneration in vitro have been published (Tomes 1985; Williams and Maheswaran 1986). Due to the importance of subculture in selection in vitro, cell suspension culture, protoplast culture, and genetic engineering, the attempt to prevent the loss of regeneration ability from long-term culture has been made.

At present, selecting embryogenic callus during every passage of subculture (Heyser et al. 1983; She et al. 1984) and optimizing the medium components and culture conditions (Kucherenko and Mammaeva 1980; Raghava and Nabors 1984, 1985; Davoyan 1986) may be the most common and effective methods for callus redifferentiation after long-term subculture. Treating callus with lower temperature (10–13° C) for 1 month before being transferred to regeneration medium increased the regeneration ability significantly (Kucherenko 1984a). One krad of gamma irradiation stimulated the morphogenesis in a cell line of IR-8 maintained for 8 months (Sharma et al., 1983), and water or salt stress could also restore the morphogenetic ability of callus (Liu and Lai 1985; Lai and Liu 1986; Kishor and Reddy 1986a). Now plants can be regenerated from callus with an in vitro history of 1500 days (Kishor and Reddy 1986b; Hildreth et al. 1987).

2.2.3 Anther, Pollen and Ovary Culture

Haploids play an important role in the rapid development of homozygous lines and in genetic manipulation, especially in isolation of desirable mutants in the very first generation. The production of new cultivars by haploid breeding reduces the breeding cycle, raises the efficiency of selection, and saves space and labor in the experimental field. For these reasons, intensive research on anther culture of rice has been carried out since 1968, when Niizeki and Oono first recovered plantlets from anther culture of rice.

So far, the regeneration frequency in *japonica* subspecies has been raised to 10% or more for some cultivars (Zhang and Yin 1985; Pulver and Jennings 1986, Taiwan Agri Cultural Research Institute 1986), some special protocols of haploid breeding and in vitro selection have been established (Zhang 1982; Chaleff and Stolarz 1982). Anther culture has been widely combined with conventional breeding for *japonica* rice. At present, about 80 varieties (or lines) have been released in China (Shen et al. 1983). With cooperation between scientists in Korea and IRRI, several anther culture-derived cold-tolerant lines have been tested in the field (Zapata et al. 1986). There are, however, still some serious barriers which hamper the wider application of this technique to breeding, especially in *indica* rice:

1. Low frequencies of callus initiation and plant regeneration. Among factors affecting androgenesis of rice (C. Chen et al. 1986), the critical factor was the genotype. The order of culture ability (percent of callus initiation × percent of regeneration) was *japonica*/*japonica* with glutinous endosperm – *japonica*/*japonica* – *indica*/*japonica* – *indica* hybrid (sterile line/restore line) – *indica*/*indica* (Shen et al. 1983). Comparing the 10% in *japonica* rice, the culture ability of *indica* is only about 0.5%. Even within one subspecies, e.g., *japonica* rice, the culture ability is different among cultivars. For this reason breeders could not use anther culture for *indica* rice improvement. In view of the fact that the culture ability may be controlled by a single block of genes (Miah et al. 1985), the transferring of these genes to those cultivars with poor culturability was suggested. Although possible technically, it is not a good way, because of the potential risk of genetic fragility resulting from narrowing the genetic background. It is necessary to develop other more universal and effective methods for anther culture.

2. High frequency of albinos. Among the regenerated plants, the average percentage of albinos was 50% (range from 5%–80%).
3. High frequency of haploids. Among the regenerated green plants, only half are spontaneously doubled diploid.

The knowledge of anther culture is accumulating, but is still not sufficient to resolve these problems. It is important to pay more attention to improving the methodology of anther culture.

Rice pollen culture was started by Kuo et al. (1977), callus being obtained from isolated pollen grains. After improving the protocol, Y. Chen et al. (1981) obtained green plantlets. For more detail about anther culture, refer to the reviews by C. Chen et al. (1986), Y. Chen (1986a,b), and Loo and Xu (1986).

Haploidy can be obtained through androgenesis as well as gynogenesis. Nishi and Mitsuoka (1969) were the first to plate unfertilized rice ovaries in vitro, and obtained diploid and tetraploid plants. Heszky and Sajo (1972) produced haploid rice plants from plated ovary-derived callus.

Asselin de Beauville (1980) and Zhou and Yang (1981) modified the technology and obtained a large number of plants from unpollinated ovaries and unfertilized embryo sacs of both *indica* and *japonica* rice, most of the regenerants being green instead of albino. This meant that not only the microspore, but also the megaspore or female gametophyte of rice can be triggered in vitro to sporophytic development. This opened a new avenue to genetic research and haploid breeding.

Lai (1986) made a cross between one *indica* variety which was semi-dwarf with good grain quality, and another which was tall with glutinous endosperm and black seed coat. The unfertilized ovaries of the hybrid were plated on modified N6 medium and diploid green plants were directly regenerated from the cultured ovaries. Some elite lines have been selected.

Although it is difficult to compare and draw conclusions about the advantages of anther and ovary culture, it seems that the main advantage of anther culture is its greater potential, largely because a much larger number of pollen are available; on the other hand, ovary culture seems to produce the majority of green regenerants.

2.2.4 *Protoplast Culture*

Research on rice protoplast culture has made considerable progress since 1985. Eight groups in the world have reported plant regeneration from rice protoplasts (Table 1). It is not surprising that each group has its own protocol. For example, most of these groups used embryogenic cell suspensions for protoplast isolation, but others isolated protoplasts from callus. Yamada et al. (1986) regenerated plants from protoplasts by the simple trick of adding calf serum to the medium, and callus could be induced from all the 26 varieties tested; however, plant regeneration occurred in only one cultivar. Cocking thought that the combination of heat shock (Thompson et al. 1987), protoplast culture in agarose (Thompson et al. 1986), and direct rapid plant regeneration from protoplast-derived callus through somatic embryogenesis was critical.

Although plants have been regenerated from the transformed or fused protoplasts of rice, it should be noted that the protoplast culture of rice, like anther culture, depends greatly on the genotype: all the protoplast-derived plants, to date, were regenerated from only a few cultivars of *japonica* and *indica* rice.

2.2.5 Cultures After Long-Term Storage

As the new improved varieties become more widely grown, more and more traditional varieties with diversified genetic bases are lost. This has resulted in a severe reduction in genetic variability for many crops. By far the best way of maintaining genetic variability in gene banks is low temperature storage of seeds. Today, more than 74,000 of old or modern rice varieties have been stored in the germplasm banks in the United States and the International Rice Research Institute. In some cases, however, storage of seeds becomes impossible because of sterility (haploid, triploid, male sterile lines etc.) or segregation of posterity (hybrid, aneuploid, etc.), so it is still necessary to maintain plants vegetatively in greenhouses or field. It is costly and laborious to protect them from disease and insects.

In vitro plant cultures can be stored by one of two approaches: slow growth and cryopreservation (Bajaj 1986b). The latter, in theory, can store materials indefinitely without periodic transfer. Recently, techniques for the preservation of rice suspension cells, mature embryos and endosperms, anther, anther-derived callus, and pollen embryos at ultra-low temperatures have been developed (Table 1). After storage for about 1 year, some cells or organs survived and even regenerated whole plantlets from induced callus. However, callus, suspension cells, or isolated protoplasts are not considered the ideal system for germplasm storage due to somaclonal variation. The development of more effective and reliable techniques for cryopreserving organs or tissues with meristems such as shoot tips may provide a valuable and genetically stable system for long-term germplasm storage.

Slow growth at a reduced temperature or in the presence of inhibitors offers the possibility of storing material for several years in vitro. This technique has been used in vegetative crops (Henshaw and O'Hara 1983). Recently, shoots of *O. sativa* L.f. *spontanea* Roschev. and other germplasm of rice have been stored in vitro without transferring for 1 year in our laboratory. Plants were then transferred to soil and grew normally. There were no significant differences either in isozyme patterns or certain agronomic traits between the stored plants and nonstored plants (Z. Sun et al. 1988a).

2.3 Somaclonal Variation

2.3.1 Variation in Callus and Regenerated Plants

Morphological differences were obvious among calli. Detailed research on the difference of callus indicated that there were two types: embryogenic (E) and nonembryogenic (NE) callus. E callus has a smooth, white, knobby appearance and is composed of small, isodiametric cells which average 31 μm in diameter; NE callus is yellow to translucent, wet, rough to crystalline in appearance and is composed of larger, elongated cells which average 52 μm in diameter and 355 m in length. Usually, E callus frequently gave rise to plants through somatic embryogenesis. NE callus occasionally gave rise to shoots or roots through organogenesis. E callus formed plants at much higher frequency than NE callus (Raghava and Nabors 1984). Recently, differences in protease activity, protein, and RNA between E and NE callus have also been found (Carlberg et al. 1984; L. Chen and Luthe 1985). Morphological differences have also been reported in rice suspension culture (Maeda 1965; Zou et al. 1986).

Variation in regenerated plants is very easy to distinguish. In the first reports on rice plantlet regeneration in vitro, all the authors described morphological and chromosomal variations (Bajaj and Bidani 1986). Albino plants existed in anther- (Niizeki and Oono 1968), tissue- (Tang and Sun 1979), ovary- (Liu and Zhou 1984), cell- (Zou et al. 1986) and protoplast- (M. Lei et al. 1986) derived plants. The frequency of albinos in anther culture was up to 50% or even higher, which limited the application of anther culture in breeding. The ploidy variation also occurred widely, which resulted in phenotype variations in sterility, grain size, and tiller number. Some other phenotypic variations, e.g., dwarf or twisted plants (Nishi et al. 1968), abnormal glumes, awns, strips (Abrigo et al. 1985), unheading (Huang et al. 1985), less leaf number, and early maturity (Z. Sun 1980; Z. Sun et al. 1981) were also observed. Oono (1978) reported that among 489 regenerated plants derived from a pure line, 58.7% showed lower fertility.

Although there were wide differences between regenerated plants and their parents, the differences were not obvious among plants originating from different parts used as explants of the same cultivar, except for the percent of albinos (Henke et al. 1978; Liu and Zhou 1984). The variation of callus or regenerated plants was reasonably considered as the result of genetic and/or environmental changes, for instance, the differences between certain E callus and NE callus were determined mainly by medium composition; however, the morphological changes in plants might be partly caused by genetic variation. Regenerated plants could not be used for rice production directly because of asynchronous development. It is also impossible to select useful variants from regenerated plant populations whose performance was greatly influenced by environment. In order to detect genetic variation and use it, much more attention should be paid to the progenies instead of to the regenerated plants.

2.3.2 Variation in Progenies of Regenerated Plants

Since Oono (1978) reported the results on the genetic behavior of plants regenerated from a pure line of rice, related data have increased enormously.

Variation of Ploidy and Chromosome. Many authors observed chromosome variations in regenerated plants, which included ploidy changes, deletion, duplication, and rearrangements. Usually, chromosome variations, especially triploid, quintyploid, and aneuploid resulted in phenotypic changes. Z. Sun et al. (1983) obtained some regenerants with thicker culms, larger leaves, awned and obviously bigger grains from immature inflorescence-derived callus. Genetic analysis proved that these plants were tetraploid. L. Zhang and Chu (1984) reported that chromosome aberrations influenced traits of regenerated plants.

Mutation in Nuclear Gene(s). Gene mutation in somaclones is very important for rice breeding. Some promising but imperfect lines could be cultured in vitro and some characters be improved. Such gene mutation was found in tomato (Evans and Sharp 1983) and rice (Z. Sun et al. 1983). In rice, among regenerants from immature inflorescence-derived callus of Tai-Zhong-Yu 39, one was 92 cm tall, quite close to its parent. In 614 SC_2 plants of this line, there were 132 dwarf mutants with a height of only 20 cm, which reached maturity normally. A X^2 test indicated that the segregation rate was in agreement with 3 normal:1 dwarf. The progenies of the dwarf mutant were all dwarf, but the F_1 of dwarf x T-Z-Y 39 was all normal in plant height, the following F_2 still segregated with the ratio of 3 normal:1 dwarf. At the same time, the progeny of normal plants in this line continuously segregated, the segregation rate fitting the expected 2:1 ratio. Thus it was confirmed that this dwarf mutant derived from tissue culture was controlled by a single recessive gene. Variation controlled by a single recessive gene was also obtained by Oono (1984).

Chloroplast Mutation. Chloroplast deficiencies, especially albino, could be found in various types of in vitro culture. Although all the albino plants died before or after transferring from the tubes, albinos continuously segregated in the progeny of green regenerants with regular frequencies. Therefore, chlorophyll deficiencies are often used as a genetic marker in the selfed generations of rice regenerants. According to Oono (1978), the rate of this mutation in SC_2 was 8.4%, comparable to X- or gamma-ray mutagenesis. Comparing the albino percentages in SC_2 and SC_3, Z. Sun et al. (1983) found that albino percentages decreased in following generations. The results suggested that chloroplast mutation in this case was controlled by cytoplasmic gene(s).

Homozygous Mutations. Reports on homozygous mutation in rice somaclones have accumulated. Once the homozygous mutation occurs, the related traits become true to type.

Oono (1978, 1981, 1984) reported in more detail this phenomenon in rice. Callus initiated from 75 seeds of a spontaneous doubled haploid derived from Norin 8, and 1121 plants regenerated. Analysis of mutations of chlorophyll, flowering, plant height, fertility, and morphology in SC_2 and SC_3 generations revealed that only 28.1% of the regenerants were normal in all these characters, and 28% of the regenerated plants carried two or more mutated traits. The results also indicated that most of SC_3 lines derived from 150 SC_2 mutants (347 out of 634) were breeding true for all of the five traits.

Z. Sun et al. (1983) also reported similar results. Among 950 immature inflorescences–derived somaclones (SC_2) from four cultivars, only 24.2% somaclones were normal in all six agronomic traits analyzed, and 36.6% of the total somaclones carried two or more varied characters. Some characteristics of the variation in all the cultivars studied were described as follows (Fig. 1):

A

B

C

D

E

1. There were obvious tendencies for plant height to become shorter, number of productive tillers to increase, and grain weight to be lighter.
2. Traits were uniform within each of more than 90% of total SC_2 lines.
3. At least up to SC_8, the traits of somaclones were true to type (Table 2) (Zhao et al. 1986).

The wide agronomic variation of somaclones at SC_2 were also described by Kucherenko (1984b) and Shen (1985). Besides plant height, tiller number, grain weight, and heading date, variations in plant type (Song et al. 1984), grain type (Zheng et al. 1989), blast resistance (Kucherenko 1984b), and cool tolerance (McKenzie and Croughan 1984) were also noted. High frequencies of somaclonal variation of agronomic characters in SC_2 were also reported by Kucherenko and Mammaeva (1979) – 38%, Davoyan (1983) – 50%–59%, Suenoga et al. (1982) and Abrigo et al. (1985) – 54% and 17.5% of SC_2 lines respectively with more than one varied trait. Reddy (1986) and Jun et al. (1986) obtained fourth and fifth generations of rice somaclones, respectively, and both obtained several breeding-true lines.

Table 2. Agronomic traits of somaclones in different generations. (Zhao et al. 1986)

Somaclone	Generation	Plant height (cm)	Panicle no.	Days to heading	1000 grain weight (g)	Flag leaf length (cm)	Panicle length (cm)	Grain No. per plant	Sterility (%)
CK_1^a		105.5 a	7.3 bc	102.7 a	21.2 d	37.8 a	19.2 a	1152.9 a	23.3 d
T-42	SC_4	92.6 b	8.4 b	95.0 c	22.1 cd	20.2 d	16.0 c	996.9 ab	32.4 c
	SC_5	94.5 b	9.3 ab	95.0 c	22.4 c	20.3 d	16.3 c	891.4 b	34.8 bc
	SC_6	94.5 b	9.5 ab	94.3 cd	22.8 c	19.9 d	16.3 c	885.1 b	38.0 bc
	SC_7	95.2 b	10.1 a	94.7 cd	22.5 c	20.6 d	16.5 c	901.9 b	32.9 c
T-7	SC_4	104.6 a	8.1 bc	93.7 d	25.1 ab	21.9 cd	17.7 b	804.3 b	38.3 bc
	SC_5	105.5 a	8.7 ab	93.7 d	25.0 b	21.7 cd	18.2 b	909.4 b	39.0 bc
	SC_6	106.3 a	8.2 bc	93.7 d	25.2 ab	21.1 cd	17.9 b	728.0 b	41.3 b
	SC_7	104.3 a	8.0 bc	93.7 d	25.1 ab	21.5 cd	17.9 b	745.9 b	47.1 ab
	SC_8	105.1 a	7.6 bc	94.0 d	25.8 a	22.5 c	18.2 b	838.6 b	49.6 a
CK_2^a		107.3 a	6.5 c	101.3 b	21.6 d	36.2 b	19.8 a	746.2 b	31.8 c

Different generations and their controls were grown in the same field and the same season, data were from randomized complete-block design with three replications. Values within a column followed by the same letter were not significantly different at 5% level of probability according to Duncan's Multiple Range Test.

[a]Both CK_1 and CK_2 were the initial variety – Tai-Zhong-Yu No. 39.

Fig. 1A-E. Somatic cell culture and somaclonal variation in rice (variety: Tai Zhong Yu 39). **A** Callus formation from immature inflorescence at early initiation stage (ca. 5–10 mm long) on N6 medium (Chu et al. 1975) supplemented with 2,4-D 1.0 mg/l, NAA 4.0 mg/l, BA 1.0 mg/l, and 3% sugar, solidified with 0.8% agar, pH 5.8 before autoclaving. **B** Plantlet regeneration from immature inflorescence derived callus on N6 medium supplemented with NAA 2.0 mg/l, BA 2 mg/l, and 2% sugar, solidified with 0.8% agar, pH 5.8. **C** Extreme dwarf mutant plant segregated in the progenies of a regenerated plant. **D** Comparison of the promising somaclone T-42 (*left*) with its parent material Tai Zhong Yu 39. The improved plant type, reduced plant height, and increased panicle number of T-42 are obvious. **E** Stability of somaclonal variation among different generations of a somaclone T-7 in the experimental field, Zhejiang Academy of Agricultural Sciences, Hangzhou, China in 1981. The D_3, D_4, and D_5 refer to SC_3, SC_4, and SC_5, respectively

In addition, similar somaclonal variations were observed in the anther culture-derived PC_2 lines (Wakasa 1982; Schaeffer 1982; Huang et al. 1985; Kishikawa et al. 1986). Yoboue (1986) found two anther-derived lines which were 20 cm shorter than the initial parent, and this character was stable after five selfed generations, genetic analysis indicating that it was recessive to normal height, showing no reciprocal effects.

The correlation between SC_1 and SC_2 (means of SC_1 progeny) was calculated at $r = 0.365$ (significant at 1% level) in seed fertility, and $r = 0.168$ (not significant) in plant height (Oono 1983). Correlation analysis between SC_1 and SC_2, and SC_2 and SC_3 was performed in our laboratory (Tables 3 and 4), the results showing that all traits except grain weight did not correlate significantly between SC_1 and SC_2; however, most traits did correlate significantly between SC_2 and SC_3 (Sun et al. 1988b).

Variation in Grain Quality. Improving grain quality is one of the goals in rice breeding. Protein and lysine content in grains are important parameters of nutritional quality. The protein and lysine content in seeds of tissue culture-derived

Table 3. Analysis of correlation between rice (Variety: Tai-Zhong-Yu 39) SC_1 and SC_2 for agronomic traits. (Z Sun et al. 1988b)

Trait	No. of somaclones	r	a	b	Sy.x	S.b	t	P%
Plant height (cm)	35	0.185	92.450	0.054	0.466	0.050	1.08	28.66
Panicle number per plant	34	-0.204	5.033	-0.018	0.097	0.015	1.18	24.58
Main panicle length (cm)	26	0.300	17.702	0.169	0.236	0.110	1.54	13.28
grain weight (g)	31	0.391	16.894	0.284	0.225	0.124	2.29	2.78

Table 4. The correlation between rice SC_2 and SC_3 plants for agronomic traits (Z-X Sun et al. 1988b)

Trait	X(SC_2)	Y(SC_3)	r	a	b	S.b	t	P%
Plant height (cm)	97.014	97.943	0.812	-73.884	1.771	0.569	3.11	2.64
Flag leaf length(cm)	37.711	40.884	0.798	7.442	0.887	0.273	3.25	1.74
2nd leaf length(cm)	46.723	47.200	0.838	1.616	0.976	0.259	3.76	0.96
Panicle number per plant	5.075	4.988	0.771	2.387	0.512	0.173	2.96	2.48
main panicle length (cm)	21.049	21.537	0.932	1.543	0.950	0.166	5.74	0.30
Total grain number per main panicle	171.157	156.057	0.629	30.786	0.732	0.405	1.81	12.93
Sterility (%)	29.981	32.201	0.702	14.354	0.596	0.247	2.41	5.11
1000 grain weight(g)	23.663	24.434	0.889	-0.771	1.065	0.224	4.76	0.36

The initial variety was Tai-Zhong-Yu 39. Both SC_2 and SC_3 plants were grown together in the same season.

plants were found to be increased (Schaeffer and Sharpe 1981). The protein content of some somaclones derived from anther culture was significantly higher than that of field control, in some at least up to the fourth generation. Sano et al. (1983) observed that the total protein in seeds of regenerants varied up to a maximum of 40%.

The amylase content, gel consistency, and alkali digestibility of several somaclones, the yields of which were higher than or equal to their respective controls, were analyzed (Z. Sun et al. 1989). Variations in cooking quality existed in somaclones and were hereditable. Mutations from nonwax endosperm to wax and, on the contrary, from wax to nonwax, were also found (Table 5). Some chalky seeds were found in the progenies of somaclones from anther or tissue culture (Schaeffer et al. 1986).

Zheng et al. (1989) investigated the grain quality of somaclones, and found that the variations in quantitative characters of all cultivars studied were consistent, grain protein contents of the somaclones were higher in one cultivar, and the variability of most traits was enhanced by combining low-dosage radiation and tissue culture.

Regulation of Somaclonal Variation. It is clear now that:

1. Somaclonal variation is relatively rare in plants regenerated through embryogenesis.
2. Prolonged subculture could increase genetic and phenotypic variation.
3. Phytohormones, especially 2,4-D, may cause somaclonal variation.

Additionally, Oono (1984, 1985) showed that high concentrations of $MgSO_4$, NH_4NO_3, and KI increased genetic variability, but $CaCl_2$, KH_2PO_4 and $MnSO_4$ decreased the variability. Base analogs such as hydroxyurea and acridine orange also increased genetic variability. Higher concentrations of BA (50 mg/l) caused genetic variation 50 times higher than lower concentration of BA (2 mg/l) did.

We know little about controlling somaclonal variation at present, but with the advance in rice genetic engineering, further search for methods to control it will be an urgent biotechnical topic.

2.4 Chromosomal Analysis and Biochemical Evidence

It is believed that there are two origins of somaclonal variation: variation preexisting in the explants, or arising during in vitro culture. In order to explain all the variability and use it, it is necessary to understand the nature of the variation. For somaclonal variation the following may be responsible: variation in chromosome number and its structure; DNA amplification and deletion, cytogenetic effects, transposons, gene inactivation and/or reactivation, and gene mutation. (Larkin and Scowcroft 1981).

Chromosome variation in both cultured cells and regenerated plants has been discussed in detail (D'Amato 1977, 1978; Constantin 1981). Recently, emphasis on this subject has moved from model plants to the most important crops such as wheat (Karp and Maddock 1984), barley (Orton 1980), corn (Edallo et al. 1981), oats

Table 5. Somaclonal variation in rice grain quality. (Z. Sun et al. 1989)

Somaclone	Generation	Alkali digestibility		Gel consistency		Amylose content		Protein
			Grade[a]	mm	Grade[b]	%	Grade[c]	%
Guang-Lu-Ai 4	CK	3.5	H	24.5	H	21.7	M	9.50
T-G 138	SC2	3.5	H	25.5	H	20.6	M	9.71
T-G 366	SC2	5.0	M	25.5	H	20.4	M	9.81
T-G 185	SC2	2.8	H	26.0	H	21.7	M	10.43
T-G 270	SC2	3.8	H	25.0	H	21.3	M	10.85
T-G 41	SC2	3.2	H	26.5	H	22.0		10.64
T-G 325	SC2	4.8	M	26.5	H	23.7	M	10.43
Dong-Xuan 4	CK	7.0	L	38.0	MH	19.27	L	8.10
T-D 3	SC2	7.0	L	36.5	MH	20.53	M	7.35
T-D 5	SC2	7.0	L	52.0	M	19.76	L	7.97
T-D 8	SC2	7.0	L	38.0	MH	19.98	L	7.77
T-D 17	SC2	7.0	L	40.0	MH	20:31	M	7.56
8046	CK	6.2	L	30.0	H	23.09	M	
T-8046 1	SC2	7.0	L	45.0	M	17.30	L	
T-8046 2	SC2	7.0	L	49.2	M	18.70	L	
T-8046 3	SC2	7.0	L	45.5	M	17.03	L	
Xiu-Shui 48	CK	7.0	L	40.7	M	20.50	M	
T-X 256	SC2	7.0	L	39.3	MH	18.35	L	
T-X 255	SC2	7.0	L	39.5	MH	20.07	M	
T-X 254	SC2	7.0	L	34.3	H	20.44	M	
T-X 253	SC2	7.0	L	36.0	MH	19.87	L	
T-X 252	SC2	7.0	L	39.5	MH	20.26	M	
T-X 251	SC2	7.0	L	43.0	M	20.50	M	
Nong-Hu 6	CK	6.9	L	51.5	M	20.11	M	
T-N 261	SC2	7.0	L	130.0	S	0	S	
T-N 262	SC2	7.0	L	114.5	S	0	S	
T-N 263	SC2	7.0	L	117.0	S	0	S	
T-N 264	SC2	7.0	L	120.0	S	1.2	S	
Ce 21	CK	7.0	L	110.0	S	1.9	S	
T-C 151	SC2	7.0	L	55.5	M	19.58	L	
Tai-Zhong-Yu 39	CK	7.0	L	37.5	MH	18.36	L	8.26
T-7	SC4	7.0	L	49.5	M	20.03	M	
T-7	SC5	7.0	L	44.8	M	18.55	L	
T-7	SC6	7.0	L	51.5	M	18.93	L	
T-7	SC7	7.0	L	48.3	M	20.08	M	
T-7	SC8	7.0	L	53.5	M	19.79	L	
T-42	SC4	7.0	L	29.5	H	22.76	M	8.10
T-42	SC5	7.0	L	30.5	H	23.08	M	
T-42	SC6	7.0	L	28.7	H	22.72	M	
T-42	SC7	7.0	L	29.3	H	22.45	M	

[a] H: high, M: medium and L: low.
[b] H: hard, MH: medium hard: M: medium and S: soft.
[c] H: high, M: medium, L: low and S: sticky.

(Mocoy et al. 1982), triticale (Jordan and Larter 1985), potato (Nelson et al. 1986), and rice (see Table 6).

In rice, since Niizeki and Oono (1968) reported that anther-derived plants were haploid or dihaploid, intensive study on cytological variation in callus and regenerated plants has been done. The chromosome number in the cells of anther-derived callus varied widely, and changed during subculture. C. Chen and Chen (1980) reported that at the first passage of anther culture, many kinds of calli consisted of different ploidy or mixed ploidies. Among them, 24% of the cultures were nonhaploid, and after 17 passages of subculture, it was found that haploid cells were eliminated from all cultures, and diploid or tetraploid cells become predominant. It was suggested that endomitosis was responsible for the changes.

The age of the callus also affected the chromosome number. The percentage of haploid in all cells analyzed decreased from 42.7% (0–3 days old) to 0 (30–33 days old); however, the percentages of diploid and polyploid increased from 39.8% to 66.7% and 17.7% to 33.3%, respectively, during the same period (C. Chen et al. 1982). She et al. (1984) analyzed about 2000 cells for each type of anther-derived callus, one of which was subcultured for 17 passages and the other for 37 passages, and obtained similar results.

The above results indicate that the chromosome number changed during in vitro culture, and suggest there is some negative relationship between redifferentiation ability and endomitosis in callus.

Variations in chromosome number in regenerated plants were also investigated. Generally about 50% of pollen plants were spontaneous diploid, mixoploid or aneuploid (Niizeki 1968; Nishi and Mitsuoka 1969; Niizeki and Oono 1971; Kharchenko and Kucherenko 1977; Huang et al. 1985). The ploidy of plants regenerated from anthers of tetraploid plants was different from that originating from diploid plants, which included diploid, trisomics, tetraploid, tetrasomics, pentasomics, and hypertetraploid (Kishikawa et al. 1986).

Different ploidies were also found in plants regenerated from ovary (Liu and Zhou 1984), seed (Oono 1978), seed embryo (Davoyan 1983), immature endosperm (Davoyan and Smetanin 1979), and immature inflorescence (Z. Sun et al. 1983). However, it seemed that the difference at chromosome level within pollen plants was greater than that within plants regenerated from other explants. Ling and his colleagues (1981) obtained 19 pollen-derived plants from an intercultivar hybrid of rice. Two of them were homologous asynaptic triploids where no chromosome pairing could be found in diplotence, diakinesis, and metaphase I. Chu et al. (1985) reported that the frequency of rice aneuploids from a total of 290 pollen plants was 10.7%. Among these the primary trisomics, tetrasomics, monosomics, and double trisomics were 1.7%, 1.3%, 1.0%, and 0.7%, respectively. Some of the primary trisomics were triplo-3, 4, 8, 10, or 12.

Variation at the chromosomal level might cause the phenotype variation. It should also be of concern that some somaclonal variations which were not associated with cytologically observable chromosome alterations might represent fine-structure changes.

The techniques of molecular biology offers an effective tool to detect somaclonal variation at the molecular level. In the early 1970's, electrophoresis patterns of the α-amylase isozyme in rice calli were studied (Saka and Maeda 1973, 1974).

Table 6. Chromosome variation in rice somaclones

Explant[a]	Chromosome no.	Analyzed cells taken from	Chromosome variation	Reference
Anther	n = 12	Regenerated plants	n = 12, n = 24	Niizeki (1968)
Anther and	n = 12	Regenerated plants	n, 2n, 3n, 4n and 5n	Nishi and Mitsuoka (1969)
ovary	n = 12	Regenerated plants	2n and 4n	
Anther	n = 12	Regenerated plants	n, 2n, 3n and 4n	Niizeki and Oono (1971)
Anther	n = 12	Regenerated plants	n, 2n, 3n, 4n and mixploid (19–30, 10–21, 18–25, 27–38)	Y. Chen et al. (1974)
Anther	n = 12	Regenerated plants	n, 2n, 3n and 4n	Kharchenko and Kucherenko (1977)
Anther (*O. perennis*)	n = 12	Regenerated plants	n and 2n	Wakasa and Watanabe (1979)
		PMC from haploid	12I, 10I + 1II, 8I + 2II	
		PMC from diploid	12II, 12I and anueploid	
Anther	n = 12	Initiated callus	n, n + 2n, n + 2n + 4n, 2n, 2n + 4n, 4n, 6n, n + 2n + 4n + 8n	C. Chen and Chen (1980)
		Subcultured callus	n, n + 2n, n + 2n + 4n, 2n, 2n + 4n, 4n, 6n, n + 2n + 4n + 8n and 3n	
Anther	n = 12	Regenerated plants	n, 2n, 3n, 4n, and 2n + 1	Oono (1981)
Anther	n = 12	PMC from regenerated plants	Asyndetic homologous triploid	Ling et al. (1981)
Anther	n = 12	Regenerated plants	n, 2n, 3n, 4n and 5n	C. Chen et al. (1982)
Anther	n = 12	Callus after different passages and with different age	n, 2n and polyploid	C. Chen (1984)
Anther	n = 12	Callus with different age	n, 2n and polyploid	She et al. (1984)
Anther	n = 12	Regenerated plants	n, 2n, n + 2n, 4n	Huang et al. (1985)
Anther	n = 12	Regenerated plants	n, 2n, Polyploid	Mercy and Zapata (1986)
Anther from 4n plant	2n = 24	Regenerated plants	2n, 2n + 1, 2n + 2, 2n + 3, 4n, 8n	Kishikawa et al. (1986)
Seed	2n = 24	Regenerated plants	2n, polyploid	Oono (1978)
Immature endosperm	3n = 24	Regenerated plants	2n	Davoyan and Smetanin (1979)
Endosperm	3n = 36	Callus, plants	Chromosome number varied from 24 to 105	Bajaj et al. (1980); Bajaj and Bidani (1986)

Immature inflorescence from haploid stem, internode	n = 12	Regenerated plants	n, 2n	Hu et al. (1983)
Immature inflorescence from n plant	n = 12	Regenerated plants	n, 2n	L. Sun and She (1982)
Seed embryo	2n = 24	Regenerated plants	2n, polyploid, aneuploid	Davoyan (1983)
Seed embryo	2n = 24	Callus, plants	n, 2n, polyploid, aneuploid	Bajaj and Bidani (1980, 1986)
Immature inflorescence	2n = 24	Regenerated plants	2n, 4n	Z. Sun et al. (1983)
Immature inflorescence	2n = 24	Regenerated plants	2n, 4n, 7II + 1X	Ling et al. (1987)
Stem, internode from n plants	n = 12	Regenerated plants	n, 2n	Reddy (1983)
Ovary	n = 12	Regenerated plants	n, 2n	Liu and Zhou (1984)
Young spikelet	n = 12	Regenerated plants	n, 2n	Heszky and Li (1984)
Immature inflorescence from n plant	n = 12	Plants regenerated from different passages	n, 2n, polyploid, aneuploid	L. Zhang and Chu (1984)
Immature inflorescence from n plant	n = 12	PMC of regenerated plants	Aneuploid, intrachromosome variation	Chu and Zhang (1985) Chu et al. (1985)

[a] All were taken from plants of *O. sativa* (2n = 24) except those noted.

Ishikawa et al. (1987) compared the electropherograms of 14 isoenzymes in calluses of 36 strains with that of original plants. It was found that the loci of some isozymes were stable during callus culture, while the electropherograms of some others were changed.

The albinism, as mentioned above (Sect. 2.2.3), is one of the most serious problems in cereal anther culture. C. Sun et al. (1974) observed that protoplastids of albino rice lacked ribosomes. Electrophoresis patterns indicated that albino plants lost the capacity to synthesize fraction 1 protein. Albino plants also lacked the 23S and 16S chloroplast ribosomal RNA's and ribulose-1,5-bisphosphate carboxylase (C. Sun et al. 1979). It is known that the large subunit of this enzyme and the chloroplast ribosomal RNA's are encoded by the chloroplast genome, and that large regions of the chloroplast genome are deleted in most wheat albino plants (Day and Ellis 1984). Thus, the next step should be research on the mechanism of plastid DNA deletion.

Oono (1986) analyzed the Southern blots of nuclear DNA of rice callus using two cloned highly repeated sequences derived from rice embryo DNA. Amplification of 100 times was found for one sequence in the callus. For another sequence, decrease in copy number was obvious. Zheng et al. (1987) cloned 11 highly repeated sequences from nuclear DNA of the suspension cells of a *japonica* variety, Roncarolo, and hybridized them with dots of nuclear DNA's from seedlings or suspension cells of the same variety. All the sequences were amplified in suspension cells, with the maximum amplification of 70 times. Such amplification appeared to relate neither to ecospecies, because similar results were obtained from an in vitro and in vivo comparison for an *indica* variety, nor to the age of the cultures, since at the time of use, one suspension was 8 years old, another 3 months. This implied that the quantitative changes of copy number of DNA repeated sequences occurred mainly during the dedifferentiation process. However, Chowdhury et al. (1988) observed only minor variation between the restriction endonuclease patterns of the mt DNA of a somaclone BL2 after long term in vitro culture and the original line PI 353705.

Proteins extracted from seedling shoots of a somaclone, T-42, (see Sect. 2.5) and its parent, Tai Zhong Yu-39, were subjected to SDS-PAGE or two- dimensional electrophoresis. The results indicated that the protein pattern of T-42 differed from that of the parent (Z. Sun and Colbert 1987). There was a dramatic increase in abundance or synthesis of a new protein of 85 kDa and pI 6.5. The results should provide a very useful basis for further evaluation of the significance of the new protein.

2.5 Utilization of Somaclonal Variation

In rice, somaclonal variation could be mainly used for the following purposes.

2.5.1 Theoretical Research

The conventional procedure for obtaining a set of trisomics is time-consuming. A certain percentage of aneuoloids occurred during anther culture (Chu et al. 1985).

It becomes less difficult to obtain a whole set of trisomics if the population for cytological analysis is large enough.

Plants derived from in vitro culture are also used for evolution research. Mercy and Zapata (1986) observed the chromosome behavior at meiosis of haploid plants derived from anther culture. Since all of the chromosomes were univalents, it is believed that the early hypothesis of rice as an ancient polyploid is unlikely.

Nowick (1986) examined meiotic chromosome pairing in a F_1 hybrid regenerated from *O. sativa* (genome AA) × *O. latifolia* (CCDD) and *O. glumaepatula* (AA) × *O. latifolia* (CCDD). The results, that chromosome pairing was found in both the hybrids, suggested there was some relationship between their respective parents.

Some regenerants with a special mutated gene such as the extreme dwarf (Z. Sun et al. 1983) could also be used as a marker for genetic research.

2.5.2 Obtaining Isogenic Line

Somaclonal variation could be used to produce new variants that retain all the favorable qualities of an existing variety while improving one trait, such as disease resistance or herbicide tolerance. Immediately after isolation of these isolines, they could be used for rice breeding, as well as rice genetic research, and in the near future, they will probably become an ideal material for gene transformation.

2.5.3 Breeding Directly

Although the mechanism of somaclonal variation is still not clear, the prospect of its application to breeding is attractive. The idea of improving rice by means of selecting somaclonal variation in the field probably first became fact at the China National Rice Research Institute, Hangzhou, China, in 1983. From immature inflorescence-derived somaclones, T42 was selected, which inherited resistance to bacterial leaf blight (BLB) from its parent, Tai Zhong Yu 39, a *japonica* variety, and combined the new traits of ideal plant type and more tillers. The average yield of T42 in the Provincial Rice Yield Trial in Zhejiang Province, China, in 1984 was 6650 kg/ha, the highest in the trial, significantly higher than the leading variety, Nong Hu No. 6. In the past few years, T42 has been tried on a large scale and accepted by the farmers (Zhao et al. 1984, 1986).

Based on these experiences, a program using somaclonal variation for rice improvement has been suggested. Referring to this program, some promising lines with ideal types and high yield, as well as better grain quality, were developed by another group in China (Song et al. 1984, 1986). Recently, these lines are being grown more widely. A somaclone derived from cv. Saturn, which was characterized by shorter stature, cool tolerance, early maturity, heavier grain, and more tillers was evaluated in replicated yield trial in Louisiana, USA (McKenzie and Croughan 1984.

Schaeffer et al. (1984) selected a line from anther culture which maintained all the parental traits but contained 10% higher lysine in seeds.

2.5.4 Selection in Vitro

One of the major advantages afforded by cell culture is that it makes possible the selection for novel phenotypes from large physiologically and developmentally uniform populations of cells grown under defined conditions. One of the prerequisites for the application of in vitro selection to rice improvement is that the reaction in the whole plant to stress correlates with that at tissue or cell level and the trait selected is heritable.

The common procedure is to add stress to the medium for callus initiation (or plant regeneration, or both), subculturing cells or callus for several passages on the medium with the stress and then regenerating plantlets from surviving calli. After that the plants are tested whether the selected trait is expressed in vivo or not, and finally, the stability of the character in sexual progenies should be evaluated. In order to increase selection efficiency, physical or chemical mutagens are usually used for the explants, callus, or suspension cells (Walther and Preil 1981). Selection in vitro has been carried out in the following areas in rice so far:

Tolerance to Saline Soil. NaCl is the principal salt in saline soils. In most of the experiments, 0.2–3% of NaCl was used as selecting agent on rice seed-derived callus (Oono 1978; Kucherenko 1980; Suenaga et al. 1982; Yamada et al. 1983), pollen-derived callus (S. Chen et al. 1983) and embryo-derived callus (Reddy and Vaidyanath 1986). Seawater was also used as selecting agent on immature embryos and immature embryo-derived callus (Yamada et al. 1983). In some cases, callus was treated with the chemical mutagen EMS before selection (S. Chen et al. 1983). Usually 1–1.5% NaCl in the medium inhibited callus growth. Calli were repeatedly subcultured on the medium with the same or increasing concentrations of NaCl. Calli with enhanced tolerance survived and grew stably. Yamada et al. (1983) suggested that four subcultures were more than enough to produce salt-tolerant callus, even though 45–50 subcultures were also reported for the selection of cell lines which could tolerate NaCl concentrations of 1–1.5% (Paul and Ghosh 1986).

Plants were regenerated on the medium with or without selecting agent. The salt tolerance of regenerated plants was higher than the control when they grew in Kimura B solution containing 17.5% seawater (Yamada et al. 1983). The selected salt-tolerant plants matured and produced good grains, but the control plantlets (one grown from normal rice seed, the other a plant regenerated from cells subcultured without seawater) did not flower.

Analyses showed that salt-tolerant rice plants absorbed more Na than K from seawater, and that much of this Na was transported to the leaves. It was possible that the salt-tolerant plants absorb excess salt, but extrude the extra salt, because sometimes many salt crystals were found on the leaf sheath, indicating that rice plants regenerated from cells tolerant to seawater possessed a salt extrusion system (Yamada et al. 1983).

Besides, Kishor and Reddy (1985) reported that calli from the Jaya cultivar tolerant to 3% NaCl were obtained by selection with 2.5 and 5% polyethylene glycol.

Resistance to Diseases. Crude phytotoxins, including culture filtrate of pathogens or direct inoculation of pathogens, were usually used as selecting agents. A pathogenic strain of *Xanthomonas oryzae* was inoculated onto calli derived from mature embryos of a susceptible variety, Nangeng 34, and a resistant variety, Nangeng 11. Based on the rates of callus growth, responses of calli to the inoculation between susceptible and resistant varieties were significantly different, the callus growth of the susceptible variety being more inhibited. Out of 365 inoculation calli of Nangeng 34, 63 calli showed sectional proliferation, from which many regenerated plants were produced. Bacterial leaf blight resistance of the regenerated plants was examined by clipping inoculation. Forty-four out of 45 regenerated plants showed resistance, and only one plant was susceptible. Evaluation of the progenies of these resistant regenerated plants showed that their resistances were stable and heritable for at least three generations (L. Sun et al. 1986).

Ling et al. (1985) reported the in vitro screening using the specific toxin (H-toxin) produced by *Helminthosporium oryzae*, based on the consistency of the in vitro and in vivo responses. Very small pieces of callus (around 1 mm^2 in size) of susceptible varieties were shaken in a toxin-enriched liquid medium for 2 days and then transferred to a toxin-free agar medium for subculture and plant regeneration. Two regenerated plants of IR-8 derived from 25% H-toxin treatment were found to be resistant to the pathogen of brown spot disease. Segregation of resistance occurred in the progenies of the resistant plant.

The pathogenicity of culture filtrate from *Pyricularia oryzae* on rice was studied in detail (Zheng et al. 1985). Using the culture filtrate of four pathogenic races of the fungus as the selecting agent, C. Li et al. (1986) screened both pollen calli and embryo calli, and green plantlets were regenerated from the resistant calli. The seedlings raised from the seeds of the regenerated plants were inoculated with the corresponding race of the pathogen and the responses were evaluated. Twelve mutants with high resistance were selected out of the regenerated plants derived from 810 resistant calli of 10 cultivars. It was also found that the somaclones of the resistant cultivars were still resistant, with widened resistant spectra, and about half of the somaclones derived from resistant calli of the susceptible cultivars became resistant.

Resistance to Amino Acids or Their Analogs. In order to improve the nutritional quality of rice, experiments were designed to recover rice cells resistant to S-aminoethyl-cystein (S-AEC), the analog of lysine, and to regenerate plants from resistant calli (Schaeffer and Sharpe 1983). The anther culture derived calli of Assam 5 were broken into small aggregates of approximately 2 mm^2 and spreaded evenly on a tissue increase medium with 10^{-3} M S-AEC. Calli that grew in the presence of S-AEC were recultured by transferring them three times in a medium with 2×10^{-3} M S-AEC to ensure resistance to the analog. Analysis of seed from the original plant regenerated from S-AEC-resistant calli showed that the lysine content expressed as percentage of total amino acids in seed protein was 3–10% higher than the control. An increase in seed protein and/or lysine consistently showed in the third through the fifth generation progeny from the regenerated plants of S-AEC-resistant callus. Protein lysine increased 3–6% in advanced generations. The mutant plants, with 10% more seed protein content than control, were found to have seed yield equal to the control. It was more interesting that the negative correlation between protein level and lysine content normally observed in rice breeding

(Beachell et al. 1972) was broken in the S-AEC-resistant variants, because the 10% increase in lysine did not produce a decrease in protein. Protein and amino acid analysis of milled and brown rice showed that the removal of the aleurone layer removed a greater percentage of the total protein from the control than from the tissue-culture mutant, indicating that increase of lysine reflected an increase in a specific protein in the endosperm.

Recently, Schaeffer (1986) reported that calli of Calrose 76 were subjected to inhibitory levels of lysine + threonine three times and of S-AEC once, and that the progeny from one resistant callus-derived plant were found to have a significantly higher percentage of lysine in seed proteins than the original cultivar. Meng and Chen (1987) found that AEC-resistant calli selected on callus induction medium with AEC was also resistant to lysine plus threonine.

In the selection for enhanced protein and lysine levels in rice seeds, other selecting agents such as 5-methyl-tryptophan (C. Chen and Chen 1979; Wakasa and Widholm 1983, 1986), L-methionine-DL-sulfoximine (Kishinami and Ojima 1980) and hydroxy-L-proline (Hasegawa and Inoue 1983) were also involved.

There were, moreover, several reports on in vitro selection for tolerance to cold (Kucherenko 1980) and drought (Kishor and Reddy 1986b,c) environments, and to aluminum (Nunez et al. 1985), and cadmium (H. Li, Chen 1987) toxicity.

Although in vitro isolation of mutants has advantages over conventional field selection, in vitro techniques for mutant isolation of agronomic significance have not been effectively used in crop breeding because of some theoretical and methodological problems. The large-scale agricultural application of in vitro mutant selection requires the cellular manifestation of the desired whole plant modification. In other words, the expression of selected traits in regenerated plants is essential for breeding application. Kucherenko (1980) provided evidence of contradictory data: no relation between salt tolerance at the organism and tissue levels on the one hand, and enhanced salt tolerance in selection-derived plants on the other. Another problem for rice is the rapid loss of totipotency after subculturing. It is believed that more progress will be made as the knowledge on molecular genetics, biochemistry, and plant physiology increases and the methodology of suspension culture and subculture improves.

3 Conclusions and Prospects

It is generally accepted that heritable variations exist in somaclones, which provides a novel approach for rice breeding. During the past 20 years, great progress has been made both in theoretical research and application of rice somaclonal variation. Some promising lines have been developed from somaclones. The increase in genome structure rearrangement during tissue culture provides a new opportunity for alien gene introgression, which will help widen the rice germplasm base, particularly in wide crosses, where crop and alien chromosomes cannot replicate through meiosis.

As plants have been regenerated from rice protoplasts, a new task, i.e., variation in protoplast-derived plants, is facing somatic genetics. Because protoplast fusion and genetic engineering of rice have been put on the agenda, it is urgent to investigate how to keep the receptor cells from variation, which is also the prerequisite to artificial seed production.

For the near future, as the foreign genes are successfully transferred to rice protoplasts and plantlets are regenerated from transformed or fused protoplasts, the in vitro and in vivo (plants and their progenies) behavior of foreign gene(s) will become an important area of somatic genetics. In brief, both somaclonal variation and somaclonal stability are going to play important roles in rice improvement.

Acknowledgments. This investigation was partially supported by funds from China National Committee of Science and Technology and by the Rockefeller Foundation International Program on Rice Biotechnology. We thank colleagues for their encouragement.

References

Abdullah R, Cocking EC, Thompson JA (1986) Efficient plant regeneration from rice protoplasts through somatic embryogenesis. Bio/Technology 4:1087–1090

Abe T, Futsuhara Y (1985) Efficient plant regeneration by somatic embryogenesis from root callus tissues of rice (*Oryza sativa* L.). J Plant Physiol 121:111–118

Abe T, Futsuhara Y (1986a) Genotypic variability for callus formation and plant regeneration in rice (*Oryza sativa* L.). Theor Appl Genet 72:3–10

Abe T, Futsuhara Y (1986b) Plant regeneration from suspension cultures of rice (*Oryza sativa* L.) Jpn J Breed 36:1–6

Abrigo WM, Novero AU, Coronel VP, Cabuslay GS, Blanco LC, Parao FT, Yoshida S (1985) Somatic cell culture at IRRI. In: Biotechnology in international agricultural research. IRRI, Manila, pp 144–158

Ameniya A, Akemine H, Toriyama K (1956) Cultural conditions and growth of immature embryo in rice plant studies on the embryo culture in rice plant. 1. Bull Natl Inst Agric Sci D 6:1–40

Anon (1966) The characteristics and cultivation technique of Ai-Jiao-Nan-Te, *Oryza sativa* L. subsp. hsien. Acta Agron Sin 5:18–22

Anon (1978a) A study of the animal and plant remains unearthed at Ho-Mu-Tu. Acta Arch Olog Sin (1):95–107

Anon (1978b) U.K. investigates virus insecticides. Nature (London) 276:548–549

Anon (1986) DNAP develops patented protected "hot" tomato. Agric Genet Rep 5(5):4

Arreghini E, Baratta G, Barbuze F, Biasini MG, Gavazzi G, Sale F, Todesco G, Tonelli C, Vecchio F (1986) Chemically induced mutagenesis versus somaclonal variation in tomato. In: Abstr 6th Int Congr Plant tissue and cell culture, Aug 3–8, 1986, Univ Min, Minneapolis, p 312

Asselin de Beauville M (1980) Obtention d'haploides in vitro à partir d'ovaires non fécondés de riz, *Oryza sativa* L. C R Acad Sci Paris D 296:489–492

Baba A, Hasezawa, Syono K (1986) Cultivation of rice protoplast and their transformation mediated by *Agrobacterium* spheroplasts. Plant Cell Physiol 27:463–471

Bajaj YPS (1980) Induction of androgenesis in rice anthers frozen at −196 °C. Cereal Res Commun 8:365–369

Bajaj YPS (1981) Growth and morphogenesis in frozen (−196 °C) endosperm and embryos of rice. Curr Sci 50:947–948

Bajaj YPS (1983) Survival of somatic hybrid protoplasts of wheat × pea and rice × pea subjected to −196 °C. Indian J Exp Biol 21:120–122

Bajaj YPS (1984) The regeneration of plants from frozen pollen embryos and zygotic embryos of wheat and rice. Theor Appl Genet 67:525–528

Bajaj YPS (ed) (1986a) Biotechnology in agriculture and forestry, vol 2. Crops I. Springer, Berlin Heidelberg New York Tokyo

Bajaj YPS (1986b) In vitro preservation of genetic resources. In: Int Symp Nuclear techniques and in vitro culture for plant improvement. IAEA, Vienna, pp 43–57

Bajaj YPS, Bidani M (1980) Differentiation of genetically variable plants from embryo-derived callus of rice Phytomorphology 30:290–294
Bajaj YPS, Bidani M (1986) In vitro induction of genetic variability in rice (*Oryza sativa* L.). Int Symp New genetical approaches to crop improvement. PIDC, Karachi, Pakistan, pp 63–74
Bajaj YPS, Saini SS, Bidani M (1980) Production of triploid plants from the immature and mature endosperm cultures of rice. Theor Appl Genet 58:17–18
Beachell HM, Khush GS, Julino BO (1972) Breeding for high protein content in rice. In: Rice breeding. IRRI, Manila, pp 419–428
Bhattacharya P, Sen SK (1980) Potentiality of leaf sheat cells for regeneration of rice (*Oryza sativa* L.) plants. Theor Appl Genet 58:87–90
Bounharmont J (1961) Embryo culture of rice on sterile medium. Euphytica 10:283–293
Bright S, Ooms G, Karp A, Foulger D, Fish N, Jones M, Evans N (1984) Somaclonal variation and genetic manipulation of potato. In: Abstr Int Symp Genetic manipulation in crops, Oct 22–26, 1984, Beijing
Carlberg I, Soderhall K, Glimelius K, Eriksson T (1984) Protease activities in non-embryogenic and embryogenic carrot cell strains during callus growth and embryo formation. Physiol Plant 62:458–464
Carlson PS (1973) Methionine sulfoximine-resistant mutants of tobacco. Science 180:1366–1368
Cella R, Colombo R, Calli MG, Nielson E, Rollo F, Sala F (1982) Freeze-preservation of rice cells: a physiological study of freeze-thawed cells. Physiol Plant 55:279–284
Chaleff RS (1975) Higher plant cells experimental organisms. In: Markham R, Davies DR, Hopwood DA, Horne RW (eds) Modification of the information content of plant cells. Elsevier/North Holland Biomedical Press, Amsterdam New York, pp 197–214
Chaleff RS (1983) Isolation of agronomically useful mutants from plant cell culture. Science 219:676–682
Chaleff RS, Stolarz A (1982) The development of anther culture as a system for in vitro mutant selection. In: Rice tissue culture planning Conf, IRRI, Manila, pp 63–74
Chang T-T (1985) Crop history and genetic conservation: rice – a case study. Iowa State J Res 59:425–455
Chang TT, Morishima H, Huang CS, Tagumpay O, Tateno K (1965) Genetic analysis of plant height, maturity and other quantitative traits in the cross of Peta × I-Geo-Tze. J Agric Assoc China 51:1–8
Chen C-C (1984) Utilization of microspore-derived plants for genetic analysis in rice. Rice Genet Newslett 1:137–138
Chen C-C, Chen C-M (1980) Changes in chromosome number in microspore callus of rice during successive subculture. Can J Genet Cytol 22:607–614
Chen C-C, Chiu W-L, Yu L-J, Ren S-S, Yu W-J (1983) Genetic analysis of anther-derived plants of rice: independent assortment of unliked genes. Can J Genet Cytol 25:324–328
Chen C-C, Tsay H-S, Huang C-R (1986) Rice (*Oryza sativa* L.): factors affecting androgenesis. In: Bajaj YPS (ed) Biotechnology in agriculture and forestry, vol 2. Crops I. Springer, Berlin Heidelberg New York Tokyo, pp 123–138
Chen C-M, Chen C-C (1979) Selection and regeneration of 5-methyltryptophan-resistant rice plants from pollen callus. Natl Sci Counc Month 7:378–382
Chen C-M, Chen C-C, Lin M-H (1982) Genetic analysis of anther derived plants of rice. J Hered 73:49–52
Chen D-F, Hsia Z-A (1987) Plant regenerated from protoplast of *Polypogon fogax* Nees ex Steud. Sci Sin B 30:54–60
Chen J-J, Chen L-J, Tsay H-S (1984) Relationship between the time of callus initiation, callus age and the ability of organ regeneration from rice anther culture. J Agric Assoc China 126:2–9
Chen L-J, Luthe DS (1985) Protein and RNA synthesis in embryogenic and non-embryogenic callus of rice (*Oryza sativa* L.) In: Abstr 1st Int Congr Plant molecular biology, Univ Georgia, USA, p 103
Chen S-L, Tian W-Z, Zhang G-H (1983) Selection of NaCl-tolerant regenerated plants from rice pollen-derived calli. Chin J Cell Biol (Spacil):29–30
Chen Y (1986a) Anther and pollen culture of rice. In: Hu H, Yang H-Y (eds) Haploids of higher plants in vitro. Science Press, Beijing, pp 3–25
Chen Y (1986b) The inheritance of rice pollen plant and its application in crop improvement. In: Hu H, Yang HY (eds) Haploids of higher plants in vitro. Science Press, Beijing, pp 118–136
Chen Y, Li L-C, Zhu J, Wang R-F, Li S-Y, Tian W-Z, Zeng S-W (1974) Investigation on the induction and genetic expression of rice pollen plants. Sci Sin 17:209–226
Chen Y, Wang R-F, Tian W-Z, Zuo Q-X, Zheng S-W, Lu D-Y, Zhang G-H (1980) Studies on pollen culture in vitro and induction of plantlets in *Oryza sativa* L. subsp. Keng. Acta Genet Sin 7:46–54
Chen Y, Zuo Q-X, Li S-Y, Zheng S-W (1981) Green plants regenerated from isolated rice pollen grains in vitro and induction factors. Acta Genet Sin 8:158–163

Chowdhury MKU, Schaeffer GW, Smith RL, Matthews BF (1988) Molecular analysis of organelle DNA of different subspecies of rice and the genetic stability of mtDNA in tissue cultured cells of rice. Theor Appl Genet 76:533–539
Chu C-C, Wang C-C, Sun C-S (1975) Establishment of an efficient medium for anther culture of rice through comparative experiments on the nitrogen sources. Asi Sin 18:659–668
Chu Q-R, Zhang L-N (1985) Cytogenetics of aneuploids derived from pollen plants of rice. Acta Genet Sin 12:51–60
Chu Q-R, Zhang Z-H, Gao Y-H (1985) Cytogenetical analysis on aneuploids obtained from pollen clones of rice (*Oryza sativa* L.) Theor Appl Genet 71:506–512
Chu Q-R, Cao H-X, Gu Y-Q, Zhang Z-H (1986) Stem node culture in vitro of twelve species of wild rice and two distant hybrids. Acta Agric Shanghai 2:39–46
Cocking EC, Abdullah R, Thompson JA (1986) Procedures for the efficient regeneration of rice plants from protoplasts. Rockefeller Found Plann Meet, Oct 1986, IRRI, Manila
Constantin MT (1981) Chromosome instability in cell and tissue cultures and regenerated plants. Environ Exp Bot 21:359–368
Coulibaly MY, Demarly Y (1986) Regeneration of plantlets from protoplast of rice. Z Pflanzenzücht 96:79–81
Croughan TP (1984) Tissue culture of US rice varieties: development of short stature variants through tissue culture. In: Abstr Int Symp Genetic manipulation of crops, Oct 22–26, 1984, Beijing, p 57
D'Amato F (1977) Cytogenetics of differentiation in tissue and cell culture. In: Reinert J, Bajaj YPS (eds) Applied and fundamental aspects of plant cell, tissue, and organ culture. Springer, Berlin Heidelberg New York, pp 343–357
D'Amato F (1978) Chromosome number variation in cultured cells and regenerated plants. In: Thorpe TA (ed) Frontiers of plant tissue culture. Univ Calgary, Can, pp 287–295
Dattee Y, San H, Andre I, Odjo A, Kouaddio (1984) Quantitative inheritance of difference between sexual self pollinated, androgenetic and gynogenetic lines in rice and barley. In: Abstr Int Symp Genetic manipulation in crops, Oct 22–26, 1984, Beijing, p 31
Davoyan EI (1981) Induction of callus and differentiation of plants from rice endosperm. Dokl BACXHUL (USSR) (6):43–45
Davoyan EI (1983) Mutagenesis in rice tissue culture and obtaining a new initial material on its basis. Genetics (USSR) 19:1714–1719
Davoyan EI (1986) On hormonal regulation of morphogenetic potency in long-term cultivated callus tissues of rice. Agric Biol (USSR) 21:85–89
Davoyan EI, Smetanin AP (1979) Callus induction and rice plant regeneration. Plant Physiol Moscow 26:323–329
Day A, Ellis THN (1984) Chloroplast DNA deletions associated with wheat plants regenerated from pollen: possible basis for maternal inheritance of chloroplasts. Cell 39:359–368
Deka C, Sen SK (1976) Differentiation in calli originated from isolated protoplasts of rice (*Oryza sativa* L.) through plating technique. Mol Gen Genet 145:239–243
Earle ED, Gracen VE (1985) Somaclonal variation in progeny of plants from tissue cultures. In: Henke RR, Hughes KW, Constantin MJ, Hollaender A (eds) Tissue culture and agriculture. Plenum, New York, pp 139–152
Edallo S, Zucchimali C, Perenzin M, Salamini F (1981) Chromosomal variation and frequency of spontaneous mutation associated with in vitro culture and plant regeneration in maize. Maydica 26:29–56
Evans DA, Sharp WR (1983) Single gene mutations in tomato plants regenerated from tissue culture. Science 221:949–951
Evans DA, Sharp WR, Madina-Filho HP (1984) Somaclonal and gamatoclonal variation. Am J Bot 71:759–774
FAO (ed) (1986) Selected indicators of food and agriculture development in Asia-Pacific region 1975–1985. RAPA, Bangkok, 1986/14
FAO Production Year Book, 1988, Vol 41
Finkle BJ, Ulrich JM (1982) Cryoprotectant removal temperature as a factor in the survival of frozen rice and sugarcane cells. Cryobiology 19:1012–1018
Finkle BJ, Ulrich JM, Tisserat B (1982) Responses of several lines of rice and date palm callus to freezing at -196°C. In: Li PH, Sakai A (eds) Plant cold hardiness and freezing stress. Academic Press, New York London, pp 643–660

Fujimura T. Sakurai M. Akagi H. Negishi T. Hirose A (1985) Regeneration of rice plants from protoplasts. Plant Tissue Cult Lett 2:74–75
Fujiwara T. Ojima (1955) Ref.: Oono K (1984) Tissue culture and genetic engineering in rice. In: Tsunoda S. Takahashi W (eds): Biology of Rice. Japan Sci Soc Press Tokyo. pp 339–358
Fukui K (1983) Sequential occurrence of mutations in a growing rice callus. Theor Appl Genet 65:225–230
Furuhashi K. Yatazawa M (1964) Indefinite culture of rice stem node callus. Kagaku (Jpn) 34:623
Furuhashi K. Yatazawa M (1970) Methionine-lysins-threonine-isoleucine interrelationships in the amino acid nutrition of rice callus tissue. Plant Cell Physiol 11:569–578
Guha S. Iyer RD. Gupta N. Swaminathan MS (1970) Totipotency of gametic cells and the production of haploids in rice. Curr Sci 39:174–176
Guha-Mukherjee S (1972) Genotypic differences in the in vitro formation of embryoids from rice pollen. J Exp Bot 24:139–144
Hasegawa H. Inoue M (1983) Induction and selection of hydroxy-L-proline resistant mutant in rice (*Oryza sativa* L.) Jpn J Breed 33:275–282
Heinz DJ. Krishnanurthi M. Nickell LG. Maretzki A (1977) Cell. tissue and organ culture in sugarcane improvement. In: Reinert J. Bajaj YPS (eds) Applied and fundamental aspects of plant cell. tissue. and organ culture. Springer. Berlin Heidelberg New York. pp 3–17
Henke RR. Mansur MA. Constantin MJ (1978) Organogenesis and plant formation from organ and seedling derived calli of rice (*Oryza sativa* L.) Physiol Plant 44:11–14
Henshaw GG. O'Hara JF (1983) In vitro approaches to the conservation and utilisation of global plant genetic resources. In: Mantell SH. Smith H (eds) Plant biotechnology. Univ Press. Cambridge. pp 219–238
Heszky LE. Sajo Z (1972) Raising haploid plants of *Oryza sativa* L. by in vitro anther and ovary culture. I. callus-tissue root and shoot induction. Agrobotenika 14:35–41
Heszky LE. Li SN (1984) Effect of callus subculture on plant regeneration capacity in somatic and haploid callus of *Oryza sativa* L. In: Novak FJ. Havel L. Dolezel J (eds) Plant tissue and cell culture application to crop improvement. Czech Akad Sci. Prague. pp 123–124
Heyser JW. Dykes TA. DeMott KJ. Nabors MW (1983) High frequency. long term regeneration of rice from callus culture. Plant Sci Lett 29:175–182
Hibberd KA. Green CE (1982) Inheritance and expression of lysine plus threonine resistance selected in maize tissue culture. Proc Natl Acad Sci USA 79:559–563
Hildreth G. Thompson M. Hanning G. Nabors M (1987) Plant regeneration from long-term cultures derived from mature seeds of Indica and Japonica rice culture. In: Proc 2nd Annu Conf International plant biotechnology network (IPBNet). Jan 11–16. 1987. Bangkok
Hu Z. Zhuang C-J. Liang H-X (1983) In vitro culture of haploid young panicle fragments of rice. Acta Bot Yunnanica (China) 5:293–300
Huang C-S. Tsay H-S. Chen C-G (1985) Genetic changes in rice variety induced by anther culture J Agric Res China 34:237–242
Huang Y-X. Liu J-L. Hu J-Y. Ke W (1960) Advances in rice breeding in Guadong Province during 1949–1959. Acta Agric Sin 11:20–29
International Rice Research Institute–IRRI (ed) (1967) Annual report for 1966. IRRI. Manila. pp 59–82
Ishikawa R. Mori K. Morishima H. Kinoshita T (1987) Expression pattern of isozyme genes in rice calli. Memoirs of the Faculty of Agriculture. Hokkaido Univ 15:371–379
Iyer RD. Govila QP (1964) Embryo culture of interspecific hybrids in the genus *Oryza*. Indian J Genet Plant Breed 24:116–121
Jena KK. Khush GS (1984) Embryo rescue of interspecific hybrid and its scope in rice improvement. Int Rice Genet Newslett 1:133–134
Jena KK. Khush GS (1986) Production of monosomic alien addition lines of *Oryza sativa* having a single chromosome of *O. officinalis*. In: Rice Genetics. Proc Int Rice Genetics Symp IRRI. Manila. Philippines. pp 199–208
Jordan MC. Larter EN (1985) Somaclonal variation in triticale (× *Triticosecale* Wittmack) cv. Carman. Can J Genet Cytol 27:151–157
Jun C. Rush MC. Croughan T (1986) The improvement of rice through somaculture. In: Abstr 6th Int Cong Plant tissue and cell culture. Aug 3–8. 1986. Univ Min. Minneapolis. p 228
Karp A. Maddock SE (1984) Chromosome variation in wheat plant regenerated from cultured immature embryos. Theor Appl Genet 67:249–255

Kawata S, Ishihara A (1968) The regeneration of rice plant, *Oryza sativa* L. in the callus derived from the seminal root. Proc Jpn Acad Sci 44:149–153
Kharchenko PN, Kucherenko LA (1977) Obtaining rice plants from anthers. Dokl BACXHUL (USSR) (4):15–16
Khush GS (1984) A lecture on rice breeding presented at CNRRI. China Natl Rice Res Inst, Hangzhou, China, pp 1–2
Khush GS, Virmani SS (1985) Some plant breeding problems needing biotechnology. In: Biotechnology in international agricultural research. IRRI, Manila, pp 51–63
Kishikawa H, Takagi Y, Egashira M, Yamashita H, Takamori Y (1986) Morphological variation in the progenies of plants regenerated from anther culture of tetraploid rice. Bull Fac Agric Saga Univ, Jpn (6):37–48
Kishinami I, Ojima K (1980) Accumulation of r-aminobutyric acid due to adding ammonium or glutamine to cultured rice cells. Plant and Cell Physiology 21:581–589
Kishor PBK, Reddy GM (1985) Resistance of rice callus tissues to sodium chloride, and polyethylene glycol. Current Science, India 54:1129–1131
Kishor PBK, Reddy GM (1986a) Regeneration of plants from long term cultures of *Oryza sativa* L. Plant Cell Rep 5:391–393
Kishor PBK, Reddy GM (1986b) Osmotic adjustment and organogenesis in long term culture of rice. In: Abstr 6th Int Congr Plant tissue and cell culture, Aug 3–8, 1986, Univ Minn, Minneapolis, p 159
Kishor PBK, Reddy GM (1986c) Improvement of rice for tolerance to salt and drought through tissue culture. Oryza 23:102–108
Kostylev PI, Yatsyna AA (1986) Use of embryo culture in rice interspecific crossings. Dokl BACXHIL (USSR) (5):11–12
Kucherenko LA (1980) Tissue culture in rice improvement: experiences in the USSR. In: Innovative approaches to rice breeding. IRRI, Manila, pp 93–102
Kucherenko LA (1984a) Conditions for plant regeneration in tissue culture. Agric Biol (USSR) (4):70–72
Kucherenko LA (1984b) Obtaining of somaclonal variation in rice tissue culture. In: Abstr Int Symp Genetic manipulation in crops, Oct 22–26, Beijing, p 53
Kucherenko LA (1986) Reproduction of non-germinating rice seed by tissue culture method. Agric Biol (USSR) (4):14–17
Kucherenko LA, Loce GD (1982) Accelerate the breeding cycle with embryo culture. Selekt Semenov (10):19–20 (in Russian)
Kucherenko LA, Mammaeva GG (1979) Study of regenerated rice plants obtained by tissue culture. Docl BACXHUL (USSR) (3):10–12
Kucherenko LA, Mammaeva GG (1980) Callus formation and organogenesis in tissue culture of different rice varieties. Agric Biol (USSR) 15:384–386
Kuo M-K, Cheng W-C, Hwang T-N, Hwang J-S, Kuan Y-L (1977) The culture of pollen grain of wheat, rice and tobacco in vitro. Acta Genet Sin 4:333–338
Kuo S-C (1982) The preliminary studies on culture of unfertilized ovary of rice in vitro. Acta Bot Sin 24:33–38
Kyozuka J, Otoo E, Shimamoto K (1988) Plant regeneration from protoplasts of indica rice: genotypic differences in culture response. Theor Appl Genet 76:887–890
Lai K-L, Liu L-F (1982) Induction and plant regeneration of callus from immature embryos of rice plants (*Oryza sativa* L.) Jpn J Crop Sci 51:70–71
Lai K-L, Liu L-F (1986) High frequency plant regeneration in water- and salt-stressed rice cell culture. In: Abstr 6th Int Cong Plant tissue and cell culture, Aug 3–8, 1986, Univ Minn, Minneapolis, p.186
Lai L-Z (1982) Plantlet derived from young preembryos of rice. Kexue Tongbao (Beijing) 27:1334–1337
Lai L-Z (1986) A promising line derived from ovary culture of rice. In: Abstr Pap Pres 3rd Natl Meet Chinese Soc Cell biology, Chengdu, China, pp 141–142
Larkin PJ, Scowcroft WR (1981) Somaclonal variation – a novel source of variability from cell culture for plant improvement. Theor Appl Genet 60:197–214
Lei M, Li X-H, Sun Y-R, Huang M-J (1986) Plantlet regeneration from rice protoplasts. Kexue Tongbao (China) 31:1729–1731
Li C-C, Gan D, Xu J (1986) New progress on screening resistant mutant of rice in vitro. Sci Agric Sin (2):93–94
Li H-M, Chen Y (1987) Selection of cadmium tolerant mutants from rice anther culture. Acta Genet Sin 14:42–48

Li H-W. Weng T-S. Chen C-C. Wang W-H (1961) Cytological studies of *Oryza sativa* L. and its related species. Bot Bull Acad Sin 2:79–86
Li SN. Heszky LE (1986) Rice tissue culture and application to breeding. I. Induction of high totipotent haploid and diploid callus from the different genotypes of rice (*Oryza sativa* L.) Cereal Res Commun 14:197–203
Ling D-H. Wang X-H. Chen M-F (1981) Cytogenetical study on homologous asyndetic triploid derived from anther culture in rice. Acta Genet Sin 8:262–268
Ling D-H. Chen W-Y. Chen M-F. Ma Z-V (1984) A study of induction and plantlet differentiation of embryogenic cluster of trihaploid rice. Acta Genet Sin 11:26–32
Ling D-H. Vidhyasekaran P. Borromeo ES. Zapata FJ. Mew TW (1985) In vitro screening of rice germplasm for resistance to brown spot disease using phytotoxin. Theor Appl Genet 71:133–135
Ling D-H. Chen W-Y. Chen M-F. Ma Z-R (1987) Chromosomal variation of regenerated plants from somatic cell culture in indica rice. Acta Genet Sin 14:249–254
Ling S-C. Min S-K. Huang Y-X. Zu D-M (1986) Cross-breeding of rice. In: Rice cultivation science in China. CAAS Agronomic Press. Beijing. pp 271–330
Liu L-F. Lai K-L (1985) High frequency plant regeneration from water-stressed rice tissue cultures. In: Abstr 1st Int Congr Plant molecular biology. Univ Georgia. p 11
Liu Z-L. Zhou C (1984) Investigation of ploidy and other characters of the gynogenic plants in *Oryza sativa* L. Acta Genet Sin 11:113–119
Loo S-W. Xu Z-H (1986) Rice: anther culture for rice improvement in China. In: Bajaj YPS (ed) Biotechnology in agriculture and forestry. vol 2. Crops I. Springer. Berlin Heidelberg New York Tokyo. pp 139–156
Lu C. Vasil V. Vasil IK (1981) Isolation and culture of protoplast of *Panicum maximum* Jacq (guinea grass): somatic embryogenesis and plantlet formation. Z Pflanzenphysiol 104:311–318
Lu J-J. Chang T-T (1980) Rice in its temporal and spatial perspectives. In: Luh BS (ed) Rice: production and utilization. AVI. Westport. Conn. pp 1–74
Maeda E (1965) Callus formation and isolation of single cells from rice seedling. Proc Crop Sci Soc Jpn 34:369–376
Maeda E (1968) Subculture and organ formation in the callus derived from rice embryos in vitro. Proc Crop Sci Soc Jpn 37:51–58
Maeda E (1971) Isolation of protoplast from seminal roots of rice. Proc Crop Sci Soc Jpn 40:397–398
McKenzie KS. Croughan TP (1984) Cool temperature seedling vigor of R2 and M2 short-staure rice (*Oryza sativa* L.) In: Abstr Int Symp Genetic manipulation in crops. Oct 22–26. 1984. Beijing. p 57
Meng Z. Chen Y (1987) Screening of rice mutant resistant to amino acid analogs. I. Screening of a rice mutant resistant to S-(2-aminoethyl)L-cysteine. Acta Genet Sin 14:100–106
Mercy ST. Zapata FJ (1986) Chromosomal behavior of anther culture derived plants of rice. Plant Cell Rep 5:215–218
Metakovsky EV. Yu Novoselskay A. Sozinov AA (1987) Problems of interpreting results obtained in studies of somaclonal variation in gliadin proteins in wheat. Theor Appl Genet 73:764–766
Miah MAA. Earle ED. Khush GS (1985) Inheritance of callus formation ability in anther culture of rice. *Oryza sativa* L. Theor Appl Genet 70:113–116
Mocoy TJ. Phillips PL. Rines HW (1982) Cytogenetic analysis of plants regenerated from oat (*Avena sativa*) tissue cultures: high frequency of parial chromosome loss. Can J Genet Cytol 24:37–50
Mori K (1986) Callus induction and protoplast culture in rice. In: Rice genetics. IRRI. Manila. Philippines. pp 781–790
Nagai C. Schnell RJ (1986) Comparison of F2 population and somaclones derived from a hybrid of *Saccharum* × *Erintus* (Ripidium). In: Abstr 6th Int Congr Plant tissue and cell culture. Aug 3–8. 1986. Univ Minn. Minneapolis. p 218
Nakajima T. Morishima H (1958) Studies on embryo culture in plants. II. Embryo culture of interspecific hybrids of Oryza. Jpn J Breed 8:105–110
Nakano H. Tashiro T. Maeda E (1975) Plant differentiation in callus tissue induced from immature endosperm of *Oryza sativa* L. Z Pflanzenphysiol 76:444–449
Nelson RS. Karp A. Bright SWJ (1986) Ploidy variation in *Solanum brevideus* plants regenerated from protoplasts using an improved culture system. J Exp Bot 37:253–261
Niizeki H (1968) Tissue culture of rice. KASEAA (Jpn) 6:743–748
Niizeki H. Kita F (1981) Cell division of rice and soybean and their fused protoplasts. Jpn J Breed 31:161–167

Niizeki H, Oono K (1968) Induction of haploid rice plant from anther culture. Proc Jpn Acad Sci 44:554-557
Niizeki H, Oono K (1971) Rice plants obtained by anther culture. Cult Tissue Plants 193:251-257
Niizeki H, Tanaka M, Saito K (1986) Response of somatic hybrid callus between rice and soybean to streptomycin. Jpn J Breed 36:75-79
Nishi T, Mitsuoka S (1969) Occurrence of various ploidy plants from anther and ovary culture of rice plant. Jpn J Genet 44:341-346
Nishi T, Yamada Y, Takahashi E (1968) Organ redifferentiation and plant regeneration in rice callus. Nature (London) 219:508-509
Nowick EM (1986) Chromosome pairing in *Oryza sativa* L. × *O. latifolia* Desv. hybrid. Can J Genet Cytol 28:278-281
Nunez Z, VM Martinez CP, Narvaez VJ, Roca W (1985) Obtaining homozygous lines of rice (*Oryza sativa* L.) tolerant of toxic Al by the use of anther culture. Acta Agron 35:7-26
Oono K (1978) Test tube breeding of rice by tissue culture. Trop Agric Res Ser 11:109-124
Oono K (1981) In vitro methods applied to rice. In: Thorpe TA (ed) Plant tissue culture, methods and application in agriculture. Academic Press, New York London, pp 273-298
Oono K (1983) Genetic variability in rice plants regenerated from cell culture. In: Cell and tissue culture techniques for cereal crop improvement. Science Press, Beijing, IRRI, Manila, pp 95-104
Oono K (1984) Regulation of somatic mutation in rice tissue culture. In: Abstr Int Symp Genetic manipulation of crops, Beijing
Oono K (1985) Tissue culture application in rice improvement. In: Int Rice research conference, IRRI, Manila
Oono K (1986) Somatic mutation in rice cultured cell. Tissue Cult Jpn 12:288-293
Oono K, Okuno K, Kawai T (1984) High frequency of somaclonal mutations in callus culture of rice. Gamma Field Symp 23:71-95
Oram R (1982) Plant breeding and genetic engineering. In: Sutton BG, Williams P (eds) Genetic engineering for agriculture – the substance behind the promise. Aust Inst Agric Occ Publ 2, pp 73-79
Orton TT (1980) Chromosomal variation in tissue cultures and regenerated plants of *Hordeum*. Theor Appl Genet 56:101-112
Ou-Lee TM, Turgeon R, Wu R (1986) Expression of a foreign gene linked to either a plant-virus or a Drosophila promotor after electroporation of protoplasts of rice, wheat, and sorghum. Proc Natl Acad Sci USA 83:6815-6819
Paul NK, Ghosh PD (1986) In vitro selection of NaCl tolerant cell cultures in *Oryza sativa* L. Curr Sci 55:568-569
Pulver EL, Jennings PR (1986) Application of anther culture to high volume rice breeding. In: Rice genetics, IRRI, Manila, pp 811-820
Qin R-Z, Wu D-Y, Chen Z-V, Song W-C, Zhang Y-H, Dao W-K (1984) Investigation on the autoteraploid rice clones derived from the elite plants of hybrid progenies. In: Abstr Int Symp Genetic manipulation in crops, Oct 22-26, 1986, Beijing
Raghava RV, Nabors MW (1984) Cytokinin mediated long-term, high-frequency plant regeneration in rice tissue culture. Z Pflanzenphysiol 113:315-323
Raghava RV, Nabors MW (1985) Plant regeneration from tissue culture of Pokkali rice is promoted by optimizing callus to medium volume ratio and by a medium-conditioning factor produced by embryogenic callus. Plant Cell Tissue Org Cult 4:241-248
Rains DW, Croughan TP, Stavarek SJ (1982) Selection of salt tolerant plants using tissue culture. In: Rains DW, Valentine RC, Hollaender A (eds) Genetic engineering of osmoregulation: impact on plant productivity for food, chemicals and energy. Plenum, New York, pp 279-292
Reddy GM (1983) Callus initiation and plant regeneration from haploid internodes in rice. In: Sen SK, Giles KL (eds) Plant cell culture in crop improvement. Plenum, New York, pp 113-118
Reddy GM (1986) Callus induction, somatic embryogenesis and plantlet regeneration from young inflorescences of rice and maize. In: Abstr 6th Int Congr Plant tissue cell culture, Aug 3-8, 1986, Univ Minn, Minneapolis, p 444
Rutger JN, Peterson ML (1976) Improved short stature rice. Cal Agric 30 (6):4-6
Saka H, Maeda E (1973) Characteristics and variated differences of amylase isozymes in rice callus tissues. Proc Crop Sci Jpn 42:307-314
Saka H, Maeda E (1974) Changes in some hydrolytic enzymes associated with the redifferentiation of shoots in rice callus tissues. Proc Crop Sci Jpn 43:207-218

Sala F, Cella R, Burroni D, Rollo F (1979) Freeze preservation of rice cells grown in suspension culture. Physiol Plant 45:170–176
Sano H, Takaiwa F, Oono K (1983) Induction of mutation tissue culture and their use for plant breeding IX. variation of seed protein and amino acid (lysine) in regenerated plants. Jpn J Breed 33(Suppl 2):84–85
Schaeffer GW (1982) Recovery of heritable variability in anther-derived doubled haploids of rice. Crop Sci 22:1160–1164
Schaeffer GW (1986) Selection of rice, *Oryza sativa*, cells from inhibitory levels of lysine plus threonine and the characterization of the progeny. In: Abstr 6th Int Congr Plant cell tissue culture, Aug 3–8, 1986, Univ Minn, Minneapolis, p 439
Schaeffer GW, Sharpe FT Jr (1981) Lysine in seed protein from S-aminoethyl-L-cysteine resistant anther-derived tissue cultures of rice. In Vitro 17:345–352
Schaeffer GW, Sharpe FT Jr (1983) Improved rice proteins in plants regenerated from S-AEC-resistant callus. In: Cell and tissue culture techniques for cereal crop improvement. Science Press, Beijing; IRRI, Manila, pp 279–290
Schaeffer GW, Sharpe FT Jr, Cregan PB (1984) Variation for improved protein and yield from rice anther culture. Theor Appl Genet 67:383–389
Schaeffer GW, Sharpe FT Jr, Carnahan HL, Johnson CW (1986) Anther and tissue culture – induced grain chalkiness and associated variants in rice. Plant Cell Tissue Org Cult 6:149–157
Sharma DR, Dawra S, Chowdhury JB (1983) Direct and indirect effects of gamma rays on stimulation of morphogenesis in longterm tissue culture of rice (*Oryza sativa* L.) Curr Sci 52:606–607
Sharma KD (1986) Induced mutagenesis in rice. In: Rice genetics. IRRI, Manila, pp 679–695
She J-M, Sun L-H, Huang M (1984) Differentiation potential and chromosome variation in subcultured callus derived from rice pollen. Hereditas (China) 6:17–19
Shen J-H, Li M-F, Chen Y-Q, Zhang Z-H (1983) Improving rice by anther culture. In: Cell and tissue techniques for cereal crop improvement. Science Press, Beijing; IRRI, Manila, pp 183–205
Shen Q-J (1985) Variations in the characteristics of plant progenies derived from somatic tissue (D2) of indica rice. Plant Physiol Commun (China) (2):35–37
Shepard JF, Bidney D, Shahin E (1980) Potato protoplasts in crop improvement. Science 208:17–24
Shimamoto K, Hayashi Y, Kyozuka J (1986) High frequency plant regeneration from protoplasts of rice (*Oryza sativa* L.) In: Abstr 6th Int Congr Plant tissue cell culture, Aug 3–8, 1986, Univ Minn, Minneapolis, p 161
Shimamoto K (1988) Genetic modifications of higher plants using cell cultures. Plant Tissue Culture Letter 5:1–5
Shome A, Bhaduri PN (1982) Response of excised embryos of rice (*Oryza sativa* L.) to X-rays. Theor Appl Genet 61:135–139
Song Z-M, Liu Z-L, Chen J-X (1984) Screening of ideotypes from somatic tissue-derived plants of rice. In: Abstr Int Symp Genetic manipulation in crops, Oct 22–26, 1984, Beijing, p 61
Song Z-M, Liu Z-L, Chen J-X (1986) Somaclonal variation in rice improvement. Anhuei Agron Sci (China) 4:34–37
Suenaga K, Abrigo EM, Yoshida S (1982) Seed derived callus culture for selecting salt-tolerant rice. I. Callus induction, plant regeneration, and variations in visible plant traits. IRRI Res Pap Ser 79:1–11
Sun C-S, Wang C-C, Chu C-C (1974) The ultrastructure of plastids in the albino pollen plants of rice. Sci Sin 17:793–802
Sun C-S, Wang C-C, Chu C-C (1979) The deficiency of soluble proteins and plastid ribosome RNA in the albino pollen plantlets of rice. Theor Appl Genet 55:193–197
Sun L-H, She J-M (1982) Induction of diploid plant from somatic cell culture in haploid plant of rice. Hereditas (Beijing) (4):9
Sun L-H, She J-M, Lu X-F (1986) In vitro selection of *Xanthomonas oryzae*-resistant mutants in rice. I. Induction of resistant callus and screening regenerated plants. Acta Genet Sin 13:188–193
Sun Z-X (1980) Observation on the regenerated plants from somatic tissue of hybrid rice. Acta Phytophysiol Sin 6:243–249
Sun Z-X, Colbert J (1987) Somaclonal variation in the electropherograms of protein of rice. In: Abstracts of the Symposium on the Plant Cell Biology, 26, Oct - 1, Nov, 1987, Dayun, Hunan
Sun Z-X, Zhao C-Z, Zheng K-L, Qi X-F, Fu Y-P (1981) Observation on the regenerated plants from somatic tissue of hybrid rice in paddy field. Acta Phytophysiol Sin 7:161–166

Sun Z-X, Zhao C-Z, Zheng K-L, Qi X-F, Fu Y-P (1983) Somaclonal genetics of rice, *Oryza sativa* L. Theor Appl Genet 67:67–73
Sun Z-X, Fu Y-P, Zhang Y-X (1988a) A preliminary research on long term storage of rice germplasm in vitro. Sci Agric Sin 21(4):9–14
Sun Z-X, Xie F-X, Rong J-P (1988b) The correlation analysis on agronomic characteristics between somaclonal generations of rice. In: Abstracts of the Papers Presented at the 55th Anniversary of the Botanical Society of China. August, 1988, Kwenming, China, p 410
Sun Z-X, Luo Y-K, Zhang Y-K, Chen Y-Y (1989) Somaclonal variation in grain quality of rice. Hereditas (Beijing), 11(5):5–8
Taiwan Agricultural Research Institute (ed) (1986) Annual report of Taiwan Agricultural Research Institute, pp 19–20
Takagi Y, Kishikawa H, Egashira Y (1982) Anther culture of tetraploid rice. Bull Fac Agric Saga Univ Jpn 52:41–47
Tamura S (1968) Shoot formation in callus originated from rice embryo. Proc Jpn Acad Sci 44:544–548
Tang Y-Y, Sun Z-X (1979) In vitro regeneration of plantlets from the tissue of hybrid rice. Acta Phytophysiol Sin 5:95–97
Terada R, Kyozuka J, Nishibayashi S, Shimamoto K (1987) Plantlet regeneration from somatic hybrids of rice (*Oryza sativa* L.) and barnyard grass (*Echinochloa oryzicola* Vasing). Mol Gen Genet 210:39–43
Thompson JA, Abdullah R, Cocking EC (1986) Protoplast culture of rice using media solidified with agarose. Plant Sci 47:123–133
Thompson JA, Abdullah R, Chen W-H, Gartland KMA (1987) Enhanced protoplast division in rice following heat shock treatment. J Plant Physiol 127:363–370
Tomes DT (1985) Cell culture, somatic embryogenesis and plant regeneration in maize, rice, sorghum and millets. In: Bright SWJ, Jones MGK (eds) Cereal tissue and cell culture. Nijoff Junk, Dordrecht, pp 175–203
Tonnelli C (1985) In vitro selection and characterization of cereal mutants. Genet Agric 39:445–470
Toriyama K, Hinatak K (1986) Plantlet regeneration from protoplasts of rice. Heredity (Jpn) 40:32–37
Toriyama K, Arimoto Y, Uchimiya H, Hinata K (1988) Transgenic rice plants after direct gene transfer into protoplasts. Biotechnology 5:1072–1074
Tsai I-S (1978) A preliminary report on callus induction and plant regeneration from the shoot apex of hybrid rice in *Oryza sativa* L. In: Proc Symp Plant tissue culture. Science Press, Beijing, pp 517–519
Uchimiya H, Fushima T, Hashimota H, Harada H, Syono K, Sugawara Y (1986) Expression of a foreign gene in callus derived from DNA-treated protoplasts of rice (*Oryza sativa* L.) Mol Gen Genet 204:204–207
Vasil V, Wan D, Vasil IK (1983) Plant regeneration from protoplasts of napier grass (*Pennisetum purpureum* Schum). Z Pflanzenphysiol 111:233–239
Vasil IK, Srinivasa C, Vasil V (1986) Culture of protoplasts isolated from embryogenic cell suspension culture of sugarcane and maize. In: Abstr 6th Int Congr Plant tissue cell culture, Aug 3–8, 1986, Univ Minn, Minneapolis, p 443
Wakasa K (1982) Application of tissue culture to plant breeding method – improvement and mutant production. Bull Natl Inst Agric Sci Ser D 33:121–200
Wakasa K (1985) Selection of nitrate reductase-deficient rice cell line by calli from microspores. Plant Tissue Cult Lett 2:33–34
Wakasa K, Watanabe Y (1979) Haploid plant of *Oryza perennis* (spontanea type) induced by anther culture. Jpn Breed 29:146–150
Wakasa K, Widholm JM (1983) In vitro selection of mutant. 4. Some characters of 5-MT-resistant mutant in rice. Jpn J Breed 33(Suppl):82–83
Wakasa K, Widholm JM (1986) 5-methyltryptophan resistant rice mutant selected in tissue culture. In: Abstr 6th Int Congr Plant tissue cell culture, Aug 3–8, 1986, Univ Minn, Minneapolis, p 440
Walther F, Preil W (1981) Mutants tolerant to low-temperature conditions induced in suspension culture as a source for improvement of *Euphoribia pulcherrima* Willd ex Klotzsch. In: IAEA (ed) Induced mutations – a tool in plant breeding. Int Atom Energ Ag, Vienna, pp 399–405
Wang G-Y, Hsia Z-A (1986) Protoplast culture of rice. In: Abstr 3rd Natl Meet Chin Soc Cell biology, Oct 30–Nov 4, 1986, Chengdu, China, pp 140–141
Wang M-S, Zapata FJ, De Castro DC (1987) Plant regeneration through somatic embryogenesis from mature seed and young inflorescence of wild rice (*Oryza perennis* Moench). Plant Cell Rep 6:294–296

Wang R-F. Zuo Q-X. Zheng S-W. Tian W-Z (1979) Induction of plantlets from isolated pollen culture in rice (*Oryza sativa* subsp. Keng). Acta Genet Sin 6:7
Wang Y-C. Klein TM. Fromm M. Cao J. Sanford JC. Wu R (1988) Transient expression of foreign genes in rice. wheat and soybean cells following particle bombardment. Plant Molecular Biology 11:433–439
WernickeW. Brettell R. Wakizuka T. Potrykus I (1981) Adventitious embryoid and root formation from rice leaves. Z Pflanzenphysiol 103:361–365
Williams EG. Maheswaran G (1986) Somatic embryogenesis: Factors influencing coordinated behaviour of cells as an embryogenic group. Ann Bot (London) 57:443–462
Wong C-K. Ko S-W. Woo S-C (1983) Regeneration of rice plantlets on NaCl-stressed medium by anther culture. Bot Bull Acad Sin 24:59–64
Wong C-K. Woo S-C. Ko S-W (1986) Production of rice plantlets on NaCl-stressed medium and evaluation of their progenies. Bot Bull Acad Sin 27:11–23
Woo S-C. Huang C-Y (1980) Anther culture of *Oryza glaberrima* Steud and its hybrids with *O. sativa* L. Bot Bull Acad Sin 21:75–79
Woo S-C. Mok T. Huang C-Y (1978) Anther culture of *Oryza sativa* L. and *O. perennis* Moench Hybrids. Bot Bull Acad Sin 19:171–178
Yamada Y. Ogawa M. Yano S (1983) Tissue culture in sea water increases salt tolerance of rice plants. In: Cell and tissue culture techniques for cereal crop improvement. Science Press. Beijing; IRRI. Manila. pp 229–235
Yamada Y. Yang Z-Q. Tang D-T (1986) Plant regeneration from protoplast-derived callus of rice (*Oryza sativa* L.). Plant Cell Rep 5:85–88
Yan C-J. Zhao Q-H (1979) Regeneration of plantlets from calli derived from leaf sheaths and rachis-branches of rice (*Oryza sativa* L.) Kexue Tongbao (China) 24:943–947
Yan C-J. Zhao Q-H (1982) Callus induction and plantlet regeneration from leaf blade of *Oryza sativa* L. subsp. Indica. Plant Sci Lett 25:187–192
Yang Z-Q. Shikanai. Yamada Y (1988) Asymmetric hybridization between cytoplasmic male-sterile (CMS) and fertile rice (*Oryza sativa* L.) protoplasts. Theor Appl Genet 76:801–808
Ye H (1984) Studies on cell suspension culture and plant regeneration in rice. Acta Bot Sin 26:52–59
Yoboue PCN (1986) Variation in genetic expression in an androgenetic progeny of rice with a high level of homozygosity. In: IAEA (ed) Nuclear techniques and in vitro culture for plant improvement. Int Atom Energ Ag. Vienna. pp 357–361
Yu R-Q. Qie Y-W (1985) Rice germplasm. In: Lin S (ed) Rice cultivation science in China. Agronomic Press. Beijing. pp 39–83
Yuan L-P (1972) An introduction to the breeding of male sterile lines in rice. In: Proc 2nd Worksh Genetics. Hainan. Guandong. China. March 1972
Yuan L-P. Virmani SS (1986) Current status of hybrid rice research and development. Pres Int Symp Hybrid rice. Oct 6–10. 1986. Changsha. Hunan. China
Yue S-X. Liu B-L. Chen S-B (1986) A preliminary study of test-tube fertilization in rice. Acta Bot Sin 28:46–49
Zao D (1980) "Yuan Feng Zao". a radiation induced mutant rice cultivar. and its performance. Appl Atom Energ Agric (4):1–6
Zapata FJ. Torrizo LB. Aldemita RR. Novero AV. Aazaredo AM. Visperas RV. Lim MS. Moon HP. Heu MH (1986) Rice anther culture: a tool for production of cold tolerant lines. In: Rice genetics. IRRI. Manila. pp 773–780
Zhang H-M. Yang H. Rech EL. Golda TJ. Davis AS. Mulligan BJ. Cocking EC. Davey MR (1988) Transgenic rice plants produced by electroporation-mediated plasmid uptake into protoplasts. Plant Cell Reports 7:379–384
Zhang L-N. Chu Q-R (1984) Characteristic and chromosome variation of somaclones and its progeny in rice (*Oryza sativa* L.). Sci Agric Sin (4):14–20
Zhang TB (1983) A study of test-tube fertilization of rice (*Oryza sativa* L.). Acta Bot Sin 25:187–191
Zhang W. Wu R (1988) Efficient regeneration of transgenic plants from rice protoplasts and correctly regulated expression of the foreign gene in the plants. Theor Appl Genet 76:835–840
Zhang Z-H (1982) Application of anther culture techniques to rice breeding. In: Rice tissue culture planning. Conf. IRRI. Manila. pp 55–61
Zhang Z-H. Yin D-C (1985) Anther culture breeding and mutation breeding. In: Lin S (ed) Rice cultivation science in China. Agriculture Press. Beijing. pp 379–418

Zhao C-Z, Zheng K-L, Qi X-F, Sun Z-X, Fu Y-P (1981) The study on the tissue and cell culture of different types of *Oryza sativa* L. Acta Phytophysiol Sin 7:287–290
Zhao C-Z, Zheng K-L, Sun Z-X, Qi X-F (1984) Somaclonal variation and rice improvement. In: Abstr Int Symp Genetic manipulation in crops, Oct 22–26, Beijing, p 57
Zhao C-Z, Sun Z-X, Zheng K-L, Qi X-F, Zhou Z-M, Fu Y-P (1986) Research on application of tissue culture to rice breeding. In: Annual reports of the China National Rice Research Institute (CNRRI) for 1983–1985, Hangzhou, China, pp 38–40
Zheng K-L, Castiglione S, Biasini MG, Biroli A, Morandi C, Sala F (1987) Nuclear DNA amplification in cultured cells of *Oryza sativa* L. Theor Appl Genet 74:65–70
Zheng K-L, Zhou Z-M, Wang G-L, Luo Y-K, Xiong Z-M (1989) Somatic cell culture of rice cultivars with different grain types: somaclonal variation in some grain and quality characters. Plant Cell, Tissue and Organ Culture 18:201–208
Zheng Z-L, Chu Q-R, Zhang C-M (1985) Pathogenecity of culture filtrate from *Pyricularia oryzae* on rice. Acta Agric Shanghai 1:85–90
Zhou C, Yang H-Y (1980) In vitro induction of haploid plantlets from unpollinated young ovaries of *Oryza sativa* L. Acta Genet Sin 7:287–288
Zhou C, Yang H-Y (1981) In vitro embryogenesis in unfertilized embryo sacs of *Oryza sativa* L. Acta Bot Sin 23:176–180
Zimny J, Lorz H (1986) Plant regeneration and initiation of cell suspensions from root-tip derived callus of *Oryza sativa* L. (rice). Plant Cell Rep 5:89–92
Zimny J, Junker B, Lorz H (1986) Root derived in vitro cultures of rice plant regeneration from callus and cell suspension, and direct gene transfer to protoplasts. In: Abstr 6th Int Congr Plant tissue and cell culture, Aug 3–8, 1986, Univ Minn, Minneapolis, p 207
Zou G-Z, Yei M-M, Jia Y-F (1986) Establishment of rice cell line and plantlet regeneration from single cell culture. Acta Agric Shanghai 2:1–8

II.3 Somaclonal Variation in Maize

E.D. EARLE[1] and A.R. KUEHNLE[2]

1 Introduction

1.1 Importance and Distribution of the Plant

In the years since Larkin and Scowcroft (1981) reviewed the literature on tissue culture-related variability and suggested that such "somaclonal variation" might serve as a resource for plant breeders, regenerated plants and progeny from a broad range of species have been examined for alterations in phenotype, physiology, cytology, and molecular characteristics. Maize (*Zea mays* L.) has been the subject of many such studies. The work with maize is easily justified by the potential value of any new approaches to improvement of this major crop plant (see next section). Moreover, maize has many other features that make it attractive material for studies of somaclonal variation (Sheridan 1982). These include the low number (2n = 20) of well-characterized chromosomes, the many nuclear gene mutants identified and mapped to particular chromosomes (Neuffer et al. 1968), and the different mitochondrial variants encoding cytoplasmic male sterility (Laughnan and Gabay-Laughnan 1983). Maize is a particularly convenient plant for genetic studies because it can readily be either self- or cross-pollinated, with production of large numbers of seeds per plant. Effects of gene dosages can be assessed by comparison of triploid endosperm and diploid somatic tissues. An additional attraction is the rapidly growing literature on the molecular genetics of maize; this encompasses both features specific to maize (e.g., storage proteins, transposable elements) and aspects of nucleic acid and protein structure common to higher plants. These considerations, coupled with increasing success in recovery of plants from tissue cultures, make studies of somaclonal variation in maize interesting to researchers from many disciplines.

Maize, one of the world's leading cereal crops, is important both as a food for human consumption and as a feed grain for livestock. Parts of maize are also found in many industrial products such as textiles, ceramics, and pharmaceuticals. In the harvest year 1984/85, the world's total harvested area of maize was close to 128 million hectares, yielding over 456 million metric tons (USDA 1986). North and Central America comprised nearly one-third of this total area and one-half of the total yield, with the USA producing more maize than any other country in the world.

[1]Department of Plant Breeding, Cornell University, 524 Bradfield Hall, Ithaca, NY 14853-1902, USA
[2]Department of Horticulture, 102 St. John Lab, University of Hawaii, Honolulu, HI 96822, USA

Biotechnology in Agriculture and Forestry, Vol. 11
Somaclonal Variation in Crop Improvement I (ed. by Y.P.S. Bajaj)

The major maize-producing countries in South America, Asia, and Africa include Argentina, Brazil, China, Egypt, India, Indonesia, the Philippines, Thailand, and South Africa. France and Italy are the principal maize producers in Western Europe, and Yugoslavia, Romania, and Hungary are the principal producers in Eastern Europe. The Soviet Union has a relatively high production of maize as well (USDA 1986).

1.2 Available Genetic Variability and Objectives for Improvement

The centers for variability in maize are in the Americas. The largest collection of Latin American and Caribbean landraces is found at the maize germplasm bank at CIMMYT, the International Maize and Wheat Improvement Center. About 12,500 entries are maintained and made available to breeders by the germplasm bank (Breth 1986). The genetic variability available for the commercial production of temperate maize, such as that grown in the USA, is extremely limited, however. Only a few of the known races are adapted to the short growing season of the northern latitudes. Most of the genetic diversity in maize is found in tropical photoperiod-sensitive germplasm from South and Central America. This material is difficult to work with in temperate regions. Two additional factors have reduced the variability in temperate maize. First, most of the world's maize is derived from the USA-based corn belt dents which originally arose from a cross between only two races of maize. Second, the dominant production of hybrid maize in the USA is based on hybrid seed derived from only a handful of inbreds, some of which were developed from the same starting population. This increase in genetic uniformity opens the way for the proliferation of diseases and pests, and limits further improvement of temperate maize through breeding.

Objectives for the improvement of maize, which can be approached not only through breeding but also through production practices, include increased yield, adaptation (to cold, heat, salt, drought, soil fertility), resistance to diseases and pests (leaf blights, downy mildew, rust, rootworms, earworms, borers, stalk, ear and root rots, etc.), improved quality (protein and oil content, protein quality, milling quality, etc.), and resistance to herbicides and insecticides. CIMMYT approaches these objectives by working at the population level, with a concentration on tropical maize. Because of the limited useful genetic diversity currently available for temperate maize, somaclonal variation may provide the needed variability not only for improving populations but also for improving elite inbreds which have been developed over many years for specific growing regions and hybrid combinations.

2 Regeneration of Maize Plants

2.1 Explants Used

Regeneration of maize plants in vitro is not difficult provided that appropriate culture systems are used. Regenerable cultures have been initiated from many different explants, including immature and mature embryos, seedling tissues (e.g., mesocotyls, shoot apices, leaf bases, and nodes) and immature reproductive structures (tassels, glumes, inflorescences, anthers). Mature vegetative tissues are notably absent from this list. Immature embryos are by far the most extensively used explants. The other tissue sources listed above have typically given low or erratic frequency of regeneration and/or very genotype-specific results, so relatively little information is available about variation among plants recovered. This review will therefore emphasize embryo-derived regenerates. Except for the initial steps, culture procedures for explants other than immature embryos (e.g., mature embryos, immature tassels) are similar to those used for immature embryos.

2.2 Regeneration from Immature Embryos

The plant regeneration protocols currently in use are derived from the one introduced by Green and Phillips (1975). The procedure begins with removal of ears from field- or greenhouse-grown maize plants when the embryos are between 1 and 2 mm long. Embryos are excised from the kernels and placed, scutellum up, on nutrient medium containing an auxin-like herbicide, usually 2,4-D. In a successful culture, embryogenic or organogenic callus develops from the scutellar portion of the embryo, usually not from the embryo itself. Such callus may either be subcultured on a similar medium or transferred to medium with little or no 2,4-D for plant regeneration. Regenerated shoots or plantlets are usually transferred to a simple medium in larger containers for continued development in vitro. Plants having several leaves and a good root system are then moved out of culture, potted, and grown to maturity in a growth chamber or greenhouse. Occasionally, regenerates are transplanted directly into the field, but there they are vulnerable to adverse environmental conditions.

2.2.1 Genotype

Genotype strongly affects culture response, with some lines producing embryogenic callus on virtually all cultured embryos, and others giving either little proliferation or only nonregenerable callus. Early studies concentrated on a limited array of genotypes, particularly A188, which regenerated readily. By now a much wider range of commercially important temperate and tropical inbred lines has been tested on a variety of media, and many of these have been successfully cultured and regenerated (e.g., Lu et al. 1982, 1983; Duncan et al. 1985; Lowe et al. 1985; Tomes and Smith 1985; Fahey et al. 1986; Hodges et al. 1986; Prioli and Silva pers. commun.). Since it is relatively easy to obtain large numbers of immature embryos, at least some plants can usually be recovered even from lines in which only a small percent of the embryos are responsive.

Response of different maize lines in culture is heritable, and crosses of regenerable lines with poorly regenerating ones often give F_1 progeny with improved culturability (Beckert and Qing 1984; Tomes and Smith 1985; Duncan et al. 1985; Hodges et al. 1986). A breeding/selection approach to improving regenerability is thus feasible. In some cases, hybrids performed better than either inbred parent (Green 1977; Tomes and Smith 1985), suggesting that specific combining ability as well as inbred performance is relevant.

Regenerable cultures can be established from lines with specific genetic features such as cytoplasmic male sterility (Gengenbach et al. 1977; Brettell et al. 1980; Earle et al. 1987), aneuploidy or monosomy (Rhodes et al. 1986), heterozygosity at specific loci (Brettell et al. 1986), sensitivity to pathogens (Gengenbach et al. 1977; Brettell et al. 1980; Wolf et al. 1987), etc. Carefully planned crosses of lines carrying the traits of interest with more regenerable lines may be required.

2.2.2 Explant

Embryos smaller than 1 mm are difficult to excise; ones larger than 2 mm usually fail to produce regenerable tissue and may instead simply swell or produce a shoot from the zygotic embryo. The days of growth after pollination until embryos are a suitable size for culture vary with genotype and plant growth conditions. Response of embryos of the same size and genotype may also differ according to time of planting or environment (e.g., Lu et al. 1983; Tomes and Smith 1985), suggesting an additional role for initial plant growth conditions in subsequent culture success.

2.2.3 Nutrient Medium

Response of a given line is also influenced by the nutrient medium supplied. Components that have been shown to be important, at least in some cases, include hormones, major salts, reduced nitrogen, and sucrose.

2,4-D-(2,4-dichlorophenoxy acetic acid) is the plant growth regulator most often used with maize tissue cultures although Dicamba, another auxin-like herbicide, has given better results in some cases (Duncan et al. 1985). In the absence of growth regulators, cultured embryos usually germinate precociously to give small plantlets. When 2,4-D, Dicamba, or other herbicidal auxins are added, growth of the root and, at higher levels, the shoot of the zygotic embryo is inhibited, and callus is formed on and around the scutellum. Levels used for induction and maintenance of callus are in the 0.25–5 mg/l range with 0.5–2 mg/l being typical. For regeneration, 2,4-D is either gradually reduced or completely omitted. Additional plant growth hormones are sometimes added to regeneration media (e.g., Lowe et al. 1985; Brettell et al. 1986), but data clearly showing that these improve results are limited. A broad range of kinetin and ABA concentrations inhibited regeneration from A188 cultures (Kamo et al. 1985). Duncan and Widholm (1988) recently showed that a brief (3–6 day) exposure to 3.5 mg/l of 6-benzylaminopurine prior to transfer to medium lacking growth regulators enhanced plant regeneration from callus of inbred Pa91.

The most commonly used major salt mixtures are MS (Murashige and Skoog 1962) and N6 (Chu et al. 1975), which differ mainly in their NO_3/NH_3 ratios [1.9 for MS and 4 for N6 (Hodges et al. 1986)]. For some genotypes, results are similar on both salt formulations, but for others growth may be markedly better on one or the other. Proline is often added to nutrient media, in part because Armstrong and Green (1985) reported that addition of proline stimulated production of friable embryogenic callus from A188 cultures grown on N6 medium. Effects of proline on other genotypes have been less clearcut. Asparagine and casein hydrolysate are other sources of reduced nitrogen sometimes included in maize culture media.

Elevation of the sucrose level in the initial medium from the usual 2–4% to 6% or even 10–12% can result in more extensive formation of embryogenic callus on cultured embryos (Lu et al. 1983; Shields pers. commun.), but the enhanced initial callus formation is not always correlated with increased plant regeneration rates (Hodges et al. 1986). Elevated sucrose concentrations (6–12%) are sometimes included in hormone-free regeneration medium to promote maturation of somatic embryos (Green 1982; Kamo et al. 1985).

Generalizations about the role of specific nutrient components are difficult, in part because tests have often compared media with multiple differences (e.g., Duncan et al. 1985; Fahey et al. 1986). Moreover, nutritional requirements for efficient regeneration vary among the many different lines tested in maize tissue culture labs (e.g., Hodges et al. 1986; Fahey et al. 1986). Such genotype by environment interaction is no more surprising in vitro than in the field, but it means that culture media may need to be optimized for each new maize line used.

2.2.4 *Types of Cultures*

Initial studies with maize cultures (Green and Phillips 1975; Springer et al. 1979) described regeneration via an organogenic pathway, but more recent work has emphasized regeneration via somatic embryogenesis (Green 1982; Lu et al. 1982; Novak et al. 1983; Kamo et al. 1985). Identification of the structures seen on scutellar explants and subcultured callus as somatic embryos often requires histological or scanning EM studies (Lu et al. 1983; McCain and Hodges 1986), and even these may not be definitive, since under certain culture regimes embryoids may coalesce or develop abnormally.

Most regenerable maize cultures, both those that are clearly embryogenic and those that produce shoots as well as embryoids, have a firm nodular texture, heterogeneous morphology and relatively slow rates of growth. Such tissue has been designated Type 1 (or I) by several authors (Green et al. 1983; Kamo et al. 1985). Regeneration of plants from freshly initiated Type 1 callus is relatively simple, but maintenance of regenerable callus over multiple transfers can be much more difficult (Lu et al. 1982; Hodges et al. 1985). Usually, morphogenetically competent areas are surrounded by nonregenerable tissue, and unselective subculture can result in a translucent rooty callus from which plant recovery is difficult or impossible. Selective transfers, in which only organized areas are subcultured, are more successful, but even with rigorous transfer regimes, the regenerability of most cultures tends to decrease sharply with time. For this reason, comparisons of different genotypes and culture media are often based on plant recovery from the initial explants themselves, not from subcultures (e.g., Lu et al. 1982, 1983; Hodges et al. 1986). The percentage of abnormal plantlets may also increase as culture period is extended (Kamo et al. 1985). However, some Type 1 cultures have remained regenerable for several years (e.g., Earle et al. 1987) or even for 7 years (Earle unpubl.).

Occasionally, sectors of more rapidly growing friable embryogenic tissue appear in Type 1 cultures, particularly in ones from specific lines such as A188 or B73 or hybrids containing these inbreds. Such tissue has also been isolated from cultures of F_1 hybrids of regenerable genotypes such as W64A or A188 with Black Mexican Sweet, a line which forms friable nonregenerable callus (Lupotto 1986; Kamo et al. 1987). Inclusion of proline in the nutrient medium can enhance growth of such cultures (Armstrong and Green 1985; Vasil and Vasil 1986). The friable embryogenic tissue, designated Type 2 (or II), tends to retain regenerability longer and can give excellent suspension cultures (Green 1982; Vasil et al. 1984; Armstrong and Green 1985; Lowe et al. 1985; Tomes and Smith 1985; Vasil and Vasil 1986; Donovan and Somers 1986; Kamo and Hodges 1986). The suspensions can be subcultured on a rapid transfer schedule or plated back on solidified medium for plant recovery. Friable regenerable suspensions are particularly valuable for in vitro selection experiments and as a source of dividing maize protoplasts. Several labs have obtained microcalli or somatic embryos from such protoplast cultures (Imbrie-Milligan and Hodges 1986; Vasil and Vasil 1987; Kamo et al. 1987). Rhodes et al. (1988) have recently recovered mature plants from protoplasts obtained from a friable regenerable cell culture. These achievements open new vistas for in vitro genetic manipulation of maize by protoplast fusion or transformation.

3 Somaclonal Variation

3.1 General Considerations

The starting material in maize regeneration programs is usually an inbred line, which would show relatively little variation in field tests. Material from cultures of such lines has by now been the subject of many studies, most of which have revealed some differences between cultured and standard material. These differences include alterations in morphology, male fertility, quantitative agronomic traits, biochemical and molecular characters, chromosome number and structure, and sensitivity to fungal toxins, insecticides, and other chemicals (Table 1). Work with hybrid lines has also revealed culture-related variations in some specific traits; however, evaluation of progeny from hybrid regenerates may be complicated by the segregation that normally occurs after selfs or crosses of hybrids.

Clearcut conclusions about the extent and nature of somaclonal variation in maize are difficult to reach because so many different genotypes, culture conditions, and experimental strategies have been used. One major variable is the level of organization considered, i.e., callus, regenerated plants, or progeny of regenerated plants. An important issue is whether alterations seen in callus or R_0[3] plants are

[3] R_0 plants: regenerated plants themselves; R_1 plants: seed progeny of R_0 plants; R_2 plants: progeny of R_1 plants, etc.

Table 1. Summary of somaclonal variation observed in maize materials[a]

A. No selection in vitro Variation	Comments	Reference[b]
Abnormalities in R_0 plants[c]	Not heritable	Gengenbach et al. (1977)
		Novak et al. (1983)
		Edallo et al. (1981)
		Hodges et al. (1985)
		Earle et al. (1987)
		Others
Qualitative morphological traits[d]	Usually recessive	Edallo et al. (1981)
		Rice (1982)
		McCoy and Phillips (1982)
		Zehr et al. (1987)
		Göbel and Lörz (1986)
		Phillips et al. (1988)
		Others
Quantitative agronomic traits[e]	Field tests	Beckert et al. (1983)
	Some hybrid trials	Novak et al. (1986)
		Zehr et al. (1987)
		Fincher (unpubl.)
		Gracen et al. (unpubl.)
Alcohol dehydrogenase mutants	Single DNA base pair changes	Brettell et al. (1986)
		Dennis et al. (1987)
Activation of transposable elements	*Ac* element activation seen in kernels	Peschke et al. (1987)
DNA methylation	R_0 plants examined	Brown and Lörz (1987)
Zein polypeptides	Recessive	Prioli and Silva (unpubl.)
Reversion to fertility and resistance to *Helminthosporium maydis* race T toxin and methomyl (cms-T maize)	Maternal inheritance	Brettell et al. (1980)
	Reversion and resistance are associated	Umbeck and Gengenbach (1983)
		Kuehnle and Earle (1987 and unpubl.)
Loss of S mtDNA plasmids (cms-S maize)	No R_0 plants regenerated	Chourey and Kemble (1982)
Reversion from cms-S to fertility	Loss of S plasmids and other changes in mtDNA	Earle et al. (1987)
	Maternal inheritance	Small et al. (1988)
Chromosome number		Green (1977)
		Edallo et al. (1981)
		McCoy and Phillips (1982)

Chromosome structure[f]	Meiotic Analysis	Benzion (1984)
		Armstrong and Phillips (1986)
		Rhodes et al. (1986)[g]
		Lee and Phillips (1987)
		Benzion et al. (1986)
Pollen sterility	Probably reflects chromosome abnormalities	Edallo et al. (1981)
		McCoy and Phillips (1982)
		Armstrong and Phillips (1986)
		Wang et al. (1987)
B. In vitro selection		
Resistance/tolerance to:		
Aminopterin	Cross resistance to methotrexate	Shimamoto and Nelson (1981)
	No plants regenerated	
Methotrexate	No plants regenerated	Tuberosa and Phillips (1986)
Lysine and threonine	Altered aspartokinase	Hibberd et al. (1980)
	Elevated free threonine in seeds	Hibberd and Green (1982)
	Single dominant gene	Miao et al. (1986)
5-methyltryptophan	Altered anthranilate synthase	Hibberd et al. (1986)
	Elevated free tryptophan in seeds	Miao et al. (1986)
	Single dominant gene	
Imidazolinone	Altered acetohydroxyacid synthase	Shaner and Anderson (1985)
	Single dominant gene	Anderson and Georgeson (1986)
	Field tests	
H. maydis race T toxin (cms-T maize)	mtDNA alterations	Gengenbach and Green (1975)
	Maternal inheritance	Gengenbach et al. (1977)
	Associated with reversion to fertility	Brettell et al. (1980)
		Dewey et al. (1986, 1987)
		Fauron et al. (1987)
Methomyl (cms-T maize)	Cross resistance to *HmT* toxin	Kuehnle and Earle (1987 and unpubl.)
	Maternal inheritance	
	Associated with reversion to fertility	
	Field tests	

[a] Except where noted, all studies listed used immature embryos as explants.
[b] In some cases, only selected and/or recent references are cited. Individual reports may not include all the types of work mentioned.
[c] Stunting, poor fertility, terminal silks or ears, etc.
[d] Dwarfs, brachytic or necrotic forms, albinos, striping and other pigment alterations, defective kernels, etc.
[e] Changes in plant height, leaf number, area or color, flowering data, ear characteristics, yield, etc.
[f] Translocations, deficiencies, duplications, etc., often near heterochromatic regions.
[g] Explants were immature tassels.

heritable or simply ephemeral responses to culture conditions. A second variable is the type of assessment done, i.e., molecular characterization at the protein or DNA level, analysis of mitotic or meiotic chromosomes, qualitative or quantitative assessment of morphology or other plant traits, performance in hybrid trials, etc. Some studies have surveyed the material for any apparent variation; others have looked only for alteration in a specific trait. Often initial observations of one type are supplemented with additional ones at a different level, i.e., cytological changes may be related to plant phenotypes, progeny may be scored for traits seen in R_0 plants, the nucleic acid basis of observed protein alterations may be determined, etc. Comparisons to similar material that has not been subjected to culture conditions are critical for assessing the significance of any variation seen in culture-derived material. It is also helpful to obtain multiple plants from a given explant, as this may confirm that the initial material was of the expected type, and may also reveal whether a novel trait appeared early or late in the culture period (Larkin 1987).

3.1.1 *Studies of Tissue Cultures*

Although ample supplies of callus are often available, choice of appropriate traits to examine for variation may be difficult. An obvious candidate is chromosome number. Unfortunately, the mitotic index in callus is usually low. Moreover, studies of mitotic chromosomes usually can reveal only alterations in number and gross rearrangements, not the more subtle differences in chromosome structure that can be detected at meiosis. Occasionally maize callus has been examined at the molecular level, e.g., assayed for presence of the S_1 and S_2 plasmids in callus from lines with the S-type of cytoplasmic male sterility (Chourey and Kemble 1982; Earle et al. 1987). Another approach with callus or suspensions is to challenge the tissue with chemicals toxic to the original material (e.g., phytotoxins or metabolic inhibitors) and check for growing resistant sectors. Recovery of resistant lines from unmutagenized callus derived from sensitive plants reveals somaclonal variation at the cellular level. Two interesting questions arise in regard to variant cells found in callus: (1) are these cells capable of regeneration?; and (2) if plants are regenerated from such cells, are the variant traits expressed and maintained?

3.1.2 *Studies of Regenerated Plants*

Recovery of plants from callus or suspension cultures permits such questions to be addressed and allows additional types of studies, including meiotic analysis and evaluation of phenotypes at the plant level. However, attention to R_0 plants alone may give misleading or incomplete results. Examination of progeny from regenerated plants is essential for most serious studies of culture-derived variation, for several distinct reasons:

1. Some novel phenotypes observed in regenerated plants are not heritable. Progeny studies are needed to reveal whether a given R_0 phenotype is stable and how it is inherited.

2. Furthermore, phenotypic changes seen in R_0 plants (and in callus) are likely to be limited to cytoplasmic or dominant ones; recessive phenotypes will usually not be apparent. Mutations to a recessive allele or gene combination will be seen in R_0's only in special cases, such as material heterozygous for a particular locus or monoploid for a chromosome or chromosome segment.
3. Many traits (e.g., days to anthesis) cannot be measured accurately on R_0 plants. Plants grown from seed provide a more clearcut starting point for such measurements.
4. Populations, preferably grown under field conditions, are required for evaluation of quantitative traits such as yield.

Progeny tests often begin with seeds from selfed R_0 plants, but it may be preferable to defer detailed analyses to later generations of progeny. This would be true if few R_1 seeds or seeds of poor quality are obtained from weak R_0 plants, if large plant populations are needed for quantitative studies, or if R_1 seeds were produced by sibbing rather than selfing so that recessives would not be revealed. Recessives may in fact be masked even in plants obtained after selfing or R_0 plants; this could occur if the R_0 plant is sectored with genetic alterations only in the tassel or only in the ear. Such "nonconcordance" could result from a multicellular origin of regenerated plants or from mutations occurring during development (McCoy and Phillips 1982). If pollen and eggs have different genetic constitutions, all R_1 plants would be heterozygous; segregation for recessives would be apparent only in the R_2 generation and then only in some R_2 progeny families.

3.2 Phenotypic Changes That Are Not Heritable

Plants regenerated from corn cultures are often stunted and/or have poor fertility, with silks in the tassel, terminal ears in place of tassels, etc. (e.g., Green 1977; Gengenbach et al. 1977; Novak et al. 1983; Hodges et al. 1985; Earle et al. 1987). These traits may be seen in either a small or a substantial part of the population; lab to lab differences in plant quality can be marked. Recovery of seeds from such abnormal plants can be difficult, but when progeny tests are feasible, they show that the abnormalities are usually not expressed in the next generation.

These variant phenotypes thus appear to be developmental aberrations induced by culture conditions, but exactly which aspect of the procedure is responsible is not yet clear. Stunting and feminization can also be seen in plants derived directly from precocious germination of zygotic embryos on MS medium lacking 2,4-D (Hodges et al. 1985; Earle and Gracen 1985); thus 2,4-D per se is not responsible for the weak growth of many R_0 plants.

Light can influence the morphology of regenerates; exposure of small plantlets to a 16- rather than 12-h photoperiod greatly increases final plant height and also improves male fertility (Earle unpubl.). This is true even if regenerates initially exposed to 12-h days are moved to standard long-day greenhouse conditions when they have about five to six leaves. These observations suggest that at least some lab to lab differences in appearance and fertility of R_0 plants may be due to differences in plantlet growth conditions rather than to earlier stages of the culture procedure.

Close attention to the environment provided to regenerates after transfer out of culture is therefore important. However, genotype differences in morphology of regenerates grown under similar regimes can also be seen, with some more sensitive to light conditions than others. In addition, some long-term cultures of inbred W182BN have yielded stunted plants even under conditions that allow vigorous growth of younger regenerates.

Abnormal morphology and fertility of R_0 plants, although usually not heritable, are still important. For example, reversion of sterile tassels to fertile ones is difficult to detect on plants with poor tassel development (Earle et al. 1987). A more general concern is that stunted, partially sterile plants are hard to pollinate, and it may be impossible to recover seeds, particularly selfed seed, from them. Since poor seed recovery makes progeny tests difficult, every effort should be made to produce vigorous fertile R_0 regenerates for somaclonal variation studies. A steady supply of seed-grown plants of the appropriate genotype can permit recovery of at least sibbed seed from regenerates with poor or asynchronous silk or pollen formation. Precautions against accidental outcrossing are important.

3.3 Changes That Are Heritable

In contrast to the transient phenotypic alterations described above, other maize variants have a more stable genetic basis. Some changes, initially detected in callus or R_0 plants, are transmitted to progeny as maternal or dominant traits. Additional changes that can be attributed to the culture procedure have been revealed in tests of progeny.

3.3.1 Changes Involving Nuclear-Encoded Traits

Variants have been categorized according to phenotype, to the pattern of inheritance, or to the mechanism underlying observed changes. Some specific reports illustrating the types of results obtained to date are summarized below.

Morphological Variants. The most easily recognized types of heritable variation are abnormal seedling, plant, or kernel phenotypes (e.g., dwarfs, necrotic forms, pigment alterations or deficiencies, striping, defective kernels, etc.). Over 50 such phenotypic variants have been reported (Phillips et al. 1988). Many of these appear similar to recessive mutations previously identified by maize geneticists (Neuffer et al. 1968), although allelism tests have rarely been carried out. Often the occurrence of such variants is merely noted, but more detailed quantitative studies have also been done. In many cases the segregation ratio in selfed R_1 or R_2 populations suggest a recessive pattern of inheritance.

Edallo et al. (1981) scored R_2 populations from W64A and S65 regenerates for seedling and endosperm mutants, counting only ones in which a 3:1 segregation was seen. The R_2 generation was used because few seeds were recovered from the weak R_0 plants. Mutation frequency was calculated as number of distinct mutant types seen in progeny of each R_0 plant. This value may be an underestimate of actual mutation rates, since each phenotype was counted only once even though it may have arisen independently several times in the growth of a culture. Even so, the rate was substantial, approximately 1 mutation/R_0. The rates were somewhat higher in W64A (regenerated after four to eight monthly transfers) than in S65 (regenerated after one to four transfers). No visible mutations were seen in control populations grown from seed.

Rice (1982) screened R_1 progeny of regenerated plants from four different lines for segregating mutations. Over 90% of 75 R_0 plants from one year old A188 and PG6 cultures showed segregation for mutant R_1 progeny; 11 and 14 different mutations were seen among the progeny of regenerates from cultures derived from single embryos of these two genotypes. Unlike Edallo et al. (1981), Rice saw some mutations in the control population. However, the mutations observed in regenerated progeny were probably due to the culture process since not all plants from a single cultured embryo carried the same mutation.

Zehr et al. (1987) reported data on qualitative variation in progeny from regenerates of four sweet corn and three dent inbred lines. Three of the sweet corn lines maintained as callus for only 25–32 days before regeneration produced progeny with no segregation for mutants but the fourth had 0.64 qualitative mutations per R_0 plant. Thus variation can, at least in some lines, occur quite early in culture. The highest frequency among the dent lines was 0.71 mutations/R_0 plant, from 130–134-day-old cultures. The lines which generated the most variants were represented by the greatest number of embryo sources and R_1 families, so broad sampling from young cultures is probably important. The percentage of plants with mutant phenotypes (in lines segregating for mutants) varied widely, from 2.4 to 100%, with most values between 10 and 30%. The deviations from a 3:1 ratio may be due to the small number of R_1 plants tested (from as few as 6 to a maximum of 99). Two separate occurrences of a striate leaf mutation present in all progeny were noted, suggesting a nonrecessive, possibly cytoplasmic, mode of inheritance. In several cases, R_2 progeny from "normal" R_1 plants segregated for variants, indicating chimerism in the original regenerate.

Prioli and Silva (pers. commun.) evaluated progeny from regenerates of three tropical maize inbred lines for morphology, male sterility, and chlorophyll deficiency, and noted several apparently recessive mutations (including brachytic form, erect leaves, lemon yellow seeds, wrinkled leaves, and male sterility). Their observations with tropical germplasm indicate that such culture-related alterations occur in a broad range of maize materials.

Quantitative Variants. Mutants of the sort described above, while easiest to detect, are of the least interest in maize improvement programs, because they are both already familiar and almost invariably deleterious. Maize breeders are more concerned with changes in agronomic traits, particularly positive ones that have been difficult to achieve via conventional breeding (Earle and Gracen 1985). A few studies have examined culture-derived material for changes in traits whose inheritance appears to be controlled by multiple "quantitative" genes.

Beckert et al. (1983) not only saw some abnormal R_0 plants and some mutants in the R_1 generation of inbreds CO158, A641 and A188 regenerated after 100 days as callus, but also carried out fairly large-scale biometrical trials with these lines.

They compared progeny of regenerates to each other and to the original genotypes in two generations of inbreeding and in hybrid trials. Quantitative traits such as leaf number, leaf area, plant height, flowering date, and grain yield components were measured. In the case of CO158, crosses of 23 progeny lines with inbred F1444 showed significant differences from the standard hybrid in many traits, but the extent of variation was not very great. Two regenerated lines were clearly different from controls. Studies with A641 and A188 showed little significant difference from the controls in the traits examined, but values for selfed regenerates were usually less than those for crosses of regenerates with the standard line. Beckert et al. concluded that variation seen as a result of culture is not greater than that normally seen in maize-breeding programs.

In field studies with progeny of regenerated inbred W182BN, Earle and Gracen (1985) saw both typical "mutant" phenotypes and plants with heritable variability in important agronomic traits. The latter included increased vigor, altered maturity (Fig. 1), increased or decreased height, darker leaf color (Fig. 2), and morphological variation. Some of these traits remained stable for at least six selfed generations.

Zehr et al. (1987) examined R_1 populations from several inbred dent lines for quantitative traits such as plant height, date to 50% anthesis, ear length, kernel rows/ear, weight/100 kernels, yield, etc. and detected some statistically significant changes, including delays in anthesis and positive and negative alterations in ear characteristics. Some of the changes observed may have relevance to breeding efforts with the lines tested. Yield per plant was not significantly altered.

Fig. 1. A maize inbred line derived from regenerated W182BN (*left foreground*) showing later maturity than the nonregenerated inbred seen at the *right* and *background*

Fig. 2. Comparison of two uniform S6 maize lines derived from one plant (N-2) regenerated from an immature embryo of inbred W182BN. The line on the *left* is taller and has darker green leaf color than the sister line on the *right*, which is similar to the standard inbred

An important further issue is whether culture-related changes in inbred lines affect performance of hybrids using such altered inbreds. Novak et al. (1986) crossed 20 regenerated lines of inbred CHI-31 with the test line $BuRO_2$ and measured hybrid yields. Thirteen lines showed significantly decreased yield and one had increased yield as compared to the hybrid produced using the initial inbred line.

In larger-scale experiments, Fincher (pers. commun. from DT Tomes) of Pioneer Hi-bred International, Inc. compared lines regenerated from B73 and other commercially important lines to comparable standard materials in inbred trials and in hybrid combinations with two testers in three locations. Characters scored were yield, moisture, lodging, staygreen, plant and ear height, i.e., ones typically rated in breeding programs. Lines regenerated from one inbred, which had been maintained in culture for only a short period, showed few differences from the standard line either in hybrid or inbred trials. Progeny of three other lines, from 14-month-old cultures, had decreased yield in all cases (see Table 2 for data from one of these lines, T66). This decrease was significant at the 0.01 level of probability in 12/26 cases (hybrid trials) and 8/12 cases (inbred trials). Staygreen ratings were also worse in all but one case, usually significantly so. A few regenerated lines showed significant differences in other traits, such as maturity and plant or ear height, but these changes were associated with decreased yield.

Field trials with progeny of regenerated inbred W182BN in several hybrid combinations (Gracen, Smith and Earle, unpubl.) have not revealed the consistent

Table 2. Differences between agronomic traits of regenerated B73 lines[a] and nonregenerated B73 checks[b]

A. Differences in testcrosses with the tester G50

Regenerated line	Yield, Bu/acre	Percent moisture	Staygreen[c]
1	-12**[d]	0.3	0.7*
2	-14**	-0.3	-1.5**
3	-14	0.1	-1.5**
4	-12**	-0.2	-0.5
6	-12**	-0.4	-0.4

B. Differences between regenerated and nonregenerated B73 inbred lines in inbred yield test

Regenerated line	Yield, Bu/acre	Percent moisture	Staygreen
1	-38**	0.1	-1.3
2	-40**	1.4	0.8
3	-27*	1.1	-1.5*
4	-11	-0.1	-1.0
5	-34**	1.2	-0.8

[a] Pioneer Hi-Bred T66 lines, regenerated from 14-month-old cultures of B73 immature embryos.
[b] Summary of data from 1986 tests at three locations (unpubl. data of R. Fincher, provided by D.T. Tomes).
[c] 1-9 scale with 9 the best score.
[d] *,** Significant at the 0.05 and 0.01 levels of probability, respectively.

yield depression observed by Fincher. Of 157 experimental hybrids involving regenerated W182BN tested in single location replicated field trials in 1984 and 1985, 48 showed significant differences in maturity from the corresponding controls (13 were earlier; 35 were later). However, only 3/157 hybrids had significantly altered yields (two higher; one lower). Table 3 shows data from one hybrid combination; significant increases and decreases in percent moisture (i.e., maturity) occur without changes in yield. Thirty of the experimental hybrids were promising enough to be advanced to multi-location trials in 1986. Of these, 12 showed significant differences from controls (two with lower yields and later maturity; three with earlier and seven with later maturity; two with higher stalk lodging). The other 18 did not differ significantly from controls in the characters measured. Any of these 18 which involve regenerated W182BN lines with superior inbred performance as well as the three earlier maturing hybrids could be useful.

These results from the various field studies demonstrate the importance of assessing overall performance of materials developed in culture. Decreased yield or quality of somaclonal lines, either at the inbred level or in crosses, will limit the attractiveness of this approach to maize improvement. Caution is clearly necessary in extrapolating from possibly desirable alterations seen at the inbred level to the applications in hybrid production. On the other hand, crosses of regenerated material with the original inbred may provide more vigorous seed parents via "sister line heterosis" (Earle and Gracen 1985; Novak et al. 1986).

Table 3. Results from testcrosses of inbred lines derived from W182BN regenerated plants with the tester (A632Ht × A641Ht)[a,b]

Regenerated plant no.[c]	Inbred no.[d]	Yield, Bu/acre	Percent moisture
N-2	17	83.1	19.9*[e]
N-2	15	74.7	20.3*
N-8	3	84.7	20.5*
N-2	20	98.9	20.6*
S-31	1	92.9	21.0*
N-2	19	76.3	21.3
N-8	2	88.0	21.9
N-2	5	73.2	22.0
N-2	18	74.4	22.4
S-26	2	67.4	23.5
N-11	6	81.8	23.6
N-11	1	66.8	24.7*
N-2	6	71.5	24.8*
N-2	7	79.4	25.9*
N-11	5	101.4	26.2*
N-11	4	73.8	26.3*
N-2	5	90.7	27.1*
N-2	10	87.6	27.3*
Control	(W182BN)	81.8	22.6
	Mean	81.5	23.4
	Standard dev.	14.2	1.0
	LSD ($P < 0.05$)	41.2	1.6

[a] The inbred W182BN crossed to this tester forms the hybrid Cornell 175.
[b] 1985 tests at Aurora, NY (unpubl. data of V.E. Gracen, M.E. Smith and E.D. Earle).
[c] Regenerated plant numbers indicate the cytoplasmic background from which the culture was established and the order number of the plant recovered (i.e., N-8 was the 8th plant recovered from a W182BN "N" cytoplasm culture). N-2, N-8 and N-11 were regenerated after 7, 19, and 20 months of callus culture, respectively; S-26 and S-31 after 15 and 16 months.
[d] Inbred numbers are arbitrary designations for inbred lines selected from selfed progeny of each regenerated plant.
[e] Significantly different from the control according to the LSD.

Molecular Variants. In a few cases, somaclonal variation at the molecular level has been identified by screening populations of regenerates for alterations in specific polypeptides. Brettell et al. (1986) and Dennis et al. (1987) found two alcohol dehydrogenase (*Adh1*) mutants by screening roots of R_0 plants regenerated from cultures of 157 embryos which were heterozygotes for *Adh1*-F (carried by A188) and *Adh1*-S (carried by Berkeley Slow). Both mutants were isolated from a population of 316 plants regenerated after only one subculture; no additional mutants were seen in 713 plants regenerated after 1–20 further subcultures. The reciprocal cross also yielded no mutants among 223 plants.

One of the mutants (*Adh1*-Usv) was found to be a slower electrophoretic variant of *Adh1*-S produced by replacement of a glutamic acid residue by a valine residue; this shift resulted from a single base change in the last codon triplet in exon 7 (Brettell et al. 1986). The other mutant was null for *Adh1*, again because of a single base change.

This change, in exon 4, converted an AAG (lysine) codon into a TAG stop codon (Dennis et al. 1987). The frequency of these culture-induced mutations (2 in 1382 plants from 218 embryos cultured) is much higher than the spontaneous mutation rate (< 1 in 10^6 seeds). The type of change seen, i.e., point mutation, is similar to that induced by other mutagenic treatments.

Prioli and Silva (pers. commun.) analyzed zein polypeptides in endosperm from 104 somaclones of tropical line Cat 100–1 by isoelectric focusing. Seeds from eight R_0 plants showed alterations in the most basic zein polypeptide. This trait segregated as a recessive in selfed progeny of one somaclone. No such variation was seen in control seeds.

Brown and Lörz (1986) have found differences in DNA methylation in regenerated maize plants and are attempting to relate these differences to observed phenotypic changes in regenerates.

Changes in Transposable Elements. Mobilization of transposable elements could lead to heritable genetic changes by a variety of different mechanisms (Larkin 1985). A convincing demonstration of transposon activation in maize cultures has recently been reported (Peschke et al. 1987). Peschke et al. tested 1200 progeny from 301 plants regenerated from 94 embryos of Ohio 43 × A188 lines for the *Activator (Ac)* transposable element. Fifty-six tests, representing ten R_0 plants from two different embryos, revealed an active *Ac* element. No active *Ac* elements were present in the noncultured controls or the explants. The frequency of culture-induced activation of transposable elements was high (about 3%) in these experiments, but the involvement of insertional events in somaclonal variation is still unclear. The stability of many alterations seen and the demonstration that some alterations are due to single base pair changes (Brettell et al. 1986; Dennis et al. 1987) do not support a primary role for transpositional changes in culture-induced variation.

Variants Recovered by in Vitro Selection. The observation that heritable variation can often be detected at the plant level in the absence of any prior selection suggests that genetic changes in cultured maize tissues are quite common. Recovery of altered cell lines or plants after in vitro selection from unmutagenized diploid callus is further evidence for extensive culture-related genetic heterogeneity. Friable embryogenic (Type 2) cultures are particularly convenient material for such studies because of their rapid growth rate and regenerative capacity; however, selections using nodular Type 1 callus have also been successful.

Several different selective agents have been used, including the amino acid combination lysine + threonine (Hibberd et al. 1980; Hibberd and Green 1982; Miao et al. 1986), the amino acid analog 5-methyltryptophan (5MT) (Hibberd et al. 1986; Miao et al. 1986), the herbicide imidazolinone (Shaner and Anderson 1985; Anderson and Georgeson 1986), and analogs of normal metabolites such as the folic acid analogs aminopterin (Shimamoto and Nelson 1981) and methotrexate (Tuberosa and Phillips 1986). In several cases, callus capable of growth in the presence of normally inhibitory concentrations of these compounds has been regenerated to plants which express the selected trait or one related to it. For example, maize selected for resistance to imidazolinone is resistant to field applications of the herbicide (Anderson and Georgeson 1986). Selection for resistance to lysine + threonine gives plants with greatly increased levels of free threonine in seeds (Hibberd and Green 1982; Miao et al. 1986). Progeny from the 5MT selections have elevated levels of tryptophan in their seeds (Hibberd et al. 1986). In these cases, inheritance of the altered trait appears to involve a single dominant gene.

The variant cell lines and plants have proven useful for analysis of the biochemical pathways affected by the selective agents. In several instances, reduced sensitivity of biosynthetic enzymes was shown to be responsible for enhanced tolerance of inhibitory compounds. Herbicide resistance of cultures growing on imidazolinone was accompanied by a reduced herbicide sensitivity of acetohydroxyacid synthase, a key enzyme in the synthesis of the branched chain amino acids (Anderson and Georgeson 1986). Callus resistant to aminopterin showed a decreased inhibition of dihydrofolate reductase, the enzyme to which the folic acid analog normally binds (Shimamoto and Nelson 1981). Another example of altered enzyme sensitivity is found in the lysine + threonine resistant lines in which the biosynthetic enzyme aspartokinase was less sensitive to inhibitory concentrations of lysine than aspartokinase from nonselected sensitive lines (Hibberd et al. 1980). The variant material obtained through in vitro selection will be valuable not only for further biochemical characterization but may provide agricultural advantages as well.

3.3.2 Changes Involving Cytoplasmic Traits

Background. Although most variants recovered from maize tissue cultures have nuclear alterations, either at the chromosomal or DNA level, cytoplasmically encoded changes have also been observed. Identification of such changes has been facilitated by the availability of several maternally inherited maize mutants, namely the male sterile maize cytoplasms. Maternally inherited failure to produce viable pollen (cytoplasmic male sterility or cms) is of great significance to hybrid seed production because it eliminates the need for costly detasselling operations. Three different types of maize cms, known as Texas (T), Charrua (C) and USDA (S), can be distinguished on the basis of the cytology and genetics of pollen abortion, restoration by nuclear restorer genes, and mitochondrial DNA (Laughnan and Gabay-Laughnan 1983). Lines carrying cms-T are sensitive to the toxin (HmT toxin) produced by race T of the fungal pathogen *Helminthosporium maydis* (also known as *Drechslera maydis* or *Bipolaris maydis*, sexual stage *Cochliobolus heterostrophus* Drechs.) (Hooker et al. 1970; reviewed in Gregory et al. 1977) and to the insecticide Lannate (active ingredient methomyl) (Humaydan and Scott 1977). Maize lines with cms-S, cms-C or male fertile (N) cytoplasms are resistant to these chemicals. Cultures of maize lines carrying the three male sterile cytoplasms have been established and studied by several laboratories.

Cms-T Variants. Fertile plants with resistance to HmT toxin or Lannate have been recovered from cms-T cultures both with and without in vitro selection. Using a crude preparation of HmT toxin, Gengenbach and Green (1975) were able to select resistant callus from cultures of a backcross of A188 with a cms-T line. In later work, Gengenbach et al. (1977) succeeded in regenerating plants from resistant callus. The plants were resistant to applications of toxin and also were as resistant to the fungal pathogen itself as N cytoplasm lines. Resistance was maternally inherited and associated with changes in mitochondrial physiology. The R_0 plants and/or their progeny had also reverted to fertility. Studies of the mitochondrial DNA of the revertants revealed that it was not identical to that seen in N cytoplasm lines but was instead a variant of the form seen in sterile cms-T lines. Thus the fertile material did not arise via selection of N cytoplasm cells or mitochondria already present in the explants. Brettell et al. (1980) recovered additional resistant male fertile plants from cms-T callus selected for resistance to HmT toxin.

Brettell et al. (1980) also recovered many male fertile toxin-resistant plants without selection with HmT toxin; of 60 plants regenerated from unselected callus cultures over a year old, 35 (58%) were revertants. Recovery of such a high frequency of spontaneous revertants from unselected cultures is particularly striking in view of the fact that no shifts from cms-T to fertility have ever been reported from field studies using unmutagenized seeds. Umbeck and Gengenbach (1983) also obtained revertants from unselected cultures but at a much lower rate; only 8/169 (5%) plants regenerated from A188 cms-T cultures over a 12-month period were fertile and toxin-resistant. Both Brettell et al. (1980) and Umbeck and Gengenbach (1983) recovered some regenerates that appeared to be cytoplasmic chimeras, as evidenced by segregation of progeny into fertile-resistant and sterile-sensitive lines.

Kuehnle and Earle (1987) used methomyl as a selective agent with cultures of W182BN, sweet corn lines, and F_1 hybrids of W182BN and sweet corn, all carrying cms-T. Many resistant male fertile plants were obtained by regenerating plants from methomyl-resistant calli, particularly when methomyl was included in the regeneration medium. Only a few (3/132 or 2.3%) fertile resistant plants were recovered from unselected calli that had been subcultured 6–8 months before regeneration. In contrast, 61% (19/31) of plants from unselected 14–16-month-old calli were fertile and resistant. The percentage of plants with spontaneous resistance varied from genotype to genotype. Methomyl resistant plants were unaffected by Lannate sprays and HmT toxin. As in the earlier studies with cms-T regenerates, the shift to fertility and resistance was maternally inherited and associated with mitochondria.

An original impetus for the cms-T work was to separate the T type of cms from sensitivity to *H. maydis* race T, cause of the devastating 1970 epiphytotic of Southern corn leaf blight in the USA. Since the shift to resistance is always accompanied by loss of the desirable cms-T, the material recovered from culture has not proved useful for breeders of hybrid field maize. Methomyl-resistant cms-T sweet corn lines carrying nuclear genes for restoration of fertility (*Rf* genes) may have agricultural value, since these lines are already used in fertile form and are particularly vulnerable to Lannate sprays used for insect control.

Culture-derived revertants have already been of great value in studies of the molecular basis of cms-T by allowing direct comparisons of mtDNA sequences in sterile and fertile cms-T lines. Such comparisons have led to identification of a specific DNA sequence, an open reading frame (ORF 13) coding for a 13-kDa protein, as the mtDNA region responsible for the T type of cms (Dewey et al. 1986, 1987; Fauron et al. 1987).

Cms-S Variants. Cultures of W182BN carrying cms-S show a high rate of reversion to fertility (Earle et al. 1987), although no such reversion has been noted in many years of field studies with this inbred. (Revertants from cms-S to fertility have been obtained from some other maize genotypes but not at the frequencies seen in vitro with W182BN). Over 100 revertants were recovered from 18 independent cultures with several different sources of cms-S. The extent of reversion increased with time; some revertants were obtained from cultures maintained as callus for only 3 months, but all cultures older than 12 months produced revertants and most failed to produce any cms plants after that time. Reversion was maternally inherited and associated with loss of the S_1 and S_2 mitochondrial plasmids characteristically seen in sterile cms-S lines. Studies of the high molecular weight mtDNA of the revertants have revealed some differences from previous revertants of other nuclear genotypes (Small et al. 1988).

Thus both cms-S and cms-T cultures show a marked trend toward reversion to fertility in vitro. With cms-T, selection using either HmT toxin or methomyl enhances recovery of revertants, particularly from young cultures, but spontaneous cms-T and cms-S revertants can be recovered even in the absence of selection. Mitochondria encoding these types of sterility appear to be less competitive under in vitro conditions than variant mitochondria associated with male fertility, but the origin of the variants remains to be clarified. As yet no revertants from cms-C have been recovered from culture (Earle et al. 1987) and no instances of in vitro shifts from male fertility to cms have been well documented.

3.4 Chromosomal Analysis

Early studies of maize callus and regenerates showed considerable cytological stability, particularly in comparison to another monocot, oats (McCoy et al. 1982; Benzion et al. 1986). Green (1977) found abnormalities in only two of 43 plants regenerated from 65–125-day-old cultures of A188 × W22 R-nj R-nj cultures. Microsporocyte samples from both were sectored: one had aneuploid cells monosomic for chromosome 5 and the other had tetraploid sectors. Pollen fertility, as measured by staining, was less than 90% in only one of 85 plants examined.

In further studies with A188 × W22 *R-nj R-nj* or the reciprocal cross, McCoy and Phillips (1982) also observed relatively few cytological changes. No more than 5% of the mitoses in 4-month-old callus cultures had a nondiploid number. Values for 8-month-old callus were similar. Some lines had no cells which were not diploid. Only five of 124 regenerated plants from these cultures were cytologically altered. Three of these had tassels which were sectored for pollen sterility. One plant had only 50% pollen fertility in all tassel branches examined and was missing most of the short arm of chromosome 10. Another had 50% small pollen but no recognizable cytological alteration.

Edallo et al. (1981) presented a detailed analysis of chromosomal constitution of regenerable maize cultures of inbreds W64A and S65. Examination of mitoses in a total of 15 callus cultures showed that most cells had the usual 2n = 20 chromosome number. Nondiploid cells were seen at variable frequencies ranging from less than 1% to over 65% in cultures from different embryos, but most calli had fewer than 10% nondiploid cells. The percentage of nondiploid cells did not increase with time, at least over eight transfers. The majority of these cells were tetraploid. Aneuploidy was also observed among some cells thought to be tetraploid with subsequent chromosome loss. Many of the plants regenerated from the W64A and S65 cultures had sterile pollen, but this sterility usually did not involve changes in chromosome numbers; only 2/110 R_0 plants had a nondiploid count (one tetraploid, one trisomic). Since the calli from which the plants were recovered were much more cytologically diverse, the process of regeneration may have selected against polyploid or aneuploid cells.

Several more recent reports from the laboratory of Phillips at the University of Minnesota have shown that cytological changes in cultured maize materials are more frequent and complex than earlier studies had revealed. These studies of meiotic chromosomes have also provided new information about the influence of genotype and cultural factors on chromosome stability.

Benzion (1984) examined meiotic chromosomes in 370 plants regenerated from 22 embryos of nine genotypes. Abnormalities were seen in 11% of the regenerates. Changes seen were mostly chromosome breakage (65% translocations and 35% deficiencies).

Lee and Phillips (1987) examined the effect of time in culture, using regenerates from Oh43 × A188 callus maintained for 3–4 or for 8–9 months. None of the 78 plants from the younger cultures showed cytological aberrations, while 91/189 (48%) of the ones from older cultures did. Again, almost all the changes were in chromosome structure, with translocations, deficiencies and duplications being the most common. The breaks were usually on chromosome arms with heterochromatic knobs.

Armstrong and Phillips (1986) compared regenerates from organogenic (Type 1) and embryogenic (Type 2) cultures of A188 or A188 derivatives of different ages. Cytological abnormalities were seen in plants from both types of cultures, 6% in Type 1 regenerates and 15% in Type 2 regenerates from 8-month-old cultures. Pollen sterility was also higher in Type 2 materials. These results suggest that cytologically abnormal cells are capable of embryogenesis and that use of embryogenic Type 2 cultures does not necessarily eliminate variation. Frequency of segregation of variant phenotypes was similar in both cases. While abnormalities were more frequent after 8 months than after 4 months, older cultures yielded a lower percentage of chimeral plants than did younger ones.

Rhodes et al. (1986) worked with cultures established from immature tassels of W22 *R r-xl* × A188 plants, some of which were identified as monosomic/aneuploid. The frequency of cytological alterations (as determined by meiotic analysis or pollen abortion) was similar in aneuploid and euploid cultures, i.e., 40% of 161 aneuploid regenerates and 49% of 115 euploid ones. Whether this high frequency of cytological change was due to use of tassels rather than embryos as explants, to genetic background (*r-xl* is a deficiency that induces instability) or to a combination of these and other factors is not clear. Most alterations involved chromosome structure (particularly translocations and heteromorphic pairs) rather than changes in number. Sectoring in the tassel was common. Since changes seen in monosomic cultures rarely involved the monosomic chromosome, it might be feasible to use monosomic cultures to select for recessive mutations of genes carried on the unpaired chromosome. No clearcut effect of time in culture on karyotypic alteration was seen, over a 3- to 17-month period.

The cytological studies done to date show that chromosome alterations occur both in callus and in regenerated plants and that changes in chromosome structure are much more common than simple changes in number. A possible mechanism for the changes observed has been suggested (McCoy et al. 1982; Benzion et al. 1986; Phillips et al. 1988). It is known that heterochromatic DNA is normally late replicating, and that such late replication may lead to formation of bridges during anaphase and subsequent breakage. Culture may accentuate this process by causing further delay in replication of heterochromatic areas. The fact that breakpoints seen in vitro often occur between the centromere and heterochromatic knobs is consistent with this interpretation. Chromosome breakage may also be involved in cultured-induced activation of transposable elements (Peschke et al. 1987).

Although cytological alterations and sectoring are common in maize regenerates, they have only rarely been related to phenotypic variations other than pollen sterility. Conversely, many observed phenotypic changes have not been linked to visible cytological changes. In some cases, cytological studies of variants have simply not been performed. In others, however, phenotypic alterations have been shown to involve only a small alteration in the base sequence of nuclear DNA (e.g., Brettell et al. 1986; Dennis et al. 1987), not a visible chromosomal rearrangement.

4 Conclusions and Prospects

Genetic alterations often occur in maize tissue cultures and in plants recovered from them. At the nuclear level, there are reports of single gene changes (recessive and dominant) affecting plant morphology, cellular physiology, and protein structure, changes in agronomic traits usually inherited in a quantitative fashion, as well as alterations in chromosome number and structure. In addition, changes in mitochondrial DNA resulting in reversion from two types of male sterility to fertility have been well documented. Cytological and genetic sectoring is common, both within callus and within regenerated plants.

Many questions remain about the factors that influence the type and frequency of changes seen. In particular, the roles of time in culture, type of culture system (embryogenic vs. organogenic), components of nutrient media, and genotype require further assessment. While some reports indicate that these factors affect stability, the evidence is often equivocal or in conflict with work from other labs with different systems. Clearly, additional well-designed experiments are required to determine the effect of culture parameters on extent and type of variation.

Information presently available suggests that culture-derived changes are not likely to be a major problem in experiments aimed at specific genetic alterations, i.e., transformation with cloned genes. While it is always important to have controls revealing the frequency of spontaneous change in the trait of interest, most regenerated plants grown under favorable conditions should give progeny essentially similar to the initial line. It is also clear that some culture-derived variants may be novel and valuable material for biochemical and molecular studies. Some traits obtained via in vitro selection may lead to breeding lines with improved nutritional quality or resistance to stress.

Less clear is the potential of somaclonal variation (without in vitro selection) as a source of desirable new traits for maize improvement. Although there has been some indication of possibly useful changes in agronomic traits, there are as yet no substantive reports of clearly superior lines incorporating traits derived from culture; some data suggest diminished overall performance. Careful field trials with good statistical treatment of data are required to assess the yield and vigor of lines incorporating variant traits. Such trials are already in progress in many parts of the world, so within a few years it should become apparent whether or not initial optimism about somaclonal variation as a component of modern maize breeding programs (e.g., Earle and Gracen 1985) was well founded.

References

Anderson PC, Georgeson MA (1986) Selection of an imidazolinone-tolerant mutant of corn. In: Somers DA, Gengenbach BG, Biesboer DD, Hackett WP, Green CE (eds) Abstr 6th Int Congr Plant tissue and cell culture. Univ Minn, Minneapolis, p 437

Armstrong CL, Green CE (1985) Establishment and maintenance of friable, embryogenic maize callus and the involvement of L-proline. Planta 164:207–214

Armstrong CL, Phillips RL (1986) Genetic and cytogenetic stability of embryogenic and organogenic maize tissue cultures. In: Somers DA, Gengenbach BG, Biesboer DD, Hackett WP, Green CE (eds) Abstr 6th Int Congr Plant tissue and cell culture, Univ Minn, Minneapolis, p 284

Beckert M, Qing CM (1984) Results of a diallel trial and a breeding experiment for in vitro aptitude in maize. Theor Appl Genet 68:247–251

Beckert M, Pollacsek, Caenen M (1983) Etude de la variabilité génétique obtenue chez le mais après callogenèse et régénération de plantes in vitro. Agronomie 3:9–18

Benzion G (1984) Genetic and cytogenetic analysis of maize tissue cultures: a cell line pedigree analysis. Ph D Thesis, Univ Minn, St Paul

Benzion G, Phillips RL, Rines HW (1986) Case histories of genetic variability in vitro: oats and maize. In: Vasil IK (ed) Cell culture and somatic cell genetics of plants, vol 3. Academic Press, New York, London, pp 435–448

Breth SA (1986) Mainstreams of CIMMYT Research: a retrospective. CIMMYT, Mexico, DF, p 12

Brettell RIS, Thomas E, Ingram DS (1980) Reversion of Texas male-sterile cytoplasm of maize in culture to give fertile, T-toxin-resistant plants. Theor Appl Genet 58:55–58

Brettell RIS, Dennis ES, Scowcroft WR, Peacock WJ (1986) Molecular analysis of a somaclonal mutant of maize alcohol dehydrogenase. Mol Gen Genet 202:235–239

Brown PTH, Lörz (1986) Methylation changes in the progeny of tissue culture derived maize plants. In: Somers DA, Gengenbach BG, Biesboer DD, Hackett WP, Green CE (eds) In: Abstr 6th Int Congr Plant tissue and cell culture, Univ Minn, Minneapolis, p 261

Chourey PS, Kemble RJ (1982) Transposition event in tissue cultured cells of maize. In: Fujiwara A (ed) Plant tissue culture 1982. Maruzen, Tokyo, pp 425–426

Chu CC, Wang CC, Sun CS, Hsu C, Yin CY, Chu CY (1975) Establishment of an efficient medium for anther culture of rice through comparative experiments on the nitrogen sources. Sci Sin 18:659–668

Dennis ES, Brettell RIS, Peacock WJ (1987) A tissue culture induced *Adh1* null mutant of maize results from a single base change. Mol Gen Genet 210:181–183

Dewey RE, Timothy DH, Levings CS III (1986) Novel recombinations in the maize mitochondrial genome produce a unique transcriptional unit in the Texas male-sterile cytoplasm. Cell 44:439–444

Dewey RE, Timothy DH, Levings CS III (1987) A mitochondrial protein associated with cytoplasmic male-sterility in the T cytoplasm of maize. Proc Natl Acad Sci USA 84:5374–5378

The literature review for this chapter was completed in October, 1987.

Donovan CM, Somers DA (1986) Genotype effects on stability of friable, embryogenic callus and suspension cultures of *Zea mays*. In: Somers DA, Gengenbach BG, Biesboer DD, Hackett WP, Green CE (eds) In: Abstr 6th Int Congr Plant tissue and cell culture, Univ Minn, Minneapolis, p 183
Duncan DR, Widholm JM (1988) Improved plant regeneration from maize callus cultures using 6-benzylaminopurine. Plant Cell Rep 7:452–455
Duncan DR, Williams ME, Zehr BE, Widholm JM (1985) The production of callus capable of plant regeneration from immature embryos of numerous *Zea mays* genotypes. Planta 165:322–332
Earle ED, Gracen VE (1985) Somaclonal variation in progeny of plants from corn tissue cultures. In: Henke RR, Hughes KW, Constantin MJ, Hollaender A (eds) Tissue culture in forestry and agriculture. Plenum, New York, pp 139–152
Earle ED, Gracen VE, Best VM, Batts LA, Smith ME (1987) Fertile revertants from S-type male-sterile maize grown in vitro. Theor Appl Genet 74:601–609
Edallo S, Zucchianali C, Perenzin M, Salamini F (1981) Chromosomal variation and frequency of spontaneous mutation associated with in vitro culture and plant regeneration in maize. Maydica 26:39–56
Fahey JW, Reed JN, Readdy TL, Page GM (1986) Somatic embryogenesis from three commercially important inbreds of *Zea mays*. Plant Cell Rep 5:35–38
Fauron CM-R, Abbot AG, Brettell RIS, Gesteland RF (1987) Maize mitochondrial DNA rearrangements between the normal type, the Texas male sterile cytoplasm, and a fertile revertant cms-T regenerated plant. Curr Genet 11:339–346
Gengenbach BG, Green CE (1975) Selection of T-cytoplasm maize callus cultures resistant to *Helminthosporium maydis* race T pathotoxin. Crop Sci 15:645–649
Gengenbach BG, Green CE, Donovan CM (1977) Inheritance of selected pathotoxin resistance in maize plants regenerated from cell cultures. Proc Natl Acad Sci USA 74:5113–5117
Göbel E, Lörz H (1986) Somaclonal variation in tissue culture derived maize plants and their selfed progeny. In: Somers DA, Gengenbach BG, Biesboer DD, Hackett WP, Green CE (eds) In: Abstr 6th Int Congr Plant tissue and cell culture, Univ Minn, Minneapolis, p 284
Green CE (1977) Prospects for crop improvement in the field of cell culture. HortSci 12:131–137
Green CE (1982) Somatic embryogenesis and plant regeneration from friable callus of *Zea mays*. In: Fujiwara A (ed) Plant tissue culture 1982. Maruzen, Tokyo, pp 107–108
Green CE, Phillips RL (1975) Plant regeneration from tissue cultures of maize. Crop Sci 15:417–421
Green CE, Armstrong CL, Anderson PC (1983) Somatic cell genetic systems in corn. In: Downey K, Voellmy RW, Fazelahmad A, Schultz J (eds) Advances in gene technology: molecular genetics of plants and animals. Miami Winter Symp Ser, vol 20. Academic Press, New York London, pp 147–157
Gregory P, Earle ED, Gracen VE (1977) Biochemical and ultrastructural aspects of southern corn leaf blight disease. Am Chem Soc Symp 62:90–114
Hibberd KA, Green CE (1982) Inheritance and expression of lysine plus threonine resistance selected in maize tissue culture. Proc Natl Acad Sci USA 79:559–563
Hibberd KA, Walter T, Green CE, Gengenbach BG (1980) Selection and characterization of a feedback-insensitive tissue culture of maize. Planta 148:183–187
Hibberd KA, Barker M, Anderson PC, Linder L (1986) Selection for high tryptophan maize. In: Somers DA, Gengenbach BG, Biesboer DD, Hackett WP, Green CE (eds) In: Abstr 6th Int Congr Plant tissue and cell culture, Univ Minn, Minneapolis, p 440
Hodges TK, Kamo KK, Becwar MR, Schroll S (1985) Regeneration of maize. In: Zaitlin M, Day P, Hollaender A (eds). Biotechnology in plant science: relevance to agriculture in the eighties. Academic Press, New York London Orlando, pp 15–33
Hodges TK, Kamo KK, Imbrie C, Becwar MR (1986) Genotype specificity of somatic embryogenesis and regeneration in maize. Biotechnology 4:219–223
Hooker AL, Smith DR, Lim SM, Beckett JB (1970) Reaction of corn seedlings with male-sterile cytoplasm to *Helminthosporium maydis*. Plant Disease Rep 54:708–712
Humaydan HS, Scott EW (1977) Methomyl insecticide selective phytotoxicity on sweet corn hybrids and inbreds having the Texas male sterile cytoplasm. HortSci 12:312–313
Imbrie-Milligan CW, Hodges TK (1986) Microcallus formation from maize protoplasts prepared from embryogenic callus. Planta 168:395–401
Kamo KK, Hodges TK (1986) Establishment and characterization of long-term, embryogenic maize callus and cell suspension cultures. Plant Sci 45:111–117

Kamo KK, Becwar MR, Hodges TK (1985) Regeneration of *Zea mays* from embryogenic callus. Bot Gaz 146:327–334
Kamo KK, Chang KL, Lynn ME, Hodges TK (1987) Embryogenic callus formation from maize protoplasts. Planta 172:245–251
Kuehnle AR, Earle ED (1987) In vitro selection for methomyl resistance in cms-T. Maize Genet Coop Newslett 61:59–60
Larkin PJ (1985) In vitro culture and cereal breeding. In: Bright SWJ, Jones MGK (eds) Cereal tissue and cell culture. Nijhoff/Junk, Dortrecht, pp 273–296
Larkin PJ (1987) Somaclonal variation: history, method, and meaning. Iowa State J Res 61:393–434
Larkin PJ, Scowcroft WR (1981) Somaclonal variation – a novel source of variability from cell culture for plant improvement. Theor Appl Genet 60:197–214
Laughnan JR, Gabay-Laughnan ST (1983) Cytoplasmic male-sterility in maize. Annu Rev Genet 17:27–48
Lee M, Phillips RL (1987) Genomic rearrangements in maize induced by tissue culture. Genome 29:122–128
Lowe K, Taylor DB, Ryan P, Paterson KE (1985) Plant regeneration via organogenesis and embryogenesis in the maize inbred line B73. Plant Sci 41:125–132
Lu C, Vasil IK, Ozias-Akins P (1982) Somatic embryogenesis in *Zea mays* L. Theor Appl Genet 62:109–112
Lu C, Vasil V, Vasil IK (1983) Improved efficiency of somatic embryogenesis and plant regeneration in tissue cultures of maize (*Zea mays* L.). Theor Appl Genet 66:285–289
Lupotto E (1986) In vitro culture of isolated somatic embryos of maize (*Zea mays* L.). Maydica 31:193–201
McCain JW, Hodges TK (1986) Anatomy of somatic embryos from maize embryo cultures. Bot Gaz 147:453–460
McCoy TJ, Phillips RL (1982) Chromosome stability in maize (*Zea mays*) tissue cultures and sectoring in some regenerated plants. Can J Genet Cytol 24:559–565
McCoy TJ, Phillips RL, Rines HW (1982) Cytogenetic analysis of plants regenerated from oat (*Avena sativa*) tissue cultures: high frequency of partial chromosome loss. Can J Genet Cytol 24:37–50
Miao SH, Duncan DR, Widholm JM (1986) Selection of lysine plus threonine and 5-methyltryptophan resistance in maize tissue culture. In: Somers DA, Gengenbach BG, Biesboer DD, Hackett WP, Green CE (eds) In: Abstr 6th Int Congr Plant tissue and cell culture. Univ Minn, Minneapolis, p 380
Murashige T, Skoog F (1962) A revised medium for rapid growth and bioassays with tobacco tissue cultures. Physiol Plant 15:473–497
Neuffer MG, Jones L, Zuber MS (1968) The mutants of maize. Crop Sci Soc Am, Madison, WI
Novak FJ, Dolezelova M, Nesticky M, Piovarci A (1983) Somatic embryogenesis and plant regeneration in *Zea mays* L. Maydica 28:381–390
Novak FJ, Hermelin T, Daskalov S, Nesticky M (1986) In vitro mutagenesis in maize. In: Horn W, Jensen CJ, Odenbach W, Schieder O (eds) Genetic manipulation in plant breeding. DeGruyter, Berlin New York, pp 563–576
Peschke VM, Phillips RL, Gengenbach BG (1987) Discovery of transposable element activity among progeny of tissue culture-derived maize plants. Science 238:804–807
Phillips RL, Somers DA, Hibberd KA (1988) Cell/tissue culture and in vitro manipulation. In: Sprague G, Dudley JW (eds) Corn and corn improvement, 3rd edition. Am Soc Agronomy, Madison, WI, pp 346–387
Rhodes CA, Lowe KS, Ruby KL (1988) Plant regeneration from protoplasts isolated from embryogenic maize cell cultures. Biotechnology 6:56–60
Rhodes CA, Phillips RL, Green CE (1986) Cytogenetic stability of aneuploid maize tissue cultures. Can J Genet Cytol 28:374–384
Rice TB (1982) Tissue culture induced genetic variation in regenerated maize inbreds. In: Proc 37th Annu Corn and sorghum research Conf, Chicago, pp 148–162
Shaner DL, Anderson PC (1985) Mechanism of action of the imidazolinones and cell culture selection of tolerant maize. In: Zaitlin M, Day P, Hollaender A (eds) Biotechnology in plant science: relevance to agriculture in the eighties. Academic Press, New York London Orlando, FL, pp 287–299
Sheridan WF (ed) (1982) Maize for biological research. Plant Mol Biol Assoc, Charlottesville, VA
Shimamoto K, Nelson OE (1981) Isolation and characterization of aminopterin-resistant cell lines in maize. Planta 153:436–442

Small ID, Earle ED, Escote-Carlson LJ, Gabay-Laughnan S, Laughnan JR, Leaver CJ (1988) A comparison of cytoplasmic revertants to fertility from different cms-S maize sources. Theor Appl Genet 76:609–618

Sprague GF (ed) (1977) Corn and corn improvement. Agron Soc Am, Madison, WI, pp 49–84

Springer WD, Green CE, Kohn KA (1979) A histological examination of tissue culture initiation from immature embryos of maize. Protoplasma 101:269–281

Tomes DT, Smith OS (1985) The effect of parental genotype on initiation of embryogenic callus from elite maize (*Zea mays* L.) germplasm. Theor Appl Genet 50:505–510

Tuberosa R, Phillips RL (1986) Isolation of methotrexate-tolerant cell lines of corn. Maydica 31:215–225

Umbeck PF, Gengenbach BG (1983) Reversion of male-sterile T-cytoplasm maize to male fertility in tissue culture. Crop Sci 23:584–588

United States Department of Agriculture (USDA) (ed) (1986) Agricultural statistics. Maize. US Gov Print Off, Washington, DC, pp 35–36

Vasil V, Vasil IK (1986) Plant regeneration from friable embryogenic callus and suspension cultures of *Zea mays* L. J Plant Physiol 124:399–408

Vasil V, Vasil IK (1987) Formation of callus and somatic embryos from protoplasts of a commercial hybrid of maize (*Zea mays* L.). Theor Appl Genet 73:793–798

Vasil V, Vasil IK, Lu C (1984) Somatic embryogenesis in long term cultures of *Zea mays* L. (Gramineae). Am J Bot 71:158–161

Wang AS, Hollingworth MD, Millic JB (1987) Mutagenesis of tissue cultures. Maize Genet Coop News Lett 61:81–83

Wolf SJ, Earle ED, Macko V (1987) Differential effects of HC toxin on susceptible and resistant corn callus. In: Abstr 38th Annu Meet Tissue culture Assoc, Washington DC

Zehr BE, Williams ME, Duncan DR, Widholm JM (1987) Somaclonal variation in the progeny of plants regenerated from callus cultures of seven inbred lines of maize. Can J Bot 65:491–499

II.4 Somaclonal Variation in Barley (*Hordeum vulgare* L.)

A. BREIMAN and D. ROTEM-ABARBANELL[1]

1 Introduction

Barley is the world's fourth most important cereal crop after wheat, maize, and rice. Three unique characteristics have enabled barley to persist as a major cereal crop through many centuries: (1) broad ecological adaptation, (2) utility as a feed and food grain, and (3) superiority of barley malt for use in brewing.

Barley is grown over a broader ecological environmental range than any other cereal and is therefore, produced in regions with unfavorable climates for the production of the other major cereals. It is grown as a crop in most parts of the world (Table 1), almost two-thirds of the world's production being in semi-humid or semi-arid regions, yet it responds well, yieldwise, to additional rainfall or irrigation water. Barley grain is used as feed for animals, malt, and as food for human consumption. The first and largest use of the grain is for animal feed. Proteins content varies from 10–15%, depending on the climate and soil conditions under which barley is grown.

The second largest use of barley is for malt. Germinating seeds produce two enzymes, alpha-amylase and beta-amylase, which hydrolyze starch to dextrose and fermentable enzymes (Peterson and Forster 1973; Dickson 1979). Barley, wheat, and rye are unique in producing both enzymes. The choice of barley for malt over wheat and rye results from a number of circumstances, including the presence of the husk on barley, which protects the acrospine during germination, the use of the husk as a filtration aid, the firm texture of the steeped barley kernel and tradition (Dickson 1979).

Finally, barley is used for human food in regions where other cereals do not grow well due to altitude, latitude, low rainfall or soil salinity (Poehlman 1985). In addition to use of the grain, the straw is used for animal bedding, and immature barley plants may be harvested for forage by grazing or by cutting for hay or silage.

1.1 Objectives for Improvement

The development of superior barley cultivars is part of a continuing dynamic process, which has achieved ever-increasing production per unit area. High-

[1]Department of Botany, The George S. Wise Faculty of Life Sciences, Tel Aviv University, Tel Aviv 69978, Israel

Biotechnology in Agriculture and Forestry, Vol. 11
Somaclonal Variation in Crop Improvement I (ed. by Y.P.S. Bajaj)

Table 1. Barley production in 1986 . (Data from FAO 1986)

Continent/country	Production 1000 mt
World	180.441
Africa	6.291
Morocco	3.563
N.C. America	28.733
Canada	15.026
S. America	824
Argentina	190
Asia	18.730
Turkey	7.000
Europe	70.163
France	10.063
Australia, New Zealand and Oceania	8.602
USSR	51.400

yielding cultivars, greater disease resistance, better straw strength, improvement of malt and feed are quality parameters of barley, as well as resistance to environmental stresses (i.e., salt tolerance, mineral stresses). These are, therefore, major objectives for barley breeding (Anderson and Reinbergs 1985).

A major effort is being invested to produce barley lines which are resistant to the four major diseases: spot blotch (*Helminthosporium sativum*), net blotch (*Puccinia graminis* f. sp. *tritici*), smuts (*Ustilago* sp.) and more recently, barley yellow dwarf virus (Kiesling 1985).

Substantial efforts are made to improve two-rowed and six-rowed malting and feed cultivars in the Pacific Northwest of the USA. Two-rowed barley with higher malt extract and enzyme levels and six-rowed cultivars with higher enzyme levels are important objectives. Development of proaonthocyanidin-free barley is being attempted to remove the need for chill-proofing in brewing (Larsen 1981). Improvements in feed barley are being sought in yield, lysine and protein content, and energy per unit area.

The breeding objectives are divided into three categories; the short-term objectives are usually to obtain high-yielding cultivars, whereas the intermediate and long-term objectives include development of germplasm that has potential in the future, i.e., synthesis of composites, incorporation of stable resistance to disease, introduction of exotic germplasm, and selection for adaptation (Dewey 1971; Schooler and Anderson 1980).

1.2 Available Genetic Variability

Spontaneous and induced genetic variation within barley has been found to be extensive (Hockett and Nilan 1985). This plant material is widely used for developing cultivars, increasing knowledge of the inheritance and genetic control of a given trait. The spontaneous variation is organized in collections of germplasm. These collections have been analyzed to determine the variation and distribution of a number of important characteristics, including disease, insect and stress resistance, yield, straw strength, early maturity, feed and malting quality components (Moseman and Smith 1985). The primary objective of any germplasm resource program is to ensure continued availability of genetically diverse germplasm with the characteristics required for developing productive and high-quality cultivars.

Although spontaneous variants have long been used for developing new and improved cultivars, induced mutants have played an increasingly important role in cultivar development (Nilan 1981b). There are several traits where induced mutations have produced alleles and form of traits not found among the spontaneous variants (Nilan 1981a,b). A striking example involves the waxy coating of the plant, where > 1300 *eceriferum* mutants have been induced. Spontaneous variation for this trait revealed six controlling loci, whereas induced variation produced many new phenotypes and revealed 77 loci, some possessing 15 to 20 alleles (Lundqvist 1976).

In addition to spontaneous and induced variation, somaclonal and gametoclonal variation are technologies that permit short-term accomplishment of breeding objectives. However, these technologies do not displace conventional plant breeding, but provide an additional and more rapid process of plant improvement (Evans et al. 1984).

2 Barley Tissue Culture

2.1 In Vitro Regeneration of Barley Plants

Plant regeneration through tissue culture has been demonstrated in many wild and cultivated barley species, as well as in inter- and intraspecific hybrids (Table 2, see also Koblitz 1986). In general, the extent of plant regeneration has been found to be affected mainly by: (1) explant, (2) genotype, and (3) in vitro conditions.

2.1.1 Effect of Explant

The physiological state of the explant appears to be a key point in regenerating plants from barley tissue culture. Immature organs and meristematic tissues have been found to be more successful than mature organs for plant regeneration (Dale and Deambrogio 1979; Goldstein and Kronstad 1986; Rotem-Abarbanell and Breiman 1989). Studies using whole immature embryos, as well as the scutellar region of the embryo, have been successful in callus induction and plant regeneration. Apical meristems, basal regions of leaves, young inflorescences and spikes, in addition to immature ovaries, have demonstrated the ability of plant regeneration (Table 2).

Whereas the regeneration capacity of calli derived from immature embryos was 100% (Lührs and Lörz 1987; Rotem-Abarbanell and Breiman 1989) plant regeneration from other immature organs, such as young inflorescences, was less abundant and did not exceed 50% (Jorgensen et al. 1986). In contrast, seeds, mature embryos, and leaves displayed limited capacity for callus induction and plant

Table 2. The effect of explant source and in vitro conditions on barley plant regeneration

Species	Explant	Pathway of regeneration	Treatment for plant regeneration Basal medium	Hormones	Additives	Reference
H. vulgare	Apical Meristem	OR/SE	MS CS	None	–	Weigel and Hughes (1983a.b. 1985)
"	"	"	CS	None	–	Cheng and Smith (1975)
"	Scutellum + apical Meristem	SE	MS	3 μM TIBA	–	Rengel and Jelaska (1986a)
"	Mesocotyl	OR	MS	3 μM TIBA or 19.5 μM 2,4,5-T	–	Jelaska et al. (1984)
"	Leaf	OR	MS	10 mg/l 2,4-D	Barley seedling extract (150 g/l)	Saalbach and Koblitz (1978)
"	Inflor IE	SE	MS	10 μM 2,4-D		Thomas and Scott (1985)
"	Immature ovaries	OR	B5	0.3 mg/l kinetin + 1 mg/l GA_3		Orton (1979)
"	ME	OR	MS	0.5–5 mg/l 2,4-D		Lupotto (1984)
"	"	"	MS (modified)	3 μM TIBA		Rengel (1987)
"	"	"	B5	10–15 μM 2,4-D		Ukai and Nishimura (1987)
"	IE	"	MS B5	0–1 mg/l 2,4-D	2 g/l yeast extract	Hanzel et al. (1985)
"	"	"	MS	None		Dale and Deambrogio (1979)
"	"	"	B5	2.5–5 mg/l Cl_3POP		Goldstein and Kronstad (1986)
"	IE whole or sections	SE	MS CC	0.05 mg/l zeatin + 1mg/l IAA		Luhrs and Lörz (1987)
"	IE	OR	1/2 MS	0.5 mg/l 2,4-D + 1 mg/l zeatin		Ahloowalia (1987)
"	Seeds	OR	B5	10–30 μM 2,4-D		Ukai and Nishimura (1987)
H. jubatum	Immature ovaries	OR	B5	0.3 mg/l kinetin + 1 mg/l GA_3		Orton (1979)
H. vulgare × *H. jubatum*	"	OR	B5	- " -	"	"

(to be continued)

Table 2. *(Continued)*

Species	Explant	Pathway of regeneration	Treatment for plant regeneration: Basal medium	Treatment for plant regeneration: Hormones	Additives	Reference
H. vulgare × *H. brachyantherum*	Young spikes	OR/SE(?)	J/25–8	None	120 mg/l L-glutamine. 50 mg/l L-aspargine 25 mg/l L-threonine 25 mg/l L-arginine 50 mg/l L-proline 25 mg/l cm	Jorgensen et al. (1986)
H. roshevitzii						
H. jubatum						
H. tetraplodium						
H. brevisubulatum						
H. lechleri						
H. procerum						
H. spontaneum	IE	OR	VKM	1 mg/l IAA + 1 mg/l zeatin	50 m/l cm	Breiman (1985)
H. bulbosum	IE	OR	VKM	- ″ -	- ″ -	Breiman (1985)
H. marinum	IE ME	OR (SE?)	MS	- ″ - or none		Rotem-Abarbanel and Breiman (1989)
H. distichum	ME	OR	MS B5	None		Katoh et al. (1986)
H. vulgare × *T. aestivum*	Inflor	SE	N6	0–0.1 mg/l 2,4-D		Chu et al. (1984)
″	IE	OR	MS (modified)	None	150 mg/l L-aspargine	Li et al. (1985)
H. vulgare × *T. crassum*	Inflor	SE	KaO (modified)	0.5 mg/l 2,3D + 0.2 mg/l GA_3	125 ml/l cm	Fedak (1985)
H. procerum × *S. vavilovii*	″	″	″	″	″	Fedak (1985)
H. procerum × *S. dalmaticum*	″	″	″	″	″	Fedak (1985)
H. pechleri × *S. cereale*	″	″	″	″	″	Fedak (1985)
H. jubatum × *S. dalmaticum*	″	″	″	″	″	″

Abbreviations

IE = immature embryo
ME = mature embryo
Inflor = inflorescence
OR = organogenesis
SE = somatic embryogenesis
MS = Murashige and Skoog (1962)

CS = Cheng and Smith (1975)
B5 = Gamborg et al. (1968)
CC = Potrykus et al. (1979)
N6 = Chu et al. (1975)
VKM = Binding (1974); Kao and Michalyuk (1975)
Kao = Kao (1977)

TIBA = 2,3,5-triiodobenzophenoxy acetic acid
2,4-D = 2,4-dichlorophenoxy acetic acid
GA_3 = gibberelic acid
Cl_3POP = 2,4,5-trichlorohenxypropionic acid
IAA = indole acetic acid
CM = coconut milk

regeneration (Saalbach and Koblitz 1978; Lupotto 1984; Katoh et al. 1986; Ukai and Nishimura 1987).

2.1.2 Genotype Effect

A relationship between plant genotype and in vitro response, particularly plant regeneration, has been reported in several studies (Hanzel et al. 1985; Goldstein and Kronstad 1986; Jorgensen et al. 1986; Rengel and Jelaska 1986a; Lührs and Lörz 1987). One such extensive study on 91 *H. vulgare* cultivars was performed in order to evaluate, not only the initiation of calli and their growth ability, but also the regeneration capacity of the different genotypes (Hanzel et al. 1985). In this study, only six genotypes demonstrated capacity for plant regeneration. Although variation in the morphogenic capability was observed among the different cultivars, the overall regeneration rate was low (15%). Table 3 illustrates some variations in regeneration rates of several *H. vulgare* cultivars, as well as of genotype combinations (interspecific hybrids of *Hordeum*). Whereas some cultivated barley genotypes, such as Dissa and Maris-Mink had a high regeneration rate (100%), interspecific hybrids yielded lower regeneration rates (up to 50%).

The relationship between genotype and regeneration rate leads to the speculation that the morphogenic capacity is regulated by specific genes. In order to establish unequivocal proof for the involvement of the genetic background in regeneration, it is necessary to introduce aneuploid and alloplasmic lines of barley

Table 3. Genotype effect on barley plant regeneration

Species (cultivar) or hybrid	Regenerative calli (%)
H. vulgare	
Dissa	100 [a]
Golden Promise	88.5[a]
Local Sebergelon	20 [a]
Maris Mink	100 [a]
Midas	79.3[a]
Ricotense	42.9[a]
H. vulgare × *H. bulbosum*	0 [b]
H. vulgare cv. Welman × *H. brachyantherum*	7 [b]
H. vulgare cv. Riso × *H. roshevitzii*	24 [b]
H. vulgare cv. Vogelsanger Gold × *H. lechleri*	50 [b]
H. vulgare cv. Riso × *H. procerum*	21 [b]
H. vulgare	
Slavonac	70 [c]
Robur	43 [c]
Alkar	69 [c]
Sokal	71 [c]

[a] Lührs and Lörz (1987) Regenerative calli: percentage of calli giving rise to plants in soil out of total calli transferred to regeneration media.
[b] Jorgensen et al. (1986) Regenerative calli: number of explants producing regenerative calli as percent of total number of explants with callus formation.
[c] Rengel and Jelaska (1986a) Regenerative calli as in [a].

and to investigate the nuclear and cytoplasmic control of regeneration. Such studies have been recently performed in *Triticum aestivum*, demonstrating that chromosomes 4B and 2BS are essential for regeneration (Mathias and Fukui 1986; Felsenburg et al. 1987)

2.1.3 *Effect of In Vitro Conditions*

Callus induction and maintenance, as well as plant regeneration, have been investigated in a variety of in vitro conditions. Several basal media, hormonal treatments, and amino acid additions have been tried, in order to optimize plant regeneration from barley cultures. Among the various media tested, the Murashige and Skoog (1962) nutrient medium has been most commonly used (Rengel and Jelaska 1986a,b; Katoh et al. 1986; Rengel 1987). Nevertheless, several other media, such as B5 (Gamborg et al. 1968), N6 (Chu et al. 1975) and C5 (Cheng and Smith 1975) have also been found to be suitable for barley cultures (Chu et al. 1984; Breiman 1985; Lührs and Lörz 1987).

A large number of growth regulators have been used in order to optimize callus and plant regeneration (Table 2). In most studies 2,4-D (2,4-dichlorophenoxy acetic acid) has been used for the initiation of calli, in concentrations ranging from 1 mg/l to 5 mg/l (Orton 1979; Lupotto 1984; Jorgensen et al. 1986). Nevertheless, other auxins and auxin-like compounds have been successfully used for initiating calli (Hanzel et al. 1985; Goldstein and Kronstad 1986; Lührs and Lörz 1987). For the induction of plant regeneration, three major hormonal treatments have been used: (1) omitting the growth regulator (2) decreasing the 2,4-D concentration, and (3) adding auxin in combination with cytokinin (Table 2). An exhaustive study on the effect of growth regulators on the induction, maintenance, and regeneration of barley has recently been published (Lührs and Lörz 1987). The authors concluded that, except for NAA (naphthalene acetic acid), all other auxins used in their study were capable of inducing embryogenic calli.

Addition of amino acids to the media was found in one case to improve somatic embryogenesis (Rengel and Jelaska 1986b), while in another study it had no effect (Lührs and Lörz 1987).

2.2 Cell Suspensions and Protoplast Culture

In general, the maceration enzymes and isolation methods of barley protoplasts are similar to those of other cereals (Vasil 1987). Enzyme concentrations for protoplast isolation from cell suspensions are approximately ten-fold higher than for their isolation from leaves.

In contrast to the ease with which dicotyledonous protoplasts isolated from leaves have been capable of regeneration, neither regeneration nor cell division have been achieved from leaf protoplasts of barley (Table 4). Only when the protoplasts have been isolated from calli or cell suspensions has cell division occurred (Ishii et al. 1987). In a recent study, transient expression of a chimeric gene has been described in barley protoplasts. In this study, 2 days after DNA uptake in the presence of PEG (polyethylene glycol), the barley protoplasts extract exhibited the expression of neomycin phosphotransferase (Junker et al. 1987).

Since cell suspensions are the only source of dividing barley protoplasts, the most critical factor in the success of plant regeneration from protoplast is the use of embryogenic suspension cultures. As the initial cell cultures used in most studies have been nonmorphogenic (Table 4), the protoplast-derived calli have also failed to form shoots or plants. In fact, there is only one study describing shoot regeneration from cell suspensions of *H. vulgare* (Kott and Kasha 1984). It is noteworthy that regeneration achieved in this embryogenic system was not from single cells, but rather from homogenized cell aggregates. Although a few faint green shoots were formed, most of the regenerated plants were albino. Other nonregenerable barley cell suspensions have been established. Both friable and watery calli have been successfully used for obtaining single cells and cell aggregates in suspension (Dale and Deambrogio 1979; Seguin-Swartz et al. 1984).

In order to achieve genetically manipulated barley plants through protoplast culture, one has to overcome the obstacle of nonregenerable cell suspensions. The success in plant regeneration from rice protoplast cultures which has been recently reported (Coulibaly and Demarly 1986; Yamada et al. 1986; Ogura et al. 1987) encourage further studies and the possibility for plant regeneration from barley protoplasts.

Table 4. Isolation and culture of barley protoplast

Species	Source of protoplasts	Cell division/ regeneration	Reference
H. vulgare	Leaves	None	Evans et al. (1972)
"	"	"	Wakasha (1973)
"	"	"	Lörz et al. (1976)
"	"	"	Hughes et al. (1978)
"	Callus	Callus	Koblitz (1976)
"	Cell suspensions	Roots	Ishii et al. (1987)
"	"	Cell division DNA uptake	Junker et al. (1987)
"	"	Colonies	Takahashi and Kaneko (1986)
"	"	Albino plantlets	Lührs and Lörz (1988)

2.3 Anther Culture

Barley haploid plants can be produced by several methods: anther (microspore) culture, (Clapham 1973; Foroughi-Wehr et al. 1976; Wei et al. 1986), the Bulbosum technique (Kasha and Kao 1970), ovary culture (San Noeum 1976; Orton 1980; Wang and Kuang 1981) and the utilization of genetic factors, i.e., hap gene (Hagberg and Hagberg 1980).

The most commonly used methods for producing barley haploid plants are anther culture and the Bulbosum method. The two methods are comparable in their efficiency and can be complementarily used (Devaux 1987). However, apparently more variation is observed among plants obtained by anther culture than by the Bulbosum method (Powell et al. 1986a; Snape et al. 1988).

There are several major factors which affect anther cultures; (1) influence of the genotype, (2) physiological state of the donor plant, (3) developmental stage of the microspore, (4) pretreatment of spikes and anthers, (5) conditions for in vitro culture.

2.3.1 *Influence of the Genotype*

A number of studies have been conducted in order to analyze the genetic components affecting the anther response in tissue culture (Foroughi-Wehr et al. 1982; Kott et al. 1986; Dunwell et al. 1987; Powell 1987; Hamachi et al. 1988). In one study, anthers of 55 different barley (*Hordeum vulgare*) and four varieties were cultured in vitro. The microspore responsiveness is a heritable character involving at least two factors: the ability of microspore within anthers to divide and form calli, and the ability of calli for morphogenesis to yield plants (Foroughi-Wehr et al. 1982).

A recent study on anther culture response revealed that the sampling of material for the anther culture plays an important role in producing variation, since there is a significant variability between spikes within plants (Dunwell et al. 1987). It was concluded that spike number on the plant affects frequency of callus production. Since two-rowed genotypes (Sabarlis) have a higher tillering capacity than six-rowed genotypes (Dissa), it may be possible to correlate the decline in productivity of anther cultured from older Dissa spikes to this phenomenon (Dunwell et al. 1987).

2.3.2 *The Physiological Status of the Donor Plant*

Critical environmental factors are light intensity, photoperiod, temperature, nutrition and CO_2 concentration. Plants grown at permanent 16–21 °C (Kasha 1987) or at 8–12 °C up to the tillering stage, and between 16–25 °C in the subsequent stages gave a good response (Foroughi-Wehr et al. 1982). Plants growing outdoors in the natural growing season are more responsive than greenhouse-grown material, and a higher frequency of callus production was obtained from plants grown at high light intensities (18–20,000 lx) (Foroughi-Wehr and Mix 1979).

2.3.3 *Developmental Stage of the Microspore*

Anthers with microspores ranging from tetrad to binucleate stage are the most responsive. Barley belongs to the class of plants whose anthers respond best when cultured after completion of meiosis, but before the first mitotic division (Sunderland et al. 1974, 1979). The induction of uninucleate microspores to form sporophytic embryoids or calli has been studied in detail in *Hordeum vulgare* (Sunderland and Dunwell 1977; Sunderland et al. 1979; Sunderland and Evans 1980). Anthers of *H. vulgare* cv. Klages bearing microspores in the mid-uninucleate stage, which are the most responsive in culture, were determined to be in the G1 of the cell cycle (Wheatley et al. 1986).

2.3.4 Pretreatment of Anthers

Certain treatments given to the whole spike or to the anthers can have a positive effect on microspore development. The most effective technique used in anther culture is cold pretreatment. A period of 21–25 days at 4°C or 14–21 days at 7°C was recommended for cut tillers (Sunderland et al. 1981). Cold pretreatment may cause an unspecific shock, resulting in the establishment of endogenous cellular conditions, which favor the development of microspores. When plants were grown at higher temperatures (16–21°C), however, the cold pretreatment was not beneficial (Kasha 1987).

2.3.5 Conditions for in Vitro Culture

Barley anthers can be grown either on solid or in liquid media (Foroughi-Wehr et al. 1976; Sunderland et al. 1979). Culture media factors are numerous and are associated with culture type and genotype.

Several defined compounds, such as the addition of glutamine and inositol, to basal media [MS (Murashige and Skoog 1962) B5 (Gamborg et al. 1968)] as well as high sucrose concentration (6–15 g/l) or starch were found to be beneficial for barley anther culture (Kao 1981; Xu and Sunderland 1981; Olsen 1987; Sorvari and Schieder 1987). Chemically undefined extracts of barley anthers or potato, sometimes named "conditioning factors", were used as stimulatory factors for anther response (Xu and Sunderland 1981; Sunderland et al. 1981).

The major advantage of liquid media is the possibility of gradual replenishment and modification of the medium to suit the changing growth requirements (Kao 1981; Xu et al. 1981; Marsolais and Kasha 1985). Addition of Ficoll to liquid media was found beneficial, since it kept the anthers floating and therefore increased their survival rate (Kao 1981). Finally, several hormones in various combinations were added to the media, 2,4-D, IAA and zeatin riboside, the 2,4-D being the most effective auxin (Sunderland et al. 1981; Marsolais and Kasha 1985).

Recently, success has been reported in culturing isolated pollen of *H. vulgare* to form calli and subsequently, to regenerate plants (Wei et al. 1986; Datta and Wenzel 1988; Balik et al. 1989). The frequency of shoot regeneration from pollen calli in the former study was 60% and the ratio of green plants to albino was 1:2.2.

3 Variation in the Callus and Regenerated Plants
(see also Vazquez and Ruiz 1986)

The term somaclonal variation was coined by Larkin and Scowcroft in 1981 to define variation among plants regenerated from cultured cells or tissue (Larkin and Scowcroft 1981). Since then, somaclonal variation has been described in a range of plant species, including cereals; maize (Green et al. 1977; Edallo et al. 1981; Peschke et al. 1987), wheat (Larkin et al. 1984; Maddock et al. 1985; Cooper et al. 1986; Breiman et al. 1987a; Ryan et al. 1987), rice (Sun et al. 1983; Ogura et al. 1987), and barley (Orton 1980; Dunwell et al. 1987; Karp et al. 1987; Breiman et al. 1987b). The main interest enhancing the research of somaclonal variation in mono- and dicots has been the claim that this phenomenon can provide an effective source for generation and recovery of genetic variability for crop improvement.

3.1 Types of Variation

The most thoroughly analyzed and familiar changes found in plants regenerated from tissue culture, except for morphological traits, are chromosomal instability and modification of ploidy. It has been suggested that aneuploidy is more readily tolerated in polyploid species. Gross chromosomal changes, such as changes in chromosome number and chromosomal re-arrangements, were reported in wheat (Karp and Maddock 1984), oats (McCoy et al. 1982) and barley (Orton 1980). In addition to these changes, subtle changes resembling single gene mutations have been detected in maize (Edallo et al. 1981) and rice (Sun et al. 1983), but confirmation of single gene mutations was reported only in tomato (Evans and Sharp 1983). Other types of changes detected by somaclonal variation are the cytoplasm genetic changes. Such changes have been detected in calli of maize plants with the cytoplasmic male sterile-T cytoplasm which were exposed to the toxin of *Drechslera maydis* in order to recover male sterile but resistant plants. The resistant maize plants recovered after selection were male fertile (Gengenbach et al. 1977) and major re-arrangements were observed in the organization of their mitochondrial DNA (Kemble et al. 1982). Recent molecular studies associated the fertility and the male sterility to specific deletions and re-arrangements of the mtDNA (Dewey et al. 1987; Wise et al. 1987; Rottmann et al. 1987). Other molecular studies demonstrated major deletions (up to 80%) in the chloroplast DNA of albino wheat and barley plants regenerated from anther culture (Day and Ellis 1984, 1985).

Transposable elements may be activated through somaclonal variation or even be responsible for the phenomenon. Maize plants containing an active Ac element were regenerated from embryogenic calli of plants without an active element (Peschke et al. 1987). Therefore, it is possible that some tissue culture variability is the result of excision or insertion of transposable elements.

3.2 Nomenclature

1. Somaclones – plants regenerated from cell cultures originated from somatic tissue (R_0).
2. Gametoclones – plants regenerated from cell cultures originated from gametic tissue.
3. Heritable somaclonal and gametoclonal variation – variation observed in the subsequent generations produced by self-fertilization of somaclones and gametoclones (R_1, R_2, R_3).

The nomenclature adopted in this review was described by Edallo (Edallo et al. 1981).

3.3 Chromosomal Analysis of Calli

Numerous chromosomal aberrations in barley cell cultures were observed (Orton 1980; Ruiz and Vazquez 1982; Singh 1986). These variations are not included in the definition of somaclonal variation and they will be discussed only briefly.

In studies of ovary-derived calli, a large array of chromosomal aberrations were detected (Orton 1980). Chromosome analysis of ten morphogenic and seven nonmorphogenic immature embryo-derived calli of *H. vulgare* cv. Himalaya revealed that morphogenic calli carried mostly a normal chromosome complement with a low frequency of numerical chromosomal aberrations, whereas in the nonmorphogenic calli numerical and structural chromosomal variation were present in the majority of the cells (Singh 1986). A general trend in multiplication of the chromosomal aberrations was correlated to the age of the cell culture. At the beginning of the cell culture, the majority of the cells were diploid, whereas after 5 months 13% of the cells were tetraploid and the tetraploids accounted for 50% of the cell population after 2 years (Ruiz and Vasquez 1982).

3.4 Somaclonal Variation

Most of the studies on somaclonal variation were done on the somaclones (R_0) and tested for chromosomal stability and occurrence of chlorophyll deficient plants and other morphological parameters.

3.4.1 Numerical and Structural Chromosome Variation

The *Hordeum vulgare* plants regenerated from immature embryo-derived calli were analyzed by chromosome counting and found to be diploids in most cases (Dale and Deambrogio 1979; Thomas and Scott 1985; Goldstein and Kronstad 1986; Ahloowalia 1987). In one study, one single tetraploid plant, out of 47 regenerants, was obtained (Dale and Deambrogio 1979).

H. vulgare, H. jubatum, and their interspecific hybrid regenerated from immature ovary-derived calli were in most cases diploids with several plants displaying aneuploid chromosome numbers (Orton 1980). The author concluded that regeneration of plants selected completely against polyploid cells and to an intermediate degree against aneuploidy. Chromosomal re-arrangements were observed only in the callus culture (Orton 1980).

H. vulgare plants regenerated from mature embryo-derived calli displayed higher numbers of tetraploid (23%) and aneuploid (17.7%) plants (Lupotto 1984) whereas *H. distichum* plants regenerated from the same explant source were all diploid (Katoh et al. 1986).

3.4.2 Albino and Chlorophyll-Deficient Plants

The albino and chlorophyll-deficient plants regenerated from somatic tissue-derived calli represent a relatively small proportion ($\leq$ 10%) of the regenerants (Dale and Deambrogio 1979; Goldstein and Kronstad 1986; Rengel and Jelaska 1986a,b; Jorgensen et al. 1986; Ahloowalia 1987; Rotem-Abarbanell and Breiman 1989). The albino plants are produced mainly from immature embryo-derived long-term cultures. Eighty-two percent albino plants were obtained from 18–24-month-old cultures (Rengel and Jelaska 1986a), whereas only 4% albino plants were reported to be present in the short-term cultures (Rengel 1987).

About 1000 *H. marinum* plants were regenerated from immature embryo-derived calli during a 5-month period. During this time 100 chlorophyll-deficient and albino plants were produced. Although chlorophyll-deficient and albino plants were formed during all the regeneration period, over 60% of these variants were produced in the 4th month. All the green plants were found to be phenotypically similar (Rotem-Abarbanell and Breiman 1989).

At least two factors can be pointed out as affecting the rate of somaclonal variation, the explant source and the age of the culture. Short-term cultures (2–3 months) derived from immature embryo explants produced the most stable plants, whereas long-term cultures derived from mature embryos caused the highest rate of albino plantlets and chromosomal aberrations. Although several barley species were analyzed, no effect of the species in somaclonal variation could be detected.

3.5 Heritable Somaclonal Variation

A recent study of heritable somaclonal variation was performed on 42 somaclones (R_0) of *H. vulgare* and their selfed progenies (R_1) (Karp et al. 1987). In this study several very sensitive assays were used in order to assess variation in the regenerated plants: (a) cytological studies, (b) isoenzyme analysis, (c) seed protein (hordein) analysis, (d) analysis of DNA restriction fragments related to B hordein, (e) analysis of ribosomal DNA spacer length polymorphism. These assays indicated that the genetic stability found on the chromosomal level of the R_0 plants was maintained in the R_1 plants, which were analyzed by all the other methods.

Two plants out of 42 plants had altered hordein patterns of SDS-PAGE. In these variants major B hordein bands wee reduced and a new band of a lower (GP1diii) or higher (B4a) molecular weights appeared (Fig. 1) (Karp et al. 1987). The barley regenerant (GP1diii) had aberrant meiosis with extensive chromosome breakage. No changes were detected in the pattern of B hordein-related DNA restriction

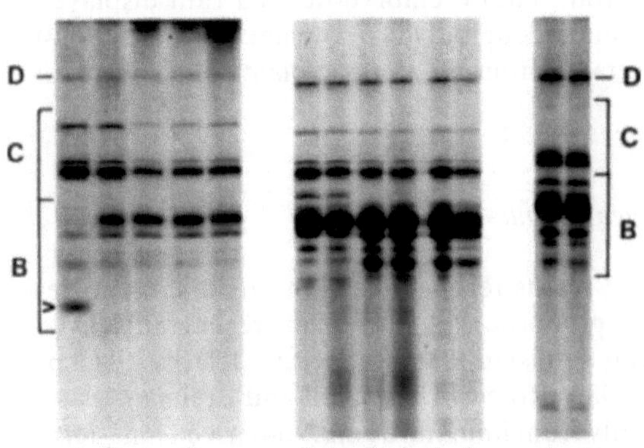

Fig. 1. SDS-PAGE of hordein fraction from half seeds of barley control and regenerated plants. Regenerant GP1(iii) (*a-b*) and Golden Promise controls (*c-e*). Regenerant B4a (*f,g*). Regenerant B4c (*h,i*). Berac controls (*j,k*). *B*, *C*, and *D* indicate the three major groups of hordein polypeptides. The *arrow* in (*a*) indicates the three major groups of hordein polypeptides. (Karp et al. 1987)

fragments of the variant, nor were any other changes observed. Similarly no variation was observed in the number or structure of the intragenic ribosomal DNA spacer units.

Heritable somaclonal variation was analyzed in *Hordeum spontaneum* and *H. marinum*. These wild barley species are diploid and were screened for their regeneration capacity (Breiman 1985) and heritable somaclonal variation of the selfed somaclones (R_1) (Breiman et al. 1987b; Rotem-Abarbanell and Breiman 1989). Twelve selfed *H. spontaneum* regenerants out of 50 which were phenotypically similar were chosen for cytological and the following molecular studies; (a) ribosomal DNA spacer length variation, (b) evaluation of B and C hordeins by SDS-PAGE and DNA restriction fragments related to B and C hordeins, (c) mitochondrial DNA organization. Although very sensitive molecular assays were used in this study, relative stability among the progenies of the *H. spontaneum* regenerated plants was observed. A single variant characterized by a different ribosomal DNA intragenic spacer pattern (Fig. 2) and a different pattern of the hordeins on the SDS-PAGE was observed. No changes in the mitochondrial DNA organization in the progenies of the regenerated plants could be observed (Fig. 3).

Uniformity in the organization of several nuclear genes (B hordein, amylase, intergenic ribosomal spacer) and mitochondrial gene was observed among 20 *Hordeum marinum* seedlings, obtained from seeds of the regenerated plants (R_1) (Fig. 4).

The major conclusion to be drawn from these studies is that somaclonal variation is very limited in barley species. Most chromosomal aberrations present in the cell culture are selected by the inability of these cells to be induced into a morphogenic pathway (Orton 1980; Singh 1986). The albino, chlorophyll-deficient and aneuploid or polyploid aberrant plants are in most cases not fertile, and therefore their genotype cannot be inherited. The R_1 plants produced by selfing of the somaclones were found to be relatively stable (Karp et al. 1987; Breiman et al. 1987b). This stability is attributed to the diploid genome, which is less buffered against variation than polyploid species, such as wheat and rice, which exhibit higher degrees of inherited somaclonal variation (Sun et al. 1983; Karp and Maddock 1984; Larkin et al. 1984; Ryan et al. 1987; Ling 1987).

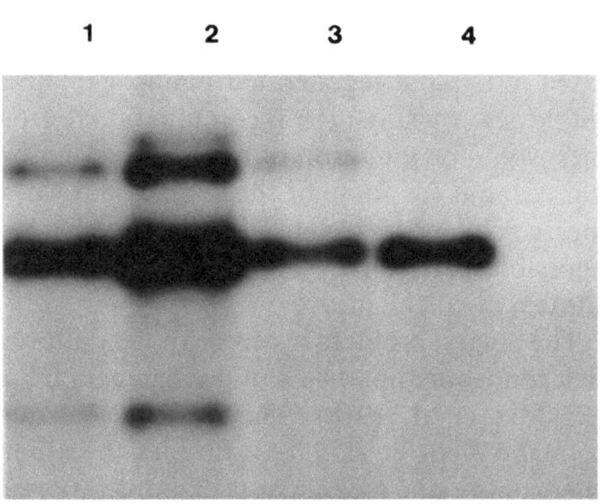

Fig. 2. Southern blot hybridization analysis of total genomic DNA from control plants and progenies of regenerated (R_1) plants of *H. spontaneum* digested with TaqI and hybridized to pHV294 (a 9 kb EcoRI fragment of barley including the 18S, 25S, and the intergenic spacer. (Gerlach and Bedbrook 1979; Breiman et al. 1987b). Lane *1* control; lanes *2–4* regenerants of accession 233

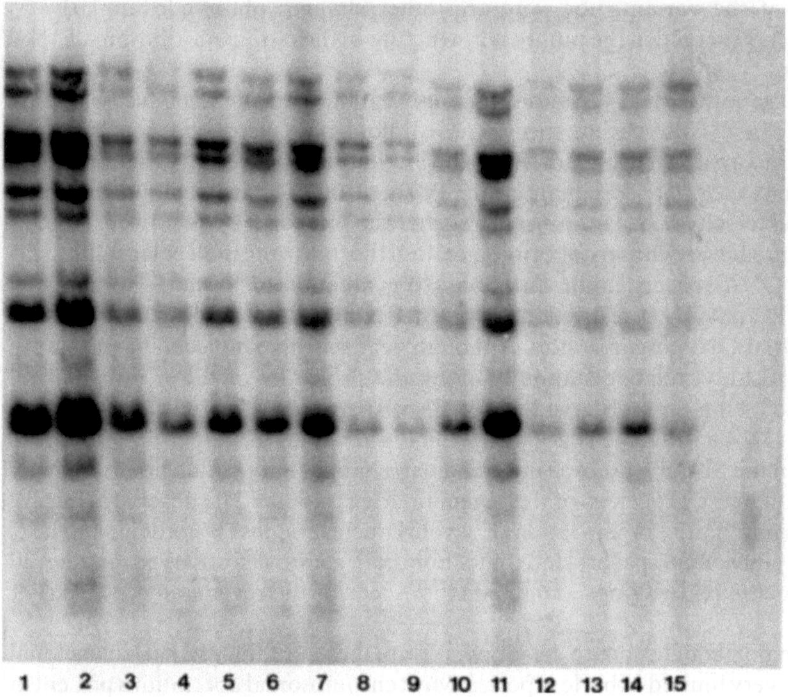

Fig. 3. Autoradiogram of Southern blots of EcoRI digest of DNA from control plants and progenies of regenerated *H. spontaneum* hybridized with a wheat mitochondorial cosmid (4A7) containing the 18S and 26S rRNA genes

3.6 Gametoclonal Variation

The types of genetic changes that are recovered in plants regenerated from cell culture are dependent on the donor material and explant source. Genetic differences exist between somaclonal and gametoclonal variation; (a) since dominant and recessive mutants induced by gametoclonal variation will be expressed in haploid regenerated plants, gametoclones can be analyzed directly to identify new variants; (b) recombinational events that are recovered in gametoclones would be the result of meiotic crossing-over, whereas mitotic crossing-over could account for some of the somaclonal variation; (c) in order to use gametoclones, colchicine treatment is needed for chromosome doubling. Therefore, it may be possible that some of the variation obtained will be due to the mutagenic treatment of colchicine.

The biggest problem in cereal anther culture is the formation of albino plantlets which contribute up to 80% of the regenerated plants (Foroughi-Wehr et al. 1976, 1982; Mix et al. 1978). This trait was demonstrated to be genotype-related (Foroughi-Wehr et al. 1982). Besides the albino plants which are obtained frm anther culture, the green plants demonstrate considerable gametoclonal variation.

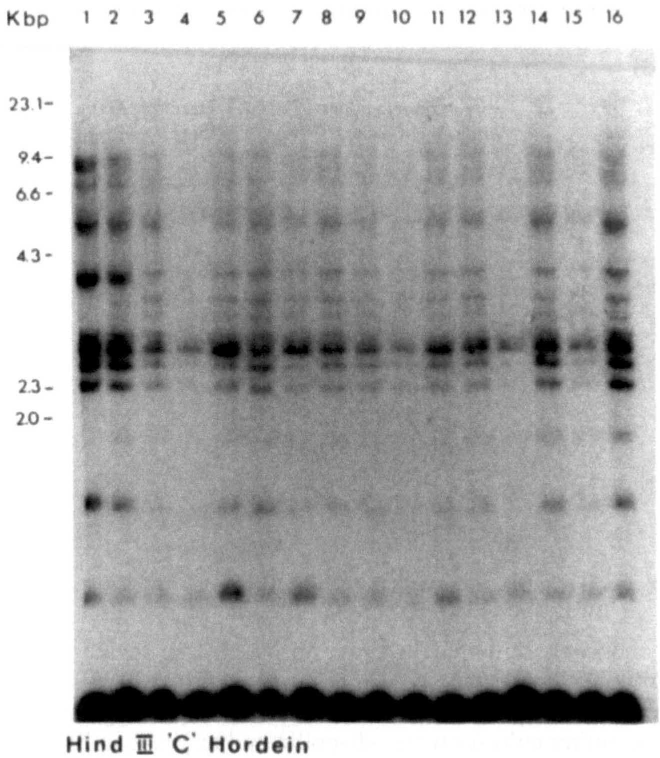

Fig. 4. Autoradiogram of Southern blots of Hind III digest of DNA from control (lane *1*) plants and progenies of regenerated *H. marinum* plants (lanes *2–16*) hybridized with a hordein cDNA clone (pcP387)

Two phenotypic categories of green plants were defined among the regenerants (Mix et al. 1978). Type A – grass-like with many tillers, narrow leaves, short stature and sterile, and Type B – of normal stature but sterile or partially fertile. Chromosome counting revealed that type A were predominantly haploids, whereas type B were mixoploid, containing aneuploid and polyploid cells. Individual phenotype B plants, with completely fertile and with partially or completely sterile ears, were encountered. A remarkable frequency of chromosomal aberrations, including ring, dicentric, and tricentric chromosomes, was detected among the regenerated plants (Mix et al. 1978). Chromosome numbers among androgenic green plants of spring and winter barley are similar (Foroughi-Wehr et al. 1982; Foroughi-Wehr and Friedt 1984). Studies on anther culture of barley were performed mainly on *H. vulgare* and the variation obtained in the regenerated plants is summarized in Table 5. A possible explanation for the observed gametoclonal variation is the nuclear fusion and endomitosis observed during microspore division (Chen et al. 1984).

Table 5. Variation in plants regenerated from anther and microspore culture

Species	Explant	Type of variation						Reference
		Gametoclonal			Heritable Gametoclonal			
		Albino	Morpho-logical	Cyto-logical	Albino	Morpho-logical	Cyto-logical	
H. vulgare	Anther	+	+	+				Mix et al. (1978)
"	"	+						Foroughi-Wehr et al. (1976, 1982)
"	"	+						Foroughi-Wehr and Mix (1979)
"	"	+		+				Foroughi-Wehr and Friedt (1984)
"	"					+		Powell et al. (1984)
"	Microspore					+	+	Powell et al. (1986ab)

3.7 Heritable Gametoclonal Variation

Studies were initiated in order to assess the degree of inherited variation of agronomic characters in a large number of lines originating from selfed, microspore-derived plants. Spontaneous double haploids produced from plants derived from anther cultures were self-pollinated to produce 74 microspore derived lines (M). The components of yield, such as tiller number, grain number and 1000-grain weight, showed significant variability in the M population, but not in the parental material (Powell et al. 1984).

The double haploid lines obtained from microspore-derived haploid plants were compared to double haploid lines derived from haploid embryos obtained by the Bulbosum technique (Powell et al. 1986a). The double haploid lines were used to monitor the segregation of five major genes; rachilla hair length, DDT susceptibility, height, C hordein polymorphism, and mildew resistance. Whereas the double haploid lines produced by the Bulbosum method segregated in the expected 1:1 ratio for four of the five genes, the lines derived from microspore products showed significant departures from the expected 1:1 ratio for three out of the five genes. In addition, the cytological examination of the M progenies revealed changes in both chromosome number and in lower chiasma frequencies. These results demonstrate a nonrandom segregation and induction of gametoclonal variation produced by this method (Powell et al. 1986a). An additional study on field performance of lines derived from haploid and diploid tissue confirms the significant variation created during the culture phase in the microspore-derived double haploid lines (Powell et al. 1986b). It can be concluded that in barley a high degree of gametoclonal variation exists at the level of the regenerated haploid plants and persists in the progenies of the selfed double haploids.

3.8 The Molecular Basis of Albino Barley Plants Produced from Anther Culture

Albinism in anther culture of cereals and grasses is one of the major factors that has impeded the use of this technique in cereal breeding programs. The percentage of albino plants produced is influenced by the genotype and physiological state of the donor plant (Bullock et al. 1982; Ouyang et al. 1983). Vaughn and co-workers (1980) have suggested that maternal inheritance of chloroplast is due to an alteration in this organelle in pollen, and that the same mechanism may be responsible for the generation of albino plants in cereal anther culture.

A molecular approach to investigate the involvement of the chloroplast genome in the albino phenotype was undertaken (Day and Ellis 1984, 1985). It was shown by Southern blotting that large regions of the chloroplast genome of wheat and barley are deleted in albino plants regenerated from separate pollen calli. Most albino plants appeared to contain heterogenous populations of chloroplast DNA molecules. Nine albino barley plants have been analyzed, and in all of them deletion of the majority of the DNA, including the inverted repeat, was demonstrated. The altered chloroplast genomes exist in a variety of physical forms and reveal that different sequence organization can exist in a single plant (Day and Ellis 1985).

A common region was retained in all albino plants, suggesting that this fragment may represent sequences sufficient for replication. The molecular studies on barley and wheat chloroplast DNA extracted from albino anther derived plants suggest that very little of the plastid genome is required for viability.

4 Application of In Vitro-Induced Variation in Barley Breeding

The gametoclonal variation arising from growing microspore-derived haploid plants was employed successfully for production of barley plants resistant to barley yellow mosaic virus (BaYMV) (Foroughi-Wehr and Friedt 1984) and to *Helminthosporium sativum* (Chawla and Wenzel 1987).

4.1 Resistance to BaYMV

A large experiment was designed in order to introduce BaYMV resistance into a high yielding cultivar (Foroughi-Wehr and Friedt 1984). Many anthers (233,445) were cultured, 159,366 of the F_1 hybrids and 74,079 of their parents. Altogether 1525 albino and 831 green plants were regenerated. The average ratio of green to albino plants was 1:1.8. The chromosome numbers revealed that the majority of plants were spontaneously diploid (69%), the remainder being triploids and aneuploids. In a single experiment from 91 androgenetic green plants, 20 plants were resistant to BaYMV. The resistance to BaYMV was combined with other important characters in a regular breeding scheme during the course of 3 years (Foroughi-Wehr and Friedt 1984; Foroughi-Wehr et al. 1986).

4.2 Resistance to *Helminthosporium sativum* by Selection

The recovery of disease-resistant plants by selection of cell and callus culture resistant to fungal culture filtrate or synthetic toxins of the pathogen can be applied for crop improvement (Wenzel 1985). In the experiment, aimed at obtaining barley plants resistant to *H. sativum* pathogens, a selection system on embryogenic callus induced from barley immature embryos was used (Chawla and Wenzel 1987). Two selection methods were employed: a continuous method in which four cycles of selection were performed one after another on toxic media, and a discontinuous method in which a pause was given after the second or third cycle of selection. The latter was superior, as it allowed the calli to regain their regeneration ability. Out of 1595 barley calli, only 6% survived the selection cycles by the first selection system, whereas from the discontinuous method 10% survived. Seventeen out of the 36 regenerant plants were resistant to the pathogen, 13 plants were intermediately resistant and three were susceptible.

5 Conclusions and Prospects

Evidence that variant characters expressed in the regenerated plant are transmitted to self-fertilized progenies and can involve single gene mutations (Evans and Sharp 1983) offer the possibility of using somaclonal variation for improvement of seed-propagated crop species.

In barley, two major factors have been found to affect the amount of variation which can be obtained through in vitro culture: the explant source and the duration in culture. The effect of the explant source can be demonstrated by comparing plants regenerated from microspore-derived calli, exhibiting the highest variability (gametoclones), with plants regenerated from immature embryos, which were relatively stable (somaclones). Long-term culture increased the amount of albino plants and plants with more chromosomal aberrations, as compared to short-term cultures.

Most of the variation demonstrated as existing in cell culture is eliminated during the regeneration process, resulting in a relatively small amount of variant somaclones. The variation is subsequently reduced by sexual reproduction, resulting in a low rate of heritable somaclonal variation in barley.

A high level of gametoclonal and heritable gametoclonal variation is present among plants regenerated from microspore-derived calli, with about 80% of the haploid plants being albino and many displaying chromosomal aberrations.

Since the morphological process involved in regeneration of plants from immature embryos and from anther culture-derived calli is very different, it may account for the genome stability or instability of the somaclones and gametoclones, respectively.

Soma- and gametoclonal variation have been assessed by various experimental approaches, including cytological, biochemical, molecular, and biometrical parameters. The choice of the experimental methodology is influenced by the type of variation in which one is interested and by the amount of material to be characterized, i.e., phenotypic variation can be measured in a large-scale field experiment, whereas molecular studies can be performed only on a very limited sample.

Further studies on somaclonal variation can be chanelled to different aims: the elucidation of molecular mechanisms involved in soma- and gametoclonal variation, and the application of the variation to specific breeding targets according to agronomic specifications.

Recent achievements in production of barley cultivars resistant to the plant pathogens, the virus BaYMV, and *Helminthosporium sativum* are encouraging in pursuing the research on somaclonal variation and in vitro selection for solving specific breeding aims.

References

Ahloowalia BS (1987) Plant generation from embryo-callus culture in barley. Euphytica 36:659–665

Anderson MK, Reinbergs E (1985) Barley breeding. In: Rasmusson DC (ed) Barley. Am Soc Agric, Crop Sci Soc Am, Soil Sci Soc Am, Madison, pp 231–267

Binding H (1974) Cell cluster formation by leaf protoplasts from axenic culture of haploid *Petunia hybrida* L. Plant Sci Lett 2:185–188

Bolick M, Sporling B, Koop HU (1989) Identification of embryogenic microspores of barley (*Hordeum vulgare*) by individual selection and culture and its potential for transformation by microinjection. XII Eucarpia Congress. 26–13 Vorträge für Pflanzenzücht 15

Breiman A (1985) Plant regeneration from *Hordeum spontaneum* and *Hordeum bulbosum* immature embryo derived calli. Plant Cell Rep 4:70–73

Breiman A, Felsenburg T, Galun E (1987a) Nor loci analysis in progenies of plants regenerated from the scutellar callus of bread-wheat. Theor Appl Genet 73:827–831

Breiman A, Rotem-Abarbanell D, Karp A, Shaskin H (1987b) Heritable somaclonal variation in wild barley (*Hordeum spontaneum*). Theor Appl Genet 74:104–112

Bullock WP, Baenziger PS, Schaeffer GW, Bottino PJ (1982) Anther culture of wheat (*Triticum aestivum* L.) Fls and their reciprocal crosses. Theor Appl Genet 62:155–159

Chawla HS, Wenzel G (1987) In vitro selection of barley and wheat for resistance against *Helminthosporium sativum*. Theor Appl Genet 74:841–845

Chen CC, Kasha KJ, Marsolais A (1984) Segmentation patterns and mechanisms of genome multiplication in cultured microspores of barley. Can J Genet Cytol 26:475–483

Cheng TY, Smith HH (1975) Organogenesis from callus culture of *Hordeum vulgare*. Planta 123:307–310

Chu CC, Wang CC, Sun CS, Hsu C, Yin KC, Chu CY, Bi FY (1975) Establishment of an efficient medium for anther culture of rice through comparative experiments on the nitrogen sources. Sci Sin 18:659–668

Chu CC, Sun CS, Chen X, Zhang WX, Du ZH (1984). Somatic embryogenesis and plant regeneration in callus from inflorescences of *Hordeum vulgare* × *Triticum aestivum* hybrids. Theor Appl Genet 68:375–379

Clapham D (1973) Haploid *Hordeum* plants from anthers in vitro. Z Pflanzenzücht 69:142–155

Cooper DB, Sears PG, Lockhart GL, Jones BL (1986) Heritable somaclonal variation in gliadin proteins of wheat plants derived from immature embryos callus culture. Theor Appl Genet 71:784–790

Coulibaly MY, Demarly Y (1986) Regeneration of plantlets from protoplasts of rice (*Oryza sativa* L.). Z Pflanzenzücht 96:79–81

Dale J, Deambrogio E (1979) A comparison of callus induction and plant regeneration from different explants of *Hordeum vulgare*. Z Pflanzenphysiol 94:65–77

Datta SK, Wenzel G (1988) Single microspore derived embryogenesis and plant formation in barley (*Hordeum vulgare*). Theor Appl Genet

Day A, Ellis THN (1984) Chloroplast DNA deletions associated with wheat plants regenerated from pollen: possible basis for maternal inheritance of chloroplasts. Cell 39:359–368

Day A, Ellis THN (1985) Deleted forms of plastid DNA in albino plants from cereal anther culture. Curr Genet 9:671–678

Devaux P (1987) Comparison of anther culture and *Hordeum bulbosum* method for the production of doubled haploid in winter barley. I. Production of green plants. Plant Breed 98:215–219

Dewey DR (1971) Synthetic hybrids of *Hordeum bogdanii* with *Elymus canadensis* and *Sitanion hystrix*. Am J Bot 58:902–908

Dewey RE, Timothy DH, Levings III CS (1987) A mitochondrial protein associated with cytoplasmic male sterility in the T cytoplasm of maize. Proc Natl Acad Sci USA 84:5374–5378

Dickson AD (1979) Barley for malting and food. In: Barley: botany, culture, winterhardiness, genetics, utilization, pests. In: USDA (ed) Agricultural handbook. US Gov Print Off, Washington DC, pp 136–146

Dunwell JM, Francis RJ, Powell W (1987) Anther culture of *Hordeum vulgare* L.: a genetic study of microscope callus production and differentiation. Theor Appl Genet 74:60–64

Edallo S, Zuccinali C, Perenzin M, Salamini F (1981) Chromosomal variation and frequency of spontaneous mutation associated with in vitro culture and plant regeneration in maize. Maydica 26:39–56

Evans DA, Sharp WR (1983) Single gene mutations in tomato plants regenerated from tissue culture. Science 221:949–951

Evans DA, Sharp WR, Medina-Filho HP (1984) Somaclonal and gametoclonal variation. Am J Bot 71:759–774

Evans PK, Keates AC, Cocking EC (1972) Isolation of protoplasts from cereal leaves. Planta 104:178–181

FAO (ed) (1986) Production yearbook 1986. FAO, Rome, pp 77–78

Fedak G (1985) Propagation of intergeneric hybrids of Triticeae through callus culture of immature inflorescence. Z Pflanzenzücht 94:1–7

Felsenburg T, Feldman M, Galun E (1987) Aneuploid and alloplasmic lines as tool for the study of nuclear and cytoplasmic control of culture ability and regeneration of scutellar calli from common wheat. Theor Appl Genet 74:802–810

Foroughi-Wehr B, Friedt W (1984) Rapid production of recombinant barley yellow mosaic virus resistant *Hordeum vulgare* lines by anther culture. Theor Appl Genet 67:377–382

Foroughi-Wehr B, Mix G (1979) In vitro response to *Hordeum vulgare* L. anthers cultured from plants grown under different environments. Environ Exp Bot 19:303–309

Foroughi-Wehr B, Mix G, Gaul H, Wilson HM (1976) Plant production from cultured anthers of *Hordeum vulgare* L. Z Pflanzenzücht 77:198–204

Foroughi-Wehr B, Friedt W, Wenzel G (1982) On the genetic improvement of androgenetic haploid formation in *Hordeum vulgare* L. Theor Appl Genet 62:233–239

Foroughi-Wehr B, Friedt W, Schuchmann R, Kohler F, Wenzel G (1986) In vitro selection for resistance. In: Semal J (ed) Somaclonal variations and crop improvement. Nijhoff, Dordrecht, pp 35–44

Gamborg OL, Miller RA, Ojima K (1968) Nutrient requirements of suspension cultures of soybean root cells. Exp Cell Res 50:151–158

Gengenbach BG, Green CE, Donovan CD (1977) Inheritance of selected pathotoxin resistance in maize plants regenerated from cell cultures. Proc Natl Acad Sci USA 44:5113–5117

Goldstein CS, Kronstad WE (1986) Tissue culture and plant regeneration from immature embryo explants of barley, *Hordeum vulgare*. Theor Appl Genet 71:631–636

Green CE, Phillips RL, Wang AS (1977) Cytological analysis of plants regenerated from maize tissue cultures. Maize Genet Coop Newslett 51:53–54

Hagberg A, Hagberg G (1980) High frequency of spontaneous haploids in the progeny of an induced mutation in barley. Hereditas 93:341–343

Hamachi Y, Komatsuda T, Nakajima K (1988) Embryoid formation and plant regeneration in anther culture of two-rowed barley cultivars in Japan. Japan J Breed 38:363–366

Hanzell JJ, Miller JP, Brinkman MA, Fendos E (1985) Genotype and media effects on callus formation and regeneration in barley. Crop Sci 25:27–31

Hockett EA, Nilan RA (1985) Genetics. In: Rasmusson DC (ed) Barley. Am Soc Agric, Crop Sci Soc Am, Soil Sci Soc Am, Madison, pp 187–229

Hughes BG, White FG, Smith MA (1978) Effect of plant growth, isolation and purification on barley protoplast yield. Biochem Physiol Pflanzen 172:67–77

Ishii C, Sato K, Okamura M, Matsuno T (1987) Protoplast isolation and plant regeneration from barley immature embryo callus. 5th Int Barley Genetics Symp, Okayama, Jpn, Oct 6–11, 1986

Jelaska S, Rengel Z, Cesar V (1984) Plant regeneration from mesocotyl callus of *Hordeum vulgare* L. Plant Cell Rep 3:125–129

Jorgensen RB, Jensen CJ, Andersen B, von Brothmer R (1986) High capacity of plant regeneration from callus of interspecific hybrids with cultivated barley (*Hordeum vulgare* L.). Plant Cell Tissue Org Cult 6:199–207

Junker B. Zimny J. Lührs R. Lörz H (1987) Transient expression of chimaeric genes in dividing and non-dividing cereal protoplasts after PEG-induced DNA uptake. Plant Cell Rep 6:329–332
Kao KN (1977) Chromosomal behavior in somatic hybrids of soybean. Mol Gen Genet 150: 225–230
Kao KN (1981) Plant formation from barley anther cultures with Ficoll media. Z Pflanzenphysiol 103:437–443
Kao KN. Michalyuk MR (1975) Nutritional requirements for growth of *Vicia hajastana* cells and protoplasts at very low density in liquid media. Planta 126:105–110
Karp A. Maddock SE (1984) Chromosome variation in wheat plants regenerated from cultured immature embryos. Theor Appl Genet 67:249–255
Karp A. Steele SH. Parmar S. Jones MGK. Shewry PR. Breiman A (1987) Relative stability among barley plants regenerated from cultured immature embryos. Genome 29:405–412
Kasha KJ (1987) Production of haploids in cereals. In: Proc IAEA/FAO Meet Use of induced mutations in connection with haploids and heterosis in cereals. Univ Guelph. Can. Dec 8–12. 1986
Kasha KJ. Kao KN (1970) High frequency haploid production in barley (*Hordeum vulgare* L.). Nature (London) 225:874–876
Katoh Y. Hasegawa T. Suzuki T. Fujii T (1986) Plant regeneration from the callus derived from mature embryos of hiproly barley *Hordeum distichum* L. Agric Biol Chem 50:761–762
Kemble RJ. Flavel KB. Brettell RIS (1982) Mitochondrial DNA analysis of fertile and sterile maize plants derived from tissue culture. Theor Appl Genet 62:213–217
Kiesling RL (1985) The diseases of barley. In: Rasmusson DC (ed) Barley. Am Soc Agric Crop Sci Soc Am. Soil Sci Soc Am. Madison. pp 269–313
Koblitz H (1976) Isolation and cultivation of protoplasts from callus cultures of barley. Biochem Physiol Pflanzen 170:287–293
Koblitz H (1986) Barley (*Hordeum vulgare* L.): establishment of cultures and the regeneration of plants. In: Bajaj YPS (ed) Biotechnology in agriculture and forestry. vol 2. Crops I. Springer. Berlin Heidelberg New York Tokyo. pp 181–203
Kott LS. Kasha KJ (1984) Initiation and morphological development of somatic embryoids from barley cell cultures. Can J Bot 62:1245–1249
Kott LS. Flack S. Kasha KJ (1986) A comparative study of initiation and development of embryogenic callus from haploid embryos of several barley cultivars. II Cytophotometry of embryos and callus. Can J Bot 64:2107–2112
Larkin PJ. Scowcroft WR (1981) Somaclonal variation – a novel source of variability from cell cultures for plant improvement. Theor Appl Genet 60:197–214
Larkin PJ. Ryan SA. Brettell RIS. Scowcroft WR (1984) Heritable somaclonal variation in wheat. Theor Appl Genet 67:433–455
Larsen J (1981) Breeding of proanthocyanidin-free malt barley. In: Asher M (ed) Barley genetics. vol 4. Proc 4th Int Barley Genetics Symp. Univ Press. Edinburgh
Li J. Zhu M. Cai T (1985) Callus induction and plant regeneration from immature barley-wheat hybrid embryos. Kexue Tongbao 30:405–407
Ling DH (1987) A quintuple reciprocal translocation produced by somaclonal variation in rice. Cereal Res Commun 15:5–12
Lörz H. Harms CT. Potrykus I (1976) Isolation of "vacuoplasts" from protoplasts of higher plants. Biochem Physiol Pflanzen 169:617–620
Lührs R. Lörz H (1987) Plant regeneration in vitro from embryogenic cultures of spring- and winter-type barley (*Hordeum vulgare*) varieties. Theor Appl Genet 75:16–25
Lührs R. Lörz H (1988) Initiation of morphogenic cell suspension and protoplast cultures of barley (*Hordeum vulgare* L.). Planta 175:71–81
Lundqvist U (1976) Locus distribution of induced eceriferum mutants in barley. In: Gaul H (ed) Barley genetics. vol 3. Proc 3rd Int Barley Genet Symp. Thieme. München
Lupotto E (1984) Callus induction and plant regeneration from barley mature embryos. Ann Bot 54:523–529
Maddock SE. Risiott R. Parmar S. Jones MGK. Shewry PR (1985) Somaclonal variation in gliadin patterns of grains of regenerated wheat plants. J Exp Bot 36:1976–1984
Marsolais AA. Kasha KJ (1985) Callus induction from barley microspores. The role of sucrose and auxin in a barley anther culture medium. Canadian J Bot 63:2209–2212
Mathias RJ. Fukui K (1986) The effect of specific chromosome and cytoplasm substitutions on the tissue culture response of wheat (*Triticum aestivum*) callus. Theor Appl Genet 71:797–800

McCoy TJ, Phillips RL, Rhines HW (1982). Cytogenetic analysis of plants regenerated from oat (*Avena sativa*) tissue cultures: high frequency of partial chromosome loss. Can J Genet Cytol 24:37–50

Mix G, Wilson HM, Foroughi-Wehr B (1978) The cytological status of plants of *Hordeum vulgare* L. regenerated from microspore callus. Z Pflanzenzücht 80:89–99

Moseman JC, Smith DH Jr (1985) Germplasm resources. In: Rasmusson DC (ed) Barley. Am Soc Agric, Crop Sci Soc Am, Soil Sci Soc Am, Madison, pp 57–72

Murashige T, Skoog F (1962) A revised medium for rapid growth and bioassays with tobacco tissue cultures. Physiol Plant 15:473–497

Nilan RA (1981a) Induced gene and chromosome mutants. Philos Trans R Soc London Ser B 292:457–466

Nilan RA (1981b) Recent advances in barley mutagenesis. In: Asher M (ed) Barley genetics, vol 4. Proc 4th Int Barley Genet Symp. Univ Press, Edinburgh

Ogura H, Kyozuka J, Hayashi Y, Koba T, Shimamoto K (1987) Field performance and cytology of protoplast-derived rice (*Oryza sativa*): high yield and low degree of variation of four Japonica cultivars. Theor Appl Genet 74:670–676

Olsen FL (1987) Induction of microspore embryogenesis in cultured anthers of *Hordeum vulgare*. The effects of ammonium nitrate, glutamine and aspargine as nitrogen sources. Carlsberg Res Commun 52:393–404

Orton TJ (1979) A quantitative analysis of growth and regeneration from tissue cultures of *Hordeum vulgare*, *H. jubatum* and their interspecific hybrid. Environ Exp Bot 19:319–335

Orton TJ (1980) Chromosomal variability in tissue cultures and regenerated plants of *Hordeum*. Theor Appl Genet 56:101–112

Ouyang JW, Zhou SM, Jia SE (1983) The response of anther culture to culture temperature in *Triticum aestivum*. Theor Appl Genet 66:101–109

Peschke VM, Phillips RL, Gengenbach BG (1987) Discovery of transposable element activity among progeny of tissue culture-derived maize plants. Science 238:804–807

Peterson GA, Forster AE (1973) Malting barley in the United States. Adv Agron 25:327–378

Poehlman JM (1985) Adaptation and distribution. In: Rasmusson DC (ed) Barley. Am Soc Agric, Crop Sci Soc Am, Soil Sci Soc Am, Madison, pp 1–17

Potrykus I, Harms CT, Lorz H (1979) Callus formation from cell culture protoplasts of corn (*Zea mays* L.). Theor Appl Genet 54:209–214

Powell W (1987) Diallel analysis of barley anther culture response. Genome 30:152–157

Powell W, Hayter AM, Wood W, Dunwell JM, Huang B (1984) Variation in the agronomic characters of microspore-derived plants of *Hordeum vulgare* cv. Sabarlis. Heredity 52:19–23

Powell W, Borrino EM, Allison MJ, Griffiths DW, Asher MJC, Dunwell JM (1986a) Genetical analysis of microspore derived plants of barley (*Hordeum vulgare*). Theor Appl Genet 72:619–626

Powell W, Caligari PDS, Dunwell JM (1986b) Field performance of lines derived from haploid and diploid tissues of *Hordeum vulgare*. Theor Appl Genet 72:458–465

Rengel Z (1987) Embryogenic callus induction and plant regeneration from cultured *Hordeum vulgare* mature embryos. Plant Physiol Biochem 25:43–48

Rengel Z, Jelaska S (1986a) Somatic embryogenesis and plant regeneration from seedling tissues of *Hordeum vulgare* L. J Plant Physiol 124:385–392

Rengel Z, Jelaska S (1986b) The effect of L-proline on somatic embryogenesis in long-term callus culture of *Hordeum vulgare*. Acta Bot Croat 45:71–75

Rotem-Abarbanell D, Breiman A (1989) Plant regeneration from immature and mature embryo explants of *Hordeum marinum*. Plant Cell Tissue & Organ Culture 16:207–216

Rottmann WH, Brears T, Hodge TP, Lonsdale DH (1987) A mitochondrial gene is lost via homologous recombination during reversion of CMS T maize to fertility. EMBO J 6:1541–1546

Ruiz ML, Vazquez AM (1982) Chromosome number evolution in stem derived calluses of *Hordeum vulgare* L. cultured in vitro. Protoplasma 111:83–86

Ryan SA, Larkin PJ, Ellison FW (1987) Somaclonal variation in some agronomic and quality characters in wheat. Theor Appl Genet 74:77–82

Saalbach G, Koblitz H (1978) Attempts to initiate callus formation from barley leaves. Plant Sci Lett 13:165–169

San Noeum LH (1976) Haploides d'*Hordeum vulgare* L. par culture in vitro d'ovaires nonfécondes. Ann Amelior Plantes 26:751–754

Schooler AB, Anderson MK (1980) Behaviour of intergeneric hybrids between *Hordeum vulgare* L. (4x) and an *Elimus mollis* type. Cytologia 45:157–162

Seguin-Swartz G, Kott L, Kasha KJ (1984) Development of haploid cell lines from immature barley *Hordeum vulgare* embryos. Plant Cell Rep 3:95–97

Singh RJ (1986) Chromosomal variation in immature embryo derived calluses of barley (*Hordeum vulgare* L.). Theor Appl Genet 72:710–716

Snape JW, Sitch LA, Simpson E, Parker BB (1988) Tests for the presence of gametoclonal variation in barley and wheat doubled haploids produced using the *Hordeum bulbosum* system. Theor Appl Genet 75:509–513

Sogaard B, von Wettstein-Knowles P (1987) Barley: genes and chromosomes. Carlsberg Res Commun 52:123–196

Sorvari S, Schieder O (1987) Influence of sucrose and melibiose on barley anther cultures. Plant Breed 99:164–171

Sun Z, Zhao C, Zheng K, Qi X, Fu Y (1983) Somaclonal genetics of rice *Oryza sativa* L. Theor Appl Genet 67:67–73

Sunderland N, Dunwell JM (1977) Anther and pollen culture. In: Street HE (ed) Plant tissue and cell culture. Blackwell, Oxford, pp 223–264

Sunderland N, Evans LJ (1980) Multicellular pollen formation in cultured barley anthers. II. The A, B and C pathways. J Exp Bot 31:501–514

Sunderland N, Collins GB, Dunwell JM (1974) The role of the nuclear fusion in pollen embryogenesis of *Datura innoxia* Mill. Planta 117:227–241

Sunderland N, Roberts M, Evans LJ, Wildon DC (1979) Multicellular pollen formation in cultured barley anthers. J Exp Bot 30:1133–1144

Sunderland N, Xu ZH, Huan B (1981) Recent advances in barley anther culture. In: Asher M (ed) Barley genetics, vol 4. Proc 4th Int Barley Genet Symp. Univ Press, Edinburgh, pp 599–703

Takahashi S, Kaneko T (1986) Isolation and culture of barley protoplast. In: 5th Int Barley Genetics Symp, Okayama, Jpn, Oct 6–11, 1986

Thomas MR, Scott KJ (1985) Plant regeneration by somatic embryogenesis from callus initiated from immature embryos and immature inflorescences of *Hordeum vulgare*. J Plant Physiol 121:159–169

Ukai Y, Nishimura S (1987) Regeneration of plants from calli derived from seeds and mature embryos in barley. Japan J Breed 37:405–411

Vasil IK (1987) Developing cell and tissue culture systems for the improvement of cereal and grass crops. J Plant Physiol 128:193–218

Vaughn KC, DeBonte LR, Wilson KG, Schaeffer GW (1980) Organelle alteration as a mechanism for maternal inheritance. Science 208:196–198

Vazquez AM, Ruiz ML (1986) Barley: induction of genetic variability through callus cultures. In: Bajaj YPS (ed) Biotechnology in agriculture and forestry, vol 2. Crops I. Springer, Berlin Heidelberg New York Tokyo, pp 204–219

Wakasa K (1973) Isolation of protoplasts from various plant organs. Jpn J Genet 48:279–289

Wang CC, Kuang BJ (1981) Induction of haploid plants from the female gametophyte of *Hordeum vulgare* L. Acta Bot Sin 23:329–330

Wei ZH, Kyo M, Harada H (1986) Callus formation and plant regeneration through direct culture of isolated pollen of *Hordeum vulgare* cv. 'Sabarlis'. Theor Appl Genet 72:252–255

Weigel RC Jr, Hughes KW (1983a) Morphogenesis in barley tissue cultures. In Vitro 20:248

Weigel RC Jr, Hughes KW (1983b) Somatic embryogenesis in barley. In Vitro 20:277

Weigel RC Jr, Hughes KW (1985) Long term regeneration by somatic embryogenesis in barley (*Hordeum vulgare* L.) tissue cultures derived from apical meristem explants. Plant Cell Tissue Org Cult 5:151–162

Wenzel G (1985) Strategies in unconventional breeding for disease resistance. Annu Rev Phytopathol 23:149–172

Wheatley WG, Marsolais AA, Kasha KJ (1986) Microspore growth and anther staging in barley anther culture. Plant Cell Rep 5:47–49

Wise RP, Pring DR, Gengenbach BG (1987) Mutation to male fertility and toxin insensitivity in Texas (T)-cytoplasm maize is associated with a frameshift in a mitochondrial open reading frame. Proc Natl Acad Sci USA 84:2858–2862

Xu ZH, Sunderland N (1981) Glutamine, inositol and conditioning factor in the production of barley pollen callus in vitro. Plant Sci Lett 23:161–168

Xu ZH, Huang B, Sunderland N (1981) Culture of barley anthers in conditioned media. J Exp Bot 32:767–778

Yamada Y, Zhi-Qi Y, Ding-Tai T (1986) Plant regeneration from protoplast-derived callus of rice (*Oryza sativa* L.). Plant Cell Rep 5:85–88

Section III Vegetables and Fruits

III.1 Somaclonal Variation in Potato

A. KARP[1]

1 Introduction

Potato (*Solanum tuberosum*) has been under cultivation by man for some 8000 years. At present, it ranks fifth amongst the most important food crops in the world. Its origin lies probably in the Andes, where wild potatoes are still widespread, although no species has yet been identified as the diploid ancestor. Most present cultivars are tetraploid ($2n = 4x = 48$) and are grown over a wide global distribution (Hawkes 1978).

The edible portion of potato (the tuber) is a modified underground stem which acts as a storage organ. The eyes on the tubers are buds, and reproduction is mostly vegetative. This is important for breeding purposes as it means that any genotype can be maintained and multiplied by clonal propagation. Indeed, most potato cultivars are highly heterozygous (Hawkes 1978). Modern cultivars originate from a narrow genetic base, but the wild gene pool is outstanding in richness and diversity. Some 180 different tuber-bearing wild species are known (Correll 1962) and many of these have been identified as useful sources of genes for crop improvement. Most potato cultivars are also capable of flowering to produce berries containing "true seed", but sterility is a serious constraint in potato breeding (Grun 1974).

Potato breeding usually involves crossing favourable parents, followed by vegetative propagation of the F_1 plants to form clones. These clones and their tuber progenies are then screened in increasing plots over several years for plants with favourable combinations of agronomic traits (Ross 1986). Current breeding aims include the need to introduce genes conferring resistance against viral- and fungal-borne diseases. These diseases significantly decrease yield each year and include those caused by viruses such as leaf roll (PLRV), leaf drop streak, or virus Y (PVY), leaf rolling mosaic, or virus M (PVM) and common mosaic, or virus X (PVX). Non-viral pathogens causing diseases include cyst nematodes such as *Globodera rostochiensis* and *G. pallida* and fungal pathogens such as late blight (*Phytophthora infestans*) and wart (*Synchytrium endobioticum*). Other breeding aims include improvement of tuber quality for specific utilization, such as crisping or the french-fry industry, and enhancement of yield.

Over the past 15 years a technology has been developing which offers new potential for potato breeding (see Bajaj 1987). At the outset, this technology should

[1]Biochemistry Department. Rothamsted Experimental Station. Harpenden. Hertfordshire AL5 2JQ. United Kingdom

Biotechnology in Agriculture and Forestry. Vol. 11
Somaclonal Variation in Crop Improvement I (ed. by Y.P.S. Bajaj)

not be viewed as a strict alternative to conventional practices, but rather as a complement. Techniques are now available for regeneration of whole plants from cultured plant tissues. Potato is very amenable to such procedures and several thousand potato plants have been regenerated from cultured explants (pieces of leaf, stem and tuber) and protoplasts (single isolated wall-less cells).

The potential applications of this technology are manifold. Tissue culture systems such as micropropagation and meristem-tip culture are already being utilized in breeding programmes for rapid multiplication of stocks, conservation of genetic resources and virus-elimination (Karp et al. 1987). Explant and protoplast regeneration provides experimental systems for mutant selection (Behnke 1979, 1980) and in addition, protoplasts can be fused together to form somatic hybrids, thus providing a mechanism for hybridization even when the desired parents are not compatible (Jones 1985). A whole new area of crop improvement is made available when regeneration systems are coupled with the advancing technology of molecular biology. Genes can be isolated, sequenced, modified and inserted back into plants in novel combinations. Transformation of potato has already been achieved by use of the natural vector system of *Agrobacterium* (Ooms et al. 1983, 1985, 1986, 1987; Yamamoto 1989).

Interesting as these developments are, they do not form the basis of this chapter. Instead, focus will be placed on an unexpected source of variability which has arisen concomitantly with the development of plant regeneration systems. This variability arises more often when plants are regenerated through a callus phase and has been called somaclonal variation (Larkin and Scowcroft 1981).

Clearly, for the production of novel plants by transformation or protoplast fusion, instability poses a serious problem which should be overcome. Conversely, observations of the nature of somaclonal variation, particularly in potato, have shown that changes can occur in traits of agronomic importance, such as yield and disease resistance. Such observations raised the issue of whether somaclonal variation could be of use for potato breeding (Shepard et al. 1980).

In this chapter the nature of somaclonal variation in potato, including the extent and the spectrum of changes observed are examined, the factors that affect the degree of variation are discussed, and a critical assessment of the potential of somaclonal variation for improvement of the potato crop is attempted.

2 In Vitro Regeneration in Potato

A wide range of tissue culture systems have been developed in potato, from micropropagation through to regeneration from cultured protoplasts. Techniques such as meristem-tip culture and micropropagation are being successfully incorporated into potato breeding programmes (Karp et al. 1987; Bajaj 1987). In their strictest form, they do not involve a stage of disorganized cell growth, or callus, and therefore are normally associated with genetic stability (Wright 1983; Denton et al. 1977). In contrast, regeneration from cultured explants and protoplasts (Fig. 1) does involve a callus phase, and it is therefore with these systems that somaclonal variation may be associated.

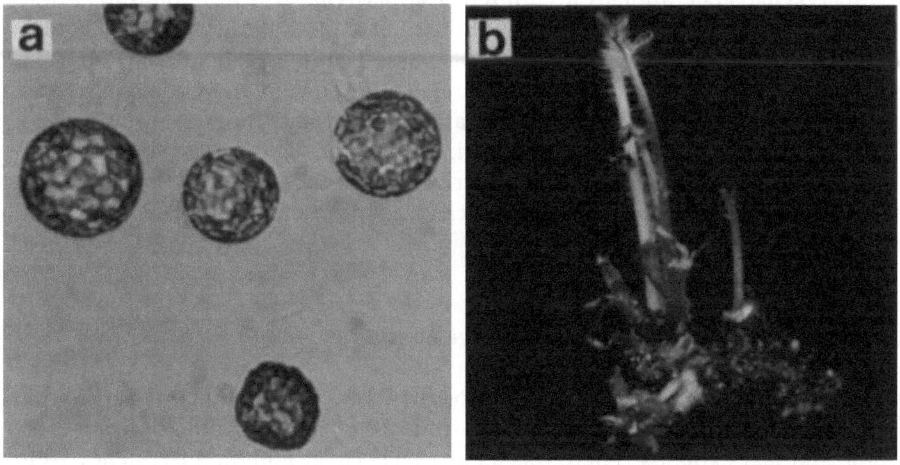

Fig. 1. (**a**) Leaf mesophyll protoplasts of potato. (**b**) Leaf explant regeneration in potato. A leaf piece has been cultured in vitro on basal medium (Murashige and Skoog 1962) containing 0.186 mg l NAA, 2.25 mg l BAP and 10 mg l GA_3. Numerous shoots have been initiated from the callus which has formed on the leaf piece

Explant regeneration is a loose term which encompasses regeneration from all complex pieces of tissue, such as leaf, stem and anther. Anther culture has a specific application, as it is a method by which haploid plants can be obtained (Sopory and Bajaj 1987). Since potato is a tetraploid, two levels of haploidy are possible. Reduction from the tetraploid gives the dihaploid (2n = 2x = 24), whilst reduction from the dihaploid gives the monohaploid (2n = x = 12). Both levels have important uses (Hermesen and Ramanna 1981) and several schemes have been devised incorporating them into potato breeding programmes (Wenzel 1981).

Production of dihaploids by pseudogamy (Hougas and Peloquin 1958; Hermesen and Verdenius 1973) is so refined a procedure that it largely outcompetes the alternative approach of tissue culture. However, advances have been made, and in a recent report, out of 20 tetraploid clones tested in anther culture, 19 produced embryoids and 90% of regenerated plants were found to be dihaploid (Johansson 1986). Monohaploids can also be produced parthenogenetically (van Breukelen et al. 1977; Jacobsen 1978), but in this case anther culture is a strong alternative (Sopory and Tan 1979; Foroughi-Wehr et al. 1977; Jacobsen and Sopory 1978).

Successful regeneration from other forms of explants has been achieved in a wide range of potato cultivars. Recent reports include cvs. Golden Wonder (Austin and Cassells 1983), Maris Bard (Karp et al. 1982; Webb et al. 1983), Cara (Ahloowalia 1982; Wheeler et al. 1985), Arran Banner, Desirée, King Edward, Majestic, Maris Piper, Pentland Crown, Record, Romano, Spunta, Up to Date, Fortyfold, Champion and Myatts Ashleaf (Webb et al. 1983; Wheeler et al. 1985). Regeneration has also been achieved from explants of monohaploid (Chwilkowska 1982; Karp et al. 1984) and dihaploid clones (Jacobsen 1981; Chwilkowska 1982; Karp et al. 1984).

Success in regeneration from cultured protoplasts has also been widespread and reports include cvs. Russet Burbank (Shepard et al. 1980), Maris Bard (Thomas 1981), Bintje (Bokelmann and Roest 1983; Sree Ramulu et al. 1983), Fortyfold (Karp et al. 1982), Majestic (Creissen and Karp 1985), Pentland Crown (Creissen unpubl. results), Desirée and King Edward (Foulger and Jones 1986) as well as numerous breeding lines (Haberlach et al. 1985).

Somaclonal variation has been described in all these systems, but most extensively in regeneration from explants such as leaf pieces, and in regeneration from leaf mesophyll protoplasts.

2.1 Somaclonal Variation in Explant Regeneration

Indications that variability might arise through tissue culture of explants were found in an experiment in cv. Desirée, devised to study the effect of X-irradiation of rachis, petiole and leaf explants. There was a clear increase in mutation frequency with increasing doses of X-irradiation, and average frequencies were 68.2% for rachis and petiole explants and 85.9% for leaf explants. Surprisingly, however, non-irradiated controls also showed a high mutation frequency. Furthermore, in the case of rachis and petiole explants, this was almost as high (50.3%) as observed in the irradiated material. The authors suggested that the "spontaneous mutation" observed amongst explant regeneration might provide a means of mutation breeding without application of mutagens (Van Harten et al. 1981).

Numerous reports since then have indicated that explant regeneration is a source of somaclonal variation in potato. Two variants were observed amongst regenerants from cultured shoot apices of cv. Cara (Ahloowalia 1982), whilst variation has also been described in explant-derived regenerants of cv. Golden Wonder (Austin and Cassells 1983) and in leaf-explant-derived clones of cvs. Desirée and Record (Wheeler et al. 1985).

Field trials of explant-derived potato clones have not only confirmed that variability is present, but also provided evidence that some of the changes could be useful. Amongst several hundred clones of cv. Desirée, regenerated from leaf, stem or rachis pieces, variation was observed in maturation time, tuber characteristics such as shape, size, number (Fig. 2) and colour, eye depth, leaf shape and size, as well as in total yield (generally lower) and ware tuber yield (generally higher). Two clones were found to have significantly high dry matter content, and whilst no change was observed in response to cyst nematode *(Globodera pallida)*, the somaclonal populations showed better response against infection by scab *(Streptomyces scabies)* compared with control clones grown in the same field plots (Evans et al. 1986).

2.2 Somaclonal Variation in Protoplast Regeneration

Amongst the first reports to draw attention to somaclonal variation in potato were those of Shepard and co-workers in cv. Russet Burbank. Over 1000 clones regenerated from leaf mesophyll protoplasts were screened in field trials lasting several

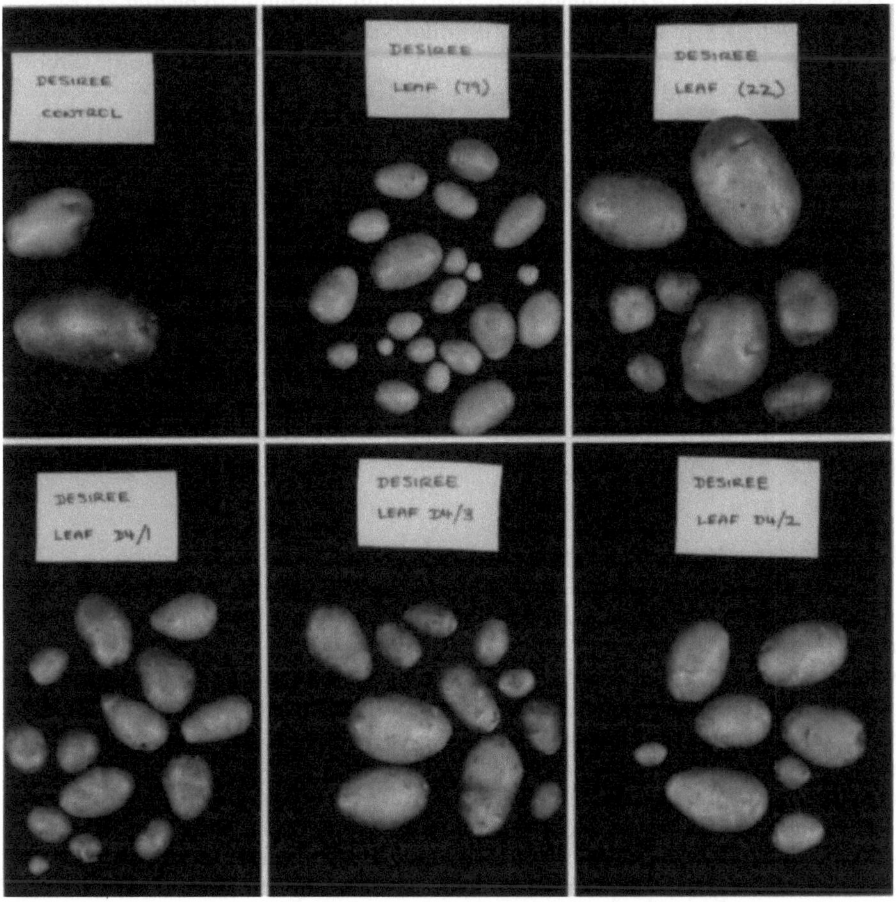

Fig. 2. Tubers of different leaf-explant derived clones of cv. Desirée, showing variation in size, number and shape. Representative tubers of a control clone are also shown (*top left*). *Numbers in brackets* and labelling *D4/1, 2, 3*, refer to the clone identities

years. Variation was extensive and observed in growth habit, tuber shape and size, skin colour, photoperiod requirements and maturation date (Shepard et al. 1980; Secor and Shepard 1981). Twenty somaclones out of 800 tested were found to show greater resistance to late blight *(Phytophthera infestans)*. It was also demonstrated that the frequency of resistant clones could be increased by isolating protoplasts from selected resistant clones and regenerating a second generation of somaclones (Ayers and Shepard 1981).

Although the field work of Shepard and co-workers was amongst the most extensive, morphological variation of the type described in cv. Russet Burbank was observed by several groups in a variety of cultivars. In one early study of cv. Maris Bard, out of 45 plants regenerated from protoplasts, none resembled the other in the ten characters scored and only one resembled the parental cultivar (Thomas et al.

1982). In later studies the frequency of normal plants was found to be higher (Fish and Karp 1986; Karp and Fish 1987). Similar variation was observed amongst protoplast-derived regenerants of cultivars Fortyfold (Karp et al. 1982), Bintje (Sree Ramulu et al. 1983) and Majestic (Creissen and Karp 1985).

More recently, replicated field trials were carried out with cvs. Feltwell and Maris Piper (Thomson et al. 1986). In Maris Piper, 13 somaclones (13% of the starting population) were selected for further trial on the basis of yield performance.

2.3 Somaclonal Variation and Disease Response

The occurrence of variation during culture, coupled with the observations that regenerated potato plants could show changes in agronomic traits, has led to an interest in the possibility of selecting disease-resistant plants through tissue culture. A number of examples will be described in which this approach has been adopted in potato.

2.3.1 Somaclonal Variation for Phytophthora infestans (Late Blight)

Phytophthora infestans, which causes late blight, is the most important fungal parasite of potato. As mentioned earlier, changes in response against *Phytophthora* have been observed amongst potato somaclones, and in cv. Russet Burbank 2% of somaclones were identified as showing significantly increased resistance (Shepard et al. 1980). The resistance in these regenerated plants did not appear to be of the major or R-gene type, but was expressed as a reduced number of infection sites, a decreased rate of pathogen development or reduced symptom intensity.

Regenerated plants expressing increased resistance to *Phytophthora* have also been obtained following selective screening. Behnke (1979) used undiluted culture filtrates of the fungus to prepare a toxic media containing the culture fluid of one pathotype. Callus initiated from leaf explants were placed on the toxic medium. The concentration of toxin resulted in 90% mortality. Sensitive calli turned brown and died within 3 weeks. After 4–6 weeks, some calli showed small growing areas. Out of more than 41,000 calli, 153 with white areas were transferred to new toxic medium, and of these 36 survived. Some of the original cultures were irradiated, and 7 out of 1400 of these survived transfer to successive toxic media. Potato plants were regenerated from the resistant calli and their disease response assessed. The number of sporangia formed was *not* significantly reduced in the somaclones, but the regenerated plants had significantly smaller lesions following mechanical inoculation with *Phytophthora* spores. The resistance did not follow a simple mode of inheritance (Behnke 1980).

In a more refined experiment, Wenzel and co-workers extracted purified toxin from the fungal growth medium by ultra-filtration. The effectiveness of the toxin was tested at different concentrations by measuring the increase in fresh weight of shoot-tip cultures. A constant concentration of the toxin was then added to defined culture medium. Dying callus sensitive to the toxin turned black, but small areas of white callus growth could be discerned in some calli. Some resistant plants have been regenerated and are now being assessed (Wenzel et al. 1987).

2.3.2 Somaclonal Variation for *Alternaria solani* (Early Blight)

Alternaria solani is the causal agent of early blight in potato. The fungus produces toxins which can be extracted and purified. Somaclones of potato (cv. Russet Burbank) regenerated from cultured protoplasts were tested for disease response by inoculation of their leaves with the purified toxins (Matern et al. 1978). Five hundred protoplast-derived plants were tested and four types of reaction were observed (1) highly sensitive (2) sensitive (3) intermediate and (4) insensitive. In highly sensitive plants the whole leaf blade turned yellow and died, whereas the other classes differed in the size of the lesions. In sensitive plants a 2–3-cm diameter lesion developed, in intermediate plants the lesion was only 1–2 cm in diameter, and in insensitive plants only a small necrotic fleck developed. Insensitive plants retained their characteristics of disease response over successive generations.

2.3.3 Somaclonal Variation for *Streptomyces scabies* (Scab)

Changes in disease response against scab *(Streptomyces scabies)* were observed in leaf-explant derived clones of cv. Desirée (Evans et al. 1986). As described later (Sect. 5.2), this change took the form of a population shift in which the scab infestation of the somaclones was generally lower than the controls over successive field trials. Two of the regenerated clones had significantly increased scab resistance over three seasons in the field. The nature of this change in response is not known.

Clones resistant against scab were also isolated from protoplast-derived regenerants of cv. Feltwell. Seed tubers of the somaclones were planted in replicated field trials in successive years. Eleven somaclones were more resistant to scab than the controls, and two of these clones were also more resistant to potato virus Y (Thomson et al. 1986).

2.3.4 Somaclonal Variation for *Fusarium* (Dry Rot)

The selective screening experiments with fungal toxins used by Wenzel and co-workers for selection of increased resistance against *Phytophtora* were also adapted for screening for resistance against *Fusarium*. Protoplast populations of four different dihaploid clones were grown and defined culture media containing exotoxin(s) produced by *F. sulphureum* or *F. solani* var. *coeruleum*. From the four susceptible clones, about 1500 protoclones resistant to the toxin at the cellular level were regenerated to give whole plants. These plants are now being tested in field trials (Wenzel et al. 1987).

2.3.5 Somaclonal Variation for Cyst Nematodes and Virus Resistance

In the protoplast-derived clones of cv. Feltwell described earlier (Thomson et al. 1986), the resistance formerly expressed by Feltwell to cyst nematode infection was unaffected by the regeneration and all clones tested so far remained resistant. In addition to the two clones resistant to scab and potato virus Y, one somaclone exceeded the potato leaf roll resistance of Feltwell and another its resistance for both virus diseases. Wenzel and Uhrig (1981) have demonstrated that anther culture can be used as a means of selecting disease-resistant plants. Exploiting the spontaneous chromosome doubling which can occur during anther culture, they isolated doubled monohaploid clones (from anther culture of dihaploid potato lines) which carried resistance to cyst nematode and to potato viruses X, Y and leaf roll. The ratio of resistant clones was relatively high, but the major limitation of the approach is the genotype-dependence of anther culture. Attempts are now underway to breed tissue culture response into valuable genotypes.

The wealth of variation present amongst protoplast- and explant-derived potato somaclones of potato held promise for potential application to potato breeding, but before such promise could be fulfilled a closer examination of the variation was required in order that the nature and stability of the changes could be ascertained.

3 The Genotypic Basis of Somaclonal Variation in Potato

Early studies of chromosome numbers in potato plants regenerated from protoplasts indicated that much of the gross morphological variation was due to aneuploidy (Table 1). In the initial experiments in cv. Maris Bard (Thomas 1981; Thomas et al. 1982) less thant 4% of the regenerants were found to have retained the normal tetraploid chromosome number of $2n=4x=48$. In cv. Fortyfold, regenerated using a different protocol, a higher frequency of tetraploids was observed (30%) and the aneuploidy was of more limited range (Karp et al. 1982). Similar aneuploid frequencies were reported for protoplast-derived regenerants of cv. Bintje (Sree Ramulu et al. 1983). In later studies, where regeneration protocols have been improved, frequencies of normal tetraploids have been increased. On average, 57% of protoplast-derived regenerants of cv. Majestic were tetraploid (Creissen and Karp 1985), whilst in a recent study in cv. Maris Bard, 60% of regenerants were normal (Fish and Karp 1986) as were protoplast-derived regenerants of cv. Russet Burbank (Gill et al. 1986) (Table 1).

These results indicate that for protoplast regeneration in potato, about 50% of regenerated plants may carry abnormal chromosome numbers. Structural chromosome aberrations have also been observed (Creissen and Karp 1985; Gill et al. 1986; Fish and Karp 1986; Sree Ramulu et al. 1986). As described later, chromosome variation is also present in explant-derived plants, but to a much lesser degree. What is the relationship between these chromosome aberrations and the morphological variation observed amongst potato somaclones?

Three classes of regenerated plants can be recognized with respect to their cytological status (Creissen and Karp 1985; Fish and Karp 1986). The first class are the tetraploids ($2n = 4x = 48$), which are morphologically similar to the controls. The second class are aneuploid at the tetraploid level (or 48 ±, e.g. $2n = 47$, $2n = 49$) (Fig. 3b). These plants are sometimes, but not always, different from the controls (Fig. 3a), depending on which chromosomes are lost/gained and how many. The final class (48 + +) are octaploid ($2n = 8x = 96$) or aneuploid at the octaploid level

Table 1. Summary of chromosome data from our studies of protoplast- and explant-derived regenerants of potato. Data given are averages – actual frequencies are genotype-dependent (see text)

Regeneration	Original ploidy level	Regenerants Ex[a]	Px[a]	Ax[a]
Protoplasts	4x	50–60	20–30	10–20
	2x (dp)[b]	25	75	–
	2x (dh)[b]	–	100	–
	x (mh)[b]	Data not available		
Explants	4x	87	–	13
	2x (dp)	50	50	–
	2x (dh)	40	60	–
	x (mh)	9	91	–

[a]Ex = euploid. Px = polyploid or aneuploid at the polyploid level. Ax = aneuploid at the original ploidy level.
[b]dp = diploid (e.g. *Solanum brevidens*). dh = dihaploid *S. tuberosum*. mh = monohaploid *S. tuberosum*.

Fig. 3a-d. Cytological classes and their representative morphology in protoplast-derived regenerants of cv. Maris Bard: (**a**) 48 ± plants, control centre; (**b**) a 48 ± cell showing 2n = 47; (**c**) a 48 + + cell showing 2n = 96; (**d**) 48 + + plants, control centre (e.g. 2n = 93, 2n = 80) (Fig. 3c). Such plants always appear different from the controls (Fig. 3d). In short, most regenerated plants with grossly abnormal morphology turn out to be in the 48 + + class. Plants with other, less severe, variations are often 48 ± plants, but not all aneuploids in this class appear abnormal.

Clearly, chromosome variation accounts for quite a large proportion of the morphological variation seen in potato somaclones. As discussed later, this is particularly the case for protoplast regeneration. However, when plants in the 48 class are examined in the field, it soon becomes clear that variation is still present. Together with the observations of structural chromosome variation, this suggests that other changes in the genome, many of which may not be detectable at the cytological level, may be occurring.

Heritable nuclear variation has been identified in explant-derived somaclones of cv. Desirée, where a change from normal red-skinned to white-skinned tubers was found to be inherited in a simple Mendelian fashion (Bright pers. commun.). Gene mutations have been described in somaclones of many other species (Karp and Bright 1985), and it can only be a matter of time before more are identified in potato. Other evidence suggesting the occurrence of genetic changes in potato include a report in cvs. Maris Piper and Feltwell, where variation was observed in the electrophoretic band pattern of tuber proteins (Smith 1986). Variation in tuber proteins amongst leaf-explant derived clones has also been observed in our own laboratory (Fig. 4).

More precise data of genomic variation has been provided by molecular studies. Two out of 12 protoplast-derived clones have been identified as carrying deficiencies in 25S ribosomal DNA (Landsmann and Uhrig 1985), whilst cytoplasmic variation has been observed in cv. Russet Burbank, where changes were detected in mitochondrial DNA (Kemble and Shepard 1984). Changes in the number and intensity of restriction fragments after probing with a ribosomal coding sequence have also been observed amongst leaf-explant-derived somaclones of cvs. Desirée and Maris Bard (Fig. 5) (Potter and Jones 1988).

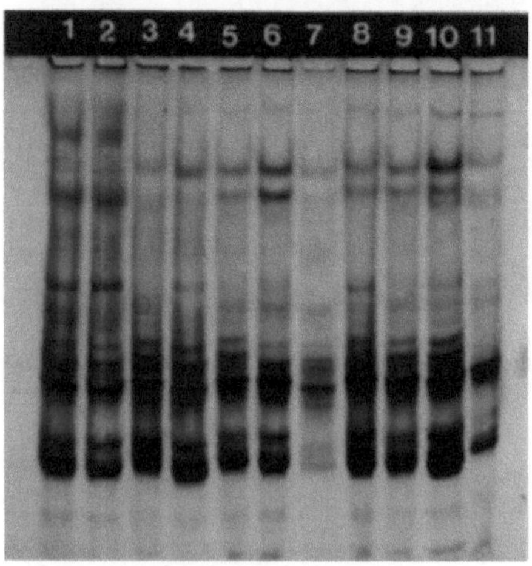

Fig. 4. Soluble tuber protein electropherogram of leaf-explant-derived regenerants of cv. Desirée showing somaclonal variation (R. Potter unpubl. results). Track *11* is the control

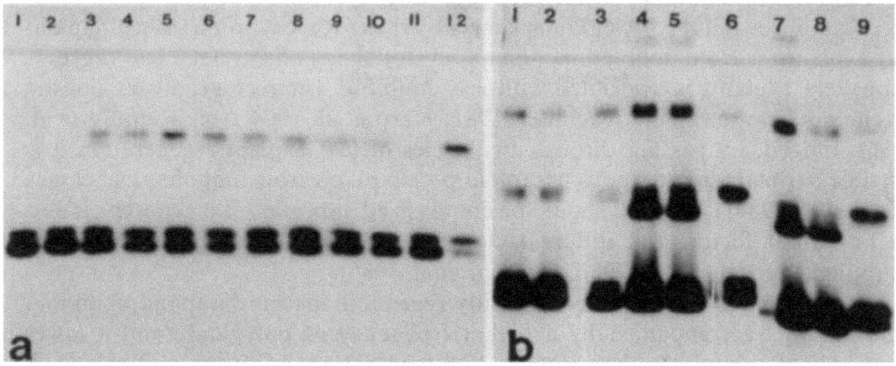

Fig. 5. **a** Eco R1 restricted DNA from leaf-explant-derived clones of cv. Desirée probed with a ribosomal coding sequence (Landsmann and Uhrig 1985). Lane *12* is the control. The *top band* is very faint in tracks *10* and *11* and is missing in tracks *1* and *2*. **b** Variation in band intensity and number in Eco R1 restricted DNA from protoplast-derived regenerants of cv. Maris Bard probed with the same probe. Track *9* is the control (R. Potter unpubl. results)

In addition to genetic mutations, some changes may occur during regeneration of potato that do not involve alterations in the informational content or quantity of DNA, but instead are related to expression. The inference comes from two sources. Firstly, not all the changes are stable. For example, a change from purple-splashed to white-skinned tubers of protoplast-derived regenerants of cv. Fortyfold was found to show reversion to varying degrees amongst tubers of different somaclones (Nelson 1984). Secondly, in maize, recent evidence indicates that changes in the degree of DNA methylation occur during regeneration of plants (Brown and Lörz 1986). Such changes have yet to be domonstrated in potato.

4 Factors Affecting Somaclonal Variation in Potato

A number of factors affect the degree to which instability is present amongst somaclones. In potato, these have been most extensively identified with respect to chromosome variation, as data on other genetic changes are somewhat limited.

One factor is the choice of regeneration system adopted, more precisely whether regeneration is from protoplasts or from explants. Explant regeneration is relatively simpler and quicker than regeneration from protoplasts, and consequently it is usually associated with much less instability. For example, a cytological survey of potato plants regenerated from leaf, stem, rachis and tuber pieces, indicated that almost 90% were normal tetraploids (Table 1), compared with 50–60% for protoplasts (Wheeler et al. 1985).

The ploidy of the source material is also, however, of importance (Table 1). Regeneration from leaf explants of monohaploid potato is associated with a high frequency of chromosome doubling (Karp et al. 1984). Cytophotometric studies have shown that doubling occurs very early during culture (Jacobsen et al. 1983; Sree Ramulu et al. 1985, 1986; Pijnacker et al. 1986a). Regeneration from dihaploid leaf explants or from doubled monohaploid (following one cycle of leaf regeneration) leaf explants is associated with less doubling. On average, 50–60% of the regenerants are tetraploid (Jacobsen 1981; Karp et al. 1984; Sree Ramulu et al. 1986). Indeed, by passing through two cycles of leaf explant regeneration it is possible to obtain homozygous tetraploid potato plants from monohaploid clones (Karp et al. 1984) (Fig. 6). It should be emphasized, however, that genotype is also an important factor, and different clones may show different frequencies of doubling (Wenzel et al. 1979; Tempelaar et al. 1985).

In conjunction with this effect of ploidy, regeneration from dihaploid, or diploid protoplasts is accompanied by a higher frequency of polyploidy and a lower frequency of aneuploidy than regeneration from tetraploid cultivars (Table 1). A cytological study of plants regenerated from protoplasts of a diploid, non-tuber-bearing wild relative of potato, *Solanum brevidens,* showed that 75% of the regenerated plants were either tetraploid (50% of the total) or aneuploid at the tetraploid level (25% of the total) (Nelson et al. 1986). Similarly, regeneration from dihaploid protoplasts results in nearly all tetraploid or higher polyploid regenerants (Wenzel et al. 1979; Sree Ramulu et al. 1986).

The source of plant cells from which regeneration is induced also affects the degree of variability observed. This is not surprising, since it has been known for a long time that changes in chromosome content, via endopolyploidy or endmitosis, may accompany the differentiation of plant cells (D'Amato 1952). In potato monohaploid and dihaploid clones, cytophotometric measurements have shown

Fig. 6. Leaves of monohaploid clone 7322 and culture-derived regenerants. The fifth youngest leaf (*left*) and an older fully expanded leaf (*right*) of: *top left* 7322 (2n = x = 12); *top right* doubled monohaploid regenerant (2n = 2x = 24); *bottom left* doubled monohaploid (2n = 4x = 48), this clone was unstable and may not have contained a balanced set of chromosomes; *bottom right* leaf from a tetraploid cultivar

that ploidy variation is present in the leaf tissue from which explants are excised and protoplasts isolated (Jacobsen et al. 1983). Also, at least in tobacco, regeneration from midveins of aged haploid leaves gives rise to more chromosome doubling than regeneration from equivalent regions of young leaves (Kasperbauer and Collins 1972). Similarly, in potato, more aneuploidy was observed in regeneration from tuber pieces than from other explants, although the sample sizes were small (Wheeler et al. 1985) and regeneration from tuber protoplasts gave rise to a high frequency of 48 + + plants (Jones et al. 1989).

Cell suspensions appear to be a particularly poor source of protoplasts as far as chromosome stability is concerned. Cell suspensions of *S. brevidens* were found to be mostly polyploid, whilst a high incidence of aneuploidy, polyploidy and chromosome structural aberrations have been observed in cell suspension of tetraploid cultivars (Karp unpubl.). Similarly, cell cultures of monohaploid and dihaploid potato clones contain a high degree of both numerical and structural chromosome variation (Pijnacker et al. 1986b). Such observations are important, as cell suspensions are often used as a source of protoplasts for protoplast fusion.

Other factors that affect the degree of chromosome variation observed amongst potato somaclones are aspects of the regeneration procedure such as length of time in culture and the hormonal composition of the media. The former can generally be summarized as a simple rule in that the longer the culture phase the greater the chance of instability, although very long periods in culture (e.g. a year) may sometimes be required for this effect to be seen. Hormonal effects are much less easily defined; they are complex and often contradictory (Karp and Bright 1985). Furthermore, they may vary depending on the genotype concerned, and even between different protoplast isolations within a genotype (Creissen and Karp 1985).

We have recently carried out a study in potato where the hormone composition of the media was changed at different stages in the regeneration procedure. This study was designed not to look at hormone effects per se but rather to attempt to increase the frequency of normal tetraploids amongst regenerants from protoplasts of potato cultivars. Numerous studies have indicated that gross genomic changes occur early during culture in both explant and protoplast regeneration systems (Sree Ramulu et al. 1984, 1985; Carlsberg et al. 1984; Pijnacker et al. 1986a,b). The first study carried out therefore investigated the effect of changing the hormone content of the initial protoplast culture medium. However, when regenerated plants were examined following a range of initial treatments, no differences were observed in the frequency of normal tetraploid regenerants (Table 2) (Fish and Karp 1986). Clearly, whatever influences the hormones may have during the initial stages were not manifested in the regenerated plants. In contrast, it had earlier been shown in cv. Bintje that the percentage of normal plants varied when different hormones were used in the regeneration medium (Sree Ramulu et al. 1983). Selection is clearly operating at this stage since, as is known from other species, more variation is present in callus than appears in regenerated plants (Orton 1980; Ogihara 1981; Browers and Orton 1982). In a second experiment in the same potato cultivar, the hormonal content of the regeneration medium was therefore altered. This time significant differences in the frequencies of tetraploids were present (Table 2) following the different hormonal treatments. These results suggest that the callus phase, when shoot regeneration is initiating, may be a critical stage in which manipulation of the

Table 2. The percentage of protoplast-derived somaclones tetraploid (48) aneuploid (48 ±) or octoploid and aneuploid at the octoploid level (48 + +) in two media experiments. In experiment 1, the initial culture medium was altered: Treatments 1 and 2 had 2 and 5 mg/l 2,4-D, respectively whilst 3 and 4 had 1 mg/l NAA and 0.5 mg/l BAP with extra salts added in 4 (Fish and Karp 1986). In experiment 2, the callus medium differed: Treatments 1* and 2* had 1 and 2 mg/l 2,4-D respectively whilst 3* and 4* had 0.1 mg/l IAA and 1 mg/l zeatin with 0.1 mg/l GA_3 added in 3 (Karp and Fish 1987)

Initial	Experiment 1			Experiment 2			Callus
treatment	48	48 ±	48 + +	48	48 ±	48 + +	treatment
1	62.2	6.7	31.9	36.4	17.2	46.4	1*
2	59.6	8.5	31.9	14.8	40.7	44.4	2*
3	62.8	9.3	27.9	36.7	24.5	38.8	3*
4	69.8	9.3	20.9	51.0	16.0	33.0	4*
	No significant differences			Significant differences			

media can influence chromosome stability. However, none of the procedures tested so far increased the frequency of stable tetraploids (consistently) above 60%, and much more work is needed in this area.

5 Somaclonal Variation and Potato Breeding

How realistic is the view that somaclonal variation can make a contribution to the breeding of new potato cultivars? In order to attempt an answer to this question, an assessment must be made firstly of the criticisms and drawbacks of somaclonal variation and secondly of what the variation has to offer.

5.1 Criticisms

One of the major criticisms of any source of variability, whether it be mutagen-induced or tissue culture-induced, is that not all of the variation is useful. This is particularly a problem in protoplast-derived potato plants where, as described earlier, 50% of the regenerants may be aneuploid at the tetraploid (48 ±) or octaploid (48 + +) levels. Aneuploidy at the octoploid level is associated with gross morphological abnormalities, including distortion of leaf shape (Fig. 3d), stunted growth and poor tuber yield. Aneuploids at the tetraploid level may also have leaf and tuber abnormalities (Fig. 3a). Clearly, none of these changes is desirable and all such plants would have to be removed in a breeding programme. The 48 + + plants are easy to distinguish at an early stage in the regeneration procedure, but this is not the case for 48 ± plants, where differences may only become apparent during tuber production (Creissen and Karp 1985; Fish and Karp 1986; Sree Ramulu et al. 1986). In close conjunction with this problem is the fact that many of the changes that occur in agronomically useful traits are changes in the wrong direction. The breeder has no use for reduction in tuber yield, for example, whatever the cause of the reduction might be.

A second major criticism is that not all of somaclonal variation is as "novel" as initially claimed. A recent study in cv. Russet Burbank has likened much of the subtle variation (not associated with aneuploidy) to that observed amongst "bolters" which routinely appear in potato field trials (Sandford et al. 1984). Others have also drawn parallels between somaclonal variants and "sports" which can occur at surprisingly high frequencies (Preil 1986). This is particularly well illustrated in ornamentals. Out of 274 azalea cultivars, for example, 144 (= 52.5%) originated as "sports" (Heursel 1977). Other examples have also been reviewed in the literature (Wasscher 1956; Horn 1968; Beauchesne 1983).

Perhaps one of the major problems of somaclonal variation, however, is unpredictability. The breeder has no guarantee that a specific desired change will occur, or even that it will take place in a desired cultivar. For example, amongst the same population of several hundred explant-derived somaclones of cv. Desirée, changes were found in response to scab (*Streptomyces scabies*) but not to cyst nematodes (*Globodera pallida*) and changes in tuber colour were observed from red to white-skinned tubers but not vice versa (Wheeler et al. 1985; Evans et al. 1986). Similarly in field trials of some 400 protoplast-derived somaclones significant increases in yield were present in 13 clones of cv. Maris Piper but not amongst clones of cv. Feltwell (Thomson et al. 1986).

Another serious problem is that when changes do occur they are not always stable. Unstable changes in tuber colour (as described earlier in cv. Fortyfold) can at least be visibly monitored, but what of such changes in disease response or yield? Evidence for unstable variation is increasing in a number of species (Oono et al. 1985; Groose and Bingham 1984; Lörz and Brown 1986), and there is no reason to suppose that potato would be free of such changes.

5.2 What Somaclonal Variation Has to Offer

There are several replies to the criticisms voiced about somaclonal variation. Accepting that not all of the variation is useful, or stable, the fact remains that some changes are. Clones with significantly increased yield were selected out of protoplast-derived plants of cv. Maris Piper (Thomson et al. 1986). Similarly, although there was an overall decrease in total yield for leaf-explant-derived regenerants of cv. Desirée, there was an increase in ware tuber yield (Evans et al. 1986). In cv. Russet Burbank, out of 500 somaclones tested for disease response to early blight, five clones were isolated with greater resistance, and out of 800 tested for response to late blight, 20 were found to be more resistant (Matern et al. 1978; Secor and Shepard 1981). In addition, amongst leaf-explant-derived clones of cv. Record, two have been identified with significantly higher dry matter contents (Bright et al. 1984). It should also be remembered that some changes which are apparently useless, for example leaf shape, could be important for establishment of a new cultivar, which after all needs to be distinct if it is to be registered as a new variety.

In addition to these successes, a second important factor is the frequency at which these changes can be found. It is difficult to place an exact figure on this but they can clearly be very high. In cv. Russet Burbank, when 65 protoplast-derived clones were tested in replicated field trials, each clone was significantly different in at least one character, and some clones were altered in 17 out of the 35 traits assessed

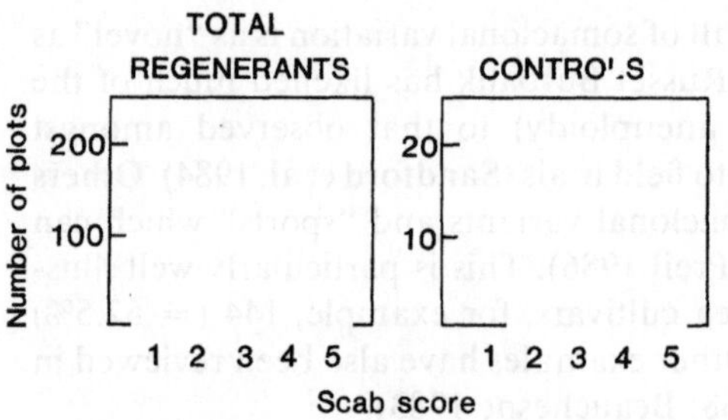

Fig. 7. The distribution of scab scores (range; *1* = 0% – *5* = 75% of surface affected) in plants grown in field plots of Broughshane, Ireland, in 1984. The plots contained explant regenerants and controls all grown at one site

(Ayers and Shepard 1981). Similar frequencies were observed in the explant-derived clones (Evans et al. 1986). Coupled with this is the observation that some of the changes take the form of whole population shifts. When the number of scab lesions per tuber were plotted for the Desirée leaf-explant somaclones, it was found that the mean value of the whole population had shifted towards better response against scab compared with the controls (Fig. 7). This shift was maintained over 3 years, and although there was not an exact correlation for individual clones between the different years, 12 out of 19 clones identified as having low scab scores in 1982 and 1983 also had low scab scores in 1984, demonstrating individual stability over the 3 years (Evans et al. 1986).

These whole population changes have been observed in other species, for example, in anther culture-derived doubled haploids of tobacco (De Paepe et al. 1982; Dhillon et al. 1983) and in sugarcane (Krishnamurthi 1974), and are a distinctive feature of somaclonal variation. Other distinctive features include a high frequency of homozygous recessive mutations, and variation in quantitative as well as qualitative traits (George and Rao 1982; Larkin et al. 1984; Karp and Bright 1985).

5.3 Weighing up the Potential of Somaclonal Variation for Improvement of the Potato Crop

It is clear from the discussion that somaclonal variation is not the complete answer to the potato breeder's problems. Realistic objectives, with careful choice of starting material and experimental procedure, are clearly required. There are numerous problems associated with the application of the variation, and extensive screening will still be needed, particularly in the early stages, to remove unwanted variants such as aneuploids. However, judging from the "successes" described in the previous section, a somaclonal programme is a chance worth taking. Breeders may prefer explant rather than protoplast regeneration, as the former show greater chromosome stability but sufficient variation in the field to be of interest. Furthermore, in the case of disease response, in vitro screens may increase chances of success, and hybrid introgression could be achieved by placing hybrids into culture. So far, the potato somaclonal programmes described here have produced clones of strong interest to the breeders. Considering that the starting population is usually of a few hundred clones, rather than the thousands used in conventional programmes, can breeders really not afford to give somaclonal variation a chance?

6 Conclusions and Prospects

Potato is amenable to a wide range of tissue culture techniques. Coupled with advancing technology of genetic manipulation, this provides a powerful tool for improvement of the potato crop in entirely novel ways. The regeneration systems are based on clonal propagation and should result in genetic stability. However, regeneration through a callus phase can be associated with considerable variability, or somaclonal variation. Such instability is a problem in techniques such as transformation and protoplast fusion, where a specific change is desired, and the causes of somaclonal variation should be elucidated. In contrast, the variation itself could be of direct use to breeders. Chromosome doubling during culture can be exploited as a means of doubling haploids, chromosome structural changes as a means of achieving introgression, and in vitro selection as a means of obtaining changes in disease response. Furthermore, extensive field trials by a number of different groups have resulted in the isolation of some clones with improved characteristics. The prospects for utilization of somaclonal variation for crop improvement are good in potato, provided a realistic approach is made and the drawbacks of the technique are clearly understood. Much more work is needed in understanding the causes and origins of the variation, so that some control can be exercised over it, and so that stability can be achieved when necessary, or variability enhanced where so required. With this control will come a greater confidence in the potential of somaclonal variation for improvement of the potato crop.

Acknowledgments. The author thanks all the members of the Biochemistry Department who have been involved in potato tissue culture for their fruitful collaboration.

References

Ahloowalia BS (1982) Plant regeneration from callus culture in potato. Euphytica 31:755–759

Austin S, Cassells AC (1983) Variation between plants regenerated from individual calli produced from separate stem callus cells. Plant Sci Lett 31:107–114

Ayers AR, Shepard JF (1981) Potato variation. Environ Exp Bot 2:379–381

Bajaj YPS (1986) Cryopreservation of potato somaclones. In: Semal J (ed) Somaclonal variations and crop improvement. Nijhoff, Dordrecht, pp 244–251

Bajaj YPS (ed) (1987) Biotechnology in agriculture and forestry, vol 3. Potato. Springer, Berlin Heidelberg New York Tokyo

Beauchesne G (1983) Appearance of plants not true to type during in vitro propagation. In: Earle D, Demarly Y (eds) Variability in plants regenerated from tissue culture. Praeger, New York, pp 268–272

Behnke M (1979) Selection of potato callus for resistance to culture filtrates of *Phytophthora infestans* and regeneration of resistant plants. Theor Appl Genet 55:69–71

Behnke M (1980) General resistance to late blight of *Solanum tuberosum* plants regenerated from callus resistant to culture filtrates of *Phytophthora infestans*. Theor Appl Genet 56:151–152

Bokelmann SJ, Roest S (1983) Plant regeneration from protoplasts of potato (*Solanum tuberosum* cv. Bintje). Z Pflanzenphysiol 109:259–265

Bright SWJ, Wheeler JA, Kirkman MA (1984) Variation in Record clones from tissue cultures. In: 9th Triennial Conf Eur Assoc Potato research, Interlaken, Abstr, pp 62–63

Browers MA, Orton TJ (1982) Transmission of gross chromosomal variability from suspension cultures into regenerated celery plants. J Hered 73:159–162

Brown PTH, Lörz H (1986) Molecular changes and possible origins of somaclonal variation In: Semal J (ed) Somaclonal variations and crop improvement. Nijhoff, Dordrecht, pp 148–159

Carlsberg I, Glimelius K, Eriksson T (1984) Nuclear DNA content during the initiation of callus formation from isolated protoplasts of *Solanum tuberosum* L. Plant Sci Lett 35:225–230

Chwilkowska B (1982) Callus formation and regeneration of plants from mono- (2n = x = 12) and dihaploid (2n = 2x = 24) *Solanum tuberosum* plants. Tsitol Genet 16:49–55

Correll DS (1962) The potato and its wild relatives. Section Tuberarium of the genus *Solanum*. Texas Res Found Contrib 4: p 606

Creissen GP, Karp A (1985) Karyotypic changes in plants regenerated from protoplasts. Plant Cell Tissue Org Cult 4:171–182

D'Amato F (1952) Polyploidy in the differentiation and function of tissues and cells in plants. A critical examination of literature. Caryologia 4:311–357

Denton IR, Westcott RJ, Ford-Lloyd BV (1977) Phenotypic variation of *Solanum tuberosum* L cv. Dr McIntosh regenerated directly from shoot-tip culture. Potato Res 20:131–136

De Paepe R, Prat D, Huguet T (1982) Heritable nuclear DNA changes in doubled haploid plants obtained by pollen culture of *Nicotiana sylvestris*. Plant Sci Lett 18:11–28

Dhillon SS, Wernsman EA, Miksche JP (1983) Evaluation of nuclear DNA content and heterochromatin changes in anther-derived dihaploids of tobacco (*Nicotiana tabacum*) cv. Coker 139. Can J Genet Cytol 25:169–173

Evans NE, Foulger D, Farrer L, Bright SWJ (1986) Somaclonal variation in explant-derived potato clones over three tuber generations. Euphytica 35:353–361

Fish N, Karp A (1986) Improvements in regeneration from protoplasts of potato and studies on chromosome stability. 1. The effect of initial culture media. Theor Appl Genet 72:405–412

Foroughi-Wehr B, Wilson HW, Mix G, Gaul H (1977) Monohaploid plants from anthers of a dihaploid genotype of *Solanum tuberosum* L. Euphytica 26:361–367

Foulger D, Jones MGK (1986) Improved efficiency of genotype-dependent regeneration from protoplasts of important potato cultivars. Plant Cell Rep 5:72–76

George L, Rao PS (1982) Yellow-seeded variants in in vitro regenerants of mustard (*Brassica juncea* Cross var. Rai-5). Plant Sci Lett 30:327–330

Gill BS, Kam-Morgan LNW, Shepard JF (1985) An apparent meiotic mutant in a mesophyll cell protoclone of the Russet Burbank potato. J Hered 76:17–20

Gill BS, Kam-Morgan LNW, Shepard JF (1986) Origin of chromosomal and phenotypic variation in potato protoclones. J Hered 77:13–16

Groose RW, Bingham ET (1984) Variation in plants regenerated from tissue culture of tetraploid alfalfa heterozygous for several traits. Crop Sci 24:655–658

Grun P (1974) Cytoplasmic sterilities that separate the group Tuberosum cultivated potatoes from its putative tetraploid ancestors. Evolution 27:633–643

Haberlach GT, Cohen BA, Reichert NA, Baer MA, Towill LE, Helgeson J (1985) Isolation, culture and regeneration of protoplasts from potato and several related *Solanum* species. Plant Sci 39:67–74

Hanneman RE Jr (1981) Strategy of the inter-regional potato introduction project JR-1. Rep Plann Conf Exploration, taxonomy and maintenance of potato germplasm III. Int Potato Centre, Lima 1979, pp 162–165

Hawkes JG (1978) The potato crop. Chapman & Hall, London, pp 1–14

Hermesen JGTh, Ramanna MS (1981) Haploidy and plant breeding. Philos Trans R Soc London Ser B 292:499–507

Hermesen JGTh, Verdenius J (1973) Selection from *Solanum tuberosum* group Phureja of genotypes combining high-frequency haploid induction with homozygosity for embryo spot. Euphytica 22:244–259

Heursel J (1977) Beschrijvende Lijst van Rhododendron simsii (*Azalea indica*) cultivars. Rijkscent Landbowk Onderzoek, Gent

Horn W (1968) Genetische Ursachen der Variation bei Zierpflanzen. Gartenbauwiss enschaften 33:317–333

Hougas RW, Peloquin ST (1958) The potential of potato haploids in breeding and genetic research. Am Potato J 35:701–707

Jacobsen E (1978) Haploid, diploid and tetraploid parthenogenesis in intespecific crosses between *Solanum tuberosum* interdihaploids and *S. phureja*. Potato Res 21:15–17

Jacobsen E (1981) Polyploidization in leaf callus tissue and in regenerated plants of dihaploid potato. Plant Cell Tissue Org Cult 1:77–84

Jacobsen E, Sopory SK (1978) The influence and possible recombination of genotypes on the production of microspore embryoids in anther culture of *Solanum tuberosum* dihaploid hybrids. Theor Appl Genet 52:119–123

Jacobsen E, Tempelaar MJ, Bijmolt EW (1983) Ploidy levels in leaf callus and regenerated plants of *Solanum tuberosum* determined by cytophotometric measurements of protoplasts. Theor Appl Genet 65:113–118

Johansson L (1986) Improved methods for induction of embryogenesis in anther cultures of *Solanum tuberosum*. Potato Res 29:179–190

Jones MGK (1985) New Plants through tissue culture. Annu Proc Phytochem Soc Eur 26:215–228

Jones H, Karp A, Jones MGK (1989) Isolation, culture and regeneration of plants from potato protoplasts. Plant Cell Reports 8:307–311

Karp A, Bright SWJ (1985) On the causes and origins of somaclonal variation. In: Miflin BJ (ed) Oxford surveys of plant molecular and cell biology, vol 2. Univ Press, Oxford, pp 199–234

Karp A, Nelson RS, Thomas E, Bright SWJ (1982) Chromosome variation in protoplast-derived potato plants. Theor Appl Genet 63:265–272

Karp A, Risiott R, Jones MGK, Bright SWJ (1984) Chromosome doubling in monohaploid and dihaploid potatoes by regeneration from cultured lead explants. Plant Cell Tissue Org Cult 3:363–373

Karp A, Jones MGK, Ooms G, Bright SWJ (1987) Potato protoplasts and tissue culture in crop improvement. In: Russell GE (ed) Biotechnology and genetic engineering reviews, vol 5. Intercept, Wimborne, Dorser pp 1–32

Kasperbauer MJ, Collins GB (1972) Reconstitution of diploids from leaf tissue of anther-derived haploids in tobacco. Crop Sci 12:98–101

Kemble RJ, Shepard JF (1984) Cytoplasmic DNA variation in a potato protoclonal population. Theor Appl Genet 69:211–216

Krishnamurthi M (1974) Notes on disease resistance of tissue culture subclones and fusion of sugar cane protoplasts. Sugar Cane Breed Newslett 35:24–26

Landsmann J, Uhrig H (1985) Somaclonal variation in potato detected at the molecular level. Theor Appl Genet 71:500–505

Larkin PJ, Scowcroft WR (1981) Somaclonal variation – a novel source of variability from cell cultures for plant improvement. Theor Appl Genet 60:197–214

Larkin PJ, Ryan SA, Brettell RIS, Scowcroft WR (1984) Heritable somaclonal variation in wheat. Theor Appl Genet 67:443–455

Lörz H, Brown PTH (1986) Variability in tissue culture derived plants – possible origins, advantages and drawbacks. In: Horn W, Jensen CJ, Odenbach W, Schieder O (eds) Genetic manipulation and plant breeding. De Gruyter, Berlin New York, pp 513–534

Matern U, Strobel G, Shepard J (1978) Reaction to phytotoxins in a potato population derived from mesophyll protoplasts. Proc Natl Acad Sci USA 4935–4939

Murashige T, Skoog F (1962) A revised medium for rapid growth and bioassays with tobacco tissue cultures. Physiol Plant 15:473–497

Nelson RS (1984) Plant regeneration from protoplasts of *Solanum tuberosum* and *Solanum brevidens*. Ph D. Thesis, Univ Bath

Nelson RS, Karp A, Bright SWJ (1986) Ploidy variation in *Solanum brevidens* plants regenerated from protoplasts using an improved culture system. J Exp Bot 37:253–261

Ogihara Y (1981) Tissue culture in Haworthia, pt 4. Genetic characterization of plants regenerated from callus. Theor Appl Genet 60:353–363

Ooms G, Karp A, Roberts J (1983) From tumour to tuber: tumour cell characteristics and chromosome numbers of crown-gall derived tetraploid potato plants (*Solanum tuberosum* cv. Maris Bard). Theor Appl Genet 66:169–172

Ooms G, Karp A, Burrell MM, Twell D, Roberts J (1985) Genetic modification of potato development using Ri-T-DNA. Theor Appl Genet 70:440–446

Ooms G, Bossen ME, Burrell MM, Karp A (1986) Genetic manipulation in potato with *Agrobacterium rhizogenes*. Potato Res 29:367–379

Ooms G, Burrell MM, Karp A, Bevan M, Hille J (1987) Genetic transformation in two potato cultivars with T-DNA from disarmed *Agrobacterium*. Theor Appl Genet 73:744–750

Oono K (1985) Putative homozygous mutations in regenerated plants of rice. Mol Gen Genet 198:377–385

Orton TJ (1980) Chromosomal variability in tissue cultures and regenerated plants of *Hordeum*. Theor Appl Genet 56:101–112

Pijnacker LP, Walch K, Ferwerda MA (1986a) Behaviour of chromosomes in potato leaf tissue cultured in vitro as studied by Brd-C-Giesmsa labelling. Theor Appl Genet 72:833–839
Pijnacker LP, Hermelink JHM, Ferwerda MA (1986b) Variability of DNA content and karyotype in cell cultures of an interdihaploid *Solanum tuberosum*. Plant Cell Rep 5:43–46
Potter RH, Jones MGK (1988) Towards molecular analysis of genetic stability of germplasm following in vitro storage of propagation. In: Dodds JH (ed) Tissue culture for conservation of plant genetic resources. Croom-Helm, Beckenham, Kent UK (in press)
Preil W (1986) In vitro propagation and breeding of ornamental plants: advantage and disadvantage of variability. In: Horn W, Jensen CJ, Odenbach W, Schieder O (eds) Genetic manipulation in plant breeding. De Gruyter, Berlin New York, pp 377–405
Ross H (1986) Potato breeding – problems and perspectives. Parey, Berlin Hamburg
Sandford JC, Weeden NF, Chyi YS (1984) Regarding the novelty and breeding value of protoplast-derived variants of Russet Burbank (*Solanum tuberosum* L.). Euphytica 33:709–715
Secor GA, Shepard JF (1981) Variability of protoplast-derived potato clones. Crop Sci 21:102–105
Shepard JF, Bidney D, Shahin E (1980) Potato protoplasts in crop improvement. Science 208:17–24
Smith DB (1986) Variation in the electrophoretic band pattern of tuber proteins from somaclones of potato. J Agric Sci Cambridge 106:427–428
Sopory SK, Bajaj YPS (1987) Anther culture and haploid production in potato. In: Bajaj YPS (ed) Biotechnology in agriculture and forestry, vol 3. Potato. Springer, Berlin Heidelberg New York Tokyo, pp 89–108
Sopory SK, Tan BH (1979) Regeneration and cytological studies of anther and pollen calli of dihaploid *Solanum tuberosum*. Z Pflanzenzücht 82:31–35
Sree-Ramulu K, Dijkhuis P, Roest S (1983) Phenotypic variation and ploidy level of plants regenerated from protoplasts of tetraploid potato (*Solanum tuberosum* L. cv. Bintje) Theor Appl Genet 65:329–338
Sree-Ramulu K, Dijkhuis P, Roest S, Bokelmann GS, De Groot B (1984) Early occurrence of genetic instability in protoplast cultures of potato. Plant Sci Lett 36:79–86
Sree-Ramulu K, Dijkhuis P, Hanisch Ten Cate Ch H, De Groot B (1985) Patterns of DNA and chromosome variation during in vitro growth in various genotypes of potato. Plant Sci 41:69–78
Sree-Ramulu K, Dijkhuis P, Roest S, Bokelmann GS, De Groot B (1986) Variation in phenotype and chromosome number of plants regenerated from protoplasts of dihaploid and tetraploid potato. Z Pflanzenzücht 97:119–128
Tempelaar MJ, Jacobsen E, Ferwerda MA, Hartogh M (1985) Changes of ploidy level by in vitro culture of monohaploid and polyploid clones of potato. Z Pflanzenzücht 95:193–200
Thomas E (1981) Plant regeneration from shoot-culture-derived protoplasts of tetraploid potato (*Solanum tuberosum* cv. Maris Bard). Plant Sci Lett 23:84–88
Thomas E, Bright SWJ, Franklin J, Lancaster V, Miflin BJ, Gibson R (1982) Variation amongst protoplast-derived potato plants (*Solanum tuberosum* cv. Maris Bard). Theor Appl Genet 62: 65–68
Thomson AJ, Gunn RE, Jellis GJ, Boulton RE, Lacey CND (1986) The evaluation of potato somaclones. In: Semal J (ed) Somaclonal variations and crop improvement. Nijhoff, Dordrecht, pp 148–159
Van Breukelen EWM, Ramanna MS, Hermesen JGTh (1977) Parthenogenetic monohaploids ($2n = x = 12$) from *Solanum tuberosum* L. and *S. verrucosum* Schl. and the production of homozygous potato diploids. Euphytica 26:263–272
Van Harten AM, Bouter H, Broertjes C (1981) In vitro adventitious bud techniques for vegetative propagation and mutation breeding for potato (*Solanum tuberosum* L.) II. Significance for mutation breeding. Euphytica 30:1–8
Wasscher J (1956) The importance of sports in some florists flowers. Euphytica 5:163–170
Webb JK, Osifo EO, Henshaw GG (1983) Shoot regeneration from leaflet discs of six cultivars of potato (*Solanum tuberosum* subsp. *tuberosum*). Plant Sci Lett 30:1–8
Wenzel G (1981) Cell culture steps in potato breeding programs. In: Proc 8th Triennial Conf Eur Assoc Potato research München, pp 178–179
Wenzel G, Uhrig H (1981) Breeding for nematode and virus resistance in potato via anther culture. Theor Appl Genet 59:333–340
Wenzel G, Schieder O, Przewozny T, Sopory SK, Melchers G (1979) Comparison of a single cell culture derived *Solanum tuberosum* L. plants and a model for their application in breeding programmes. Theor Appl Genet 55:49–55

Wenzel G, Debnath SC, Schuchmann R, Foroughi-Wehr B (1987) Combined application of classical and unconventional techniques in breeding for disease resistant potatoes. In: Ellis GJ, Richardson DE (eds) The production of new potato varieties: technological advances. Univ Press, Cambridge, pp 277–288

Wheeler VA, Evans NE, Foulger D, Webb KJ, Karp A, Franklin J, Bright SWJ (1985) Plant regeneration from explant cultures of fourteen potato cultivars and study of cytology and morphology of regenerated plants. Ann Bot (London) 55:309–320

Wright NS (1983) Uniformity among virus-free clones of ten potato cultivars. Am Potato J 60: 381–388

Yamamoto M (1989) Transformation in potato. In: Bajaj YPS (ed) Biotechnology in agriculture and forestry, vol 9. Protoplasts and genetic engineering II. Springer, Berlin Heidelberg New York, pp 122–139

III.2 Somaclonal Variation in Tomato

M. BUIATTI[1] and R. MORPURGO[2]

1 General Account

Lycopersicon species originated in the Andean region of South America. Domestication of the tomato began in Mexico, where wild populations of *L. esculentum* var. *cerasiforme* are still found. From this region the cultivated tomato spread first to Europe in the Mediterranean region and then to North America in the 18th century. The common Mexican origin (Rick and Forbes 1975) of all domesticated tomatoes is demonstrated by the fact that variation in allozyme patterns in the cultivated tomato is extremely poor. In fact, the majority of the temperate cvs. and the great majority of the Mesoamerican ancestors var. *cerasiforme* proved to be monomorphic for all tested APS, EST, GOT, and PRX loci. In wild and Andean cultivated tomato, a greater degree of variability exists between individuals, populations, and races. Maximum variability was found in Peruvian and Ecuatorian coast regions.

A complete range of mating systems is found in *Lycopersicon* species, from the almost autogamous *L. cheesmanii* and *parviflorum* to the obligately outcrossed, self-incompatible biotypes of *L. chilense, hirsutum,* and *S. pennelli.* Self-fertility and various degrees of facultative outcrossing are found in *L. chmielewskii, pimpinellifolium, esculentum,* and self-compatible biotypes of *L. hirsutum* and *S. pennellii.* Domestication of *L. esculentum* was accompanied by a transition from exerted to inserted stigmas and a consequent obligate autogamy.

Hybridization tests between all the possible species combinations have shown the existence of two distinct groups, one formed by *L. chilense* and *peruvianum*, and the other containing the remaining seven species. The barrier between these two groups can be broken only by the application of embryo culture, which succeeds only when *esculentum* group members are used as female parents. Reciprocal crosses succeed only if highly inbred lines of *L. peruvianum* are used (Hogenboom 1972).

F_1 hybrids of all the successful combinations display a normal or nearly normal chromosomal behavior, and the resultant bivalents segregate normally at the anaphase of the two meiotic divisions. It is therefore evident that speciation occurred mainly as a result of gene mutation and to a very minor extent by chromosomal differentiation (Rick 1979). However, as far as hybrid fertility is concerned, a wide range of situations are encountered, from an almost completely

[1] Department of Animal Biology and Genetics, Via Romana 17, Florence, Italy
[2] Department of Physiology, International Potato Center (CIP), P.O. Box 5969, Lima, Peru

Biotechnology in Agriculture and Forestry, Vol. 11
Somaclonal Variation in Crop Improvement I (ed. by Y.P.S. Bajaj)

normal situation in *L. esculentum* × *L. pimpinellifolium* to various degrees of F_1 and F_2 sterility, reduced recombination, and modified segregation ratios in *L. esculentum* × *L. pennellii*.

According to FAO (1985), 102 countries are interested in tomato cultivation, and world production of tomatoes reached 60,825,000 metric tons in that year. The ten top tomato producers were at that time the USA (7,836,000 t), the USSR (6,900,000 t), Italy (6,115,000 t), China (5,260,000 t), Turkey (4,900,000 t), Egypt (2,800,000 t), Greece (2,512,000 t), Spain (2,200,000 t), Brazil (1,932,000 t), and Mexico (1,665,000 t), but new production areas are developing in Africa, the Near East, and Eastern Europe.

1.1 Breeding Objectives

In recent years, tomato breeding has concentrated on increased yields, improved plant quality, alteration of plant growth habit, and pest resistance. Wild species have been used to introduce valuable traits, particularly resistance to several bacterial, fungal, and viral diseases. These include resistance to bacterial canker, bacterial wilt, leaf mold, anthracnose, fusarium and verticillium wilts, late blight, and tobacco mosaic virus.

Introgression of useful genes from wild species in the cultivated *Lycopersicon* appears particularly useful for breeding new cultivars adapted to the new geographical regions characterized by constrained environments. In spite of the fact that stress-tolerant traits are known in the tomato collections, nonstress-tolerant cvs. have also been developed till now, due to the lack of efficient screening procedures. Therefore there is a need for improved breeding techniques able to by-pass the problems arising from the incompatibility barriers shown by the tomato.

On the other hand, in countries with a well-established tomato cultivation, future research is aimed at developing new cvs. with improved color, flavor, solid content, and nutritional value.

1.2 Micropropagation and Regeneration

Tomato plants can be easily propagated from meristems (Kartha et al. 1977; Novak and Maskova 1979; Lopez Peralta et al. 1978) and from hypocotyls. Cotyledons are also amenable to shoot and plant regeneration with little or no callus formation, but to a lesser extent (Feher 1979; Gunay and Rao 1980). Embryo culture is fairly easy and is currently used mainly to overcome incompatibility barriers in introgression experiments (Vorob'eva and Prikhod'ko 1980; Neal and Topoleski 1983, 1985; Uralets 1984; Bonito 1985). Extensive data on the basic problems of tomato culture can be found in the review by Sink and Reynolds (1986).

Regeneration of *L. esculentum* is, on the other hand, generally a difficult task in the case of callus formation, and particularly if cultures are kept in an undifferentiated state (callus, cells in liquid culture) for a long time. Thus, the source of the explant and the duration of proliferation prior to differentiation play a critical role, along with genotypes. Nevertheless, plants have been differentiated in a few cases also from long-term cultures and/or protoplast, although only in part of the genotypes tested.

As shown in Table 1, regeneration has been successfully attempted from various plant parts and found to be particularly easy when young explants were cultured for very short periods of time. However, also in these cases, plant genotypes and environmental factors (composition of the medium, light etc.) have been found to play a significant role. Behki and Lesley (1976) observed regeneration from leaf discs in only 12 out of 15 tomato mutant clones. Fedeeva et al. (1979a,b,c) analyzed root formation from cotyledons in a series of genotypes, F_1's and F_2's, and observed a direct dependence on parental cultivars, the direction of the cross, and the age of the plant at the time of excision. Similar data were obtained by Ohki et al. (1978), while Frankenberger et al. (1979, 1981a,b) carried out a thorough genetic analysis of the genetic variability for shoot-forming capacity of leaf disc explants from 21 varieties. Besides observing the effects of environment and age of the plant, these authors, in a diallel cross of six cultivars, obtained high heritability and general combining ability estimates, while specific combining ability was also found to be significant. Similarly, Kurtz and Lineberger (1983) observed genotypic differences in the ability to regenerate shoots and in the number of shoots, although in this case the genotype-hormone interaction was particularly clear. Finally, Zelcer et al. (1984), in analyzing the regenerating response of leaf discs of 100 cvs., found that only one quarter of them, but not the F_1's, displayed relatively good shoot-forming capacity. It must be pointed out, however, that regeneration cannot be predicted

Table 1. Some cases of successful regeneration from short-term cultures

Source of the explant	Reference
Leaf disc or section	Behki and Lesley (1976)
	Bharati et al. (1978)
	Coleman and Greyson (1977a,b)
	Frankenberger et al. (1981a, 1979)
	Kurtz and Lineberger (1983)
	Pence and Caruso (1984)
	Tal et al. (1977)
	Evans and Sharp (1983)
Stem	Cassells (1979)
	Vnuchova (1975)
Cotyledons	Fedeeva et al. (1979a,b,c)
	Gavazzi et al. (1987)
	McCormick et al. (1986)
	Buiatti et al. (1985)
	Sibi (1980)
Hypocotyls	Heideveld et al. (1984)
	Ohki et al. (1978)
	Seeni and Gnanam (1981)
	Zelcer et al. (1984)
	Caruso et al. (1986)
	Asano and Sekioka (1983)

from the plant habit, as proven by Frankenberger et al. (1979) and Tal et al. (1977), who failed to show a correlation between in vivo-altered hormonal balance and in vitro behavior in a series of mutants.

Regeneration can be altered also by introgression of "regeneration" genes from other species. Koorneef et al. (1987) obtained regenerating genotypes by transferring regeneration capacity from *L. peruvianum* to *L. esculentum*. In these experiments, regeneration from established callus cultures was also shown to be controlled by two dominant genes, while callus growth from primary explants, callus growth in long-term cultures, and shoot regeneration from explants, all had high heritabilities. Moreover, regeneration from cotyledonary explants was also induced on a medium without hormones by co-cultivation with *Agrobacterium tumefaciens* strains harboring T plasmids inactivated in auxin-synthesizing genes (Bogani et al. 1987).

As mentioned earlier, the source of the explant seems to be important for the regeneration of the culture. Generally speaking, it can be said that young leaf and cotyledonary explants seem to be best suited, hypocotyls ranking third, and stems being less endowed with morphogenetic potential.

No general conclusion can be drawn for the effect of different phytohormones, the optimal hormone balance being strictly dependent on other factors, and probably highly specific for changes in differentiation patterns (Nacmias et al. 1987). As can be seen from Table 2, regeneration was possible only in a few cases also from long-term callus cultures, and generally only with a few genotypes. Thus, Koblitz and Koblitz (1982) and Morgan and Cocking (1982) only regenerated plants from cv. Lukullus (out of 12 cvs. tested), Meredith (1979) only from VFNT Cherry out of four tested, Niedtz et al. (1985) in six cvs. out of eight, while Muhlbach and Zapata (1980) and Herman and Haas (1978) failed to obtain regeneration at all. Only Shahin (1985) claimed regeneration in 14 cultivars on a single basal medium supplemented with phytohormones (GA_3, BA, 2-iP, various auxins) at concentrations dependent on the cultivars utilized. All the experiments mentioned except

Table 2. Some regeneration experiments from long-term cultures

Source of the callus	Result	Reference
Leaf	+	Behki and Lesley (1976)
	–	Herman and Haas (1978)
Hypocotyl	+	Heideveld et al. (1984)
	+	Imanishi and Hiura (1976)
	+	Seeni and Gnanam (1981)
Protoplast	+	Shahin (1985)
	+	Niedtz et al. (1985)
	+	O'Connell et al. (1986)
	[a] +/–	Koblitz and Koblitz (1982)
	[a] +/–	Meredith (1979)
	[a] +/–	Morgan and Cocking (1982)
	–	Muhlbach and Zapata (1980)

[a] Positive in only one cultivar.

that by Herman and Haas (1978) were carried out on protoplast-derived callus cultures. Protoplasts are, in fact, easily obtained and cultured and offer good material for fusion experiments with other species. Also in this field, tomato research has led to pioneering results like that of Melchers et al. (1978) in obtaining somatic tomato-potato hybrids from which several lines were later obtained differing in the relative contents of the two nuclear and extrachromosomal genomes, called "pomato" when containing potato plastids and "topato" when the plastids were of tomato origin (Roddick and Melchers 1985). Fusion hybrid plants have also been obtained from *Solanum nigrum* × *L. esculentum* (Jain et al. 1985), *L. esculentum* × *L. peruvianum* (Kinsara et al. 1986), *L. esculentum* × *S. lycopersicoides* (Sink et al. 1986, Handley et al. 1986).

2 Somaclonal Variation

Lycopersicon esculentum certainly is one of the better-known plants from the genetic point of view, as it is self-fertilized and has a relatively short life cycle and a good morphology. This has allowed the construction of a detailed genetic map, the establishment of collections of mutants, spontaneous or recovered after treatment with chemical and physical mutagens, and extensive breeding work which has led to obtaining a very wide range of cultivars.

Tomato is thus a particularly suitable plant for the analysis of genetic variability in cultured cells, regenerated plants, and their progenies, its only drawback being the relatively small size of its mitotic chromosomes. This is probably the reason why the data on cytological variability in cultured cells and regenerated plants are rather scanty if compared with those obtained on other species (D'Amato 1985). Some results are, however, summarized in Table 3.

Mixoploidy and the differentiation of cell clones with specific ploidy (triploid and aneuploid) have been described by Kiforak et al. (1977) and Levenko et al. (1977), who also showed (Levenko and Kiforak 1977) that the presence of fluorophenilalanine in the medium led to selective proliferation of cells with lower chromosome numbers and reduced the frequency of chromosome aberrations. Changes in ploidy levels were also observed in plants regenerated from anther calli (Zagorska et al. 1982), protoplast (Benton and Bastian 1986) and, with a lower frequency, cotyledons (Evans and Sharp 1983; Evans et al. 1984; Buiatti et al. 1985). Polyploidy and mixoploidy were also described by Handley et al. (1986) in *Lycopersicon esculentum* × *Solanum lycopersicoides* fusion hybrids, while the importance of the nature of the source of regenerants was confirmed by O'Connell et al. (1986) in *Lycopersicon esculentum* × *L. pennelli* sexual hybrids. In this case, plants derived from leaf slices were 90% diploid, while those regenerated from protoplasts were mostly tetraploid or octoploid. *Lycopersicon esculentum* is, with maize (Edallo et al. 1981), to our knowledge the first plant where genetic analysis of the variability present in regenerated plants has been carried out with success. The first report of mutations in progenies of plants regenerated from tomato cotyledons seems to be that of Sibi (1980, 1986) followed by others (Evans and Sharp 1983; Evans et al. 1984; Buiatti et al. 1984; 1985; Zagorska et al. 1982, Gavazzi et al. 1987). All these

Table 3. Chromosome numbers in cells and regenerants from *L. esculentum* and crosses with related species

Plant material	Chromosome no.	Reference
L. esculentum calli	60–160	Meredith (1978)
L. esculentum regenerants from cotyledons	4n (3/35)	Buiatti et al. (1985)
L. esculentum. L. peruvianum regenerants from cells	4n (1/19)	Thomas and Pratt (1982)
L. esculentum R_3 plants	2n (13/13)	Shahin and Spivey (1986)
L. esculentum regenerants from cotyledons	3 aneuploids/70	Sibi (1986)
L. esculentum regenerants from anthers	12,24,36,48,mix.	Zagorska et al. (1982)
MO393, MO504 regenerants from leaves	4n (2/280)	Zhuchenko et al. (1981)
L. esculentum. L. pennelli regenerants from protoplasts	4n (76/100), 8n (13/100)	O'Connell et al. (1986)
L. esculentum. L. pennelli regenerants from leaves	4n (3/26)	O'Connell et al. (1986)

experiments consistently showed some peculiarities in the genetic variability observed after varying periods of culture in vitro.

Firstly, mutation frequencies were generally very high, ranging from 4–5% (Evans et al. 1984) to 20–30% in other experiments. Secondly, the mutation spectrum seems generally to be different both from that covered by spontaneous mutation and that observed in "traditional" mutagenesis experiments. In all the experiments analyzed from this point of view, some particular mutations appeared more frequently than others. This was the case for male sterility, indeterminate, and other morphological mutants in Evans and Sharp's work (1983), developmental mutants in Gavazzi et al.'s experiments (1987), and xantha-like mutants in our work (Buiatti et al. 1985).

Complementation tests carried out by crossing somaclonal mutants with known mutants obtained from the collection of C.M. Rick (University of California, Davis) showed the novelty of a fruit variant, tangerine (Evans et al. 1984) and xantha (Tognoni, Lipucci and Buiatti, unpubl.). Moreover, striking differences in mutation spectra were observed when they were compared in progenies of plants regenerated from tomato cotyledons and derived from seeds or mature pollen treated with ethyl methane sulfonate (EMS) (Gavazzi et al. 1987). Particularly, three mutant types, all affecting plant development, were very frequently recovered in progenies from selfed regenerated plants (early ripening, potato leaf, and reduced number of lateral shoots). On the other hand, while loss of germination was also common in culture-derived plants, a high percentage of seedling death was induced by mutagenic treatments. It should be finally stressed that the abnormally high frequency of mutation affecting development was not correlated with a similar number of changes in isozyme patterns when they were carefully studied in two interspecific hybrids (one sexual and one derived from protoplast fusion) (O'Connell et al. 1986; Handley et al. 1986). In both cases, in fact, no apparent mutation was observed in regenerated plants, except for the lack of one glutamate oxaloacetate transaminase-4 band in the sexual hybrid *Lycopersicon esculentum* x *L. pennellii* (O'Connell et al. 1986).

The reasons for both the appearance of novel mutants and the difference in mutation spectra from those found spontaneously and after traditional mutagenesis experiments are still obscure. One of the possible causes, as suggested by Gavazzi et al. (1987) and Caruso et al. (1986), may be the selective advantage in vitro of cells carrying mutations which would be selected against by diplotic selection under normal ontogenetic constraints. Direct support to this hypothesis is given by the work of Caruso et al. (1986), who attributed the high in vitro reversion rate of homozygous La/La (lanceolate) mutants to a selective effect of the medium, as suggested by the observed increased regenerative capacity of La/La + heterozygotes. A surprising feature of the genetic structure of regenerants is the frequent observation of nonsegregating (true breeding) progenies in selfing experiments. Sibi (1980) observed four true breeding progenies out of 30 analyzed, all differing from control for leaf color and plant and leaf shape. A diallel cross with these lines and a control one (derived from selfing of a normal seed plant) revealed a high additive component for several quantitative characters and, in several cases, asymmetrical transmission effects.

Evans et al. (1984) thoroughly analyzed a series of mutants whose phenotype was already evident in regenerated plants and bred true in their progeny. This was the case of "jointless", "molted", and a series of other morphological mutants. In the first two cases, complementation tests showed that both mutants were distinct from any of the previously reported tomato mutants. Similar results were also obtained by Gavazzi et al. (1987) for a series of developmental mutated types. Although the origin of true breeding mutants has still to be ascertained, some authors suggested that mitotic crossing over may play a role in their determination (Evans et al. 1984). Somatic crossing over was also suggested as one of the putative causes of the loss of the marker "potato leaf" in F_2 of regenerants from leaves of F_1 plants, from crosses between the mutants Mo393 and Mo504 treated with 2,3,6-TBA and Picloran (Zhuchenko et al. 1981).

The extent and nature of genetic variability in tomato tissue cultures and regenerants seem to offer a useful tool for breeding purposes in a plant which, due to the very intensive selection work, certainly would profit from a further widening of genetic diversity. Up to now, to our knowledge, attempts to utilize this new source of variability have been mainly directed toward obtaining cultivars resistant to biotic and abiotic stresses.

2.1 Somaclonal Variation for Resistance to Biotic Stresses

Resistance to pathogens is a very complex phenomenon which may involve both passive and active defense responses by the plant. Genes for resistance to several pathogens are present in *L. esculentum*, or may be obtained from wild relatives.

However, their range does not cover the present needs and, moreover, introgression is a lengthy process, particularly when interspecific crosses are involved. It would then be of real help if novel variability for resistance to biotic stresses could be derived from regenerated plants, and possibly selected in vitro prior to regeneration. The possible use of in vitro cultures for these purposes is suggested by two series of considerations.

Firstly, a correlation between in vitro behavior and in vivo response has been shown in several host-parasite systems (Buiatti and Scala 1984; Buiatti 1987) and also in the tomato-*Fusarium* interaction (Scala et al. 1985). In this case, cells from a resistant line were shown to produce more phytoalexins (substances with antifungal action generally synthesized by the plant during the hypersensitive reaction) than susceptible genotypes, when challenged with fungal cell wall components. Secondly, field scoring of progenies of regenerated plants has often revealed the presence of novel resistant genotypes (Buiatti 1987). This has been the case for resistance to *Verticillium alboatrum* found in a rather high number of progenies from plants regenerated from tomato cotyledons. Similarly, six out of 370 somaclones derived from leaf discs of a tomato cultivar susceptible to TMV were found to show varying degrees of resistance to the virus (Barden et al. 1986).

The two series of results just mentioned, taken together, seem to open the way for the establishment of techniques of direct in vitro selection of resistant genotypes based on parameters easily screened in vitro, which correlate well with in vivo resistance to pathogens. In other words, the rationale of such an approach would be to dissect the biochemical steps which lead to incompatible reactions, and consequently devise methods to easily screen cells in vitro for those which more clearly discriminate between resistant and susceptible genotypes.

The most immediate application of such an approach has been the use of media containing fungal toxic filtrates, crude or more or less purified, as selective agents. As shown in Table 4, this approach has been successfully used in selecting tomato cells or plants resistant to *Alternaria solani* (Shepard 1986), *Phytophthora infestans* (Illig and Dall'Acqua 1986), *Fusarium oxysporum* (Shahin and Spivey 1986). It should be noted, however, that these results, like others obtained on other host-parasite systems, have been challenged on the basis that nonspecific toxins are not

Table 4. Some successful experiments which resulted in obtaining: cells (*), plants (**), and progenies from regenerated plants (***) resistant to pathogens

Resistance to pathogen	Selection agent	Reference
*A. solani****	Culture filtrate	Shepard (1986)
*Ph. infestans***	"	Illig and Dallacqua (1986)
*F. oxysporum**	"	Toyoda et al. (1984a)
*F. oxysporum**	Fusaric acid	Toyoda et al. (1984b)
*F. oxysporum****	"	Shahin and Spivey (1986)
*F. oxysporum***	Culture filtrate	Scala et al. (1984)
*F. oxysporum**	"Elicitors" and BUdR	Buiatti et al. (1987)
*V. alboatrum****	None	Gavazzi et al. (1987)
TMV***	None	Barden et al. (1986) Smith and Murakishi (1986)

thought to play a major role in disease development, and moreover, that in some cases the frequencies of resistant plants in progenies from selected and unselected cells do not seem to differ significantly (Ingram and Mac Donald 1986).

In tomato, Shepard (1986) claimed to have found increased susceptibility to *Alternaria* and no resistant plants in progenies of regenerants from tissues grown on basal medium while recovering resistant genotypes in plants derived from cv. Petomech callus cultures grown on toxic filtrates. In Shahin and Spivey's experiments (1986), seven single gene mutants for resistance to *Fusarium* were recovered from a total of 73 calli which survived on 20 to 30 μM fusaric acid, while six such mutants were also found scoring plants derived from 126 regenerating calli not submitted to selection. In this case moreover, all mutants behaved as single dominant mutations. These experiments, therefore, although suggesting an easier recovery of resistant genotypes from selected material, confirm the presence of spontaneous relatively high-frequency mutants also in unselected cell populations. Although selection on toxic filtrates or toxins has been by far the most frequently used selection method other techniques are being established which are based on attempts to screen cells for processes which result in active defense, from host-parasite recognition to the induction of proteins and enzymes involved in the hypersensitive reaction. Using this approach, Buiatti et al. (1987) treated tomato cv. Red River suspension cultures with *Fusarium oxysporum f.sp. lycopersici* heat-released cell wall components (elicitors) with or without the addition of BUdR. As hypersensitive cell growth should be stopped by elicitor treatments, the experiments should allow the selection of nonhypersensitive cells in the absence of BUdR and the recovery of genotypes with increased hypersensitivity when the BUdR negative selection technique (Shimamoto and King 1983) was used. The results were encouraging, as cell lines were obtained which produced significantly more or less phytoalexins than control when treated with *F. oxysporum* elicitor. Moreover (Buiatti et al. unpubl.), highly hypersensitive cells were shown to respond to fungal cell wall components with increased peroxidase activity and to inhibit fungal growth in dual culture experiments (Fig. 1). These results also open the way to the use of early peroxidase activity, easily observed through a guayacol staining method developed in our laboratory or β-glucan content scored through in vivo calcofluor staining as markers for resistance (Storti et al. submitted).

2.2 Somaclonal Variation for Resistance to Abiotic Stresses

Resistance to abiotic stresses would be of great interest in tomato breeding, as it may broaden areas of cultivation of the crop and allow the use of low-cost agronomic techniques. Reports in this field, however, seem less conclusive than in the case of resistance to pathogens, mainly because the mechanism of stress itself is less known, and hence, the development of selection techniques is less easy. Moreover, in several cases, regeneration after in vitro selection has proven to be difficult, due probably to the long time needed to habituate cells to the stresses. The only case of obtaining tomato plants and lines resistant to an abiotic stress agent is, to our knowledge, the isolation of Paraquat-tolerant plants carried out by Thomas and Pratt (1982) working on a 75% *L. peruvianum*, 25% *L. esculentum* genotype. In this case, tolerant

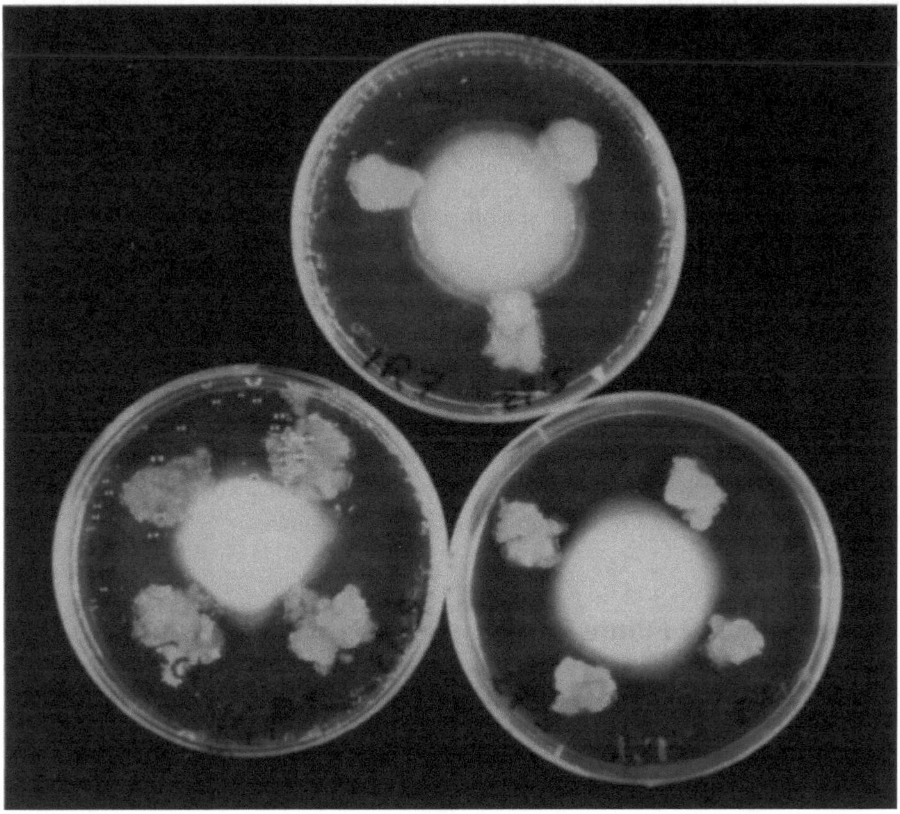

Fig. 1. A typical dual culture experiment where tomato cell lines selected for increased response to *Fusarium oxysporum* elicitors (*lower dishes*), or unselected (*upper dish*) are challenged with the pathogen

cell lines were isolated at a frequency of 5×10^{-8}. Plants were then regenerated which showed a 30-fold increase over wild-type tolerance. Progeny analysis showed that most selected genotypes were due to single, dominant mutations.

Selection for resistance to water stress has been mainly attempted by growing cells on media containing different concentrations of polyethylene glycol (with different water potentials) (A.K. Handa et al. 1982, 1983). In this case, a considerable variation for tolerance among the clones was found (A.K. Handa et al. 1983) which was maintained through further subcloning. However, a more careful analysis showed that tolerant subclones were still mixed populations of cells, and that the PEG method did not seem to be adequate for discriminating between adapted and putative mutant cells. The utility of this selection method seems therefore under question.

Better results seem to have been obtained on the other hand, although only at the cell level, in attempts to select aluminum-, cadmium-, and sodium chloride-resistant variants. Meredith (1978) isolated aluminum-resistant cell clones which

Table 5. Examples of obtaining through in vitro selection: cells (*), plants (***) or progenies (***) resistant to abiotic stresses

Resistance	Selection agent	Reference
Paraquat***	Paraquat	Thomas and Pratt (1982)
Sodium chloride**	Sodium chloride	Rahman et al. (1986)
Cadmium*	Cadmium sulfate	Scheller et al. (1986)
Aluminum*	Aluminum	Meredith (1978)
Water stress*	Polyethylenglycol	Handa A.K. et al. (1983).

occurred at the estimated frequencies of 2.03×10^{-7} in callus and 8.32×10^{-6} in suspension cultures. Most of these variants were found to be stable and highly polyploid, and did not differ from susceptible controls in aluminum uptake. Similarly, cells resistant to cadmium at concentrations up to 2000 mM were reported to have been selected by Scheller et al. (1986), while plants were regenerated from tissues grown on media supplemented with up to 80 mM NaCl, although no data are available on the behavior of regenerants or their progenies.

3 Conclusion and Future Prospects

The amount and nature of genetic variability in in vitro cultures of tomato regenerants and their progenies certainly open interesting avenues for both basic and applied research. Firstly, the mechanisms of origin and molecular basis of in vitro-released mutations still have to be elucidated in this, as in other plants, with the relative advantage, already mentioned, that tomato is a particularly well-suited plant for genetic analysis. Particularly the relative roles of events like activation of transposable elements, DNA amplification, translocation and rearrangements of structural genes, mutations in regulatory sequences, and somatic crossing-over (Buiatti 1977, 1988) in determining genome instability in culture have to be analyzed. Unfortunately, and somewhat surprisingly, molecular data on tomato genome in culture are almost absent in the literature. However, as tomato genomic libraries are already available, fast progress in this field is expected in the near future.

Moreover, the high number of developmental mutants recovered in progenies from regenerated plants may offer a new important tool for studying the physiological and genetic basis of plant development, also given the interesting morphology of the tomato plant. In other words, in vitro mutations may offer genetic tools for a work comparable to that carried out with the aid of homeotic mutants in *Drosophila*, based on the dissection of developmental processes.

The peculiar features of genome fluctuation in in vitro conditions may be of help also from the applied point of view, in that, at high frequencies, regenerated plants seem to contain novel mutations never recorded in other experiments. Of particular interest is the fact, again, that these mutations seem to be generally highly pleiotropic and to affect the general physiological phenotype of the plant. For this reason, they may be utilized both for creating cultivars resistant to biotic and abiotic stresses (changes in metabolism leading to resistance are indeed very complex) and for "modeling" plant development according to the needs of lower costs and better suitability for consumption and transformation of the crop. The main drawback for practical utilization of culture-released variability remains the fact that not all cultivars are easily regenerated from cell and protoplast and thus suitable for the application of direct in vitro selective techniques. Hope to overcome this drawback comes from the already mentioned work by Koornneef et al. (1987) on the introduction of "regeneration" genes from *Lycopersycon peruvianum* into *L. esculentum*, and the possibility of changing regenerating capacity through the introduction of *Agrobacterium tumefaciens* genes for auxin and cytokinin synthesis. In fact, recombinant DNA techniques will certainly be in the near future of utmost importance for solving both theoretical and practical problems connected with the study and application of the genetic changes in tomato tissue cultures.

It should be noted that an efficient transformation technique leading to the production of good numbers of transgenic plants has only recently been established (McCornick et al. 1986). In this work, transformation was easily achieved through co-cultivation of tomato leaf explants or cotyledon/hypocotyl section with *A. tumefaciens* strains harboring binary or co-integrate T-DNA vectors. Three hundred transgenic plants from eight tomato cultivars were thus obtained, transformation being more efficient with co-integrate than with binary vectors.

This technique has already been applied by the same group to the insertion of insect control protein gene from *Bacillus thuringensis* var. Kurstaki HD-1 which produced resistance to *Manduca sexta, Heliothis virescens,* and *Heliothis zea* in transgenic plants (Fischoff et al. 1987), by Bogani et al. (1987) to the production of plants containing genes for auxin and cytokinin synthesis from *A. tumefaciens.*

Recombinant DNA techniques may thus now be currently utilized both for the insertion of alien sequences and for the study of their interaction with normal and in vitro-modified tomato genomes.

Acknowledgment. Published and unpublished research carried out by the authors was supported by IPRA-CNR and a contract within the frame of the Biotechnology Action Programme (EEC-BAP).

References

Asano WY, Sekioka TT (1983) Auxin effects on the growth, ethylene production and differentiation of hypocotyl calli from three tomato varieties. Hort Sci 18:(II) 566

Barden K, Schiller A, Smith S, Murakishi H (1986) Regeneration and screening of tomato somaclones for resistance to tobacco mosaic virus. Plant Sci 45:209–213

Behki RM, Lesley SM (1976) In vitro plant regeneration from leaf explant of *Lycopersicon esculentum* (tomato). Can J Bot 54(21):2409–2414

Benton WD, Bastian KL (1986) Karyotypic variation among protoplast-derived plants. 6th Int Congr Plant tissue and cell culture, Univ Minn, Minneapolis p 216

Bharati O, Ramakrishnan T, Vaidyanathan CS (1978) Regeneration of hybrid tomato plants from leaf callus. Curr Sci 47:458–460

Bogani P, Caffio FMR, Preiti A, Bettini P, Nacmias B, Pellegrini MG, Buiatti M (1987) Regeneration from tomato cotyledons through infection with *Agrobacterium tumefaciens* carrying different oncogene complements. In: Modern trends in tomato genetics and breeding EUCARPIA Tomato Work Group. Salerno. 2–6 Sept. 1987, p 123

Bonito R di (1985) Embryo culture in tomato: a technique to accelerate the breeding for disease resistance. Genet Agr 39(3):321–322

Buiatti M (1977) DNA amplification and tissue culture. In: Reinert J, Bajaj YPS (eds) Applied and fundamental aspects of plant cell, tissue, and organ culture. Springer, Berlin Heidelberg New York, pp 358–374

Buiatti M (1987) I nuovi metodi di miglioramento genetico e la resistenza a stress biotici. Agric Ric 77:33–42

Buiatti M (1988) Information flux and constraints in development and evolution: a critical view. In: Chaos and complexity, World Scientific 331–349

Buiatti M, Scala A (1984) In vitro analysis of plant pathogen interactions and selection procedures. In: Int Symp Plant tissue culture application to crop improvement. Czech Acad Sci, Prague, pp 331–340

Buiatti M, Tognoni F, Lipucci M, Marcheschi G, Pellegrini G, Grenci FC, Scala A, Bettini P, Bogani P, Bett G (1984) Genetic variability in tomato *(Lycopersicon esculentum)* plants regenerated from in vitro cultured cotyledons. Genet Agr 38:325

Buiatti M, Marcheschi G, Tognoni F, Collina Grenci F, Martini G (1985) Genetic variability induced by tissue culture in the tomato *(Lycopersicon esculentum)*. J Plant Breed 94:162–165

Buiatti M, Simeti C, Vannini S, Marcheschi G, Scala A, Bettini P, Bogani P, Pellegrini MG (1987) Isolation of tomato cell lines with altered response to *Fusarium* cell wall components. Theor Appl Genet 75:37–40

Cappadocia M, Sree Ramulu K (1980) Plant regeneration from in vitro culture of anthers and stems internodes in an interspecific hybrid, *Lycopersicon esculentum* × *L. peruvianum* Mill. and cytogenetic analysis of the regenerated plants. Plant Sci Lett 20:157–164

Caruso JL, Franco PI, Snider JA, Pence VC, Hangarter RP (1986) Evidence for a revertant at the *la* locus in regenerating hypocotyl segments in tomato. Theor Appl Genet 72:240–243

Cassells AC (1979) The effect of 2,3,5-triiodobenzoic acid on caulogenesis in callus cultures of tomato and *Pelargonium*. Physiol Plant 46:159–164

Còleman WK, Greyson RI (1977a) Analysis of root formation in leaf discs of *Lycopersicon esculentum* Mill cultured in vitro. Ann Bot (London) 41:307–320

Coleman WK, Greyson RI (1977b) Promotion of root initiation by gibberellic acid in leaf discs of tomato *(Lycopersicon esculentum)* Mill cultured in vitro. New Phytol 78:47–54

Collina Grenci F, Lipucci M, Tognoni F, Buiatti M (1981) Sdifferenziazione e differenziazione in vitro di specie, varieta e linee di *Lycopersicum*. Riv Ortoflorofrutt It 65:178–187

D'Amato F (1985) Cytogenetics of plants cell and tissue cultures and their regenerates. CRC Crit Rev Plant Sci 3:73–112

Edallo SC, Zucchinali C, Perenzin M, Salamini F (1981) Chromosomal variation and frequency of spontaneous mutation associated with in vitro culture and plant regeneration in maize. Maydica 26:39–56

Evans DA, Sharp WR (1983) Single gene mutation in tomato plants regenerated from tissue culture. Science 221:949–951

Evans DA, Sharp WR, Medina-Filho HP (1984) Somaclonal and gametoclonal variation. Am J Bot 71:755–774

Fadeeva TS, Kozyreva OG, Lutova LA (1979a) Plant regeneration as a genetic characteristic. Issl Genet 8:160–170

Fadeeva TS, Lutova LA, Kozyreva OG (1979b) 3-ya Vses Konf Kul'tura kletok Rast, Abovyan, 1979. Tez Dokl, Abovyan, Armenian SSR, pp 119–120

Fadeeva TS, Kozyreva OG, Lutova LA (1979c) 3-ya Vses Konf Kul'tura kletok Rast, Abovyan, 1979. Tez Dokl, Abovyan, Armenian SSR, pp 118–119

FAO (ed) (1985) Production yearbook 1985. FAO Rome

Feher T (1979) Tomato propagation from parts of sterile seedlings. Kertgazdag 11(5):59–64

Fischoff DA, Bowdish KS, Perlak FJ, Marrone PG, McCormick SM, Niedermeyer JG, Dean DA, Kusano-Kretzmer K, Majer EJ, Rochester DE, Rogers SG, Fraley RT (1987) Insect tolerant transgenic tomato plants. Biotechnology 5:807–813

Frankenberger EA, Hasegawa PM, Tigchelaar EC (1979) Analysis of shoot-forming capacity from leaf discs of selected tomato varieties. HortSci 14(3) III:478

Frankenberger EA, Hasegawa PM, Tigchelaar EC (1981a) Influence of environment and developmental state on the shoot-forming capacity of tomato genotypes. Z Pflanzenphysiol. 102(3):221–232
Frankenberger EA, Hasegawa PM, Tigchelaar EC (1981b) Diallel analysis of shoot-forming capacity among selected tomato genotypes. Z Pflanzenphysiol 102(3):233–242
Gavazzi G, Tonelli C, Todesco G, Arreghini F, Raffaldi F, Sala F, Biasini MG, Vecchio F, Barbuzi F (1987) Chemically induced mutagenesis versus somaclonal variation in tomato. Theor Appl Genet 74:733–739
Gunay AL, Rao PS (1980) In vitro propagation of hybrid tomato plants (*Lycopersicon esculentum* L.) using hypocotyl and cotyledons explants. Ann Bot (London) 45(2):205–207
Handa AK, Bressan RA, Hasegewa PM (1982) Characteristics of cultured tomato cells after prolonged exposure to medium containing polyethyleneglycol. Plant Physiol 69:514–521
Handa AK, Bressan RA, Handa S, Hasegawa PM (1983) Clonal variation for tolerance to polyethylene glycol-induced water stress in cultured tomato cells. Plant Physiol 72:645–653
Handa S, Bressan RA, Handa AK, Carpita NC, Hasegawa PM (1983) Solutes contributing to osmotic adjustment in cultured plant cells adapted to water stress. Plant Physiol 73:834–843
Handley LW, Nickels RL, Cameron MW, Moore PP, Sink KC (1986) Somatic hybrid plants between *Lycopersicon esculentum* and *Solanum lycopersicoides*. Theor Appl Genet 71:691–697
Heideveld M, Jansen CE, Kool AJ (1984) Regeneration of tomato cultivars. In: A new era in tomato breeding. Synopses, 9th Meet EUCARPIA Tomato Work Group, 22–24 May 1984, Wageningen, Neth, p 118
Herman EB, Haas GJ (1978) Shoot formation in tissue cultures of *Lycopersicon esculentum* Mill. Z Pflanzenphysiol 89(5):467–470
Hogenboom NG (1972) Breaking breeding barriers in *Lycopersicon* 5. The inheritance of the unilateral incompatibility between *L. peruvianum* Mill. and *L. esculentum* Mill. and the genetics of its breakdown. Euphytica 21:405–414
Illig RD, Dallacqua AN (1986) In vitro selection for resistance to culture filtrates of *Phytophthora infestans* in tomato by callus culture growth in the presences of pathogen culture filtrate and regeneration of resistance lines. 6th Int Congr Plant tissue cell culture, Univ Minnesota, Minneapolis, p 376
Imanishi S, Hiura I (1976) Organogenesis from long-term callus culture from hypocotyl in *Lycopersicon esculentum* L J Yamagata Agric For Soc 33:35–38
Ingram D, Mac Donald MV (1986) In vitro selection of mutants. In: Nuclear techniques and in vitro culture for plant improvement. In: Proc Int Symp IAEA, 19–23 Aug 1985, Vienna, pp 241–258
Jain SM, Shahin EA, Sun S (1985) Somatic hybridization between atrazine resistant *Solanum nigrum* and tomato. In vitro 21:25–27
Kartha KK, Champoux S, Gamborg OL, Pahl K (1977) In vitro propagation of tomato by shoot apical meristem culture. J Am Soc Hortic Sci 102(3):346–349
Kiforak OV, Levenko BA, Chernetskii VP, Semenyuk DV (1977) Cytogenetic analysis of cell tomato callus tissue after treatment with kinetin riboside. Tsitol Genet 11:310–313
Kinsara A, Patnaik SN, Cocking EC, Power JB (1986) Somatic hybrid plants of *Lycopersicon esculentum* Mill. and *Lycopersicon peruvianum* Mill. J Plant Physiol 125:225–234
Koblitz H, Koblitz D (1982) Experiments on tissue culture in the genus *Lycopersicon* Miller. Shoot formation from protoplast of tomato long term cell cultures. Plant Cell Rep 1(4):147–150
Korneef M, Hanhart CJ, Martinelli L (1987) A genetic analysis of cell culture traits in tomato. Theor Appl Genet 74:633–641
Kurtz SM, Lineberger RD (1983) Genotypic differences in morphogenic capacity of cultured leaf explant of tomato. J Am Soc Hortic Sci 108(5):710–714
Kut SA, Bravo JE, Evans DA (1984) Tomato. In: Sharp WR, Evans DA, Ammirato PV, Yamada Y (eds) Handbook of plant cell culture. Vol. Crop Species Macmillan, New York 3:247–289
Levenko BA, Kiforak OV (1977) The effect of fluorophenilalanine on tissue culture in tomato and *Haplopappus* Eksperim. Genet Rast Kiev Ukranian SSR, pp 138–142
Levenko BA, Kunakh VA, Yurkova GN, Legeida VS, Kiforak OV, Alpatova LK (1977) Formation of strains of plant callus tissue during prolonged culture in vitro. 3-ii s"ezd Vses ob-va genetikov i selektsionerov im. NI Vavilova, Leningrad, USSR, p 292
Lopez Peralta C, Castaneda GC, Salceda SVM (1978) A study on the development of cell cultures of *Phaseolus vulgaris* L. and *Lycopersicon esculentum* in media supplied with mead and coconut water. 2. Growth and differentiation of the apical meristems of the shoot. Agrociencia 31:75–81

McCormick Sh M, Nidermeyer J, Fry J, Barnason A, Horsh R, Fraley R (1986) Leaf disc transformation of cultivated tomato (*Lycopersicon esculentum*) using *A. tumefaciens*. Plant Cell Rep 5:81–84
Melchers G, Sacristan MD, Holder AA (1978) Somatic hybrid plants of potato and tomato regenerated from fused protoplasts. Carlsberg Res Comun 43(4):203–218
Meredith CP (1978) Selection and characterization of aluminium resistant variants from tomato cell cultures. Plant Sci Lett 12:25–34
Meredith CP (1979) Shoot development in established callus cultures of cultivated tomato (*Lycopersicon esculentum* Mill). Z Pflanzenphysiol 95(5):401–411
Morgan A, Cocking EC (1982) Plant regeneration from protoplasts of *Lycopersicon esculentum* Mill. Z Pflanzenphysiol 106:97–104
Muhlbach HP, Zapata FJ (1980) Different regeneration potentials of mesophyll protoplasts from cultivated and a wild species of tomato. Planta 148(1):89–96
Nacmias B, Ugolini S, Ricci MD, Pellegrini MG, Bogani P, Bettini P, Inze D, Buiatti M (1987) Tumor formation and morphogenesis in different *Nicotiana* sp. and hybrids induced by *Agrobacterium tumefaciens* T-DNA mutants. Dev Genet 8:61–71
Neal CA, Topoleski LD (1983) Effects of the basal medium on growth of immature embryos in vitro. J Am Soc Hortic Sci 108(3):434–438
Neal CA, Topoleski LD (1985) Hormonal regulation of growth and development of tomato embryos in vitro. J Am Soc Hortic Sci 110(6):869–873
Niedtz RP, Rutter EM, Handley LW, Sink KC (1985) Plant regeneration from leaf protoplasts of six tomato cultivars. Plant Sci 39(3):199–204
Novak FJ, Maskova I (1979) Apical shoot tip culture of tomato. Sci Hortic 10(4):337–344
O'Connell MA, Hosticka LP, Hanson MR (1986) Examination of genome stability in cultured *Lycopersicon*. Plant Cell Rep 5:276–279
Ohki S, Bigot C, Mousseau J (1978) Analysis of shoot forming capacity in vitro in two lines of tomato (*Lycopersicon esculentum* L.) and their hybrids. Plant Cell Physiol 19(1):27–42
Pence VC, Caruso JL (1984) Effects of IAA and four IAA conjugates on morphogenesis and callus growth from tomato leaf discs. Plant Cell Tissue Org Cult 3:101–110
Rahman MM, Kochhar TS, Kaul K (1986) Isolation of sodium chloride resistant callus of tomato – salt tolerance. 6th Int Congr Plant tissue cell culture, Univ Minn, Minneapolis, p 379
Rick CM (1979) Biosystematic studies in *Lycopersicon* and closely related species of *Solanum*. In: Hawkes JG, Lester RN, Skelding AD (eds) The biology and taxonomy of the solanaceae. Academic Press, New York London, pp 325–346
Rick CM, Forbes JF (1975) Allozyme variation in the cultivated tomato and closely related species. Bull Torrey Bot Club 102:376–384
Roddick JG, Melchers G (1985) Steroidal glycoalkaloyd content of potato, tomato and their somatic hybrids. Theor Appl Genet 70:655–660
Scala A, Bettini P, Buiatti M, Bogani P, Pellegrini G, Tognoni F (1984) In vitro analysis of the tomato-*Fusarium oxysporum* system and selection experiments. In: Novak FJ, Havel L, Dolezel J (eds) Plant tissue and cell culture application to crop improvement. Czech Acad Sci, Prague, pp 361–362
Scala A, Bettini P, Bogani P, Pellegrini MG, Tognoni F (1985) Tomato-Fusarium interaction: in vitro analysis of several possible pathogenic factors. Phytophatol Z 113:90–94
Scheller HV, Huang B, Goldsbrough PB (1986) Selection and characterization of cadmium-tolerant cell lines in *Lycopersicon esculentum* – tomato. 6th Int Congr Plant tissue cell culture, Univ Minn, Minneapolis, p 77
Seeni S, Gnanam A (1981) In vitro regeneration of chlorophyll chimeras in tomato (*Lycopersicon esculentum*). Can J Bot 59:1941–1943
Shahin EA (1985) Totipotency of tomato protoplasts. Theor Appl Genet 69(3):235–240
Shahin EA, Spivey R (1986) A single dominant gene for *Fusarium* wilt resistance in protoplast-derived tomato plants. Theor Appl Genet 73:164–169
Shepard SLK (1986) Selection for early blight disease resistance in tomato: use of tissue culture with *Alternaria solani* culture filtrate – callus culture. 6th Int Congr Plant tissue cell culture, Univ Minn, Minneapolis, p 211
Shimamoto K, King PJ (1983) Isolation of a histidine auxotroph of *Hyoscyamus muticus* during attempts to apply BUdR enrichment. Mol Gen Genet 189:69–72
Sibi M (1980) Heritable epigenetic variations from in vitro tissue culture of *Lycopersicon esculentum* (var.

Montalbo). In: Earle D. Demarly Y (eds) Variability in plants regenerated from "in vitro" tissue cultures. Praeger, New York, pp 228–244
Sibi M (1986) Non-mendelian heredity – genetic analysis of variant plants regenerated from "in vitro" culture: epigenetics and epigenic. In: Semal J (ed) Somaclonal variations and crop improvement. Nijhoff, Dordrecht, pp 53–54
Sink KC, Reynolds R (1986) Tomato (*Lycopersicon esculentum* L.). In: Bajaj YPS (ed) Biotechnology in agriculture and forestry, vol 2. Crops I. Springer, Berlin Heidelberg New York Tokyo, pp 319–362
Sink KC, Handley LW, Niedtz RP, Moore PP (1986) Genetic manipulation in plant breeding. In: Horn W, Jensen C, Odenbach W, Schieder O (eds) Proc Int Symp Eucarpia, Sept 8–13, 1985, Berlin (West). De Gruyter, Berlin New York, pp 405–413
Smith SS, Murakishi HH (1986) Nature of resistance of tomato somaclones to tomato-mosaic virus and tobacco-mosaic virus. 6th Int Congr Plant tissue cell culture, Univ Minn, Minneapolis, p 326
Tal M, Dehan K, Heikin H (1977) Morphogenetic potential of cultured leaf sections of cultivated and wild species of tomato. Ann Bot (London) 41(175):937–941
Thomas BR, Pratt D (1982) Isolation of paraquat tolerant mutants from tomato cell cultures. Theor Appl Genet 63:169–176
Toyoda H, Tanaka N, Hirai T (1984a) Effects of the culture filtrate of *Fusarium oxysporum f. sp. lycopersici* on tomato callus growth and selection of resistant callus cells to the filtrate. Ann Phytopathol Soc Jpn 50:53–62
Toyoda H, Hayashi H, Yamamoto K, Hirai T (1984b) Selection of resistant tomato calli to fusaric acid. Ann Phytopathol Soc Jpn 50:538–540
Uralets LI (1984) In vitro embryo culture in tomato. Fiziol Biokim Kul't Rest 16(5):495–499
Vnuchova VA (1975) Study of cultivation conditions of the tomato species *Lycopersicon esculentum* Mill., *L. peruvianum* Mill. and *L. hirsutum* H. et B. in tissue culture. Tezisy Dokl Konf Selektsiya i Genet Ovoshch Kul'tur 3, Kishinev, Moldavian SSR, pp 59–60
Vorob'eva GA, Prikhod'ko NI (1980) Use of in vitro embryo culture to obtain interspecific tomato hybrids. Tr Priklad Bot Genet Selekt 67(3):64–74
Zagorska NA, Abaddjiava MD, Georgiev HA (1982) Inducing regeneration in anther cultures of tomatoes (*Lycopersicon esculentum* Mill.) C R Acad Bulg Sci 35:97–100
Zelcer A, Soferman O, Izhar S (1984) An in vitro screening for tomato genotypes exhibiting efficient shoot regeneration. J Plant Physiol 115(3):211–215
Zhuchenko AA, Chernova LK, Sedov GI, Kolos VP (1981) Effect of factors inducing recombination on the occurrence of genetic deviations in in vitro regenerated plantlets of interspecific hybrids. Ekol Genet Rast i Zhivotnykh Tez Dockl Vses Konf 2, Kishinev, Moldavian SSR, pp 212–213

III.3 Somaclonal Variation in Eggplant (*Solanum melongena* L.)

R. ALICCHIO[1]

1 Introduction

1.1 Importance and Distribution of the Plant

Eggplant (*Solanum melongena* L., n = 12), is cultivated in all the Mediterranean and in Asian countries, from where most of the production derives. Northeastern India and, secondarily, China are believed to be the centers of diffusion of this plant (Vavilov 1951). The species is well adapted to tropical weather, requiring a long hot season to mature fruits. In temperate climates, plants are often started in hotbeds and transplanted outdoors in late spring.

The fruits contain many seeds, which retain the capacity for germinating through 6–7 years. Cultivars are classified mostly by the shape of the fruits, which may be oblong, round, or pear-shaped, and by the color which may be dark, purple, white, or green.

Annual production in the world is about 4.5×10^6 mt. (Hinata 1986).

It is an important food plant especially in Japan and other parts of the Orient, but it is also used in French and Italian cookery, both fresh and pickled.

1.2 Objectives for Improvement

Most of the breeding work on eggplant is reported from India, Italy, and Japan. The conventional breeding of eggplant, as of many other vegetable crops, involves essentially selection programs and mostly hybrid breeding (Pearson 1983) to obtain varieties with valuable gene combinations and corresponding useful agronomic phenotypic traits, such as precocity of flowering time, quality of fruits (shape, size, color), and obviously, yield.

Commercial eggplant varieties are highly susceptible to diseases caused by *Verticillium* ssp., *Fusarium* ssp., and, in tropical countries, by nematodes (*Meloidogyne* ssp.) and bacteria *Pseudomonas*. These pests and pathogens are so destructive in Japan, that 20% of commercially produced eggplants are grafted onto resistant rootstocks (Yamakawa 1982). Eggplant suffers also from too dry conditions and excessive flooding. It is therefore of great importance to obtain new gene

[1]Department of Evolutionary and Experimental Biology, University of Bologna, Via F. Selmi 1, 40126 Bologna, Italy

Biotechnology in Agriculture and Forestry, Vol. 11
Somaclonal Variation in Crop Improvement I (ed. by Y.P.S. Bajaj)

combinations which combine both yield characteristics and resistance traits to pathogens and to extreme environmental conditions.

1.3 Available Genetic Variability

Eggplant is essentially an autogamic plant: this modality of reproduction implies that commercial cultivars are highly homozygous. Numerous attempts have been made to obtain F_1 hybrids displaying heterosis for desirable traits. Usually, the more the parental cultivars differ in morphology, physiology, and geographical provenance, the more the hybrids display heterosis for yield, phenotypic stability, and adaptation (Kakizaki 1931).

The possibility of using F_1 hybrids showing more than 50% heterosis for yield was first reported by Kakizaki (1931) and later by Dharme (1977), who found heterotic effect in the number of fruits and fruit weight, both components of yield.

Many authors reported heritability estimates for several yield components (plant height, days to flower, number of fruits per plant). The high heritability (estimates ranging from 44% to 81%) represents a good basis for selection in varietal improvement (Eldin 1967; Silvetti and Brunelli 1970; Peter and Singh 1976).

However, as Pearson (1983) pointed out, when a breeding program based on selection or hybridization is aimed at obtaining the desired fruit shape, transgressive selection for yield and vigor may occur in eggplant, as in other Solanaceae, because not all combinations will be desirable.

As regards pathogen resistance, hybrids do not always show traits improved with respect to their parents. Nicklow (1983) found seven introductions showing tolerance to *Verticillium* manifested as a 7% survival when infected as seedlings with the fungus. F_1 hybrids were all susceptible, indicating that different tolerance genes were involved. In efforts to achieve an increased tolerance to *Verticillium*, a recurrent selection program on selfed progenies was realized; the percentage of survivors increased, but resistance was never achieved.

2 Somaclonal Variation

2.1 Need for Variation in Eggplant

The success of conventional breeding programs depends on the provision of an adequate pool of genetic variation through selection and intraspecific and interspecific crosses. To be effective in developing new varieties, selection requires a long time and much space. The production of hybrid eggplant seed is labor-intensive, and natural outcrossing seldom occurs. Furthermore, hybrids from intraspecific crosses do not always inherit the agronomic characteristics of the parents; artificial hybridization at specific or generic levels encounters barriers to crossability, and in eggplant it has been reported (Narasimha Rao 1979) that barriers have developed even between varieties.

Taking into account that progress in genetic improvement may be hindered by several problems, plant breeders have recognized the importance of introducing new genetic techniques as an adjunct to conventional breeding (see Larkin and Scowcroft 1981 and Gunn and Day 1986) for recovering useful genetic variants. The advantages offered by in vitro methods are: (a) micropropagation for preserving desirable traits, (b) in vitro evaluation of the plant material to assess genotypic responses; (c) in vitro selection of cells resistant to pathogens, chemical and physical treatment, (d) production of haploids through anther cultures to early release of varieties, and (e) somatic hybridization for recombining genomes of sexually incompatible species.

Hinata (1986) envisaged an increased role of tissue culture techniques in the breeding work of eggplant, stressing the importance of in vitro regeneration, haploid production, and protoplast fusion. In particular, in vitro selection of cells resistant to toxins and physicochemical treatments, early screening of somaclonal variants, and somatic hybridization between *Solanum melongena* and sexually incompatible wild species resistant to pests and pathogens, represent a powerful option for eggplant improvement.

2.2 Brief Review of In Vitro Studies

Many studies are available concerning the in vitro culture of different explants such as hypocotyls, cotyledons, leaves, stem, roots tissues, and embryos in eggplant. Callus was induced from all these explants on LS or MS media supplemented with different amounts of auxins (NAA, IAA, 2, 4-D) (see Hinata 1986). Regeneration from callus was obtained through embryogenesis and organogenesis, depending on the hormones added to the inducing medium. Regeneration was also achieved from cultured protoplasts and anthers.

2.2.1 Regeneration of Plants from Segments and Micropropagation

Since the appearance of the first paper on regeneration from *Solanum melongena* callus by Yamada et al. (1967), many reports have accumulated on in vitro culture of different explants and plant regeneration via shoot buds and/or embryogenesis.

Segments excised from hypocotyls were often used as explants to obtain plant regeneration with or without a previous callus induction. Kamat and Rao (1978) studied the effects of various hormones on the capacity of inducing adventitious shoot buds directly on excised hypocotyl segments; they found the maximum frequency of plants when IAA (9.6 mg/l) and Z (11.6 mg/l) were added suppressed shoot regeneration.

The effects of various combinations of NAA and BA6 on the induction of callus in hypocotyls and on the development of embryoids and shoot buds were analyzed by Matsuoka and Hinata (1979). The authors demonstrated that the two patterns of regeneration via shoot buds or via embryogenesis were under the control of hormonal balance and were inversely related. NAA at high concentrations (8 mg/l) seemed effective in eggplant for embryogenesis. The addition of BA (0.2225 mg/l) enhanced shoot bud formation while it suppressed embryogenesis.

These observations were in agreement with the results reported by Gleddie et al. (1982, 1983); embryogenesis from leaf explants was specifically induced by high concentrations of auxins like NAA (10 mg/1), and totally inhibited by various cytokinins when supplied at 1 mg/1. Shoot organogenesis was induced by cytokinins in amounts ranging from 0.5 to 10 mg/1, while it was suppressed by auxins like NAA. From these results it can be inferred that the potential for embryogenesis is inversely related to the potential for organogenesis; this conclusion was supported by the fact that cultivars with high potential for embryoid induction had a lower potential for shoot production and vice versa (Matsuoka and Hinata 1979).

Shoot regeneration was achieved also in root cultures by Zelcer et al. (1983), who used MS without hormones as inducing medium.

Micropropagation through bud cultures can be easily obtained by cutting the stem of sterilized seed plantlets in pieces containing at least one node; placing them on LS or MS media without hormones (Rotino unpubl.), or supplemented with BA (Hisajima 1982) a new plant originates in a few days as growth of a single axillary bud (Fig. 1).

2.2.2 Plants from Callus and Suspension Cultures

Plant regeneration was achieved also in calli excised from the explant and subcultured for many passages on solid and on liquid media. Yamada et al. (1967) first obtained embryoid formation on the surface of embryo callus which was subcultured for 2 years on White's medium supplemented with 1 mg/l NAA. Embryogenesis was later obtained by Gleddie et al. (1982, 1983) in leaf cell suspensions; even after 18 months of growth in liquid medium with 2, 4-D, transfer of cells to MS medium-supplemented with 10 mg/1 NAA induced high-frequency embryogenesis.

Matsuoka and Hinata (1983) obtained embryoid formation in MS with 8 mg/l NAA from callus which was subcultured on low NAA medium (2 mg/l) for 19 passages. This NAA specificity for embryogenesis is unique for eggplant.

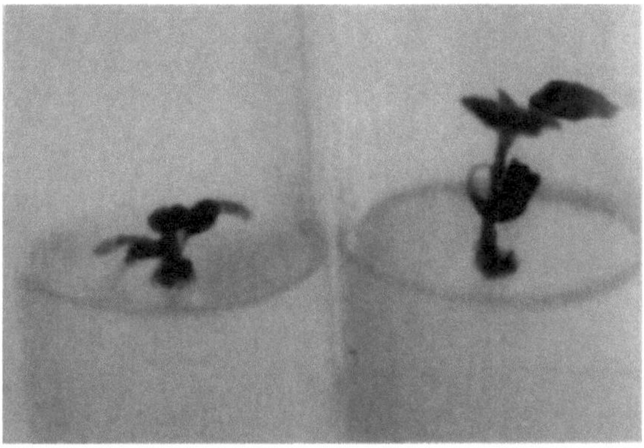

Fig. 1. Micropropagation of eggplant by in vitro bud cultures (Courtesy of Dr. G.L. Rotino unpubl.)

Shoot organogenesis was induced by Alicchio et al. (1982b), in calli excised from hypocotyls, leaves, and cotyledons subcultured for many passages on LS + 2, 4-D 0.4 mg/l and then placed onto inducing media (LS without hormones, LS + kinetin 2 mg/l, LS + gibberellic acid 2 mg/l). The best results in terms of shoot formation were obtained in LS medium (Fig. 2A), the addition of kinetin induced the formation of plantlets with compact nodular callus at the base (Fig. 2B), whereas the addition of gibberellin gave very thin shoots (Fig. 2C).

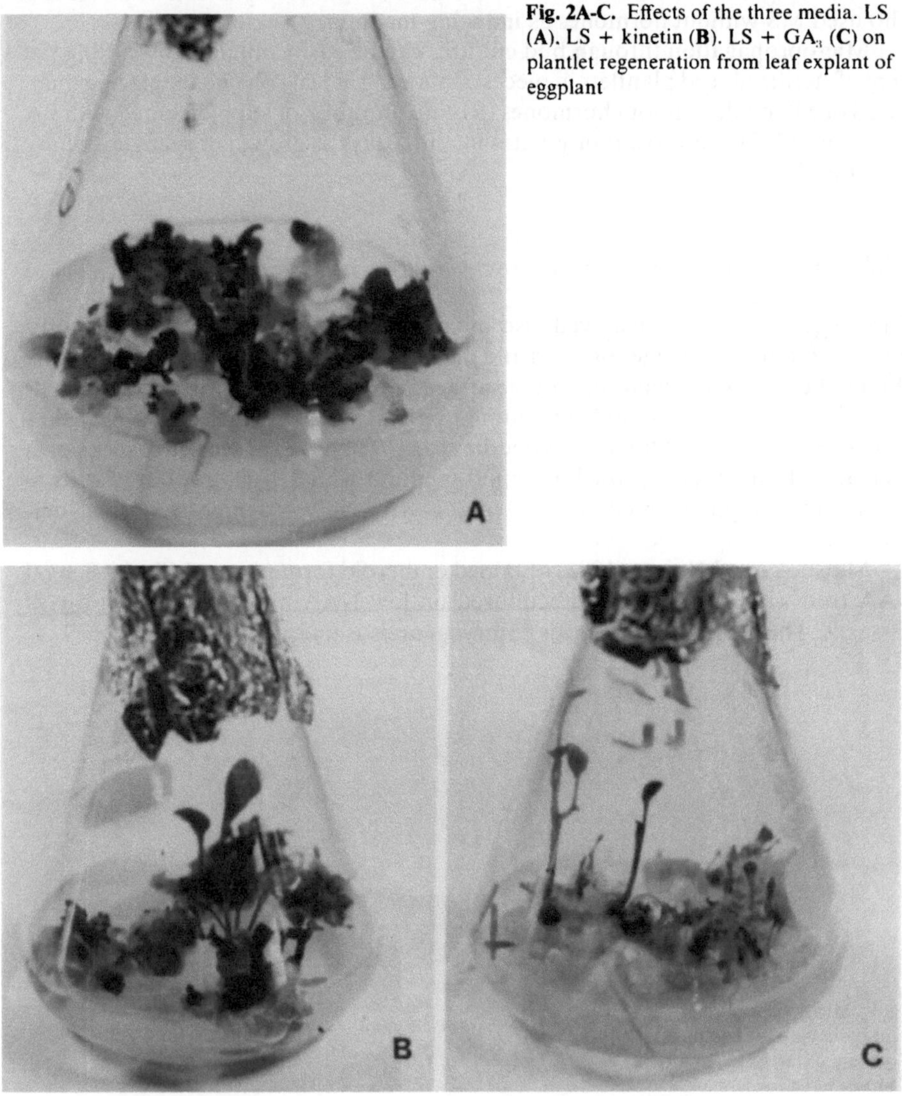

Fig. 2A-C. Effects of the three media. LS (**A**). LS + kinetin (**B**). LS + GA_3 (**C**) on plantlet regeneration from leaf explant of eggplant

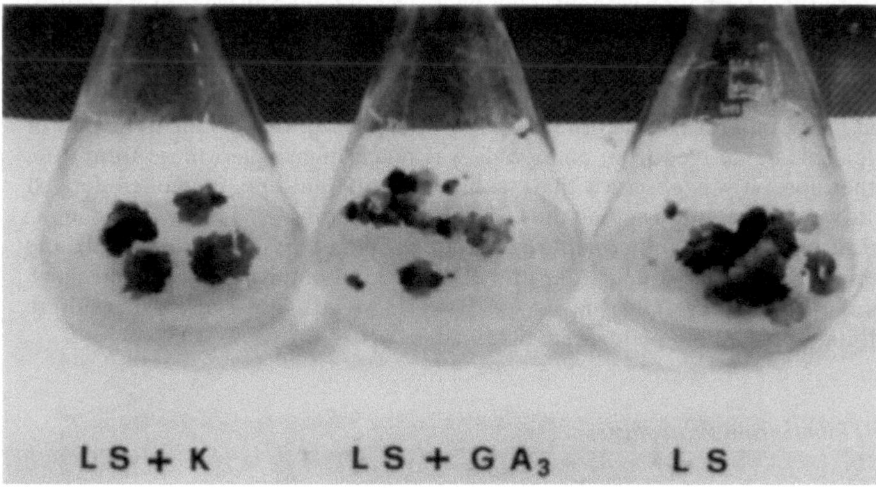

Fig. 3. Pieces of green callus from hypocotyl explant of eggplant

The authors found quantitative and qualitative differences among the calli from the three explants in regard to their regenerative potential; in fact the regeneration from hypocotyl callus was observed only in hybrids, but not in inbred cultivars, where only pieces of green callus grew (Fig. 3). It was suggested that the dedifferentiation from different explants gives rise to cell populations which may keep the ability of producing different exogenous growth regulators concerned with differentiation.

Many experiments proved that the regenerative potential is genetically controlled; Gleddie et al. (1982, 1983) observed quantitative differences in frequency of embryogenesis among seven cultivars, and Alicchio et al. (1982b) demonstrated that hybrids, but not the parental lines, have the capacity of regenerating plants from late subcultures.

2.2.3 *Plants from Anthers*

Anther culture, with consequent haploid production, is of great importance in plant breeding, both in studies on the induction of mutations and in the production of homozygous plants which are obtained spontaneously in vitro or by chemical induction of chromosome doubling in haploids. The critical step in anther culture is the choice of buds containing anthers with uninucleate microspores; at this stage most anthers are embryogenetic.

Raina and Iyer (1973) first observed differentiation of shoot buds from anther callus on a medium with high content of IAA and kinetin (2 mg/l). Since all the plants recovered were diploid, the authors suggested that the diploidization was the result of in vitro chromosome duplication.

Embryoid formation without callus phase was obtained by Isouard et al. (1979) on a basal medium supplemented with NAA (0.1 mg/l) and BAP (0.2 mg/l); the few plants regenerated were mostly haploid. A higher number of plants were obtained from anther cultures through embryoid formation or shoot induction in callus by the Research Group for Haploid Breeding in Peking (1978). Usually the frequency of haploid plants derived from embryoids was much higher than those from calli. Furthermore, it was observed that, within each original line, the pollen-derived plants were more uniform than those derived from callus. An improvement in the yield of plants from anther culture was later obtained by Dumas de Vaulx and Chambonnet (1982) by incubating anthers at + 35 °C in the dark during the first 8 days of in vitro culture. On using, IAA 0.01 mg/l instead of 2,4-D in the first culture medium, only haploid plants were obtained.

2.2.4 Plants from Protoplasts

The technique of isolation and culture of protoplast in eggplant has been reported in many protocols differing in the concentration of mannitol (from 0.3 M to 0.7 M) and in the concentration of enzymes (Bhatt and Fassuliotis 1981; Jia and Potrykus 1981; Saxena et al. 1981; Gleddie et al. 1982; 1986b). Usually the protoplasts were incubated overnight in the dark or dim light at 25 °C for the digestion. Optimal plating density ranged from 5×10^4 to 1×10^6 protoplasts/ml. The plating efficiency observed in these experiments never exceeded 1%.

Guri and Izhar (1984) described a method to improve the efficiency of plating and regeneration of plants. The presence of a reservoir medium containing 1% charcoal (Shepard et al. 1980) adjacent to the culture medium, and the incubation in the dark for 4 days immediately after plating, stimulated the initial divisions. Plating efficiency, calculated after 3 weeks, was 15%. The best results in shoot regeneration were obtained in about 1 month in a medium MS + zeatin 2 mg/l.

2.3 Variations in Callus and Regenerated Plants

2.3.1 Morphological Changes

In spite of the large amount of experimental data concerning the various techniques of tissue cultures, little information is available on the characteristics of regenerated plants. Gleddie et al. (1982) obtained embryoids from leaf callus which presented various abnormalities like fused cotyledons, abnormal radicles, shoot apices, and callus proliferation. Other data relative to regeneration from protoplasts after a very short callus phase showed that plants were similar to the parental lines, developed fertile flowers, and set seeds (Jia and Potrykus 1981; Guri and Izhar 1984).

Plantlets with irregular leaves were sometimes regenerated from callus (Fig. 4). Furthermore, when plantlets were transferred to soil, only a few of them survived (Alicchio et al. 1984). Plants were morphologically normal, but the karyotypic instability caused high sterility; in fact after self-pollination they produced small berries, many of them were without seeds.

Fig. 4. A plantlet regenerated from eggplant leaf callus showing irregular leaves. (Alicchio unpubl.)

2.3.2 Enzymatic Changes

To understand the kind of transformations implied by the process of in vitro regeneration, it is necessary to analyze the role of different explants and the consequences of dedifferentiation in the callus. Del Grosso and Alicchio (1981) made an accurate analysis of the biochemical patterns of peroxidases, both in organized (explants) and unorganized tissues (callus). In the callus from different organs (hypocotyl, cotyledon, leaf) more isozymes are expressed than in the original explants (Fig. 5), indicating that the dedifferentiation to callus involves the activation of genes normally repressed in the differentiated tissues. Differences between shoots and callus tissue in the peroxidase pattern of *Sinapis alba* were also observed by Bajaj et al. (1973).

Moreover, it was observed that the recovery of activity is different among the three calli both in the number of isozymes and in the substrate affinity (Fig. 5). These results, along with the data concerning the regenerative potential of hypocotyl, cotyledon, and leaf (Alicchio et al. 1982b), and the differential growth of calli from the three explants in toxic medium (Alicchio et al. 1982c), confirm that dedifferentiation follows different pathways to callus, depending on the explant.

Further studies by Del Grosso et al. (1982) verified that the biochemical differences associated with calli of various origins, persist during the regeneration process and are particularly evident in leaf and root of the regenerated plants (Fig. 6). On the basis of these comparative analyses, the authors concluded that the cell populations obtained from the three explants retain a different gene expression in spite of the dedifferentiation process.

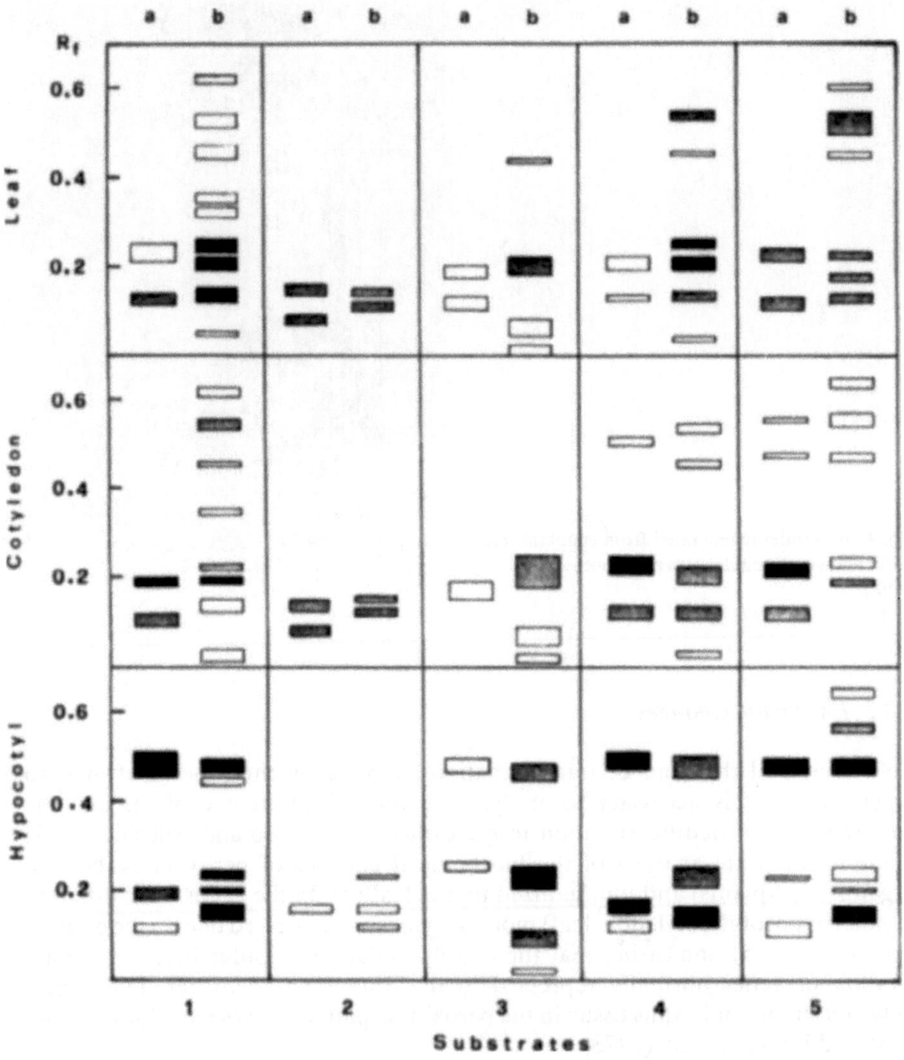

Fig. 5. Isozymatic patterns of peroxidases in hypocotyl, cotyledon and leaf of *Solanum melongena*; *a organized tissues. b* callus. *Black, gray, white bands* represent decreasing degrees of enzyme activity. Substrates: *1* 0-dianisidine; *2* DOPA; *3* Ferulic acid, *4* eugenol; *5* cathecol. (Del'Grosson and Alicchio 1981)

2.3.3 Chromosomal Analysis

Relatively few analyses are available on the chromosome constitution of regenerated plants. Matsuoka and Hinata (1983) checked the chromosome number of plantlets regenerated from hypocotyl callus, and found that the karyotypic in-

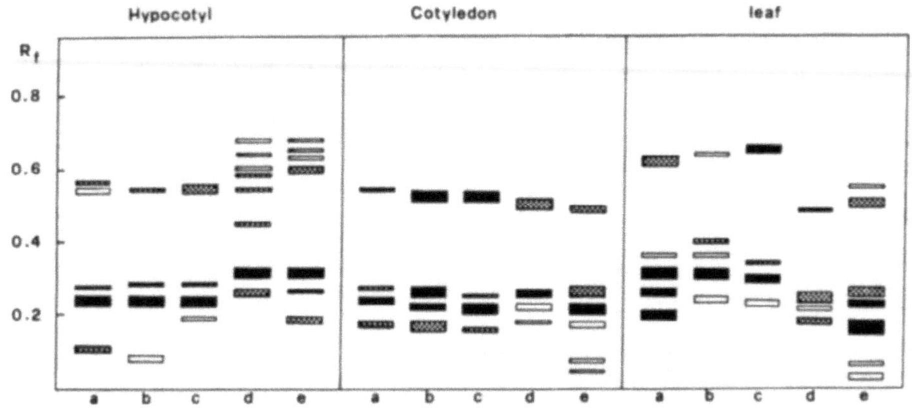

Fig. 6. Isozymatic patterns of peroxidases at different stages of plant differentiation from hypocotyl, cotyledon and leaf callus of eggplants. *a* callus; *b* shoot; *c* stem; *d* leaf; *e* root. *Black, gray, white bands* represent decreasing degrees of enzyme activity. (Del Grosso et al. 1982)

stability increased with the number of callus subcultures. In fact, while plants regenerated from early subcultures were all diploid except one, those obtained from later passages (up to 19 passages) were mostly tetraploid. This observation seems to confirm that the degree of polyploidy in regenerated plants increases progressively with increasing age of the callus, and it is consequent to the mitosis disorders proper of the callus phase such as endoreduplication and spindle fusion in binucleate or multinucleate cells (D'Amato 1977).

On the other hand, the age of callus and the variation in ploidy do not seem to reduce the morphogenetic capacity of the culture. When regeneration is achieved from a short-term culture as in callus from protoplasts (Guri and Izhar 1984), plants were obtained which have the same chromosome number and look similar to the parental cultivar.

Unfortunately, few cytological observations accompany the data on plant regeneration and investigations of somaclonal variation; it is therefore difficult to draw conclusions about a possible correlation between morphological modification and changes in chromosome number. However, it is clear that aneuploidy is by far more detrimental than polyploidy because it results in genic unbalance. Nuti Ronchi et al. (1973), working on the stem pith of *Nicotiana glauca*, observed a very active nuclear fragmentation (amitosis) followed by mitoses. The consequence of this abnormal cell division cycle in the callus is the production of cells with different chromosome numbers. The multicellular origin of in vitro regenerated shoots and roots may cause the presence in plants of mixoploidy and aneuploidy proper of the callus.

Alicchio et al. (1982a, 1984) reported the cytological analysis made on root tips of plants regenerated after six subcultures of callus; they found that the condition of aneuploidy and mixoploidy was the rule (Fig. 7). These results agree with the observations of various authors who found that regeneration frequently produces a mixoploid condition in root apical meristems.

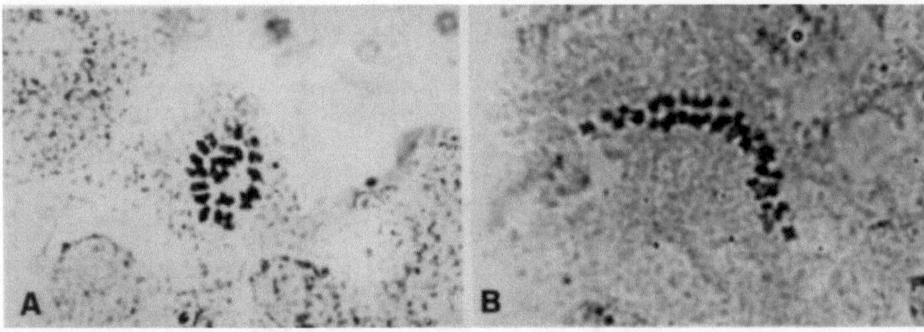

Fig. 7A,B. Karyotypic variability in plants regenerated from leaf callus of *Solanum melongena* (LVR cv.). **A** hypodiploid cell. **B** Hyperdiploid cell

Two genotypes (LVR and P.I. 23327) and their F_1 hybrid were analyzed; the distribution of root tip cells with different chromosome numbers are reported in Fig. 8A; the modal karyotype of plants regenerated from callus is diploid (2n = 24), but plants with cells hypodiploid and hyperdiploid are present. The same analysis made on plants regenerated from callus grown on medium containing toxic substances in the filtrate culture of *Verticillium dahliae* revealed a greater heterogeneity of chromosome numbers (Fig. 8B), and in particular a higher frequency of hyperdiploid cells in LVR cv. and in the hybrid.

The authors were able to discard the hypothesis that filtrate culture of *Verticillium* may act as a polyploidizing agent, and advanced the hypothesis that the toxic medium acts as a selective factor, allowing hyperdiploid cells to divide more rapidly than diploid ones. Furthermore, the differences among genotypes suggest that the toxic medium may affect the chromosomal number by a selective process involving genes controlling the mitosis.

The persistence of mosaicism observed in the first selfed progeny of LVR cv., even if associated with a partial diplontic selection, (Fig. 9), conveys the hypothesis that toxic substances induced an altered genic control of the cell division cycle which results in a stabilization of the cytological instability.

2.4 Utilization of Somaclonal Variation

2.4.1 *Resistance to Verticillium*

As we have reported in the precedent sections, a lot of work has been done to obtain and improve the systems of regeneration from callus, pollen, and protoplasts. Nevertheless, very few researchers have reported observations and data on the appearance of somaclonal variants in regenerated plants. At the most, the experiments deal with obtaining plants which, as the authors describe, are fertile and set seeds, while it would be of great importance in the breeding of eggplant to obtain new variants which, besides maintaining the most important characteristics of commercial cultivars, present resistance traits to diseases and environmental stress.

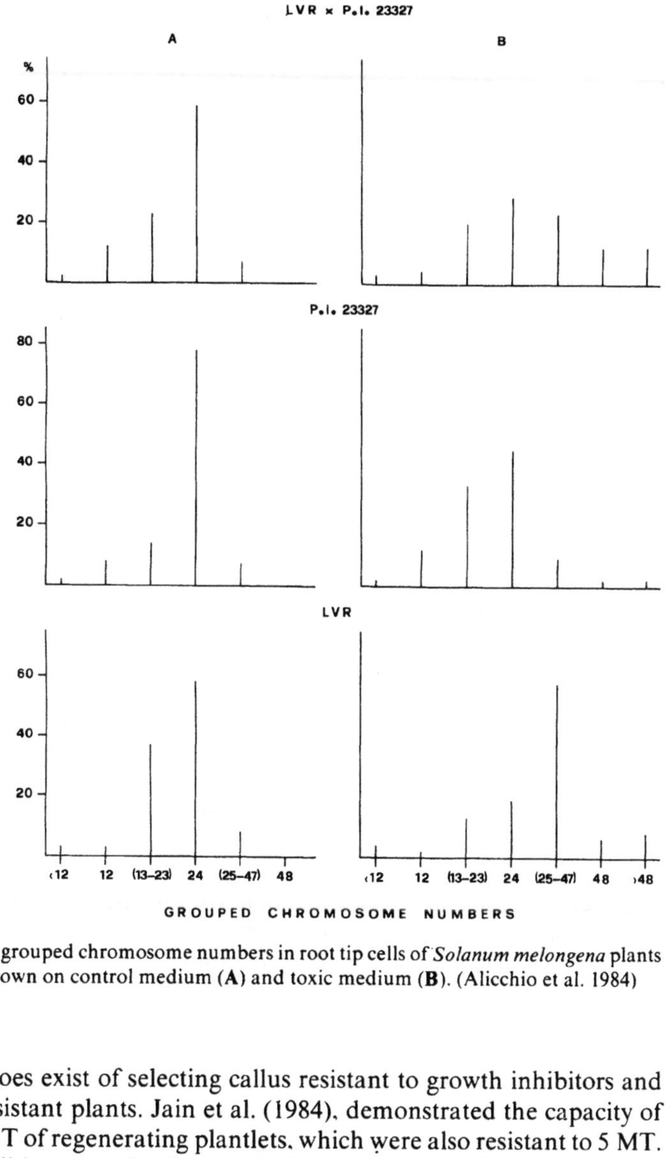

Fig. 8A,B. Distribution of grouped chromosome numbers in root tip cells of *Solanum melongena* plants regenerated from callus grown on control medium (**A**) and toxic medium (**B**). (Alicchio et al. 1984)

The possibility does exist of selecting callus resistant to growth inhibitors and able to regenerate resistant plants. Jain et al. (1984), demonstrated the capacity of callus resistant to 5 MT of regenerating plantlets, which were also resistant to 5 MT. Vice versa, it was possible to transfer characteristics from the organized tissues to the callus (Mitra et al. 1981); in fact, callus cultures were grown from both healthy and little leaf (a mycoplasma disease)-infected tissue and callus from diseased tissues regenerated 50% of plantlets diseased.

The possibility of selecting, from cell populations of susceptible cultivars, clones resistant to phytotoxins produced by *Verticillium dahliae* and regenerating somaclones exhibiting pathogen tolerance, was reported by Alicchio et al. (1982c, 1984 and unpubl.). Wilt of eggplant caused by *Verticillium* is one of the most widespread and destructive diseases of *Solanum melongena*. The recognition of

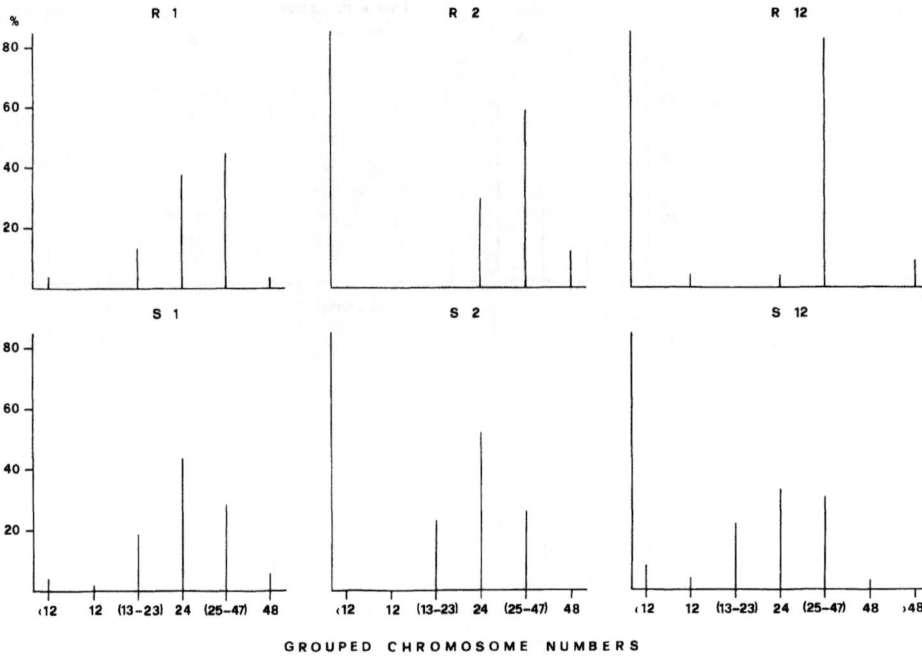

Fig. 9. Distribution of grouped chromosome numbers in root tip cells of three eggplant regenerated plants (*R*) and of their first selfed progeny (*S*). LVR cv. (Alicchio et al. 1984)

Verticillium as a major pathogen in eggplant refers to this species, being apparently the preferential host of the fungus with neither spontaneous nor induced resistant varieties recovered.

Several authors have suggested that toxemia, besides vascular occlusion, may play a role in determining wilting symptoms following *Verticillium* infection. The fungus elaborates, in liquid culture, phytotoxic metabolites acting as growth inhibitors (Pegg 1965), capable of inducing wilt in healthy shoots of lucerne (Stoddart and Carr 1966) and cotton (Keen et al. 1970). More recently, Nachmias et al. (1985) devised immunodiffusion and immunofluorescence techniques which revealed that the low molecular weight toxin was antigenically related to a substance present in *Verticillium*-infected potato tissue. Taking these reports into account, it seemed reasonable to employ tissue cultures as a valuable material from which cell lines resistant to phytotoxins might be selected. It is obvious that the success and utilization of this technique in breeding need the fulfilment of the following conditions: (a) the correlation between the pathogenic effect of fungus with the pathogenic effect of its toxin; (b) the similarity between the effects of toxin in vivo and in vitro; (c) the possibility of obtaining plants from cultured cells which retain resistance. Observations by Alicchio et al. (1982c, 1984) made it clear that the prerequisites needed for successful use of tissue culture in breeding for increased disease resistance were satisfied. In fact, both the pathogen and the cell-free culture filtrate caused a darkening of the central vessel in sectioned stems and wilting

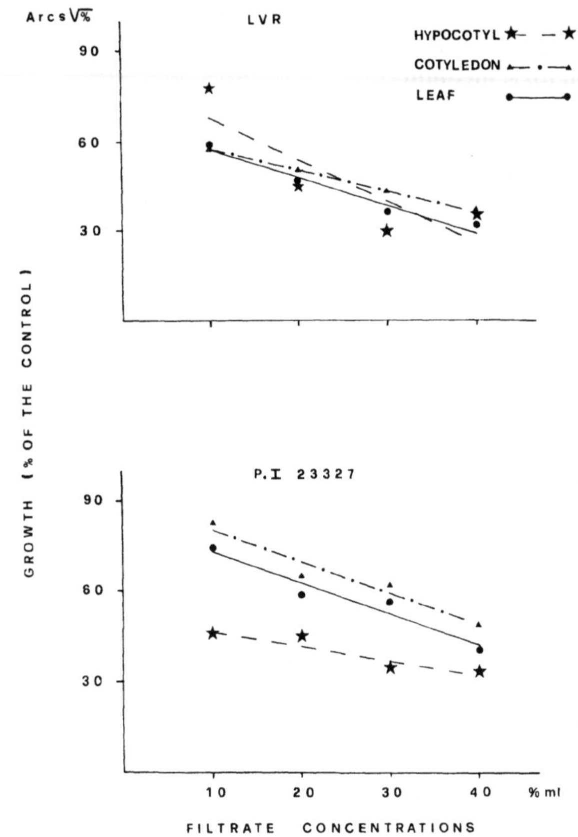

Fig. 10. Regression of eggplant callus growth percentages on concentration of toxin-containing filtrate in the basal medium. Calli from hypocotyl, cotyledon, and leaf. (Alicchio unpubl.)

symptoms in leaves on plants in vivo. Therefore the authors assumed that the toxic substances contained in the culture filtrate of *Verticillium* played a major role in determining wilting disease in eggplant. Furthermore, the addition of culture filtrate to the basic medium (LS) determined an inhibition of the callus growth which was linearly dependent on the concentration of filtrate (Fig. 10). In one of the two genotypes tested, (P.I. 23327), callus resistance to *Verticillium* toxin seemed to be affected by differences among calli from different explants; these results again support the hypothesis that the dedifferentiation process follows different pathways to callus, depending on the genotype and the explant (Del Grosso and Alicchio 1981; Alicchio et al. 1982b).

The pieces of callus exposed to the effect of the toxin for 1 month were alternatively transferred on basic medium (LS) and toxic medium (LS + filtrate). The calli which did not grow on toxic medium ("sensitive") did not grow at all when placed subsequently on basic medium; this would indicate that high concentrations of filtrate caused their death. "Resistant" calli, after one transfer on basic medium, increased their growth percentages when again placed in the toxic medium. The response followed different patterns in the two genotypes; in fact, while a second transfer on toxic medium was sufficient to sort out all the "resistant" cells in P.I. 23327, the other genotype (LVR) showed a gradual response to successive transfers on toxic medium.

These results confirmed that the cell resistance to toxin is differently determined in the two genotypes; in fact, if a mono or oligo-factorial determination for resistance in P.I. 23327 can be assumed, in the other genotype (LVR) the pattern of response to selection could be explained with genetic and chromosomic heterogeneity, which may give rise to somatic selection and/or adaptive effects. The latter was highly unlikely because the resistant calli were alternatively transferred onto medium without toxin, which should have removed the adaptive effects determined by fungal toxin on the callus growth.

Plants were regenerated from LVR callus grown in the toxic medium (20% filtrate in the basic medium); only five plants survived when planted in soil. They were allowed to self-pollinate and produced many berries with few seeds (from 1 to 300 per plant). The progeny of the first selfing generation (S) was tested for resistance to *Verticillium dahliae*. At the first two-nomophylle stage, plantlets were infected by inoculating roots with broken tips for 20 min in a distilled water suspension of agar in which cultures of different strains of *V. dahliae* isolated from eggplant were grown. The disease symptoms were scored 30 days after the inoculation as wilting in the leaves (Fig. 11). The results of inoculating plants are reported in Table 1. The progeny of the five selfed plants gave a heterogeneous response to fungus infection. In fact, only a low percentage of plants were recovered without symptoms of disease. The same test made on plantlets regenerated from LVR callus not selected for resistance to filtrate culture of *Verticillium* gave 100% diseased plants. It is therefore evident that the conditions of the experiments need to be improved to obtain a higher efficiency of resistant plants; this may be reached both by increasing the

Fig. 11a-c. Effects of inoculation with *Verticillium dahliea* on the selfed progeny of a plant regenerated from eggplant callus. **a** Plant not infected. **b** Diseased plant. **c** Plant without disease symptoms. (Alicchio unpubl.)

Table 1. Effects of inoculation with *Verticillium dahliae* on the selfed progeny of *Solanum melongena* plants regenerated from callus grown on toxic medium

Selfed progeny	No. of inoculated plants	No. of resistant plants
S1	81	3
S2	31	1
S12	20	2
S14	25	2
S15	20	3

selection pressure and regenerating plants from callus grown on medium with higher concentration of filtrate. Furthermore, the technique of regenerating plants after a shorter callus phase from protoplasts exposed to toxic medium could overcome the disadvantage of the high karyotypic instability which, in this experiment, may be partly responsible for the heterogeneous response of plants to fungus infection.

2.4.2 Resistance to Nematodes

The southern root-knot nematode (*Meleidogyne incognita*) is a major pest of eggplant, while some wild species of *Solanum*, like *S. sisymbriifolium*, are resistant to this pest (Fassuliotis and Bhatt 1982). While successful crosses of eggplant with related species have become more numerous throughout the world (Yannick 1985), unsuccessful attempts have been made in interspecific crosses between *S. melongena* and *S. sisymbriifolium* Lam (Fassuliotis 1975; Seck 1983).

Progress in fusion of protoplasts isolated from related species may represent a useful approach to bypass sexual incompatibility barriers to interspecific hybridization and to transfer resistance genes and desirable agronomic traits from wild to cultivated species of *Solanum*. Gleddie et al. (1986a) were successful in fusing protoplasts of *S. melongena* and *S. sisymbriifolium*. A double selective system was devised to identify somatic hybrids; colonies were selected for their ability to grow in the presence of 6-azauracil (which inhibited the growth of *S. sisymbriifolium* cells but not the growth of azauracil-resistant eggplant lines used in the experiment) and of zeatin, which stimulated anthocyanin production only in shoots of *S. sisymbriifolium* .Twenty-six potential somatic hybrids were regenerated from 37 selected colonies out of 740 screened colonies. Several morphological traits of hybrids were intermediate to those of the parents. Usually, regenerants were less vigorous than parental lines. Most of them died after transplantation; those which survived flowered prematurely and formed numerous but small flowers. Isozymatic analysis confirmed the hybrid constitution of eight regenerants. Hybrids were revealed to be mostly mixoploid, and this cytological instability is probably the cause of their sterility.

Preliminary tests showed that hybrids were resistant to root-knot nematodes (Gleddie et al. 1985). The inoculation of plants with eggs of *Meleidogyne incognita* produced only occasional galls in *S. melongena*, while nematode reproduction was observed in eggplant roots. Only sexually undifferentiated larvae were found in roots of hybrids indicating a resistance similar to *S. sisymbriifolium*. .

The practical application of these hybrids in breeding programs depends on the possibility of overcoming their sterility. So far hybrids have been propagated by cutting and by in vitro micropropagation, but now efforts have to be made to restore partial fertility in hybrids. Possible approaches, as Gleddie et al. (1986a) suggested, include the backfusion of hybrid protoplasts to eggplant protoplasts with the objective of eliminating more *S. sisymbriifolium* chromosomes and shortening the callus phase prior to using cells in hybridization experiments.

3 Conclusions and Prospects

To date much work has been reported on the improvement of the various culture techniques in *Solanum melongena*, but unfortunately very few studies deal with the appearance and the exploitation of somaclonal variation.

The experimental work so far realized demonstrates that somaclonal variation may be exploited for the improvement of this crop in conjunction with conventional breeding programs, particularly in the achievement of varieties resistant to common eggplant pests and to environmental stresses.

For the future, it would be of great importance to obtain plants regenerated after a short callus phase from protoplasts exposed to selective agents (chemical or physical agents). This procedure allows selecting a high number of cells and regenerating plants from a single selected cell. The short callus phase would also eliminate the more detrimental cytological disorders.

Protoplast fusion between commercial cultivars and wild species of *Solanum* resistant to common eggplant pests, along with the selection of somaclonal variants, represents a useful approach for the release of new varieties.

Acknowledgments. The author would like to thank Mrs. M. Bioni and Mr. E. Boschieri for their technical assistance.

References

Alicchio R, Antonioli C, Palenzona DL (1982a) Chromosome numbers of plantlets of *Solanum melongena* regenerated from callus culture. Atti AGI 28:15–18

Alicchio R, Del Grosso E, Boschieri E (1982b) Tissue cultures and plant regeneration from different explants in six cultivars of *Solanum melongena*. Experientia 38:449–450

Alicchio R, Del Grosso E, Cavina R, Palenzona DL (1982c) Effects of filtrates of *Verticillium dahliae* on the growth of *Solanum melongena* callus tissue. Genet Agr 36:156

Alicchio R, Antonioli C, Palenzona DL (1984) Karyotypic variability in plants of *Solanum melongena* regenerated from callus grown in presence of culture filtrate of *Verticillium dahliae*. Theor Appl Genet 67:267–271

Bajaj YPS, Bopp M, Bajaj S (1973) Patterns of peroxidases and differentiation in *Sinapis alba*. Phytomorphology 23:43–52

Bhatt DP, Fassuliotis G (1981) Plant regeneration from mesophyll protoplasts of eggplant. Z Pflanzenphysiol 104:81–89
D'Amato F (1977) Cytogenetics of differentiation in tissue and cell cultures. In: Reinert J, Bajaj YPS (eds). Applied and fundamental aspects of plant cell, tissue, and organ culture. Springer, Berlin Heidelberg New York, pp 343–350
Del Grosso E, Alicchio R (1981) Analysis in isozymatic patterns of *Solanum melongena*: differences between organized and unorganized tissues. Z Pflanzenphysiol 102:467–470
Del Grosso E, Alicchio R, Borgatti S (1982) Analysis of isozymatic pattern in different stages of in vitro regeneration of *Solanum melongena* plantlets. Atti AGI 28:139–142
Dharme GMV (1977) Genic analysis of yield and yield components in brinjal. Mysore J Agric Sci 11:426
Dumas de Vaulx R, Chambonnet D (1982) Culture in vitro d'anthères d'aubergine (*Solanum melongena* L.): stimulation de la production de plantes au moyen de traitements à + 35 °C associés à de faibles teneurs en substances de croissance. Agronomie 2:983–988
Eldin SAB (1967) The inheritance of certain quantitative characters in eggplants. Diss Abstr 28:491–B
Fassuliotis G (1975) Regeneration of whole plants from isolated stem parenchyma cells of *Solanum sisymbriifolium*. J S Am Soc Hortic Sci 100:636–638
Fassuliotis C, Bhatt DP (1982) Potential of tissue culture for breeding root-knot nematode resistance into vegetables. J Nematol 14:10–14
Gleddie S, Keller WA, Setterfield G (1982) In vitro morphogenesis from tissues, cells, and protoplasts of *Solanum melongena* L. In: Fujiwara A (ed) Plant tissue culture 1982. Maruzen, Tokyo, pp 139–140
Gleddie S, Keller WA, Setterfield G (1983) Somatic embryogenesis and plant regeneration from leaf explants and cell suspensions of *Solanum melongena* (eggplant). Can J Bot 61:656–666
Gleddie S, Fasssuliotis G, Keller WA, Setterfield G (1985) Somatic hybridization as a potential method of tranferring nematode and mite resistance into eggplant. Z Pflanzenzücht 94:348–351
Gleddie S, Keller WA, Setterfield G (1986a) Production and characterization of somatic hybrids between *Solanum melongena* L. and *Solanum sisymbriifolium* Lam. Theor Appl Genet 71:613–621
Gleddie S, Keller WA, Setterfield G (1986b) Somatic embryogenesis and plant regeneration from cell suspensions derived protoplasts of *Solanum melongena* L. Can J Bot 64:355–361
Gunn RE, Day PR (1986) In vitro culture in plant breeding. In: Withers LA, Alderson PG (eds) Plant tissue culture and its agricultural applications. Butterworths, London, pp 313–336
Guri A, Izhar S (1984) Improved efficiency of plant regeneration from protoplasts of eggplant (*Solanum melongena* L.). Plant Cell Rep 3:247–249
Hinata K (1986) Eggplant (*Solanum melongena* L.) In: Bajaj YPS (ed) Biotechnology in agriculture and forestry, vol 2. Crops I. Springer, Berlin Heidelberg New York Tokyo, pp 363–370
Hisajima S (1982) Microplant propagation through multiple shoot formation from seeds and embryos. In: Fujiwara A (ed) Plant tissue culture 1982. Maruzen, Tokyo, pp 141–142
Isouard G, Raquin C, Demarly Y (1979) Obtention de plantes haploides et diploides par culture in vitro d'anthères d'aubergine (*Solanum melongena* L.). C R Acad Sci Paris Ser D 288:987–989
Jain RK, Chowdhury JB, Sharma DR, Chowdhury VK (1984) Selection and characterization of brinjal (*Solanum melongena* L) cell cultures resistant to tryptophan and phenylalanine analogues. Indian J Exp Biol 22:589–591
Jia J, Potrykus I (1981) Mesophyll protoplasts from *Solanum melongena* var depressum Bailey regenerate to fertile plants. Plant Cell Rep 1:71–72
Kakizaki Y (1931) Hybrid vigor in eggplants and its practical utilization. Genetics 16:1–25
Kamat MG, Rao PS (1978) Vegetative multiplication of eggplants (*Solanum melongena*) using tissue culture techniques. Plant Sci Lett 13:57–65
Keen NT, Long M, Erwin DC (1972) Possible involvement of a pathogen-produced protein-lipopoly-saccharide complex in *Verticillium* wilt of cotton. Physiol Plant Pathol 2:317–331
Larkin PJ, Scowcroft WR (1981) Somaclonal variation – a novel source of variability from cell cultures for plant improvement. Theor Appl Genet 60:197–214
Matsuoka H, Hinata K (1979) NAA-induced organogenesis and embryogenesis in hypocotyl callus of *Solanum melongena* L. J Exp Bot 30:363–370
Matsuoka H, Hinata K (1983) Factors affecting embryoid formation in hypocotyl callus of *Solanum melongena* L. Jpn J Breed 33:303–309
Mitra DK, Gupta N, Bhaskaran S (1981) Development of plantlets from brinjal (*Solanum melongena*) stem tissue infected with little leaf, a mycoplasma disease. Indian J Exp Biol 19:1177–1178

Nachmias A, Buchner V, Burstein Y (1985) Biological and immunochemical characterization of a low molecular weight phytotoxin isolated from a protein-lipopolysaccharide complex produced by a potato isolate of *Verticillium dahliae*. Kleb. Physiol Plant Pathol 26:43–55
Narasimha Rao N (1979) The barriers to hybridization between *Solanum melongena* and some other species of *Solanum*. In: Hawkes JG, Lester RN, Skelding AD (eds) The biology and taxonomy of the Solanaceae. Academic Press, New York London, pp 605–614
Nicklow CW (1983) The use of recurrent selection in efforts to achieve *Verticillium* resistance in eggplant. HortSci 18:600
Nuti Ronchi V, Bennici A, Martini G (1973) Nuclear fragmentation in dedifferentiating cells of *Nicotiana glauca* pith tissue grown in vitro. Cell Differ 2:77–85
Pearson OH (1983) Heterosis in vegetable crops. In: Frankel R (ed) Heterosis – reappraisal of theory and practice. Springer, Berlin Heidelberg New York, pp 138–188
Pegg GF (1965) Phytotoxin production by *Verticillium albo atrum* Reinke et Berthold. Nature (London) 208:1228–1229
Peter KV, Singh RD (1976) Combining ability, heterosis, and analysis of phenotypic variation in brinjal. Indian J Agric Sci 44:393–399
Raina SK, Iyer RD (1973) Differentiation of diploid plants from pollen callus in anther cultures of *Solanum melongena* L. Z Fflanzenzücht 70:275–280
Research Group of Haploid Breeding in Peking (ed) (1978) Induction of haploid plants of *Solanum melongena* L. In: Proc Symp Plant tissue culture. Pitman, London; Science Press, Beijing, pp 227–232
Saxena PK, Gill R, Rashid A, Maheshwari SC (1981) Plantlet formation from isolated protoplasts of *Solanum melongena* L. Protoplasma 106:355–359
Seck A (1983) Amélioration génétique de l'aubergine: étude des crossements de *Solanum melongena* L. par *S. torvum* et *S. sisymbriifolium*. Mem le Annee. Thesis, Montpellier
Shepard JF, Bidney D, Shahin E (1980) Potato protoplasts in crop improvement. Science 208:17–24
Silvetti E, Brunelli B (1970) Analysis of a diallel cross among some eggplant varieties and of their F_1 progeny. Genetica Agraria 258–268
Stoddart JL, Carr AHJ (1966) Properties of wilt-toxins produced by *Verticillium albo-atrum*. Reinke et Berth. Ann Appl Biol 58:81–92
Vavilov NI (1951) The origin, variation, immunity and breeding of cultivated plants. Chron Bot 13:1–364
Yamada T, Nakagawa H, Sinoto Y (1967) Studies on the differentiation in cultured cells. I. Embryogenesis in three strains of *Solanum* callus. Bot Mag (Tokyo) 80:68–74
Yamakawa K (1982) Use of rootstocks in Solanaceous fruit-vegetable production in Japan. I. Jpn Agric Res Quant 15:175–179
Yannick H (1985) Resistance comparée de 9 espèces du genre Solanum au flétrissement bactérien (*Pseudomonas solanacearum*) et au nematode *Meleidogyne incognita*. Interêt pour l'amélioration de l'aubergine (*Solanum melongena* L.) en zone tropicale humide. Agronomie 5(1):27–32
Zelcer A, Soferman O, Izhar S (1983) Shoot regeneration in root cultures of Solanaceae. Plant Cell Rep 2:252–254

III.4 Somaclonal Variation in Cucurbits

V. MORENO and L.A. ROIG[1]

1 Introduction

1.1 Importance, Origin and Distribution of Cucurbitaceae

Cucurbits were among the first plants used by man, and the bottle gourd may be the only plant known in both the New and Old Worlds in early prehistoric times. Species of *Cucurbita* were major crops in the agriculture of the Aztec, Inca, and Mayan civilizations in Central and South America. Cucumbers have been cultivated in India for perhaps 3000 years (Esquinas-Alcazar and Gulik 1983).

Presently, some species of the Cucurbitaceae (see Table 1a) are major market crops in Latin America, North America, North Africa, southern Europe, and tropical and temperate Asia. In addition, there is also large commercial production of cucumbers in a number of more northern countries. Table 2 shows the world production of major Cucurbitaceae crops. Other species (see Table 1b) have lower global economical interest, although some of them have certain local interest, mainly in developing countries.

The origin and centers of diversity of the cultivated species of this family have been discussed by Esquinas-Alcazar and Gulik (1983). Cultivated *Cucurbita* originated in two centers in the New World: central Mexico is the center of origin for *Cucurbita pepo, C. moschata*, and *C. mixta* and southern Peru, Bolivia, and northern Argentina are the centers of origin for *C. maxima*. Today the area of diversity for *C. pepo* is mainly northern Mexico. That for *C. mixta* extends from central Mexico and the Yucatan Peninsula to Costa Rica. The center of diversity for *C. moschata* extends from Mexico city south through Central America into northern Colombia and Venezuela and for *C. maxima* it is in northern Argentina, Bolivia southern Peru, and northern Chile. The center of origin for melon is still not clear, although the evidence points to Africa, where wild species of *Cucumis* with the same basic chromosome number (x = 12) frequently occur. However, domestication may have occurred independently in Southeast Asia, India, and East Asia. Today the primary center of diversity is in Southwest and Central Asia. There are also secondary centers of diversity in China, Korea, and the Iberian Peninsula. Cucumber is thought to have originated in India, where its wild relatives of

[1] Plant Tissue Culture Laboratory, Department of Biotechnology, Escuela Técnica Superior de Ingenieros Agrónomos, Universidad Politecnica de Valencia, C/ Camino de Vera, 14, 46020 Valencia, Spain

Biotechnology in Agriculture and Forestry, Vol. 11
Somaclonal Variation in Crop Improvement I (ed. by Y.P.S. Bajaj)

Table 1. Cultivated species and chromosome number of the Cucurbitaceae family (Esquinas and Gulik 1983)

1a Species of global interest

Cucumis melo L. (melon; 2n = 24)
Cucumis sativus L. (cucumber; 2n = 14)
Citrullus lanatus (Thunb.). Matsum. & Nakai (watermelon; 2n = 22)
Cucurbita maxima Duch. ex Lam. (pumpkin, winter squash; 2n = 40)
Cucurbita mixta Pang. (pumpkin, winter squash, cushaw; 2n = 40)
Cucurbita moschata (Duch. ex Lam.) Duch. ex Poir (winter squash; 2n = 40)
Cucurbita pepo L. (summer squash, winter squash, pumpkin, gourd; 2n = 40)

1b Species of local or regional interest

Cucumis anguria L. (West Indian gherkin; 2n = 24)
Citrullus colocynthis (L.) Schrad. (colocynth; 2n = 22)
Cucurbita ficifolia Bouché (fig-leaf or Malabar gourd; 2n = 40)
Cucurbita foetidisima H.B.K. (buffalo gourd; 2n = 40)
Benincasa hispida (Thunb.) Cogn. (wax or white gourd; 2n = 24)
Lagenaria siceraria Mol. Standl. (bottle gourd; 2n = 22)
Luffa acutangula (L.) Roxb. (angled loofah; 2n = 26)
Luffa cylindrica M.J. Roem. (smooth loofah, loofah gourd; 2n = 26)
Momordica charantia L. (bitter gourd; 2n = 22)
Momordica dioica Roxb. ex Willd. (bitterless bitter gourd; 2n = 28)
Sechium edule (Jacq.) Sw. (chayote; 2n = 24)
Coccinia indica Wight & Arn. (little gourd; 2n = 24)
Cucumeropsis mannii Naud. (egusi melon; 2n = 22)
Cyclanthera pedata (L.) Schrad. (2n = 32)
Hodgsonia macrocarpia (Bl.) Cogn. (chinese lard fruit)
Praecitrullus fistulosus Pang.
Sicana odorifera (Vell.) Naud. (cassabanana)
Telfairia occidentalis Hook. F. (fluted gourd)
Telfairia pedata (Sm. ex Sims) Hook. (oysternut)
Trichosanthes cucumerina L. (snake gourd; 2n = 22)
Trichosanthes dioica Roxb. (pointed gourd; 3n = 33)

chromosome number (2n = 14) are found at present. Today the center of diversity is in this area. Watermelon originated in Africa, and was probably introduced into cultivation in Egypt and spread to India in very early times. The centers of diversity of cultivated forms are in India and in Tropical and Subtropical Africa. A more exhaustive discussion of the origin, evolution, distribution, and phylogenetic relationships of cucurbits can be found in several articles (Whitaker and Davis 1962; Whitaker and Bemis 1964, 1976; Dane 1983; Perl-Treves et al. 1985; Perl-Treves and Galun 1985; Mallick and Masui 1986).

1.2 Breeding Objectives and Available Genetic Variability

It is difficult to state succintly which are the breeding objectives of the major Cucurbitaceae crops, as naturally, each one presents particular problems, and the objectives change with time according to market demands. However, some generalizations may be made.

Table 2. Statistics in harvested area, yield and production of major Cucurbitaceae crops, in different part of the world, in 1985 (FAO 1985)

Parts of the world	Area (x 1000 ha)	Yield (kg/ha)	Prod. (x 1000 mt)
	Pumpkins, squash and gourds		
Africa	69	14,189	980
North and Central America	41	6,970	288
South America	75	9,841	741
Asia	202	12,368	2,498
Europe	155	76,977	1,190
Oceania	12	12,352	151
World	555	10,548	5,852
	Cucumber and gherkins		
Africa	23	16,234	368
N. and C. America	70	13,797	970
Soutn America	3	16,602	47
Asia	420	15,501	6,512
Europe	123	19,331	2,370
Oceania	1	13,612	18
USSR	161[a]	9,006	1,450[a]
World	801	14,653	111,734
	Watermelons		
Africa	120	18,404	2,200
N. and C. America	117	14,100	1,649
South America	120	9,342	1,118
Asia	970	16,617	16,125
Europe	135	20,682	2,787
Oceania	5	14,556	79
USSR	390	9,744	3,880[b]
World	1,857	14,950	27,757
	Melons		
Africa	48	15,899	756
N. and C. America	79	16,116	1,280
South America	26	13,648	351
Asia	317	12,880	4,083
Europe	121	14,406	1,737
Oceania		13,527	2
World	590	13,904	8,209

[a] Unofficial figure.
[b] FAO estimates.

Quality and productivity improvements are common objectives. To obtain Fl hybrids in cheaper ways is another important objective in view of the acceptance and the prices that the hybrid seed is reaching on the market. As far as the cucumber is concerned, the genetics of sex expression and its modification through growth regulators have been extensively studied and it is now common to use gynoecious lines for the production of Fl hybrid seed. As regards melon, three genes conditioning for male sterility have been described (Bohn and Whitaker 1949; Bohn and Principe 1964; Mc Creigth and Elmstrom 1984). These genes, if introduced into commercial varieties, would allow the cheap and quick production of Fl hybrids. Other sources of male sterility could be introduced by means of interspecific crosses.

As for watermelon, considerable effort has been made to produce triploid lines with seedless fruits. The problem is to obtain quickly and easily the tetraploid plants which will then be crossed with the normal diploid plants to attain the triploid lines.

The introduction of resistance to environmental stress, above all resistance to salinity, may be of special interest. In fact, in some countries or in specific locations, water sea seeping into continental aquifers after long drought periods leads to their salinization, and irrigation with water whose salt content is high induces a steep fall of productivity. It should be mentioned, however, that in some species (e.g., melon) irrigation with slightly salinized water produces fruits of greater quality. Therefore, it is likely that obtaining salinity-resistant lines, besides avoiding important losses in the productivity, would also improve the quality.

Besides these points, another aspect has to be stressed. Currently, the general problem affecting the major Cucurbitaceae crops is that of the abundant yearly losses caused by their susceptibility to an assortment of pests and diseases. As the seriousness of the problem requires, a considerable effort has already been carried out to develop pest- and disease-resistant lines but, unfortunately, the problem is far from solved. As regards melon, the problem has even worsened in the southeastern mediterranean area with the appearance of new diseases (e.g., yellowing and muerte súbita), that are producing high yearly losses.

Cucurbitaceae are open pollinated but self-compatible. Taking into account the fact that we are dealing with allogamous (though not strictly so) species, one should expect to find a high degree of potential intraspecific genetic variability. This high degree of variability in fact exists, so that melon is a highly polymorphic species and *Cucurbita maxima* is noted for its extremely rich diversity, some authors suggesting that it has more cultivated forms than any other cultivated plant (Esquinas-Alcazar and Gulick 1983). In fact, the exploitation of this rich intraspecific genetic diversity has already allowed some important breeding achievements, as already stated.

In spite of this, we are facing an acute problem in the high degree of genetic erosion, which is narrowing the margins of the potential variability. Taking the example of *C. maxima*, The Vegetables of New York listed, in 1935, about 200 cultivars of it. Today there are no more than 20 cultivars available on the market, although there may still be about 100 of the old cultivars available in breeders' collections (Esquinas-Alcazar and Gulick 1983). Another problem is posed by the fact that for some pests and diseases for anyone given species, it has not been possible to find sources of resistance inside the same species itself. Taking this into account, considerable effort has been devoted to increasing the gene pool of the cultivated species through interspecific hybridization, with the chief aim of introducing pest- and disease-resistant genes from wild species (see den Nijs and Visser 1985; Custers and den Nijs 1986). Table 3 shows an incomplete list of wild species, where accessions with resistance to plagues and diseases have been found. Through the conventional methods of sexual crossings in species of *Cucurbita*, some partial successes have been achieved (e.g., *C. pepo* × *C. ecuadorensis*, Dumas de Vaulx and Pitrat 1980; *C. maxima* × *C. ecuadorensis*, Cutler and Whitaker 1969; Weeden and Robinson 1986). However, with *Cucumis*, in spite of the many studies made (e.g., Deakin et al. 1971; Fassuliotis 1977; Niemierowicz-Szczytt and Kubicki 1979;

Table 3. Wild species of *Cucumis* and *Cucurbita* where resistance to pests and/or diseases have been described in some accessions

Species	Resistance[a]	References
Cucumis metuliferus Naud.	PM_1	Clark et al. (1972)
	Aphids	Clark et al. (1972)
	PMV	Provvidenti and Robinson (1974)
	WMV	Provvidenti and Robinson (1974)
	Meloydogine spp.	Fassuliotis (1967)
Cucumis africanus Lindley	BSM	Kroon et al. (1979)
	CGMMV	Knipping et al. (1975); de Ponti (1978); Kroon et al. (1979)
	PM_1	Lebeda (1984)
Cucumis anguria L. var. *anguria*	PM_1 and PM_2	Lebeda (1984)
Cucumis anguria L. var. *longipes*	BSM	Knipping et al. (1975); de Ponti (1978); Kroon et al. (1979)
	PM_2	Lebeda (1984)
	CGMMV	Kroon et al. (1979)
	Meloydogine spp.	Kroon et al. (1979)
Cucumis ficifolius A. Rich	CGMMV	Kroon et al. (1979)
	PM_1 and PM_2	Lebeda (1984)
	Meloydogine spp.	Kroon et al. (1979)
Cucumis myriocarpus Naud.	BSM	Knipping et al. (1975); de Ponti (1978)
	CGMMV	Kroon et al. (1979)
	PM_1	Lebeda (1984)
Cucumis angolensis Hook.	WF	Kovalewsky and Robinson (1977)
Cucumis asper Cogn.	WF	Kovalewsky and Robinson (1977)
Cucumis dinteri Cogn.	WF	Kovalewsky and Robinson (1977)
	PM_1 and PM_2	Lebeda (1984)
Cucumis sagittatus Peyr.	PM_2	Lebeda (1984)
Cucumis leptodermis Schweik.	PM_1	Lebeda (1984)
Cucumis dipsaceus Ehrenb. ex Spack	WF	Esquinas-Alcazar and Gulik (1983)
Cucumis hardwickii Royle	CGMMV	Esquinas-Alcazar and Gulik (1983)
	Meloydogine spp.	Esquinas-Alcazar and Gulik (1983)
Cucumis heptadactylus Naud.	CGMMV	Esquinas-Alcazar and Gulik (1983)
	Meloydogine spp.	Esquinas-Alcazar and Gulik (1983)
Cucumis hookeri Naud.	CGMMV	Esquinas-Alcazar and Gulik (1983)
	Meloydogine spp.	Esquinas-Alcazar and Gulik (1983)
Cucurbita ecuadorensis Cuttler and Whitaker	PM_1	Dumas de Vaulx and Pitrat (1980)
	CMV	Dumas de Vaulx and Pitrat (1980)
	WMV	Dumas de Vaulx and Pitrat (1980)
	ZYMV	Provvidenti et al. (1984); Provvidenti and Gonsalves (1984)
Cucurbita martinezii Bailey	CMV	Dumas de Vaulx and Pitrat (1980)
	PM_2	Dumas de Vaulx and Pitrat (1980)
	WMV	Esquinas-Alcazar and Gulik (1983)
Cucurbita lundelliana Bailey	PM	Esquinas-Alcazar and Gulik (1983)
Cucurbita okeechobeensis Bailey	PM	Esquinas-Alcazar and Gulik (1983)
Citrullus colocynthis (L.) Schrad	ZYMV (tolerant)	Provvidenti and Gonsalves (1984)
Lagenaria siceraria Mol. Standl	ZYMV	Provvidenti and Gonsalves (1984)

[a] PM_1: *Sphaeroteca fugilinea* (Schlecht ex Fr.) Pollacci (powdery mildew).
PM_2: *Erisiphe cichoracearum* DC ex Merat (powdery mildew).
PMV: Pumpkin Mosaic Virus.
WMV: Watermelon Mosaic Virus type 1.
BSM: *Tetranychus urticae* Koch. (bean spider mite).
CGMMV: Cucumber Green Mottle Mosaic Virus.
WF: *Trialeurodes vaporariorum* Westw. (white fly).
CMV: Cucumber Mosaic Virus.
ZYMV: Zucchini Yellow Mosaic Virus.

Kroon et al. 1979; Dane et al. 1980; Kho et al. 1980), the cross of any of the wild species mentioned with one of the two cultivated species of *Cucumis* (melon and cucumber) has proven to be impossible. Norton (1969) reported the hybridization of a feral melon (PI 140471) with *C. metuliferus*; however, he himself indicated the impossibility of reproducing the cross (Norton and Granberry 1980). The problems existing in this field have been recently reviewed (den Nijs and Visser 1985; Custers and den Nijs 1986).

Today, the development of methods for plant regeneration from melon protoplasts (Moreno et al. 1986; Roig et al. 1986) and the updating of fusion methods which are adequate for cucurbit species (Roig et al. 1986) may make it feasible to apply somatic hybridization to overcome the barriers of sexual incompatibility. At present, a program of somatic hybridization between melon and several wild species of cucurbits is being carried out in our laboratory.

Another way to increase genetic variability in the cultivated cucurbits is to take advantage of somaclonal variation. The results already achieved in this field have certainly been impressive in the last few years (Shepard et al. 1980; Larkin and Scowcroft 1981, 1983; Scowcroft 1984; also see Bajaj 1986). However, the prerequisite for taking advantage of the somaclonal variation originating from cell culture is the availability of methods to regenerate whole plants from primary calli or from permanent or established cell lines, suspension cultures or, above all, from protoplasts. In the following section a brief review of the current status as regards cucurbits will be attempted.

2 Plant Regeneration in in Vitro Cultures of Cucurbits

In Tables 4 and 5, a brief review of the studies on morphogenesis in cell cultures of cucurbits is made. Some of these works carried out with major cucurbitaceous crops, particularly stressing the systems set by our work group for obtaining plant regeneration from explant-derived calli, cell suspension cultures, and protoplast of melon are summarized. In this discussion no reference will be made to the works carried out with species of the genus *Cucurbita*, as a recent review has been devoted to that aspect (see Jelaska 1986; in Vol. 2 of this Series).

2.1 Morphogenesis in Explant-Derived Calli

2.1.1 Melon

The induction of unorganized growth and the formation of calli from root segments was first described by Fadia and Mehta (1973, 1975, 1976). However, no morphogenetic response was reported in these works. In 1980, we reported obtaining whole plants from calli derived from cotyledon segments of cv. Cantaloup Charentais by organogenesis (Moreno 1980). Later we published the results of a study that aimed at an assessment of the influence of diverse IAA and kinetin concentrations on the organogenesis in calli derived from the same kind of explants of cv.

Table 4. Morphogenesis in different cell cultures of cucurbits

Species	Variety, cultivar or line	Plant material	Morphogenetic response	Reference
		A. In Explant-derived calli		
Cucumis melo L.	Cantaloup Charentais	Cotyledon	Shoots and plants	Moreno (1980)
Cucumis melo L.	var. *utilissimus* Duthie & Fuller	Hypocotyl and cotyledon	Buds and plantlets	Halder and Gadgil (1981, 1982)
Cucumis melo L.	Banano, Dixie Jumbo, Morgan, Saticoy Hybrid, Planters Jumbo.	Cotyledon and leaf	Shoots	Blackmon and Reynolds[a] (1982)
Cucumis melo L.	Amarillo Oro	Cotyledon and hypocotyl	Shoots and plants, embryoids	Moreno et al. (1985)
Cucumis melo L.	Cantaloup Doublon, Piboule, Cantaloup Charentais T, CM 17187, Ogon No. 9	Cotyledon	Shoots and plants	Bouabdallah and Branchard[b] (1986)
Cucumis melo L.	Tokyo Early, Sakata's Sweet, Valenciano Pinyonet, Perlita, Tendral Verde Tardío, Ogon, Temprano Roget, Piel de Sapo, Cantaloup Doublon, Meloncio, Cantaloup Charentais, Cantaloup Westland, Honey Dew	Cotyledon	Shoots and plants	Orts et al.[c] (1987)
Cucumis sativus L.	Hokus	Hypocotyl and cotyledon	Buds	Custers et al. (1980)
Cucumis sativus L.	85 cultivars and lines	Hypocotyl Cotyledon	Roots but no shoots Roots and shoots	Wehner and Locy (1981)
Cucumis sativus L.	Skierniewickii, Borszczagowski and GY-3	Leaf	Embryoids and plants	Malepszy and Nadolska-Orczyk (1983); Nadolska-Orczyk and Malepszy (1984)
Cucumis sativus L.	14 cultivars and lines	Cotyledon and root	Embryoids and plants	Trulson and Shahin (1986)
Citrullus lanatus (Thumb.) Matsum & Nakai	Charleston Gray 133.	Cotyledon	Shoots	Blackmon and Reynolds (1982).

ariety, cultivar or line	Plant material	Morphogenetic response	Reference
rad XIV-2 (2n), Arad XIV (4n), rad VII-1001 (4n), Sweet (2n), impuriu de Canada (2n), Baby ugar (2n), Graybelle (2n).	Hypocotyl and cotyledon	Shoots and plants	Anghel and Rosu (1985)
ar. *longipes* Meeusse	Leaf	Shoots and plants	Garcia-Sogo et al. (1986)
	Cotyledon	Buds and plantlets	Halder and Gadgil (1981, 1982).
	Fleshy pericarp wall	Embryoids and plantlets	Schroeder (1968)
	Hypocotyl and cotyledon	Embryoids and plantlets	Jelaska (1972, 1974, 1980)
	B. In anther cultures		
MR-58	Anther	Embryoids and plantlets	Lazarte and Sasser (1982)[d]
	C. In suspension cultures		
antaloup Charentais	Cell suspension from hypocotyl or cotyledon	Embryoids	Moreno (1980)
uperpickle	Cell suspension from hypocotyl	Embryoids, shoot-buds plants.	Rajasekaran et al. (1983)
Y-3	Cell suspension from leaf	Embryoids and plantlets	Malepszy et al. (1986)
elilah	Cell suspension from hypocotyl	Embryoids and plantlets	Ziv and Gadasi (1986)[e]

rphogenetic response.
hout morphogenetic response.
Honey Dew gave practically no morphogenetic response.
etophytic or sporophytic origin was not verified.
os was accomplished by culturing them in a double layer: agar (lower)/liquid (upper) media. For more information
ro see Jelaska 1986.

Table 5. Isolation, culture and plant regeneration from protoplasts in cucurbits

Species	Variety, cultivar or line	Source of protoplasts	Response of Culture	Reference
Cucumis melo L.	Cantaloup Charentais	Suspension cultures and leaves	Callus	Moreno et al. (1980)
Cucumis melo L.	Cantaloup Charentais, Pinyonet Piel de Sapo, Valenciano Pinyonet.	Leaves (axenic plants)	Callus	Moreno et al. (1984)
Cucumis melo L.	Cantaloup Charentais	Leaves (axenic plants)	Embryoids	Moreno and Zubeldia (1984)
Cucumis melo L.	Cantaloup Charentais	Leaves (axenic plants)	Embryoids, shoots and plants	Moreno et al. (1986)
Cucumis melo L.	Cantaloup Charentais	Cotyledons	Shoots and plants	Roig et al. (1986)
Cucumis anguria L.	var. *longipes* Meeusse	Precultured cotyledons	Callus	Roche et al. (1987)
Cucumis metuliferus Naud	–	Precultured cotyledons	Callus	Roche et al. (1987)
Cucumis ficifolius A. Rich	–	Precultured cotyledons and leaves (ax. plants)	Callus	Roche et al. (1987)
Cucurbiba martinezzi Bayley	–	Precultured cotyledons	Callus	Roche et al. (1987)
Cucumis sativus L.	Ashley, China	Leaves	Callus	Coutts and Wood (1975)
Cucumis sativus L.	Ashley, China	Cotyledons and leaves	Callus and roots	Coutts and Wood (1977)
Cucumis sativus L.	GY-3	Leaves (axenic plants)	Embryoids and plantlets	Orczyk and Malepszy (1985) Malepszy et al. (1986)
Cucumis sativus L.	F_1 hybrid 37-1Gx78–50	Cotyledons	Embryoids and plantlets	Jia et al. (1986)
Cucumis sativus L.	Straight Eight	Cotyledons	Embryoids and plants	Trulson and Shahin (1986)

Amarillo Oro (Moreno et al. 1985). In Figs. 1, 2, and 3 the results (unpubl.) of other complementary studies carried out with the same cultivar are presented. Figure 1 shows the results of the influence of different mineral solutions on the organogenesis in cotyledon-derived calli. The best results were obtained with Murashige and Skoog (1962) solution. Under the same cultural conditions, B5 (Gamborg et al. 1968) solution allowed the organogenesis in only half of the assayed explants. All the rest of the mineral solutions were clearly deficient as far as the organogenic response is concerned.

In another study, where the influence of different concentrations of IAA and kinetin was analyzed (see Fig. 2), the following conclusions could be drawn: (1) generally speaking, for a given concentration of auxin, as the concentration of exogenous kinetin goes up, so does the organogenic response; (2) initially, for each kinetin concentration, as the IAA concentration increases so do the responses, which reach their highest points at 0.5 mg/l IAA (with 3 mg/ kinetin), 1 mg/l IAA (with 4.5 mg/l kinetin) and 1.5 mg/l IAA (with 6 mg/l kinetin), but from this point onward, further auxin increases are detrimental; (3) although the percentage of calli with morphogenetic response depends on the exogenous concentrations of IAA and kinetin, organogenesis was achieved in media with very different concentrations of both regulators. Additionally, in Fig. 3 it can be seen that the tissues from cotyledon segments of melon need no exogenous auxin in order to undergo morphogenetic response. In this study, where different kinetin concentrations were tried, it was possible to verify that the type of container had a bearing on the percentage of organogenic calli obtained.

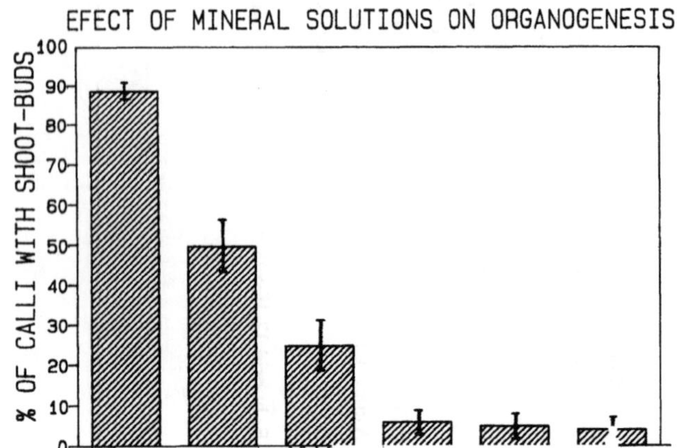

Fig. 1. Effect of different mineral solutions on the organogenic response in cotyledon-derived calli of melon cv. Amarillo Oro. *MS* Murashige and Skoog (1962); *B5* Gamborg et al. (1968); *SH* Schenk and Hildebrandt (1972) modified to contain the same level of iron as MS; *H* Heller (1953); *NN* Nitsch and Nitsch (1956). The experiment was carried out under the same cultural conditions as previously described (Moreno et al. 1985), using the IK1560 as basal medium and changing only the corresponding mineral solution. More than 50 replicates per medium were scored, the *bars* indicating the S.E. of each result

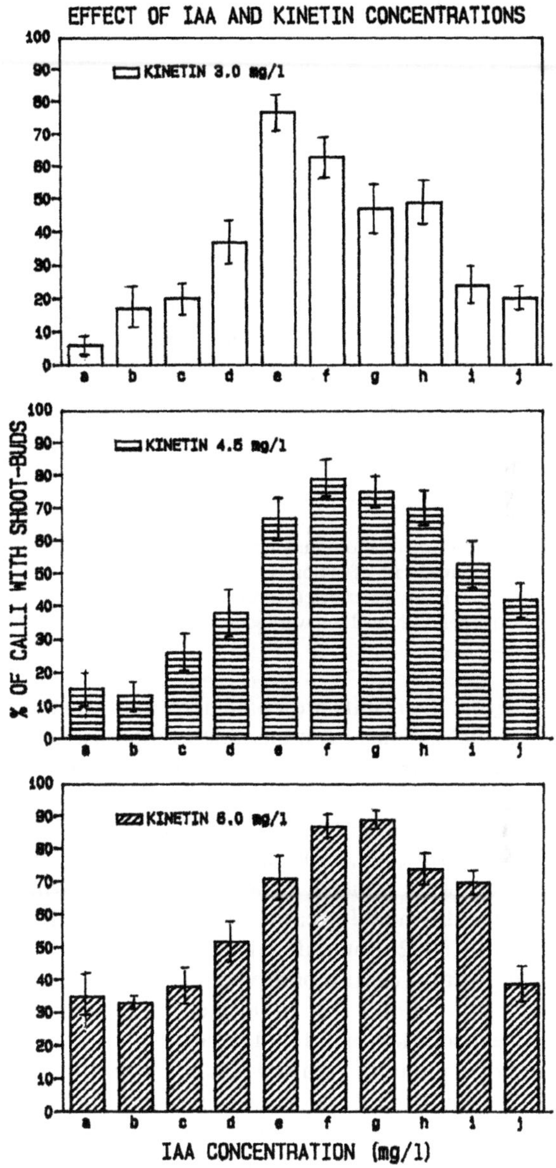

Fig. 2. Effect of different concentrations of IAA and Kinetin on the organogenic response in cotyledon-derived calli of melon cv. Amarillo Oro. IAA concentrations: *a* 0; *b* 0.01; *c* 0.05; *d* 0.10; *e* 0.50; *f* 1.00; *g* 1.50; *h* 3.00; *i* 4.50; *j* 6.00 (mg/l). This experiment was realized under the same cultural conditions as previously described (Moreno et al. 1985) and some data are taken from that paper (columns *a, g, h, i* and *j*). The total number of calli scored per each medium was always higher than 50, and the *bars* indicate the S.E. of each result

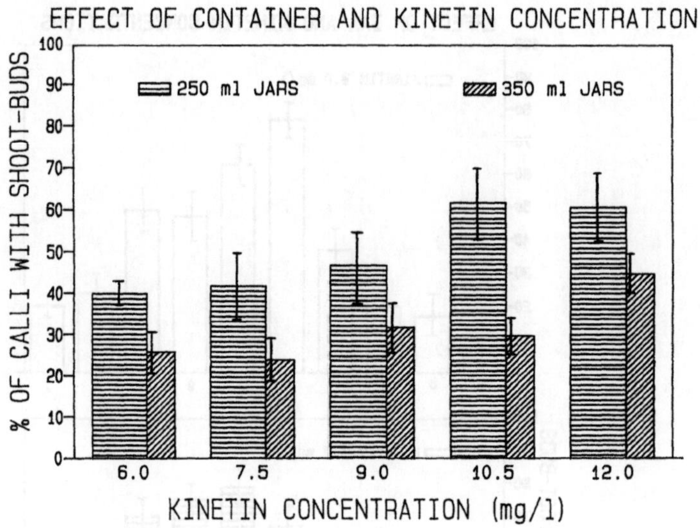

Fig. 3. Effect of type of container and kinetin concentrations on the organogenic response in cotyledon-derived calli of melon cv. Amarillo Oro growing in absence of auxin. The cultural conditions were as previously described (Moreno et al. 1985) and the basal medium used was MB3. More than 50 replicates per each medium were scored and the *bars* indicate the S.E. of the data

From the results presented here and from those previously published (Moreno et al. 1985), it has been possible to establish a protocol (see later) which permits obtaining calli with shoot-buds in more than 90% of the explants for the cv. Amarillo Oro.

We have recently applied this protocol to 13 other melon cultivars (Orts et al. 1987). In this study, important differences in the organogenic response of the various cultivars were noticed. In work carried out with different melon cultivars, other authors also noted great differences in morphogenetic response (Blackmon and Reynolds 1982; Bouabdallah and Branchard 1986). Additionally, it is interesting to point out that we also observed important differences in one and the same cultivar (Orts et al. 1987) that indicated correspondingly important genetic differences between seed lots of different accessions, initially classified as the same cultivar. The kind of explant is also a key factor. Under the described culture conditions we had no morphogenetic response (or only with low frequency in cultivars of high response) in calli derived from hypocotyl segments (Moreno et al. 1985; Orts et al. 1987). The importance of the epigenetic factor (i.e., kind of explant) has also been referred to by other authors (Bouabdallah and Branchard 1986).

2.1.2 *Cucumber*

Custers et al. (1980) obtained buds from calli of hypocotyl and cotyledon segments, while Wehner and Locy (1981), working with 85 cultivars and lines, found that only 28 of them formed shoots from cotyledon explants, and no shoots were formed from hypocotyl pieces in any of the 85 cultivars and lines.

Under very different culture conditions, Malepszy and Nadolska-Orczyk (1983) reported obtaining embryoids and whole plant regeneration from leaf-derived calli of three cultivars. The described system implied a primary culture in a medium containing a phenoxy-derived (2,4-D or 2,4,5-T) and a cytokinin (BAP or 2-iP) followed by one or several subcultures in a secondary medium without auxin or with reduced concentration both of auxin and cytokinin. Whole plant regeneration was achieved by subsequently transferring the embryoids to the same medium without growth regulators.

Recently, Trulson and Shahin (1986) have reported plant regeneration from cotyledonary and root explants via somatic embryogenesis: the explants were cultured in a semi-solid medium containing 2,4-D, NAA, and BAP, embryo maturation was achieved in the same medium without 2,4-D, and plant regeneration in the hormone-free basal medium. In that work, all 14 cultivars or lines assayed produced embryoids in cotyledon-derived calli, differing in the number of genotypes from each line or cultivar which produced immature embryos, mature embryos, and shoots. As regards embryogenesis in root-derived calli, the authors tested four out of the 14 cultivars and attained plant regeneration with big differences in the number of shoots/seedling produced by each cultivar (3, 3, 1 and 100 respectively). The authors also tested this protocol using other cucurbits (two cultivars each of muskmelon, watermelon, and squash) and found only sporadic plant regeneration (two plants in one cultivar and three in the other) in cotyledonary tissue of muskmelon, with none in watermelon and squash.

On the other hand, the results obtained by Gross et al. (1981) should be mentioned. They proved that cucumber tissues, unlike those of the other plant species, are able to grow and form calli in media with galactose or galactose-containing olygosaccharides. Curiously, in secondary cultures, stachyose and rafinose proved to be the most efficient sugars as far as calli growth is concerned. It would be of great interest to verify whether these sugars are also used by the morphogenetic callus cultures and whether other species of the same family can also use these sugars.

2.1.3 Watermelon

Plant regeneration from watermelon cotyledon-derived calli has also been described (Blackmon and Reynolds 1982; Anghel and Roşu 1985). In the latter work, organogenic response was obtained in different cultivars, diploids as well as tetraploids and, as previously indicated for melon, the kind of explant proved to be a key factor, since under the same culture conditions, hypocotyl-derived calli could not produce shoot buds.

2.2 Morphogenesis in Anther Cultures

To the best of our knowledge, anther culture in only one species of the family, i.e., cucumber, has been reported (Lazarte and Sasser 1982). In this case obtaining embryoids and plantlets was described but, unfortunately, no chromosome counting of the regenerated plants was carried out and so, as the authors themselves say, it is impossible to determine whether they were of gametophytic or sporophytic origin. We have carried out experiments with melon anther cultures, though not systematically, and in some cases we obtained calli and then, after subculturing them, shoots and later plantlets, but in all cases these plants were diploids and not haploids.

2.3 Morphogenesis in Suspension Cultures

2.3.1 *Melon*

In 1980, we described the methods and conditions to set suspension cultures and, subsequently, to obtain somatic embryogenesis from them in cv. Cantaloup Charentais. We improved the protocol further and Table 6 shows (unpubl.) the number of somatic embryos obtained when calli from hypocotyl and cotyledon segments of this cultivar were transferred to an MB3 basal medium (Moreno et al. 1985) supplemented with a complex extract. As can be seen, the number of embryos is high, but apparently the maturation process is defective, as only rarely are whole normally developed plants produced. We are currently carrying out intensive experiments which aim both at obtaining somatic embryogenesis in a totally defined medium and at solving the problems thus far found in embryo maturation. Alternatively, the regeneration of whole plants through organogenesis of cell-derived calli and cell aggregates of cv. Cantaloup Charentais suspension cultures may be feasible if the methods and conditions for the regeneration of whole plants from protoplast-derived calli (see below) are followed.

2.3.2 *Cucumber*

Rajasekaran et al. (1983) described the establishment of suspension cultures and obtaining somatic embryogenesis as well as organogenesis from calli incubated in agitated liquid media. Seemingly, in this case, morphogenesis was achieved by way of embryogenesis or organogenesis, according to the concentration of 2,4-D used. These authors stressed the in vitro flowering of plants regenerated by both procedures.

Recently, Ziv and Gadasi (1986), in a study aimed at minimizing the frequency of abnormal embryoids usually obtained through somatic embryogenesis, and especially at increasing the number of embryoids capable of further development to give whole plants, have indicated that some treatments (such as increased osmolarity, the addition of ABA, zeatin, or activated charcoal) which are used to improve the morphological quality of the embryoids in other species, had little effect on cucumber. Apparently only the use of double-layer cultures containing activated charcoal (0.5%) in the lower agar layer and ABA (0.4 μM) with elevated calcium in the upper liquid phase was found to prevent the abnormal embryogenesis and to increase, although only slightly, the number of plantlets obtained from the embryoids. Nevertheless, none of the treatments prevented vitreous leaf formation.

Table 6. Number of embryoids[a] obtained in suspension cultures of melon after subculturing in E19D5[b] liquid medium two kind of explant-derived calli grown in C[c] solid medium

Source of calli	Days of culture in C medium	
	30 days	60 days
Hypocotyl	91 ± 9	126 ± 38
Cotyledon	23 ± 3	103 ± 18

[a] Expressed as mean of 10–15 replicates ± S.E. Each replicate is the number of embryoids appeared in a 250 ml Erlenmeyer flask containing 40 ml of E19D5 liquid medium after 20 days of incubation under the same cultural conditions previously described for embryogenesis in protoplast-derived calli of melon (Moreno et al. 1986).
[b] E19D5 liquid medium: MB3 medium (Moreno et al. 1985) additioned with 1 g/l of a cell line extract originating from a permanent cell culture of tobacco in C solid medium.
[c] C solid medium: MB3 medium plus 2.5 mg/l NAA and 1.0 mg/l BA.

Leaf growth, and subsequent plant survival after transplanting to soil was improved by the hardening of the plants through their partial desiccation with calcium sulphate under aseptic conditions. Flowering in vitro was also observed with as many as 25 flower buds per plant, although it could be reduced and delayed in plants regenerated in the double layer cultures and hardened by desiccation.

2.4 Isolation, Culture and Plant Regeneration from Protoplasts

2.4.1 Melon

We have described methods for isolation (Moreno et al. 1980), culture and callus formation (Moreno et al. 1984), and plant regeneration (Moreno et al. 1986; Roig et al. 1986) from cotyledon protoplasts and from mesophyll protoplasts of axenically grown plants. The medium used for growing the plants under axenic conditions proved to be very important for the mitotic activity of the protoplast-derived cells when mesophyll from axenic plants was used as starting material (Moreno et al. 1984). Later, Shahin (1985), working with tomato, has also stressed the importance of the cultural conditions of the maternal plants in obtaining viable protoplasts endowed with high mitotic activity.

In our experience, the culture medium used for growing the protoplasts was less important than the plant culture medium and incubation conditions. Thus, Table 7 (unpubl.) shows the results of a study where melon mesophyll protoplasts were incubated in two series of media: firstly, ZEPC medium (Moreno et al. 1984) and two simplified ZEPC's; secondly, M medium (see formulation in Table 7) and three simplified M's. As can be noticed, the frequencies of cells undergoing cell divisions is high in all the assayed media. It can thus be stated that melon protoplast culture is feasible in different, and relatively simple, media provided that the protoplasts have been isolated from plants grown under specific conditions.

Morphogenesis in protoplast-derived calli has been achieved by somatic embryogenesis (Moreno et al. 1986) as well as through organogenesis (Moreno et al.

Table 7. Mitotic activity[a] of melon (cv. Cantaloup Charentais) mesophyll protoplasts after growing 20 days[b] in diverse culture media[c]

Protoplast culture medium[c]	Mitotic activity[a]
ZEPC	83.0 ± 6.7
$ZEPC_{vit}$-	77.5 ± 9.8
$ZEPC_{naa}$-	78.1 ± 9.2
M	92.2 ± 3.6
M_{cm}-	89.9 ± 3.3
M_{rt}-	87.3 ± 7.4
M_{cm}-$_{rt}$-	84.0 ± 6.5

[a] Expressed as percent of cells that undergo cell division with regard to the cells regenerated from protoplasts. Each number is the average of 20 replicates ± S.E.
[b] The experiment was carried out under the same cultural conditions as previously described (Moreno et al. 1984).
[c] ZEPC medium: Moreno et al., 1984
ZEPC's modifications as follows:
$ZEPC_{vit}$-: without vitamins $ZEPC_{naa}$-: without NAA.
M medium: MB3 medium (Moreno et al. 1985)+RT vitamins (Staba 1969)+2% (v/v) coconut milk+2.5 mg/l NAA + 1 mg/l BA.
M's modifications:
M_{cm}-: without coconut milk M_{rt}-: without RT vit.
M_{cm}-$_{rt}$-: both products were removed.
All media were osmotically estabilized with 0.6 M-mannitol.

1986; Roig et al. 1986). As regards embryogenesis (Fig. 4A-D), a great number of embryoids can be obtained by transferring the microcalli to E19D5 agitated liquid medium (see formulation in Table 6), but the number of these embryoids capable of undergoing normal further development and becoming whole plants with vigorous growth is too low. Therefore, an efficient and reproducible protocol which leads to obtaining protoclones by somatic embryogenesis is a problem still to be solved. Fortunately, it was possible to establish a reproducible protocol to obtain protoclones via organogenesis, and thus whole plants could be obtained following these methods.

Recently, we have notably improved the efficiency of this protocol in order to increase the number of whole plants regenerated from melon protoplasts (see Fig. 5A-E). After a number of experiments carried out with the aim of studying the effect of different metallic ions (which in the standard media are supplied as trace elements) on the organogenic response of explant-derived calli, we could verify that the addition of relatively high concentrations of copper ion (1-5 mg/l SO_4 Cu) to culture medium noticeably increased not only the frequency of calli with shoot buds but also the number of shoot buds per callus. The addition of these concentrations of copper sulfate to the media employed to obtain organogenesis in protoplast-derived calli (see media formulations in Table 8) also allows not only a greater frequency of organogenic calli but also a quicker production of well-developed shoots as well as a greater number of them. Thus, at present, with this modification

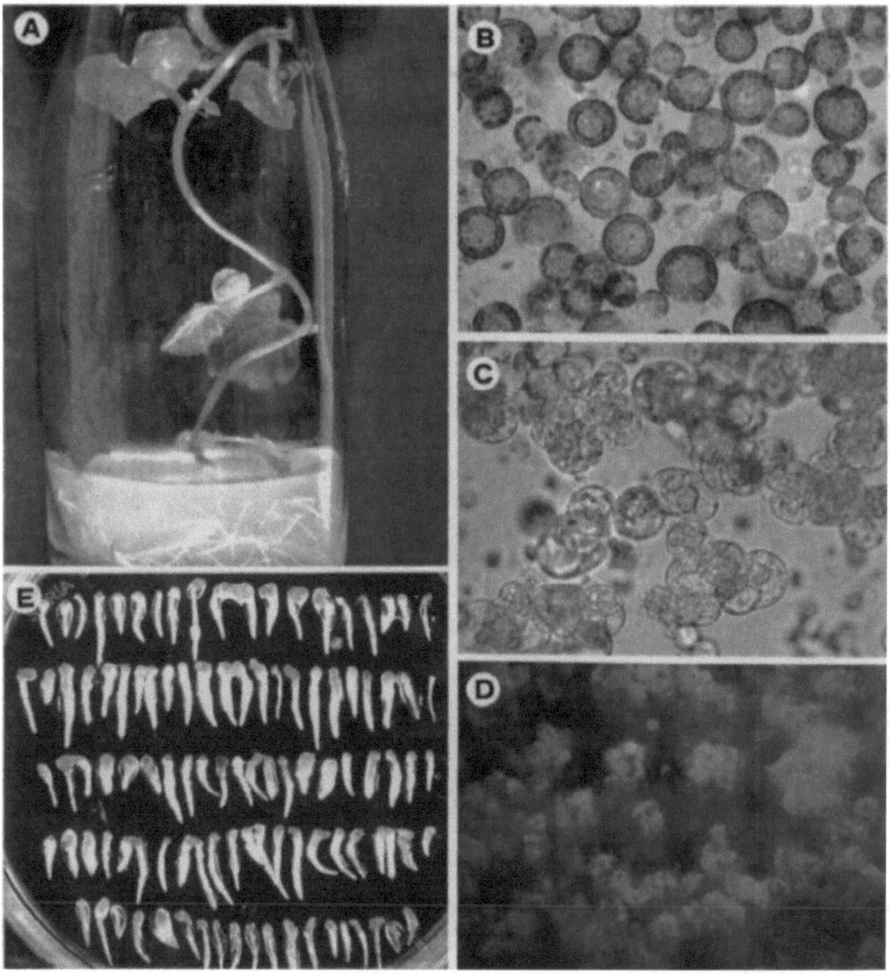

Fig. 4A-E. Embryogenesis in protoplast-derived calli of melon cv. Cantaloup Charentais. **A** Axenic plant growing on MEL solid medium (Moreno et al. 1984). **B** Mesophyll protoplasts just isolated from these plants. **C** Cells and cell clusters from protoplasts after 20 days in culture in ZEPC 0.6 M-mannitol liquid medium. **D** Microcalli obtained after another 20 days in ZEPC 0.3 M-mannitol liquid medium. **E** Embryoids appeared in the suspension cultures of those microcalli in E19D5 liquid medium

and some methodological changes (see Protocols), we have developed an improved and quicker method in order to regenerate plants from protoplasts.

2.4.2 Cucumber

Coutts and Wood (1975, 1977) were the first to report the isolation, culture, and formation of calli from protoplasts of this species. However, they could obtain only

Fig. 5 A-E. Organogenesis in protoplast-derived calli of melon cv. Cantaloup Charentais. **A** Microcalli in CEN solid medium. **B** Shoot buds in calli growing in IKO160 + 1 mg/l copper sulfate solid medium. **C** Shoots and shoot buds in the first subculture in $NB00101_s$ + 1 mg/l copper sulfate solid medium. **D** Well-developed shoots in calli after the third subculture in the same medium. **E** Plantlets rooted in R solid medium

Table 8. Composition of the media (mg l^{-1}) used for culture and plant regeneration of melon protoplasts

Components[a]	Culture medium					
	ZEPC(0.6)[b]	ZEPC(0.3)[c]	CEN[d]	IK0160[e]	NB00101[f]$_s$	R[g]
Mineral solution	B5	B5	MS	MS	MS	MS
Sucrose	10,000	10,000	30,000	30,000	30,000	20,000
Glucose	5,000	5,000	–	–	–	–
Myo-inositol	100	100	100	100	100	–
Thiamine-HCl	10	10	1	1	1	–
Nicotinic acid	1	1	–	–	–	–
Pyridoxin-HCl	1	1	–	–	–	–
RT-vitamins	No	No	No	No	Yes	Yes
Casein hydrolysate	–	–	100	100	100	100
IAA	–	–	0.1	0.1	–	0.1
NAA	0.5	0.5	–	–	0.01	–
2,4-D	1	1	–	–	–	–
Kinetin	–	–	10	6	–	–
BA	0.5	0.5	–	–	0.1	–
Mannitol	99,000	44,000	–	–	–	–
Agar	–	–	6,000	8,000	8,000	8,000

[a] All components from Sigma Chemical Co., except of Agar from Oxoid (Technical No. 3). The amounts are in milligrams per liter. RT-vitamins from Staba (1969). Mineral solutions: B5 from Gamborg et al. 1967. MS from Murashige and Skoog (1962). The final pH of all media was adjusted to 5.7.
[b] Medium for protoplast culture (Moreno et al. 1984).
[c] Medium for protoplast-derived cell aggregates (Moreno et al. 1984).
[d] Medium for plating the cell aggregates.
[e] Medium for callus growth and adventitious buds formation.
[f] Medium for subculturing the organogenic calli several times to obtain well-developed shoots.
[g] Medium for rooting the plantlets.

roots but not whole plants from the calli. In 1985, Orczyk and Malepszy reported obtaining cucumber plants from protoplasts of the cv. GY-3. They indicated that the addition of 0.1 M glycine (an osmoprotector of the cell membrane) to the pre-plasmolitic, maceration, and washing media was essential for the survival and subsequent divisions of the cucumber mesophyll protoplasts. Plant regeneration through somatic embryogenesis was achieved from the protoplast-derived calli by repeatedly subculturing them in the same media as was used to bring about the same process in explant-derived calli. Jia et al. (1986) have additionally reported obtaining calli from cotyledon protoplasts of six cultivars, three gynoecious lines and two Fl hybrids. In this case, the addition of glycine as osmoprotector was not necessary. However, plant regeneration was only achieved from calli of one of the assayed Fl hybrids.

Trulson and Shahin (1986), working with cotyledons of the cv. Straight Eight, obtained high yields of viable and vigorously growing protoplasts. In this work, the degree of morphogenetic response and the number of regenerated plants are clearly stated: 5–10% of the minicalli produced embryoids and less than 1% of these embryoids continued to develop into normal mature embryos that could be grown to mature plants. A total of 11 plants were regenerated, six of which were transplanted to the greenhouse, looked phenotypically normal, flowered, and produced seeds.

2.4.3 *Watermelon and Other Species*

In recent works we have been able to achieve cell wall regeneration, cell division and aggregate formation from cotyledon protoplasts of watermelon (unpubl.). The formation of calli from protoplast of different wild species of cucurbits has also been achieved using the standard method developed for melon protoplasts (Roche et al. 1987).

3 Somaclonal Variation

Few studies have been carried out on the existence of chromosomal variation in somaclones of cultivated cucurbits. However, there is some evidence of the existence of a high degree of variation. Thus, in an experiment with 42 calliclones regenerated from melon cotyledon segments incubated in a medium consisting of MB3 basal medium additioned of 0.77 mg/l IAA and 2.56 mg/l kinetin, we found that 26 of the plants were diploid, whereas 16 were tetraploid. In this small plant sample no triploids were detected. However, in another experiment with melon, Bouabdallah and Branchard (1986) indicated the appearance of triploid and tetraploid plants, besides diploids, among the plants regenerated from cotyledon-derived calli subcultured twice (the number of plants of each type was not mentioned.)

In a greenhouse assay with 22 leaf-derived calliclones, 85 cotyledon-derived calliclones, and 11 protoclones of melon, we have observed great phenotypic variation. Some of the noted variant characters in the somaclonal plants (SC_1 according to the nomenclature from Scowcroft 1984), were: growth pattern (Fig. 6A), size and shape of the leaf, and other vegetative traits, flower maturation pattern (Fig. 6B), and number of flowers per plant, size and weight of the fruit, size of the blossom scar (Fig. 6C), size and shape of the seeds (Fig. 6D), and contents in soluble solids per plant. The phenotypic variation degree was also high. Table 9 shows the percentage of somaclones SC_1 that present phenotypic variation for each of the six fruit characters studied. The highest percentage of variants was in the protoplast-derived somaclones (protoclones), followed by the cotyledon-derived calliclones, in which the percentage of variants was considerably higher than in the leaf-derived calliclones. Figure 7 shows the percent of somaclones presenting from 0 to 6 variant characters per plant, for the six fruit characters mentioned. The average of variant characters per plant observed was, respectively, of 1.36 in the leaf-derived calliclones, 3.27 in the cotyledon-derived calliclones and 4.80 in the protoclones. Definitive corroboration of whether the variant characters observed in individual plants have a genetic basis must wait for the subsequent culture of self-pollinated progeny (SC_2) and backcrosses of these plants. However, and through the experiments carried out in past years, we already have clear evidence which indicates the sexual transmission of some of these variant characters. Thus, in three families of SC_2 plants from three calliclones, which showed remarkable differences in the SC_1 for several traits, sexual transmission for some of these characters was observed: the first family was variant for size of the blossom scar, fruit weight, and seed shape; the second and third families were variant for size of the blossom scar and seed shape.

Fig. 6A-D. Some aspects of the somaclonal variation in *Cucumis melo* cv. Cantaloup Charentais. **A** Dwarf and rosette-shaped protoclone with absence of apical dominance and normal root development. **B** Normal male flowers (*upper raw*) from the cultivar and variants (*lower raw*) from another protoclone. **C** Fruit from protoclone 8C showing the unusually big blossom scar. **D** Normal seeds (*upper group*) from the cultivar and variants (*lower group*) from protoclone 11E

Table 9. Percentage of somaclones[a] SC_1 phenotypically variants for each of the six fruit characters scored

Character	Leaf-derived calliclones	Cotyledon-derived calliclones	Protoclones
Blossom scar size	27	64	82
Longitudinal diameter	9	47	91
Ecuatorial diameter	5	26	73
Length-to-width ratio	5	33	55
Seed shape	45	72	91
Fruit weight	45	67	91

[a] Total number of somaclones scored: 22 leaf-derived, 85 cotyledon-derived and 11 protoclones.

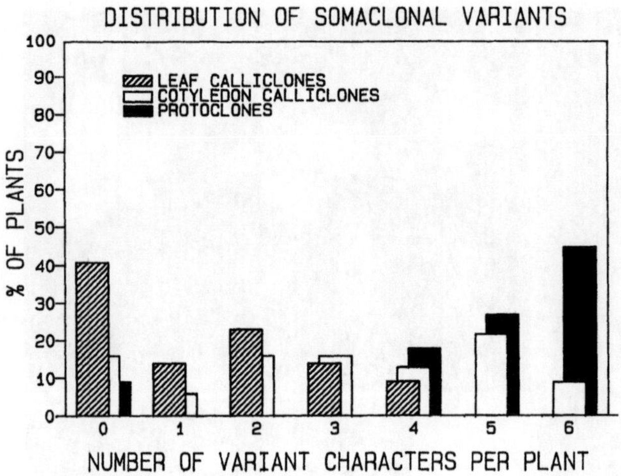

Fig. 7. Distribution of percent of plants showing different number of variant characters in somaclones SC_1 originating from explant-derived calli (calliclones) and from protoplast-derived calli (protoclones) of melon cv. Amarillo Oro. The total number of somaclones assayed was: 22 leaf-derived calliclones, 85 cotyledon-derived calliclones and 11 protoclones

Other authors have described differences in the weight of the fruits of SC_1 plants. Nadolska-Orczyk and Malepszy (1984) indicated that 15 somaclones of cucumber cv. Borszczagowski had fruits of a weight which was significantly higher than that of plants originating from seed. There is also a good deal of interest in the fact that these authors pointed out that the somaclones put flowered more intensely than the control plants, as we have been able to observe the same phenomenon in many of the melon somaclones studied.

It is necessary to wait for the availability of date from the character evaluation, in the SC_2, of a greater number of calliclones and protoclones before safely discussing the extension of somaclonal variation in melon. However, we do already have reliable data on other species; specifically, with *Cucumis anguria* var. *longipes* we obtained a somaclonal variant (CASRI) which, according to the analysis of clonally propagated plants from this variant cultivated over two years and in two different localities, showed variation for at least 14 characters. These include fruit characters (less weight, shorter longitudinal and equatorial diameters, post-maturing white-yellowish instead of green-yellowish skin), plant color (light green, because of its chlorophyll deficiency; see Fig. 8A), flowers (shorter length of the female flowers, shorter female flower petiole, abortive male flowers; see Fig. 8B) vegetative characters (different branching pattern, shorter internodal length, shorter leaf diameter and length, less stem thickness), and fertility (lower in the fruits originating from the backcross with the original line).

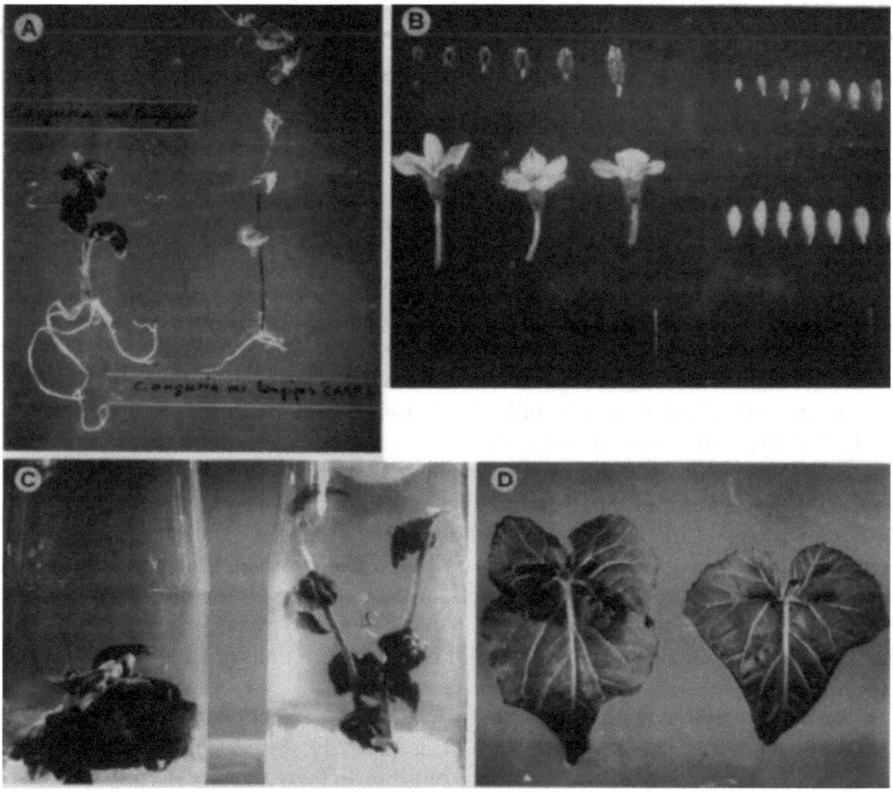

Fig. 8A-D. Some variant characters found in *Cucumis anguria* var *longipes* somaclones. **A** Normal *C. anguria* plant (*left*) and the CASRI calliclone (*right*). Note the light-green color due to chlorophyll deficiency of the somaclone. **B** Male flowers in progressive maturation states of normal plants (*left*) and of CASRI variant (*right*). Note the different pattern leading to give abortive flowers in the variant. **C** Dwarf and rosette-shaped CASR5 variant (*left*) versus normal *C. anguria* plant (*right*). **D** Differences in leaf shape between normal *C. anguria* plants (*right*) and the rosette-shaped CASR5 calliclone (*left*)

4 Final Remarks and Outlook

Although to date the species of the Cucurbitaceae family have been comparatively less studied than those belonging to other families, the results obtained by different groups have allowed the setting up of efficient and reproducible methods to regenerate whole plants from in vitro cultures. Especially remarkable are the results obtained with protoplasts. Indeed, the work of our group has made it possible to develop highly efficient methods to obtain melon protoclones and those of other researchers have proved the feasibility of obtaining regenerated plants from cucumber protoplasts.

With other crop species, not only a high degree of somaclonal variation has been detected but, above all, somaclonal variants which are genetically improved with regards to the original plants have been obtained (see reviews of Larkin and Scowcroft 1981, 1983; Scowcroft 1984). Since Shepard's work with potato (Shepard 1980; Shepard et al. 1980), the published papers on this subject are ever-increasing, and the most important point is that taking advantage of somaclonal variation has yielded practical results in a very short time, if compared with that necessary to achieve similar results when conventional genetic improvement techniques are used.

The updating of methods to obtain calliclones in various cultivated cucurbits as well as melon and cucumber protoclones lends credibility to the assertion that in a not too distant future the somaclonal variation will be used in the genetic improvement of these highly valuable species. Thus, for instance, some preliminary experiments of ours and of other authors indicate that, even in plants regenerated from explant-derived calli, a remarkable degree of chromosomal variation takes place. As regards melon, work to assess the advantages and disadvantages of the triploid and tetraploid plants was published a long time ago (Dumas de Vaulx 1973, 1974), but the conventional methods used to obtain these plants were not too efficient. Cell culture techniques are much more advisable to find out the possibilities of using autopolyploidy from a practical point of view in diverse cultivars. Particularly, obtaining tetraploids may be of great interest to evaluate, quickly and easily, the characters of the triploid plants obtained by crossing them with diploids. In some crop species, triploids are more advantageous than diploids and, moreover, the possibility of heterosis that some triploids possess, having been obtained through crossings of diploids and tetraploids of different lines or cultivars, could be exploited.

With watermelon, the use of triploids which give seedless fruits is spreading in different countries. The problem here is, as already stated, that obtaining the tetraploid parents is difficult and not too profitable. In this respect, cell culture seems to offer a more advantageous alternative.

The characters to be sampled to make use of somaclonal variation can differ, according to the particular problems of each one of the cultivated species of the family. However, one objective is common to all of them: the search for tolerance or resistance to pests and/or diseases is one of the most pressing problems to be addressed. At present, as regards melon, we are carrying out assays in this respect, using the methods already developed.

Lastly, occasionally, somaclonal variation arising from cell culture may have no direct relevance for genetic improvement, but certainly an indirect one. This is the case of the somaclonal variant CASRI of *Cucumis anguria* var. *longipes* that we obtained, and to which we referred above. In this species, valuable resistance to be used for melon improvement has been reported, but the somaclonal variant not only demonstrates genetic markers (e.g., chlorophyll deficiency) which make it a very useful parent in experiments on somatic cell hybridization, but also desirable characters, especially male sterility, the introduction of which into melon cultivars could be of a great practical interest.

5 Protocols for Plant Regeneration from in Vitro Cultures of Melon

Here we describe the most favorable methods and cultural conditions to obtain somaclones from melon cell cultures of different origin. For a more detailed description see the original papers referred to in each section.

5.1 Calliclones from Explant-Derived Calli of Melon (Moreno et al. 1985; Orts et al. 1987)

1. Seeds are surface sterilized by immersion in 20% commercial bleach (equivalent to 10 g/l of active chlorine) for 20 min, followed by three successive rinses with sterile distilled water.
2. Seeds are aseptically germinated in MG medium, consisting of MS mineral solution supplemented with 1% sucrose and 0.8% agar (Technical No. 3, Oxoid). Seed germination as well as incubation of all the further cell cultures are carried out in a culture room at 27 ± 1 °C with a photoperiod of 16 h light (2000 lx, provided by 36W Gro-lux cool fluorescent tubes, Sylvania, equivalent to 90 $\mu E/s/m^2$).
3. Cotyledons from 11–13-day old seedlings, just before the first leaf appears, are used as source of explants. Each cotyledon is cut in halves and the edges (1 mm width) are removed before they are placed in the culture medium.
4. Explants are placed, with the back side toward the medium, in 250 ml jars (twist-off type) containing 40 ml of IK1560 solid medium, consisting of MS mineral solution supplemented with 3% sucrose, 1 mg/l thiamine-HCl, 100mg/l myo-inositol, 1.5 mg/l IAA, 6 mg/l kinetin and 0.8% agar. The jars are covered with metallic or plastic lids and incubated in the culture room for 25–30 days. When varieties of high morphogenetic response are used, 90% or more of explants give rise to callus with shoot buds, but the number of well-developed shoots in these primary calli was usually to be low, although it depends on the line utilized.
5. In order to increase this number, subculture of the organogenic calli is suitable, transferring fragments with shoot buds (removing the unorganized growth areas) to $NB00101_s$ solid medium (see formulation in Table 8). After 15–25 days of incubation, new adventitious shoot buds and well-developed shoots can be rooted.
6. To obtain calliclones, the well-developed shoots are excised from the organogenic calli and transferred to 250 ml bottles containing 30 ml of R solid medium (see Table 8) and covered with cotton plugs. It is feasible to increase the number of plants by subculturing pieces of callus with not entirely developed shoots in this R medium to proceed with their development. When the shoots can be easily manipulated they are transferred again to the same solid medium for rooting.
7. When the calliclones have a good root system they are transferred to dried peat containers (Jiffypots, Denmark Ryomgaard) with a mixture of peat-substrate-sand (1:1:1), taking care of removing the culture medium attached to the roots. The plants are grown in a growth chamber at 30 ± 1 °C day and 24 ± 1 °C night, 80% R.H. and 16 h light photoperiod (12,000 lx provided by 215 W Sylvania VHO cool white fluorescent tubes, equivalent to 540 $\mu E/s/m^2$) until they reach active growth and are transplanted to soil.

5.2 Calliclones from Organogenic Cell Lines of Melon (Moreno et al. 1985)

1. It is easy to establish organogenic cell lines growing indefinitely and able to give shoot buds continuously. Organogenic pieces of callus have to be subcultured into fresh medium as described above. The success of this procedure depends on the time elapsed between transfers and the way the subcultures are made. Although the time to make the transfers is variable for each line or variety, it must be short (usually every 15 days), and for each subculture (specially the first ones) all the unorganized areas of the organogenic piece of callus must be eliminated.
2. Different culture media can be used to establish these lasting organogenic cell lines. Nevertheless, in our case, the most suitable medium for our different lines and varieties was the $NB00101_s$ medium.
3. Calliclones from these cell lines can be obtained following steps 6–7 of preceding protocol.

5.3 Plant Regeneration from Protoplasts of Melon

This protocol for obtaining melon protoclones is an improved modification of those previously reported (Moreno et al. 1986; Roig et al. 1986) and was initially developed for the cv. Cantaloup Charentais.

1. The most suitable plant material to be used as source of protoplasts capable of regenerating plants, are the following:
   a) Cotyledons from axenic seedlings (like those described above in 5.1) are cut in halves and precultured in Petri dishes (9 cm in diameter, 10 cotyledon segments per plate) containing 20 ml of C solid medium (same composition as IKI560 but changing the growth substances for 3 mg/l NAA and 1 mg/l BAP) for 3 days at 27°C in darkness.
   b) Leaf mesophyll from axenic plants obtained by subculturing the apical buds of axenic seedlings into 250 ml bottles containing 40 ml of MEL solid medium (MS mineral solution supplemented with 3% sucrose and 0.1% Difco Yeast Extract) covered with cotton plugs. Protoplast isolation is made when the plant tip reaches the cotton plug. Although in some varieties the isolation can be done directly from the excised leaves, in other varieties it is necessary to preculture the leaves before protoplast isolation, following the method described for cotyledons.
2. The source of protoplasts is cut into 1–2 mm wide strips and then placed in 100 ml Erlenmeyer flasks containing 8 ml of a filter sterilized enzyme solution consisting of 2% cellulase Onozuka R-10 (Yakult Pharmaceutical Ind. Ltd.), 0.6 M-mannitol, 50 mM MES in MS mineral solution, at pH 5.7. One milliliter of enzyme solution is used for each 0.1 g of plant tissue. The incubation is performed in the dark at 28°C for 5 h in a reciprocal shaker at 100 strokes/min (amplitude 20 mm). Alternatively, the maceration can be done for 12–14 h under the same conditions, but lowering the cellulase content to 1%.
3. The resulting crude protoplast suspension is poured through nylon mesh (100 μm pore size) and purified by centrifugation at 50 g in a double phase system:

protoplasts remain floating in the interphase between the enzyme solution and a 28% sucrose solution. Protoplast ring is collected with a Pasteur pipette and resuspended in washing solution (same as maceration solution without enzyme). After three rinses by centrifugation at 100 *g* the last pellet is resuspended in the protoplast culture medium at a final concentration of 1×10^5 protoplast/ml.
4. Although several media can be used to culture the protoplasts (see Table 7) we are routinely using the ZEPC 0.6 M-mannitol (see Table 8). Incubation is carried out in 6 cm plastic Petri dishes (4 ml per dish) at 27°C in the dark for 20 days. Cell wall regeneration is visible after 48 h incubation and the mitotic activity of the cells is high (see Table 7).
5. Cells and small cell clusters originating from protoplasts are subcultured in fresh ZEPC 0.3 M-mannitol liquid medium for another 20 days under the same cultural conditions.
6. The resulting microcalli are collected by centrifugation and resuspended in 5 ml of fresh CEN, then mixed with 5 ml of semi-solid CEN medium (kept melted at 45°C) and finally plated over a layer of 15 ml solid CEN medium, in plastic Petri dishes of 9 cm diameter. Henceforth the incubation of cultures is made in the culture room with photoperiod described above (see 5.1.2).
7. The small calli (see Fig. 5A) are transferred to solid IKO160 medium (see formulation in Table 8). After an initially unorganized growth, shoot bud containing zones are formed in this medium (see Fig. 5B). The addition of 1 mg/l copper sulfate to this medium accelerate the appearance of such organogenic areas in the calli and, especially, increases the frequency of organogenic calli as well as the degree of organogenic response (number of shoots/callus).
8. The bud-containing areas of the calli are subcultured two or three times in the solid medium $NB00101_s$. When 1 mg/l copper sulfate is added to this medium, calli with a large number of shoot buds can be obtained in the first subculture yet (see Fig. 5C), and well-developed shoots appear in the following subcultures (see Fig. 5D). This morphogenic response can be kept for a long time by continuous transfers to fresh medium as indicated in 5.2.
9. Rooting of protoclones is made in R solid medium (see Fig. 5E) and the plants thus obtained can be transplanted to soil following the procedure already described (5.1.7).

By means of these protocols it is possible to obtain sufficient number of calliclones and protoclones to undertake a study aimed at the practical exploitation of the somaclonal variation.

Acknowledgements. The authors express their appreciation to CAICYT (Comisión Asesora Investigación en Ciencia y Tecnologia, Ministry of Education and Science, Spanish Government) for financial support to the studies on morphogenesis in cell cultures of melon (Project No. 3021/83). Our special thanks are due to Ms. Carmen Hernandez Pino for carrying out the characterization of melon somaclones and for her permission to use some unpublished data from these studies. We also acknowledge the permission given by IBPGR, FAO and Martinus Nijhoff/Dr. W. Junk Publishers to include in this article the corresponding data expressed in the text.

We extend our acknowledgement to all our work group namely M.V. Roche, B. Garcia-Sogo, A. Salvador, M.C. Orts and I. Granell for their effort in the realization of all the experiments.

References

Anghel I, Rosu A (1985) In vitro morphogenesis in diploid, triploid and tetraploid genotypes of watermelon, *Citrullus lanatus* (Thumb.) Mansf. Rev Roum Biol Biol Veget 30:43–55

Bajaj YPS (ed) (1986) Biotechnology in agriculture and forestry, vol 2. Crops I. Springer, Berlin Heidelberg New York Tokyo

Blackmon WJ, Reynolds BD (1982) In vitro shoot regeneration of *Hibiscus acetosella*, muskmelon, watermelon, and winged bean. HortSci 17:588–589

Bohn GW, Principe JA (1964) A second male-sterility gene in the muskmelon. J Hered 55:211–215

Bohn GW, Whitaker TW (1949) A gene for male sterility in the muskmelon (*Cucumis melo* L.). Proc Amer Soc Hort Sci 53:309–314

Bouabdallah L, Branchard M (1986) Regeneration of plants from callus cultures of *Cucumis melo* L. Z Pflanzenzücht 96:82–85

Clark RL, Jarvis JL, Braverman SW, Dietz SM, Sowell G Jr., Winters HF (1972) A summary of reports on the resistance of plant introductions to diseases, nematodes, insects and mites. *Cucumis sativus* L. and *Cucumis spp.* US Dep Agric New Crop Res Branch, Beltsville, MD (mimeographed pamphlet)

Coutts RHA, Wood KR (1975) The isolation and culture of cucumber mesophyll protoplasts. Plant Sci Lett 4:189–193

Coutts RHA, Wood KR (1977) Improved isolation and culture methods for cucumber mesophyll protoplasts. Plant Sci Lett 9:45–51

Custers JBM, Den Nijs APM (1986) Effects of aminoethoxyvinylglycine (AVG), environment, and genotype in overcoming hybridization barriers between *Cucumis* species. Euphytica 35:639–647

Custers JBM, Van Ee G, Den Nijs APM (1980) Tissue culture methods and interspecific hybridization in the breeding of cucumber (*Cucumis sativus* L.). In: Eucarpia, Sect Legumes (ed) Application de la culture in vitro à l'amélioration des plantes potagères. INRA, CNRA, Versailles, pp 132–136

Cutler HC, Whitaker TW (1969) A new species of *Cucurbita* from Ecuador. Ann Missouri Bot Garden 55:392–396

Dane F (1983) Cucurbits. In: Tanskley SD, Orton TJ (eds) Isozymes in plant genetics and breeding, pt. B. Elsevier, Amsterdam, pp 369–390

Dane F, Denna DW, Tsuchiya T (1980) Evolutionary studies of wild species in the genus *Cucumis*. Z Pflanzenzücht 85:89–109

De Ponti OMB (1978) Resistance in *Cucumis sativus* L. to *Tetranychus urticae* KOCH. I. Search for sources of resistance. Euphytica 27:167–176

Deakin JR, Bohn GW, Whitaker TW (1971) Interspecific hybridization in *Cucumis*. Econ Bot 25:195–211

Den Nijs APM, Visser DL (1985) Relationships between African species of the genus *Cucumis* L. estimated by the production, vigour and fertility of F_1 hybrids. Euphytica 34:279–290

Dumas De Vaulx R (1973) Some results on production of triploid seeds. Eucarpia, Sect Hortic (ed) Premières journées d'étude sur la sélection du melon, pp 73–76

Dumas De Vaulx R (1974) Etude des possibilités d'utilisation de la polyplïdie dans l'amélioration du melon (*Cucumis melo* L.). Ann Amelior Plantes 24:389–403

Dumas De Vaulx R, Pitrat M (1980) Application de la culture d'embryos immatures à la réalisation de l'hybridation interspécifique entre *Cucurbita pepo* et *Cucurbita ecuadorensis*, F_1 et BC_1. In: Eucarpia, Sect Legumes (ed) Application de la culture in vitro à l'amélioration des plantes potagères. INRA, CNRA, Versailles, pp 126–131

Esquinas-Alcazar JT, Gulik PJ (1983) Genetic resources of Cucurbitaceae -a global report. IBPGR Secr, Rome

Fadia VP, Mehta AR (1973) Tissue culture studies on cucurbits: pt III Growth and nutrition of *Cucumis melo* L. callus cultures. Indian J Exp Biol 11:424–426

Fadia VP, Mehta AR (1975) Tissue culture studies on cucurbits: factors limiting growth of *Cucumis melo* L. tissue grown as batch cultures. Indian J Exp Biol 13:590–591

Fadia VP, Mehta AR (1976) Tissue culture studies on cucurbits: chlorophyll development in *Cucumis* callus cultures. Phytomorphology 26:170–175

FAO (ed) (1985) Production yearbook. FAO, Rome

Fassuliotis G (1967) Species of *Cucumis* resistant to the root-knot nematode, *Meloidogyne incognita acrita*. Plant Disease Rep 51:720–723

Fassuliotis G (1977) Self-fertilization of *Cucumis metuliferus* NAUD and its cross-compatibility with *C. melo* L. J Am Soc Hortic Sci 102:336–339

Gamborg OL, Miller RA, Ojima K (1968) Nutrient requirements of suspension cultures of soybean root cells. Exp Cell Res 50:151–158

Garcia-Sogo M, Granell I, Garcia-Sogo B, Roig LA, Moreno V (1986) Plant regeneration from explant-derived calli of *Cucumis anguria* L. var *longipes*. Cucurbit Genet Coop Rep 9:108–109
Gross KC, Pharr DM, Locy RD (1981) Growth of callus initiated from cucumber hypocotyls on galactose and galactose-containing oligosaccharides. Plant Sci Lett 20:333–341
Halder T, Gadgil VN (1981) Morphogenesis in some plant species of the family Cucurbitaceae. In: Rao AN (ed) Proc COSTED Symp Tissue culture of economically important plants, Singapore, pp 98–103
Halder T, Gadgil VN (1982) Shoot bud differentiation in long-term callus cultures of *Momordica* & *Cucumis*. Indian J Exp Biol 20:780–782
Heller R (1953) Recherches sur la nutrition minérale de tissus végétaux cultivés in vitro. Ann Sci Nat Bot Biol Veg 14:1–223
Jelaska S (1972) Embryoid formation by fragments of cotyledons and hypocotyl in *Cucurbita pepo*. Planta 103:278–280
Jelaska S (1974) Embryogenesis and organogenesis in pumpkin explants. Physiol Plant 31:257–261
Jelaska S (1980) Growth and embryoid formation in *Cucurbita pepo* callus cultures. In: Eucarpia, Sect Legumes (ed) Application de la culture in vitro à l'amélioration des plantes potagères. INRA, CNRA, Versailles, pp 172–178
Jelaska S (1986) Cucurbits. In: Bajaj YPS (ed) Biotechnology in agriculture and forestry, vol 2. Crops 1. Springer, Berlin Heidelberg New York Tokyo, pp 371–386
Jia SR, Fu YY, Lin Y (1986) Embryogenesis and plant regeneration from cotyledon protoplast culture of cucumber (*Cucumis sativus* L.). J Plant Physiol 124:393–398
Kho YO, Den Nijs APM, Franken J (1980) Interspecific hybridization in *Cucumis* L. species. 2. An investigation of "in vivo" pollen tube growth and seed set. Euphytic 29:661–672
Knipping PA, Patterson CG, Knavel DE, Rodriguez JG (1975) Resistance of cucurbits to twospotted spider mite. Environ Entoml 4:507–508
Kovalewsky E, Robinson RW (1977) White fly resistance in *Cucumis*. Bull IOBC/WPRC 77/3
Kroon GH, Custers JBM, Kho YO, Den Nijs APM, Varekamp HQ (1979) Interspecific hybridization in *Cucumis* L. 1. Need for genetic variation, biosystematic relations and possibilities to overcome crossability barriers. Euphytica 28:723–728
Larkin PJ, Scowcroft WR (1981) Somaclonal variation: a novel source of variability from cell cultures for plant improvement. Theor Appl Genet 60:197–214
Larkin PJ, Scowcroft WR (1983) Somaclonal variation and crop improvement. In: Hollaender A, Kosuge T, Meredith C (eds) Genetic engineering of plants. An agriculture perspective. Plenum, New York, pp 289–314
Lazarte JE, Sasser CC (1982) Asexual embryogenesis and plantlet development in anther culture of *Cucumis sativus* L. HortSci 17:88
Lebeda A (1984) Screening of wild *Cucumis* species for resistance to cucumber powdery mildew (*Erysiphe cichoracearum* and Sphaerotheca *fugilinea*. Sci Hortic 24:241–249
Malepszy S, Nadolska-Orczyk A (1983) In vitro culture of *Cucumis sativus* I. Regeneration of plantlets from callus formed by leaf explants. Z Pflanzenphysiol 111:273–276
Malepszy S, Nadolska-Orczyk A, Orczyk W (1986) Systems for regeneration of *Cucumis sativus* plants in vitro. In: Horn W, Jensen J, Odenbach W, Schieder O (eds) Genetic manipulation in plant breeding. De Gruyter, Berlin New York, pp 429–432
Mallick MFR, Masui M (1986) Origin, distribution and taxonomy of melons. Sci Hortic 28:251–261
McCreigth JD, Elmstrom GW (1984) A third muskmelon male-sterile gene. HortSci 19:268–270
Moreno V (1980) Aislamiento y cultivo de células y protoplastos de melon (*Cucumis melo* L.). PhD Thesis, Univ Valencia, Spain
Moreno V, Zubeldia L (1984) Primeros resultados en torno a la identificación y selección de células hibridas *Cucumis melo* (x) *Cucumis metuliferus* obtenidas mediante fusión de protoplastos. In: Proc 1st Cong Natl SECH, Valencia (Spain), pp 399–407
Moreno V, Roig LA, Garcia-Sogo B (1980) Isolation and culture of protoplasts of melon (*Cucumis melo* L.). In: Eucarpia, Sect Legumes (ed) Application de la culture in vitro à l'amélioration des plantes potagères. INRA, CNRA, Versailles, pp 166–171
Moreno V, Zubeldia L, Roig LA (1984) A method for obtaining callus cultures from mesophyll protoplasts of melon (*Cucumis melo* L.). Plant Sci Lett 34:195–201
Moreno V, Garcia-Sogo M, Granell I, Garcia-Sogo B, Roig LA (1985) Plant regeneration from calli of melon (*Cucumis melo* L. cv. Amarillo Oro). Plant Cell Tissue Org Cult 5:139–146
Moreno V, Zubeldia L, Garcia-Sogo B, Nuez F, Roig LA (1986) Somatic embryogenesis in protoplast-derived cells of *Cucumis melo* L. In: Horn W, Jensen J, Odenbach W, Schieder O (eds) Genetic manipulation in plant breeding. De Gruyter, Berlin New York, pp 491–493

Murashige T, Skoog F (1962) A revised medium for rapid growth and bioassays with tobacco tissue cultures. Physiol Plant 15:473–497
Nadolska-Orczyk A, Malepszy S (1984) Cucumber plant regeneration from leaf explants-selected characteristics. Bull Pol Acad Sci 32:425–428
Niemierowicz-Szczytt KAW, Kubicki B (1979) Cross fertilization between cultivated species of genera *Cucumis* L. and *Cucurbita* L. Genet Pol 20:217–224
Nitsch JP, Nitsch C (1956) Auxin-dependent growth of excised *Helianthus tuberosus* tissues. Am J Bot 43:839–851
Norton JD (1969) Incorporation of resistance to *Meloidogyne incognita acrita* into *Cucumis melo*. Proc Assoc S Agric Workers 66th Ann Conv, p 212
Norton JD, Granberry DM (1980) Characteristics of progeny from an interspecific cross of *Cucumis melo* with *C. metuliferus*. J Am Soc Hortic Sci 105:174–180
Orczyk W, Malepszy S (1985) In vitro culture of *Cucumis sativus* L. V. Stabilizing effect of glycine on leaf protoplasts. Plant Cell Rep 4:269–273
Orts MC, Garcia-Sogo B, Roche MV, Roig LA, Moreno V (1987) Morphogenetic response of calli derived from primary explants of diverse cultivars of melon. HortSci 22:666
Perl-Treves R, Galun E (1985) The *Cucumis* plastome: physical map, intrageneric variation and phylogenetic relationships. Theor Appl Genet 71:417–429
Perl-Treves R, Zamir D, Navot N, Galun E (1985) Phylogeny of *Cucumis* based on isozyme variability and its comparison with plastome phylogeny. Theor Appl Genet 71:430–436
Provvidenti R, Robinson RW (1974) Resistance to squash mosaic virus I in *Cucumis metuliferus*. Plant Disease Rep 58:735–738
Provvidenti R, Gonsalves D, Humaydan HS (1984) Occurence of zucchini yellow mosaic virus in cucurbits from Connecticut, New York, Florida, and California. Plant Disease 68:443–446
Rajasekaran K, Mulins MG, Nair Y (1983) Flower formation in vitro by hypocotyl explants of cucumber (*Cucumis sativus* L.). Ann Bot (London) 52:417–420
Roche MV, Orts MC, Zubeldia L, Roig LA, Moreno V (1987) Aislamiento y cultivo de protoplastos de especies silvestres de *Cucumis* y *Cucurbita*. In: Actas 2nd Cong Natl SECH, Cordoba (Spain), pp 871–880
Roig LA, Zubeldia L, Orts MC, Roche MV, Moreno V (1986) Plant regeneration from cotyledon protoplasts of *Cucumis melo* L. cv. Cantaloup Charentais. Cucurbit Genet Coop Rep 9:74–77
Schenk RV, Hildebrandt AC (1972) Medium and techniques for the induction and growth of monocotyledonous and dicotyledonous plant cell cultures. Can J Bot 50:199–204
Schroeder CA (1968) Adventive embryogenesis in fruit pericarp tissue in vitro. Bot Gaz 129:374–376
Scowcroft WR (1984) Genetic variability in tissue culture: impact on germplasm conservation and utilization. IBPGR Rep 152
Shahin EA (1985) Totipotency of tomato protoplasts. Theor Appl Genet 69:235–240
Shepard JF (1980) Mutant selection and plant regeneration from potato mesophyll protoplasts. In: Rubenstein I, Gegenbach B, Phillips RL, Green CE (eds) Emergent techniques for the genetic improvement of crops. Univ Minnesota Press, Minneapolis, pp 185–219
Shepard JF, Bidney D, Shahin E (1980) Potato protoplasts in crop improvement. Science 208:17–24
Staba EJ (1969) Plant tissue culture as a technique for the phytochemist. In: Seikel MK, Runcekles VC (eds) Recent advances in phytochemistry, vol 2, Appleton-Century-Crofts, New York, pp 75–196
Trulson AJ, Shahin EA (1986) In vitro plant regeneration in the genus *Cucumis*. Plant Sci 47:35–43
Weeden NF, Robinson RW (1986) Allozyme segregation ratios in the interspecific cross *Cucurbita maxima* (x) *C. ecuadorensis* suggest that hybrid breakdown is not caused by minor alterations in chromosome structure. Genetics 114:593–609
Wehner TC, Locy RD (1981) In vitro adventitious shoot and root formation of cultivars and lines of *Cucumis sativus* L. HortSci 16:759–760
Whitaker TW, Bemis WP (1964) Evolution in the genus *Cucurbita*. Evolution 18:553–559
Whitaker TW, Bemis WP (1976) Cucurbits. In: Simmonds NW (ed) Evolution of crop plants. Longman, London New York, pp 64–69
Whitaker TW, Davis GN (1962) Cucurbits. Botany, cultivation and utilization. (World Crops Books). Hill, London; Wiley Interscience, New York
White PR (1963) The cultivation of animal and plant cells, 2nd edn. Ronald, New York
Ziv M, Gadasi G (1986) Enhanced embryogenesis and plant regeneration from cucumber (*Cucumis sativus* L.) callus by activated charcoal in solid/liquid double-layer cultures. Plant Sci 47:115–122

III.5 Somaclonal Variation in Sugarbeet

J.W. SAUNDERS[1], W.P. DOLEY[2], J.C. THEURER[1], and M.H. YU[3]

1 Introduction

1.1 Importance and Distribution of Sugarbeet

The sugarbeet (*Beta vulgaris* L.) is an important world food crop in temperate and subtropical climates. On the international trade market its principle product, sugar, is an extremely important commodity, and it is also a significant industrial crop in some countries. Sugarbeets are cultivated as a warm or cool season crop in Europe, North and South America, northern Africa, and much of Asia where climate permits.

The sugarbeet is a naturally cross-pollinated herbaceous dicotyledon of the family Chenopodiaceae. It is of biennial growth habit and normally completes its life cycle in 2 years. During the first year the plant produces a rosette of closely spiraled leaves and a large fleshy tap root. The root is generally of an elongated cone shape with two vertical grooves or sutures down the sides. In the second year, after a low temperature exposure (4–7 °C for 80–110 days will suffice under artificial conditions), the plant initiates a seedstalk from its apex. Sessile flowers develop on the terminal end of the main axis and on profuse lateral branches. The flowers are perfect, consisting of a tricarpellate pistil surrounded by five stamens and a perianth of five narrow sepals. Petals are absent. Each flower is subtended by a slender green bract. The fruit and seed may be single (monogerm) or an aggregate where two to six flowers are fused together (multigerm). Pollination is effected by the wind.

Sugarbeet is a relatively new crop, having been developed within the past 200 years. Initial domestication of the species from wild progenitors occurred in the Mediterranean area more than 2000 years ago. The beet plant was a common part of the diet in Egypt during the building of the pyramids, although its sweetener value was not discovered until the mid 1700's. Leaves had been used for food, but the roots were used only for medicinal purposes (de Bock 1986).

[1] US Department of Agriculture, Agricultural Research Service, Crops and Entomology Research Unit, Box 1633, East Lansing, MI 48826-6633, USA
[2] Department of Crop and Soil Sciences, Michigan State University, East Lansing, MI 48824, USA
[3] US Department of Agriculture, Agricultural Research Service, Sugarbeet Production Research Unit, 1636 East Alisal St. Salinas, CA 93905, USA

Abbreviations: MS, Murashige-Skoog medium (1962); mMS, modified Murashige-Skoog medium; BA, 6-benzyladenine; 2,4-D, 2,4-dichlorophenoxyacetic acid; GA_3, gibberellic acid; IAA, indole-3-acetic acid; NAA, 1-naphtaleneacetic acid; TIBA, 2,3,5-triiodobenzoic acid; Z, zeatin.

Biotechnology in Agriculture and Forestry, Vol. 11
Somaclonal Variation in Crop Improvement I (ed. by Y.P.S. Bajaj)

While the main product of the sugarbeet today is sugar, it also has application in food processing, chemical and other industries, including the production of liquid fuel. Sugar and the molasses remaining after sugar extraction can be used for alcohol production. Using microbiological processes it has been possible to produce citric acid, L-lysine, L-glutamic acid, aconitic acid, vitamin B_{12}, and many other by-products from the molasses. Leaves, molasses, and dried root pulp also have value as animal feed. Atanassov (1986a) has estimated that the head (crown) and leaves of sugarbeet on a fresh weight basis are equal to about 50% of the root yield. Hence, additional feed value from one ton of sugarbeet is equal to 450 kg corn or 498 kg barley, and the digestable protein is equal to 450 kg dry alfalfa hay (Atanassov 1986a).

Sugarbeets are grown today on about 8.7 million ha in 41 countries. The Soviet Union is the largest producer with about 3.5 million ha. France, Poland, and China each produce about 0.5 million ha, whereas the USA, the Federal Republic of Germany, and Turkey produce about 0.4 million ha. Almost 95 million mt of beet sugar are produced annually. By 1980 almost 42% of the world sugar supply came from sugarbeet.

1.2 Objective for Genetic Improvement

The increasing world population will soon need more sugar, in spite of the present surplus on the world market and the financial stress that some processors are currently experiencing. Domestic agricultural policies will continue to demand that sugarbeet remain the main source of sucrose for countries in the temperate zone, regardless of competitive availability of high fructose corn syrup and noncaloric sweeteners. Thus our overall objective must be to improve efficiency for all aspects of sugarbeet production to achieve the maximum economic yield of sucrose.

Maximum sugar yield depends upon a regular full stand of plants in the field, elimination of weed competition, and control of diseases, insects, and other pests. More effective harvesting and better storage ability are needed to help decrease the 10% loss of beet incurred during and after harvest. Better-yielding cultivars with more efficient photosynthesis and respiration, rapid early plant development, improved sucrose transport, and reduced nonsucrose solubles, viz. free amino acids, betaine, sodium, and potassium, should be sought. Smooth nonsutured root types would improve the efficiency of harvest and storage by lessening soil retention and bruising during harvest, thus reducing respiration and storage rot losses. Greater herbicide tolerance in cultivars would reduce dependence on critical weather conditions at application.

Development of hybrid cultivars depends on the use of nuclear cytoplasmic male sterility (cms), which in sugarbeet has been limited to a single source of sterile cytoplasm (Bosemark 1979). New sources of sterile cytoplasm and diagnostic methods to identify maintainer genotypes (O-type) are needed for breeding efficiency. Monogermness (one true seed per seedball) and resistance to certain endemic diseases are a necessity. Germplasm resistant to the curly top virus is required in beets grown in semi-arid climates. Virus yellows resistance is needed for parts of northwestern Europe and the western US. Cercospora leaf spot resistance must be bred into cultivars for southern Europe and the eastern half of the US sugarbeet hectarage. Blackleg (root rot complex) resistance incited by *Pythium, Phoma, Rhizoctonia,* and *Aphanomyces* needs to be enhanced for northern Europe, the eastern US, and Chile. Rhizomania, the beet necrotic yellow vein virus disease transmitted by the fungus *Polymyxa betae* Keskin, is a devastating worldwide disease with special attention needed in much of Europe, Japan, the US, and Canada. Yellow wilt, caused by a rickittsia-like organism, is presently found only in Argentina and Chile, and is potentially one of the most serious sugarbeet diseases of the world.

Traditional methods of breeding sugarbeet are slow. New selection and screening techniques must be developed to more rapidly and accurately identify and isolate superior genotypes with good combining ability. Genetic engineering has the potential to overcome breeding barriers among species, with protoplast hybridization and recombinant DNA techniques likely to contribute to future cultivar development. Shoot culture cloning, in vitro selection or screening with toxins or challenging agents, genetic transformation, and somaclonal variants may also expedite the goal of developing cultivars with higher sucrose and better quality without sacrificing root yield.

1.3 Available Genetic Variability

Domestication of cultivated forms of *Beta* probably occurred from ancestral maritime populations. *Beta* spp. show a wide natural distribution in the eastern Mediterranean region, eastward as far as India, and northward along the coasts of the British Isles, Denmark, and Sweden (de Bock 1986).

The beet was first fully domesticated as a leaf beet similar to Swiss chard by 500 B.C. (Bosemark 1979). Subsequently, selection was made for swollen root types and the mangel-wurzel became a part of the mixed farming systems of Europe by 1800 (de Bock 1986).

Sugarbeet is believed to have originated from a relatively limited germplasm range of fodderbeet. There is some belief that spontaneous hybridization with leaf types and with *B. maritima* contributed to genetic variation (reviews by Bosemark 1979; de Bock 1986). However, the genetic base of the crop is probably narrower than most other cultivated crops.

In 1747, the German chemist Marggraf discovered that the sweet substance found in the beet was a crystallizable sugar identical to cane sugar. About 50 years later, Achard began historic experiments to extract sugar from beets. He recovered about one-half of the 6% available sugar (on a fresh weight basis) (de Bock 1986; Smith 1987). Achard used mass selection to increase sugar content, and the contemporary von Koppy family continued selection work until 1830. Their research resulted in White Silesian, which many consider the source of modern sugarbeets. In contrast, Bosemark (1979) suggested that *Beta imperialis*, with 11–12% sugar content, developed by the German breeder Knauer, was considered the source of the sugarbeet. It can be argued that the intense selection for high sugar content resulted in gradually narrowing the original gene pool.

About the year 1900, European breeders began to select and market cultivars with different combinations of sugar content and root yield (Bosemark 1979). Ernte (E) type cultivars were selected for high root yield, Zucker (Z) type cultivars were bred for high sucrose percentage, and Normal (N) type cultivars were bred for a balance of tonnage and sugar content. This marketing strategy probably conserved more genetic variation than would have occurred with only a single type of beet.

Attempts have been made to broaden the genetic base of sugarbeet by introgression of specific characteristics, especially pest resistance, from *Beta* spp. However, many of these attempts have not been of sufficient intensity or duration to be of practical or recognized value.

The genus *Beta* is taxonomically composed of 15 species in four sections, viz. Vulgares, Corollinae, Patellares and Nanae (Table 1). The sugarbeet and other species in the section Vulgares are diploids with 2n = 18 chromosomes. Other *Beta* spp. may have 18, 36, or 54 chromosomes. Sugarbeet readily hybridizes with the four other Vulgares species. Wide variation is noted for many characters in segregating generations among these species, which can be considered to have great potential value for sugarbeet improvement. Dale et al. (1985) recently assessed a selection of germplasm of the section Vulgares for some characteristics important in sugarbeet breeding and observed considerable favorable variation for resistances to downy mildew, virus yellows, and aphids, and for nuclear cms maintainer alleles. Many of our current Cercospora leaf spot-resistant cultivars can be traced back to resistance genes introgressed from *B. maritima* by Munerati as early as 1932. Genes for round monogerm seedballs, regularly shaped roots, partial resistance to the beet cyst nematode, and low temperature germination have been identified in undomesticated germplasm of the section Vulgares. However, a major problem associated with the introduction of favorable genes has been that of overcoming the weedy and usually annual character of the wild forms (de Bock 1986).

Species of section Corollinae have been used in hybridization attempts with sugarbeets (Smith 1987) to introduce genes for apomixis, monogermness, and

Table 1. Sections and species of *Beta* and their normal chromosome numbers[a]. (Smith 1987)

Section	Species	Chromosome no.
Vulgares	*B. vulgaris* L.	2n = 2x = 18
	B. maritima L.	2n = 2x = 18
	B. macrocarpa Guss.	2n = 2x = 18
	B. patula Ait.	2n = 2x = 18
	B. atriplicifolia Rouy	2n = 2x = 18
Patellares	*B. patellaris* Moq.	2n = 4x = 36
	B. procumbens Chr. Sm.	2n = 2x = 18
	B. webbiana Moq.	2n = 2x = 18
Corollinae	*B. macrorhiza* Stev.	2n = 2x = 18
	B. trigyna Wald. et Kit	2n = 4x = 36, 5x = 45, 6x = 54
	B. foliosa (sensu Haussk.)	2n = 2x = 18
	B. lomatogona Fisch. et Mey.	2n = 2x = 18, 4x = 36
	B. corolliflora	2n = 4x = 36
Nanae	*B. nana* Bois. et Held.	2n = 2x = 18

[a] According to Coons (1975). *Beta trigyna*, as collected in Hungary and the Crimea and distributed to various herbaria, is a hexaploid. The plants collected in the Caucasus and having 2n = 36 were named *corolliflora* by Zossimovitch, but considered by Coons to be *B. triygyna*. A pentaploid (5x) form of *B. trigyna* found to be apomictic has also been identified.

resistance to curly top virus, virus yellows, drought, and cold temperature. Hybrids occur at a low frequency, little chromosome homology is observed, and the hybrids are usually sterile (de Bock 1986). Thus, introgression of genes from the section Corollinae into sugarbeet has not been recognized.

Section Patellares is thought to contain the most distantly related *Beta* spp. to sugarbeet. Characteristics such as monogermness and resistance to Cercospora leaf spot, curly top virus, beet cyst nematodes, and aphids are recognized in these species (de Bock 1986). Hybridization and intensive selection by Savitsky (1975) succeeded in producing diploid nematode-resistant recombinants from sugarbeet crossed to *B. procumbens*. Resistant diploids, however, performed no better than susceptible cultivars in intensely infested fields, but the nematodes failed to reproduce in resistant germplasm (McFarlane et al. 1982).

Plants of *B. nana* are small perennial cold-resistant types which have not been studied relative to disease reaction or possibilities of hybridization with sugarbeets (de Bock 1986; Smith 1987).

Monogermness is now an essential character for the economical production of sugarbeet because it eliminates the thinning requirement due to plant crowding obtained with multigerm seedballs. A single source, released by the US Department of Agriculture (Savitsky 1950) to the sugarbeet industry as SLC 101, is the basis for all monogerm cultivars worldwide. SLC 101 was self-fertile, which facilitated the development and maintenance of inbred lines, but at the same time reduced the genetic variability of monogerm source populations from which superior lines could be developed.

In spite of the diversity of germplasm in some of the early open-pollinated cultivars, disease epidemics resulted in reducing the available genetic variation. For example, curly top virus-resistant cultivars had to be developed in the 1930's–1940's or sugarbeet production in the western US would have ceased. Resistant cultivars were developed by mass selection from US 1, the original US Department of Agriculture release. The original Cercospora leaf spot synthetics were also produced from limited genetic sources.

In the early 1940's, Owen (1945) discovered a source of nuclear-cms which revolutionized the sugarbeet industry because it made possible large-scale commercial hybrid seed production. Nuclear genetic variability in the resulting hybrids was somewhat maintained by the search for O-types (double homozygous recessives) among diverse germplasm or by backcrossing to incorporate the nuclear maintainer factors into other genetic backgrounds. However, international use of the single nuclear-cms system has greatly reduced variability in cytoplasmic factors of hybrid cultivars compared to that existing in older cultivars. Sugarbeets are thus potentially vulnerable to effects of cytoplasm-related diseases, as occurred with the southern corn leaf blight that was found to be associated with cms-T of maize. Oldemeyer (1957), Kinoshita (1977) and Bosemark (1979) have reported other sources of cms, but these have either been determined to be similar to Owen's cms source, or have not been developed sufficiently to be of use in commercial hybrids.

Cytoplasmic genetic diversity may be as important as nuclear genetic diversity. Restriction endonuclease digestion analyses of sugarbeet mitochondrial and chloroplast DNA's have shown polymorphic differences for normal, Owen's cms and other cms sources (Powling 1981; Powling and Ellis 1983; Mikami et al. 1984,

1985). However, functional or RNA transcriptional differences need to be established before this variation can be causally related to cms.

Many genetic as well as environmental factors affect extractability and final crystallization of sucrose. Wide variation has been demonstrated in selection experiments with major nonsucrose constituents, e.g., sodium, potassium, betaine, and amino acids (Smith 1987).

It is our feeling that the present gene pool of sugarbeet is not dangerously narrow, with the exception of the cytoplasmic uniformity. New collections of *B. maritima* are now being made by D.L. Doney and wild species in the section Vulgares offer excellent potential for widening the genetic base of the crop (Dale et al. 1985). Most breeders would concur with Bosemark (1979), that broadening of the cytoplasmic as well as nuclear genetic variation within sugarbeet would be desirable.

A recently available source of genetic variation that warrants examination is somaclonal variation, which includes all variation, whether genetic or epigenetic, that is derived through in vitro procedures (Larkin and Scowcroft 1981). The most valuable somaclonal variation will be that which is not available in *Beta*, and which is of agronomic value. This might include tolerance to diseases, insect vectors, salt, excessive moisture, herbicides, and other stresses. Variants that modify cytoplasmic factors conditioning male sterility should be sought. Most likely, useful variation will be derived from somatic cell selection using toxic agents or environmental shock treatments. Additionally, if chromosome breakage and interchanges occur in beet callus and suspensions, these mechanisms might be used to enhance gene transfer in interspecific hybrids, as originally proposed by Larkin and Scowcroft (1981).

2 In Vitro Regeneration of Plants

2.1 Shoot Culture and Direct Adventitious Shoot Regeneration

Vegetative propagation of sugarbeet genotypes by in vitro shoot culture is used extensively in applied breeding programs. Seedling apices (Hussey and Hepher 1978; Atanassov 1980), inflorescence tips (Coumans-Gilles et al. 1981), and flower buds (Margara 1977; Atanassov 1980; Miedema 1982) are all commonly used as starting explants. Genotype and cultural details have been compared by Atanassov (1986b). In our laboratory, axillary buds from floral stalks are disinfested and placed on mMS + 0.25 mg/l BA (Saunders 1982b). The same medium is used for shoot multiplication with at least a tripling of shoot number per 2-week cycle. Shoots are rooted on mMS + 3 mg/l NAA under moderate light intensity, and resulting plantlets are easily grown in the glasshouse if strong light is initially avoided. Cold treatment of the donor tissue (Atanassov 1980) and addition of GA_3 to the medium (Coumans-Gilles et al. 1981) are reported to enhance the multiplication process. Such in vitro clonal multiplication permits the indefinite preservation of elite breeding clones. It also facilitates precise estimates of genotypic and environmental effects since, like inbred lines, there is no genetic variance among ramets of a clone. Because shoot culture multiplication is based on the hormonal release of preexisting axillary buds from apical dominance, the genetic fidelity of the ramets is rarely questioned.

Adventitious buds are known to arise on the petioles and leaves of shoot cultures (Hussey and Hepher 1978; Saunders 1982b), and adventitious buds arising on various explants have been suggested as a means of clonal propagation (Miedema 1982; Rogozinska and Goska 1978). Since adventitious buds are a source of somaclonal variation in some species (e.g., Van Harten et al. 1981), it may also be true for sugarbeet. The only study which has addressed the genetic fidelity of sugarbeet plants derived from adventitious buds arising in shoot cultures concluded that the risk of somaclonal variation is not high (Abdel-Latif and Saunders 1985). Slavova et al. (1982) found low levels of triploids and tetraploids in plants regenerated directly from blade and petiole explants of diploid donor plants. Diploid plants regenerated directly from explants of tetraploid plants have also been reported (Slavova et al. 1982; Zagorska and Atanassov 1985).

2.3 Androgenesis and Gynogenesis

Because the haploid nature of gametophytic tissue allows identification of recessive mutations, regeneration of sugarbeet plants from such tissue would be an invaluable source of somaclonal variation. However, regeneration of haploid plants from microsporogenous tissue or pollen has not yet been successful in beet. Rogozinska et al. (1977) induced callus from sugarbeet anthers, and one callus produced a number of regenerant shoots. Unfortunately, all of the regenerants were diploid and probably arose from somatic tissue, although chromosome doubling was not ruled out. Van Geyt et al. (1985) obtained proembryoids from microspore culture, but plant regeneration was not achieved.

Regeneration of haploid sugarbeet plants through gynogenesis seems more promising than through androgenesis. Using media containing various types and levels of auxin and cytokinin, Hosemans and Bossoutrot (1983) obtained haploid plants from excised ovules, but the recovery frequency was only 0.23%. In a second report (Bossoutrot and Hosemans 1985), the recovery rate was only 0.17% when using the best medium from the previous report. Plants derived from ovules originated via embryogenesis (Bossoutrot and Hosemans 1985). Goska (1985), using a medium with BA and NAA, obtained haploid plants from excised ovules at a frequency of 2.5%. This tenfold variability in recovery rate may be partially attributable to the limited amount of germplasm sampled. When sampling several sugarbeet populations and two related *Beta* spp., Van Geyt et al. (1985) and Speckmann et al. (1986) obtained proembryoids from microspore callus, but plant regeneration was not achieved. Van Geyt et al. (1987) reported recovery rates ranging from 0 to 2.2%. D'Halluin and Keimer (1986) also noted pronounced genotypic effects, with a maximum response frequency of 26%.

Haploid and spontaneously derived diploid plants have been regenerated from callus obtained from unfertilized ovules (Smith pers. commun.). Regenerant plants have been typically mixoploid, with root tips showing twice the chromosome complement of shoot tips. Callus line OVl, from an unfertilized ovule, was maintained in the dark for over a year, and readily regenerated shoots when transferred to the light. About 20% of the regenerant shoots were chimeric green and white. The frequency of somaclonal variation in products of gynogenesis may determine how useful the haploids will be in breeding applications.

Ovary culture may lead to haploid sugarbeet plants more efficiently than culture of excised ovules. Van Geyt et al. (1987) reported recovery frequencies up to 6.1% when ovaries were cultured on various levels of BA, NAA, and 2,4-D. The presence of charcoal in the medium reduced callus frequency, but stimulated plant regeneration after transfer to medium without 2,4-D. Possible reasons for the superiority of ovary culture were thought to be the sensitivity of excised ovules to manipulation and to light. The developmental state of the ovule was less critical in ovary culture. Plants derived from ovaries are also thought to originate via embryogenesis.

2.4 Regeneration from Protoplasts

With the advent of leaf disc transformation systems, regeneration of plants from protoplasts now seems less critical for the successful application of in vitro gene transfer techniques. Since leaf disc transformation of sugarbeet has not been reported, efforts to regenerate sugarbeet from protoplasts are still needed. Cultural conditions influencing beet protoplast isolation and plating efficiency have been reported for protoplasts derived from suspension cultures (Szabados and Gaggero 1985; Bhat et al. 1985) and for mesophyll protoplasts derived from shoot culture leaves (Bhat et al. 1986).

Bhat et al. (1986) succeeded in regenerating roots from protoplast calli of three diverse beet genotypes, but shoot regeneration was not achieved. One report of somaclonal variation in sugarbeet includes information on plants regenerated from protoplasts of the tetraploid line 68–3 (Steen et al. 1986), but the procedure for regeneration of sugarbeet plants from protoplasts has yet to be published.

2.5 Regeneration from Callus and Suspension Cultures

Despite difficulties encountered in early efforts to induce shoot regeneration in sugarbeet, several recent reports indicate that procedures for plant production from callus are now available for a wide range of germplasm (Van Geyt and Jacobs 1985; Saunders and Doley 1986; Saunders and Shin 1986; Tetu et al. 1987). These reports differ in the extent to which the data permit generalization about models of in vitro response to variations in growth-regulator additives in the medium. It is clear, however, that sugarbeet, unlike most other species, possesses a shoot regeneration system based on cells autonomous for growth regulators.

The simplest and best-characterized system appears to involve only cytokinins as inducers of shoot regeneration (Saunders and Doley 1986; Saunders and Shin 1986). Leaf disc explants from shoot cultures (Saunders 1982b) or from intact plants are placed one per 35 ml of medium (mMS + 1 mg/l BA) in a Petri dish and incubated in darkness or dim light at 28–32 °C. Callus is first seen no sooner than 3 weeks after inoculation of the plates. Callus arises from one or more sites on the leaf discs, but growth is favored for those calli arising in contact with the medium but not overly covered by the explant. Callus is usually loose, friable, and easy to separate from the explant. Genotype determines prolificacy of callus initiation, speed of shoot formation (if any) from callus, and prolificacy of shoot formation. Shoots can be excised from the callus and multiplied on mMS + 0.25 mg/l BA. Shoots regenerated directly from callus, or then multiplied in shoot culture, are rooted on mMS + 3 mg/l NAA.

Suspension cultures can be easily obtained for some genotypes by placing 3–4-week-old primary callus, obtained as described above, into mMS + 1 mg/l BA without agar in flasks on an illuminated shaker (Saunders unpubl.). After one to three weekly subcultures, a well-dispersed culture is obtained. For somatic cell selection, these suspension cultures have been sieved and then plated on mMS + 1 mg/l BA + 0.9% agar in the presence of the selective agent Chlorsulfuron. Using a genotype with good shoot regeneration ability, cells surviving the challenge grew into a colony, which quickly regenerated (Fig. 1).

The callus produced on leaf explants on mMS + 1 mg/l BA has always proven to be hormone-autonomous when tested for subsequent growth on hormone-free mMS for two or more monthly culture periods (Saunders and Doley 1986). The lag time of at least 3 weeks between first exposure of the leaf piece to the medium and first appearance of callus is significant, and probably indicates the amount of time it takes one or more cells to develop the mechanism for hormone autonomy. This type of callus also differs from callus most commonly obtained with a combination

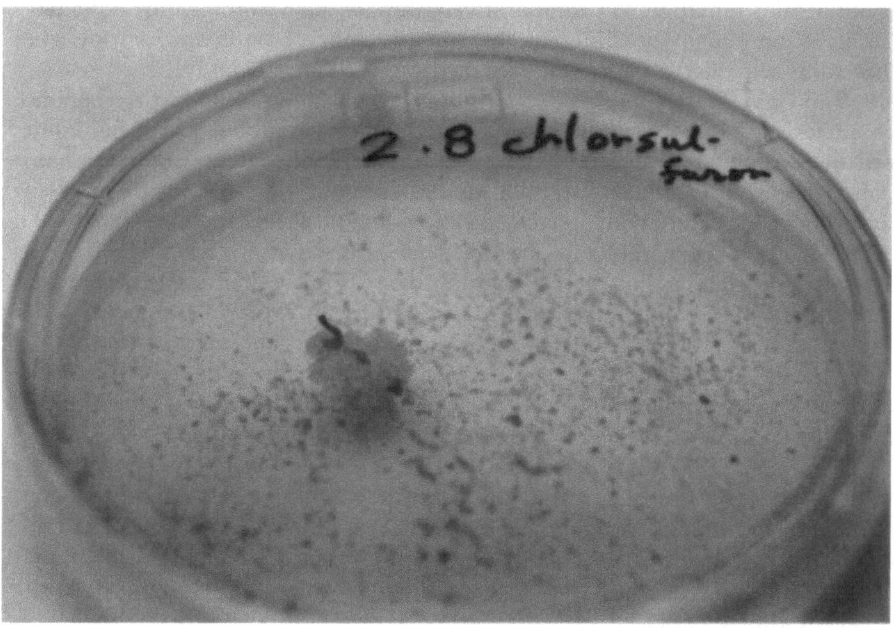

Fig. 1. Single regenerant shoot on surviving callus after plating suspension culture onto mMS + 1 mg/l BA + 2.8 nM Chlorsulfuron

of auxin and cytokinin in that callus initiation can be limited to as few as one randomly located site per explant. When the proportion of explants callusing is at a minimum, such as with explants from more expanded leaves (Saunders and Doley 1986), those explants with callus commonly have only one or two callus initiation sites, each of which may correspond to a single cell on the explant. When the proportion of explants callusing is high, as when partially expanded leaves are used as explant donors, numerous callus initiation sites are evident.

That shoots could be regenerated from hormone-autonomous callus using BA as the only growth regulator was demonstrated for a range of germplasm using callus arising on shoot culture leaf pieces (Saunders and Shin 1986) and whole plant leaf pieces (Saunders and Doley 1986). Although, at different times, we have used both sources of leaf pieces to obtain regenerating callus (on mMS + 1 mg/l BA) from several genotypes, it has not been determined whether there are response differences between these two sources of explant when other media are used.

A cytokinin additive is not required in order to obtain callus on leaf explants from shoot culture (Saunders 1982a), but it must be remembered that the leaf tissue could be "charged" with cytokinin from the shoot culture medium. The degree of charging may depend in part on the age of the culture. Thus, callus arising from shoot-culture leaf pieces on hormone-free medium may have been induced by cytokinin carried over in the explant. On the other hand, leaf pieces from partially expanded leaves of intact plants of many genotypes can also give rise to callus on hormone-free mMS (Doley and Saunders, in press). In the study of 74 plants from 16 germplasm sources, 49% of the explants representing 14 of the sources initiated callus. Callus initiation was comparatively slow, with an average time of 96 days. Callus of some genotypes differentiated somatic embryos, buds, and/or roots when this callus remained on hormone-free medium without subculture.

Van Geyt and Jacobs (1985) concluded that hormone autonomy is a common characteristic in sugarbeet cell suspensions after they were able to obtain habituated cell suspensions from different genotypes and explant sources. Cytokinin habituation occurred spontaneously, but auxin habituation occurred with successive lowering of the external auxin concentration. Shoot regeneration was obtained in the form of self-regenerating lines when "young regenerating shoots from primary explants and small callus pieces from the bases of the primary regenerant were subcultured" on a liquid induction medium. Self-regenerating lines were defined as "autonomously growing cultures which retained their regeneration capacity" (Van Geyt and Jacobs 1985). Plantlets were produced after approximately five subcultures of the self-regenerating lines. The development of the lines was very similar to the single example described by De Greef and Jacobs (1979).

Tetu et al. (1987) have reported different intensities of shoot regeneration or somatic embryogenesis depending on the growth regulator regimes and medium sequences. Low intensity shoot morphogenesis was obtained from petiole and root explants of 4-week-old axenic plants when callus was initiated on MS with different levels of BA and either NAA or IAA, and then placed on MS + 1 mg/l NAA + 0.5 mg/l BA after 4 weeks.

To obtain higher intensity shoot morphogenesis, the same authors initiated callus with MS + 1 mg/l NAA + 5 mg/l BA, and transferred it to MS or PGo medium (De Greef and Jacobs 1979) + 1 mg/l TIBA with 1 or 3 mg/l BA or Z. Callus derived from petioles, roots, floral buds, or shoot tip explants yielded shoots, but cotyledon callus did not.

Tetu et al. (1987) also described a procedure that led to somatic embryogenesis. Callus induced on petiole explants on PGo + 1 mg/l NAA was subcultured on PGo + 1 mg/l NAA + 1 mg/l BA. After two or three subcultures, the friable and green calli thus obtained were transferred to hormoneless PGo for 4 to 6 weeks. After a further subculture on PGo + 1 mg/l NAA + 1 mg/l BA, first in darkness for 1–2 weeks, then in light for 17 h per day, somatic embryos appeared.

It would be interesting to determine what similarity the shoot regeneration system described by Tetu et al. (1987) has with the hormone autonomous callus system described by Saunders and Doley (1986) or with the self-regenerating lines of Van Geyt and Jacobs (1985). Tetu et al. (1987) did not discuss whether callus initiated in their procedure was hormone-autonomous. It is apparent, however, that self-regenerating lines, as defined by Van Geyt and Jacobs (1985), have not been seen in the work reported by Saunders and Shin (1986) and Saunders and Doley (1986).

3 Somaclonal Variation Without Selection

3.1 Genetic and Procedural Considerations

Somaclonal variation includes all variation, whether genetic or epigenetic, that is derived through in vitro procedures. It can be treated according to whether or not artificial selection is applied to cells or shoots prior to plantlet production. Somatic cell selection is the term given to the deliberate application of some chemical or physical agent or environment to cultured cells with the intention of favoring the survival or multiplication of certain variant cells in the cell population. In some instances of somatic cell selection, chemical or radiological mutagens and/or haploid cells are used, but increasingly, the callus, suspension, or protoplast culture procedure itself is being used as the mutagen.

In the absence of artificial selection, the population of regenerant plants should display a random array of genetic variation characteristic of that produced in the undifferentiated culture. Exception to this generality must be made for the bias attributable to natural selection for cellular vigor and for the ability to differentiate into a plant. In addition to this selection for maintenance of competency, there may also be a natural selection in vitro for variant cells more capable of regenerating shoots.

3.2 Variability in Vitro Without Selection

Gross cytogenetic variation among cells in callus or suspensions is well documented. In some species this would seem to be involved in the loss of organ-forming capacity in long-term cultures (Smith and Street 1974; Chandler and Dodds 1983). In sugarbeet, polyploidy and aneuploidy have been detected soon after the initiation of callus (Atanassov et al. 1978).

Fig. 2. Chlorophyll-deficient and normal regenerants derived from the same donor genotype and maintained on mMS + 0.25 mg/l BA

In callus cultures of red beet, Girod and Zryd (1987) observed phenotypic variation for betalain pigment distribution. When green-habituated callus was transferred from low to high light intensity, red-colored spots appeared on the surface. The genetic nature of this variability was not ascertained.

Plant regeneration is a prerequisite for determining the genetic behavior of variants. Chlorophyll-deficient variant shoots, such as the example in Fig. 2, are generally not capable of autotrophic growth. Therefore, inheritance studies of these variant traits are difficult. Grafting them to normal shoots in vitro might be one solution to this dilemma.

3.3 Variability in Regenerated Plants Without Selection

The original self-regenerating line described by De Greef and Jacobs (1979) produced plants which were diploid (ten plants were sampled). These plants had acquired the unusual property that when leaf pieces of regenerated plants were used as explants, they formed regenerating callus on hormone-free medium.

In contrast, when a reproducible procedure for isolating self-regenerating lines was developed (Van Geyt and Jacobs 1985), leaf pieces from regenerated plants required the presence of auxin and cytokinin in the medium for callus growth to occur. These results demonstrated that, for most regenerant plants, hormone habituation was not due to spontaneous mutations, but was seen as inducible and reversible.

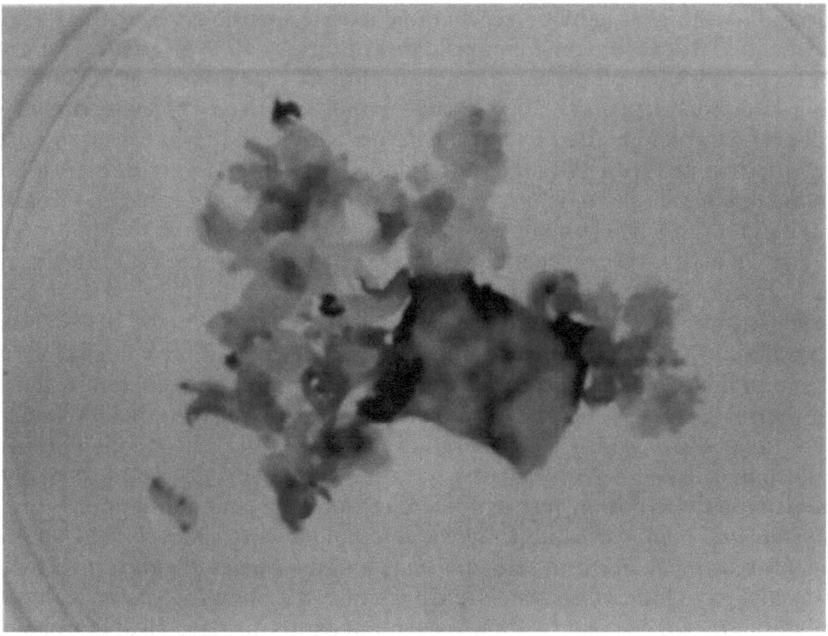

Fig. 3. Prolific regeneration from leaf disk callus in the somaclonal variant N in response to mMS + 1 mg/l BA. (Saunders and Doley 1986)

A similar conclusion was reached with plants regenerated from callus capable of growing on hormone-free medium (Saunders and Doley 1986). Only one regenerant plant, N, of four tested in that study, had an enhanced shoot regeneration capacity compared to the control source plant, and produced habituated callus on hormone-free medium (Fig. 3). In more recent unpublished work, leaf discs from 24 regenerant plants of three different genotypes had in vitro leaf disc responses indistinguishable from those of discs of respective source plants. Plant N thus appears to be a unique somaclonal variant for in vitro behavior. This variant was a less vigorous plant with smaller, narrower leaves.

Whereas the most obvious conclusion in the above two reports is that habituation is reversible, the callus (considered habituated because growth is sustained on hormone-free medium) could be a mixture of habituated and nonhabituated cells, with the latter obtaining needed hormones from the former by diffusion. This possibility could be tested with clonal analysis (Meins and Binns 1978), using calli from protoplasts derived from shoot-producing habituated callus to test the hormone autonomy on a cell by cell basis. Factors which could invalidate any conclusion are differential protoplast survival/regrowth for habituated vs. nonhabituated cells, and continued initiation of hormone autonomy.

With reliable shoot regeneration methods available only recently, comprehensive reports on the scope of somaclonal variation in beets are lacking. In the largest study available (Steen et al. 1986), 1760 plants were regenerated from 16 different donor plants via somatic embryogenesis. Within genotypes the degree of variation was higher in plants regenerated from suspension cultures than in plants regenerated from callus, and the variation appeared to be even greater in the plants regenerated from protoplasts. Doubling of chromosome number was common with diploid donor plants, and diploid regenerants were sometimes recovered from callus of tetraploid plants (Table 2). Isozyme banding patterns for seven isozymes were determined with 190 regenerant plants, revealing 25 variant plants. Loss of preexisting bands and gain of new bands were both found (Table 3). Variation in leaf and reproductive structures was also found in regenerated plants.

When plants were regenerated from the self-regenerating line Pr3 after growth on hormone-free medium for 3 years, no variation was found in four isozyme systems (phosphoglucomutase, isocitrate dehydrogenase, malic dehydrogenase, peroxidase) (Van Geyt pers. commun.). All whole-plant morphological variation scored in the field could be explained by assuming that the different regenerants were not at the same state of development when transferred to the field. When shoots were isolated from the self-regenerating line (De Greef and Jacobs 1979) grown autotrophically for 8 years, rooting was difficult to achieve, and most regenerant plants brought to flower were male sterile (Van Geyt pers. commun.).

Yu (1987) induced callus from leaf discs and in vitro shoots on MS + 1.0 or 2.0 mg/l BA, using plants from 12 *Beta* species and subspecies. Plants from three other *Beta* spp. gave little or no response from leaf discs. Shoot-producing callus came only from *B. vulgaris* plants. Some regenerant plants displayed leaf morphological and pigmentation differences. Plants of 36 and 38 chromosome number were

Table 2. Variation in chromosome number within plants regenerated from four genotypes. (Steen et al. 1986)

Donor/ genotype	Chromosome no. in donor	Culture	No. of reg. plants	Chromosome no. in reg. plants	% Reg. plants with chromosome number diff. from donor
145000	18	Callus	123	18	
		Callus	18	36	12.8
		Cell-susp.	42	18	
		Cell-susp.	10	36	19.2
145018	18	Callus	426	18	
		Callus	93	36	17.9
		Cell-susp.	37	18	
		Cell-susp.	9	36	19.6
62/2	18	Callus	167	18	
		Callus	40	36	19.3
		Cell-susp.	184	18	
		Cell-susp.	89	36	32.6
68–3	36	Callus	12	18	14.8
		Callus	69	36	
		Protoplast	2	18	8.0
		Protoplast	23	36	

0.8% aneuploids were found, but the chromosome numbers of these are unstable.

Table 3. Variation in seven isozyme systems within plants regenerated from four genotypes. (Steen et al. 1986)

Donor/ genotype	No. of plant		*Isozyme systems* PGM	GPI	IDH	MDH	AK1	AK2	ACO	DIA
145000		44	x	x	1	1	x		x	2
		3	x	x	1	1	x		1	2
		2	x	x	x	1	x		1	2
		1	x	1	1	1		1	x	2
	Total	50								
145018		36	1	1	x	1	1		2	x
		1	x	1	1	1	x		x	x
		1	x	x	x	x	x		2	x
	Total	38								
62/2	Total	71	1	x	x	x		1	2	2
68-3		14	x	x	1	x	x		x	x
		11	x	x	1	x	x		x	2
		4	2	x	1	x	x		x	x
		1	x	x	1	x		1	x	2
		1	x	x	1	x		1	x	x
	Total	31								

1 and 2 = homozygote alleles. x = heterozygote alleles. PGM = phospho-gluco-mutase. GPI = glucose-phosphate-isomerase. IDH = iso-citrate-dehydrogenase. MDH = malate-dehydrogenase. AK1 = adenylate-kinase. AK2 = adenylate-kinase. ACO = aconitase. DIA = diaphorase.

obtained from 18 and 19 chromosome plants, respectively. Mixoploid plants containing roots with either doubled or undoubled chromosome number were also obtained.

Subsequent work (Yu unpubl.) has examined regenerant plants from five individuals of population 2488, which consists of diploid and monosomic addition entities and bears a *B. procumbens* hybridization background in its remote ancestry. Approximately 20% of the 136 plants regenerated, representing four of the five monosomic addition donor plants sampled, displayed a leaf intumescence characterized by a frothy pockmark appearance of the leaf surface (Fig. 4A,B,C). In one case, the intumescence feature did not arise until a second cycle of regeneration, i.e., in regenerants that were derived from callus of primary regenerants. Thus leaf intumescence appears to be a recurring trait that has arisen independently multiple times. As such it warrants further investigation into the mechanism(s) involved.

In the same study, the most common type of gross chromosomal variation detected was doubling of chromosome number. Polyploid regenerants usually developed wider leaf blades, hence smaller leaf length to width ratios, as compared with diploids (Fig. 4A,D). In one experiment where chromosome complements of 116 regenerated plants were determined, 61% contained the somatic chromosome number of the donors, 31% had doubled chromosomes, and the remaining 8% were mixoploids having sectors of diploidy and tetraploidy in the same regenerated plants (Table 4), based on root-tip chromosome counts. Some chimeric plants with

A B C D

Fig. 4A–D. Leaf characteristics of regenerated plants derived from callus of leaf sections from sugarbeet germplasm line 2488 induced on medium containing 1 mg/l BA. **A** Normal appearance of a regenerated plant that resembles its explant parent. **B** and **C** Adaxial and abaxial surfaces, respectively, of an intumescent leaf in which the protuberances on the upper surface are indented from beneath, or vice versa. Note the two intumescences at the edge of the leaf that are shaped like small cups. **D** Leaf from a regenerated plant with doubled number of chromosomes. Note that the length to width ratio of the blade in **D** is considerably decreased as compared to that in **A**. (Yu unpubl.)

Table 4. Chromosome composition of the regenerated plants derived from callus of sugarbeet leaf or shoot explants. (Yu unpubl.)

Source sugarbeet	Callus origin[a]	Ploidy level[b] 2x	4x	2x/4x
H10A	Leaf	6	2	1
H10D	Leaf	5	4	1
H10D	Regenerated shoot	4	4	3
61851	Leaf		2	1
61936	Leaf	3	1	
TA3	Leaf	8	5	
TA3	Shoot	7	6	1
TA5	Leaf	6		
T192	Leaf	8		
T193	Leaf		11	2
T194	Leaf	6		
T194	Regenerated shoot	18	1	
Total		71	36	9

[a] Leaf and shoot explants were cultured on a MS medium containing 1 and 0.25 mg/l BA, respectively, for callus induction; regenerated shoots were rooted in medium containing 1.5 mg/l NAA without BA.
[b] Based on available root-tip chromosome counts.

roots of either doubled or undoubled chromosome number were also recovered among control plants produced through shoot culture. This phenomenon has been reported in other instances where roots are induced on beet shoots in vitro, and appears to be an artifact of that procedure (Abdel-Latif and Saunders 1985; Bossoutrot and Hosemans 1985; Van Geyt et al. 1987).

A study of somaclonal variation has been initiated utilizing as source germplasm a series of full sib genotypes bred for shoot regeneration capacity in combination with floral annualism and self-fertility (Saunders unpubl.). The latter two characters will facilitate inheritance studies of variant features. These genotypes are all heterozygous at the B, M, and R loci (Smith 1980) and some are heterozygous at the phosphoglucomutase locus. Use of multiple heterozygosis should permit detection of somaclonal change at these loci.

In order to quantify the frequency of different categories of genetic change detected in the population of regenerants, care was taken to assign each regenerant shoot a unique number, including a reference to the particular leaf disc from which it was derived. This coding was done to insure that two variant regenerant plants had independent origins, and to preclude inflation of frequency values. Regenerant shoots were multiplied prior to root induction (Saunders 1982b) so that multiple copies were available for comparison with copies of source genotypes in the glasshouse.

Examination of these regenerant plants has so far focused on foliar appearance. A narrow leaf morphology is typical of several regenerants from different source genotypes. Interestingly, three instances of chimerism in regenerated shoots or their derivative shoots have been detected. In one case a narrow leaf plant and a normal leaf plant were derived from the same regenerant shoot through shoot culture multiplication, which involves outgrowth of axillary buds, and have maintained their differences. In another instance, a single plant displayed bilateralism (normal vs. atrophied appearance) (Fig. 5). A third case involved plants derived from an axillary bud taken from a regenerant plant. The regenerant had displayed several small protuberances, but otherwise appeared normal. An attempt to propagate it was made so that it could be examined with multiple copies and in different environments. When the propagation cycle was completed and multiple ramets were grown in the glasshouse, all were alike but quite different from the regenerant from which the axillary bud was taken. All propagules had narrow leaves and a confused mixture of short and long internodes on a short flower stalk. All were completely both male and female sterile, in contrast to good fertility seen on the source regenerant plant.

Two heritable foliar traits have also been observed. One was seen on the regenerant shoot and involved large patches and spots of intense red color. This pigmentation pattern diminished greatly during rooting of the shoots, but has been maintained in shoot cultures. Pigmentation of these vegetative plants reverted to a type nearly identical with the normal source genotype in the glasshouse, but many leaves on floral stalks have been partially red-pigmented. Outcrossing of these plants has yielded progeny with speckled seedling leaves (Fig. 6), indicating a dominant mode of inheritance. This trait has a few similarities with the dominant trout character for speckled leaves reported by Owen and Ryser (1942), and may be

Fig. 5. Chimeric regenerant plant displaying bilateralism

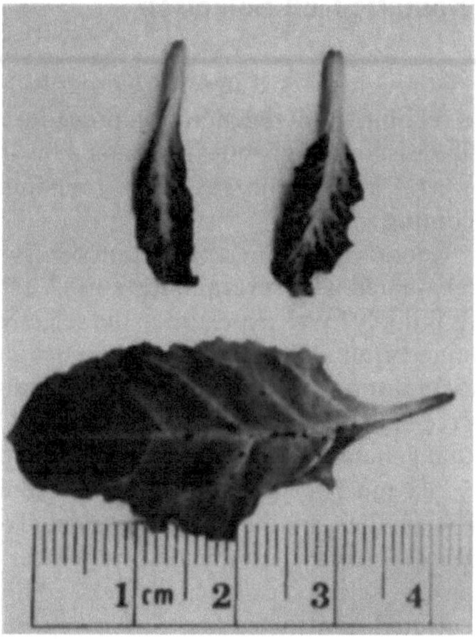

Fig. 6. Leaves of red patch somaclonal variant. *Top* intense red pigmentation on shoot culture leaves of this regenerant. *Bottom* speckled leaf from F_1 progeny of this regenerant

Fig. 7. Necrotic patch somaclonal variant. *Left, center* Necrotic patch leaves from F_1 plant. *Right* normal leaf from normal F_1 segregate

similar to an unstable anthocyanin variant recovered from alfalfa callus (Groose and Bingham 1986).

The second heritable foliar trait noticed in the regenerant population involved necrotic patches on leaves of regenerant plants (Fig. 7). The trait was seen again after selfing and outcrossing, suggesting dominance. The necrotic patch segregants are impaired in their growth rate. The necrotic patch and red patch variants could both be examples of variegation due to transposable elements, which have been implicated in some types of somaclonal variation in other species (e.g., Groose and Bingham 1986; Peschke et al. 1987).

4 Somatic Cell Selection

Even though beet tissue culture progress has been limited until recently by the lack of a reliable shoot regeneration procedure, there has been success in selecting beet cells resistant to chemical agents. This area of endeavor should prove to be of powerful application to future breeding goals now that plant regeneration is becoming routine.

Reports about the application of cell selection techniques to plant cells should be evaluated with several questions clearly in mind. At what ploidy level were the cells initially? Was exposure to the selective agent chronic or acute? Was resistance retained after a period of growth in the absence of the selective agent? Answers to these questions, and indeed design details of future selection efforts, relate to the degree of expression of the resistance in whole plants, its heritability in subsequent sexual generations, and its value in crop genetic improvement.

Pua and Thorpe (1986), addressing the need for Na_2SO_4 tolerance in sugarbeets grown on the Canadian prairies, selected callus cultures for growth on Na_2SO_4 with stepwise increase in concentration over at least 10 months. After this chronic exposure, selected callus also tolerated higher levels of NaCl (560 mM) that killed unselected control callus. Shoot regeneration from selected cultures was not reported.

Increased callus tolerance to the pathotoxin cercosporin produced by the causal agent of leaf spot, *Cercospora beticola*, was reported after callus from a self-regenerating line was exposed to 25 ppm cercosporin for ten successive culture periods (Viseur 1984). At least one regenerant plant was obtained that displayed major morphological alterations, but the reaction of tissue from the plant to cercosporin or to the disease inoculum was not reported.

Herbicide resistance is a character of general interest to crop biotechnologists. In sugarbeet, sensitivity to sulfonylurea herbicides applied in preceding cereal crops has become an economic concern. Because the sulfonylurea herbicides inhibit an enzymatic step in amino acid biosynthesis (Chaleff and Mauvais 1984) that most likely would be operative in whole plants, selection for resistance to the sulfonylurea herbicide chlorsulfuron was initiated (Saunders unpubl.). The philosophy was to minimize the time spent in the callus and suspension phase so as to maximize the probability of obtaining shoot regeneration from surviving colonies and minimize the amount of background somaclonal variation. The exposure to the herbicide was acute, with the intent of obtaining resistance that would be simply inherited, and no artificial mutagen was used.

The starting genotype was REL-1, a diploid annual self-fertile regenerator clone which is available to the public. All suspension cells plated on mMS + 1 mg/l BA + 2.8 nM chlorsulfuron died except one surviving colony which promptly produced a shoot (Fig. 1). Other shoots were later regenerated from the surviving callus, given unique identification numbers, and treated as separate isolates in the event that additional somaclonal variation would occur. Subcultured surviving callus demonstrated tolerance to 2.8 nM Chlorsulfuron, whereas source callus died. When shoots of the different isolates and the source genotype were challenged with 28 nM Chlorsulfuron incorporated into the rooting medium (mMS + 3 mg/l NAA), no visible damage was seen on isolate shoots, whereas control shoots quickly died

Fig. 8. Shoot cultures from unselected source (*left*) and solitary surviving callus (*right*) on mMS + 3 mg/l NAA + 28 nM chlorsulfuron

(Fig. 8). Regenerant plants of the isolates are now being self- and cross-pollinated to test for genetic transmission of the resistance. Because one isolate was found to be male sterile, and others have necrotic foliar patches, the decision to handle the multiple regenerant shoots on a pedigree basis may prove useful.

Tolerance to the herbicides 2,4-D and Picloram has been obtained in callus derived from protoplasts (Senanayake and Jacobs pers. commun.). No plants were regenerated, but transfer of the picloram resistance in fusions with *Nicotiana plumbaginifolia* nitrate reductase-deficient protoplasts indicated a dominant or co-dominant mode of inheritance.

5 Tolerating Somaclonal Variation

In most instances, somaclonal variants can be either utilized or discarded. Exceptions to this occur where concomitant genetic changes of a detrimental nature interfere with recovery of whole plants or progeny of desired variants. A well-documented case in potato somatic cell selection exemplifies this problem. Only 21 of 67 hydroxyproline-resistant cell lines from one source clone regenerated shoots within a year after transfer to regeneration conditions. Shoots of only 14 lines could be rooted. Of these, plantlets of only four lines survived transfer to the greenhouse, and all were completely sterile (Van Swaaij et al. 1987). These authors concluded that more attention should be paid to reducing somaclonal variation, probably by reducing the length of the cell culture phase.

The recovery of three stable plant types (normal fertility and morphology, male sterile, and necrotic leaf) from a single sugarbeet colony surviving Chlorsulfuron exposure (Saunders unpubl.) may be a lesser example of this problem. Recovery of fertile regenerant plants from somatic cell selection efforts will probably succeed when exposure to the toxic agent and the shoot regeneration medium occur before excessive genetic changes occur.

The risk of somaclonal variation has generally precluded use of shoot regeneration from callus as a means of clonal propagation. However, regenerant plants could be useful in genotypic evaluations such as for multiple disease screenings or where simply inherited traits such as cytoplasmic male sterility maintenance need to be ascertained by test crosses. Time would be gained by plant breeders in these situations if shoots are obtained from leaf discs of young field plants. The probability of somaclonal variation affecting any single locus or small number of loci is probably low. If employed in this way, regenerant plants should probably not be used to perpetuate the source genotype.

6 Prospects for Utilization of Somaclonal Plants

Variation derived from in vitro culture has been detected in sugarbeet plants regenerated from callus, cell suspensions, and protoplasts. This variation includes apparent halving of chromosome number (Zagorska and Atanassov 1985; Steen et al. 1986), high frequency variation in isozyme patterns (Steen et al. 1986), recurrence of a novel intumescence character in regenerants from different donor plants (Yu unpubl.), and a heritable pigment variant of possibly unstable nature (Saunders and Theurer unpubl.). Most of these variants have not yet been well characterized. There have been no reports of somaclonal variation in sugarbeet for quantitative agronomic traits.

Somaclonal variant sugarbeet plants have yet to be incorporated into applied breeding programs. It seems unlikely that the variation produced and randomly sampled in tissue culture will make a significant contribution to yield or quality, the quantitative traits for which improvement is most desired. Such random variation, as with products of chemical or radiological mutagenesis, is more often than not deleterious.

More likely, useful simply inherited variation for qualitative traits will be derived through somatic cell selection with judicious choice of target traits and imaginative choice of selective agents or environments. Close cooperation between tissue culturists and plant breeders is essential to insure that selection efforts are directed toward traits of economic value. The most useful somaclonal variants will be those for which natural variation is not available or is too complexly inherited.

Although somaclonal variants in beet have yet to be utilized, there are some considerations for the future. Assessments of the incidence of somaclonal variation provide underestimates because only a subset of morphological or biochemical parameters can be monitored. Given the recovery of a desirable character in a regenerated plant, assurance that statistically significant somaclonal variation for reductions in vigor, economic yielding ability, or combining ability is absent would require extensive and expensive testing. In a regeneration system and germplasm background similar to that used by Steen et al. (1986), for example, where 21% of regenerants were variant in at least one isozyme system, the frequency of multiple genetic changes in individual regenerants is probably high. For novel traits to be utilized in the most risk-free manner, they will have to be crossed out of their regenerant germplasm and into an acceptable breeding background, probably via a backcross program. The initial steps of this procedure provide information on the mode of inheritance of the trait and would be part of normal genetic characterization. Use of monogenic annualism would accelerate the backcross procedure in an otherwise biennial species.

The alternative to backcrossing in sugarbeet would be to select an elite regenerable clone based on agronomic performance or combining ability from a valuable breeding population. Multiple isolates of independent origin from somatic cell selection efforts, for example, would then be tested for vigor and combining ability to choose the isolate with the least background somaclonal variation. This isolate could then be used directly in hybrid production after clonal propagation. Thus, the alternative to backcrossing involves considerable testing to identify both the elite clone for subsequent in vitro use, and the least debilitated isolate after regenerants are recovered. Because beet breeding and seed production are overwhelmingly population- and seed propagation-oriented, the seed industry will probably find backcrossing to be the most acceptable way to incorporate somaclonal variants into commercial cultivars. Furthermore, the same risk of background somaclonal variation will be present when genetic transformation becomes available as a breeding tool in beet.

For basic research, an argument can be made for utilizing a single clone to produce genetic and cytogenetic stocks of somaclonal or transgenic origin in a common genetic background. Variants of a given clone with a single recognized genetic change represent near-isogenic lines of the founder clone that cannot otherwise be produced in beet because of inbreeding depression. In this respect, beet is similar to alfalfa (Bingham and McCoy 1986).

It is likely that many of the novel traits arising from somaclonal variation will be products of somatic cell selection, as this is an efficient way to screen for products of somaclonal variation. Because it is likely that each resistant isolate will represent only one sample from a population of mutations conferring resistance to any one toxic agent, it will be important to maximize the number of isolates so as to have available a variety of resistance mechanisms, inheritance modes and possibly pleiotropic effects such as cross-resistance.

The several instances of chimerism seen in regenerant plants (Saunders unpubl.) suggest a multicellular origin of the regenerating shoots. The possibility of chimerism in regenerated plants used for inheritance studies cautions against reliance on segregation ratios seen in immediate progeny of regenerant plants or their vegetative propagules (McCoy and Phillips 1982; Buiatti et al. 1985). The ease of shoot culture procedures in beet permits vegetative propagation to maintain and to provide multiple copies of variant regenerants or sexual progeny for phenotypic stability studies, and can allow chimerism to be expressed.

The greatest prospects for economic impact of somaclonal variation in sugarbeet lie in the area of somatic cell selection. Disease resistance, particularly where naturally occurring variation is limited, and herbicide tolerance are likely to be incorporated into commercial hybrids in the near future. Tolerance to environmental stresses is another promising area for the application of somatic cell selection. More challenging to the imagination will be the use of select metabolic inhibitors in somatic cell selection systems to heritably alter the chemical composition of the beet root for more efficient sugar extraction in the processing operation, for example by modifying membrane composition. By acute selection in diploid or tetraploid cells, it should be possible to recover resistant mutants with monogenic dominant inheritance.

Somatic cell selection could become a complement to genetic transformation. Whereas transformation offers the precision of known nucleotide sequences with expected phenotypes, somatic cell selection could generate a range of resistance mechanisms with demonstrated expression, with less expertise required. Finally, variant characters obtained from somatic cell selection in sugarbeet could provide isolated genes for transgenic insertion into other crop species.

References

Abdel-Latif TH, Saunders JW (1985) Genetic fidelity of *Beta vulgaris* L. ramets derived from adventitious buds on shoot culture petioles. Agron Abstr Am Soc Agron, Madison, WI, p 130

Atanassov AI (1980) Method for continuous bud formation in tissue cultures of sugar beet (*Beta vulgaris* L.). Z Pflanzenzücht 84:23–29

Atanassov AI (1986a) Sugar beet. In: Evans DA, Sharp WR, Ammirato PV (eds) Handbook of plant cell culture, vol 4. MacMillan, New York, pp 652–680

Atanassov AI (1986b) Sugarbeet (*Beta vulgaris* L.). In: Bajaj YPS (ed) Biotechnology in agriculture and forestry, vol 2. Crops I. Springer, Berlin Heidelberg New York Tokyo, pp 462–470

Atanassov AI, Kikindonov T, Antonova G (1978) Cytological changes in permanent sugar beet tissue cultures cultivated in vitro. In: Stoilov M (ed) Proc Int Symp on experimental mutation of plants. Bulg Acad Sci, Sofia, pp 309–317

Bhat SR, Ford-Lloyd BV, Callow JA (1985) Isolation of protoplasts and regeneration of callus from suspension cultures of cultivated beets. Plant Cell Rep 4:348–350

Bhat SR, Ford-Lloyd BV, Callow JA (1986) Isolation and culture of mesophyll protoplasts of garden, fodder and sugar beets using a nurse culture system: callus formation and organogenesis. J Plant Physiol 124:419–423

Bingham ET, McCoy TJ (1986) Somaclonal variation in alfalfa. Plant Breed Rev 4:123–152

Bosemark NO (1979) Genetic poverty of the sugarbeet in Europe. In: Zeven AC, Van Harten AM (eds) Proc of the Conf on broadening the genetic base of crops. Pudoc, Wageningen, pp 29–35

Bossoutrot D, Hosemans D (1985) Gynogenesis in *Beta vulgaris* L.: from in vitro culture of unpollinated ovules to the production of doubled haploid plants in soil. Plant Cell Rep 4:300–303

Buiatti M, Marcheschi G, Tognoni F, Lipucci di Paola M, Collina Grenci F, Martini G (1985) Genetic variability induced by tissue culture in the tomato (*Lycopersicon esculentum*). Z Pflanzenzücht 94:162–165

Chaleff RS, Mauvais CJ (1984) Acetolactate synthase is the site of action of two sulfonylurea herbicides in higher plants. Science 224:1443–1445

Chandler SF, Dodds JH (1983) Adventitious shoot initiation in serially subcultured callus cultures of *Solanum laciniatum*. Z Pflanzenphysiol 111:115–121

Coons GH (1975) Interspecific hybrids between *Beta vulgaris* L. and the wild species of *Beta*. J Am Soc Sugar Beet Technol 18:281–306

Coumans-Gilles MF, Kevers C, Coumans M, Ceulemans E, Gaspar Th (1981) Vegetative multiplication of sugarbeet through in vitro culture of inflorescence pieces. Plant Cell Tissue Org Cult 1:93–101

Dale MFB, Ford-Lloyd BV, Arnold MH (1985) Variation in some agronomically important characters in a germplasm collection of beet (*Beta vulgaris* L.). Euphytica 34:449–455
De Bock ThSM (1986) The genus *Beta*: domestication, taxonomy, and interspecific hybridization for plant breeding. Acta Hortic 182:335–343
De Greef W, Jacobs M (1979) In vitro culture of the sugarbeet: Description of a cell line with high regeneration capacity. Plant Sci Lett 17:55–61
D'Halluin K, Keimer B (1986) Production of haploid sugarbeet (*Beta vulgaris*) by ovule culture. In: Horn W, Jensen CJ, Odenbach W, Schieder O (eds) Genetic manipulation in plant breeding. De Gruyter, Berlin, pp 307–309
Doley WP, Saunders JW Hormone-free medium will support callus production and subsequent shoot regeneration from whole plant leaf explants in some sugarbeet (*Beta vulgaris* L.) populations. Plant Science Letters (in press)
Girod P-A, Zryd J-P (1987) Clonal variability and light induction of betalain synthesis in red beet cell cultures. Plant Cell Rep 6:27–30
Goska M (1985) Sugar beet haploids obtained in the in vitro culture. Bul Acad Pol Sci Biol Sci 33:31–33
Groose RW, Bingham ET (1986) An unstable anthocyanin mutation recovered from tissue culture of alfalfa (*Medicago sativa*). I. High frequency of reversion upon reculture. Plant Cell Rep 5:104–107
Hosemans D, Bossoutrot D (1983) Induction of haploid plants from in vitro culture of unpollinated beet ovules (*Beta vulgaris* L.). Z Pflanzenzücht 91:74–77
Hussey G, Hepher A (1978) Clonal propagation of sugar beet plants and the formation of polyploids by tissue culture. Ann Bot (London) 42:477–479
Kinoshita T (1977) Genetic relationship between pollen fertility restoring genes and cytoplasmic factors in the male sterile mutants of sugar beets. Jpn J Breed 27:19–27
Larkin PJ, Scowcroft WR (1981) Somaclonal variation – a novel source of variability from cell cultures for plant improvement. Theor Appl Genet 60:197–214
Margara J (1977) La multiplication vegetative de la betterave (*Beta vulgaris* L.) en culture in vitro. CR Acad Sci Paris Ser D 285:1041–1044
McCoy TJ, Phillips RL (1982) Chromosome stability in maize (*Zea mays*) tissue cultures and sectoring in some regenerated plants. Can J Genet Cytol 24:559–565
McFarlane JS, Savitsky H, Steele A (1982) Breeding for resistance to the sugarbeet nematode. J Am Soc Sugar Beet Technol 21:311–323
Meins F, Binns AN (1978) Epigenetic clonal variation in the requirement of plant cells for cytokinins. In: Subtelny S, Sussex IM (eds) The clonal basis for development. Academic Press, New York London, pp 185–201
Miedema P (1982) A tissue culture technique for vegetative propagation and low temperature preservation of *Beta vulgaris*. Euphytica 31:635–643
Mikami T, Sugiura M, Kinoshita T (1984) Molecular heterogeneity in mitochondrial and chloroplast DNAs from normal and male sterile cytoplasms in sugar beets. Curr Genet 8:319–322
Mikami T, Kishima Y, Sugiura M, Kinoshita T (1985) Organelle genome diversity in sugarbeet with normal and different sources of male sterile cytoplasms. Theor Appl Genet 71:166–171
Murashige T, Skoog F (1962) A revised medium for rapid growth and bioassays with tobacco tissue culture. Physiol Plant 15:473–497
Oldemeyer RK (1957) Sugar beet male sterility. J Am Soc Sugar Beet Technol 9:381–386
Owen FV (1945) Cytoplasmically inherited male-sterility in sugar beets. J Agric Res 71:423–440
Owen FV, Ryser GK (1942) Some mendelian characters in *Beta vulgaris* and linkages observed in the Y-R-B group. J Agric Res 65:155–171
Peschke VM, Phillips RL, Gengenbach BG (1987) Discovery of transposable element activity among progeny of tissue culture-derived maize plants. Science 238:804–807
Powling A (1981) Species of small DNA molecules found in mitochondria from sugarbeet with normal and male sterile cytoplasms. Mol Gen Genet 183:82–84
Powling A, Ellis THN (1983) Studies on the organelle genomes of sugarbeet with male-fertile and male-sterile cytoplasms. Theor Appl Genet 65:323–328
Pua E-C, Thorpe TA (1986) Differential response of nonselected and Na_2SO_4-selected callus cultures of *Beta vulgaris* L. to salt stress. J Plant Physiol 123:241–248
Rogozinska J, Goska M (1978) Induction of differentiation and plant formation in isolated sugar beet leaves. Bull Pol Acad Sci Biol Sci 26:343–345

Rogozinska JH, Goska M, Kuzdowicz A (1977) Induction of plants from anthers of *Beta vulgaris* L. cultured in vitro. Acta Soc Bot Pol 46:471–479

Saunders JW (1982a) Cytokinin effects on formation of high frequency habituated callus and adventitious buds in sugarbeet (*Beta vulgaris* L.). In: Fujiwara A (ed) Plant tissue culture 1982. Maruzen, Tokyo, pp 153–154

Saunders JW (1982b) A flexible in vitro shoot culture propagation system for sugarbeet that includes rapid floral induction of ramets. Crop Sci 22:1102–1105

Saunders JW, Doley WP (1986) One step shoot regeneration from callus of whole plant leaf explants of sugarbeet lines and a somaclonal variant for in vitro behavior. J Plant Physiol 124:473–479

Saunders JW, Shin K (1986) Germplasm and physiologic effects on induction of high-frequency hormone autonomous callus and subsequent shoot regeneration in sugarbeet. Crop Sci 26:1240–1245

Savitsky VF (1950) Monogerm sugar-beets in the United States. Proc Am Soc Sugar Beet Technol 6:156–159

Savitsky H (1975) Hybridization between *Beta vulgaris* and *B. procumbens* and transmission of nematode (*Heterodera schachtii*) resistance to sugarbeet. Can J Genet Cytol 17:197–209

Slovova Y, Zahariev A, Antonova G (1982) Improving and applying the method of sugar beet vegetative propagation in vitro. Plant Physiol (Sofia) 8(1):95–99

Smith GA (1980) Sugarbeet. In: Fehr WR, Hadley HH (eds) Hybridization of crop plants. Am Soc Agron, Madison, WI, pp 601–613

Smith GA (1987) Sugar beet. In: Fehr WR (ed) Principles of cultivar development. MacMillan, New York, pp 577–625

Smith SM, Street HE (1974) The decline of embryogenic potential as callus and suspension cultures of carrot (*Daucus carota* L.) are serially subcultured. Ann Bot (London) 38:223–241

Speckmann GJ Jr, Van Geyt JPC, Jacobs M (1986) The induction of haploids of sugarbeet (*Beta vulgaris* L.) using anther and free pollen culture or ovule and ovary culture. In: Horn W, Jensen CJ, Odenbach W, Schieder O (eds) Genetic manipulation in plant breeding. De Gruyter, Berlin, pp 351–353

Steen P, Keimer B, D'Halluin K, Pedersen HC (1986) Variability in plants of sugar beet (*Beta vulgaris* L.) regenerated from callus, cell-suspension and protoplasts. In: Horn W, Jensen CJ, Odenbach W, Schieder O (eds) Genetic manipulation in plant breeding. De Gruyter, Berlin, pp 633–634

Szabados L, Gaggero C (1985) Callus formation from protoplasts of a sugarbeet cell suspension culture. Plant Cell Rep 4:195–198

Tetu T, Sangwan RS, Sangwan-Norreel BS (1987) Hormonal control of organogenesis and somatic embryogenesis in *Beta vulgaris* callus. J Exp Bot 38:506–517

Van Geyt JPC, Jacobs M (1985) Suspension culture of sugarbeet (*Beta vulgaris* L.). Induction and habituation of dedifferentiated and self-regenerating cell lines. Plant Cell Rep 4:66–69

Van Geyt J, D'Halluin K, Jacobs M (1985) Induction of nuclear and cell divisions in microspores of sugarbeet (*Beta vulgaris* L.). Z Pflanzenzücht 95:325–335

Van Geyt J, Speckmann GJ Jr, D'Halluin K, Jacobs M (1987) In vitro induction of haploid plants from unpollinated ovules and ovaries of the sugarbeet (*Beta vulgaris* L.). Theor Appl Genet 73:920–925

Van Harten AM, Bouter H, Broertjes C (1981) In vitro adventitious bud techniques for vegetative propagation and mutation breeding of potato (*Solanum tuberosum* L.). II. Significance for mutation breeding. Euphytica 30:1–8

Van Swaaij AC, Nijdam H, Jacobsen E, Feenstra WJ (1987) Increased frost tolerance and amino acid content in leaves, tubers and leaf callus of regenerated hydroxyproline resistant potato clones. Euphytica 36:369–380

Viseur J (1984) Selection in vitro de cals de betterave tolerants a la cercosporine, une toxine fongique extraite de *Cercospora beticola* Sacc. Ann Gembloux 90:39–44

Yu MH (1987) Observations on callus induction and somaclonal variation in beet species. Genetics 116:s17

Zagorska N, Atanassov A (1985) Somaclonal variation in tobacco and sugar beet breeding. In: Henke RR, Hughes KW, Constantin MJ, Hollaender A (eds) Tissue culture in forestry and agriculture. Plenum, New York, p 371

III.6 Somaclonal Variation in Chicory

S. GRAZIA[1], G. ROCCHETTA[1], and E. PIERAGOSTINI[2]

1 Introduction

1.1 Importance and Distribution

Chicory (*Cichorium intybus* L., n = 9) is a species of the Compositae family that is concentrated in the warm temperate zone of the northern hemisphere, particularly in Europe and around the Mediterranean Basin. According to Vavilov (1951), the Mediterranean is the primary center of distribution of this plant. Chicory grows well in every type of soil and does not have particular climatic requirements, although the best qualitative and quantitative results have been obtained when it is cultivated in fresh well-drained soils. However, chicory needs all nutrient elements, above all phosphates (Antoniani et al. 1975).

Chicory leaves are principally utilized as fresh or cooked food, even though they are of little nutritional value. In Italy, chicories with red and/or variegated leaves are known as "radicchio" and belong to different cultivars, i.e., Rosso di Verona, Rosso di Chioggia, Rosso di Treviso, and Variegato di Castelfranco. These cultivars are principally grown in the northern regions (Lombardia and Veneto) and represent 30.7% of the total Italian chicory production. An appreciable amount of "radicchio" is annually exported to Austria, Switzerland, Belgium, France and the FRG. Some cultivars with large roots, i.e., Brunswick chicory, Magdeburg chicory, Brussells chicory, are grown as coffee substitutes or additive plants and for the production of the "witloof", that is a very common vegetable in Belgium and France. In recent years, experiments have been performed to investigate the possibility of using this plant as fodder (Mittal et al. 1983; Maddaloni et al. 1985) and as an energy crop (Douglas and Poll 1986).

1.2 Objectives for Improvement

As for all horticultural plants, the improvement of chicory is mainly aimed at obtaining a product better appreciated by consumers and with a total yield suited to market demand. For this reason, the major cultivated types have been obtained

[1]Department of Evolutionary and Experimental Biology, University of Bologna, Via Belmeloro 8, 40126 Bologna, Italy
[2]Department of Animal Production, University of Bari, Via Amendola 165 A, 70100 Bari, Italy

Biotechnology in Agriculture and Forestry, Vol. 11
Somaclonal Variation in Crop Improvement I (ed. by Y.P.S. Bajaj)

by farmers through the simple method of mass selection for morphological and organoleptic traits. In Italy, selection for physiological traits has produced cultivars with different production times, so that chicory is available all year round (Pimpini 1978). A second important objective for improvement is to obtain resistance to the principal diseases affecting this plant, and papers are available on resistance to *Alternaria* (Garibaldi and Tesi 1971), blotch virus (Pozdena et al. 1980), and yellow mottle virus (Piazzolla et al. 1978).

1.3 Available Genetic Variability

Cichorium intybus L. is a typically allogamous species and entomophilous pollination accounts for much of the genetic variation in this plant. Applying the Mather's model to F_1 and segregating generations (F_2 and backcrosses) of two populations of chicory, Olivieri (1972a,b) analyzed the variability of the most important agronomic traits. Additive genetic variation was found for increment of leaf number, plant diameter, and time of floral stalk development, while dominance and nonallelic interactions did not seem to be important. These results, confirmed by Olivieri and Parrini (1974), suggested that selection processes could be utilized in varietal improvement. On the other hand, Parrini (1969) and Bannerot and Corninck (1970, 1976) suggested that the total yield of this plant could be increased by the use of F_1 hybrids. The use of F_1 hybrids for improvement is correlated with the possibility of obtaining inbred lines. In chicory, some degree of self-incompatibility makes it difficult to utilize this method. Major studies have been made on this subject by Eenink (1979, 1981, 1984), in particular, on the spread of self-compatibility between and within populations, the specific analysis of the incompatibility system, and the methodology to overcome it. Selection programs and F_1 hybrids represent the tools by which genetic variation within populations can be exploited. Until some years ago, the sexual cycle and mutation represented the only way to obtain new genetic variability. Cell and tissue manipulations have revolutionized these traditional methods, allowing genetic variation irrespective of the sexual process.

2 In Vitro Culture and Somaclonal Variation

Most investigations on in vitro cultures of *Cichorium intybus* concerned the root and leaf segment culture for the study of the flowering process, since this plant is one of the few known species able to form inflorescences in vitro. Apical and axillary meristems have often been used to establish vegetative multiplication in vitro and regeneration from embryos or root callus through organogenesis or embryogenesis has been obtained. Studies on the isolation and culture of protoplasts are the most recent. For general information on various aspects of in vitro culture of chicory also see Schoofs and de Langhe (1988) and Chupeau (1988).

2.1 Regeneration of Plants from Segments and Meristems

Many experiments have been made on in vitro cultures of root explants obtaining buds and/or callus in different environmental conditions. The first studies on regeneration from root segments of *Cichorium intybus* have shown that this plant is one of the few known species able to form inflorescences in vitro. Paulet and Nitsch (1964a,b) and Nitsch (1966) have obtained normal flowers from root tissues cultured using Knop and Murashige and Skoog modified media. They observed that vernalization of the root followed by a long day was necessary for flower initiation. Moreover, they found that the addition of IAA to the culture medium prevented the formation of flower bud even at concentrations which did not inhibit the initiation and development of vegetative buds; gibberellic acid (10^{-7}, 10^{-6} M) promoted flowering. Phenolic compounds, such as p-coumaric acid (10^{-5} M) and coumarin (10^{-6} M) increased the percentage of flower buds.

Flowering was induced by Margara et al. (1966), Margara and Rancillac (1966), Buoniols and Margara (1968) on MF×2 medium supplemented with sucrose (45 g/l) and agar (12 g/l) at 24°C in continuous light. In subsequent years, Margara et al. chose *Cichorium intybus* as a model for studying in vitro floral induction of long-day biennial plants requiring vernalization. They characterized three successive phases in floral induction of neoformed buds obtained from root segment culture: the preinduction phase, the photoinduction phase, and the initiation of floral primordia (Margara and Touraud 1968; Margara 1973b, 1977).

The preinduction phase, which runs during the first culture week and is independent of the photoperiod, is characterized by the acquisition of the aptitude to flower (photoinducibility). It can be achieved at relatively high temperatures (24°C) even without vernalization (Margara 1965, 1973a). The factors which promote preinduction are principally the relative dehydration of tissues, the nature (liquid or solid) of nutrient media (Margara and Bouniols 1967, 1968; Bouniols and Margara 1971a; Bouniols 1974), and nutritional factors such as glucides, N, Z (Margara et al. 1966; Margara 1973a,b). The choice between the vegetative and the flowering state was also determined by growth regulators (Margara 1972). Bouniols and Margara (1971b) observed that the addition of proline to the medium promotes flower induction.

The photoinduction phase needs a long day and starts during the second week of culture (Margara and Touraud 1968); the application of a short photoperiod in this phase results in loss of flower aptitude (Margara 1970).

The initiation of floral primordia is independent of photoperiod. This phase begins after the second week of culture, and Rancillac (1973) observed that at culture day 14 the buds are still vegetative.

Floral induction can be reversed; reversion factors act mostly in the preinduction phase (Margara and Touraud 1968; Margara 1974, 1975), but reversion was also observed after complete photoinduction (Margara 1970).

Many authors have studied in vitro flowering regulation of root fragments of *Cichorium*, analyzing endogenous variations of phenolic compounds, free cytokinins and cytokinins in tRNA's during thermic or photoperiodic treatments (Paulet 1979). The culture method used in these experiments is basically that described by Paulet and Nitsch (1964a,b).

Analyzing the phenolic compounds during cold treatment Mialoundama and Paulet (1975a,b) saw that flowering ability was related to the conversion of chlorogenic acid into isochlorogenic acid.

Verger (1978) and Verger et al. (1978) showed that cold treatment increases free cytokinins and cytokinins in tRNA's in the roots. The addition of zeatin riboside (10^{-6} M) to the culture medium increases flowering ability in the explants (Joseph 1976).

Martin-Tanguy et al. (1984) studied the relation between phenolamides and floral induction in different culture conditions (thermic, photoperiodic, nutritional). They found that accumulation of these complexes was closely linked to in vitro flowering. In fact, the phenolamide levels were much lower under the conditions of flower inhibition than under the conditions favorable to flower induction.

Different photoperiodic treatments were analyzed by Badila et al. (1985), who observed that the period of highest sensitivity to light conditions of root explants is between the 8th and 16th days of culture (photoinductive period).

Badila and Paulet (1986) found that hydroxyciannamic esters (especially chlorogenic acid) increase during the photoinductive period, largely as a result of increased photosynthetic activity in long-day conditions.

Many reports deal with regeneration induced on excised root segments without flower induction. Vasser and Roger (1983) found that in a Heller medium supplemented with 10^{-6} M kinetin the root explants produced only adventitious buds; syntheses of nucleic acids and proteins during the bud induction were studied.

The effects of various combinations of NAA and kinetin were analyzed by Sene et al. (1983), who obtained root production on explants in Heller nutrient agar medium supplemented with NAA (10^{-6} M), kinetin (10^{-7} M) and glucose (10 g/l). Light and GA (10^{-6}, 10^{-4} M) also promoted root formation. These authors, and Vasser et al. (1986) observed that rooting capacity is influenced by the size of the initial cultured explant.

The effects of different hormonal treatments on plantlet regeneration were analyzed by Profumo et al. (1985). They demonstrated again that Heller medium supplemented with kinetin was very effective for shoot induction; NAA induced both shoot differentiation and root formation. Shoot regeneration was also achieved by Bennici (1985) using MS medium containing IBA.

Vasser and Dubois (1985) reported that mineral nutrition strongly affects the proliferation of root explants. They compared various nutrient solutions based on the formulas of Gautheret, White, Margara, Murashige and Skoog, Gamborg, and Heller; all the media contained 5×10^{-5} M NAA, 10^{-7} M kinetin and vitamins. The best proliferation was obtained in a modified MS solution with a high ionic strength characterized by the absence of ammonium salts and a high level of Ca and K ions.

Toponi (1963) made a great contribution to the in vitro culture of leaf segments by analyzing the combined effect of kinetin and IAA on the organogenesis of leaf fragments. Nitsch and Nitsch (1964) studied the flowering capacity of leaf segments cultured in vitro on a Paulet and Nitsch (1964a,b) medium. Flower buds were observed when leaves were vernalized, and flowering was greatest after 3–4 weeks of cold treatment. Flower buds were also obtained without vernalization, using leaves of plants kept for 5 weeks at 22–24°C with a photoperiod of 9 h. These results did not confirm those of Hartman (1956) and Michniewcz and Kamienska (1964).

indicating that short day and a relatively high temperature (22–24 °C) could have a vernalization effect.

Bouriquet (1966) studied the effect of light on regeneration from leaf segments. The culture technique was that described in previous works (Bouriquet and Vasseur 1966a,b); the explants were put on a half-strength Knop medium at a temperature of 22 °C. Two stocks were made: one was kept in the dark, the other in the light for 12 h. They observed that light was not necessary for the formation of buds, but darkness reduced the proliferative capacity of explants, as well as the number of buds formed. Moreover, the photoperiod affected in vitro flowering, as already observed by Paulet and Nitsch (1964b), so that the explants produced flowers only if exposed to the light for 18 h.

Proliferation and root formation were also obtained by Profumo (1967) and Profumo and Dameri (1969) from leaf fragments grown in vitro in a medium supplemented with IAA or NAA (10^{-8}, 10^{-5} M) at a temperature of 22–23 °C and a photoperiod of 12 h. The explants were taken from different leaf areas, and their distal or proximal part planted in the substratum or else laid horizontally on it; the frequency of proliferation was influenced only by the position of the fragments in relation to the substratum. Histological variations connected with the initiation and development of callus and roots were studied.

Bouriquet and Vasseur (1973) also observed that proliferation, organogenesis, and flowering are more important in the youngest tissues. Regeneration took place above all in the callus originating from the basal part of the explants, confirming the results of Vasseur (1965, 1966a,b).

The same authors (Vasseur and Bouriquet 1973, Vasseur 1979) pointed out that the formation of buds on fragments of etiolated leaf is linked to the formation of callus which cannot be suppressed without concomitant inhibition of bud formation. Maleic hydrazide reduced the number of buds formed without modifying callus development. A first phase of RNA synthesis preceded cell division and apparently regulated its extent, while a second phase, when the bud primordia arose, appeared to regulate bud formation.

Legrand (1974, 1975) observed that a dark period of 3 days at the beginning of culture otherwise conducted in continuous light stimulated bud regeneration. Conversely, a 3-day light pretreatment before culture in darkness inhibited bud regeneration. The tissues grown in continuous light showed lower peroxidase activity than those grown in darkness. Moreover, he verified that iron, chelated or not, stimulated bud formation in chicory explants, and the effect was more marked in light than in darkness.

Mix (1985) obtained regeneration from in vitro culture of leaf vein segments using a modified MS medium. Liebert and Tran Thanh Van (1972) studied the organogenetic potential of thin layers of leaf cells. The mineral medium was supplemented with coconut milk 1%, thiamine (0.4 mg/l), mesoinositol (100 mg/l), IAA (10^{-4}, 10^{-6} M), kinetin (10^{-6} M), sucrose (30 g/l). IAA at 10^{-4} M promoted a higher rate of cell division and a greater root proliferation than 10^{-6} M IAA. Buds developed quicker in a liquid than on solid medium, and regeneration was supported by replacement of kinetin with BAP (10^{-4}, 10^{-5} M); at a concentration of 10^{-4} M there was greater bud formation than at 10^{-5} M; further bud development was increased by this last concentration of BAP.

Dorchies and Rambour (1985) measured nitrate reductase activity in roots and leaves produced by *Cichorium intybus* seedlings or by in vitro grown root or leaf explants. This activity was always higher in young roots. In field grown plants the activity decreased as the root became tuberous and increased in the green leaves of the rosette. Tuberous roots or etiolated leaves were unable to reduce nitrate in vivo. When tissue explants from these organs were cut and cultured in vitro, nitrate reductase activity appeared.

Apical and axillary meristems have often been used to establish vegetative multiplication in vitro. Venturi and Rossi (1979) aseptically cut axillary meristems of different Chicory cultivars and placed them in a MS medium supplemented by citric acid (10 mg/l), NAA (0.05 mg/l), BAP (0.4 mg/l) and GA (19 mg/l). In vitro multiplication without callus production was obtained by transferring the shoots to MS medium in which the growth regulators concentration and/or combination was modified; rooting was established by placing the plantlets in a medium with NAA (0.1 mg/l), GA (0.1 mg/l).

Fortunato et al. (1983) cultured vegetative apices on a basal medium containing MS macroelements, Nitsch and Nitsch microelements, meso-inositol (10^{-1} g/l), thiamine-HCl (4×10^{-4} g/l), sucrose (20 g/l), agar (6 g/l) and growth regulators. The best growth regulator combination for the proliferation stage was 2-iP (2 mg/l), IAA (0.1 mg/l), GA_3 (0.1 mg/l); roots were promoted on a medium containing 2 mg/l IAA.

Pieragostini et al. (1981) and Grazia et al. (1985) obtained plantlets of *Cichorium* by micropropagation on a modified MS medium containing 0.3 mg/l GA, 0.01 mg/l NAA and 0.2 mg/l BAP. The medium of Venturi and Rossi (1979) was used for rooting.

Vegetative multiplication was also achieved by Theiler-Hedtrich and Badoux (1986) through in vitro culture of apical or axillary meristems. MS medium supplemented with IAA (0.5 mg/l), BAP (1 mg/l) was used; root formation was achieved by removing BAP.

2.2 Callus and Suspension Cultures

Successful attempts to initiate callusing in *Cichorium endivia* were made by Vasil et al. (1964), growing mature embryos on a D-nutrient medium (Hildebrandt et al. 1964) supplemented with coconut milk, 2,4-D, NAA; this callus tissue was regularly subcultured on a C medium (same as D medium minus 2,4-D). Differentiation was obtained on MS medium supplemented with thiamine HCl (0.1 mg/l), nicotinic acid (0.5 mg/l), pyridoxine HCl (0.5 mg/l), glycine (2 mg/l), meso-inositol (100 mg/l), IAA (10 mg/l) and kinetin (0.04 mg/l). The callus, grown on a shaker in the liquid medium, formed a fine suspension of cells and cell groups. By further cell division and growth, embryoids were formed which subsequently developed roots and leaves so that normal plantlets were formed.

These observations were in agreement with the results reported by Vasil and Hildebrandt (1966a,b), who obtained only in MS liquid medium embryoids from single cells which first developed roots and then shoots. Moreover, studying the behavior of embryo callus tissue in different agar and liquid nutrient media, they observed variation in regenerative potential: fresh and dry weight percentage, and chlorophyll content; incorporation of yeast extract or high concentration of inositol, kinetin, or casein hydrolysate improved growth and organ formation.

Gwozdz (1973) and Gwozdz et al. (1974) analyzed the effect of IAA on growth, organogenesis and RNA content of root callus. They observed that IAA induced intensive callus growth accompanied by an increase in the RNA content and simultaneous decrease in RNAase activity. Fractionation of RNA by column chromatography showed that the auxin-induced increase in RNA synthesis mainly concerned the ribosomal fraction.

Somatic embryogenesis was also induced by Heirwegh et al. (1985) using callus obtained from root explants. An MS semi-solid basal medium, supplemented with 2,4-D and IBA, and a liquid MS medium devoid of growth regulators were used respectively for induction of callus and embryoids. Plantlets were regenerated from embryoids in semisolid MS medium.

Shoot regeneration from callus obtained by root segment was achieved by Caffaro et al. (1982) and Profumo et al. (1985) on modified Heller medium supplemented with 2,4-D.

Vasseur and Sene (1984) obtained only callus proliferation when they cultured root explants on modified Heller medium containing a high concentration of IAA (5×10^{-5} M), kinetin (10^{-7} M), casein hydrolysate (1 g/l) and vitamins (100 mg/l inositol, 0.5 mg/l nicotinic acid, 0.5 mg/l pyridoxine, 0.1 mg/l thiamine, and 0.05 mg/l biotin).

2.3 Protoplasts

The first paper on isolation and culture of protoplasts in *Cichorium* was published by Binding et al. (1981). The protoplasts were isolated as already described by Binding and Nehls (1980) and the sources were shoot tip regions and leaves. Protoplasts were plated in 0.7 ml of liquid media at densities of 10^3 to 2×10^4 in Petri dishes 35 mm in diameter and were kept in a growth chamber at 26°C under continuous illumination by white fluorescent light of 1–2 klx. Plating efficiency was 60%. The suspension of regenerated cells and cell clusters was diluted by equal volumes of liquid or soft agar media at 2–8-day intervals depending on the growth rates. They were later plated into high osmotic agar media and finally transferred to low osmotic agar media.

Protoplast culture media were: V-KM (Binding and Nehls 1977) containing 6-BA (2.5 μM), NAA (5 μM) and 2,4-D (0.5 μM); V-KM soft agar (0.3% agar) and V-KM solid agar (0.8% agar). Protoplasts from the apical regions regenerated equally as well as those from leaves, but mitotic activities and browning of the cell clusters ceased soon afterward. These disadvantages could be partially overcome using shoot tips as protoplast sources, plating at high densities (2×10^4) and diluting the regenerated cell clusters every 2 days. Moreover, the callus had to be grown to a considerable size on V-KM solid agar medium about 3–4 months after isolation of the protoplasts. Protoplast-derived calluses about 2 mm in diameter were transferred into a low osmotic medium with 0.8% agar. Shoot formation was obtained in a modified NT (Nagata and Takebe 1971) medium supplemented with

6-BA (2.5 μM), NAA (10 μM); root formation requires a medium without growth regulators. Crépy et al. (1982) tried to establish whether necrosis and mitotic arrest of the small *Cichorium intybus* colonies were due to toxicity of one of the components of the culture medium, and monitored the dose effect of each component of the medium. For protoplast isolation young sterilized leaves were cut into 0.8 mm wide strips. About 500 mg of leaf material were incubated in Petri dishes with 10 ml of protoplast isolation medium prepared by mixing 9 ml of culture medium pH 5.5 and 1 ml of a concentrated filter sterilized enzyme solution containing 1% Cellulase Onozuka R-10, 0.5% Driselase, and 0.2% Macerozyme R-10. The resulting protoplast suspension was filtered through an 80 μ mesh stainless steel sieve and sedimented by centrifugation (5 min, 70 *g*) and washed twice by repelleting in 20 ml of a simple saline medium: KCl 2.5%, $CaCl_2$ 0.2% H_2O 0.2%, pH 5.5. Protoplasts were plated in 50 × 13 mm Petri dishes with 3 ml liquid media contained macroelements derived from MS medium, microelements from Heller solution, vitamins from Morel and Wetmore, NAA (3 mg/l) and 6-BA (1 mg/l). Incubation was made at 27 °C and 70–90% relative humidity in the dark for 48 h, then exposed to a 16-h photoperiod (1500 lx). Colony formation and survival (40–50%) was optimal for plating densities between 5×10^3 and 5×10^4, but very few cell colonies survived in this medium

Examination of the role of each macro-ion revealed that toxicity was due to nitrate. Using glutamine as the sole nitrogen source, plating efficiency and cell colony survival were improved. Variation of the concentration of microelements and vitamins and the type and level of sugars did not result in any significant improvement of colony growth. On the contrary, replacement of NAA by IAA (1 mg/l) and 6-BA by zeatin (1 mg/l) enhanced both plating efficiency and colony survival, so that up to 70% of the plated protoplasts divided in the first week.

Early transfer (before the second week of culture), low NAA level (0.1 mg/l) and osmotic pressure equivalent to 0.5 M mannitol were essential for development of microcalli. Under these conditions, 2% of the transferred colonies gave rise to small calli in about 1 month; this phase of protoplast culture appears once again to be a limiting step in regenerating plants. 20% of protoplast-derived calli, transplanted to agar medium with IAA (0.5 mg/l) and 6-BA (1 mg/l) produced buds which differentiated within a month. Then the low frequency of shoot formation was mainly due to necrosis of the transplanted tissues since all the growing calli differentiated.

The removal of nitrates from the medium proved most suitable for growth also for Saski et al. (1986), who observed that mesophyll protoplasts showed substained division in a medium containing mineral salts, vitamins, NAA (1 mg/l), BA (1 mg/l), sucrose (0.5%), mannitol (8%), and glutamine as the sole nitrogen source (750 mg/l). Bud induction required partial or complete substitution of glutamine by NO_3^- reduced levels of growth regulators (0.05 mg/l NAA and 0.5 mg/l BA) and no mannitol (also see Chupeau 1988).

2.4 Variation in the Regenerated Plants

Many studies have been reported concerning various culture techniques, but very few analyze somaclonal variation. In the first experiments of Grazia, Rocchetta, and

Pieragostini on clones of *Cichorium intybus* obtained through micropropagation by different maternal plants, they observed that the clones could be classified as H clones or L clones, depending on the high or low variability expressed in the first clonation by the "rooting time" trait (number of days needed for the root spring).

These two clone types also behaved differently in the following clonations. Hierarchical analysis of variance made at the fourth clonation (Table 1) shows that the variance between primary ramets within clones was significant in the H clones but not in the L clones. Moreover, the H clone plantlets were smaller, slower in rooting, and had closer internodes than the L clones (Table 2). Since changes in peroxidase activity have been observed during plant development (Van Huystel and Cairns 1982), and this variation is also evident during the rooting process (Gaspar et al. 1975), peroxidase activity was determined on plantlets of H and L clones. The results show that, as well as for morphological characters, these two types of clones are different also for this biochemical parameter. In fact, the quantity of protein and the enzymatic activity are lower in H clones: moreover, in these clones, the peroxidase activity shows a greater variability than in L clones (Table 3). These first results seemed to suggest that the variability shown in the first clonation could be utilized to indicate a particular type of intraclonal variation.

In order to investigate the origin of the somaclonal variation shown by the H clones, an experiment of two-way selection for the trait-rooting time was carried out for eight subsequent cycles of clonation (Pieragostini et al. 1981). The early line (Plus

Table 1. Hierarchical analysis of variance for the trait-rooting time on the H and L clones (Grazia et al., unpublished data)

	Source of variation	d.f.[a]	MS	F
H Clones	Between clones	5	62.01	4.95 $p<0.05$
	Between primary ramets within clones	6	12.51	3.14 $p<0.01$
	Within primary ramets	241	3.99	
L Clones	Between clones	5	26.37	13.12 $p<0.01$
	Between primary ramets within clones	6	2.009	0.94 n.s.
	Within primary ramets	158	2.127	

[a] Degrees of freedom.

Table 2. Mean values of rooting time, plantlet height, internodal distance (expressed as the ratio of the length of the highest leaf on the plantlet height) for the H and L clones and Student's t comparisons (Grazia et al., unpublished data)

	H Clones		L Clones		
	n	$\bar{x} \pm$ S.E.	n	$\bar{x} \pm$ S.E.	t
Rooting time	74	7.98 ± 0.191	62	6.95 ± 0.181	3.89 $p<0.01$
Plantlet height	74	4.47 ± 0.162	62	5.87 ± 0.197	5.47 $p<0.01$
Internodal distance	74	0.74 ± 0.018	62	0.65 ± 0.034	2.30 $p<0.05$

Table 3. Protein content and peroxidase activity in the H and L clones: comparisons between the mean values (t test) and the variance estimates (F test) (Grazia et al., unpublished data)

		Protein content (mg/ml)	Peroxidase activity (units/mg protein)
	n	21	21
H Clones	$\bar{x} \pm$ S.E.	0.43 ± 0.022	1.57 ± 0.442
	s2	0.01054	4.105
	n	20	20
L Clones	$\bar{x} \pm$ S.E.	0.58 ± 0.018	2.51 ± 0.328
	s2	0.00680	2.159
t test		5.21 $p < 0.01$	1.70 n.s.
F test		1.55 n.s.	1.90 $p < 0.10$

line) exhibited no variation either in rooting time or in variability during selection. On the contrary, the later line (Minus line) showed a larger variability when compared with the Plus line, and a progressive increase in rooting time up to the third generation, after which the divergence between the two selection lines kept quite constant (Figs. 1, 2). Single plantlets of the two selection lines showed also a different multiplication rate; in Fig. 3 it is evident that the Plus line produces a greater number of propagules than the Minus line. Moreover, the two lines became morphologically different after 3 weeks in the rooting medium; in fact, plantlets of the Minus line were smaller and had closer internodes than those of the Plus line (Figs. 4A,B). These two lines responded in different ways when transferred onto culture media with different concentrations of NAA (N1 = 0.04 mg/l, N3 = 0.07 mg/l, N5 = 0.10 mg/l, N7 = 0.13 mg/l, N9 = 0.16 mg/l). In particular, only the Minus line showed a decrease in rooting time correlated with the increased NAA concentration as well as the original population (Fig. 5). The above results suggested the presence in the micropropagated clone of a certain amount of heterogeneity exploited by selection, and emphasized by changes in culture media composition.

Grazia et al. (1985) performed an experiment where the variability of some morphological traits was analyzed on H clones but varying the length of the propagation period. Two clone types were obtained by the same genotypes; A clones (short cycle = clonation every 14 days), B clones (long cycle = clonation every 28 days). The analysis of the variability shown by average number of leaf, leaf length, plantlet height, and rooting time was made for the A and B clones after the first and the sixth cycles of clonation (Table 4).

The results obtained showed that the variance between primary ramets within the clone of the A and B clones was always significant, both at the first and the sixth clonation, yet the trend of this variance was different in A and B clones. In A clones this variance for all the traits decreased from the first to the sixth clonation, while in B clones the various characters examined exhibited neither an increase in variance nor any modifications. Moreover, after the six cycles, the B clones showed a larger phenotype and an accelerated rooting in comparison with the corresponding A clones (Table 5). Summing up the results, the most impressive datum of this

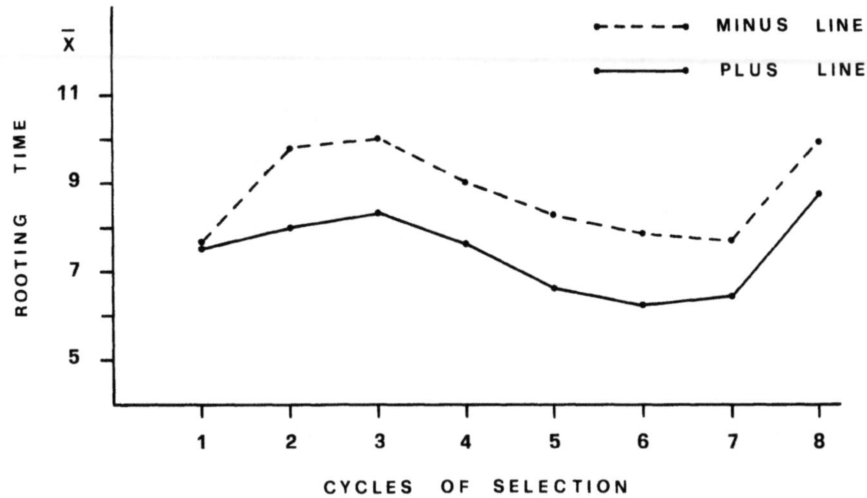

Fig. 1. Mean values of rooting time in the eight subsequent cycles of selection in the Plus and Minus lines

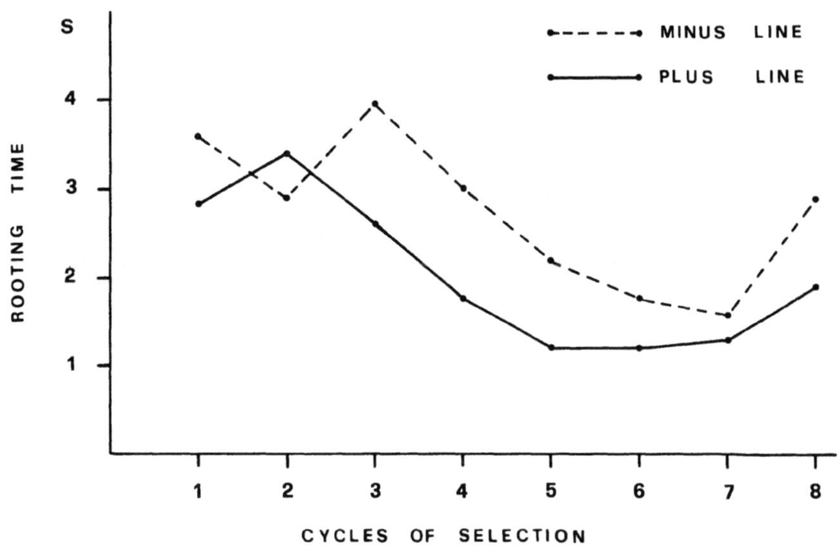

Fig. 2. Standard deviations of rooting time in the eight subsequent cycles of selection in the Plus and Minus lines

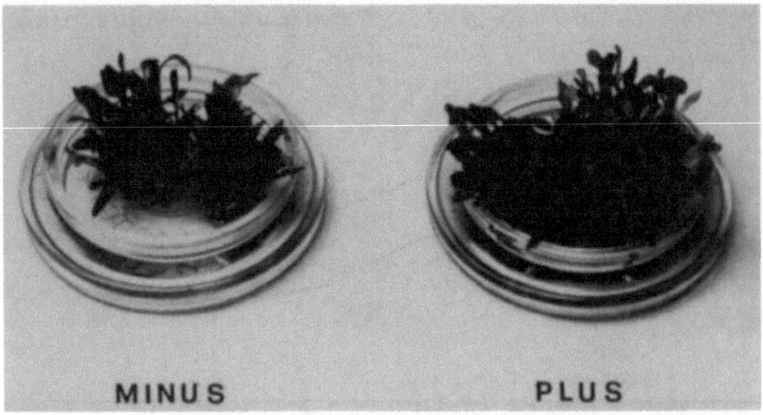

Fig. 3. Plantlets of the Plus and Minus lines after 28 days on multiplication medium (modified MS with 0.3 mg/l GA. 0.01 mg/l NAA. and 0.02 mg/l BAP)

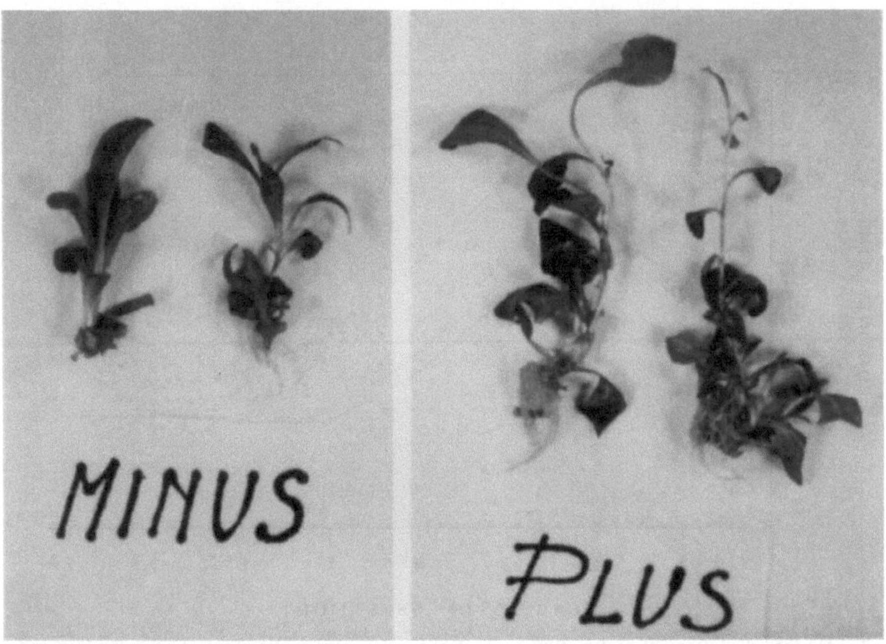

Fig. 4. A Plantlets of the Minus line after 21 days on rooting medium (modified MS with 0.1 mg/l NAA. 0.1 mg/l GA). **B** Plantlets of the Plus line after 21 days on rooting medium (modified MS with 0.1 mg/l NAA. 0.1 mg/l GA)

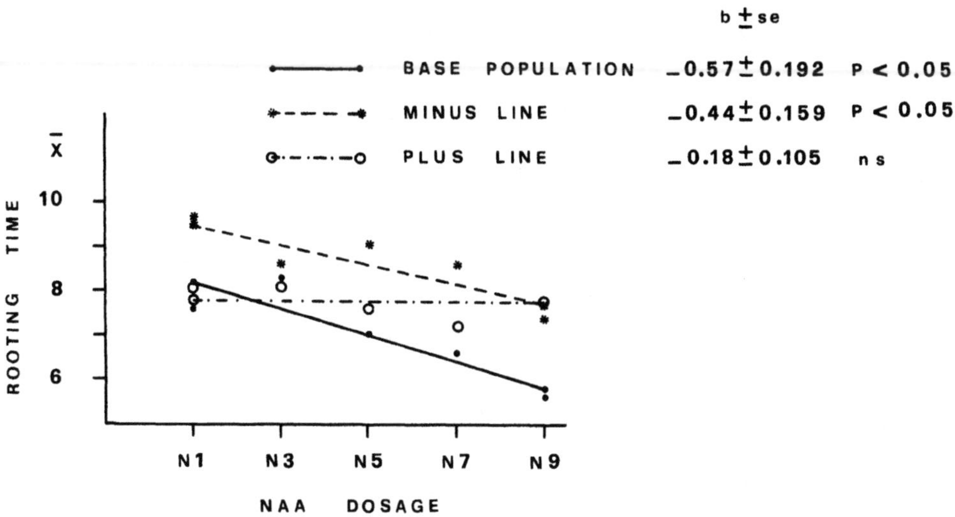

Fig. 5. Linear regression of rooting time on NAA dosage in the base population and in the Plus and Minus selected lines

Table 4. Hierarchical analysis of variance for A and B clones after the first (A_1, B_1) and sixth cycle (A_6, B_6) of cloning: s_a^2 = variance between primary ramets within clone; s_b^2 = variance within primary ramets; $d.f._a$ = degrees of freedom of s_a^2; $d.f._b$ = degrees of freedom of s_b^2 (Grazia et al. 1985)

	$d.f._a$	s_a^2	$d.f._b$	s_b^2	F
Average number of leaves per plantlet					
A_1	18	0.563	24	0.758	0.74
A_6	18	0.812	47	0.482	1.68
B_1	18	0.726	24	0.506	1.43
B_6	18	4.351	24	1.305	3.33**
Plantlet height					
A_1	18	565.064	452	181.092	3.12**
A_6	18	235.995	681	131.867	1.79**
B_1	18	406.838	448	152.598	2.66**
B_6	18	371.120	477	138.596	2.68**
Leaf length					
A_1	18	839.02	2079	127.420	6.58**
A_6	18	492.821	2976	99.705	4.94**
B_1	18	297.950	1935	91.700	3.25**
B_6	18	1077.228	2762	110.114	9.78**
Rooting time					
A_1	18	13.985	533	4.029	3.47**
A_6	18	8.813	538	4.563	1.93**
B_1	18	9.697	517	3.348	2.89**
B_6	18	8.036	517	2.195	3.66**

**$p < 0.01$.

Table 5. Mean values and SE of the four traits analyzed on the A and B clones after the first and sixth cycles of cloning. (Grazia et al. 1985)

Genotypes	Average no. of leaves per plantlet				Plantlet height (mm)				Leaf length				Rooting time (days)			
	A_1	A_6	B_1	B_6	A_1	A_6	B_1	B_6	A_1	A_6	B_1	B_6	A_1	A_6	B_1	B_6
1	5.2 ± 0.32	4.1 ± 0.22	4.7 ± 0.34	4.7 ± 0.27	28.9 ± 1.30	17.9 ± 0.76	29.4 ± 1.50	31.3 ± 1.20	18.2 ± 0.51	12.8 ± 0.32	18.6 ± 0.62	18.6 ± 0.48	7.6 ± 0.18	8.1 ± 0.22	6.4 ± 0.14	7.7 ± 0.18
2	4.5 ± 0.26	5.0 ± 0.23	4.2 ± 0.22	4.3 ± 0.29	29.7 ± 0.99	28.8 ± 0.75	26.2 ± 1.05	32.3 ± 1.02	19.4 ± 0.51	17.1 ± 0.31	15.7 ± 0.45	18.4 ± 0.44	7.2 ± 0.12	8.0 ± 0.23	6.9 ± 0.12	7.5 ± 0.10
3	3.19 ± 0.24	4.6 ± 0.24	3.7 ± 0.28	6.4 ± 1.16	38.3 ± 2.07	32.2 ± 1.28	22.9 ± 150	35.4 ± 1.22	25.8 ± 0.85	20.6 ± 0.54	15.5 ± 0.57	20.9 ± 0.45	8.6 ± 0.20	8.0 ± 0.15	8.15 ± 0.29	7.9 ± 0.14
4	4.3 ± 0.22	3.8 ± 0.24	4.0 ± 0.37	5.3 ± 0.50	35.1 ± 1.66	25.8 ± 1.08	23.2 ± 1.33	36.3 ± 1.61	22.7 ± 0.71	17.9 ± 0.48	15.4 ± 0.60	18.6 ± 0.49	7.6 ± 0.17	8.9 ± 0.25	7.4 ± 0.22	6.7 ± 0.14
5	4.2 ± 0.31	3.9 ± 0.18	4.3 ± 0.14	5.4 ± 0.30	34.9 ± 1.72	28.5 ± 1.20	30.3 ± 2.00	36.5 ± 2.00	19.6 ± 0.72	16.0 ± 0.47	15.5 ± 0.58	15.6 ± 0.51	7.5 ± 0.25	8.8 ± 0.31	7.1 ± 0.25	8.1 ± 0.19
6	4.8 ± 0.37	3.9 ± 0.21	4.0 ± 0.23	6.0 ± 0.37	38.3 ± 1.50	30.9 ± 1.15	27.1 ± 1.14	42.5 ± 1.37	25.4 ± 0.69	20.3 ± 0.52	17.3 ± 0.61	22.4 ± 0.52	7.3 ± 0.12	8.3 ± 0.22	6.8 ± 0.14	7.7 ± 017
Student's t-test for paired comparison	d.f.	t	d.f.	t	d.f.	t	d.f.	t	d.f.	t	d.f.	t	d.f.	t	d.f.	t
	5	0.39	5	2.76*	5	4.89**	5	4.32**	5	8.82**	5	2.92**	5	2.56*	5	1.48

*$p < 0.05$; **$p < 0.01$.

experiment was represented by the differential response of the same genotype to the micropropagation on varying the length of the cloning period.

In conclusion, it is our opinion that the variability expressed in first clonation by the time of rooting trait can be a suitable parameter to select undesirable phenotypes sensitive to environmental stresses such as clonation, changes in culture media composition, and length of propagation cycle.

2.5 Chromosomal Analysis

Few papers report chromosomal analysis of the regenerated plants. Caffaro et al. (1982) observed that root explants of a diploid (2n = 18) cultivar of chicory on modified Heller's medium produced callus followed by shoots; cytological analysis on 18 shoot apices showing mitoses revealed that the majority of plantlets had chromosome number mosaics, i.e., aneusomatic (Table 6). Whereas some plantlets seemed to consist exclusively or essentially of diploid cells (2n = 18), others were characterized mostly by a wide range of chromosome numbers from haploid, hypodiploid, to diploid and hyperdiploid. Tetraploid or hypertetradiploid cells were not found. Fifteen root apices, collected from several explants, showed only diploid mitoses. Cytological analyses in developing primary calli showed a high incidence of amitotic phenomena, which were confirmed by DNA cytophotometry (Profumo et al. 1985).

Table 6. Chromosome counts in apices of *Cichorium intybus* L. regenerated shoots (2n = 18) (Caffaro et al. 1982)

Shoot no.	Distribution of mitoses with chromosome numbers as grouped below				
	9(n)	10–17[a]	18(2n)	19–35[a]	36(4n)
1	–	2(11), 1(13), 4(14) 1(15)	5	–	–
2	2	6(12), 2(14), 4(15)	8	1(21)	–
3	1	1(11), 1(12), 5(14), 3(15), 3(16), 1(17)	19	3(19), 6(20), 1(21), 3(22)	–
4	–	1(15), 1(16)	3	1(19)	–
5	–	1(15), 2(16)	2	1	–
6	1	6.1(10), 1(11), 4(14)	6	1(19)	–
7	–	1.3(14)	4	–	–
8	–	–	2	–	–
9	–	2.1(14)	24	7.1(23)	–
10	–	1	8	1.1(20)	–
11	–	–	6	–	–
12	–	–	1	1(24)	–
13	–	4.1(15), 1(16)	15	1	–
14	–	4.1(12)	2	–	–
15	–	1.2(14), 2(15), 1(16)	8	1	–
16	–	1.1(11), 2(12), 1(15)	6	1(20)	–
17	6	2.5(14), 1(16)	5	1(19)	–
18	2	6.1(10)	13	–	–

[a] Aneuploid chromosome numbers which could be counted exactly are reported in parenthesis.

Fortunato et al. (1983) reported that plantlets of *Cichorium intybus* obtained by in vitro micropropagation had root apices with normal chromosome number.

3 Summary and Conclusions

In chicory, regeneration has been obtained from root, leaf, embryo, and protoplast cultures. Moreover, vegetative multiplication in vitro can be obtained by apical and axillary meristem cultures. At present, very few studies treat the application of these in vitro techniques for the resolution of important problems such as maintenance of varietal standards, incompatibility, selection for different types of resistance (diseases, herbicides, environmental stresses). Moreover, little information is available on the somaclonal variation of the regenerated plants. It is our opinion, also in respect to our experimental results, that in vitro culture techniques will have greater role to play in future plant improvement of *Cichorium* if they are exploited as an important source of new genetic variability.

Acknowledgment. We would like to thank Professor D. L. Palenzona for his support of this work and for many helpful discussions.

References

Antoniani C, Panizzon A, Schiavio G (1975) Prova di concimazione su radicchio Variegato di Clioggia. Comun Nuova Note Inf 6:3–7

Badila P, Paulet P (1986) Hydroxycinnamic acid esters and photoperiodic flower induction in vitro in *Cichorium intybus*. Physiol Plant 66(1):15–20

Badila P, Lauzac M, Paulet P (1985) The characteristics of light in floral induction in vitro of *Cichorium intybus*. The possible role of phytochrome. Physiol Plant 65(3):305–309

Bannerot H, Corninck B (1970) L'utilisation des hybrides F1: nouvelle méthode d'amélioration de la chicoriée de Bruxelles. Int. Symp Gembloux, pp 99–118

Bannerot H, Corninck B (1976) Heterosis in witloof chicory at the forcing stage. In: Proc Eucarpia Meet, Wageningen, March 1976, vol 1, pp 4–7

Bennici A (1985) In vitro clonal multiplication of chicory (*Cichorium intybus L.*). Riv Ortofluor It 69(4):235–239

Binding H, Nehls R (1977) Regeneration of isolated protoplasts to plants in *Solanum dulcamara*. Z Pflantzenphysiol 85:279–280

Binding H, Nehls R (1980) Regeneration of isolated protoplasts of *Senecio vulgaris*. Z Pflantzenphysiol 99:183

Binding H, Nehls R, Kock K, Tinger J, Mordhorst G (1981) Comparative studies of protoplast regeneration in herbaceous species of the Dicotyledoneae class. Z Pflantzenphysiol 101:119–130

Bouniols A (1974) Neoformation de bourgeons floraux in vitro à partir de fragments de racine d'endive *Cichorium intybus L.*: influence du degré d'hydratation des tissus, et ses conséquences sur la composition en acides amines. Plant Sci Lett 2:363–371

Bouniols A, Margara J (1968) Recherches expérimentales sur la néoformation de bourgeons inflorescentials ou végétatifs in vitro à partir d'explantats d'endivie. Ann Physiol Veg 10(2):69–81

Bouniols A, Margara J (1971a). Influence de la nature du milieu liquide ou gelose sur la composition en acides amines des tissus de racine d'endivie. CR Acad Sci Paris 273:1104–1107

Bouniols A, Margara J (1971b) Influence de l'adjonction d'acides amines au milieu de culture sur l'induction florale des bourgeons de *Cichorium intybus L.* néoformes in vitro. Photochem Photobiol 5:391-395

Bouriquet R (1966) Action de la lumière sur le développement des tissus de feuilles d'endive cultivés in vitro. Photochem Photobiol 5:391-395

Bouriquet R, Vasseur J (1966a) Sur la croissance et le bourgeonnement des tissus de feuilles d'endive cultivés in vitro. In: Juornees Int Etudes Phytohormones et organogenèse, Liége

Bouriquet R, Vasseur J (1966b) Action comparée de la kinétine, des bases puriques et pyrimidiques et du lait de coco sur le développement in vitro des tissus de feuilles d'endive. Bull Soc Bot N Fr 19:139-147

Buoriquet R, Vasseur J (1973) Croissance et bourgeonnement des tissues de feuilles d'endive en function de l'age et du lieu de prélèvement des explantats. Bull Soc Bot Fr 120(1/2):27-32

Caffaro L, Dameri RM, Profumo P, Bennici A (1982) Callus induction and plantlet regeneration in *Cichorium intybus L.*: 1. A cytological study. Protoplasma 111(2):107-112

Chupeau Y (1988) Regeneration of plants from chicory (*Cichorium intybus* L.) protoplasts. In: Bajaj YPS (ed) Biotechnology in agriculture and forestry, vol 8. Plant protoplasts and genetic engineering I. Springer, Berlin Heidelberg New York Tokyo, pp 206-216

Crépy L, Chupeau MC, Chupeau Y (1982) The isolation and culture of leaf protoplasts of *Cichorium intybus* and their regeneration into plants. Z Pflanzenphysiol 107(2):123-131

Dorchies V, Rambour S (1983) Activité de la nitrate reductase mesurée in vitro chez *Cichorium intybus* (var. Witloof) au cours du forcage. Physiol Veg 21(4):705-713

Douglas JA, Poll JTK (1986) A preliminary assessment of chicory (*Cichorium intybus*) as an energy crop. NZ J Exp Agric 14(2):223-225

Eenink AH (1979) Study of witloof chicory (*Cichorium intybus*) for the production of inbred lines and F1 hybrids. Zaadbelangen 33(9):260-266

Eenink AH (1981) Compatibility and incompatibility in witloof, chicory (*Cichorium intybus L.*). 2. The incompatibility system. Euphytica 30(1):77-85

Eenink AH (1984) Compatibility and incompatibility in witloof chicory (*Cichorium intybus L.*). 4. Formation of self-seeds on a self-incompatible and a moderately self-compatible genotype after double and triple pollinations. Euphytica 33(1):161-167

Fortunato IM, Dellacecca V, Mancini L (1983) First experiments on the technique of in vitro propagation for chicory cv. Catalogna (*Cichorium intybus L.*). Riv Ortofluor It 67(6):445-453

Garibaldi A, Tesi A (1971) Resistenza ad *Alternaria porri f.* sp. cichorii in Cichorium sp. e sua ereditarieta. Riv Ortoflouro It 4:350-355

Gaspar T, Penel C, Greppin H (1975) Peroxidase and isoperoxidases in relation to root and flower formation. Plant Biochem J 2:33-37

Grazia S, Rocchetta G, Pieragostini E (1985) Somatic variation in micropropagated clones of *Cichorium intybus*. Experientia 41:393-395

Gwozdz E (1973) Effect of IAA on growth, organogenesis and RNA metabolism during the development of *Cichorium intybus* root explants cultured in vitro. Acta Soc Bot Pol 42(3):493-506

Gwozdz E, Wozny A, Szweykowska A (1974) Induction by auxin of polyribosomes and granular endoplasmic reticulum in the callus tissue of *Cichorium intybus*. Biochem Physiol Pflanzen 165(1/2):82-92

Hartman TA (1956) After-effects of low temperature on leaf morphology of *Cichorium intybus L.* Proc Koningl Akad Wetensch Amsterdam 59:667-684

Heirwegh KMG, Banerjee N, Nerum K, Langhe E (1985) Somatic embryogenesis and plant regeneration in *Cichorium intybus L.* (witloof, Compositae). Plant Cell Rep 4(2):108-111

Hildebrandt AC, Wilmar JC, Johns H, Riker AJ (1964) Iron nutrition for growth and chlorophyll development of some plant tissue culture. Nature (London) 202:1235-1236

Joseph C (1976) Contribution à l'étude de la néoformation de bourgeons inflorescetiels in vitro à partir d'explantats de racine de *Cichorium intybus L.* These 3me Cycle Univ Orleans

Legrand B (1974) Influence des conditions d'éclairement sur la néoformation des bourgeons par les tissus de feuilles d'endive cultivés in vitro et sur l'activité peroxydasique de ces tissus. CR Hebd Acad Sci Ser D 278(19):2425-2428

Legrand B (1975) Action du fer et de l'EDTA sur la néoformation des bourgeons par les fragments de feuilles d'endive cultivés in vitro. CR Hebd Acad Sci Ser D 280(19):2215-2218

Liebert J, Tran Thanh Van M (1972) Progrès de la multiplication clonale intense chez les plantes cultivées: cas de l'endive. Etudes de la capacité néoformatrice des couches minces de cellules de type épidermique de *Cichorium intybus L.* CR Acad Agric Fr 58(7):472–477

Maddaloni J, Bertin OD, Josifovich J (1985) *Cichorium intybus* and meat production. Inf Tech Est Exp Reg Agric Pergamino 196:9

Margara J (1965) Sur la néoformation in vitro chez certaines lignées précoces d'endive (*Cichorium intybus L.*) en l'absence de toute vernalisation, de bourgeons inflorescentiels anormaux à partir d'explantats de racine. Bull Soc Bot Fr 112:113–118

Margara J (1970) Sur la reversibilité de l'aptitude à l'induction florale photopériodique des bourgeons de *Cichorium intybus L.* néoformés in vitro. CR Acad Sci Paris 270:1104–1107

Margara J (1972) Effets de régulateurs de croissance sur l'initiation florale des bourgeons de *Cichorium intybus L.* néoformés in vitro. In: 97th Congr Soc Sav, Nantes

Margara J (1973a) Analyse expérimentale en culture in vitro du besoin de vernalisation chez *Cichorium intybus L.* CR Acad Sci Paris 276:2373–2376

Margara J (1973b) Interaction de facteurs nutritionnels sur l'initiation florale des bourgeons de *Cichorium intybus L.* néoformés in vitro. CR Acad Sci Paris 277:2673–2676

Margara J (1974) Vernalisation in vitro des bourgeons de *Cichorium intybus L.* ayant perdu l'aptitude à l'induction florale photopériodique. CR Acad Sci Paris 278:2283–2286

Margara J (1975) Conditions de la néoformation in vitro de bourgeons inflorescentiels chez *Cichorium intybus L.*. La reversion de l'aptitude à fleurir. 100th Congr Soc Sav, Paris

Margara J (1977) Schema du developpement en culture in vitro de *Cichorium intybus L.*, exemple type de plante bisannuelle de jour long a besoin de vernalisation. Bull Soc Bot Fr 124:491–501

Margara J, Bouniols A (1967) Comparison in vitro de l'influence du milieu liquide ou gelose sur l'initiation florale chez *Cichorium intybus L.* CR Acad Sci Paris 264:1166–1168

Margara J, Bouniols A (1968) Id VI Mise en évidence de l' influence de l'hydratation des tissus. Ann Physiol Veg 10:69–81

Margara J, Rancillac M (1966) Observations préliminaires sur le rôle du milieu nutritif dans l'initiation florale des bourgeons néoformés in vitro chez *Cichorium intybus L.* CR Acad Sci Paris 263:1455–1458

Margara J, Touraud G (1968) Id VL'induction photoperiodique. Ann Physiol Veg 10:41–56

Margara J, Rancillac M, Bouniols A (1966) Recherches expérimentales sur la néoformation de bourgeons inflorescentiels ou végétatifs in vitro à partir d'explantats d'endive *Cichorium intybus L.* III. Etude critique de la méthode. Ann Physiol Veg 8:285–305

Martin-Tanguy J, Margara J, Martin C (1984) Phénolamides et induction florale de *Cichorium intybus* dans différentes conditions de culture en serre ou in vitro. Physiol Plant 61(2):257–262

Mialoundama F, Paulet P (1975a) Variations de la teneur en acides chlorogénique et isochlorogénique au cours du traitement de vernalisation des racines de *Cichorium intybus L.*, en relation avec leur aptitude à la floraison in vitro. CR Acad Sci Paris 280:1385–1387

Mialoundama F, Paulet P (1975b) Evolution des teneurs en acide chlorogénique et isochlorogénique au cours du traitement de vernalisation des racines de *Cichorium intybus*. Physiol Plant 35:39–44

Michniewicz M, Kamienska A (1964) Flower formation induced by kinetin and vitamin E treatment in cold requiring plant (*Cichorium intybus L.*) grown under non inductive conditions. Naturwissenschaften 51:473–497

Mittal GK, Barsaul CS, Maheshwari ML (1983) Use of kasni seeds (*Cichorium intybus L.*) as a concentrate for sheep. Vet Res J 6(1/2):23–26

Mix G (1985) Regeneration in vitro of plants from leaf vein segments of chicory (*Cichorium intybus L.*). Landbauforsch Volkenrode 35(2):59–62

Nagata T, Takebe I (1971) Plating of isolated tobacco mesophyll protoplasts on agar medium. Planta 99:12–20

Nitsch JP (1966) Phytohormones et genèse des bourgeons végétatifs et floraux. In: Journees Int Etudes, phytohormones et organogenèse, Liège

Nitsch JP, Nitsch C (1964) Néoformation de boutons floraux sur culture in vitro de feuilles et de racines de *Cichorium intybus L.*. Existence d'un état vernalisé en l'absence de bourgeons. Soc Bot Fr 111:299–304

Olivieri AM (1972a) Individuazione di alcuni parametri genetici in un incrocio di *Cichorium intybus L.* Riv Agron 3:171–174

Olivieri AM (1972b) Alcune considerazioni sulla frequenza di incrocio in popolazioni di Cichorium. Riv Agron 4:235–242

Olivieri AM, Parrini P (1974) Analisi di un incrocio diallelico tra quattro popolazioni di *Cichorium intybus L.*. Riv Agron 1:39–46
Parrini P (1969) Manifestazioni di eterosi nell incrocio interspecifico *Cichorium intybus L.* × *Cichorium endivia* var. latifol. um. Riv Agron 2/3:70–76
Paulet P (1979) Sur la régulation de la néoformation florale in vitro. Physiol Veg 17(3):631–641
Paulet P, Nitsch JP (1964a) Néoformation de fleurs in vitro sur des cultures de tissus de racines de *Cichorium intybus L.*. CR Acad Sci Paris 258:5952–5955
Paulet P, Nitsch JP (1964b) La néoformation de fleurs sur cultures in vitro de racines de *Cichorium intybus L.*: étude physiologique. Ann Physiol Veg INRA 6:333–345
Piazzolla P, Gallitelli D, Vovlas C, Quacquarelli A (1978) New data on chicory yellow mottle virus. Phytopathol Med 17(3):149–152
Pieragostini E, Grazia S, Rocchetta G, Palenzona DL (1981) Analisi della variabilita intraclonale in *Cichorium intybus* mediante selezione in vitro su diversi terreni di coltura. Atti SIGA 35:77–79
Pimpini F (1978) I radicchi veneti: tipi, tecniche colturali, miglioramento genetico. Inf Agric 26:2201–2222
Pozdena J, Cech M, Smrz J, Filigarova M, Branisova H (1980) Host range and some properties of chicory blotch virus. In: Tag Akad Landbauwissenschaft der DDR, pp 45–51
Profumo P (1967) Osservazioni sulla coltura in vitro di foglie e frammenti fogliari di *Cichorium intybus L.* Gior Bot It 101:311
Profumo P, Dameri RM (1969) Proliferazione e rizogenesi in frammenti fogliari di *Cichorium intybus L.*: osservazioni istologiche. Gior Bot It 103:301–324
Profumo P, Gastaldo P, Caffaro L, Dameri RM, Michelozzi GR, Bennici A (1985) Callus induction and plantlet regeneration in *Cichorium intybus L.*: II. Effect of different hormonal treatments. Protoplasma 126(3):215–220
Rancillac M (1973) Néoformation de bourgeons et induction florale en culture de tissus: étude histophysiologique. Thèse Doct, Univ Paris
Saski N, Dubois J, Millecamps JL, Vasseur J (1986) Régéneration de plantes de chicorée Witloof cv. Zoom à partir de protoplastes: influence de la nutrition glucidique et azotée. CR Acad Sci Paris Ser 3 302(5):165–170
Schoofs J, De Langhe E (1988) Chicory (*Cichorium intybus* L.). In: Bajaj YPS (ed) Biotechnology in agriculture and forestry, vol 6. Crops II. Springer Berlin Heidelberg New York Tokyo, pp 294–321
Sene A, Vasseur J, Lefebvre R (1983) Sur l'aptitude de petits explantats racinaires de *Cichorium intybus L.* (var. Witloff) à produire des racines adventives en culture in vitro. CR Acad Sci Paris Ser 3 297:81–86
Theiler-Hedtrich R, Badoux S (1986) Using in vitro multiplication in a program for breeding red bitter chicory (*Cichorium intybus L.*). Rev Suis Vit Arbor Hortic 18(5):271–275
Toponi L (1963) Sur la culture de feuilles d'endive (*Cichorium intybus L.*). CR Acad Sci Paris 257:3212–3215
Van Huystel RB, Cairns WL (1982) Progress and prospects in the use of peroxidase to study cell development. Phytochemistry 21:1843–1847
Vasil IK, Hildebrandt AC (1966a) Growth and chlorophyll formation in plant callus tissue grown in vitro. Planta 68:69–82
Vasil IK, Hildebrandt AC (1966b) Variation of morphogenetic behaviour in plant tissue cultures. I. *Cichorium endivia*. Am J Bot 53:860–869
Vasil IK, Hildebrandt AC, Riker AJ (1964) Endive plantlets from freely suspended cells and cell groups grown in vitro. Science 146:76–77
Vasseur J (1965) Sur les conditions de culture in vitro des tissus de feuilles d'endive. Bull Soc Bot N Fr 18:205–212
Vasseur J (1966a) Action de quelques facteurs de croissance sur le développement in vitro des tissus de feuilles d'endive. Bull Soc Bot N Fr 19:37–43
Vasseur J (1966b) Sur la polarité des tissus de feuilles d'endive cultivés in vitro. Bull Soc Bot N Fr 19:188–194
Vasseur J (1979) Action de l'acide indolyl-acétique, de la kinétine et de l'hydrazide maléique sur la néoformation des bourgeons et la synthèse d'ARN observées au cours de la culture in vitro de fragments de feuilles étiolées d'endive. CR Hebd Sci Acad Sci 289(2):93–96

Vasseur J, Bouriquet R (1973) Interaction de l'hydrazide maléique et des bases pyrimidiques ou puriques sur le bourgeonnement in vitro de fragments de feuilles d'endive. CR Hebd Sci Acad Sci 277(7):641–644
Vasseur J, Dubois J (1985) Influence de l'alimentation minérale sur la prolifération de petits explantats racinaires de *Cichorium intybus L.* (var. Witloof) cultivés in vitro. CR Acad Sci Paris Ser 3 300(20):717–720
Vasseur J, Roger V (1983) Synthèses d'acides nucléiques et de protéines au cours de l'initiation de bourgeons adventifs sur des explantats de *Cichorium intybus* cultivés in vitro. Physiol Plant 57:485–491
Vasseur J, Sene A (1984) Influence de quelques régulateurs de croissance sur la prolifération de petits explantats racinaires de *Cichorium intybus L.* (var. Witloof) cultivés in vitro. CR Acad Sci Paris Ser 3 298:371–374
Vasseur J, Lefebvre R, Backoula E (1986) Sur la variabilité de la capacité rhizogène d'explantats racinaires de *Cichorium intybus* (var. Witloof) cultivés in vitro: influence de la dimension des explantats initiaux et de la durée de conservation des racines au froid. Can J Bot 64(1):242–246
Vavilov NI (1951) The origin, variation, immunity and breeding of cultivated plants. Chron Bot 13:1–364
Venturi V, Rossi V (1979) Comportamento in vitro di alcune varieta di cicoria. Atti Incontro su "tecniche di culture in vitro". Pistoia 59–64
Verger A (1978) Variation du taux de cytokinines des t-RNAs de la racine de *Cichorium intybus L.* au cours du traitement par le froid. Thèse 3me Cycle, Univ Orleans
Verger A, Joseph C, Paulet P (1978) Cytokinins in t-RNAs from *Cichorium intybus L.* root: isolation and changes during flower induction. Fed Eur Soc Plant Physiol Inaugural Meet, Edinburgh Abstr, pp 548–549

III.7 Somaclonal Variation in Strawberry

K. NIEMIROWICZ-SZCZYTT[1]

1 Introduction

1.1 Importance and Distribution

The genus *Fragaria* can be divided into about a dozen species which can be grouped into four classes with respect to their ploidy level, taking seven chromosomes as the basic number (Staudt 1953; Darrow 1966; Fadeeva 1975). The first group comprises diploid species (2n = 2x = 14), among which *F. vesca* L. is the most widely distributed and appears throughout Europe, North Asia, and America, as well as in northern Africa. Basically, there are two cultivated strawberry species, namely, *Fragaria* × *ananassa* Duch. and *F. vesca* var. *semperflorens*; the latter is grown by home gardeners only, since it produces minute and fragile fruits difficult to transport.

A species characteristic for the second ploidy class is tetraploid (2n = 4x = 28) *F. orientalis* Los., which is native in Western Siberia, Mongolia, Manchuria, and Korea. It is a wild species which has not been used for creating new varieties so far. The third class includes hexaploids and is represented by *F. moschata* Duch. (2n = 6x = 42) which occurs in Northern and Central Europe, and a few of its varieties with perfect flowers and musk aroma are cultivated in the USSR (Darrow 1966). The modern octoploid strawberry *F.* × *ananassa* Duch. (2n = 8x = 56) originated from two octoploid *Fragaria* species *F. chiloensis* (L.) Duch. and *F. virginiana* Duch. During the last 50 years it has become an important world crop. *F.* × *ananassa* does not occur as a wild species, but sometimes it is found escaped from cultivation. This species is easily adapted to various climatic conditions, because a great diversity of varieties has been created. It is possible to distinguish three main types of varieties: the June-bearing single-crop type which needs short days and cool temperatures for flower formation, the ever-bearing type which can produce flowers during either long or short periods of daylight, and the day-neutral one, which can be programmed to produce fruit approximately 3 months after planting (Bringhurst and Voth 1980). In hot climatic conditions strawberry is cultivated as an annual crop and in temperate climate as a perennial for up to 3 years. At present, all modern varieties are characterized by perfect flowers, but it is still possible to find some old clones with

[1]Department of Genetics and Horticultural Plant Breeding, ul. Nowoursynowska 166, 02-766 Warsaw, Poland

Biotechnology in Agriculture and Forestry, Vol. 11
Somaclonal Variation in Crop Improvement I (ed. by Y.P.S. Bajaj)

Table 1. Strawberry production (FAO 1985)

Producers	Production (1000t)
World total	1948.6
Continents	
Africa	3.9
N.C. America	554.8
South America	15.2
Asia	293.1
Europe	953.2
Oceania	8.3
Australia	3.1
Countries	
USA	462.1
Poland	212.0
Japan	201.8
Italy	160.7
USSR	120.0
France	92.0
Spain	85.0
Mexico	58.9
UK	51.0

female flowers. Female plants are highly productive, but they require pollinators, which makes their cultivation rather difficult.

Strawberry production in different parts of the world and in selected countries obtaining the best results is shown in Table 1.

1.2 Objectives for Improvement

Due to its hybrid nature and high ploidy level, allooctoploid strawberry offers a complex subject for investigation. Among the objectives for studies it is possible to distinguish theoretical ones, concerning genetic and systematic investigations, and applied breeding problems. The first group includes germplasm resources and preservation, long-distance crosses, and better understanding of the relationships among species of *Fragaria*, polyploid genetics (sexual and somatic polyploidization, haploidization, detection of ploidy differences), and gene mutations. The main objective in the second group is breeding cultivars adapted for regional environmental conditions and resistant to diseases. Yield, vigor, and fruiting habit are the most important traits to be considered in breeding programs. They mostly depend on the country, time of ripening, winter hardiness, high-temperature tolerance, length of rest period, concentrated ripening, and adaptability to machine harvesting. Favorable fruit characters are also included in these objectives (Scott and Lawrence 1975).

Up to now two principal breeding methods have been used for cultivar improvement, namely, outcrossing and inbreeding. The first method is more often applied, and consists in the crossing of plants derived from clones or seedlings. Interspecific crossing can also be a source of transferring desirable characters such as winter hardiness, resistance to red stele, root rot and virus diseases, and day-neutral fruiting traits.

1.3 Available Genetic Variability

As has already been mentioned, *F.* × *ananassa* is an allooctoploid species, whose genomic designation is AAA'A'BBBB (Senanayake and Bringhurst 1967). On the basis of the above, it is possible to say that octoploid strawberries have mixed polyploidy (Sanford 1983). Due to high heterozygosity and the possibility of distant crosses with wild species, the variability of traits has been considerable. However, there is still need for new desirable traits and therefore to widen the existing range of variability. Such required characters are among others:

- concentrated ripening connected with fruit firmness, which would facilitate the mechanical harvest of strawberries;
- resistance to diseases, among others to variegation (June yellows);
- homozygosity of breeding material, which can partly be obtained by anther culture;
- high multiplication rate in meristem culture determined by proper reaction to growth substances;
- low probability of producing off-type plants in tissue culture;
- meristem culture fitness for long storage at low temperatures;
- ability to sustain high concentrations of various chemical substances, such as antibiotics.

More intensive studies on the fusion of *Fragaria* species protoplasts might bring about the possibility of plant regeneration and in consequence, ploidy manipulation. It would be of great interest to study the role of a single genome in high polyploidy *Fragaria* species.

2 Somaclonal Variation

2.1 Brief Review of in Vitro Studies

In vitro studies on strawberry have developed during the last two decades. After Boxus (1974) determined the proper concentration of benzylamino purine (BAP) necessary to obtain a great number of shoots from meristem tips, it became possible to multiply plants on an industrial scale. Earlier investigations aimed at the formation of fruits by cultivating pollinated flowers (Nitsch 1950; Bajaj and Collins 1968) and the induction of healthy virus-free plants from meristems. The full literature review on this subject is given by Boxus et al. (1977) and Jungnickel (1988). Further works dealt with various aspects of micropropagation, since this technique has been applied in many countries.

2.2 In Vitro Regeneration of Plants

2.2.1 Meristems, Embryos, and Segments

Investigations on meristem culture from virus-infected plants gave good results (Belkengren and Miller 1962; Vine 1968; Nishi and Ohsawa 1973; Boxus 1974; Popov 1974; Mullin et al. 1974; McGrew 1980; Sobczykiewics 1980; Swartz et al. 1981; Converse and Tanne 1984; Kondakova and Kacharmazov 1986). Similar studies are being continued in order to show that strawberry plants infected by other diseases, e.g., *Phytophthora fragariae* can be freed from pathogens by this method (Seemuller and Merkle 1984).

A series of articles was also devoted to the influence of different macro- and microelements and growth substances, especially BA, on shoot formation and plant development in meristem culture (Damiano et al. 1979; Waithaka et al. 1980; Maliarcikova 1981; Kondakova et al. 1983; Hunter et al. 1984; Marcotrigiano et al. 1984; Atkinson et al. 1986). As a result, it was found that response to medium composition and BA concentration depended on the variety. In some cases, high concentration of BA and multiple subcultures brought about the regeneration of off-type plants (Swartz et al. 1981; Anderson et al. 1982; Kondakova et al. 1983). Tissue culture appeared to be particularly favorable for ever-bearing varieties, because they produce a small number of runners. In vitro-regenerated Tribute plants were characterized by fewer flower clusters and produced more runners (Scott et al. 1985). Other factors determining maximum growth and development, such as temperature and light intensity, were also studied, and it was found that for var. Cambridge Favourite the optimum temperature was 28°C and light intensity of from 4000 lx for explant establishment and growth to 6000 lx for maximum propagule growth, and 6000–7000 lx for shoot development and root initiation (Hunter et al. 1983). At present, some investigations are being conducted to compare plants obtained from meristem culture or their daughter plants from runners with traditionally propagated plants (Swartz et al. 1981; Damiano et al. 1982, 1983; Maliarcikova and Okossyova 1984; Thuesen 1984; Cameron et al. 1985; Neumann and Seipp 1986; Navatel et al. 1986; Cameron and Hancock 1986). These studies confirm that June-bearing varieties micropropagated in vitro needed some time and low temperature for further typical development.

Such plants usually have smaller fruits with a longer ripening period, but they produce more runners, which is important for nurseries. Vegetative progeny of micropropagated strawberry plants show increased vigor and productivity.

Strawberry can also be regenerated from meristem tips previously frozen to -196°C (Sakai et al. 1978; Kartha et al. 1980).

Shoot meristems of *F. vesca* were also cultured in order to make them free from viruses (Belkengren and Miller 1962) and to induce a considerable number of identical individuals for experimental purposes (Niemirowicz-Szczytt and Malepszy 1980). *F. vesca* meristems were irradiated in vitro with γ rays (Popov and Ravakin 1977) and treated with colchicine (Niemirowicz-Szczytt et al. 1984). The mutagenic effect of γ irradiation brought about the development of somatic and morphological mutants. About 50% of the colchicine-treated meristems which regenerated plants were tetraploids, and the rest varied in their ploidy level.

Immature zygotic embryo culture was conducted to induce somatic embryogenesis (Wang et al. 1984). Embryogenic tissue initiation was highest in the presence of 5 mg/l 2,4-D, 0.5 mg/l BA and 500 mg/l casein hydrolysate. Only a few somatic embryos continued to develop and formed more advanced structures. It is worthwhile mentioning that the embryogenic tissue which turned into soft callus finally lost its morphogenic potential after two subcultures on a medium containing 2,4-D and BA. It appeared possible to regenerate plantlets from mature embryos by transferring them to a medium containing 0.5 mg/l of BA and 0.1 mg/l of NAA. Using this medium, Ilieva and Zagorska (1986) obtained somatic embryos and regenerants from embryogenic callus. Germinating seeds also proliferated quickly when placed on Boxus medium (1974) and formed many additional shoots. The results obtained by Izsak and Izhar (1983) revealed no variation among plants originating from a single seed.

Regeneration of shoots from leaf explants was reported by Niemirowicz-Szczytt and Malepszy (1980) and petiole culture by Lee and Campbell (1982). This method can be useful for the micropropagation of ever-bearing varieties. In explant culture new shoots regenerated in wounded margins only and the best results, e.g., six to eleven plants per explant were obtained on LS medium (Linsmaier and Skoog 1965) enriched with 0.5 mg/l BA, 0.2 mg/l NAA and 1000 mg/l of meso-inositol. Lee and Campbell divided petiole into proximal and distal segments and cultured them on MS medium (Murashige and Skoog 1962) containing 2 mg/l BA or 0.02 mg/l 2,4-D, or their combination. Of the proximal segments 55% differentiated shoots on a medium containing BA and 2,4-D.

Recently, first experiments in the inoculation of strawberry vegetative parts and various tissues with *Agrobacterium tumefaciens* have been reported (Jelankovic et al. 1986). Actively growing inoculated runner segments developed large callus, while non-inoculated ones showed much weaker reaction. The test of callus for the presence of opines was positive. It is supposed that at least some of the callus cells were transformed by the Ti plasmid. There was no information about regenerated plants.

2.2.2 Callus Culture

As a result of cultivating flower pedicels, the receptacle, or tissue discs taken from the receptacle, Bajaj and Collins (1968) were able to induce callus on medium with the addition of kinetin and auxin. Nishi and Ohsawa (1973) obtained callus using meristem tips as explants. Callus capable of plant regeneration was induced only when the medium was enriched with 10^{-5} or 10^{-6} M IAA or NAA in the presence of 10^{-5} M BAP, while the addition of 2,4-D inhibited plant formation. Callus was also obtained in meristem culture on different media supplemented with 1 mg/l NAA (Waithaka et al. 1980) or in leaf explant culture when the medium contained 0.5 mg/l of BAP combined with 0.5 mg/l 2,4-D and possibly 1 mg/l IAA (Niemirowicz-Szczytt and Malepszy 1980). It is also possible to develop callus in anther or pistil culture, as will be discussed in the next section.

2.2.3 Anther Culture

In anther culture, plant regeneration also occurred through the callus stage by applying growth substances at different concentrations (Table 2).

Callus obtained from anther culture (Fig. 1) was then transferred to regenerating media enriched with: 1 mg/l BAP (Niemirowicz-Szczytt and Zakrzewska 1980; Barrientos-Pérez and Murillo 1986), 2.2 mg/l BAP (Niizeki and Fukui 1983) or 3.4 mg/l BAP and 2 mg/l kinetin (Laneri and Damiano 1980). The number of regenerated shoots was variable (Fig. 2).

In order to reduce the chromosome number Niizeki and Fukui (1983) added PFP (P-fluorophenylalanine) to the medium, increasing its concentration at the rate of 10 mg/l, so that it finally reached 85 mg/l. The best development of callus was observed at 5 and 15 mg/l, and above 55 mg/l an apparent reduction of growth occurred.

2.2.4 Cells and Protoplasts

The method of cell and protoplast isolation and culture is still being elaborated for strawberry. Sarwar (1984) obtained mesophyll cells from fully expanded healthy leaves by hand homogenizer. One g of leaves yielded 10^7 cells. Only the agar plating technique gave some positive results, but the cell cluster percentage was below 10%. Viable strawberry protoplasts were obtained by Oosawa and Takayangai (1984) with the use of the combination 0.1% Pectolyase Y-23, 1% Driselase and 2% Cellulose Onozuka R.S. in 0.6 M of Mannitol solution at pH 5.8. There was no evidence of cell division on any of the media applied.

Table 2. Composition of growth substances applied to media for anther callus induction

Medium	Growth substances in mg/l					Reference
	kin	BA	IAA	NAA	2,4-D	
Linsmaier and Skoog 1965	–	2.2	1.7	or 1.8	–	Nishi and Oosawa (1973)
LS	0.2	–	2.0	–	0.4	Rosati et al. (1975)
Gresshoff and Doy 1972	5.0 or 1.0	–	–	2.0	–	Laneri and Damiano (1980)
LS	–	0.1	2.0	–	0.4	Niemirowicz-Szczytt and Zakrzewska (1980)
LS	–	2.2	–	0.2	–	Niizeki and Fukui (1983)
GD 1972	–	1.0	0.1	or 0.1	–	Velchev and Milanov 1984)
Murashige and Skoog 1962	–	–	–	–	1.0	Barrientos-Pérez and Murillo (1986)

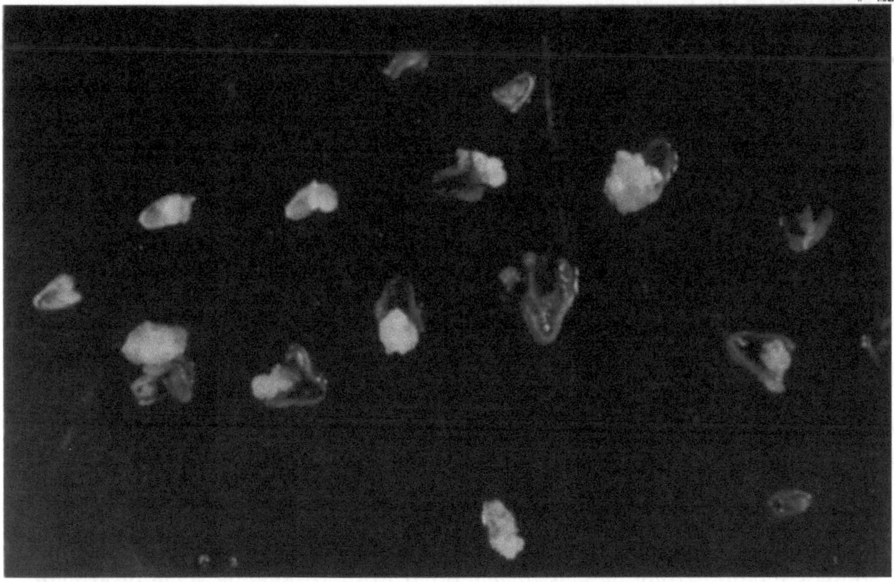

Fig. 1. Initiation of callus from different parts of anthers: second week of anther culture

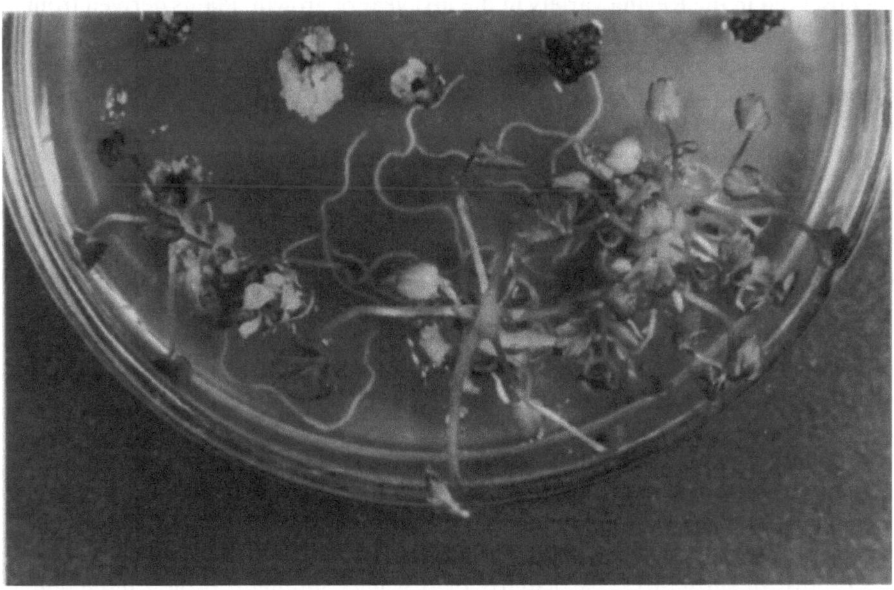

Fig. 2. Plantlets differentiated from anther callus; some degenerating calli observable

2.3 Variation in the Regenerated Plants

Some of the plants regenerated in vitro from meristems are off-types, which is determined by medium composition, time of culture, and number of subcultures (Shaeffer et al. 1980). Some off-type characters of plants disappear after 1–2 years' growth under typical cultivation conditions (Damiano et al. 1979; Damiano 1980), while others are permanent. One of the first research studies on phenotypic stability (Swartz et al. 1981) reported three off-type variants in meristem culture, namely: white-streaked leaf chlorosis, an irregular sectorial yellow chlorosis, and dwarf-type plants. The dwarf plants were characterized by reduced leaf and petiole sizes, lack of runnering, lower yields, and smaller size of fruits. Off-type plants were found in Earliglow and Redchief varieties. Other tissue culture discrete variants included compact trusses as well as runnerless and female sterile plants.

In other papers (Anderson et al. 1982) multi-apexing (bud-off) plants were noted. The multi-apexing persisted in the main shoot during 2 years' observation. Variants similar to those mentioned above such as chlorosis, fasciated flower bearing, and female sterility were detected in those varieties (Kondakova et al. 1983; Kondakova and Kacharmazov 1986). Anther callus-regenerated plants also showed some morphological changes, e.g., varied size of plants (Fig. 5). Some individuals having strong defects degenerated in their juvenile stage of growth. Numerous teratological changes such as tuft leaves, plantlets capable of rooting at the tip of the fruit (Fig. 3), seeds on the fruit germinating to seedlings, fasciate and pinnate runner fruits (Fig. 4), and inflorescences (Niemirowicz-Szczytt et al. 1983) were observed in the Regina variety in the first year of growth. Plants derived from anther callus gave rather low yields as compared with their outset varieties (Table 3). No plants even reached the lowest yielding standard variety level. It is evident from the table that as many as 91% and 93% of the plants derived from Regina and Red Gauntlet respectively yielded less than the Ostara standard variety.

On the basis of the data previously presented by the author (Niemirowicz-Szczytt 1987) concerning fruit- and seed-bearing, it is possible to state that out of 85 tested plants 93% bore fruits, but among them about 16% produced inviable seeds after open pollination. About 50% of these fruiting plants showed very low fertility, and they can practically be considered as nonfertile. Niizeki and Fukui (1983) reported that plants regenerated from anther callus did not show considerable differences in leaf shape, flowering time, or type of plants. Only those plants which were characterized by a very high chromosome number ($2n = 98$) differed from the original variety or from regenerated plants. They had smaller leaves and petioles, but their flowers became larger.

In summary, it is worthwhile mentioning that plants regenerated both from meristems and from anther callus can undergo morphological changes. Some of the changes are durable, while others disappear under typical growing conditions. The percentage of untypical plants is relatively low. High unbalanced concentrations of growth substances, particularly of BA, may stimulate variant induction. Anther callus-regenerated plants, although morphologically unchanged, can differ from their outset varieties in yield per plant and seed viability but also in cytological traits.

Fig. 3. Two plantlets on the top of a fruit of Regina tetrahaploid as a result of teratological changes (photo T. Plodowski)

Fig. 4. Fasciate fruits of a tetrahaploid strawberry. This type of change frequently occurs in plants regenerated from anther callus (photo P. Gajewicz)

Table 3. Yield characteristics of plants developed in anther culture in comparison with outset varieties. (Niemirowicz-Szczytt 1987)

Outset variety	No. of plants	Fruit yield per plant (g)		No. of cropping plants	
		Average	Range	Below 110 g	Percent
1	2	3	4	5	6
Red Gauntlet	31	52	3–121	29	93
Regina	41	37	2–137	39	91
Climax	7	35	16–94	7	100
Ostara	2	103	90–116	1	50
		$\bar{x}$ = 57			$\bar{x}$ = 91.5
Total	81			76	
1	2	3	4	5	6
Standard (outset variety)					
Red Gauntlet		450	390–470		
Regina		360	320–380		
Climax		250	220–275		
Ostara		220	205–257		

2.4 Cytological Analysis

Cytological investigations have mainly concerned anther callus-regenerated plants and only occasionally meristem-regenerated ones (Niemirowicz-Szczytt et al. 1983; Niizeki and Fukui 1983; Fukui and Niizeki 1983; Niemirowicz-Szczytt 1987). Anther callus used for plant regeneration in the above experiments was developed on two media (see Table 2). The determination of chromosome numbers gave the following results: 15 tetrahaploids (Fig. 6), 1 hexaploid, and 9 octoploids (Niemirowicz-Szczytt and Zakrzewska 1980); 67 tetrahaploids, 1 hexaploid, and 17 mixoploids (Fig. 5) or aneuploids (Niemirowicz-Szczytt 1987); 12 heptaploids, 4 octoploids, and 6 mixoploids or others (Niizeki and Fukui 1983). The last group of plants regenerated from callus developed on a medium with low PFP concentration. Unexpected results were obtained by Niizeki and Fukui (1983), who among meristem callus-regenerated plants detected 3 heptaploids, 14 octoploids, and 2 other ploidy individuals in control combination (without PFP). When PFP concentration was as high as 75 mg/l, the number of heptaploidal plants reached ten. At the concentration of 15 mg/l, two plants of $2n = 98$ were determined. Thus Niizeki and Fukui concluded that the induction of plants with reduced chromosome numbers seemed to be independent of the concentration of PFP. Mïcroscope observations of stomata indicated that their sizes ranged in the sequence of $2n = 98 > 2n = 56 > 2n = 49$. (Niizeki and Fukui 1983; Fukui and Niizeki 1983).

Pollen were analyzed with respect to the ploidy level of plants. (Niemirowicz-Szczytt 1987; Table 4). On the basis of pollen grain diameter measurements, a group of plants characterized by homogeneously small pollen (17–20 μm) was

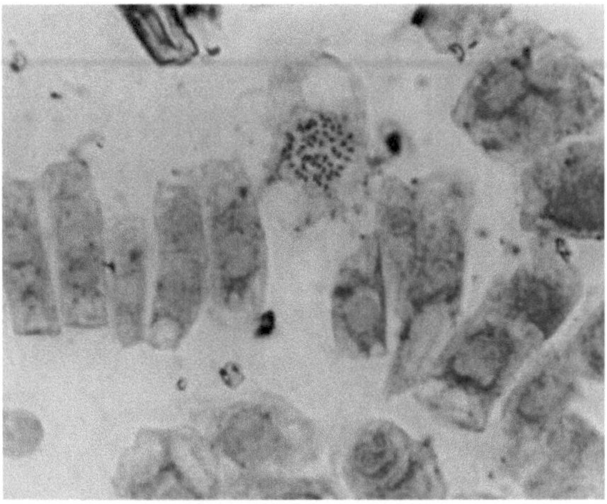

Fig. 5. Chromosome number reduced to 49; lack of one genome. Such a chromosome number is sometimes observed in cells of mixoploids

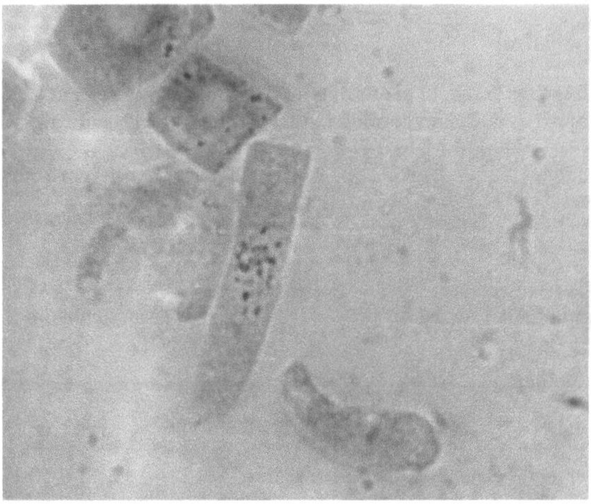

Fig. 6. Cell showing a reduced chromosome number 2n = 28, typical of tetrahaploids

distinguished (Fig. 7). Such pollen grains are typical of *Fragaria* species showing lower ploidy levels. The two other groups determined were: one characterized by large pollen grains (22.5–25 μm), typical of octoploid strawberry, and the other with diverse pollen grain sizes: large and small. Tetrahaploid plants produced different sized pollen grains but also homogeneously small ones (which are presumed to be nonfertile) and homogeneously large pollen grains, typical of octoploids.

On the basis of the experiments discussed above, it is possible to conclude that anther callus can be a source of induction of tetrahaploids and other ploidy plants, e.g., heptaploids, mixoploids, and aneuploids. However, in some experiments, no differences in chromosome number were revealed (Rosati et al. 1975; Laneri and Damiano 1980; Velchev and Milanov 1984). It is supposed that an addition of PFP or BA to a medium is responsible for changes in chromosome number.

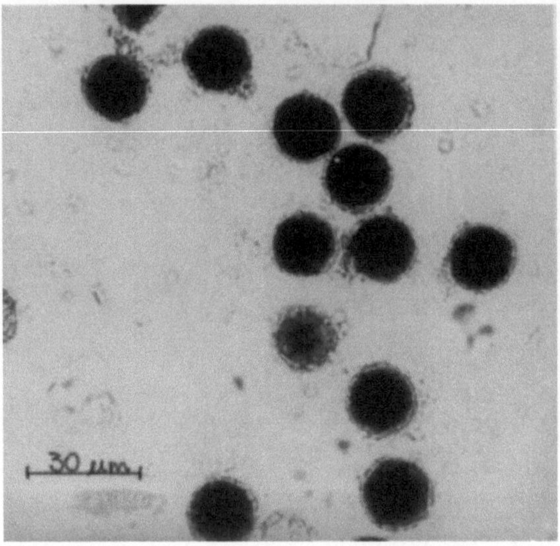

Fig. 7. Small regular pollen grains typical of lower ploidy species

Table 4. Number of plants obtained in strawberry anther culture classified in size and pollen stainability groups[a] as well as according to their ploidy level (Niemirowicz-Szczytt 1987)

Ploidy level	No. of plants	No. of plants with different kinds of pollen								
		Small			Mixed			Large		
		A	B	Total	A	B	Total	A	B	Total
Tetraploids	64	11	21	32	8	21	29	1	2	3
Mixoploids and others	17	6	1	7	4	4	8	2	–	2
Total	81	17	22	39	12	25	37	3	2	5

[a] Percentage of stainable pollen grains: A – 0.1–50.0; B – 50.1–100

2.5 Utilization of Somaclonal Plants

Variants regenerated in standard meristem culture, as described in Section 2.3, have always been treated as off-type plants. In the publications available there have been no examples of experiments to obtain the next generation by in vitro multiplication or by seed. Further investigations concerned only plants regenerated from anther or meristem callus. The next generation was obtained by self- and open pollination of such plants (Niemirowicz-Szczytt 1987). In the work quoted above, the following problems were taken into consideration: the possibility of sexual reproduction on a tetrahaploid level, the development of unreduced gametes by tetrahaploids and their return to an octoploid level, increasing or decreasing the range of variability in the generative progeny of tetrahaploids, and the possibility of utilizing the variability of these plants in breeding practice. The progeny obtained was tested for the following traits: pollen size, stainability and germination, meiotic process, chromosome number in mitotic divisions, seed germination, and fruit yield per plant. It was found that tetrahaploids could develop unreduced gametes, and thus octoploids and other ploidy plants with more than four genomes could be obtained. In the progeny of tetrahaploids a considerable variability of pollen grain stainability and diameter was observed. Some plants produced giant pollen grains, (30–38 μm), not detected before in outset varieties or tetrahaploid plants (Fig. 8). Pollen grains variable in size and stainability developed as a result of disturbances of particular phases of meiosis (Figs. 9–12). The yield of fruits per plant in the tested offspring of tetrahaploids varied, and the average yield of plants developed after selfing was usually lower than that of plants developed after open pollination. It is possible to propose a strawberry breeding program with the application of haploids. In the first stage of such a program, anther culture of several valuable varieties of strawberry would be a source of haploid development. In the second stage, it would be necessary to obtain progeny of tetrahaploids by various pollination methods, especially cross-pollination. The third stage would be the evaluation of the progeny obtained. Then the best individuals could be used for comparative clonal experiments (stage 4). Additionally, on the basis of the best progeny, it would be possible to select the parental forms with the highest combined abilities.

Plants of changed chromosome number can be subjected to further in vitro experiments (Fukui and Niizeki 1983). As a result of repeated callus induction from meristems of heptaploid plants on a medium containing 30 mg/l of PFP, 300 plants were regenerated, among which 7 were hexaploids, 85 heptaploids, 18 octoploids, and the rest belonged to other ploidy groups. These plants showed a greater range of variability (in pollen size, fruit shape, and stomata size) than Hōkōwase, the outset variety.

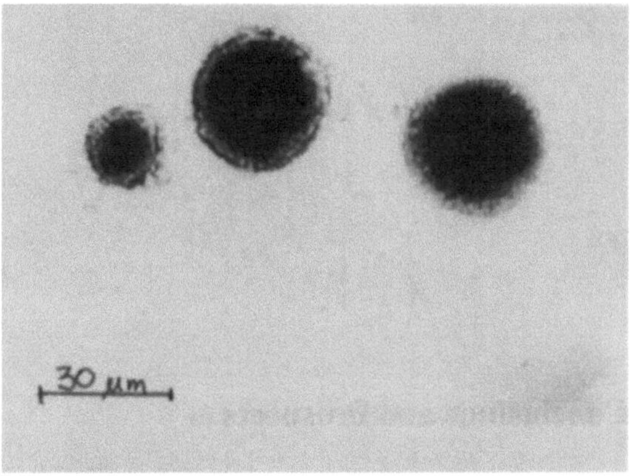

Fig. 8. Giant pollen grains, larger than octoploid strawberry pollen

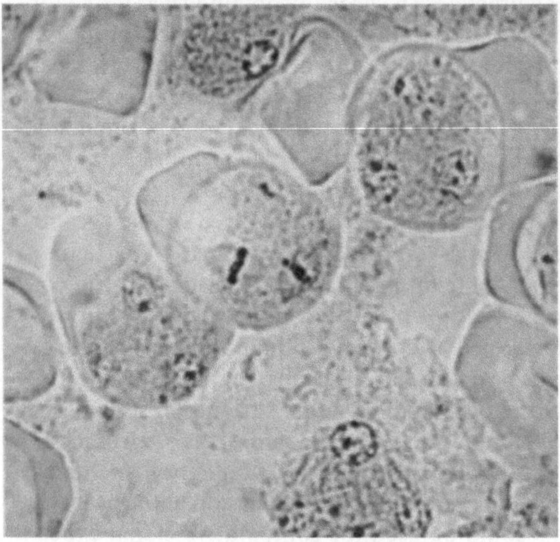

Fig. 9. Disturbances in meiotic divisions; minute spindle

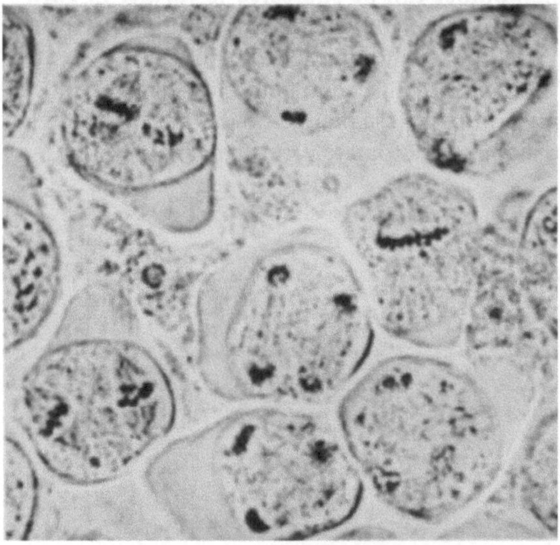

Fig. 10. Spindle fusion characteristic for the process of unreduced gamete

3 Conclusions and Prospects

There are two tendencies in in vitro strawberry culture. The first aims at the micropropagation of strawberry varieties on a large scale to produce healthy, well-developed planting material, while the other is intended to obtain morphologically diversified plants of different ploidy levels.

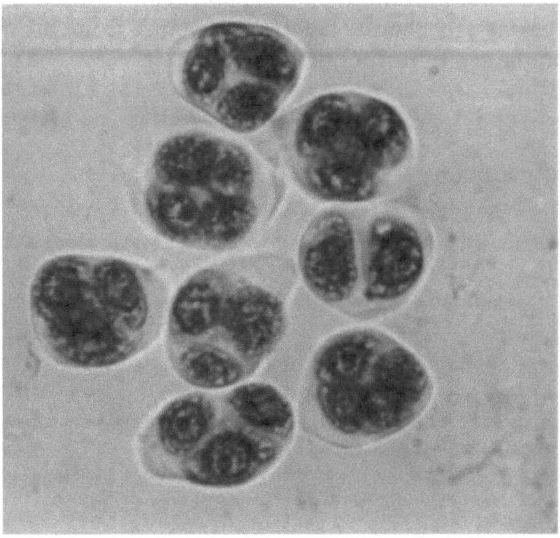

Fig. 11. Diad and some tetrads. The presence of a diad may speak for the fact that there was no second meiotic division

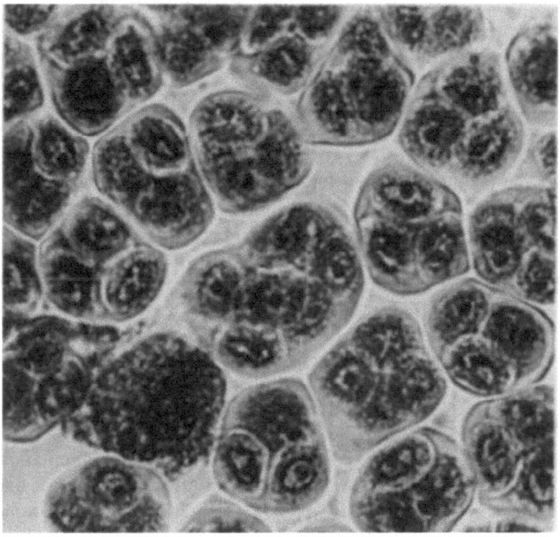

Fig. 12. Polyad and several tetrads. Polyad formation is connected with the disturbances in meiotic divisions

In commercial micropropagation, off-type individuals are undesirable and are thrown away. It seems, however, that the selection of the varieties showing the greatest somaclonal variation would be of interest.

So far the best source of plants with the greatest range of chromosome number has been anther callus culture. Plants of lower ploidy, as compared with their octoploid outset variety, show inconsiderable morphological changes and therefore require time-consuming cytological tests (Niizeki and Fukui 1983; Niemirowicz-

Szczytt 1987). Lower ploidy plants can be used in breeding programs, but it is necessary to choose the best varieties from the point of view of traits to be maintained in further breeding.

References

Anderson HH, Abbot AJ, Wiltshire S (1982) Micropropagation of strawberry *Fragaria × ananassa* plants. In vitro effect of growth regulators on incidence of multi-apex abnormality. Sci Hortic 16:331–342

Atkinson D, Crisp CH, Wiltshire SE (1986) The effect of medium composition on the subsequent initial performance of micropropagated strawberry plants. Acta Hortic 179:877–878

Bajaj YPS, Collins WB (1968) Some factors effecting the in vitro development of strawberry fruits. Proc Am Soc Hortic Sci 93:326–333

Barrientos-Pérez F, Murillo E (1986) Seedling selection in strawberry plants from anther culture. Hort Sci 21/3, 2:684

Belkengren RO, Miller PW (1962) Culture of apical meristems in *Fragaria vesca* strawberry plants as a method of exluding latent A virus. Plant Disease Rep 46:119–121

Boxus P (1974) The production of strawberry plants by in vitro micropropagation. J Hortic Sci 49:209–210

Boxus P, Quorin H, Laine JM (1977) Large scale propagation of strawberry plants from tissue culture. In: Reinert J, Bajaj YPS (eds) Applied and fundamental aspects of plant cell tissue, and organ culture. Springer, Berlin Heidelberg New York, pp 130–143

Bringhurst RS, Voth V (1980) Six new strawberry varieties released. Cal Agric 34.2:12–15

Cameron JS, Hancock JF (1986) Enhanced vigour in vegetative progeny of micropropagated strawberry plants. HortSci 21.5:1225–1226

Cameron JS, Hancock JF, Nourse TM (1985) The field performance of strawberry nursery stock produced originally from runners or micropropagation. Adv Strawberry Prod 4:56–58

Converse RH, Tanne E (1984) Heat therapy and stolon apex culture to eliminate mild yellow-edge virus from Hood strawberry. Phytopathology 74,11:1315–1316

Damiano C (1980) Strawberry in vitro culture. Ann Ist Sper Frutt 11:33–42

Damiano C, Laneri V, Arias EJ (1979) Different effects of nitrate ammonium jointly with citric-acid and succinic acid in strawberry *Fragaria × ananassa* micropropagation. Ann Ist Sper Frutt 10:43–52

Damiano C, Faedi W, Cobianchi D (1982) Nursery runner production fruiting and behaviour of micropropagated strawberry plants. Ann Ist Sper Frutt 13:69–78

Damiano C, Faedi W, Cobianchi D (1983) Nursery runner plant production fruiting and behaviour of micropropagated strawberry plants. Acta Hortic 131:193–200

Darrow GM (1966) The strawberry. Holt, Reinhart & Winston, New York

Fadeeva TS (1975) Genetika ziemlaniki. Izd Univ Leningrad

FAO (ed) (1985) Production yearbook. FAO, Rome 39:202

Fukui K, Niizeki H (1983) Chromosome engineering in somatic cells of plants with special references to the chromosome elimination by PFP. Bull Natl Inst Agric Sci Ser D 34:113–185

Hunter SA, Foxe MJ, Hennerty MJ (1983) The influence of temperature and light intensity on the in vitro propagation of the strawberry (*Fragaria × ananassa* Duch.) cv. Cambridge Favourite. Acta Hortic 131:153–161

Hunter SA, Hanon M, Foxe MJ, Hennerty MJ (1984) Factors affecting the in vitro production of strawberry (*Fragaria × ananassa* Duch.) meristems (cv. Cambridge Favourite). J Life Sci; R Dubl Soc 5:13–1

Ilieva I, Zagorska N (1986) Somatic embryogenesis in the strawberry (*Fragaria × ananassa* Duch.). Genet Selek 19,5:446–448

Izsak E, Izhar S (1983) Rapid micropropagation of strawberry plantlets from seeds for breeding purposes. Acta Hortic 131:101–103

Jelankovic G, Chee-Kok Chin, Billings S (1986) Transformation studies of *Fragaria × ananassa* Duch. by Ti plasmids of *Agrobacterium tumefaciens*. HortSci 21(3):695

Kartha KK, Leung NL, Pahl K (1980) Cryopreservation of strawberry *Fragaria* × *ananassa* cultivar Redcoat meristems and mass propagation of plantlets. J Am Soc Hortic Sci 105,4:481–484

Kondakova V, Kacharmazov V (1986) Tissue cultures and selection of disease-free material in strawberry. Genet Selek 19,5:386–388

Kondakova V, Kacharmazov V, Khristov R (1983) Effect of different nutrient media on organogenesis in strawberry tissue culture. Gradin Loz Nauka 20,6:61-64

Laneri V, Damiano C (1980) Strawberry anther culture. Ann Ist Sper Frutti 11:43–48

Lee YB, Campbell WF (1982) Effects of benzylamino purine and 2,4-3 on shoot and root formation from petiole segments of strawberry *Fragaria* × *ananassa* plant in the in vitro culture. HortSci 17/3, 2:483

Linsmaier EM, Skoog F (1965) Organic growth factor requirements of tobacco tissue culture. Physiol Plant 18:100–127

Maliarcikova V (1981) Obtaining strawberry plants by the method of meristem culture. Ved Prace OPSEMEX VUOOD Bojniciach 3:111–115

Maliarcikova V, Okossyova M (1984) The importance of comparative testing of the efficiency of strawberry clones bred by the explant method. Ved Prace Vysk Ustavu Ovoc Okras Drevin Bojniciach 5:77–88

Marcotrigiano M, Swartz HJ, Gray SE, Tokarcik D, Popenoe J (1984) The effect of benzylamino purine on the in vitro multiplication rate and subsequent field performance of tissue-culture propagated strawberry. Adv Strawberry Prod 3:23–25

McGrew JR (1980) Meristem culture for production of virus-free strawberries. In: Proc Conf Nursery production of fruit plants through tissue culture – applications and feasibility. USDA, SEA, APR-NE-11, pp 80–85

Múllin RH, Smith SH, Frazier NW, Schlegel DE, McCall SR (1974) Meristem culture frees strawberries of mild yellow edge, pallidossis, and mottle diseases. Phytopathology 64, 11:1425–1429

Murashige T, Skoog F (1962) A revised medium for rapid growth and bioassay with tobacco tissue cultures. Physiol Plant 15:473–497

Navatel JC, Carachaud G, Rondeillac P, Bardet A (1986) Agronomic comparison of strawberry plants from micropropagated mother plants and from traditional material. Inf Agrar 42,6:79–82

Neumann WD, Seipp D (1986) Cropping performance of the progeny of strawberry mother plants from tissue culture/meristem plants compared with conventionally propagated plant material. Obstbau 11,1:16–19

Niemirowicz-Szczytt K (1987) Strawberry (*Fragaria* × *ananassa* Duch.) haploids and their generative progeny. Induction and characteristics. Warsaw Agric Univ Press, Warszawa, pp 1–64

Niemirowicz-Szczytt K, Malepszy S (1980) Micropropagation from the meristems and young leaves of *Fragaria*. Bull Pol Acad Sci 28,5:335–340

Niemirowicz-Szczytt K, Zakrzewska Z (1980) *Fragaria* × *ananassa* anthers culture. Bull Pol Acad Sci 5, 341–347

Niemirowicz-Szczytt K, Zakrzewska Z, Malepszy S, Kubicki B (1983) Characters of plants obtained from *Fragaria* × *ananassa* anther culture. Acta Hortic 131:231–237

Niemirowicz-Szczytt K, Ciupka B, Malepszy S (1984) Polyploids from *Fragaria vesca* L. meristems, induced by colchicine in the in vitro culture. Bull Pol Acad Sci 32,1-2:58–63

Niizeki H, Fukui K (1983) Elimination of somatic chromosomes in strawberry *Fragaria ananassa* plants by treatment with P-fluorophenylalanine. Jpn J Breed 33,1:55–61

Nishi S, Oosawa K (1973) Mass production method of virus-free strawberry plants through meristem callus. Jap Agric Res Qrt 7,3:189–194

Nitsch JP (1950) Growth and morphogenesis of strawberry as related to auxin. Am J Bot 37:211–215

Oosawa K, Takayangai K (1984) Protoplast approach in vegetable breeding I. Isolation and culture conditions of protoplast from mesophyll cells of vegetable crops. Bull Veg Ornam Crops Res Ser A 12:12–19

Popov Yu G (1974) Meristem culture of strawberry affected by virus diseases. Sk Biol 9,5:694–697

Popov Yu G, Ravakin AS (1977) Effect of irradiation on isolated meristems of strawberry. In: Ispolz. biofiz. metodov v genet. – selektsion. experymente. Kishiniev Moldavian SSR, pp 43–44 after Ref Zh 1978, 5:65–92

Rosati P, Devreux H, Laneri V (1975) Anther culture of strawberry. HortSci 10:119–120

Sakai A, Yamakawa M, Sakata D, Harada T, Yakuwa T (1978) Development of a whole plant from an excised strawberry runner apex frozen to -196°C. Inst Low Temp Sci 36:31–38

Sanford JC (1983) Ploidy manipulation. In: Moore JN, Janick J (eds) Methods in fruit breeding. Prudue Univ Press, West Lafayette, Indiana, pp 100–123

Sarwar M (1984) The effect of different media and culture techniques on plating efficiency of strawberry mesophyll cells in culture. Physiol Plant 60,1:57–60

Scott DH, Lawrence FJ (1975) Strawberries. In: Janick J, Moore JN (eds) Advances in fruit breeding. Prudue Univ Press, West Lafayette, Indiana, pp 71–97

Scott DH, Galetta GJ, Swartz HJ (1985) Tissue culture as aid in the propagation of "Tribute" everbearing strawberry. Adv Strawberry Prod 4:59–60

Seemuller E, Merkle F (1984) Elimination of *Phytophthora fragariae* by meristem culture. Gartenbauwissenschaft 49,5/6:227–230

Senanayake YDA, Bringhurst RS (1967) Origin of Fragaria polyploid I. Cytological analysis. Am J Bot 54,2:221–228

Shaeffer GW, Damiano C, Scott DH, McGrew JR, Krul WR, Zimmerman RH (1980) Transcription of a panel discussion on the genetic stability of tissue culture propagated plants. In: Proc Conf Nursery production of fruit plant through tissue culture – applications and feasibility. USDA-ARS, pp 64–79

Sobczykiewicz D (1980) Heat treatment and meristem culture for the production of virus-free strawberry plants. Acta Hortic 95:79–82

Staudt G (1953) Die geographische Verbreitung der Gattung Fragaria und die Bedeutung für die Phylogenie der Gattung. Ber Dtsch Bot Ges 66,6:237–239

Swartz HJ, Galetta GJ, Zimmerman RH (1981) Field performance and phenotypic stability of tissue culture propagated strawberries *Fragaria ananassa*. J Am Soc Hort Sci 106,5:667–673

Thuesen A (1984) Yield of fruit from meristem propagated strawberry plants. Tiddskr Plant 88,1:75–80

Velchev V, Milanov E (1984) Regeneration of callus cultures from anther and pistils in strawberry. Gradin Loz Nauka 21,2:29–35

Vine SJ (1968) Improved culture of apical tissues for production of virus-free strawberries. J Hortic Sci 43:293–297

Waithaka K, Hildebrandt AC, Dama MN (1980) Hormonal control of strawberry *Fragaria* × *ananassa* axillary bud development in vitro. J Am Soc Hortic Sci 105,3:428–430

Wang D, Wergin WP, Zimmerman RH (1984) Somatic embryogenesis and plant regeneration from immature embryos of strawberry *Fragaria* × *ananassa*. HortSci 19,1:71–72

III.8 Somaclonal Variation in Peach

F. HAMMERSCHLAG[1]

1 Introduction

For several thousand years, conventional breeding techniques have been used to genetically improve peaches. Now we have tissue culture and genetic engineering technologies at our disposal that will enable us not only to work at the whole plant level, but also at the organ, tissue, cell, protoplast, organelle, chromosome, and gene levels in our efforts to improve this species. In the future, the specific approach chosen will depend on the problem to be solved and the amount of groundwork laid as far as the basic biology and the specific technology is concerned. Peach breeders should welcome these new technologies as a means of providing much needed variability, for the germplasm base is quite narrow for most commercial cultivars in the United States (Okie et al. 1985; Scorza et al. 1985; Arulsekar et al. 1986). In addition, many cultivars of North American origin are being used in intensive breeding efforts in many countries (Hesse 1979). A number of reports have surfaced, indicating a scarcity of peach germplasm with resistance to *Botryosphaeria dothidea* (Okie and Reilly 1983), *Cytospora leucostoma* (Scorza and Pusey 1984), *Pseudomonas syringae* (Petersen 1975), and the stem pitting virus (Cochran 1975), all important pathogens of peach. A recent report by Werner et al. (1986) indicated that several *commercial cultivars* previously rated highly resistant to bacterial leaf spot (Fogle et al. 1974) were only moderately resistant. This report was corroborated by in vitro screening (Hammerschlag 1988a).

Another trait impacted by the scarcity of germplasm variability is cold tolerance. Recurrent crop losses due to inadequate flower bud hardiness emphasize the need for cultivars that can withstand low temperatures. In the United States in 1985, extremely low temperatures destroyed 100% of the peach crop in ten states (Anon. 1985).

One approach for generating variation that has received a good deal of attention is to obtain somaclonal variants generated by the tissue culture cycle (Larkin and Scowscroft 1981). The screening of somaclonal variants can be carried out at either the callus, cell, or protoplast level if the screening agent, such as a toxin, acts directly at the cellular level. If not, then screening must be carried out at the whole plant level. Several recent reviews have extensively described screening for somaclonal variants (Flick 1983; Reisch 1983; Hammerschlag 1984b; Maliga 1984; Daub 1986).

[1]Tissue Culture and Molecular Biology Laboratory, USDA/ARS/BA, Beltsville, MD 20705, USA

Biotechnology in Agriculture and Forestry, Vol. 11
Somaclonal Variation in Crop Improvement I (ed. by Y.P.S. Bajaj)

It is clear from many studies, that the type of variation that can result, i.e., chromosomal changes, chromosomal rearrangements, gene rearrangements, etc. (Larkin and Scowcroft 1981; Evans and Sharp 1983; D'Amato 1985), and the degree of variation will depend on the tissues being cultured (Murashige 1974; D'Amato 1975; Skirvin and Janick 1976; Barbier and Dulieu 1980), and will increase with the length of time that the cells or tissues are maintained in vitro (Skirvin and Janick 1976; Barbier and Dulieu 1980). More studies on more crops need to be conducted to determine the specific genetic events that occur in vitro and how these events are influenced by the explant being cultured, the cultural conditions, and the environmental conditions. Not all types of variation are welcomed by the plant breeder, and certainly no variation is welcomed by the plant propagator who intends to use tissue culture technology for propagation.

Several reviews (Skirvin 1984; Hammerschlag 1986a,b) have indicated the specific tissue culture techniques being used to propagate and improve peaches. This chapter will discuss phenotypic and genotypic variation in peach plants either propagated from shoot tips in vitro or regenerated from cell cultures. Genotypic variation in peach cell cultures will also be discussed.

2 Variation in Micropropagated Plants

Very few reports are available on field performance of tissue-cultured plants or on genetic analysis of peach plants. Three self-rooted tissue cultured peach cultivars were shown to produce fruit less than 2 years from in vitro culture, and all yielded as well as controls (Martin et al. 1983). Rosati and Gaggioli (1987) reported that there were no differences in the cumulated production (1983–1985) of the scions on tissue-cultured rootstocks GF 677 and GF 43 compared with the conventionally produced rootstocks. In this latter example, the tissue culture technique was truly being put to the test, because the same plant material, propagated by two different methods, was being compared. From a scientific standpoint this type of comparison is best; however, from a commercial standpoint, it is also important to compare self-rooted, tissue cultured scions with scions budded onto commercially produced rootstocks. Currently, the author and coworkers are comparing four commercially important self-rooted, tissue-cultured cultivars with the same cultivars budded onto Lovell rootstocks. Data collected thus far (Hammerschlag 1987a) indicate that fruit production of the tissue-cultured trees exceeds that for the bud-grafted trees. Genetic analysis of the tissue-cultured plants indicated that all the plants were diploid $2x = 2n = 16$ (Hammerschlag et al. 1987).

3 Variation in Embryo-Derived Callus and the Regenerated Plants

There is only one detailed report in the literature on regeneration of peach plants from callus derived from embryos (Hammerschlag et al. 1985). In that study, plants were regenerated from callus derived from immature embryos (Fig. 1). Only green

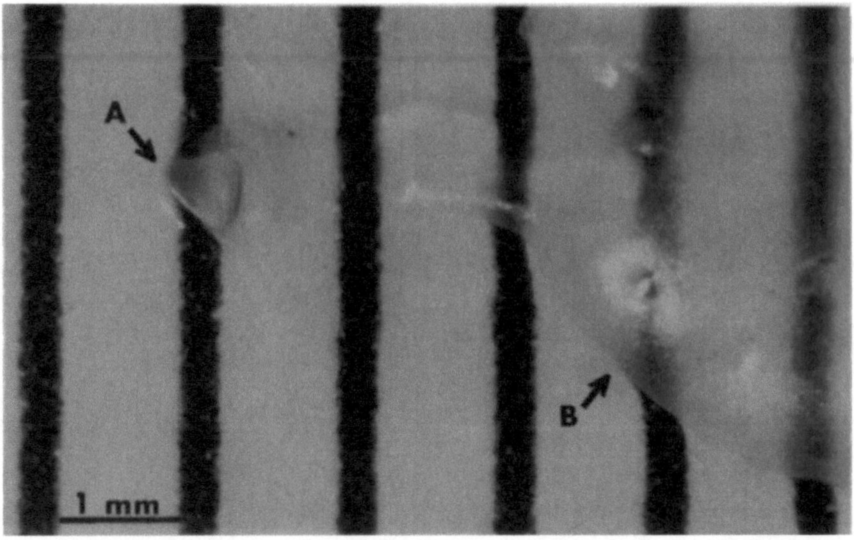

Fig. 1. Immature peach embryo (*A*) dissected from endosperm (*B*) that was removed from surface sterilized seed 54 days after full bloom

Table 1. Percentage of friable callus, derived from peach embryos at different stages of development, that produced nodular callus. Stage of development is indicated by PF_1 value which equals $\frac{\text{embryo length}}{\text{seed length}} \times 100$. Friable callus from a minimum of 12 embryos per stage of development per seed source were plated (Hammerschlag et al. 1985)

Seed source	PF_1 Value					
	3		28		100	
Seed source	White nodules	Green nodules	White nodules	Green nodules	White nodules	Green nodules
Sunhigh (primary callus)	45	55	10	90	0	100
Suncrest (primary callus)	33	67	10	90	10	90

nodular callus formed from friable callus derived from mature embryos (Fig. 2a and Table 1). Only primary callus from immature embryos could be induced to form a highly regenerative white nodular callus (Fig. 2b and Table 1) from which embryoid-like structures (Fig. 2c) and then plants (Fig. 2d) could be regenerated. Chromosome counts (Fig. 2e) and DNA determinations using a cytofluorometric technique (Hammerschlag 1983) revealed that regenerated plants were diploid, $2n = 2x = 16$. Further studies (Hammerschlag and Bauchan 1984) revealed that primary callus from immature and mature embryos was diploid which suggested that gross cytological changes were not responsible for the lack of morphogenetic

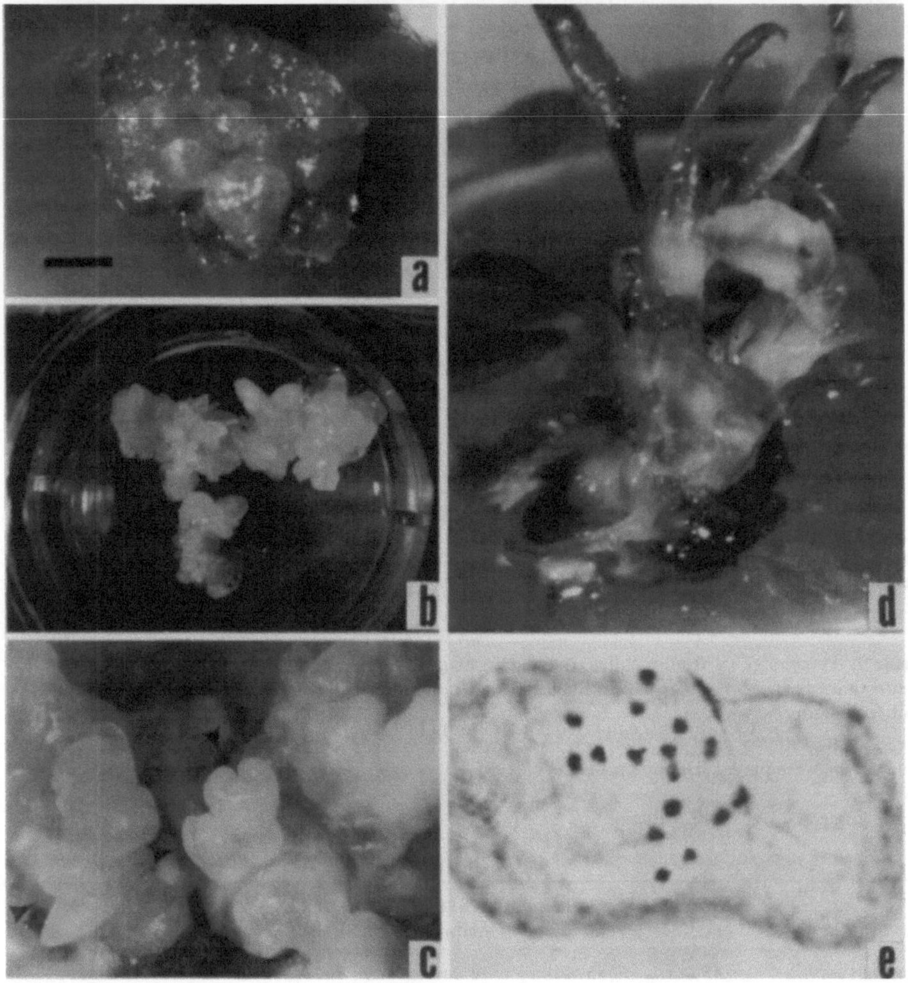

Fig. 2. **a** Green nodular callus formed from friable callus derived from mature peach embryos. **b** White highly regenerative nodular callus formed from friable callus derived from immature peach embryos. **c** Embryoid-like structures (*arrow heads*) produced from highly regenerative callus. **d** Shoots regenerated from highly regenerative callus. **e** Feulgen-stained root tip chromosomes in *Prunus persica* (2n = 2x = 16) regenerated from highly regenerative callus. *Bar*: **a** = 1 mm; **b** = 4.5 mm; **c** = 1 mm; **d** = 1.5 mm; **e** = 5 μm. (Hammerschlag et al. 1985)

potential in primary callus derived from mature embryos. Preliminary in vitro and greenhouse evaluations of these plants have revealed substantial variation in response to the bacterial leaf spot pathogen, *Xanthomonas campestris* pv. *pruni* (Hammerschlag 1987b). Phenotypic evaluation of these plants, grown under field conditions, is now in progress. In addition, studies are in progress to evaluate peach regenerants for resistance to *Meloidogyne incognita* (root-knot nematode) using an in vitro screening method developed by Huettel and Hammerschlag 1986), and to *Pseudomonas syringae* (bacterial canker).

4 Variability in Anther-Derived Callus

To date, plants have not been regenerated from anther callus. Stiles et al. (1980) obtained callus from peach anthers and reported that among the 232 mitotic cells observed in squashed samples of 25 calli, 18% were haploid, 78% were diploid, 2% were polyploid, and 2% were aneuploid. They suggest that both diploid and polyploid cells may have originated in somatic tissues. Hammerschlag (1983) removed white nodular callus from within the anther (Fig. 3), and to provide enough callus for genetic analyses, she cultured this callus for 4 months. Ploidy levels were determined by first measuring DNA levels in interphase nuclei by a cytofluorometric technique (Hammerschlag 1983) in which fluorescence measurements were made of nuclei stained with 4,6-diamidino-2-phenylindole (DAPI) (Fig. 4), a dye specific for DNA. DNA measurements were then compared with a standard. A DNA standard for cells from a diploid plant was established by

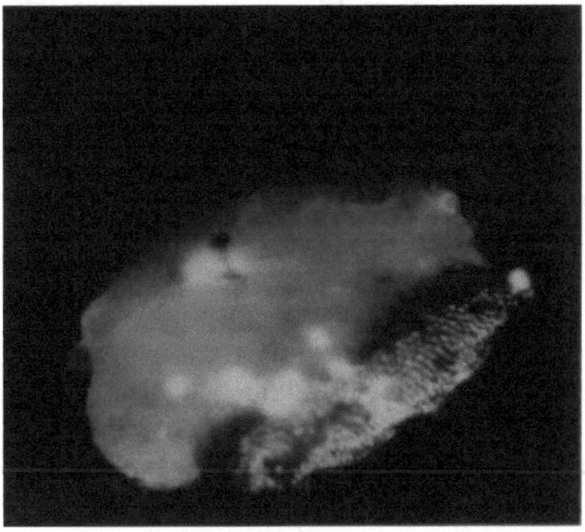

Fig. 3. Sunhigh peach anther callus produced by serial float culture. (Hammerschlag 1983)

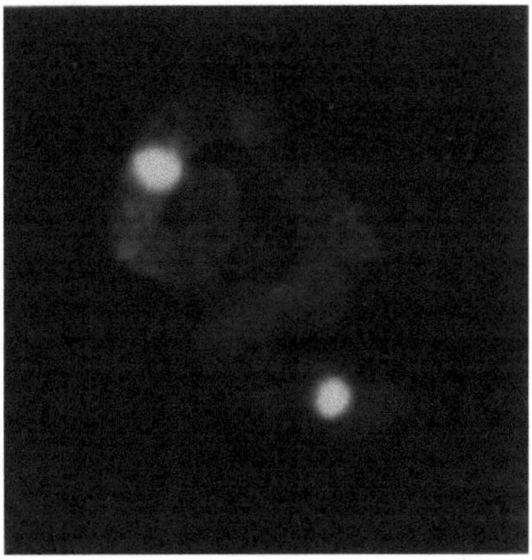

Fig. 4. Peach callus cells stained with the fluorochrome 2,6-diamidino-2-phenylindole (DAPI)

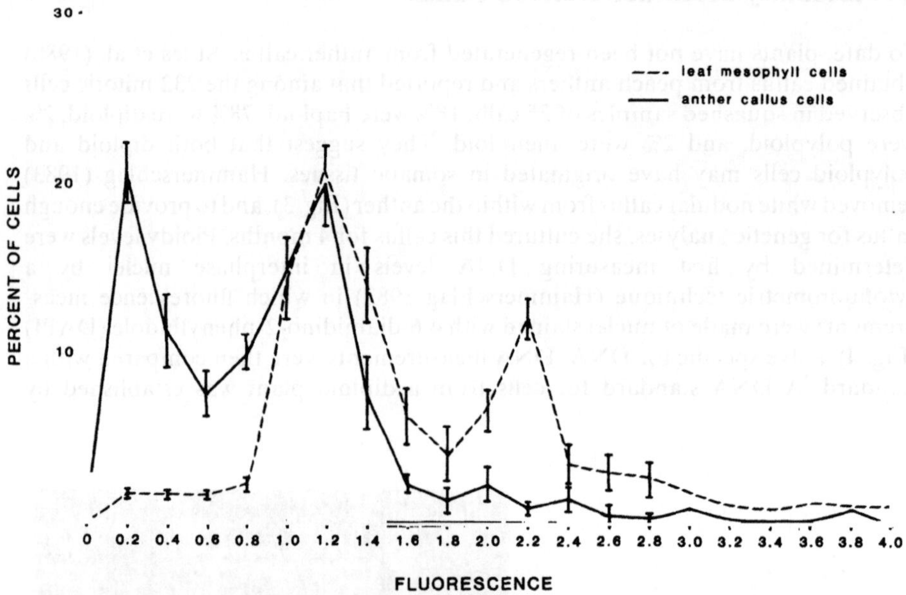

Fig. 5. The fluorescence of nuclei of leaf mesophyll cells from diploid Jerșey Queen peach and nuclei of Compact Redhaven peach anther callus cells following exposure to the fluorochrome 4.6-diamidine-2-phenylindole (DAPI). Values expressed ± SE. (Hammerschlag 1983)

counting chromosomes in leaf mesophyll cells (2n = 2x = 16) and then measuring fluorescence of DAPI stained nuclei. A comparison of DNA measurements of anther callus cells with the standard indicated that most of the anther callus cells were haploid (Fig. 5). Thus, anther callus cells are genetically stable, with regard to ploidy, for at least the first four months of culture.

5 Variability Through in Vitro Mutant Selection

In vitro selection for somaclonal variants was alluded to earlier in this chapter. This approach has been used successfully to obtain plants with useful agricultural traits not present in the parent plants from which the cell cultures were derived. For this approach to work, and to be of use agriculturally, the following criteria must be met: (1) a selective agent that acts directly at the cellular level must be identified, (2) the response of the cells in vitro to the selective agent must correlate with the response of cells in the intact plant to the selective agent, and (3) a reliable system for regenerating plants from cell cultures must be available. With these criteria in mind, Hammerschlag (1984a) demonstrated that a toxic metabolite produced by the peach leaf spot bacterium (*X. campestris* pv. *pruni*) was involved in disease development and acted at the cellular level. Next, a reliable system for regenerating peach plants from embryo callus was developed (Hammerschlag et al. 1985) and 400 calli, derived from embryos of the highly susceptible cultivar Sunhigh were exposed to the toxic metabolite and four plants were regenerated from the two surviving calli. These clones were micropropagated (Hammerschlag et al. 1987) and both an in vitro wick bioassay (Hammerschlag 1988a) and a detached leaf bioassay (Hammerschlag 1988b) were used to determine if the clones were resistant to leaf spot. Two of the four clones were significantly more resistant to *X. campestris* pv. *pruni* than the cultivar Sunhigh and one was significantly more resistant than the moderately resistant cultivar Redhaven. Currently, these plants are being evaluated for spot resistance under field conditions.

The in vitro screening approach has definite advantages over screening whole plants. First, very little space is required and second, much larger numbers can be screened. The 400 calli used by the author were composed of approximately 3 million cells each of which theoretically could have been regenerated into a whole plant. To screen that number of whole plants in the greenhouse would have been impossible. Because large numbers can be screened at one time, the length of time required to obtain resistant phenotypes is much less than that required by conventional breeding.

6 Summary and Conclusion

To date, there is some evidence that indicates that peach plants derived from axillary shoot cultures are morphologically and cytogenetically like the parent plants from which the shoot cultures were derived. On the other hand, some phenotypic variability is seen in peach plants derived from callus cultures. Chromosome counts reveal that this variability is not due to changes in chromosome number. These results support earlier studies (Murashige 1974; D'Amato 1975) which claim that axillary shoots are more likely to be identical to the parent plant than adventitious shoots, especially if the latter arise from callus. Whether the variability in peach regenerants is due to epigenetic or genetic events remains to be seen. Numerous genetic changes have been shown to occur in cultured cells which could account for the variability (Larkin and Scowcroft 1981; Evans and Sharp 1983; D'Amato 1985).

It is clear from these studies that the explant of choice for propagation should be an axillary shoot, whereas for the induction of variability, any explant that yields highly regenerative callus is desirable. It should be noted that once a reliable system is devised for regenerating plants from callus, numerous in vitro selection studies can be performed, as long as selective agents continue to be identified.

References

Anon (1985) The 1985 peach crop. Peach Times 30:1, 4

Arulsekar S, Parfitt DE, Beres WH, Hansche PE (1986) Genetics of malate dehydrogenase isozymes in peach (*Prunus persica* L. Batsch). J Hered 77:49–51

Banks-Izen MS, Polito VS (1980) Changes in ploidy level in calluses derived from two growth phases of *Hedera helix* L., the English ivy. Plant Sci Lett 18:161–167

Barbier M, Dulieu HL (1980) Effets génétiques observés sur des plantes de Tabac régénerées à partir de cotyledons par culture in vitro. Ann Amelior Plantes 30:321–344

Cochran LC (1975) Viruses. In: Childers NF (ed) The peach. Horticultural Publ, New Brunswick, pp 363–366

D'Amato F (1975) The problem of genetic stability in plant tissue and cell cultures. In: Frankel OH, Hawkes JG (eds) Crop genetic resources for today and tomorrow. Univ Press, Cambridge, pp 333–348

D'Amato F (1985) Cytogenetics of plant cell and tissue cultures and their regenerates. Crit Rev Plant Sci 3:73–112

Daub ME (1986) Tissue culture and the selection of resistance to pathogens. Annu Rev Phytopathol 24:159–186

Evans DA, Sharp WR (1983) Single gene mutations in tomato plants regenerated from tissue culture. Science 221:949–951

Flick CE (1983) Isolation of mutants from cell cultures. In: Evans DA, Sharp WR, Ammirato PV, Yamada Y (eds) Handbook of plant cell culture, vol 1. Techniques for propagation and breeding. Macmillan, New York, pp 393–441

Fogle HW, Keil HL, Smith WL, Mircetich SM, Cochran LC, Baker H (1974) Peach production. USDA Agric Handb 463

Hammerschlag FA (1983) Factors influencing the frequency of callus formation among cultured peach anthers. HortSci 18:210–211

Hammerschlag FA (1984a) Optical evidence for an effect of culture filtrates of *Xanthomonas campestris* pv. *pruni* on peach mesophyll cells. Plant Sci Lett 34:295–304

Hammerschlag FA (1984b) In vitro approaches to disease resistance. In: Collins GB, Petolino JG (eds) Applications of genetic engineering to crop improvement. Nijhoff/Junk, Dordrecht, pp 453–490

Hammerschlag FA (1986a) Peach (*Prunus persica* L. Batsch). In: Bajaj YPS (ed) Biotechnology in agriculture and forestry, vol 1. Trees I. Springer, Berlin Heidelberg New York Tokyo, pp 170–183

Hammerschlag FA (1986b) Temperate fruits and nuts. In: Zimmerman RH, Griesbach RJ, Hammerschlag FA, Lawson RH (eds) Tissue culture as a plant production system for horticultural crops. Nijhoff, Dordrecht, pp 221–236

Hammerschlag FA (1987a) Field performance of in vitro-propagated, own-rooted peaches under field conditions. HortSci (Abstr) 22:1067

Hammerschlag FA (1987b) Somaclonal variation in peach plants regenerated from callus derived from immature embryos. HortSci (Abstr) 22:1117

Hammerschlag FA (1988a) Screening peaches in vitro for resistance to *Xanthomonas campestris* pv. *pruni*. J Am Soc Hortic Sci 111:164–166

Hammerschlag FA (1988b) Selection of peach cells for insensitivity to culture filtrates of *Xanthomonas campestris* pv. *pruni* and regeneration of resistant plants. Theor Appl Genet 76:865–869

Hammerschlag FA, Bauchan G (1984) Genetic stability of callus cells derived from peach embryos. HortSci (Abstr) 19:554

Hammerschlag FA, Bauchan G, Scorza R (1985) Regeneration of peach plants from callus derived from immature embryos. Theor Appl Genet 70:248–251

Hammerschlag FA, Bauchan GR, Scorza R (1987) Factors influencing in vitro multiplication and rooting of peach cultivars. Plant Cell Tissue Org Cult 8:235–242

Hesse CO (1979) Peaches. In: Janick J, Moore NJ (eds) Advances in fruit breeding. Purdue Univ Press, W Lafayette, pp 285–335

Huettel RN, Hammerschlag FA (1986) Influence of cytokinin on in vitro screening of peaches for resistance to nematodes. Plant Disease 70:1141–1144

Larkin PJ, Scowcroft WR (1981) Somaclonal variation – a novel source of variability from cell cultures for plant improvement. Theor Appl Genet 60:197–214

Maliga P (1984) Isolation and characterization of mutants in plant cell culture. Annu Rev Plant Physiol 35:519–542

Martin C, Carre M, Vernoy R (1983) La multiplication végétative in vitro des végétaux ligneux cultivés: cas des arbres fruitiers et discussion générale. Agronomie 3:303–306

Murashige T (1974) Plant propagation through tissue culture. Annu Rev Plant Physiol 25:135–166

Okie WR, Reilly CC (1983) Reaction of peach and nectarine cultivars and selections to infection by *Botryosphaeria dothidea*. J Am Soc Hortic Sci 108:176–179

Okie WR, Ramming DW, Scorza R (1985) Peach, nectarine and other stone fruit breeding by the USDA in the last two decades. HortSci 20:633–641

Petersen DH (1975) Bacterial canker of peach. In: Childers NF (ed) The peach. Horticultural Publ, New Brunswick, pp 358–359

Reisch B (1983) Genetic variability in regenerated plants. In: Evans DA, Sharp WR, Ammirato PV, Yamada Y (eds) Handbook of plant cell culture, vol 1. Techniques for propagation and breeding. Macmillan, New York, pp 748–769

Rosati P, Gaggioli D (1987) Field performance of micropropagated peach rootstocks and scion cultivars of sour cherry and apple. Acta Hortic 212:379–390

Scorza R, Pusey PL (1984) A wound freezing inoculation technique for evaluating resistance to *Cytospora leucostoma* in young peach trees. Phytopathology 74:569–572

Scorza R, Mehlenbacher SA, Lightner GW (1985) In breeding and coancestry of freestone peach cultivars of the Eastern United States and implications for peach germplasm improvement. J Am Soc Hortic Sci 110:547–552

Skirvin RM (1984) Stone fruits. In: Ammirato PV, Evans DA, Sharp WR, Yamada Y (eds) Handbook of plant cell culture, vol 3. Crop species. Macmillan, New York, pp 402–452

Skirvin RM, Janick J (1976) Tissue culture-induced variation in scented *Pelargonium* spp. J Am Soc Hortic Sci 101:281–290

Stiles HD, Biggs RH, Sherman WB (1980) Some factors affecting callus production by peach anthers. Proc Fla State Hortic Soc 93:106–108

Werner DJ, Ritchie DF, Cain DW, Zehr EI (1986) Susceptibility of peaches and nectarines, plant introductions and other *Prunus* species to bacterial spot. HortSci 21:127–130

Section IV Ornamentals and Forage Plants

IV.1 Somaclonal Variation in *Pelargonium*

A. BENNICI[1]

I Introduction

The genus *Pelargonium*, l'Her. Storksbill (Geraniaceae) comprises plants of various habit, often succulent: the leaves, either digitately or pinnately veined, are entire, lobed, or dissected; the flowers, of many colors, usually have five sepals and petals, the two upper petals mostly larger, the lower mostly narrow and occasionally very small; there are ten stamens, some of them without anthers; fruits with five valves. Nearly all pelargoniums are from South Africa (Bailey 1964).

There are more than 230 species and very many well-marked hybrids also exist. *Pelargonium* is distinguished from the genus *Geranium* by technical characters. In most cases, the flowers of *Geranium* are regular, but those of *Pelargonium* are irregular. The most constant difference between the two genera is the presence in *Pelargonium* of a nectar tube, extending from the base of one of the sepals and adherent to the side of the calyx tube or pedicel. Because of the great number of variable and even confusing wild-type species and varieties, the genus offers exceptional advantages and problems. Since the nineteenth century, many species have been in cultivation in Europe, and experiments in hybridizing and breeding became common. Few classes of plants could have more interest to the amateur, because the species are numerous and varied, the colors mostly very attractive, the habit of the plant interesting, and the foliage often with a pleasing fragrance (Bailey 1963).

Most of the cultivated forms of *Pelargonium* may be grouped into four general horticultural classes: (1) the zonal types, known to gardeners as "geraniums", which seems to be derived from *P. zonale* and *P. inquinans*; (2) the ivy-leaved geranium, products largely of *P. peltatum*; (3) the "show" or fancy type, known to gardeners as "pelargonium"; all these types are used as ornamentals and are very popular in many countries, especially in Europe; (4) various scented-leaved geraniums, known mostly as "rose geraniums". These latter are of several species, with their hybrids and derivatives (for example *P. graveolens*, *P. radula*, *P. odoratissimum*, or *P. fragrans*) (Bailey 1963). The last types are very important as aromatic plants for the production of commercial geranium oil utilized in perfumery (main components of

[1]Plant Biology Department, Faculty of Agriculture, University of Florence, Italy. Address for correspondence: Prof. A. Bennici, Dipartimento di Biologia Vegetale, P. le delle Cascine 28, 50144 Firenze, Italy

Biotechnology in Agriculture and Forestry, Vol. 11
Somaclonal Variation in Crop Improvement I (ed. by Y.P.S. Bajaj)

this oil are citronellol, geraniol, iso-menthone, citronelyl-formate, and geranyl-formate). All these compounds (monoterpenes) are localized in essential oil glands which are distributed on the surface of the leaves.

2 Somaclonal Variation

2.1 Brief Review of In Vitro Studies

In this ornamental and commercial plant, somaclonal variation, displayed through in vitro plant regeneration, may be of great importance for the genetic improvement of the crop (see Sect. 2.5).

The first researches showed that in calli of *Pelargonium* shoot formation occurred on MS medium (Murashige and Skoog 1962) with the addition of high kinetin (10 mg/l) and low auxin (0.03 mg/l) (Pillai and Hildebrandt 1969). A similar medium, used by Twyford Laboratories, enabled multiple production of buds, although a lower concentration of cytokinin (0.5 mg/l) was advantageous (Holdgate 1977). Pillai and Hildebrandt (1969) and Holdgate (1977) reported that a major problem with *Pelargonium* was that hormonal requirements varied from variety to variety, and even using apparently optimal hormone concentrations, variations in multiplication rates still existed. Furthermore, callus progressively lost its regenerative capacity, but if regenerating callus was continuously removed from the rest of the callus-containing regenerating tissue, then organogenesis could be maintained over long periods, although the rates were slow and expensive. It was also found that media containing low concentrations of auxin (e.g., 0.1 mg/l) were beneficial because under these conditions, nonregenerating callus grew slowly. At the Twyford Laboratories no genetic variation was observed in plants produced in culture. (Holdgate 1977), and also Pillai and Hildebrandt (1968, 1969) made no mention of any variation. Cytological research of Bennici et al. (1968) revealed that when haploid *Pelargonium* internodes were cultured in vitro, prolonged culture led to the production of nuclei with DNA contents (16C and 32C) greater than those occurring in vivo (8C). Later, Bennici (1974) showed that the loss of the stability of the chromosome number in the haploid *Pelargonium* callus (high level of diploid and tetraploid mitoses) was reflected in regenerated roots, shoots, and plants. Moreover, a chimeric condition was observed in some roots and, in a higher degree, in shoots. Skirvin and Janick (1976a) systematically compared plants regenerated from the callus of five cultivars of *Pelargonium* spp. They found that plants obtained from stem cuttings in vivo were uniform, whereas plants from in vivo root and petiole cuttings and plants regenerated from callus were quite variable. Chen and Galston (1967) induced callus formation from stem pith explants on White's medium containing 1 mg/l IAA and 1.5 mg/l BAP, and having subcultured it to MS medium with the same growth regulators, obtained caulogenesis on a semi-solid medium without either auxin or cytokinin. Harney (1982) similarly reported obtaining regenerative stem tip callus on MS medium plus 0.5 mg/l NAA and 0.5 mg/l kinetin. Holdgate (1977) stated that shoot formation was induced on a modified MS medium with 0.1 mg/l NAA and 0.5 mg/l BAP, and that it could be prolonged in a commercial situation by continuously removing nonregenerative callus from the regenerative tissue (cf. George and Sherrington 1984). Geranium leaf petiole segments (about 2 cm in length), placed on MS medium with 3% sucrose and 1 mg/l zeatin, gave rise to adventitious shoot directly, after a small amount of callus formation, or from green nodules formed in callus. The first-formed shoots prevented the development of others unless 0.1 mg/l TIBA was also added to the medium (Cassells 1979; George and Sherrington 1984). On a modified Linsmaier and Skoog's (LS) medium, maximum growth of callus from stem explants taken from 2-week-old *Pelargonium* plants occurred in presence of 5 mg/l NAA plus 5 mg/l kinetin, but these concentrations inhibited callus formation from similar explants taken from 2-year-old plants. Optimum callus proliferation from the latter occurred with 0.2 mg/l of both regulators. The formation of meristemoids, and eventually shoots and roots, occurred only on the young explants cultured on 0.2 mg/l NAA and 0.2 mg/l kinetin (Hammerschlag and Bottino 1981). Shoot tip cultures of *Pelargonium* have also been established with success. In seeking to obtain virus-free stocks, Pillai and Hildebrandt (1968) and Beauchesne et al. (1977) were able to grow meristem tip explants directly into single shoots which could be rooted. Later, as reported above, Pillai and Hildebrandt (1969) induced differentiation of callus into shoots and roots. Cultures capable of producing 3–4 axillary shoots in 7 weeks from *P. hortorum* shoot tips (in 10 weeks from meristem tips) were described by Debergh and Maene (1977). Their medium included 1 g/l casein hydrolysate, 10 mg/l kinetin and 0.5 mg/l IAA. Cassells et al. (reported by George and Sherrington 1984) cultured meristem tips of *P. domesticum*, *P. peltatum*, and *P. zonale* on MS agar medium (3% sucrose) with additional ammonium nitrate (825 mg/l), $NaH_2PO_4 \cdot 2H_2O$ (150 mg/l), meso-inositol (100 mg/l) and with 1 g/l casein hydrolysate. The growth regulators IAA (2 mg/l), kinetin (4 mg/l), adenine sulfate (50 mg/l) and GA_3 (1 mg/l) were incorporated. Axillary shoot proliferation occurred when the cultures were allowed to develop without subdivision. Harney (1982) obtained proliferating shoot tip cultures of a number of *Pelargonium* taxa on MS basal medium supplemented with 0.5 mg/l NAA and 0.5 mg/l kinetin.

In a study made to determine the effects of selected high intensity monochromatic radiations on the growth of pith callus tissue of *Pelargonium zonale* var. "Enchantress Fiat", Ward and Vance (1968) found that the mean increase in fresh weights of tissues was significantly greater in the blue and full spectrum (white) treatments than in the red, green, or dark treatments. Tissues grown in the dark yielded mean fresh weight increases significantly lower than tissues grown under blue, red, and white light. In this work procedures similar to those of Chen and Galston (1965) were used to initiate pith callus growth. Light treatments consisted of monochromatic radiation obtained by filtering incandescent light from General Electric Reflector Flood Lamps through layered plexiglass filters purchased from Carolina Biological Supply. A 300 W lamp was used for the red and green treatments. This research is very interesting because up to now little is known about the influences of visible radiation on the growth and development of plant callus tissue.

The effects of exogenous monoterpenes on cell viability and growth of fine suspension cultures of *Pelargonium fragans* were investigated by Brown et al. (1987). A wide range of monoterpenes showed some toxicity, but there appeared to be no correlation between toxicity and both the structure of the monoterpene or its solubility in the growth medium. Cultures in the later stages of the growth cycle were generally more tolerant than those in the early stages. Adventitious regenerants, obtained by Cassells and Carney (1987) in stem and petiole tissue cultures of the cv. "Grand Slam", were up to 16% variants, depending on the explant origin. No variants were detected in control stem cuttings or in vitro from nodal culture. The authors suggested that genome instability in Grand Slam might produce a useful variation but they mitigated against the use of adventitious regeneration in micropropagation.

2.2 In Vitro Regeneration of Plants

2.2.1 Meristems and Stem Tips

Pillai and Hildebrandt (1968) described aseptic culture of stem tip meristems to raise geranium (*P. hortorum* Bailey) plants. From stem tips (3 mm length), cultured on two different MS media (MS1 containing 0.04 mg/l kinetin and 10 mg/l IAA or MS2 with 10 mg/l kinetin plus 1 mg/l NAA), shoot induction and growth and root formation were obtained in the explants within 3 to 4 weeks when the cultures were incubated in the greenhouse at 26° C. On the contrary, on the same media, shoot tips developed callus instead of growing into plantlets (see Sect. 2.2.2). A technique which resulted in the production of four to eight shoots per culture, depending on the cultivar, was developed. Terminal shoot tips were taken from greenhouse-grown geraniums and disinfected by soaking in 0.9% sodium hypochlorite. Thus multiple shoots were obtained from *P. hortorum*, *P. domesticum*, and *P. peltatum* (Harney 1982). Welander (1982), in a series of experiments also studied the possibility of using micropropagation as a method for propagating *P. zonale* hybr. on a large scale, using apical shoot meristems with one or two leaf primordia and various treatments with mineral salts, cytokinins (zeatin, kinetin, BAP), and TIBA. She found that meristems forming callus were frequent on media containing a high level of macronutrients when the concentration of kinetin was low (0.04 mg/l) and IAA was high (10 mg/l). The formation of one shoot or several shoots per meristem was affected by the composition of the macronutrients. In addition, more shoots were formed per meristem on media containing BAP (2 mg/l). The addition of TIBA (0.01 mg/l) did not affect the number of shoots per meristem. When shoots formed from the meristems were transferred to new growth medium, further shoot formation was induced from the transferred shoots. When meristems were taken from stock plants grown in a growth chamber under artificial light, the number of shoots per meristem was higher compared to meristems from stock plants grown in the glasshouse. Most of the shoots raised in vitro had no roots. To stimulate root formation, some shoots were placed on media with various IAA concentrations, or in a mixture of perlite and soil. Neither of the ways was better than the other and only about 25% of the plants survived after being planted in soil. Among the plantlets raised by micropropagation 5–30% were malformed. Cassells and Minas (1983a) carried out an extensive study with the goal of advancing knowledge relevant to the development of commercially viable micropropagation procedures for *Pelargon*-

ium cultivars. In this work the plant material consisted of 27 cultivars which were grown throughout the year under glass with a minimum temperature of 15°C. Bud tips, consisting of the apical dome and first pair of leaf primordia, were cultured on the medium of Hamdorf (1976) and modified by Cassells et al. (1980). Moreover, the rooting ability of axillary shoots (2 cm in height) was studied. Among the different aspects investigated the more important were: (1) culture responses of a range of cultivars on a single defined medium; (2) photoperiod effects for plant growth in relation to bud-tip culture responses; (3) effect of removal of the apical bud on the culture response of lateral bud tips; (4) development of a rooting medium for adventitious shoots. The zonal cultivars (*Pelargonium* × *hortorum*) showed high rates of establishment with 20–70% of the cultures forming buds. The regal hybrids were less successful, with proliferation (of new buds) occurring in 2–50% of the cultures. The ivy-leaf cultivars (*Pelargonium peltatum*) were more uniform with 20–37% of the cultures responding. Semi-ivy-leaf and scented cultivars showed very high proliferation frequencies. Exposure to high temperatures (20–30°C) of the cultivar "Irene Red Purple" resulted in lower establishment rates and proportionately decreased proliferation potential. Highest establishment and proliferation were observed in bud-tip cultures from 10°C pretreated donor plants. In the same cultivar mentioned above the removal of the apical bud did not influence the subsequent culture response of the lateral bud tips and, moreover, when the plants were grown in the glasshouse in darkness, in 8- and 16-h photoperiods and in continuous light for up to 4 weeks, the cultures of bud tips from plants from dark or short-day pretreatments were shown to have the highest establishment and proliferation rates. Moreover, a rooting medium was designed for Irene Red Purple to give a multiroot system resistant to damage during the transfer to soil.

In another paper Cassells and Minas (1983b) reported an investigation on abnormal *Pelargonium* types in relation to their culture behavior to determine the basis of this abnormality, i.e. whether it was due to chimeras or caused by pathogenic organisms. They found that in 36 progeny plants obtained by bud-tip culture of cultivars affected by the virus PPSA, none of the plants showed petal streak symptoms. Similarly, no net vein symptoms developed in 48 progeny plants produced from "Crocodile". The chimeral cultivars gave normal patterns of variegation (often variable and unstable) in the leaves of the progeny plants. Explant cultures of the chimeral cultivars "Mme Salleron" and "Mrs Cox" gave rise to albino and all-green progeny. No variegated plants were produced.

2.2.2 *Callus and Suspension Cultures*

Pillai and Hildebrandt (1968), culturing stem tips of *P. hortorum* on media different from those used for the production of plants directly from the apex meristem, developed mainly callus. Callus pieces transferred to MS2 medium (see Sect. 2.2.1) and incubated at 26°C in the greenhouse or growth chamber with a photoperiodic cycle of 16 h light and 8 h darkness differentiated shoots in 8–10 weeks. Light periods of 12 to 16 h alternating with the corresponding dark periods induced shoot formation. If the shoots were kept on the same medium, roots developed in another 8 to 10 weeks, but if the shoots were transferred to MS1 medium (see Sect 2.2.1), roots developed in 5 to 6 weeks. If they were grown for 20 to 25 days on MS1 medium and then transferred to White's medium, roots developed more rapidly. The same authors were able to induce shoot differentiation also from geranium (*P. hortorum*) callus formed from internode pith explants. If this callus, produced on media with or without coconut milk and 2,4-D, was subcultured on MS medium supplemented with different concentrations and combinations of NAA, kinetin, IAA, GA, and incubated at different photoperiods light/dark cycle, shoots were induced in 8–10 weeks and roots in another 8–10 weeks. The best combinations of NAA plus kinetin for plantlets regeneration were 0.1, 0.03, 0.05 mg/l, and 10 mg/l, respectively. Among photoperiods that induced differentiation, 15/9 and 16/8 h were most effective (Pillai and Hildebrandt 1969).

Organogenesis in in vitro cultures of a haploid *Pelargonium* was induced by Bennici (1974) with the aim of obtaining data on chromosome number in relation to organogenesis (see Sect. 2.3). First, callus was obtained from petals excised from young buds of the haploid *Pelargonium* cv. Kleiner Liebling, grown in the greenhouse. The petals, after sterilization with a commercial hypochlorite solution (18% for 20 min), were sown on LS basic medium supplemented with thiamine 400 mg/l, meso-inositol 100 mg/l, glutamine 1 mg/l, NAA 0.5 mg/l, kinetin 1 mg/l, agar 0.8%, and sucrose 3%, transfers were made every 20–30 days. The cultural conditions for organ differentiation were the same as used for callus induction and growth. Under these conditions, callus regenerated roots and shoots: while the agar-containing (solid) medium increased the number of buds, the liquid medium increased root formation. Callus transferred into liquid medium after 2 and 9 months of solid culture produced roots profusely. Liquid cultures were kept at 25 ± 1 °C under continuous light (500 lx) on a New Brunswick rotatory shaker (75 rpm) for aeration. Regenerated buds and shoots were isolated from cultures grown on solid medium for 2–3 months at 4000 lx light intensity and at 25 ± 1 °C (Fig. 1). In liquid culture,

Fig. 1. Regenerated *Pelargonium* plantlet from callus grown on solid medium for 2–3 months at 4000 lx light intensity (at 25 ± 1 °C) on LS basic medium supplemented with thiamine 400 mg/l, myo-inositol 100 mg/l, glutamine 1 mg/l, NAA 0.5 mg/l, kinetin 1 mg/l, agar 0.8%, and sucrose 3% (Bennici 1974)

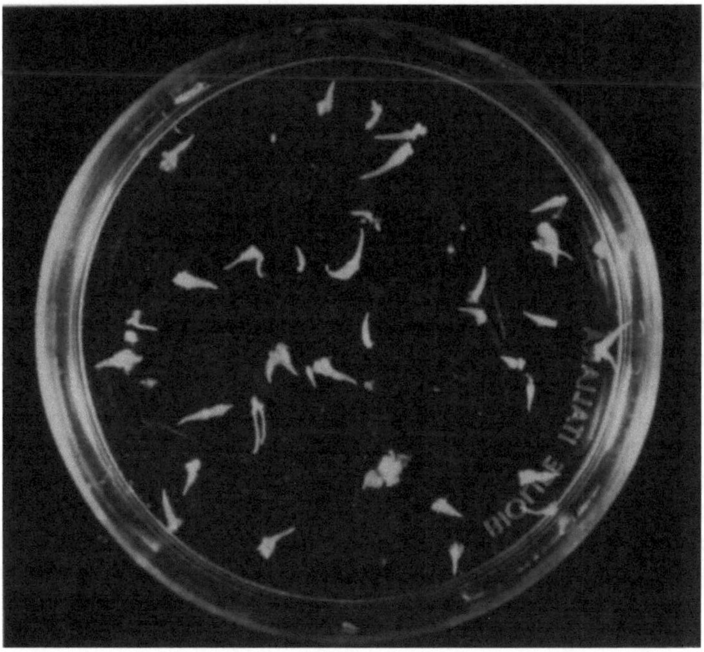

Fig. 2. Roots with cell aggregates at their basal end, in liquid culture obtained 8–10 days after transfer of callus maintained from 2 to 9 months in solid culture. Liquid cultures were kept at 25 ± 1°C under continuous light (500 lx) on a rotatory shaker (75 rpm). Composition of the liquid culture medium as in Fig. 1. (Bennici 1974)

8–10 days after transfer from solid media, the formation of numerous roots, which showed at their basal end cell aggregates, was observed (Fig. 2). The presence of these aggregates at the base of the roots might indicate either: (1) a detachment of roots or root primordia from a callus and subsequent callus formation on the organized structures, or (2) dissociation of larger calli into small aggregates, giving rise on the one end to roots and on the other to further undifferentiated cell proliferation.

Skirvin and Janick (1976a) investigated the variability associated with tissue culture of various clones of geranium (see Sect. 2.3). They used techniques for producing plants from callus cultures which were modified from Pillai and Hildebrandt (1969). Disinfection was best obtained by soaking explants in 0.525% sodium hypochlorite (10% Clorox) for 20 min. Callus production was induced in the dark at 25°C on MS medium supplemented with NAA (1 mg/l), kinetin (5 mg/l), adenine (4 mg/l), and vitamin supplement, i.e., cyanocobalamin (1.5 μg/l), folic acid (0.5 mg/l), riboflavin (0.5 mg/l), biotin (1 mg/l), choline chloride (1 mg/l), Ca pantothenate (1 mg/l), thiamine-HCl (1 mg/l), nicotinamide (2 mg/l), and pyridoxine-HCl (2 mg/l). Although there were differences between cultivars, shoot production was best with Pillai and Hildebrandt's (1969) modification of MS medium supplemented with NAA (0.1 mg/l) and kinetin (10 mg/l) under diffuse light (16-h light) at about 25°C. Rooting occurred when young shoots were transplanted to White's medium without hormones, but supplemented with iron (37.3 mg/l NA_2EDTA and 27.8 mg/l $FeSO_4 \cdot 7H_2O$). Calliclones grew rapidly in the greenhouse. There were large differences in the ability of *Pelargonium* species and cultivars to produce calliclones. None of the ten cultivars of *P. hortorum* Bailey formed plantlets. However, *P. domesticum* Bailey cv. Country Girl and five of the six scented geraniums *P. graveolens* Thunb. (cv. Rober's Lemon Rose), *P. domesticum* × *P. denticulatum* Poir (cv. Clorinda), *P. capitatum* L. (cv. Attar of Rose), *P. graveolens* Thunb. (cv. Old Fashioned Rose), *P. adcifolium* (cv. Snowflake), *P. crispum* L'Her. (cv. Lemon) produced calliclones readily. Recently, long-term cultures have been established from stem tissue of 27 variants of scented pelargoniums (Brown and Charlwood 1985), in order to investigate many aspects of monoterpene production in vitro. For callus initiation stem explants (6 cm in length and with leaves and axillary buds removed), surface sterilized with sodium hypochlorite (3% aqueous solution) for 3 min, were used. Stem segments (1.5 cm long), each of which was divided longitudinally, were placed on initiation medium (containing kinetin 0.2 mg/l plus 2,4-D 1 mg/l) with the cut section in contact with the agar. Incubation was carried out at 26°C in a light regime of 16 h light (photosynthetic photon flux density 3 $\mu E\ m^2\ s^1$) and 8 h dark. Initiated calli were transferred after 14 to 21 days to maintenance medium of Murashige and Skoog, containing 3% sucrose, 1% agar, and supplemented with BAP (5 mg/l) and NAA (1 mg/l). The calli were subcultured into fresh maintenance medium at 28-day intervals. Callus material (ca. 0.4 g fresh weight) from the species *P. australe, P. citriodorum, P. crispum, P. graveolens, P. filifolium,* and *P. tomentusum* and from the variants Duke of York, Miss Australia, and Prince of Orange, which had been subcultured four times on maintenance medium, was used to investigate the effects on growth and differentiation of various concentrations of growth regulators. The calli were incubated on maintenance medium supplemented with appropriate levels of cytokinin and auxin (BAP plus NAA respectively 0.05 + 0.05, 0.05 + 0.50, 0.05 + 1.00, 0.25 + 2.00, 0.50 + 0.05, 0.50 + 0.15, 0.50 + 0.50, 1.00 + 1.50, 1.50 + 0.15, 1.50 + 1.00, 1.50 + 2.25, 2.25 + 1.50, 5.00 + 1.00 mg/l) under the temperature and light regimes described above, and observed daily over a period of 4 weeks. At the end of this period, root and shoot formation was assessed quantitatively. In order to determine whether the ability to regenerate shoots was general, newly initiated callus material from each variant was transferred to, and maintained on, regenerating medium (maintenance medium with BAP 0.5 mg/l and NAA 0.05 mg/l). After initiation, individual shoots were removed from the surface of the callus and recultured on regeneration medium. Prior to root induction, plantlets were grown to a minimum height of 5 cm. Segments of callus material from the two species (i.e., *P. tomentusum* and *P. australe*) that were most rapid in their regeneration under the above conditions were transferred to regeneration medium after each 4-weekly subculture on maintenance medium to determine whether ability to regenerate was impaired with time. All 27 variants produced callus after 5–22 days. When the cytokinin level was maintained below 1 mg/l, a decrease in auxin level resulted in an increase in shoot production. Root formation was observed only at low cytokinin (0.25 mg/l and below) and higher auxin levels. The growth responses at BAP (0.5 mg/l) and NAA (0.05 mg/l) indicated that such a supplementation should provide a medium for shoot regeneration for most variants. All calli (except for *P. echinatum*) exhibited shoot regeneration when placed on regeneration medium. The calli did not lose their ability to regenerate even after 19 subcultures. Moreover, shoot morphogenesis was more prolific in liquid than in corresponding solid medium.

2.2.3 *Anther*

With the aim of producing pathogen-free geraniums, *P. hortorum* plants were induced from anther callus cultures by Abo-El-Nil and Hildebrandt (1971). They used anthers at different stages of development (with tetrads, uninucleate immature pollen, and binucleate mature pollen grains). For callus induction, the anthers were cultured on a modified White's agar medium supplemented with each of parachlorophenoxyacetic acid (P-CPA), NAA, kinetin or 2,4-D in concentration of 0.5, 1, 2, 2.5, 5 mg/l, with or without 15% coconut milk. Anthers with uninucleate immature pollen formed vigorous white or green calli on the medium supplemented with 150 ml/l coconut milk, and 0.1–2.5 mg/l NAA, and 2.5 mg/l kinetin. Direct production of embryoids from tetrads, uninucleate, or mature pollen was never observed. Calli produced from anthers were transferred to MS medium supplemented with 0.1, 0.2, 0.3, 0.4, 0.5, 1, 2, or 2.5 mg/l NAA, and 0.1, 0.5, 1, or 2.5 mg/l kinetin in various combinations. Calli grown on a medium supplemented with 0.5 mg/l NAA and 2.5 mg/l kinetin differentiated embryoids after 1-month incubation under a 16-h light photoperiod, but not with continuous illumination. Plantlets developed from the embryoids during the next 2 months were transferred to White's agar medium for root induction. These plants, transplanted in soil, were free from the virus-like leaf symptoms present in the parent plants. Squash preparation of root tips showed that the induced plants were tetraploid. However, in later research, haploid plants were obtained from anther cultures (Abo-El-Nil and Hildebrandt 1973). The differentiation of the tetraploid plants may have been from callus not originatng from pollen, or may have resulted from chromosome endoreduplication of the pollen callus cells or endomitosis. Also the haploid geranium plants produced were without virus symptoms.

2.2.4 *Protoplasts*

Very recently, plant regeneration from cultured cell-derived protoplasts of several *Pelargonium* species has been reported (Yarrow et al. 1987). Callus was induced from leaf and petiole material of *P. hortorum*, *P. peltatum*, and *P. aridum* on MS medium containing 3% sucrose, 0.2 mg/l NAA, and 0.5 mg/l BAP. After 5–6 weeks, callus was well enough established to be subcultured. This callus was used to obtain cell suspensions from which protoplasts were produced with an enzyme mixture consisting of 2% Rhozyme HP150, 4% Meicelase, 13% Mannitol, and inorganic salts. Protoplasts were grown on a culture medium which contained a mixture of two appropriate liquid media. Small protoplast-derived colonies were transferred to the surface of MS medium with 0.1 mg/l NAA, 0.5 mg/l BAP, 3% sucrose, and 4.5% Mannitol (MS2). These colonies were further subcultured on Mannitol-free medium, prior to the initiation of plant regeneration. In order to initiate plant regeneration, it was necessary to change the culture conditions as follows: for *P. aridum* calli, MS medium was used with 0.05 mg/l NAA, 0.5 mg/l BAP, 3% sucrose, and 0.8% agar (MS3); *P. peltatum* calli were placed in MS2 without Mannitol liquid medium; the culture of *P. hortorum* calli was similar to that for *P. peltatum* (MS2 liquid medium with 0.5 mg/l BAP as the sole phytohormone). On agar solidified regenerative medium, such calli underwent plant regeneration at frequencies approaching 100% for *P. aridum* and 10% for *P. hortorum*. Under similar conditions, shoot primordia developed in 5% of *P. peltatum* calli, but these never developed into normal shoots. However, following a liquid shake culture regime, whole plants were induced in 20% of *P. peltatum* and in 60% of *P. hortorum* calli.

2.3 Variations in Callus and Regenerated Plants

Despite the numerous studies on *Pelargonium* in vitro mentioned above, very little research has been done on the genetic (chromosomal) status of the callus and regenerated plants. The only reports on these subjects are those of Bennici et al. (1968), Bennici (1974), Skirvin and Janick (1976a), Janick et al. (1977) and Tokumasu and Kato (1979). Tissues of the haploid *Pelargonium* cv. Kleiner Liebling (n = 9) were chosen by Bennici et al. (1968) for an extensive cytological analysis of the cell nuclei in vivo and in callus obtained in vitro. For callus formation internode explants of 5×3×4 mm size, previously sterilized in 15% commercial hypochlorite for 30 min, were sown on MS medium with the addition of thiamine 0.5 mg/l, pyridoxine 0.5 mg/l, nicotinic acid 0.5 mg/l, meso-inositol 100 mg/l, sucrose 3%, 2,4-D, and kinetin both at 10^{-6} M. Samples for cytological and DNA cytophotometric analyses were collected 6 and 15 days after the primary explant, and 12 and 19 days after the first transfer: this was made after 15 days of primary culture. The collected material, together with shoot tips and control internodes, after a colchicine pre-treatment, was fixed in alcoholacetic acid 3:1 and stained by the Feulgen technique. Measurement of nuclear DNA content was carried out with a Deeley-type cytophotometer GN2 (produced by Barr and Stroud, UK). In agreement with the cytological observation of Daker (1966), all mitoses in the shoot apices of Kleiner Leibling were found to be haploid. In Fig. 3, the results of a DNA cytophotometric analysis of 90 interphase nuclei and ten mitoses in a shoot apex are shown. In this, and all other graphs, the DNA values, in arbitrary units, are reported as logarithms to the base 2; therefore, distances of 1 unit correspond to successive duplications of DNA contents (1 = 1C; 6 = 32C). Figure 3 shows that the ten mitoses examined had the expected 2C DNA content (haploid) and 72 out of the 90 interphases (80%) were in the various stages of the DNA cycle in haploid nuclei, from the pre-DNA synthetic phase or G_1 (1C) to the post-DNA synthetic phase or G_2 (2C). Of the remaining 18 interphases (20%), two were 4C and the remaining had 2C–4C intermediate DNA contents, a condition demonstrating the tendency to chromosome endoreduplication in the material. In the internodes of Kleiner Liebling the few mitoses present were all haploid (probably cambium cells); the interphase nuclei showed a much greater variation in volume as compared to those of shoot tips. In two internodes, the cells of which were analyzed with the cytophotometer (Figs.

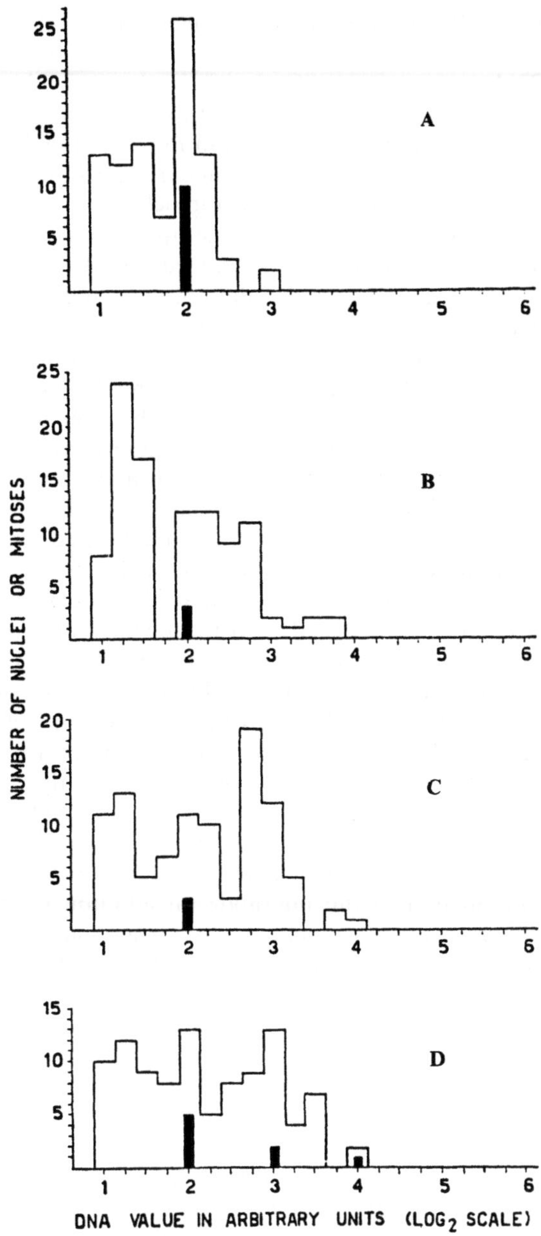

Fig. 3A-D. Nuclear conditions in haploid *Pelargonium*. **A** Shoot apex. **B,C** Internodes in vivo. **D** Internode grown in vitro for 6 days

3B,C), 61% and 47% of the interphase nuclei fell in the range 1 to 2C; the rest had undergone, or were in the process of undergoing, chromosome endoreduplication (DNA contents up to 8C). In internodes grown in vitro for 6 days, the cytological picture did not differ from that of control internodes, apart from the occurrence, in addition to haploid mitoses, and mitoses of endopolyploid nuclei. Since, at that time, many of these nuclei were undergoing the first mitosis post-endoreduplication, mitotic stages with either nine diplochromosomes (4-chromatid chromosomes) or nine chromosomes with a more complex structure (quadruplochromosomes?) were seen. DNA cytophotometry indeed revealed the occurrence of 2C, 4C and 8C mitoses (Fig. 3D). After 15 days of in vitro growth, most mitoses showed normal 2-chromatid chromosomes (monochromosomes). Although no extensive analysis was made, all well-analyzable mitoses gave chromosome counts in the euploid series 9, 18, and 36. Also the cytophotometer readings of mitoses gave very uniform values of 2C, 4C and 8C. Another difference between 6 and 15 days of culture was seen in the composition of the population of interphase nuclei. In the sample illustrated in Fig. 4, the proportion of haploid nuclei had decreased to 40% and a 16C nucleus was measured. Twelve days after the first transfer (Figs. 4B and C), the proportion of nuclei with DNA contents higher than 8C had further increased, being ca. 20% in the sample of Fig. 4C, in which a rather low frequency of haploid (1C to 2C) nuclei was manifest. This cytological situation did not seem to have changed after one more week in culture (Figs. 4D and E); the maximum degree of polyploidy in dividing cells was still 4n, probably indicating that 16C and 32C nuclei were hardly induced to undergo division by the conditions prevailing in the culture. The *Pelargonium* Kleiner Liebling belongs to the vast category of plants in which chromosome endoreduplication occurs as a concomitant of tissue differentiation. Chromosome endoreduplication already starts in the shoot tip, probably at the base of the apical meristem, an area in which an active endopolyploidization is known to occur from studies on several species (D'Amato 1965). The endopolyploidy course increased, in both degree and extent, during the formation of the anatomical structure of the stem; so that in the internode, in addition to the haploid (meristematic) cell population (DNA contents 1C to 2C), a population of endopolyploid nondividing cells was established (DNA contents up to 8C). When the internodes were cultured in vitro, differentiated cells were stimulated to divide; a population of diploid and tetraploid mitoses was then added to the preexistent meristem cell population (haploid). In the course of time, diploid and tetraploid mitoses continued to be present in the callus, whilst haploid mitoses decreased in number and eventually disappeared. Nineteen days after the first transfer, a few calli showed no mitoses at all. The mitoses analyzed had the chromosome numbers 9, 18, and 36, and the mitotic DNA values measured were correspondingly 2C, 4C, and 8C. Considering, however, the limited number of mitoses studied, the presence of aneuploidy in this material could not be excluded; although, if occurring as all, it was rather rare. Under the experimental conditions used, growth in vitro for 2 weeks or more led to the production of nuclei with DNA contents (16C and 32C) not found in internodes in vivo; these values demonstrated the stimulation of one and two additional DNA replications beyond the limit attained in vivo (8C). In *Pelargonium* calli from which haploid mitoses had disappeared (Fig. 4C-E), the cytophotometric analysis still revealed the presence of cells of the meristematic, haploid, line (1 to 2C DNA

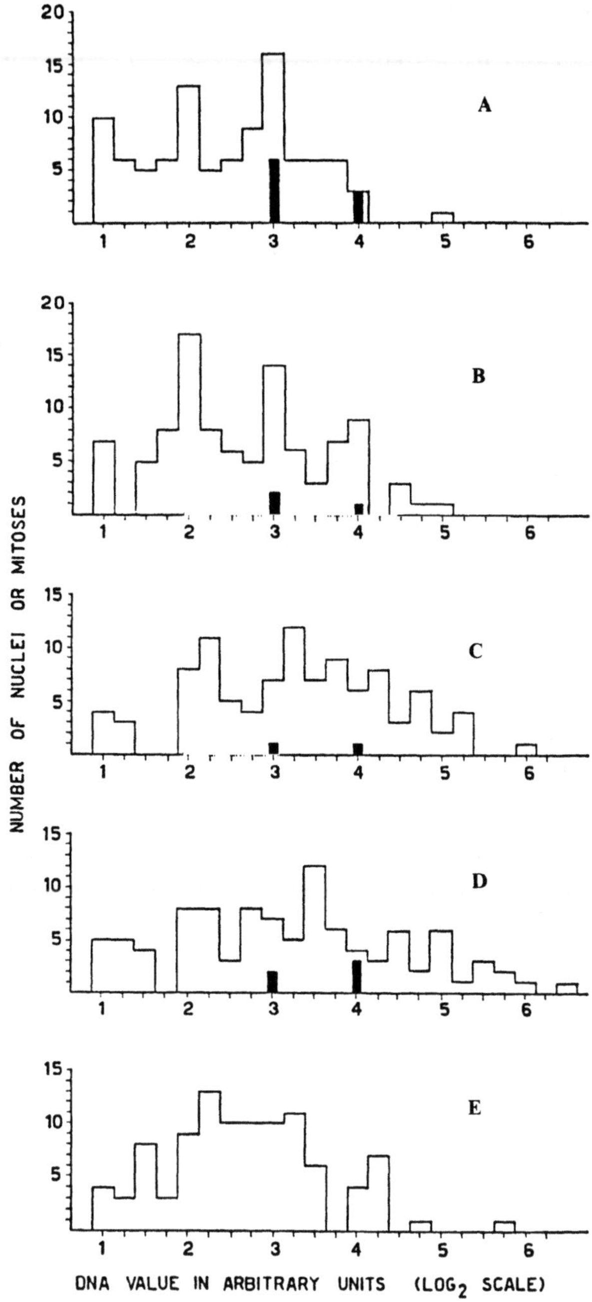

Fig. 4A-E. Nuclear conditions in haploid *Pelargonium*. **A** Callus from internode grown in vitro for 15 days. **B,C** Calli 12 days after the first transfer. **D,E** Calli 19 days after the first transfer. The first transfer was made after 15 days of primary culture values). This showed that in *Pelargonium* haploid cells might be carried along in culture without active division.

Later, Bennici (1974) used the same material with the purpose of investigating organogenesis in vitro by comparing caulogenesis with rhizogenesis in relation to chromosome number in order to assess the possible chimeric structure of the regenerated organs. For the cytological analysis of regenerated organs (see Sect. 2.2.2) root and/or shoot meristems and calli were collected and fixed in a 3:1 (v/v) alcohol/acetic acid mixture, stained by the Feulgen technique and squashed. In some cases, a 0.05% colchicine pre-treatment for 5–6 h at room temperature was used to improve resolution in chromosome counts. The results of the cytological analysis of root and cell aggregates obtained in liquid culture are reported in Tables 1 and 2. Table 1 shows ploidy differences between root apices derived from cultures of different age (2 and 9 months old). It appears that old cultures produce tetraploid roots in high frequency (44.82%), whilst young cultures give rise predominantly to diploid roots (71.42%). The percentages of haploid structures, however, do not seem to differ significantly in the two experiments. This result seemed to be in accord with the data on the cytology of callus development in the haploid *Pelargonium* mentioned before, where an increase in ploidy with age of culture was observed. Moreover, the cytological analysis of initial callus development from petals showed haploid, diploid, and tetraploid mitoses in percentages of 5, 60 and 36 respectively. Table 1 also shows chimeric [9 and 18 (n and 2n); 27 and 36 (3n and 4n)] and triploid root meristem (17.24%). Most probably, triploidy resulted from fusion of haploid and diploid nuclei (D'Amato 1972). From the cytological analysis reported in Table 2, it is seen that out of 12 structures analyzed, six had the same chromosome number in the root meristem and the basal cell aggregate, whereas four showed a higher ploidy level in the cell aggregate, and only one in the root. The cytological analysis of young shoot apices (Table 3) revealed in most cases the coexistence in the same meristem of different levels of ploidy (chimeric condition). In a total of 191 mitoses observed, 61.78% were diploid and only 4.18% tetraploid. Aneuploid mitoses were frequent especially at higher chromosome numbers (18–36). An analysis of regenerated adult plants (Fig. 2) had in most cases shown apical shoot meristems with the diploid chromosome number only. In one plant, aneuploid (9 ÷ 36 chromosomes) and triploid mitoses had been observed. A point worth noting was the presence of diplo-chromosomes (= 4-chromatid chromosomes) in some te-

Table 1. Cytological conditions in apical meristems of roots regenerated from old and young calli of haploid *Pelargonium* (n = 9) in liquid culture. (Bennici 1974)

Ploidy levels	Roots from old callus %	Roots from young callus %
n	10.34	19.04
n and 2n	13.79	–
2n	13.81	71.42
3n	17.24	–
3n and 4n	–	4.76
4n	44.82	–
Aneuploids (n ÷ 4n)	–	4.76

Table 2. Nuclear conditions in meristems of roots and in cell aggregates at the base of each root in liquid cultures of haploid *Pelargonium* (n = 9). (Bennici 1974)

No.	Root meristems		Basal cell aggregates	
	No. of mitoses observed	Chromosome no.	No. of mitoses observed	Chromosome no.
1	3	9	25	9
2	8	9	10	18
	10	18	13	Aneupl.[a]
3	20	18	7	18
	1	9	5	Aneupl.[a]
4	7	36	8	18
	3	Aneupl.	3	Aneupl.
5	1	18	30	18
6	4	9	25	9
7	6	27	8	27
8	6	18	7	18
			5	Aneupl.
9	1	9	2	9
			2	18
10	7	36	10	36
11	4	36	6	36
12	7	36	8	36
			1	54

[a] Chromosome number from 18 to 36.

Table 3. Chromosome counts in cells of 20 shoots apices regenerated in in vitro cultures of haploid *Pelargonium* (n = 9). (Bennici 1974)

Shoot apex no.	No. of mitoses observed	Chromosome no.						
		9 (n)	9 ÷ 18	18 (2 n)	27 (3 n)	18 ÷ 36	36 (4 n)	η 36
1	3	–	–	3	–	–	–	–
2	13	–	1	12	–	–	–	–
3	21	1	7	7	3	2	–	1
4	10	6	4	–	–	–	–	–
5	35	–	–	27	–	5	3	–
6	4	–	–	–	–	2	2	–
7	4	–	–	3	–	1	–	–
8	3	–	–	3	–	–	–	–
9	5	–	–	5	–	–	–	–
10	4	–	–	2	–	–	–	2
11	5	–	–	3	–	2	–	–
12	5	–	–	3	–	2	–	–
13	15	–	–	9	–	5	1	–
14	12	–	–	6	–	3	1	2
15	3	–	1	2	–	–	–	–
16	18	–	–	11	–	7	–	–
17	13	–	–	11	1	1	–	–
18	6	–	–	4	–	2	–	–
19	5	–	2	3	–	–	–	–
20	7	–	–	4	1	1	1	–
Total	191	7	15	118	5	33	8	5
%		3.66	7.85	61.78	2.61	17.27	4.18	2.61

traploid (2n diplos), diploid (n diplos) and aneuploid mitoses both in roots and in shoots, particularly in the younger organs. These results raised two points of interest: (1) Shoot and, in a lower degree, roots showed a chimeric condition in many cases. Two hypotheses could be made about this condition, one suggesting a concrescence of two or more heteroploid cells including cells already endoreduplicated during meristem differentiation, and hence a number of initials higher than one, the other implying that higher chromosome numbers derived from somatic polyploidization after meristem organization. In this case, the meristems would originate from single cells. In favor of the second hypothesis was the occurrence of mitoses with diplochromosomes, which demonstrated that endoreduplication occurred after organ initiation. In this context, it was noted that the structures analyzed were still exposed to hormonal effect, a treatment which is known to stimulate division of preexisting endoreduplicated cell and to induce chromosome endoreduplication as well. (2) A low frequency of haploid cells was found in the callus originating from petals of haploid *Pelargonium*. That endopolyploid nuclei were already present in the differentiated cells of explants was documented by the frequency of diploid mitoses (60%) found in the callus just formed and the occurrence of diplochromosome mitoses. Moreover, the presence of tetraploid nuclei with 2n diplochromosomes in the callus clearly showed the tendency to chromosome doubling during callus growth. When the nuclear cytology of a callus and of the organs derived therefrom is considered, it is seen that the nuclear conditions of the regenerated organ generally reflect those occurring in callus culture. In the experiments above, ploidy levels higher than haploid were found in roots, shoots, and plants. The tendency to selection for the diploid rather than haploid condition could be seen in shoots (although a chimeric state existed) and roots originating from young callus (Table 1). Stabilization in chromosome number in apical meristems is also documented by the data in Table 2 showing that high ploidy levels and heterogeneity in chromosome numbers are more frequent in cell aggregates than in organized structures. Skirvin and Janick (1976a) compared calliclones (see Sect. 2.2) with plants derived from stem, root, and petiole cuttings of five cultivars (Tables 4, 5, 6). Plants from stem cuttings of all cultivars were uniform and identical to the parent clone. Plants from root and petiole cuttings were more variable, with the amount of variation dependent upon cultivar. High variability was associated with calliclones. Aberrant types included changes in plant and organ size, leaf and flower morphology, essential oil constituents, fasciation, pubescence, and anthocyanin pigmentation. Calliclone variation was dependent upon clone and age of callus. These changes were attributed to segregation of chimeral tissue, polyploid changes, and unknown heritable changes, possibly individual chromosome aberrations or gene mutations. The polyploid changes associated with calliclones included chromosome doubling, as well as reduction (haploidy) of calliclone polyploids. The stabilization of some of the aberrant calliclone phenotypes in a second cycle of tissue culture indicated a fixation of genetic changes. The decrease in variability from calliclone cycles suggested cellular variability associated with the original clones.

Tokumasu and Kato (1979) examined the variation of anther-derived plants with respect to chromosome numbers and essential oil components. They used as sources of cultured anthers diploid and tetraploid plants of *P. roseum* Willd. cv. Bourbon. The tetraploid was induced by colchicine treatment. Anthers were

Table 4. Phenotypes and leaf measurements of propagules of the parental clone (lobed) of Rober's Lemon Rose scented geranium. (Skirvin and Janick 1976a)

Propagule Phenotype	n	Leaf measurements: Size (mm^2) (l×w)	Shape (l/w)	Form (w/waist)	Stomate size (μ)
Stem cuttings					
Lobed	15	36.2 ab[a]	0.69 cd	10.4 c	18.0 b
Root cuttings					
Dentated	29	58.1 b	0.61 a	6.8 b	16.9 a
Petiole cuttings					
Dentated	29	39.9 ab	0.64 ab	6.4 b	17.5 b
Hairy	2	35.6 ab	0.66 ab	4.0 a	23.0 ef
Calliclones					
Lobed	1	34.9 ab	0.68 bcd	12.1 d	16.7 a
Dentated	139	53.5 b	0.65 ab	6.6 b	18.4 c
Hairy	21	58.7 b	0.64 ab	4.4 a	24.7 f
Hairy dwarf	1	38.9 ab	0.65 abc	3.7 a	21.7 e
Dwarf	3	16.3 a	0.74 e	3.6 a	19.2 d
Flat leaf	1	60.6 b	0.73 de	4.0 a	19.0 cd

[a] Mean separation within columns by LSD, 5% level.

Table 5. No. of propagules significantly different from either lobed stem cuttings or dentated root cuttings of Rober's Lemon Rose scented geranium. (Skirvin and Janick 1976a)

Propagule phenotype	n	No. plants significantly different (5%): Leaf size	Leaf shape	Leaf form	Stomate size	Total plants
			from lobed stem cuttings			
Root cuttings						
Dentated	29	10	8	1	0	16
Leaf cuttings						
Dentated	29	0	1	1	0	2
Hairy	2	0	0	2	2	2
Calliclones						
Lobed	1	0	0	0	0	0
Dentated	139	19	0	5	5	29
Hairy	21	2	0	17	21	21
Hairy dwarf	1	0	0	1	1	1
Dwarf	3	0	0	3	0	3
Flat leaf	1	0	0	1	0	1
			from dentated root cuttings			
Petiole cuttings						
Dentated	29	0	0	0	0	0
Hairy	2	0	0	0	2	2
Calliclones						
Lobed	1	0	0	1	0	1
Dentated	139	1	0	4	5	10
Hairy	21	1	0	0	21	21
Hairy Dwarf	1	0	0	0	1	1
Dwarf	3	1	1	0	0	2
Flat leaf	1	0	0	0	0	0

Table 6. Comparison of phenotypes derived from propagules of Rober's Lemon Rose scented geranium. (Skirvin and Janick 1976a)

Variable	Parental (lobed)	Calliclones					
		Lobed	Dentated	Hairy	Hairy dwarf	Dwarf	Flat leaf
Chromosome number	72	72	72	144	–	72	72(?)
Stomate length (μ)	18.0	16.7	18.4	24.7	21.7	19.2	19.0
Flower							
Relative size	Normal	Normal	Normal	Large	–	Normal	Normal
No. stigma parts	2	2	5	5	–	5	5
Pollen							
Volume ($\mu^3 \times 1000$)	363	371	296	742	–	280	180
Viability (%)	0	0	0	56	–	0	0
Hair length (mm)	0.53	0.49	0.54	0.66	0.69	0.47	0.51
Gland size (μ)							
height	27	–	30	34	–	27	25
width	14	–	15	16	–	14	16
No. major peaks by GLC	3	3	3	3	–	3	2
Plant height (cm)	–	–	92	97	54	60	70

cultured according to the method first mentioned of Abo-El-Nil and Hildebrandt (1971); 26 plants were obtained from anthers of the diploid and 23 plants from those of the tetraploid. Chromosome counts showed that the derivatives of the diploid plants remained diploid but those of the tetraploid were "haploid" (diploid). In order to examine the variation of oil components, only three main compounds were considered: iso-menthone, citronellol, and geraniol. With respect to citronellol and geraniol, 4x-derivatives always showed higher values in the coefficients of variation than 2x-derivatives throughout all seasons. This suggested that 4x-derivatives were composed of genetically different plants. Therefore, these plants probably originated from pollen grains.

2.4 Objectives for Improvement and Utilization of Somaclonal Plants

It is known that the basic chromosome numbers of the genus *Pelargonium* are 8, 9, 10, and 11. *Pelargonium* plants can be propagated very easily by stem cuttings or leaf bud cuttings, and these are the methods more commonly used by gardeners for commercial purposes. Although seed-propagated geraniums have achieved considerable popularity, they lack certain horticultural attributes such as the semi-double florets and nonshattering inflorescences found in the vegetatively propagated cultivars (Harney 1982). The vegetative propagation of this crop is complicated by viral pathogens which very often infect the plants. It is very important, therefore, to have or to produce pathogen-free geraniums by tissue culture to use as mother plants for mass clonal propagation (see Sect. 2.2.1). New cultivars may be produced by traditional breeding methods, i.e., clonal selection or cross-fertilization and selection trials or, if possible, by cross between different species, followed by backcross with one of the parent species. The breeder's genetic pool is frequently extremely large and the variations possible are enormous, but the combination of genes in the desired manner by repeated backcrossings may take an excessively long time. The application of pollen culture for the production of haploid plants which may be converted to the fertile homozygous diploid (or polyploid) forms to use in breeding programs has obvious potential benefits. However, the use of this method also requires a certain time (Holdgate 1977; Reisch 1983).

The creation of polyploids in *Pelargonium* can be very useful in some cases. For example, hybridization between two important species cultivated for geranium oil production in Japan, *P. roseum* and *P. denticulatum*, is impossible because they are male sterile and cannot produce seeds, while the artificially induced tetraploids of these two species became fertile and produced good seeds. Thus, it is also possible to obtain hybrid seeds when these different species are crossed (Tamai et al. 1963). Genetic variations can be produced by culturing plants treated with various mutagenic agents, but these changes are random and the resulting plants are often chimeral.

Crop improvement by traditional methods can be complemented by tissue culture techniques. Phenotypic variation has been reported in a great number of plant species regenerated using tissue culture (Larkin and Scowcroft 1981). This genetic variability may be agriculturally useful when integrated into existing breeding programs. In fact, exploiting cellular variation (either preexisting or tissue culture-induced) through tissue culture may provide a new means for plant improvement. This can be achieved by imposing selection on the variable populations of cells in culture that may arise naturally or be induced by mutagenic agents (Skirvin and Janick 1976a). Variants associated with in vitro culture may be of different types: (1) physical and morphological changes in undifferentiated callus; (2) chromosomal and true genetic changes manifested in differentiated plants. Many plants obtained via bud formation (or via somatic embryogenesis) will be of single-cell origin and, hence, of pure mutant (variant) type, avoiding the traditional chimerism phenomena of mutation breeding studies (Skirvin 1977). No adequate explanation has ever been advanced for the occurrence of culture variation. The culture environment itself may be mutagenic, since stable variants arise from it. Moreover, besides genetic mutation, karyotype changes, chromosome rearrangements, transposable genetic elements, gene amplification, and depletion may be responsible for genetic variation in cultured cells (Larkin and Scowcroft 1981; D'Amato 1985). Therefore the main objective to reach for the improvement of *Pelargonium* species is the induction of somaclonal variation through tissue culture methods (also using mutagenic treatments, protoplast fusion, or DNA uptake) and the recovery of this variability in regenerated plants (calliclones). The study of Skirvin and Janick (1976a) may be considered a model system for work in intraclonal plant improvement. In fact, following their extensive investigations of somaclonal variation in *Pelargonium* sp. an improved scented geranium that they named Velvet Rose was developed. They proposed that the variability among regenerated plants might be useful, especially with highly polyploid, sterile, asexually propagated clones. Plant variation generated by the use of the tissue culture cycle may be a pool on which selection can be imposed. However, it may be necessary and useful to increase the percentage of variability in callus and calliclones by techniques such as mutagenic agents or selection applied to single cell clones for stress conditions and/or ability to resist or utilize specific metabolites (Skirvin and Janick 1976a).

Molecular biology could aid the introduction of alien variation by making possible the isolation of particular genes from an alien source, not necessarily a related species, and their incorporation into the species of interest, either by a vector, or possibly by direct injection (Austin et al. 1986). These new biotechnologies might be particularly useful for the introduction of new perfume genes into traditional scented *Pelargonium* species, avoiding the sexual hybridization methods for the introgression of desirable alien genes. Moreover, many male sterile *Pelargonium* species, after regeneration in vitro, are tetraploid and male fertile, making it possible to use them in hybridization programs.

References

Abo-El-Nil MM, Hildebrandt AC (1971) Differentiation of virus-symptomless geranium plants from anther callus. Plant Disease Rep 55:1017–1020

Abo-El-Nil MM, Hildebrandt AC (1973) Origin of androgenetic callus and haploid geranium plants. Can J Bot 51:2107–2109

Austin RB, Flavell RB, Henson IE, Lowe HJB (1986) Molecular biology and crop improvement. Univ Press, Cambridge

Bailey LH (1963) The Standard cyclopedia of horticulture, vol 3. MacMillan, New York

Bailey LH (1964) Manual of cultivated plants. MacMillan, New York

Beauchesne G, Albony J, Morand JC, Daguenet J (1977) Clonal propagation of *Pelargonium* × *hortorum* and *Pelargonium* × *peltatum* from meristem culture for disease-free cuttings. Acta Hortic 78:397–402

Bennici A (1974) Cytological analysis of roots, shoots and plants regenerated from suspension and solid in vitro cultures of haploid *Pelargonium*. Z Pflanzenzücht 72:199–205

Bennici A, Buiatti M, D'Amato F (1968) Nuclear conditions in haploid *Pelargonium* in vivo and in vitro. Chromosoma 24:194–201

Brown JT, Charlwood BV (1986) The control of callus formation and differentiation in scented pelargoniums. J Plant Physiol 123:409–417

Brown JT, Hegarty PK, Charlwood BV (1987) The toxicity of monoterpenes to plant cell cultures. Plant Science 48:195–201

Cassells AC (1979) The effect of 2,3,5-triiodobenzoic acid on caulogenesis in callus cultures of tomato and *Pelargonium*. Physiol Plant 46:159–164

Cassells AC, Carney BF (1987) Adventitious regeneration in *Pelargonium domesticum* Bailey. Acta Hort 212:419–425

Cassells AC, Minas G (1983a) Plant and in vitro factors influencing the micropropagation of *Pelargonium* cultivars by bud-tip culture. Scientia Hortic 21:53–65

Cassells AC, Minas G (1983b) Beneficially-infected and chimeral *Pelargonium*: implications for micropropagation by meristem and explant culture. Acta Hort 131:287–297

Cassells AC, Minas G, Long R (1980) Comparative tissue culture studies of pelargonium hybrids using meristem and explant culture: chimeral and beneficially infected varieties. In: Ingram DS, Helgeson JP (eds) Tissue culture for plant pathologists. Blackwell Scientific, Oxford, pp 125–130

Chen HR, Galston AW (1967) Growth and development of *Pelargonium* pith cells in vitro. II. Initiation of organized development. Physiol Plant 20:533–539

Daker MG (1966) "Kleine Liebling". A haploid cultivar of *Pelargonium*. Nature (London) 211:549–550

D'Amato F (1965) Endopolyploidy as a factor in plant tissue development. In: White PR, Grove AR (eds) Proc Int Conf Tissue Culture. McCutchan, Berkely, pp 449–462

D'Amato F (1972) Significato teorico e pratico delle colture in vitro dei tessuti e cellule vegetali. Acc Naz Lincei 173:1–55

D'Amato F (1985) Cytogenetics of plant cell and tissue cultures and their regenerates. CRC Crit Rev Plant Sci 3:73–112
Debergh P, Maene L (1977) Rapid clonal propagation of pathogen-free *Pelargonium* plants starting from shoot tips and apical meristems. Acta Hortic 78:449–454
George EF, Sherrington PD (1984) Plant propagation by tissue culture. Exegetics, Eversley Basingstoke, Hants, England
Hamdorf G (1976) Propagation of pelargonium varieties by stem-tip culture. Acta Hortic 59:143–151
Hammerschlag F, Bottino P (1981) Effect of plant age on callus growth, plant regeneration, and anther culture of geranium. J Am Soc Hortic Sci 106(1):114–116
Harney PM (1982) Tissue culture propagation of some herbaceous horticultural plants. In: Tomes DT, Ellis BE, Harney PM, Kasha KJ, Peterson RL (eds) Application of plant cell and tissue culture to agriculture and industry. Univ Press, Guelph, pp 187–208
Holdgate DP (1977) Propagation of ornamentals by tissue culture. In: Reinert J, Bajaj YPS (eds) Applied and fundamental aspects of plant cell, tissue, and organ culture. Springer, Berlin Heidelberg New York, pp 18–43
Janick J, Skirvin RM, Janders RB (1977) Comparison of in vitro and in vivo tissue culture system in scented geranium. J Hered 68:62–64
Larkin PJ, Scowcroft WR (1981) Somaclonal variation – a novel source of variability from cell cultures for plant improvement. Theor Appl Genet 60:197–214
Murashige T, Skoog F (1962) A revised medium for rapid growth and bioassays with tobacco tissue cultures. Physiol Plant 15:473–497
Pillai SK, Hildebrandt AC (1968) Geranium plants differentiated in vitro from stem tip and callus cultures. Plant Disease Rep 52:600–601
Pillai SK, Hildebrandt AC (1969) Induced differentiation of geranium plants from undifferentiated callus in vitro. Am J Bot 56:52–58
Reisch B (1983) Genetic variability in regenerated plants. In: Evans DA, Sharp WR, Ammirato PV, Yamada Y (eds) Handbook of plant cell culture, vol 1. Techniques for propagation and breeding. Macmillan, New York, pp 748–769
Skirvin RM, Janick J (1976a) Tissue culture induced variation in scented *Pelargonium* spp. J Am Soc Hortic Sci 101(3):281–290
Skirvin RM, Janick J (1976b) Velvet Rose *Pelargonium*, a scented geranium. HortSci 11:61–62
Skirvin RM (1977) Natural and induced variation in tissue culture. Euphytica 27:241–266
Tamai T, Tokumasu ST, Yamada K (1963) Studies on the breeding of *Pelargonium* species used for the essential oil production. II. Artificially induced tetraploid plant in *Pelargonium denticulatum*. Jpn J Breed 13(3):7–12
Tokumasu S, Kato M (1979) Variation of chromosome numbers and essential oil components of plants derived from anther culture of the diploid and the tetraploid in *Pelargonium roseum*. Euphytica 28:329–338
Ward HB, Vance BD (1968) Effects of monochromatic radiations on growth of *Pelargonium* callus tissue. J Exp Bot 58:119–124
Welander T (1982) Mikroförökning av *Pelargonium zonale* hybr. Sver Lantbruksuniv Rap 21 (Alnarp)
Yarrow SA, Cocking EC, Power JB (1987) Plant regeneration from cultured cell-derived protoplasts of *Pelargonium aridum*, *P.* x *hortorum* and *P. peltatum*. Plant Cell Rep 6:102–104

IV.2 Somaclonal Variation in *Fuchsia*

J. BOUHARMONT and P. DABIN[1]

1 Introduction

1.1 Importance and Distribution of the Plant

In the wild, the genus *Fuchsia* (Onagraceae) comprises about 100 species, living mainly in the mountains of South and Central America, with a small section in the Pacific. Some species were introduced into Europe at the end of the 18th century as garden plants.

In nature, the flowers are generally pollinated by hummingbirds, and the coevolution of the partners has led to a broad array of flower forms and colors. Nevertheless, most species are not reproductively isolated; many interspecific hybridizations have been performed by professional and nonprofessional horticulturists. Some 6000 cultivars have been obtained in Great Britain, the United States, Holland, France, and Germany over the past 150 years, but only slightly more than 2000 of them are still currently in cultivation (Ewart 1982). Nevertheless, the number of wild species involved in the breeding programs is rather small. Other cultivars are spontaneous mutations derived from older forms.

The cultivation of the fuchsias has become very popular in Britain and France, and different cultivars have been adapted for gardens, greenhouses, and homes. The foundation of the American Fuchsia Society in 1929 and the British Fuchsia Society in 1938 started a worldwide chain of fuchsia societies, embracing Holland, Belgium, France, the FRG, Denmark, South Africa, Zimbabwe, Australia, and New Zealand, each country with its own climatic problems to contend with, but all succeeding in making the fuchsia more popular today (Ewart 1982). Later, fuchsia was extended through most temperate regions, mainly Western Europe and North America. Some cultivars are permanently planted in gardens in areas with humid climate and mild winters. Many others are maintained for only one season inside or outside.

Fuchsias produce a wealth of color, whether grown in pots in the greenhouse or outdoors in the border or rockery, as climbers against a wall, in hanging baskets, or as house plants. All flower profusely in sun or shade (Proudley and Proudley 1975).

According to Berry (1982), the genus *Fuchsia* is divided into nine sections: *Quelusia, Fuchsia, Ellobium, Hemsleyella, Kierschlegeria, Schufia, Jimenezia, Encliandra,* and *Skinnera*, with a total of 97 species.

[1]Laboratoire de Cytogénétique, Université Catholique Louvain, Place Croix du Sud 4, 1348 Louvain-la-Neuve, Belgium

Biotechnology in Agriculture and Forestry, Vol. 11
Somaclonal Variation in Crop Improvement I (ed. by Y.P.S. Bajaj)

For North (1979), the fuchsia cultivars principally come from three wild species: *F. magellanica* (*Quelusia*), *F. fulgens*, and *F. macrostigma* (*Fuchsia*). Moreover, some hybrid cultivars called "triphylla" combine three species of the section *Fuchsia*.

The basic chromosome number for the genus is 11 and the somatic numbers reported for the wild species are 22, 33, 44 (Darlington and Wylie 1955). For the garden forms, the higher number would be 88.

1.2 Genetic Variability and Objectives

In spite of the rather limited number of wild species involved in their origin, fuchsia cultivars display a broad diversity. The size, form, and color of the flowers are the most polymorphic characters. The wild species and some cultivars have four petals, but many hybrids and mutants have a double corolla and sometimes a large number of petals. The shape of the plant is generally bushy, but some forms are climbing or hanging, and used to make up baskets.

In spite of the available combinations of red, lilac, rose, and white colorations in the sepals and petals, it has not yet been possible to obtain completely white flowers. Another missing character is the yellow pigmentation. The hardiness of the cultivars is variable, but improving cold and frost tolerance is an important goal in many countries. Adaptation of the plant shape to different cultivation patterns is another objective. Finally, as in other ornamental plants, many morphological anomalies are welcome.

Some desirable characters can be obtained through further crossing between cultivars or species, but mutation is often a more appropriate way because it can preserve the general phenotype of a good commercial variety and modify only one character, like the shape of the plant, the color of flower parts, the leaf pigmentation, or the cold-hardiness. Mutations can be induced by irradiations or treatments by chemical mutagens, but somaclonal variation represents a substitute or a complementary means. Many somaclonal variants have been reported in polyploid plants vegetatively propagated (potato): the same results could be expected in fuchsias, since most cultivars are polyploids derived from hybridizations.

2 In Vitro Culture in *Fuchsia*

2.1 Micropropagation

Fuchsias are normally propagated by cuttings without difficulty, but in vitro culture has been proposed for some cultivars, when cutting propagation is not efficient, as well as for obtaining virus-free plants. Nevertheless, little work has been carried out on these plants. A summary of in vitro culture is given in Table 1.

Stevenson and Harris (1980) obtained an efficient multiplication in *Fuchsia hybrida* (cv. Swingtime) every 4 weeks through shoot culture on a medium containing the major and minor salts, sugar, and vitamins according to Gamborg et al.

Table 1. In vitro culture of *Fuchsia*

Cultivar	Inoculum	Medium (mg/l)	Growth Response	Reference
Swingtime	Shoot tips,	B5	Growth induction	
	petioles,	B5 + BAP (3)	Buds and plants	Stevenson and Harris (1980)
	leaves,	B5 + 2-iP (5)	Buds and plants	
	roots			
Swingtime,	Ovaries	B5 + IAA (1)	Callus	
Rose Van Den Berg.		B5 + 2,4-D (1)	Callus	
		B5 + NAA (1)	Callus	
		B5 + NAA (1) + IAA (1)	Callus	Dabin and Beguin (1987)
		B5 + BAP (1)	Adventitious buds	
		B5 + IAA (1)	Embryos and plants	Bouharmont and Dabin (1986a,b)
		B5 + kinetin (1)	Embryos and plants	
Constance,	Petals,	B5 + IAA (1)	Callus	
Swingtime,	sepals,	B5 + 2,4-D (1)	Callus	Dabin and Bouharmont (1984)
	ovaries,	B5 + BAP (1)	Buds and plants	
	filaments	B5 + IAA (1)	Embryos and plants	Bouharmont and Dabin (1986a,b)
		B5 + kinetin (1)	Embryos and plants	
Swingtime	Leaves,	B5 + IAA (1 to 5)	Callus	Dabin and Vaerman (1985)
Constance,	veins	B5 + 2,4-D (1 to 5)	Callus	Bouharmont and Dabin (1986a,b)
		B5 + BAP (1)	Buds and plants	
Swingtime	–	–	Buds and plants	Harris and Mason (1983)
F. hybrida	Stem segments	MS/2	Buds and plants	Yang (1981)
	Stem tips	MS + NAA (1)	Roots	
F. hybrida	Petiole segments	Modified MS	Callus and roots	Klimaszewska (1981)
F. hybrida (18 cv.)	Shoot tips,	MS + BAP (1) + IAA	Axillary buds,	
	nodes	or NAA or IBA (1)	plants	Kevers et al. (1983)
		MS + IBA (1)	Roots	
Constance,				
Symphony,				
Koralle,		B5	Induction	Dabin and Choisez-Givron (1982)
P.M. Slater,		B5 + BAP (3)	Axillary buds	
Viva Ireland,	Axillary buds	B5	Elongation	
La Campanella,		B5	Roots	
Mrs. W. Rundle,				Choisez-Givron and Dabin (1983)
E. Ott,				
S. Cash.				

(1968) and BAP (3 mg/l) or 2-iP (5 mg/l). The shoot tips were cultured for 2 weeks on agar medium and then transferred to the same liquid medium with BAP or 2-iP for proliferation. The proliferated shoots elongated when the cytokinin was eliminated and rooted on the same medium solidified with agar before being transferred in greenhouse.

Dabin and Choisez-Givron (1982) and Choisez-Givron and Dabin (1983) used the same culture medium with BAP (3 mg/l) for the propagation of nine other cultivars. In vitro behavior of the explants, evolution of the axillary buds, and multiplication efficiency showed large differences according to the cultivars but, for some of them, the physiological state of the mother plants were probably also involved. Later, some other cultivars and species were successfully propagated in the same conditions.

According to the genotype, 0 to 23 viable plants were obtained from each explant and the same multiplication rates were observed every 35 days starting from shoots elongated in vitro.

Rapid development of axillary buds from shoot tips and nodes of 18 cultivars was also obtained on solid Murashige and Skoog medium (MS) with 1 mg/l BA + 0.1 mg/l of an auxin (NAA, IAA, IBA); NAA was the most effective auxin. Vegetative shoots were subsequently isolated and they developed up to 15 supplementary axillary shoots on the same solid medium. Agitated and nonagitated liquid media of the same composition were less effective. The rooting is done in MS medium with 1 mg/l IBA before transfer in soil (Kevers et al. 1983). Yang (1981) and Klimaszewska (1981) have cultured various explants of *Fuchsia hybrida* on modified MS medium to regenerate whole plants. Harris and Mason (1983) have described two methods for in vitro commercial propagation of plants, including *Fuchsia hybrida* cv. Swingtime.

2.2 Callus Culture and Plant Regeneration

For callus induction, various explants of two cultivars (Constance and Swingtime) were inoculated on the same medium supplemented with IAA or 2,4-D (1 mg/l) and cultivated in the dark (Dabin and Vaerman 1985). All ovary fragments produced calli after 1 month. Callogenesis was also highly efficient for leaf parenchyma, petals, sepals, and filaments, while calli were less frequent and their proliferation was slower for leaf veins and stem segments. Table 2 gives the percentages of callogenesis on B5 medium with IAA or 2,4-D and adventitious budding on B5 medium with BAP (Dabin and Bouharmont 1984). After transfer to a medium containing BAP (1 mg/l), some calli regenerated plantlets through adventitious meristems.

The origin of the calli has been observed microscopically on explants fixed a few days after inoculation. Cell proliferation always starts in parenchyma, after dedifferentiation of the cells next to the vascular bundles. The growth of the calli and the way of further differentiation depend on the explant origin and on the culture conditions. These calli contain generally irregular clumps of tracheids. In some conditions, they produce a number of adventitious roots: meristems differentiate from groups of meristematic cells located in the inner parts of the calli, and the roots

Table 2. Culture of flower fragments

Explant	Auxins	Number	cv. Constance calli		Buds	Number	cv. Swingtime calli		Buds
			1 month	2 months			1 month	2 months	
	0	20	0%	0%	0%	16	0%	0%	0%
Petals	IAA	16	94%	100%	31%	12	58%	92%	8%
	2.4-D	25	96%	100%	80%	12	100%	100%	25%
	0	20	0%	0%	0%	16	0%	0%	0%
Sepals	IAA	22	100%	100%	72%	20	87%	87%	42%
	2.4-D	16	100%	100%	81%	20	100%	100%	17%
	0	10	40%	40%	40%	4	100%	100%	0%
Ovaries	IAA	10	100%	100%	80%	6	100%	100%	67%
	2.4-D	10	100%	100%	100%	6	83%	100%	17%
	0	25	0%	0%	0%	16	0%	0%	0%
Filaments	IAA	32	84%	91%	31%	20	20%	20%	0%
	2.4-D	32	78%	78%	0%	20	35%	35%	0%

grow through the parenchymatic tissues. In contrast, the origin of the apical meristems is superficial: a large number of parenchymatic cells are transformed in irregular meristems proliferating at the surface of the calli. They give rise to more or less regular foliaceous knobs, but few of them develop true apical meristems and shoots.

Different experiments have been performed in order to increase the efficiency of plant regeneration. The best hormonal complement is composed of IAA (1 mg/l) and kinetin (1 mg/l).

For conservation and growth of undifferentiated calli, the solid or liquid culture medium contains a higher vitamin concentration (five times the original Gamborg's medium), and increased amount of sucrose (30 g/l), tryptone (1 g/l) and 2,4-D (1.3 mg/l). These calli retain their organogenetic potential when they are transferred to an appropriate medium.

2.3 Somatic Embryogenesis

Two types of somatic embryogenesis have been observed according to the superficial or internal location of the primordia. In general, when calli induced in the dark are transferred on a solid medium containing IAA or kinetin (1 mg/l) and cultivated under 16 h daily illumination, somatic embryos differentiate on their surface besides adventitious meristems. Some embryos are similar to zygotic ones, but most have poorly differentiated cotyledons and discrimination between true embryos and adventitious shoots is not always easy (Figs. 1, 2).

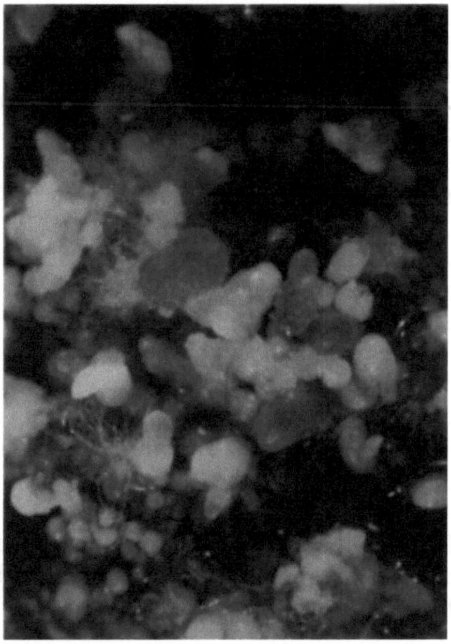

Fig. 1. Somatic embryogenesis in *Fuchsia*. Proliferation of embryos on a callus of the cv. Rose Van den Berg on B5 medium with IAA and kinetin (1 mg/l)

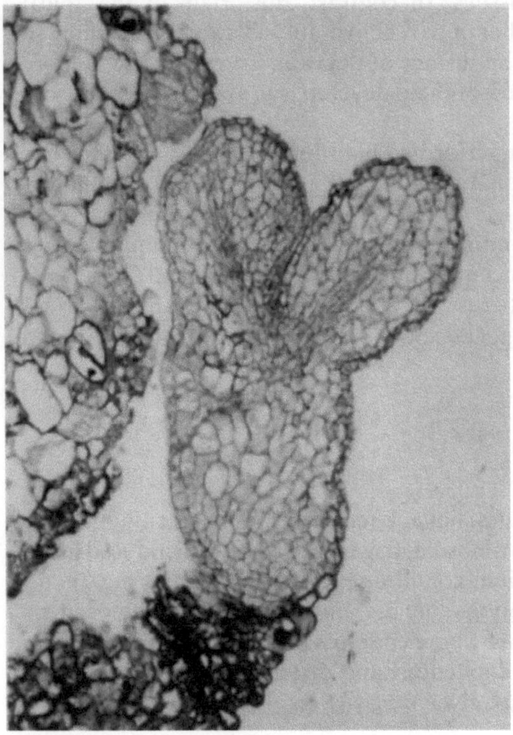

Fig. 2. Somatic embryogenesis in *Fuchsia*. Section in a superficial somatic embryo

The origin of these superficial embryos is a small group of callus cells producing a globular or elongated bud with a regular epidermis. A small apical meristem and more or less regular cotyledons differentiate at one end. At the other end, a typical radicle is rarely apparent, but several adventitious roots develop from the base of the short hypocotyl. These embryos separate easily from the callus before or after their germination (Bouharmont and Dabin 1988).

Some calli or callus parts become friable and translucent. Isolated internal cells divide and produce small white meristematic globules growing into typical bipolar embryos with cotyledons, hypocotyl, and radicle. Compared to superficial ones, these embryos show a more regular form, and they appear to derive from single cells.

Several modifications of the culture medium have been tested in order to increase the number of somatic embryos and to improve their germination. A sucrose concentration of 50 g/l instead of the original 20 g/l generally gave a higher number of embryos in the calli in a shorter time. Better results were obtained on a medium containing the high vitamin concentration (five times), 30 g/l of sucrose, IAA (1 mg/l), kinetin (1 mg/l), and tryptone (100 mg/l); under these conditions, the differentiation of the embryos is more frequent and uniform. Nevertheless, germination rates are low: thus it seems advisable to transfer calli, after embryo differentiation, on a last medium containing only 5 g/l of sucrose for germination.

3 Somaclonal Variation

3.1 Morphological Variation

The first regenerated plants were derived from a single callus induced on a leaf fragment of the cultivar Swingtime. After multiplication and culture in the greenhouse for 2 years, several clones showed modifications in flower size or pigmentation. The first observed variants were mosaic or chimeral plants: some branches were generally different from the others. After several propagations through cuttings, some clones became uniform, but other plants remained heterogeneous (Fig. 3).

Swingtime and many other cultivars are tetraploid hybrids with 44 chromosomes. The first variant was characterized by large flowers, stout sepals and petals; the root tips showed about 88 chromosomes. Thus the large flowers of that variant may be explained by a chromosome doubling induced by callus proliferation.

A second variant is also probably polyploid; its flowers are large, sepals are stout and broader than in the first form, with green apex, and supplementary short petals are located outside the normal corolla. Another form preserved the original flower size, but the number of petals was increased. The root tips of another variant revealed 44 somatic chromosomes, as in the mother plant. A clone is characterized by very small flowers. Some variations are not well characterized: the flowers may

Fig. 3. Variation of the flower size and shape on a plant regenerated from a leaf callus of the cv. Swingtime (somaclonal variation)

be very dark to pale pink on the same plant, sepals remain scarcely expanded or they are more or less fully curved back (Bouharmont and Dabin 1986a).

3.2 Isolation of an Embryogenic Cell Line

Such a line appeared in a suspension culture derived from an ovary of the cultivar Rose Van den Berg. After induction in the dark, calli have been propagated for 19.5 months in different solid and liquid media. The calli were then transferred in a liquid medium containing the high vitamin concentration (five times), sucrose (30 g/l), 2,4-D(0.65 mg/l) and tryptone (1 g/l). After 2 months in a roller bottle, some calli break up into a large number of microcalli. These microcalli were maintained in suspension, and their proliferation was continued indefinitely in the same culture medium or with a higher 2,4-D concentration (1.3 mg/l). Very friable calli also proliferate after plating of the suspension on the same solid medium.

When microcalli are transferred in an embryogenic liquid solution or plated on the same agar medium, they differentiate, and give a very large number of somatic embryos after only 4 to 6 weeks. These embryos germinate on the same medium and regenerate normal plants. The total number of plantlets obtained in this way is high. Nevertheless, the proportion of germinating embryos is low and must be increased for an efficient application in selection.

Variants have not been observed among the plants recovered from these somatic embryos, but the size of the sample is at present still limited.

4 Possible Use of Induced Variation

Several mutagenic treatments have been applied in order to increase the genetic diversity and the chance of finding new interesting characters (Bouharmont and Dabin 1986b). One of the objectives was the complete elimination of flower pigments; such mutations are possible through irradiation and would be welcomed by horticulturists. Chimeral plants with different pigmentation levels in the cell layers of the petals or leaves do not seem present in the available cultivars; mutations could induce such chimeral structures in the apical meristems.

Treatments by irradiation or chemical mutagens have been applied on shoots of some fuchsias, and morphological modifications have been observed in the plants, but the induced characters were unstable or unfavorable.

On the other hand, explants have been treated with several mutagens before callogenesis: these mutagens were methyl nitro-nitrosoguanidine (MNNG), ethyl methane-sulfonate (EMS), sodium azide, and hydrazine. Soaking of the explants (ovaries) in aqueous solutions of mutagens, for 1 hour, does not significatively reduce callus proliferation on B5 medium with 2,4-D for the following concentrations: MNNG: 80 mg/l, EMS: 0.5%, NaN_3: 0.1 M, N_2H_4: 0.05 M.

After 1 to 6 months on B5 medium with kinetin or IAA + kinetin (1 mg/l), differentiation is still possible and the percentages of differentiated calli comprise between 32% for MNNG and 35% for EMS.

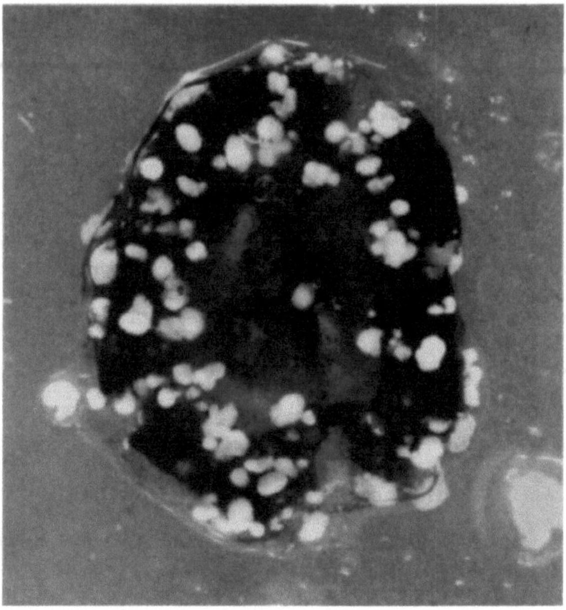

Fig. 4. Differentiation of somatic embryos on a callus induced on an ovary fragment of the cv. Rose Van den Berg cultivated at low temperature (10°C) on B5 medium with kinetin (1 mg/l)

Induced calli were also treated by MNNG (80 mg/l) and EMS (1%). Few results were obtained from these treatments; the soaking in EMS solution causes the death of all calli; while those treated by MNNG develop normally, the differentiation of somatic embryos sometimes occurred after 2 months on B5 medium with kinetin (1 mg/l). No variation was observed in the 500 regenerated plants.

Some plants of the cv. Rose Van Den Berg recovered through somatic embryogenesis showed leaves with a white margin: they are probably chimeras resulting from a chlorophyll mutation in the epidermis, induced by MNNG.

In order to select cell lines and plants more tolerant to low temperatures, explants and calli were cultivated for several months in cool rooms (7–13°C). In these conditions, callus induction was slower, but the growth of the calli was not inhibited. The induction of embryogenesis was earlier and the number of somatic embryos was higher at the lowest temperatures (Fig. 4).

References

Berry PE (1982) The systematics and evolution of *Fuchsia* sect. *Fuchsia* (Onagraceae). Ann Missouri Bot Garden 69.1:1–68

Bouharmont J, Dabin P (1986a) Somaclonal variation in some cultivars of *Fuchsia*. In: Semal J (ed) Somaclonal variations and crop improvement. Nijhoff, Dordrecht, pp 257–260

Bouharmont J, Dabin P (1986b) Application des cultures in vitro à l'amélioration du *Fuchsia* par mutation. In: Nuclear techniques and in vitro culture for plant improvement. IAEA, Vienna, pp 339–347

Bouharmont J, Dabin P (1988) Origine et évolution des embryons somatiques chez les fuchsias. Cellule 74:309–316

Choisez-Givron D, Dabin P (1983) Application de la culture in vitro à sept variétés cultivées de *Fuchsia* et comparaison des taux de multiplication. Bull Soc R Bot Belg 16:25–30

Dabin P, Beguin F (1987) Somatic embryogenesis in *Fuchsia*. Acta Hortic 212:725–726

Dabin P, Bouharmont J (1984) Use of in vitro cultures for mutation breeding in *Fuchsia*. In: Novak J, Havel L, Dolezel J (eds) Plant tissue and cell culture, application to crop improvement. Acad Sci Prague, Czech, pp 451–452

Dabin P, Choisez-Givron D (1982) Culture in vitro de deux variétés de *Fuchsia* (Constance et Symphony) et comparaison avec les méthodes traditionnelles. Bull Soc R Bot Belg 115:27–32

Dabin P, Vaerman AM (1985) Callogenèse et organogenèse chez deux cultivars de *Fuchsia* (Constance et Swingtime). Bull Soc R Bot Belg 118:172–178

Darlington CD, Wylie AP (1955) Chromosome atlas of flowering plants. Allen & Unwin, London, 519 pp

Ewart R (1982) Fuchsia lexicon. Blandford, Poole, Dorset, 336 pp

Gamborg OL, Miller RA, Ojima K (1968) Nutrient requirements of suspension cultures of soybean root cells. Exp Cell Res 50:151–158

Harris RE, Mason EBB (1983) Two machines for in vitro propagation of plants in liquid media. Can J Plant Sci 63.1:311–316

Kevers C, Coumans-Gilles M-F, Coumans M, Gaspar T (1983) In vitro vegetative multiplication of *Fuchsia hybrida*. Sci Hortic 21(1):67–71

Klimaszewska K (1981) Plant regeneration from petiole segments of some species in tissue culture. Acta Agrobot 34(1):5–28

North C (1979) Plant breeding and genetics in horticulture. McMillan, London, 150 pp

Proudley B, Proudley V (1975) Fuchsias in colour. Blandford, Poole, Dorset, 206 pp

Stevenson JH, Harris RE (1980) In vitro plantlet formation from shoot-tip explants of *Fuchsia hybrida* cv. Swingtime. Can J Bot 58:2190–2192

Yang NB (1981) In vitro clonal propagation of twelve plant species. Acta Bot Sin 23(4):284–287

IV.3 Somaclonal Variation in Carnations

B. Leshem[1]

1 Introduction

The carnation (*Dianthus caryophyllus* L.) is a member of the Caryophyllaceae family. It is a semi-hardy perennial with branching stems, bearing linear, glaucous leaves in opposite and decussate pairs. Each stem forms a terminal flower and the inflorescence is a loose cyme. The flowering shoot can be treated for marketing in one of two forms: either the flower buds formed on short lateral shoots arising from the axils of the upper leaves are removed to leave one large, terminal flower on a long leafy stem ("standard" type), or the terminal flower bud is removed at an early stage to encourage more even development of the lateral flowers, which then produce a multiple-flowered stem ("spray" or "miniature" types). Special cultivars have been selected for spray production.

Modern carnation cultivars are perennial plants and are grown for cut flower production in heated glasshouses in northwestern Europe and in the northern USA, and in unheated plastic structures in the Mediterranean area. They are also grown in equatorial regions if the altitude is sufficiently high to provide low air temperatures, e.g., in Colombia and Kenya. Propagation is by stem cuttings, 10 to 15 cm long, from disease and virus-free mother plants. Progeny testing is required to avoid multiplication from mutants having flowers of poor shape, low productivity, or showing reversion to types that flower only in long day. Rooted cuttings are planted in beds, where they are allowed to produce a succession of flowers for one (miniature types) or more years (standard types) (Bunt and Cockshull 1985).

The glasshouse carnation is of great commercial interest. It has gradually become one of the most important cut flower crops since the development in the USA of the cultivar William Sim in 1938, and particularly since an impressive number of sports and high quality strains were obtained from this cultivar by selection and stabilization within the genetic variation that arose spontaneously. Besides these strains, spray (miniature) type carnations have more recently attracted attention. Carnation is among the three most important flowers on an international scale, and it is produced in the largest numbers. The most important countries to produce carnation are (listed according to their importance) Italy, Mexico, Columbia, Holland, Japan, and Israel. The supply of carnation (both standard and spray) in Europe was 1150 million pieces in 1985, and increased to 1300 million

[1]Department of Plant Genetics and Breeding, ARO, The Volcani Center, P.O. Box 6, Bet Dagan 50-250, Israel

Biotechnology in Agriculture and Forestry, Vol. 11
Somaclonal Variation in Crop Improvement I (ed. by Y.P.S. Bajaj)

pieces in 1986. Imports of carnation in the USA were 800 and 980 million pieces in 1985 and 1986, respectively (Erler and Siegmund 1986; Hoogendoorn 1987).

As a cut flower crop with such worldwide importance, breeding efforts are constantly aimed at extending the range of colors and shape of the flowers by exploiting the variation within the genus *Dianthus* in combination with modern horticultural and tissue culture techniques (Sparnaaij and Demmink 1983).

2 The Chimeral Nature of Carnation – A Source of Variation

Carnation is indeed bred for new varieties; however, a vast majority of the existing cultivars are sports which are constantly being formed spontaneously. The chimeral nature of the sports was first shown by Richter and Singleton (1955). They observed induced color changes due to somatic mutations which were induced by radiation, as well as histological changes, which were ascribed to the genetic replacement of the epidermal layer by deeper-lying tissue. Sagawa and Mehlquist (1957), by treating shoots with 2.5 or 5 krad X-irradiation, showed that the sports White Sim and Pink Sim are stable periclinal chimeras of the red-flowered William Sim.

Chimeras are defined as organisms containing two or more genetically distinct cell types. This situation usually evolves from changes in the shoot apex. Two major zones in the shoot apex of higher plants can be recognized: (1) the tunica layer, usually two layers thick (LI and LII), where cell divisions are mainly in the anticlinal plane, and (2) the corpus (LIII), where cell divisions are both anticlinal and periclinal. Mutation in the initial cells of the LI layer, which form the epidermis, will result in a "periclinal" chimera, an arrangement which is usually stable. There are also "mericlinal" and "sectorial" chimeras, when a particular cell layer contains more than one genotype. As flower colors are developed in the epidermal layer alone, it becomes clear that mutation in color-determining genes could form a new cultivar, which would be a periclinal chimera. Mericlinal and sectorial chimeras are not stable, and thus of little horticultural importance.

Genetic variability in carnation can thus be achieved by conventional breeding, induced mutations, and changes in the order of the cell layers at the shoot apex. Induced mutations by irradiation of carnation shoot cuttings were carried out by many workers (Farestveit 1969; Dommergues and Gillot 1973; Pereau-Leroy 1974; Sparnaay et al. 1974). Pereau-Leroy found that, in addition to color mutations, there probably was a physiological interaction between the genetically distinct cell layers to form new colors. As flower color is determined by various alleles of several genes (Mehlquist and Geissman 1947), there is ample scope for many small deviations in colors, which are discovered after irradiation and can be propagated and marketed as new cultivars. Still, low doses are relatively ineffective in inducing flower color changes, and a high dose is necessary. However, the color mutations which are thus obtained are always accompanied by many unfavorable mutations, so that the benefits are outweighed by the drawbacks.

In addition, it was shown (Farestveit 1969) that irradiation often destroyed plugs of tissue, including the epidermis, and extending several cells deep. Replacement came from the division of internal cells from LIII, regenerating a new epidermis, and sometimes even a reconstituted whole apex, thus accounting for changes in periclinal chimeras. The topic is reviewed in detail by Broertjes and Van Harten (1978) and by Tilney-Bassett (1986).

3 Tissue Culture – A Source of Somaclonal Variation

Historically, plant tissue culture was seen essentially as a method of cloning a particular genotype. From this standpoint it was assumed that all the plants arising from a tissue culture should be exact copies of the parent plant. However, phenotypic variants have been observed frequently amongst regenerated plants. In addition, spontaneous variation among subclones of particular parental cell lines has occurred. Thus, it would seem that tissue culture per se, may be a source of variability. More recently, it has been recognized that this method, which introduces changes into crop plants, could be used to develop new breeding lines. Larkin and Scowcroft (1981) used the term "somaclonal variation" for the variation detected in plants generated by the use of tissue culture. The subject has been reviewed by Evans et al. 1984 (see also Chap. I.1, this Vol.).

The mechanisms by which culture variation occurs are not completely clear. The culture environment may be considered mutagenic in the broad sense, since stable variants can arise from it. However, the frequency of these events is too great to be accounted for by simple base changes or deletions caused by some chemical mutagenetic component of the culture medium. In addition, the degree of variation can be influenced by varying components in the medium, particularly the auxin concentration. Some of the possibilities for the mechanisms by which culture variation occurs include karyotype changes, chromosome rearrangements, and gene amplification and deletion events, which have been reviewed (George and Sherrington 1984).

4 In Vitro Regeneration of Carnation

4.1 Virus Elimination by Meristem Tip Culture

Carnation tissue culture was initially used to obtain virus-free plants from clones of vegetatively propagated cultivars which were heavily infected with viruses. Stone (1963) reported carnation plants which were grown from cultured meristem tips 0.2–0.5 mm long, and subsequently (Stone 1968) on the elimination of viruses from carnation by meristem tip culture. At the same time, Phillips and Matthews (1964) described the development of the meristem tip in vitro. Since then, meristem tip culture has established its credibility as an efficient technique which is in regular practice (Hartmann and Kester 1983).

However, because of the chimeral nature of carnation, micropropagation techniques can cause loss or change in desirable chimeral characteristics of cultivars. In practice, it is often very difficult to control the growth of the original meristem tip, and many shoots may arise adventitiously either directly or indirectly from basal callus. Conditions in vitro may also favor a more rapid division of cells of one layer, leading to an alteration in chimerical characteristics. Therefore strict horticultural as well as virus-indexing should be done to ensure true-to-type and virus-free mother plants.

Nevertheless, the fact that practically all cultivars undergo meristem tip culture and in most cases retain their original chimeral nature, shows that if properly cultured, the integrity of the apical meristem is not disturbed.

4.2 Propagation from Shoot Tips

Shoot tips have been the most accepted explants for carnation plant propagation (Table 24, pp. 401–402 in George and Sherrington 1984). Such explants normally developed in vitro into whole plants, but they sometimes turned into callus, from which organs were regenerated. These two modes of development differ widely in their chances of producing somaclonal variation.

The development of shoot tips in vitro was followed by Phillips and Matthews (1964), who found that excised, 0.2-mm-long shoot tips had new leaf and root primordia in 11 days. Afterwards, stem growth and leaf elongation were not restricted unless an abundance of callus was produced at the base. Shabde and Murashige (1977) reported on the hormonal requirements of excised carnation shoot meristem, 0.1 mm long and 0.2 mm wide at the base. Both IAA (indole-3-acetic acid), and kinetin were indispensable for their development, and after 7 days in the IAA-kinetin medium, a bipolar shoot-root axis had been established. Exogenous gibberellin (GA) and abscissic acid did not show any promotive effect on growth.

Shoot tips of carnation have been used also for vegetative propagation through adventitious shoot formation. Plants originated adventitiously may be a promising source for somaclonal variation, as was shown for chrysanthemum (Bush et al. 1976; Sutter and Langhans 1981). Shoot apices of carnation cv. Scania were cultured on Murashige and Skoog's (MS) solid medium containing 0.5 mg/l BAP (6-benzylamino purine), which induced multiplantlet formation (Weryszko and Hempel 1979). Histological studies showed that adventitious buds had formed from the epidermis of leaves and from the epidermis and subepidermal parenchyma of the shoot apex base. Many axillary adventitious buds (70–80 per plantlet) developed in plantlets cultured from carnation shoot apices on high BAP (0.5 mg/l) solid medium (Leshem unpubl.). Buds of exogenous or adventive origin might have apices with a new order of cell layers, which differs from that of the original chimera.

Hackett and Anderson (1967) propagated shoot tips of cv. White Sim in vitro. Most of the regenerated plants appeared normal, except that some reverted to the red flower color. This indicates that the outer layer of cells of the meristem sometimes broke down, allowing regeneration of the outer layer from inner cells. Adventitious buds were used by Dommergues and Gillot (1973) to develop a solid shoot apex from a chimeral one. In order to avoid the disadvantages of irradiation, these researchers induced adventitious buds by application of hormones to the exposed surface of cultured cut stems of shoot tips of cv. White Sim. This treatment resulted in a proportion of the regenerated plants having atypically pigmented flowers. Johnson (1980) also experimented with meristem cultures and with cultures of physically macerated shoot tip explants, which he compared with irradiation. Meristem culture of the chimera S. Arthur Sim gave rise to 7% unchanged chimeral flowers, 27% flowers derived from LI and 66% from LII.

Propagation from shoot tips cultured in liquid medium was tried by Earle and Langhans (1975). Shoot tips of carnations (cv. CSU white Pikes Peak) formed multiple shoots on agar medium containing 0.5 mg/l kinetin and 0.1 mg/l NAA (α-naphthalenacetic acid). Shootlets were transferred to liquid medium, where they could grow, and then subcultured onto fresh medium or rooted and grown to flower. All the plants flowered and had the normal color, which indicates that the chimeral arrangement had not been disturbed by the culture procedure. Multiple adventive shoots developed also from shoot tips cultured on medium containing 2 mg/l kinetin (Jelaska and Sutina 1977) or from irradiated nodal stem explants cultured on medium with 1 mg/l BAP (Roest and Bokelmann 1981). Unfortunately, in both studies, no tests for true-to-type propagation were made.

4.3 Callus as a Source of Organogenesis

A lack of uniformity amongst the plants regenerated from callus has been described in a wide range of plant species. Although it could sometimes have had a physiological origin, in most cases the somaclonal variation will have resulted from plants being regenerated from genetically altered cells. Some regenerated plants may be grossly abnormal; others may show useful changes in general plant morphology, yield attributes, flower color, or disease resistance (Larkin and Scowcroft 1981).

In carnation, shoot tips develop the greatest amount of callus when cultured on a proper medium. Various other explants generate less callus (Kakehi 1970). Callus obtained from carnation shoot tips developed rapidly on MS solid medium, could be continuously subcultured, and regenerated only roots (Kakehi 1972). This callus contained normal and abnormal cells; the latter were very large, with various shapes, and multinucleated. There were loosely connected cells, meristematic cells with an abnormal number of chromosomes in most cases, and scattered vessels. Callus growth was dependent mainly on division of normal cells.

Petru and Landa (1974) reported stem regeneration from callus isolated from hypocotyl segments, which was cultured on a modified MS solid medium. On the other hand, Engvild (1972) cultured callus of carnation from stem pith on both solid and liquid media, but, except for regenerated roots, all attempts to induce formation of shoots gave negative results. Similarly, shoot tips of carnation on MS solid medium containing 2 mg/l 2,4-D (2,4-dichlorophenoxyacetic acid) developed into friable callus, which grew rapidly and was easily subcultured. When auxin level was reduced, the callus regenerated only roots, and no buds were ever seen (Earle and Langhans 1975). On liquid medium, callus suspension cultures grew into lumpy, firm callus. Anatomical observation of this callus's interior did not reveal any differentiation of tissue or organ (Kakehi 1975). Mesophyll protoplasts obtained from young carnation leaves in enzyme solutions grew within 3 months into compact colonies, which regenerated only roots. Shoot formation was not observed in any media tested (Mii and Cheng 1982).

4.4 Regeneration from Anthers and Petals

From all the flower organs, only anthers and petals appeared as a source of plants with interesting characteristics. Villalobos (1981) studied flower formation in excised anthers of carnation cultured in vitro. Anthers containing unicellular microspores were removed from young flower buds (0.5 cm in diameter) of the cv. White Sim, and cultured on solid MS medium supplemented with IAA (1 mg/l), kinetin (1 mg/l), and coconut milk (10% v/v).

Small green leaf primordia were first observed on the surface of the anthers 16 weeks after planting. Regeneration of small shootlets took place after 32 weeks in about one-third of the cultured anthers. Within 52 weeks the plantlets in the tubes rooted and developed flowers. Their color was always white, as in the original variety, but they differed in size and had a reduced number of petals and no anthers at all. No attempt was made to transfer the flowering plantlets into pots.

Induction of redifferentiated plants from petal tissue was first reported by Kakehi (1979). Petals of cv. Scania were cultured on MS solid medium containing NAA (1 mg/l) and BAP (1 mg/l). A large number of leafy shootlets was formed, but most of them remained dwarf. If GA (2 mg/l) was added, many of them elongated into normal shootlets within 50 days. Rooting occurred in MS medium with NAA (1 mg/l) after 30 days. The differentiated plants developed normally and showed no remarkable difference in external morphology and flower color when compared with normal carnation. All of them produced a number of shoots with many flowers during 7 months after the beginning of culture.

Gimelli et al. (1984) also succeeded in obtaining adventitious buds from petals of cvs. Rubino and Lisa. The petals were excised from young lateral flower buds (10 mm in diameter) which were cultured on solid MS medium supplemented with NAA (2 mg/l) and BAP (0.5 mg/l). Regeneration occurred on only 20% of the explants, and both normal and flowering shootlets developed. All were rooted on solid MS medium with IBA (indolebutyric acid 0.5 mg/l) and then transferred into pots for growth in the greenhouse. The normal shootlets grew as normal tall plants, whereas the flowering shootlets developed into short plants with thin stems and flowering habitus.

Leshem (1985, 1986) reported on three types of plants which developed from cultured petals. The basal half of petals of young flower buds of cv. Cerise Royalette was cultured on solid MS medium with NAA (1 mg/l) and BAP (1 mg/l). Direct contact of the cut base of the petal with the medium appeared to promote the formation of adventitious buds. New leaves were seen after 3 weeks, but transplantable shoots, 15 mm long, were obtained only after 50–60 days. Altogether, 110 petals were planted, and 58 new adventitious shootlets appeared. Of the 40 shoots that were transplanted to the hormone-free rooting medium, 13 formed roots, grew well in the greenhouse, and produced light pink flowers. Plants from petals grew into one of three types (Fig. 1):

1. normal – with long flowering stalks (90–100 cm high) bearing 14–17 pairs of leaves (Fig. 1A);
2. intermediate – with a single flower stalk (50–60 cm high) bearing only seven or eight pairs of leaves (Fig. 1B);

Fig. 1A-D. Types of plants which grew from buds regenerated on petals of carnation cv. Cerise Royalette. **A** Normal type with long (90–100 cm high) flowering stalks, each bearing 14–17 pairs of leaves. **B** Intermediate type with a single flower stalk (50–60 cm high) bearing only seven or eight pairs of leaves. **C** Short type with flower stalks 4–5 cm high, each bearing three or four pairs of leaves. **D** Short type with elongated internodes of the flower stalks (5–10 cm high). Petals of flowers on the short type have been removed to show the ovary

3. short – with flower stalks (4–10 cm high) bearing only three or four pairs of leaves (Fig. 1C, D)

Type (3) seems similar to the plants with the flowering habitus described by Gimelli et al. (1984).

Of the 13 plants which were transferred to the greenhouse, three were normal, three of type (2), and seven of type (3). The future development of plantlets in the tubes could not in any way be predicted. Plants of type (2) bore one to three flower buds on each flower stalk, and those of type (3) only one. The small round leaves of the latter were completely different from the linear, glaucous carnation leaves. All the plants were kept from March until November, during which time they kept producing new stems with the same growth habitus and color of flowers.

Since apical vegetative shoot tips, from which carnation plantlets are generally regenerated, did not differentiate flowers in vitro, then apparently types (2) and (3) could result from partial or full determination of the meristems developed on the petals to the flowering state. Similar development has been described for tobacco (Tran Thanh Van 1973). The results of Gimelli et al. (1984) indicate that determination might be reversible.

4.5 Regeneration from Vitrified Plantlets

Vitrification is attributed to an abnormal development of cultured shoot tips of many species (Vieitez et al. 1985). In carnation, as many as 50% of the cultured shoot tips grow into vitrified plantlets (Leshem 1983a,b; Ziv et al. 1983). These are stunted, bushy succulent plantlets, with no apical dominance, a short succulent stem, and thick fragile leaves (Fig. 2). The vitrified plantlets are unable to survive outside culture conditions. However, in the tubes, they maintain their abnormal habit through many subcultures (Leshem and Sachs 1985). The availability of both unlimited water and high auxin in the medium tends to increase the proportion of abnormal plants (Debergh et al. 1981; Leshem 1983a; Ziv et al. 1983; Leshem and Sachs 1985). However, normal shoots do appear on these plants, although not as a direct continuation of the shoot tips that had already produced abnormal tissues. Normal shoots were found on plants that had been cultured without transfer for relatively long periods of 6 to 9 weeks. Their occurrence and location were not regular or predictable. The normal shoots produced healthy plants once they were rooted and grown in a greenhouse.

Lateral buds from virus-free plants (cv. Cerise Royalette) were cultured on solid MS medium with NAA (1 mg/l) and BAP (0.05 mg/l). About half the apices grew as abnormal vitrified plants. The apical 2 to 3 mm of the shoots of vitrified plants were subcultured at monthly intervals, and the abnormal growth was maintained for 20 months. About 70 normal shoots were taken from the 9th and 10th transfers of the abnormal plants and rooted on half-strength solid MS with no hormones. Half of the plantlets formed roots and were transferred subsequently to pots and grown in a greenhouse; 19 of these plants have flowered. The color of the petals varied from the dark pink of the present cultivar to pink and even to the light pink typical of Silvery Pink.

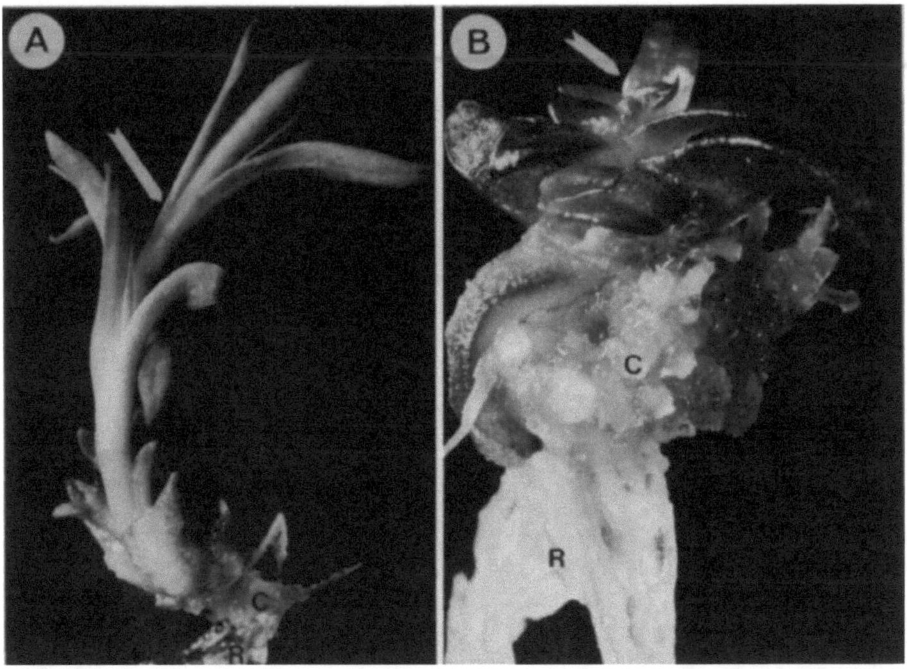

Fig. 2A,B. *Dianthus* plantlets growth for 21 days on standard media. **A** normal; and **B** "vitrified" or teratomatous plantlet. *Arrows* show the location of the shoot apex. *C* callus tissue; *R* roots. (Leshem and Sachs 1985)

The changes in flower colors could be due to mutation. However, it has been shown (Leshem 1983a; Werker and Leshem 1987) that the apical meristems of vitrified shoots are of the mantle-core type, and the core cells are large and prominently vacuolated (Fig. 3). When new normal meristems are regenerated in the vitrified plantlet, the internal layers are sometimes not represented, or may play a lesser role than in the parental cultivar, thus leading to disruption of the chimeral structure.

Buds from Callus Tissues. The initiation of adventitious shoots from callus tissue was achieved when the concentration of cytokinin in the medium was greatly increased. When axillary buds of Cerise Royalette were placed on 0.1 mg/l NAA and 3 mg/l BAP, callus developed at the base during the first 10 days of culture, buds did not elongate, and leaves did not grow. Elongation of the main apex and about ten laterals followed this initial period. New adventitious buds appeared directly on the callus in 17 of the 60 buds studied, about 18–20 days after the cultures were started. These buds grew mainly as abnormal, vitrified plants, but when transferred with some callus onto the low-cytokinin medium, they formed some normal shoots after three or more passages, rooted, and showed normal growth under greenhouse conditions.

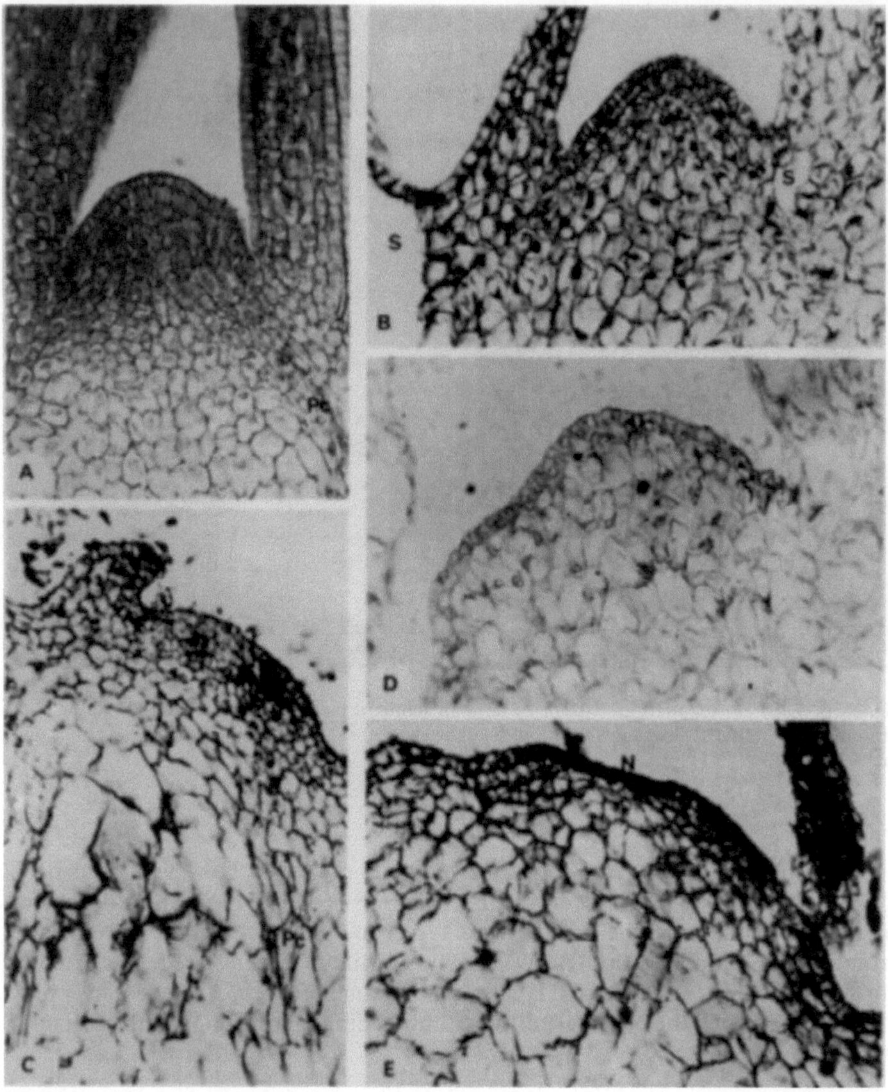

Fig. 3A-E. Median longitudinal sections of shoot apices. **A** Normal plantlet. ×240. **B-E** Vitrified plantlets at various stages of shoot apex determination. **B** Rib and peripheral meristems are absent. ×240. **C** Cell enlargement has proceeded almost to the central mother cells. ×240. **D** Partial meristematic traits seen only in the outermost tunica layer. × 100. **E** Cell necrosis (*N*) at the summit of the shoot apex; the two tunica layers are still meristematic. *PC* procambrium; *S* schizo-lysigenous space. (Werker and Leshem 1987)

5 Conclusions

There is increasing awareness that for those species which are amenable to tissue culture, somaclonal variation may provide an important source of variability for plant improvement. In carnation, natural variability is constantly maintained by the frequent regeneration of new sports from the known cultivars. Besides conventional breeding, great efforts have been directed at changing the chimeral structure of the carnation shoot apices, aiming at inducing solid cultivars or increased variability. Irradiation, which was used for this purpose, was sometimes beneficial but in other cases caused deleterious mutations.

Carnation has been cultured in vitro since the 1960's, first to obtain virus-free mother plants, and subsequently for propagation. The shoot tip proved to be quite stable and mostly has kept its original order of the cell layers of the periclinal chimeras. Plant regeneration from callus has always been considered a promising source for somaclonal variation. In carnation, however, most of the efforts to regenerate plants from callus have not met with much success.

The vitrified plantlets of carnation have been shown to be easy to propagate in vitro for long periods and constantly to regenerate normal shootlets, which can be rooted and raised to normal plants. The substantial changes in the shoot meristem structure may increase the probability of inducing new variability amongst the regenerated plants.

Acknowledgments. This chapter is dedicated to the memory of my nephew, Yuval Harel. Contribution from the Agricultural Research Organization, The Volcani Center, Bet Dagan, Israel, No. 2188-E, 1987 series.

References

Broertjes C, van Harten AM (1978) Application of mutation breeding methods in the improvement of vegetatively propagated crops. Elsevier, Amsterdam, pp 156–162

Bunt AC, Cockshull KE (1985) *Dianthus caryophyllus*. In: Halevy AH (ed) Handbook of flowering, vol 2. CRC, Boca Raton, Fl, pp 433–440

Bush SR, Earle E, Langhans RW (1976) Plantlets from petal segments, petal epidermis and shoot tips of the periclinal chimera, *Chrysanthemum morifolium* Indianapolis. Am J Bot 63:729–737

Debergh P, Harbaoui Y, Lemeur R (1981) Mass propagation of globe artichoke (*Cynara scolymus*). Evaluation of different hypotheses to overcome vitrification with special references to water potential. Physiol Plant 53:181–187

Dommergues P, Gillot J (1973) Obtention de clones génétiquement homogènes dans toutes leurs couches ontogéniques à partir d'une chimère d'oeillet américain. Ann Amelior Plantes 23:88–93

Earle E, Langhans RW (1975) Carnation propagation from shoot tips cultured in liquid medium. HortSci 10:608–610

Engvild KC (1972) Callus and cell suspension cultures of carnation. Physiol Plant 26:62–66

Erler R, Siegmund I (1986). Yearbook of the international horticultural statistics. Int Assoc Hortic Prod. Inst Gartenbauökon Univ Hannover 34:45

Evans DA, Sharp WR, Medina-Filho HP (1984) Somaclonal and gametoclonal variation. Am J Bot 71:759–774

Farestveit B (1969) Flower colour chimeras in glasshouse carnations, *Dianthus caryophyllus* L. Yearb R Vet Agric Coll Copenhagen 1969:19–33

George EF. Sherrington PD (1984) Plant propagation by tissue culture. Exegetics. Eversley Basingstoke. England. pp 74–83
Gimelli F. Ginatta G. Venturo R. Positano S. Buiatti M (1984) Plantlet regeneration from petals and floral induction in vitro in the Mediterranean carnation (*Dianthus caryophyllus* L.). Riv Ortoflorofrutt It 68:107–121
Hackett WP. Anderson JM (1967) Aseptic multiplication and maintenance of differential carnation shoot tissue derived from shoot apices. Proc Am Soc Hortic Sci 90:365–369
Hartmann HT. Kester DE (1983) Plant propagation. Prentice Hall. New Jersey. pp 527–532
Hoogendoorn C (1987) International developments in production and consumption of carnations. In: Book of abstracts. 3rd Int Symp Carnation. Noordwijkerhout. Neth. p 19
Jelaska S. Sutina R (1977) Maintained culture of multiple plantlets from carnation shoot tips. Acta Hortic 78:333–337
Johnson RT (1980) Gamma irradiation and in vitro – induced separation of chimeral genotypes in carnation. HortSci 15:605–606
Kakehi M (1970) Studies on the tissue culture of carnation. I. Relationships between the growth and histodifferentiation of callus and each tissue of aseptic culture. Bull Hiroshima Agric College 4:40–49
Kakehi M (1972) Studies on the tissue culture of carnation. II. Cytological studies on cultured cells. J Jpn Soc Hortic Sci 41:72–75
Kakehi M (1975) Studies on the tissue culture of carnation. IV. Callus suspension culture. Bull Hiroshima Agric College 5:133–137
Kakehi M (1979) Studies on the tissue culture of carnation. V. Induction of redifferentiated plants from the petal tissue. Bull Hiroshima Agric College 6:159–166
Larkin PJ. Scowcroft WR (1981) Somaclonal variation – a novel source of variability from cell cultures for plant improvement. Theor Appl Genet 60:197–214
Leshem B (1983a) Growth of carnation meristems in vitro: Anatomical structure of abnormal plantlets and the effect of agar concentration in the medium on their formation. Ann Bot (London) 52:413–415
Leshem B (1983b) The carnation succulent plantlet – a stable teratological growth. Ann Bot (London) 52:873–876
Leshem B (1985) Carnation plants from vitrified plantlets. callus and petals – a possible source of somaclonal variation. Hassadeh 66:534–536 (Hebrew with English summary)
Leshem B (1986) Carnation plantlets from vitrified plants as a source of somaclonal variation. HortSci 21:320–321
Leshem B. Sachs T (1985) Vitrified *Dianthus* – teratomata in vitro due to growth factor imbalance. Ann Bot (London) 56:613–617
Mehlquist GAL. Geissman TA (1947) Inheritance in the carnation (*Dianthus caryophyllus*). III. Inheritance of flower color. Ann Missouri Bot Garden 34:39–74
Mii M. Cheng SM (1982) Callus and root formation from mesophyll protoplasts of carnation. In: Fujiwara A (ed) Plant tissue culture 1982. Maruzen. Tokyo. pp 585–586
Pereau-Leroy P (1974) Genetic interaction between the tissues of carnation petals as periclinal chimeras. Radiat Bot 14:109–116
Petru R. Landa Z (1974) Organogenesis in isolated carnation plant callus tissue cultivated in vitro. Biol Plant 16:450–453
Phillips DJ. Matthews GJ (1964) Growth and development of carnation shoot tips in vitro. Bot Gaz 125:7–12
Richter A. Singleton WR (1955) The effect of chronic gamma radiation on the production of somatic mutations in carnations. Proc Natl Acad Sci USA 41:295–300
Roest S. Bokelmann GS (1981) Vegetative propagation of carnation in vitro through multiple shoot development. Sci Hortic 14:357–366
Sagawa Y. Mehlquist GAL (1957) The mechanism responsible for some X-ray induced changes in flower color of the carnation. *Dianthus caryophyllus*. Am J Bot 44:397–403
Shabde M. Murashige T (1977) Hormonal requirements of excised *Dianthus caryophyllus* L. shoot apical meristem in vitro. Am J Bot 64:443–448
Sparnaaij LD. Demmink JF (1983) Carnation of the future. Acta Hortic 141:17–23
Sparnaay LD. Demmink JF. Garretsen F (1974) Clonal selection in carnation. In: Eucarpia Meet Ornamentals. Frejus Inst Hortic Plant Breed. Neth. pp 39–50
Stone OM (1963) Factors affecting the growth of carnation plants from shoot apices. Ann Appl Biol 52:199–209

Stone OM (1968) The elimination of four viruses from carnation and sweet william by meristem tip culture. Ann Appl Biol 62:119–122

Sutter E, Langhans EW (1981) Abnormalities in chrysanthemum regenerated from long term cultures. Ann Bot (London) 48:559–568

Tilney-Bassett RAE (1986) Plant chimeras. Arnold, London, pp 122–136

Tran Thanh Van M (1973) Direct flower neoformation from superficial tissue of small explants of *Nicotiana tabacum*. Planta 115:87–92

Vieitez AM, Ballester A, San Jose MC, Vieitez E (1985) Anatomical and chemical studies of vitrified shoots of chestnut regenerated in vitro. Physiol Plant 65:177–184

Villalobos V (1981) Floral differentiation in carnation (*Dianthus caryophyllus* L.) from anthers cultured in vitro. Phyton 41:71–75

Werker E, Leshem B (1987) Structural changes during vitrification of carnation plantlets. Ann Bot (London) 59:377–385

Weryszko E, Hempel M (1979) Studies on in vitro multiplication of carnations. II. The histological analysis of mutliplantlets formation. Acta Hortic 91:323–331

Ziv M, Meir G, Halevy AH (1983) Factors influencing the production of hardened glaucous carnation plantlets in vitro. Plant Cell Tissue Org Cult 2:55–65

IV.4 Somaclonal Variation in *Haworthia*

Y. OGIHARA[1]

1 Introduction

1.1 Characteristics of the *Haworthia* Plant

Plants of the genus *Haworthia* belonging to the tribe Aloineae of the Liliaceae are distributed in Southeast Africa and Madagascar; they have bimodal chromosomes, namely large (L) and small (S) chromosomes (2n = 14; 8L + 6S, Sato 1942). The large chromosomes (L_1 to L_4) can be identified by their respective homologs, whereas the identification of the small individual chromosome (S_1 to S_3) is not possible with any certainty. This karyotype is widespread throughout the Aloineae, several of which are cultivated for medical use, and have consistent stability, suggesting that strong selection might be operating for the maintenance of this basic karyotype. However, plants with chromosomal anomalies such as aneuploids, chromosomal rearrangements including deletions, duplications, paracentric and pericentric inversions, interchanges, and meiotic anomalies, e.g., E-type bridges, subchromatid aberrations, and U-type exchange (Riley and Mukerjee 1962; Riley and Majumder 1966; Brandham 1969, 1970, 1976; Vig 1970) can be found in natural populations and they occasionally occupy a certain niche (Brandham 1976; Brandham and Johnson 1977), probably because *Haworthia* plants are outbreeding and can be propagated vegetatively. This accumulation of knowledge on chromosome variants of *Haworthia* provides the opportunity to investigate the nature of chromosomal variations of plants regenerated from cultures and found in nature.

1.2 *Haworthia* for the Study of Somaclonal Variation

Haworthia species are suitable for both tissue culture and cytological work, since induction of de- and redifferentiation is easily carried out in vitro (Majunder 1970; Kaul and Sabharwal 1972), and cytological studies of cultured tissue are also not difficult (Yamabe and Yamada 1973). From these points of view, *Haworthia setata* was used to clarify somaclonal variations, especially chromosomal modifications in cultures and regenerants of plants (see Ogihara 1981).

[1]Section of Cytogenetics, Kihara Institute for Biological Research, Yokohama City University, Nakamura-cho 2-120-3, Minami-ku, Yokohama 232, Japan

Biotechnology in Agriculture and Forestry, Vol. 11
Somaclonal Variation in Crop Improvement I (ed. by Y.P.S. Bajaj)

2 Establishment of Callus Cultures and the Regeneration of Plants

Kaul and Sabharwal (1972) succeeded in chemical control of callus induction and root/shoot regeneration in *Haworthia*. Calli were induced from flower buds of *Haworthia setata* Haw. (2n = 14; 8L + 6S) and subcultured for 2 years on RM-1964 agar medium (Linsmaier and Skoog 1965) containing 5 mg/l NAA, by transferring to the new medium every 2 months. The stock calli were maintained in 100-ml Erlenmeyer flasks placed under continuous light at 25° C. Under these conditions they grew vigorously and doubled in fresh weight after 7 days (Ogihara and Tsunewaki 1978).

In order to investigate callus growth and the regeneration of plants, 64 combinations of auxins (IAA, NAA, and 2,4-D) and kinetin were set up. From the same stock callus mentioned above, 576 callus pieces, ca. 25 mg fresh weight, were prepared. These were divided equally into blocks, each supplemented with IAA, NAA, or 2,4-D. Each auxin was added in one of seven concentrations (0, 0.1, 0.5, 1, 5, 10, and 50 mg/l), along with kinetin in one of three concentrations (0, 0.2, and 2 mg/l) to the RM-1964 basal medium in all possible combinations, giving 21 plots for each auxin. Nine samples were prepared for each plot. The weight of the callus was measured just before and after 42 to 50 days of inoculation. Growth rate (r) was calculated according to the formula $r = \ln (Wt/Wo) + 1$, where Wo and Wt are the initial and final fresh weight, respectively. At the time of measuring final fresh weight, formation of shoots and/or plantlets was recorded.

The growth rate of the callus increased in most cases, with increasing auxin concentration, reaching a maximum, and then decreasing at higher concentrations. The best callus growth was obtained with 0.1–5 mg/l of auxin, depending on the kind of auxins and the concentration of kinetin supplied along with it. The effect of kinetin was also detectable in three auxin blocks. This is mostly attributable to better growth of the callus with 0.2 mg/l of kinetin than with 2 mg/l or without kinetin. With 5–10 mg/l 2,4-D callus growth was more or less reduced, irrespective of the concentrations of kinetin. The best growth of the callus was obtained with 1–5 mg/l of IAA, or NAA plus 0.2 mg/l of kinetin, and with 0.1–1 mg/l of 2,4-D plus 0.2 mg/l kinetin. It has been generally observed that the level of the three auxins necessary for callus growth is the highest in IAA, followed by NAA and 2,4-D in this order.

Figure 1 shows regeneration rate of calli in response to different combinations of auxins with kinetin. The highest percentage of the calli showing shoot or plantlet regeneration was obtained in the medium containing 0.1 or 0.5 mg/l IAA without kinetin (regeneration rate was 66 and 56%, respectively). Strikingly, this result suggests that addition of kinetin suppresses shoot formation of the callus in the presence of IAA. Concerning NAA, the highest concentration of 50 mg/l suppressed the regeneration of shoots more strongly than lower concentrations. Surprisingly, kinetin (2 mg/l) did not inhibit plant regeneration when 50 mg/l NAA was present. 2,4-D, in general, inhibited shoot formation, but the lowest concentration of 2,4-D induced formation of some plantlets from callus (5 shoots in 27 callus pieces). Kinetin was not effective in plant formation. The regenerated plantlets in the 2,4-D-containing media (total five plants) all grew poorly and were less green than those in the IAA- and NAA-containing media. Similarly, regenerated plants in the media containing high concentrations of auxins (10 mg/l or more) showed

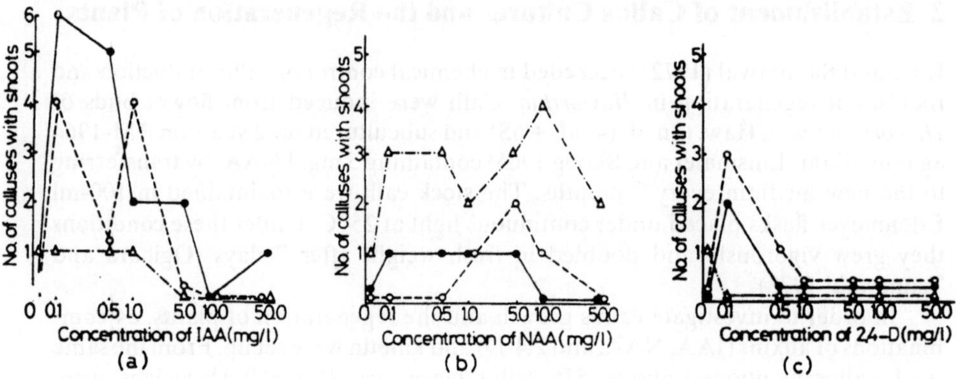

Fig. 1a-c. Number of calli with shoot(s) in each auxin block. Each medium includes nine callus pieces. (**a**) IAA, (**b**) NAA, (**c**) 2,4-D —●— 0 mg/l; --○-- 0.2 mg/l; --△-- 2.0 mg/l kinetin. (Ogihara 1979)

generally poor growth and were only slightly green. All these plants withered when they were transplanted into sterilized soil. It is generally accepted that the balance between auxin and kinetin concentration operates the de- and redifferentiation of cultured tissues (Skoog and Miller 1957; Kaul and Sabharwal 1972; Nishi et al. 1973). The result, using *H. setata* callus, indicates that this scheme is not strictly applicable to any cultured tissues. Efforts to examine the optimal condition of callus growth and regeneration to target plant species are required.

3 Somaclonal Variation in Callus Cultures and Regenerants

3.1 Chromosomal Variations in Callus Culture

It is well known that epigenetic (Braun and Wood 1976; Maliga et al. 1976; Meins and Binns 1977) as well as genetic (Marton and Maliga 1975; Gengenbach et al. 1977; Brettel et al. 1986; Ryan and Scowcroft 1987) alterations of many plant species occur during tissue and/or cell culture cycle. Some regenerants transmit these alterations into offspring through gametes (Maliga 1984). This indicates that cultured tissues are able to produce novel chromosomal, genic, and physiological variations caused by mitosis, which are different from those through the meiotic cell cycle (somaclonal variation). Data that show less range of variations of regenerated plants compared to cultured tissues (Ogihara 1981) show that the selection pressure of maintaining a certain karyotype, mainly euploid, might operate during the course of regeneration. One must look for mutant plants via tissue culture by accidental chance in the case of the ordinary successive subculture method. The rate of chromosomal mutant regenerants is approximately less than a few percent by this method. On the other hand, if one can replace the model karyotype of culture with a new one, the possibility of producing regenerants with new genetic components (Sacristan 1975; Ogihara and Tsunewaki 1979; Ogihara 1981) may increase. In this sense, "callus cloning" in which successive callus division is performed, is considered to be a powerful tool.

From the stock callus induced from flower bud and maintained on RM-1964 agar medium supplemented with 5 mg/l NAA and 0.1 mg/l kinetin (first culture generation as the start) by four successive callus clonings, each of which was referred to as the second to the fourth culture generation to amplify the variation of the callus. The term "clone" is used here to indicate a group of cells. After the fifth culture generation when callus cloning was stopped, each callus was divided into two halves; one of them was transferred to the subculture medium and the other to the regeneration medium, that is the RM-1964 agar medium supplemented with 0.1 mg/l IAA. A scheme of the establishment of callus clones is presented in Fig. 2. As early as in the third and fourth generations, variations in some morphological and physiological characters were clearly noticed (Ogihara and Tsunewaki 1979). Several typical variants obtained during successive subcultures are shown in Fig. 3. These variations were increased in early culture generations when callus cloning was carried out. Furthermore, these variations showed no drastic increase during later simple successive subcultures. Similar tendencies were observed in chromosomal variations. The major karyotype observed among the cells of the stock callus (the first culture generation) was 8L + 6S, its frequency being 55.5%. The frequency of hypodiploid cells of various karyotypes was 28.9%. The tetraploids (16L + 12S) and hypotetraploid cells were in minor fractions (3.5 and 8.1%, respectively). Hyperdiploid and hypertetraploid cells were scarcely observed (4 and 0%).

Strikingly, a new variant karyotype, 7L + 1M + 6S, in which a part of the long arm of the L_2 chromosome was deleted, was first observed in the third culture generation. In this generation, the variant karyotype constituted the modal class in three of the four primary clones. The normal karyotype (8L + 6S) was scarcely found in these three primary clones, although it remained as the model type in the fourth primary clone. No more variant karyotypes constituting the modal class appeared in later culture generations. The diploid cells, including those of the 7L + 1M + 6S

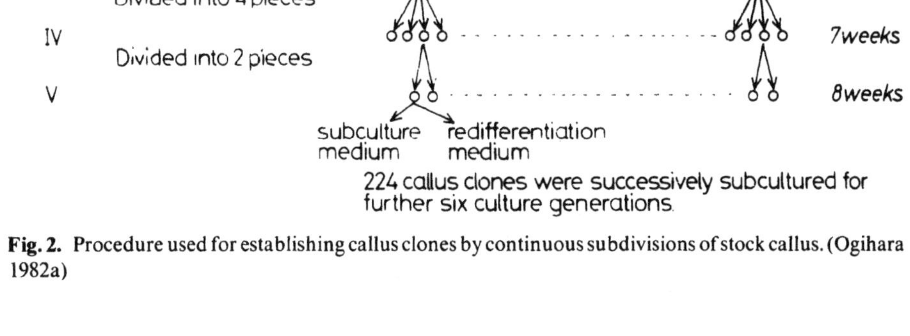

Fig. 2. Procedure used for establishing callus clones by continuous subdivisions of stock callus. (Ogihara 1982a)

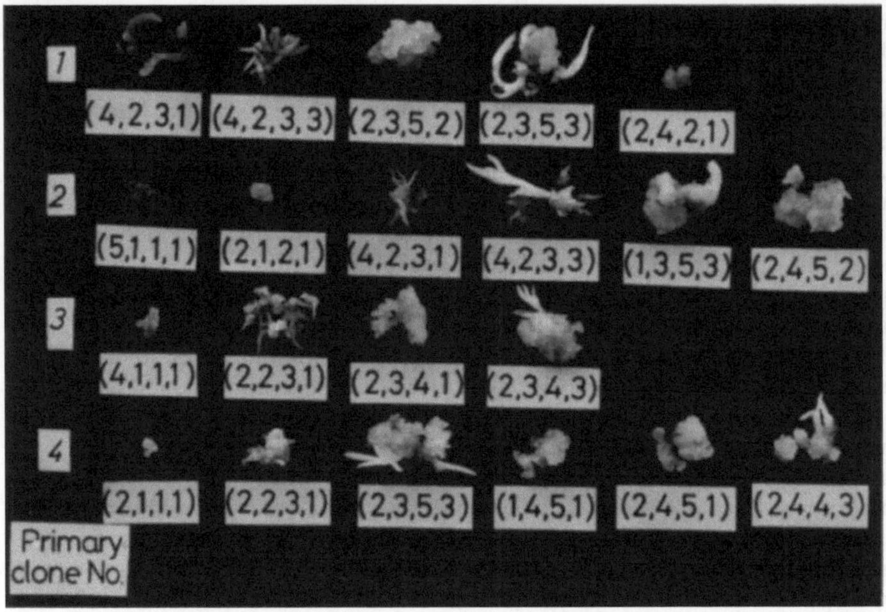

Fig. 3. Calli with various characteristics (about four characters); calli were given different grades depending on each of four characters. For details see Ogihara and Tsunewaki (1979)

karyotype, constituted the modal class in all generations (about 50%). Hyperdiploids, hypo-, and hypertetraploids remained as minor fractions in all culture generations, their frequencies being 1–4% for hyperdiploids and hypertetraploids and 5–9% for hypertetraploids. Hypodiploid cells were found at a moderate frequency, although somewhat decreasing in their frequency from high (40% in the third culture generation) to low (12% in the tenth culture generation). Tetraploid cells, including 14L+2M+12S cells, increased in their frequency in the tenth culture generation. These replacements could never be observed during ordinary successive subcultures so far examined for 3 years, except the tetraploidization. Since this karyotype appeared at almost the same time in three primary clones, it seems to have existed in the stock callus in a latent state. Rapid replacement of the normal cells by cells of an aberrant karyotype was also reported by Sacristan (1975) in the tumorous callus culture of *Crepis capillaris*. She ascribed this phenomenon to the selective advantages of this karyotype, because karyotypic alterations of callus associated with a change in growth rate were observed in *Crepis capillaris*. In *Haworthia*, such a relationship could not be detected (Ogihara and Tsunewaki 1979). Therefore, the change of modal karyotype in the callus might be due to directed selection by callus cloning. Once such a variant callus line has been established, further alterations would scarcely occur by simple successive subcultures. The established callus line would, at certain frequencies, continue to produce some variant cells of lower selective values. Thus, the counteracting forces of producing variants and selecting them will result in the maintenance of the model cell type of a higher selective value at an almost constant frequency under the equilibrated condition. Although the aberrant karyotypes with the deletion of one L_2 chromosome occupied the largest number of cells in three of four primary clones, in the fourth primary clone, in which the cells with the normal karyotype occupied the mode, a deletion of the long arm of one L_2 chromosome was also produced in the tenth culture generation. Its breaking position was similar to that of the deletion observed in the other three primary clones. In addition, dicentric chromosomes and acentric fragments were observed, as shown in Fig. 4. Including these deletions, 14 chromosome breakages in total were detected in large chromosome group; two of them were in L_1, six in L_2, one in L_3, and five in L_4 chromosome. All of them occurred

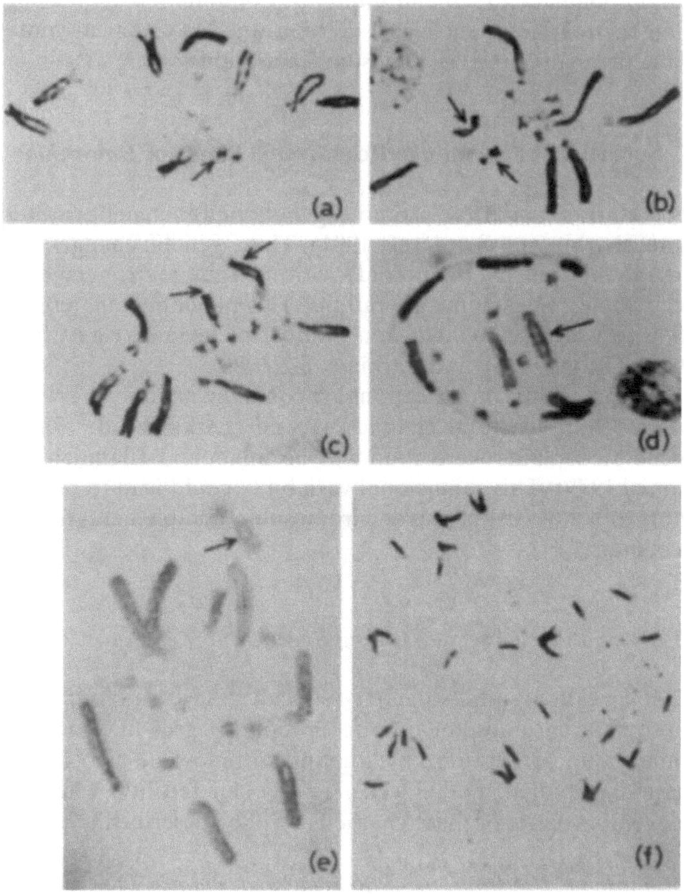

Fig. 4a-f. Variant metaphase cells in calluses of *Haworthia setata*. *Arrows* indicate: **a** 8L + 5S, in which two small chromosomes (S_1 and S_2) were fused. **b** 6L + 2M + 6S, in which two large chromosomes (L_1 and L_2) were deleted of a part of their long arms. **c** Nonreciprocal translocation from the long arm of L_3 to the long arm of L_4 chromosome. **d** A dicentric chromosome. **e** Acentric fragment. **f** Octoploid (32L + 24S). (Ogihara 1982)

in the long arm of the respective chromosome. These facts suggest that there might be fragile positions of the chromosome in cultured tissues. Singh et al. (1972) reported that some homologous pairs of chromosomes were more stably retained than others in *Vicia hajastana* cultured cells. Similar tendencies, not statistically followed, were observed in *Haworthia* cultures.

Structural changes of chromosomes in cultured cells are often reported. These are dicentric chromosome, acentric fragment, deletion, and interchanges (Norstog et al. 1969; Kao et al. 1970; Sacristan and Wendt-Gallitelli 1971, 1973; Singh et al. 1972; Bayliss 1980; Orton 1980; Murata and Orton 1983). In most cases, chromosomes had been broken in the interstitial region, and not in the centromeric region. However, radiation treatment of calli and supplement of PFP (para-fluorophenylalanine) into the culture medium caused centromeric breakage and reunion (Ogihara 1982b and unpubl.). Combinations of the culture method with mutagen treatment, such as radiation and/or chemical mutagens, may allow producing many novel chromosomal variations.

3.2 Somaclonal Variation in Regenerated Plants of *Haworthia*

The culture of tissues also promotes genetic and epigenetic variations of regenerated plants (Larkin and Scowcroft 1981). These genetic changes, so far observed, are genome multiplicity, aneuploidy, chromosome rearrangements, gene mutations, and cytoplasmic genome alterations. This phenomenon, namely somaclonal variation, is well documented in a number of reviews, e.g., Chaleff (1983), Maliga (1984), Orton (1984), and Larkin et al. (1985).

Chromosomal modifications are common, not only the change of chromosome number, but structural changes as well (Larkin et al. 1985). In certain cases, regenerated plants were found to be mixoploid (Bennici and D'Amato 1978; Bennici 1979). This somaclonal variability could lead to production of beneficial plants with novel genotypes or chromosomal complements for basic study and plant breeding.

3.2.1 Variations Observed Among Regenerants

From the calli of simply or cloned subcultures were examined for eight characters, i.e., somatic chromosome number in root tips, growth vigor, leaf shape, leaf color, stomata number, esterase zymogram, chromosome association at meiotic metaphase I in pollen mother cells, and pollen fertility. Characterization of these regenerants on six of these characters will be described in the following section.

Chromosome Constitution of Regenerated Plants. The root tips of some plants showed chromosome mosaicism in which hypodiploid cells were frequently observed, as occurred in the callus cells. In these cases, their modal karyotype was regarded as a representative karyotype. As presented in Table 1, more than 80% of the regenerants from the simply subcultured calli were diploid, while only 42% of those originating from the cloned calli were diploid. The frequency of tetraploid

Table 1. Frequency in percent of the regenerants with different karyotypes. (Ogihara 1981)

Culture methods	No. of plants observed	Karyotype						
		7L+6S	7L+1M+6S	1LL+6L +1M+6S	8L+6S	15L+12S	14L+2M +12S	16L+12S
Simple	241	0.4	0	4.1	89.2	0	0	6.2
Cloning	147	0	20.4	0	42.2	2.0[a]	9.5	25.8
Total	388	0.3	7.7	2.6	71.4	0.8	3.6	13.7

[a] One hypotetraploid (16L+11S) is included.

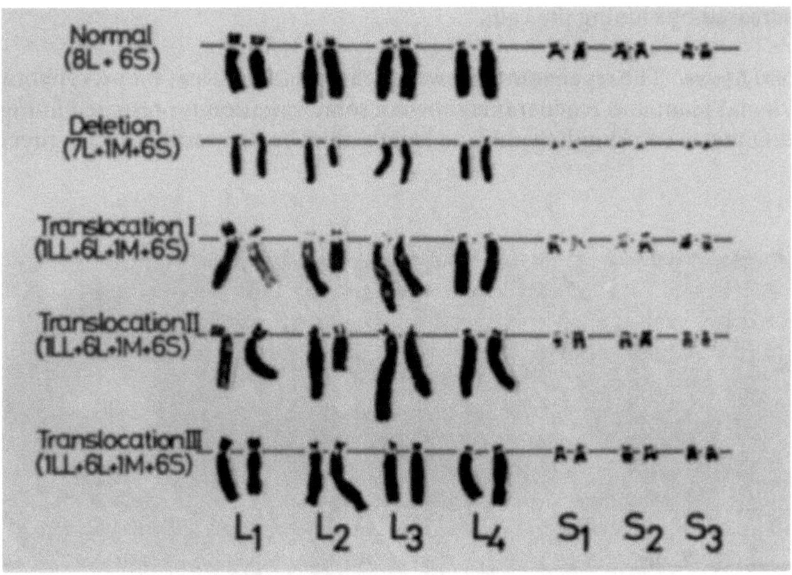

Fig. 5. Karyotypes found among regenerated plants. *From top to bottom* : 8L+6S (normal), 7L+1M+6S (deletion of a segment of the long arm of L_2), 1LL+6L+1M+6S (translocation I from L_2 to L_3), 1LL+6L+1M+6S (translocation II from L_2 to L_3), 1LL+6L+1M+6S (translocation III from L_1 to L_2). (Ogihara 1981)

plants (16L+12S, 13.7% in total) was much higher than that of tetraploid cells (3%) in the stock callus. Several new karyotypes were found among the regenerants. Four of these are shown in Fig. 5. These karyotypes had a similar chromosome constitution 1LL+6L+1M+6S, which had never been seen in the stock callus. One extra-large (LL) and one medium (M) chromosome in the karyotypes were produced by the following translocation between two large (L) chromosomes: (I) translocation of a segment of the long arm of L_2 to the long arm of L_3 chromosome, (II) similar translocation of a larger segment than the first case, of the long arm of L_2 to that of L_3, and (III) translocation of a segment of the long arm of L_1 to the long arm of the L_2 chromosome. Among the regenerants from cloned calli, more new karyotypes, i.e., 7L+1M+6S, 15L+12S, 16L+11S, and 14L+2M+12S were observed, all of which are already known to occur in the cloned calli. The M chromosome(s) included in these karyotypes was derived from deletion of an L_2 chromosome.

Growth Vigor. In total regenerants, 71% of the plants showed vigorous growth, while 28% grew fairly well and 1% grew poorly. About 75% of the plants belonging to the normal diploid karyotype (8L + 6S) grew vigorously, while only 59% of the plants with aberrant karyotypes showed vigorous growth. The difference between the two classes of karyotype (normal diploid vs. others) and growth vigor was statistically significant by a chi-square test at the 1% level. The two methods of subculture (simple vs. cloning) seemed to have affected differences in the growth of the regenerants. This indicates that the proportion of subnormal regenerants was increased by cloning the calli.

Leaf Shape. The regenerants showed great variation in leaf shape. A photograph of original plant and regenerants showing some variation is presented in Fig. 6. Two leaf indices, i.e., length/width and length/thickness were adopted in order to clarify

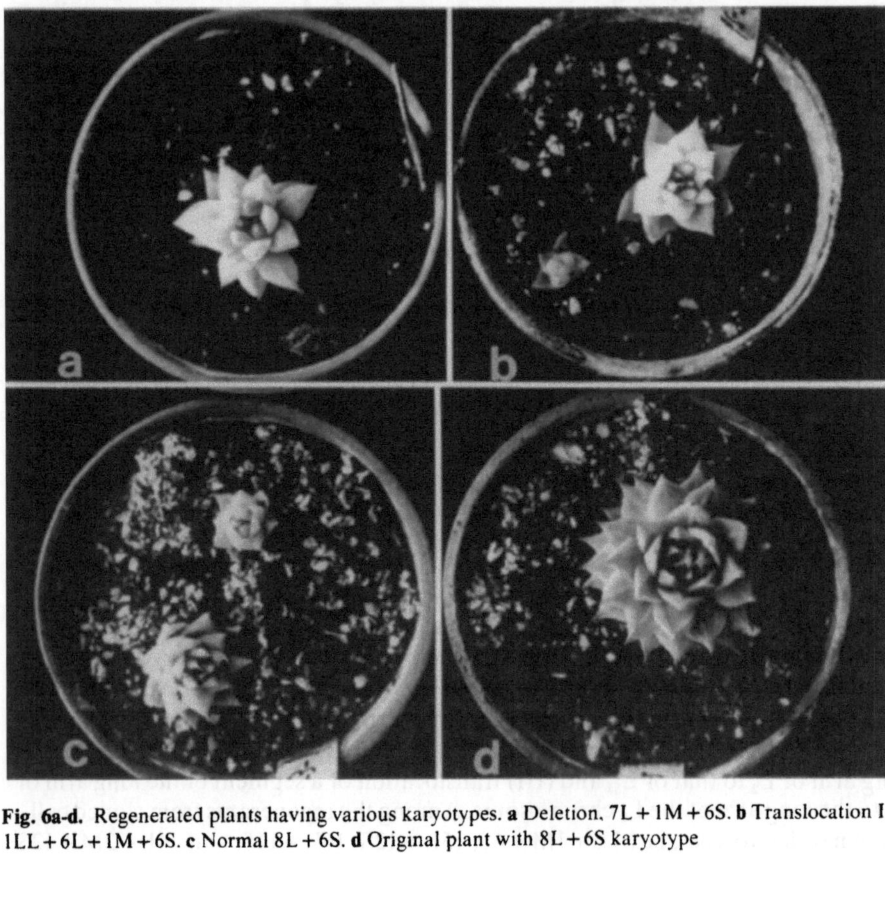

Fig. 6a-d. Regenerated plants having various karyotypes. **a** Deletion, 7L + 1M + 6S. **b** Translocation I, 1LL + 6L + 1M + 6S. **c** Normal 8L + 6S. **d** Original plant with 8L + 6S karyotype

these leaf variations. Leaf shape indices of the karyotypic aberrants of the diploid (1LL + 6L + 1M + 6S and 7L + 1M + 6S) could be studied in comparison with those of the normal diploid (8L + 6S). No effect of chromosomal aberration was detected in the diploid level. Tetraploids (16L + 12S and 14L + 2M + 12S) had broader and thicker leaves than diploids (8L + 6S and 7L + 1M + 6S) in the two clones. Analysis of variance indicated that their differences are significant at the 1% level. The diploids from simple culture had slightly more slender and thinner leaves than those from cloning culture. These differences in both leaf shape indices were significant statistically, indicating effects of the subculture methods on leaf shape.

Esterase Zymogram. Esterase zymograms observed in the leaves and roots of the regenerants are shown in Fig. 7. Five types (I–V) of zymograms were found in the leaves. The type III zymogram, that had four highly active bands (3–6), four minor bands (10–13) and four faint bands (2, 7–9), was obtained from the parental plant and most of the regenerants. The type II zymogram lacked these middle faint bands (7–9). The type I zymogram had only four major bands. Type IV zymogram had one additional faint band to type I, and type V had one more additional band (14) to type III. The most frequent of the regenerants from the simply subcultured calli was type III; type I occurred the most frequently among the regenerants from the cloned calli. No effect was attributed to their ploidy level and their karyotype difference. Comparison of the zymogram frequencies between plants derived from simply subculture and cloning revealed that the effect of subculture methods was statistically significant.

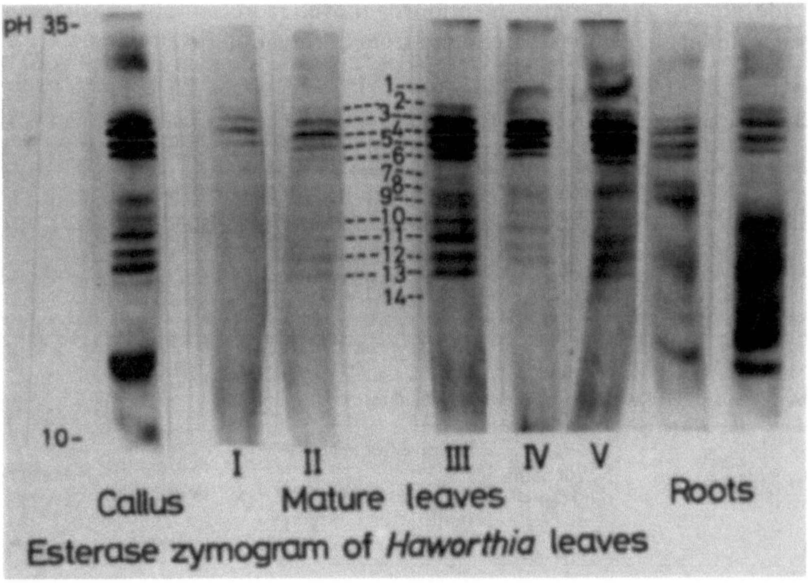

Fig. 7. Esterase zymograms observed in leaves and roots of the regenerants, compared to that of the stock callus (*left*) (Ogihara 1981)

Chromosome Association at Metaphase I in PMC's. *Haworthia* plants occasionally show chromosomal aberrations at meiosis (Brandham 1976). However, the original plant of *H. setata* showed no cytological abnormality in either somatic or meiotic cells, and formed seven normal bivalents at MI of PMC's. On the other hand, regenerated plants revealed a wide variation of chromosome association as shown in Table 2. In euploids (8L + 6S), most PMC's (92%) formed seven bivalents, but some showed univalents (ca. 0.4 univalents/PMC). A few cells formed multivalents; namely, 0.03 trivalents and 0.1 quadrivalents per PMC. Plants with translocation and tetraploids formed univalents and multivalents more frequently than expected. A chi-square test revealed that the differences on chromosome association between the two classes of karyotypes, i.e., euploid (8L + 6S) vs. eutetraploid (16L + 12S), and normal diploid (8L + 6S) vs. translocation heterozygote (1LL + 6L + 1M + 6S) were both significant at the 1% level.

Pollen Fertility. More than 90% of the pollen grains were morphologically normal in the original plant (Table 3), but some regenerated plants became partially sterile (60–90% pollen fertility) or partially fertile (40–60%). Even in the eudiploids (8L + 6S), there was some variation in pollen fertility. All translocation heterozygotes (1LL + 6L + 1M + 6S) showed varying degrees of pollen sterility. It is noteworthy that tetraploids had a pollen fertility similar to that found in the diploids.

3.2.2 Plants with Chromosomal Aberrations in Pollen Meiosis

In addition to the variants mentioned above, more variant regenerants with chromosomal aberrations were found from observations of PMC.

Table 2. Mean chromosome association at MI of PMC's observed in regenerants. (Ogihara 1981)

Karyotype of root tip cell	No. of plants observed	No. of cells observed	Mean chromosome association								Freq. of PMC's showing 7″
			I	II	III	IV	V	VI	VII	VIII	
8L + 6S	60	1890	0.43	6.58	0.03	0.09	0.0				73.3
16L + 12S	10	138	1.78	4.62	0.46	3.30	0.06	0.10	0.03	0.06	---
1LL + 6L + 1M + 6S	3	32	0.81	5.56	0.19	0.41					21.9

Table 3. Pollen fertility of the regenerants. (Ogihara 1981)

Karyotype observed	No. of plants fully fertile	Relative frequency (%)			Mean pollen
		Partially sterile (90%)	Partially fertile (60–90%)	Fertility (%) (40–60%)	
8L + 6S	98	54	35	11	84.9
16L + 12S	13	69	31	0	91.1
1LL + 6L + 1M + 6S	4	0	50	50	50.0

Reciprocal Translocation. Three plants with the nonreciprocal translocation belonging to the first type of translocation mentioned above had observable pollen meiosis and pollen fertility. They formed quadrivalents, 0.29, 0.29, and 0.64 per cell, and their pollen fertilities were 62, 40, and 37%, respectively. Therefore, diploid plants with more quadrivalents per cell than 0.30 were regarded as having reciprocal translocation because no aberration was detected in their root tip cells. In total, three plants with reciprocal translocation were regenerated. They formed quadrivalents, 0.63, 0.33, and 0.38 per cell and showed 79, 58, and 58% of pollen fertility.

Paracentric Inversion. As shown in Fig. 8, many aberrations were observed in anaphase I of pollen meiosis, for example, bridge, bridge and fragment, lagging chromosomes, and abnormal chromosome segregation at AI. Three plants showed bridge and fragment at AI of PMC's. They were regarded as plants carrying paracentric inversion, not U-type exchange, because the length of their fragments was almost constant at AI.

Sub-Chromatid Aberration. Some plants formed only bridges which were not fragments at anaphase I of PMC's. Six plants in total were considered to show sub-chromatid aberrations (Brandham 1970; Vig 1970).

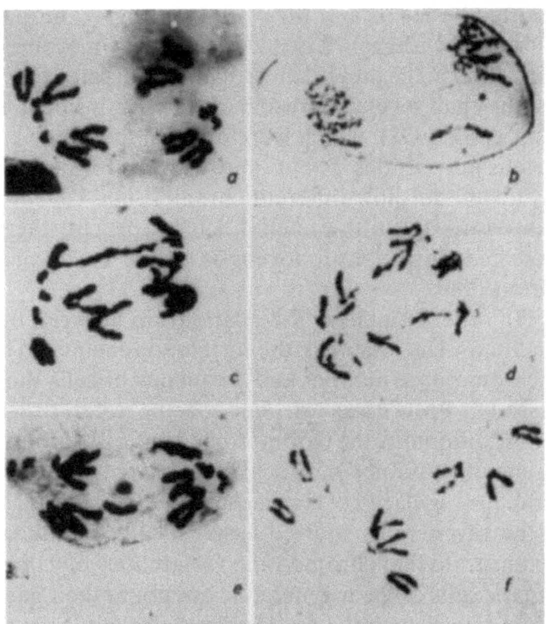

Fig. 8a-f. Chromosome segregation at anaphse I of diploid pollen meiosis. **a** Normal segregation (7-7). **b** Bridge and fragment. **c** Bridge. **d** Double bridge in one cell. **e** Laggard. **f** Irregular 8-6 chromosome segregation. (Ogihara 1981)

3.2.3 Chromosome Variability in Regenerated Plants

Studies on chromosome variability among regenerated plants showed that regenerated plants exhibited less chromosome variabilities than the original callus cells (Sacristan and Melchers 1969; Orton 1980). It is, therefore, evident that selection in favor of specific chromosome constitution(s) is operating during the course of plant regeneration from calli. The intensity of the selection seems to vary depending on the materials used; very strong selection for euploids in maize, common wheat, and carrot (Shimada et al. 1969; Mok et al. 1976; Gengenbach et al. 1977), moderate selection in tobacco, durum wheat, and potato (Novak and Vyskot 1975; Bennici and D'Amato 1978; Shepard et al. 1980; Gill et al. 1986), and almost no selection in tobacco (Ogura 1976).

In the present material, selection for euploids (diploids and tetraploids) was rather strong, although a few plants with abnormal karyotypes (deletions and translocations) were produced. However, chromosome chimerism was detected in many regenerants of *Haworthia*, that is, a few tetraploid or aneuploid cells were observed among many eudiploid cells in the root tips of regenerants. Thus, the relationship between chromosome variabilities of the calli and the regenerated plants was examined. The correlation coefficient between the standard deviations of their chromosome number was significant at the 1% level, indicating that chromosome variation observed in the root tips of the regenerants is closely related to that of the callus from which they originated. Such a relationship was previously reported by Heinz and Mee (1971) in sugarcane. The following possibility can be considered with regard to the origin of chromosome variability of regenerants. The callus used in the present investigation consisted of a heterogeneous cell population, and they had several meristematic pockets inside them (Kaul and Sabharwal 1972; King et al. 1978), which were also composed of karyotypically different cells in various proportions. Shoots and/or roots formed from multicellular initials might have come to show different chromosomal constitutions, including chromosome chimerism, from the modal karyotype. Furthermore, new karyotypes were produced during morphogenesis in the meristematic tissue of the regenerants (Orton 1980).

In order to analyze the relationship between the chromosome variabilities of somatic and meiotic cells, the correlation coefficient between the standard deviation of chromosome number and frequency of cells showing seven bivalents at metaphase I of PMC's was calculated, being not statistically significant.

Additionally, the chromosome variabilities observed in the root tip cells of the regenerants did not statistically correlate with their pollen fertility. On the other hand, the regularity of chromosome association at MI of PMC's and pollen fertility of the regenerants showed a statistically significant correlation. All these facts demonstrate that chromosome variabilities of callus cells were carried over to the somatic cells of the regenerants, but not to their generative cell lines. This result is contradictory with the findings obtained with tobacco (Zagorska et al. 1974).

3.2.4 Morphology of the Regenerants in Relation to Their Chromosome Constitutions

There have been a number of reports on the phenotypic variations of regenerated plants from cultures with respect to their chromosome constitutions. Some of these examples are listed in Table 4. *Crepis capillaris* seems to be the most remarkable case, where a rearrangement of chromosomes was closely associated with abnormal plant morphology (Sacristan and Wendt-Gallitelli 1971).

In the present material, the alterations of chromosome constitutions somewhat affected the morphology of regenerated plants. Changes in ploidy level caused alteration in growth vigor, leaf morphology, and of course, chromosome association at MI in PMC's, but no effects were detected on leaf color esterase, isozyme pattern, and pollen fertility. On the other hand, changes in chromosome structure at the diploid level (deletion and translocations) brought about no alterations in morphology of vegetative organs, but a marked effect was exerted on chromosome association at MI in PMC's and pollen fertility, as expected. Aneuploids and chromosomally chimeric plants did not reveal any special anomalies, which is in contrast to the phenotypic anomalies observed in chromosomally chimeric regenerants from tobacco (Ogura 1976). Therefore, it can be said that *Haworthia* plants are relatively stable in plant morphology when chromosome constitution is changed. However, it should be emphasized that, for example, Shepard et al. (1980) selected many regenerants showing anomalies in several horticultural characters from protoplasts of potato associated with and without chromosomal disturbance. Alterations of regenerated plant morphology are considered as the results of epigenetic and/or polygenic, particularly genic (e.g., homeotic genes in *Drosophila*, Gehring and Hiromi 1986) modification during the course of plant development from cultures (Chaleff 1983). Consequently, although effects of chromosomal alterations to the plant morphology might be secondary, chromosomal mutants such as aneuploidy, deletion, translocation, and inversion occurring in cell culture must provide many possibilities to study the position and/or gene dosage effects of genes in the plant genome, and contribute to plant breeding strategy.

3.2.5 Chromosome Variations Derived from Callus Culture and Those Found in Nature

Field collections of *Haworthia* contain many chromosomal variants (Brandham and Johnson 1977). This is mainly due to vegetative propagation and occasional outbreeding in this plant. Examples of these chromosomal variants so far reported are presented in Table 5, in which those derived from callus culture of *Haworthia setata* are also tabulated. From the comparison of variant plants, some remarkable differences between the regenerants from culture and the plants collected in nature are illustrated. The first is a higher frequency of diploid plants carrying the deletion of a whole chromosome (one per 388 plants) or chromosome segment (30 per 388) in the regenerants. No such frequency was found among 597 diploid plants collected in nature (Brandham 1976). This might show that diploid plants carrying such chromosome deletions have some selective disadvantage in a natural environment.

Table 4. Chromosomal changes associated with modifications in plant character

Plant species	Starting tissues of culture	Chromosomal changes	Modifications in plant character	Reference
Allium sativum	Shoot tips	Rearrangement	Plant morphology	Novak (1980)
Apium graveolens	Petioles	Monosomics Rearrangement	Plant morphology PGM and SDH isożyme	Orton (1983)
Crepis capillaris	Tumor	Rearrangement	Plant morphology	Sacristan and Wendt-Gallitelli (1971)
Haworthia *setata*	Flower bud	Deletion Translocation Inversions	Meiosis and fertility	Ogihara (1981)
Hordeum hybrid	Immature ovary tissue	Interchanges	Plant morphology Meiosis	Orton (1980)
Lolium hybrid	Embryo	Aneuploidy Deletion Interchange Inversion	Plant morphology	Ahloowalia (1976)
Lycopersicum pervianum	Stem and anther	Aneuploidy	Plant morphology	Sree Ramulu et al. (1976)
Nicotiana sylvestris	Seedlings	Rearrangement	Fertility	Maliga et al. (1979)
Nicotiana tabacum	Tumor	Aneuploidy Rearrangement	Plant morphology	Sacristan and Melchers (1969)
Oryza sativa	Anther Ovary	Aneuploidy Polyploidy	Plant morphology	Oono (1985) Nishi and Mitsuoka (1969)
Saccharum sp.	Parenchyma tissue	Aneuploidy	Plant morphology Peroxidase pattern	Heinz and Mee (1971)
Solanum tuberosum	Mesophyll	Aneuploidy Polyploidy	Plant morphology	Gill et al. (1986)
Triticale	Immature embryo	Interchange Deletion	Plant morphology	Nakamura and Keller (1982) Lapitan et al. (1984)
Triticum aestivum	Immature embryo	Isochromosome Translocation	Plant morphology ADH zymogram	Davies et al. (1986)
	Immature embryo	Isochromosome Deletion U-type chromatid exchange		Karp and Maddock (1984)
Zea mays	Immature embryo	Aneuploids	Fertility	Gengenbach et al. (1977)

Table 5. Comparison between chromosome variants of *Haworthia* regenerated from callus culture and those found in nature

Source of variation	No. of plants observed in field collection[a]	Reference	Regenerated plants (388 plants)
Aneuploid			
3n–2	*H. limifolia*	Riley and Mukerjee (1962)	4 (1.0%)
4n–2	*H. reinwardtii*	Riley and Majunder (1966)	
6n–2	*H. tesselata*		
Deletion			
Diploid	0/597 (0.0%)		30 (7.7%)
Polyploid	10/449 (2.2%)		
Duplication	1/2234		0 (0.0%)
Translocation			
Diploid	8/597 (1.3%)		13 (3.3%)
Polyploid	54/2234(2.4%)		0 (0.0%)
Inversión			
Paracentric	10/253 (4.3%)		2 (0.5%)
Pericentric	9/2234(0.4%)		0 (0.0%)
Meiotic anomalies			
E-type bridge	*H. icosiphylla*	Brandham (1969)	0 (0.0%)
Sub-chromatid aberrations	*H. attenuata*	Vig (1970)	6 (1.5%)
U-type exchange	*H. nigra*	Brandham (1970)	0 (0.0%)

[a] Frequency of variant plants found in nature was calculated according to Brandham 1976, and other frequencies were not given in the relevant literature.

The second is the breaking point of chromosomes in the case of deletion and interchanges. Brandham (1976) critically analyzed the 51 interchanges found in the Aloineae. Of 102 breaks in these interchanges, 30 were at the centromere, and 72 were at other positions, which indicated a large departure from randomness in favor of the centromere breaks. A similar tendency was reported in *Haplopappus* (Jackson 1965) and *Gibasis* (Jones 1974). In the present study, all breaks found in the cultured regenerants were located in the chromosome arms, and none in the centromeric region. Translocations reported by Sacristan and Wendt-Gallitelli (1971) in *Crepis capillaris* and Maliga et al. (1979) in *Nicotiana sylvestris* seem to have breaks in noncentromeric region. Recently, there have been some reports that chromosome breakage took place at the centrome and the resultant iso-chromosomes were observed in celery cultures (Murata and Orton 1983), regenerants from lily culture (Sheridan 1974), and regenerants from wheat cultures (Davies et al. 1986). We may conclude, based on these findings, that chromosomal breakages through the culture cycle occur at the late-replicating heterochromatic region (Larkin et al. 1985). Clarification of mechanism(s) and the possibility of chromosomal variation in cultures and regenerants require further investigations.

4 Conclusions and Prospects

Somaclonal variations in *Haworthia*, especially focused on chromosomal modifications in culture and regenerants, have been described by comparing them with chromosome mutants found in nature. From the finding that *Haworthia* calli showed characteristic responses to three auxins and kinetin on callus growth and shoot regeneration different from tobacco, the optimum culture condition of the objective plant should be set up before full-scale culture experiments. In *Haworthia* flower buds as explant, a combination of 5 mg/l NAA with 0.2 mg/l kinetin as growth regulators gave the best results for callus induction and establishment of callus cultures, respectively; 0.1 mg/l of IAA without kinetin was the optimal condition for regeneration of plants from callus cultures. Callus cloning, which means successive callus mass dividing, not single-cell origin, was very convenient to induce variation and screen certain chromosome aberrants, and an ordinary method of subculture was able to keep these stable as callus lines. Furthermore, regenerated plants were obtained from these variant calli. Regenerants from calli established by callus cloning showed a wide range of variations on chromosomal as well as physiological and morphological characters, compared to those derived from simply subcultured calli. Chromosome variabilities of callus cells were carried over to the somatic cells of regenerants, but not to their generative cell line.

Although most of these chromosomal variants had already been found in plants collected in nature, a few novel variants could be screened. Since the supplement of chemicals to the culture medium (amino acid analog, in this case) induced new chromosomal aberrations never found in the ordinary cultures, combinations of tissue culture technique with mutagen treatment such as chemical mutagens and irradiation may lead to the production of plants with novel chromosome constitutions. Especially the use of chemicals which disturb the cell cycle and/or DNA replication is very promising, because the proposal has been made that the chromosome may break at the late-replicating heterochromatic region during cultures and regeneration (McCoy et al. 1982; Larkin et al. 1985).

These rearrangements of chromosomes produce new linkage groups among genes. This will make it possible to study the position and dosage effects of genes which are involved in the changed chromosome segment (Davies et al. 1986).

There have now been a number of reports on somaclonal variation associated with molecular modifications such as gene amplification, activity of transposable element, point mutations, and alteration of multigene family expression (Larkin et al. 1985; Brettel et al. 1986; Cooper et al. 1986; Davies et al. 1986; Ryan and Scowcroft 1987). These provide impulses for plant genetics and contribute to new breeding strategies (Shepard et al. 1980). In fact, there are many examples of somaclonal variants which could soon be released for practical breeding programs. Several mechanisms might operate to produce somaclonal variation. Further investigations at the molecular and cellular level are required to clarify these mechanisms and the potential application of somaclonal variations.

References

Ahloowalia BS (1976) Chromosomal changes in parasexually produced ryegrass. In: Jones K, Brandham PE (eds) Current chromosome research. North Holland, Amsterdam, pp 77–87

Bayliss MW (1980) Chromosomal variation in plant tissues in culture. Int Rev Cytol 11A:113–143

Bennici A (1979). A cytological chimera in plants regenerated from *Lilium longiflorum* tissues grown in vitro. Z Pflanzenzücht 82:349–353

Bennici A, D'Amato F (1978) In vitro regeneration of durum wheat plants. 1. Chromosome numbers of regenerated plants. Z Pflanzenzücht 81:305–311

Brandham PE (1969) Chromosome behaviour in the *Aloineae*. 1: the nature and significance of E-type bridge. Chromosoma 27:201–215

Brandham PE (1970) Chromosome behaviour in the *Aloineae*. 3: correlations between spontaneous chromatid and subchromatid aberrations. Chromosoma 31:1–17

Brandham PM (1976) The frequency of spontaneous structural change. In: Jones K, Brandham PE (eds) Current chromosome research. North Holland, Amsterdam, pp 77–87

Brandham PE, Johnson MAT (1977) Population cytology of structural and numerical chromosome variants in the *Aloineae (Liliaceae)*. Plant Syst Evol 128:105–122

Braun AC, Wood HN (1976) Suppression of the neoplastic state with the acquisition of specialized functions in cells, tissues, and organ of crown gall teratomas of tobacco. Proc Natl Acad Sci USA 73:496–500

Brettel RIS, Dennis ES, Scowcroft WR, Peacock WJ (1986) Molecular analysis of a somaclonal mutant of maize alcohol dehydrogeneae. Mol Gen Genet 202:235–239

Chaleff RS (1983) Considerations of developmental biology for plant cellular genetics. In: Kosuge T, Meredith CP (eds) Genetic engineering of plants. Plenum, New York, pp 257–270

Cooper DB, Sears RG, Lookhart GL, Jones BL (1986) Heritable somaclonal variation in gliadin proteins of wheat plants derived from immature embryo callus culture. Theor Appl Genet 71:784–790

Davies PA, Pallota MA, Ryan SA, Scowcroft WR, Larkin PJ (1986) Somaclonal variation in wheat: genetic and cytogenetic characterization of alcohol dehydrogenase 1 mutants. Theor Appl Genet 72:644–653

Gehring WJ, Hiromi Y (1986) Homeotic genes and the homeobox. Annu Rev Genet 20:147–173

Gengenbach BG, Green CE, Donovan CM (1977) Inheritance of selected pathotoxin-resistance in maize plants regenerated from cell cultures. Proc Natl Acad Sci USA 74:5113–5117

Gill BS, Kam-Morgan LNW, Shepard JF (1986) Origin of chromosomal and phenotypic variation in potato protoclones. J Hered 77:13–16

Heinz OJ, Mee GWP (1971) Morphologic, cytogenetic, and enzymatic variation in *Saccharum* species hybrid clones derived from callus tissue. Am J Bot 58:257–262

Jackson RC (1965) A cytogenetic study of a three-paired race of *Haplopappus gracilis*. Am J Bot 52:946–953

Jones K (1974) Chromosome evolution by Robertsonian translocation in *Gibasis (Commelianaceae)*. Chromosoma 45:353–368

Kao KN, Miller RA, Gamborg OL, Harvey BL (1970) Variations in chromosome number and structure in plant cells grown in suspension cultures. Can J Genet Cytol 12:297–301

Karp A, Maddock SE (1984) Chromosome variation in wheat plants regenerated from cultured immature embryos. Theor Appl Genet 67:249–255

Kaul K, Sabharwal PS (1972) Morphogenetic studies on *Haworthia*: Establishment of tissue culture and control of differentiation. Am J Bot 59:377–385

King PJ, Potrykus I, Thomas E (1978) In vitro genetics of cereals: Problems and perspectives. Physiol Veg 16:381–399

Lapitan NLV, Sears RG, Gill BS (1984) Translocations and other karyotypic structural changes in wheat × rye hybrids regenerated from tissue culture. Theor Appl Genet 68:547–554

Larkin PJ, Scowcroft WR (1981) Somaclonal variation – a novel source of variability from cell culture for plant improvement. Theor Appl Genet 60:197–214

Larkin PJ, Brettel RIS, Ryan SA, Davies PA, Pallota MA, Scowcroft WR (1985) Somaclonal variation: impact on plant biology and breeding strategies. In: Day P, Zaitlin M, Hollaender A (eds) Biotechnology in plant science. Relevance to agriculture in the eighties. Academic Press, New York London, pp 83–100

Linsmaier EM, Skoog F (1965) Organic growth factor requirements of tobacco tissue cultures. Physiol Plant 18:100–127

Majumdar SR (1970) Production of plantlets from the ovary wall of *Haworthia turgida* var. pallidifolia. Planta 90:212–214
Maliga P (1984) Isolation and characterization of mutants in plant cell culture. Annu Rev Plant Physiol 35:519–542
Maliga P, Lazar G, Svab Z, Nagy F (1976) Transient cycloheximide resistance in a tobacco cell line. Mol Gen Genet 149:267–271
Maliga P, Kiss ZR, Dix PJ, Lazar G (1979) A streptomycine resistant line of *Nicotiana sylvestris* unable to flower. Mol Gen Genet 172:13–15
Marton L, Maliga P (1975) Control of resistance in tobacco cells to 5-bromodeoxyuridine by a simple mendelian factor. Plant Sci Lett 5:77–81
McCoy TJ, Phillips RL, Rines HW (1982) Cytogenetic analysis of plants regenerated from oat (*Avena sativa*) tissue cultures: High frequency of partial chromosome loss. Can J Genet Cytol 24:37–50
Meins F Jr., Binns A (1977) Epigenetic variation of cultured somatic cells: Evidence for gradual changes in the requirement for factors promoting cell divisions. Proc Natl Acad Sci USA 74:2928–2932
Mok MC, Gabelman WH, Skoog F (1976) Carotenoid synthesis in tissue cultures of *Daucus carota* L. J Am Soc Hortic Sci 101:442–449
Murata M, Orton TJ (1983) Chromosome structural changes in cultured celery cells. In vitro 19:83–89
Nakamura C, Keller WA (1982) Callus proliferation and plant regeneration from immature embryo of hexaploid *Triticale*. Z Pflanzenzücht 88:137–164
Nishi T, Mitsuoka S (1969) Occurrence of various ploidy plants from anther and ovary culture of rice plant. Jpn J Genet 44:341–346
Nishi T, Yamada Y, Takahashi E (1973) The role of auxins in differentiation of rice tissue cultured in vitro. Bot Mag (Tokyo) 86:183–188
Norstog K, Wall WE, Howland GP (1969) Cytological characteristics of ten-year old ryegrass endosperm tissue cultures. Bot Gaz 130:83–86
Novak FJ (1980) Phenotype and cytological status of plants regenerated from callus cultures of *Allium sativum* L. Z Pflanzenzücht 84:250–260
Novak FJ, Vyskot B (1975) Karyology of callus cultures derived from *Nicotiana tabacum* L. haploids and ploidy of regenerants. Z Pflanzenzücht 75:62–70
Ogihara Y (1979) Tissue culture in *Haworthia*. 2. Effects of three auxins and kinetin on greening and redifferentiation of calluses. Bot Mag (Tokyo) 92:163–171
Ogihara Y (1981) Tissue culture in *Haworthia*. 4. Genetic characterization of plants regenerated from callus. Theor Appl Genet 60:353–363
Ogihara Y (1982a) Tissue culture in *Haworthia*. 5. Characterization of chromosomal changes in cultured callus cells. Jpn J Genet 57:499–511
Ogihara Y (1982b) Characterization of chromosomal changes of calluses induced by callus cloning and para-fluorophenyl-alanine (PFP) treatment and regenerates from them. In: Fujiwara A (ed) Plant tissue culture 1982. Maruzen, Tokyo, pp 439–440
Ogihara Y, Tsunewaki K (1978) Tissue culture in *Haworthia*. 1. Effects of auxins and kinetin on callus growth. Bot Mag (Tokyo) 91:83–91
Ogihara Y, Tsunewaki K (1979) Tissue culture in *Haworthia*. 3. Occurrence of callus variants during subcultures and its mechanism. Jpn J Genet 54:271–293
Ogura H (1976) The cytological chimeras in original regenerates from tobacco tissue cultures and in their offsprings. Jpn J Genet 51:161–174
Oono K (1985) Putative homozygous mutations in regenerated plants of rice. Mol Gen Genet 198:377–384
Orton TJ (1980) Chromosomal variability in tissue cultures and regenerated plant of *Hordeum*. Theor Appl Genet 56:101–112
Orton TJ (1983) Spontaneous electrophoretic and chromosomal variability in callus cultures and regenerated plants of celery. Theor Appl Genet 67:17–24
Orton TJ (1984) Somaclonal variation: theoretical and practical considerations. In: Gustafson JP (ed) Gene manipulation of plant improvement. Plenum, New York, p 427
Riley HP, Majumder SK (1966) Cytogenetic studies of an aneuploid species of *Haworthia* with special reference to its origin. J Cytol Genet 1:46–54
Riley HP, Mukerjee D (1962) Two aneuploid plants of *Haworthia*. J Hered 53:105–109
Ryan SA, Scowcroft WR (1987) A somaclonal variant of wheat with additional b-amylase isozymes. Theor Appl Genet 73:459–464

Sacristan MD (1975) Cloned development in tumorous cultures of *Crepis capillaris*. Naturwissenschaften 62:139–140

Sacristan MD, Melchers G (1969) The karyological analysis of plants regenerated from tumorous and other callus cultures of tobacco. Mol Gen Genet 105:317–333

Sacristan MD, Wendt-Gallitelli MF (1971) Transformation to auxin autotrophy and its reversibility in a mutant line of *Crepis capillaris* callus culture. Mol Gen Genet 110:355–360

Sacristan MD, Wendt-Gallitelli MF (1973) Tumorous cultures of *Crepis capillaris*; Chromosome and growth. Chromosoma 43:279–288

Sato D (1942) Karyotype alteration and phylogeny in *Liliaceae* and allied families. Jpn J Bot 12:58–161

Shepard JF, Bidney D, Shahin E (1980) Potato protoplasts in crop improvement. Science 208:17–24

Sheridan WMF (1974) Long-term callus cultures of *Lilium*. Relative stability of the karyotype. J Cell Biol 63:625

Shimada T, Sasakuma T, Tsunewaki K (1969) In vitro culture of wheat tissues. 1. Callus formation, organ redifferentiation and single cell culture. Can J Genet Cytol 11:294–304

Singh BD, Harvey BL, Kao KN, Miller RA (1972) Selection pressure in cell population of *Vicia hajastana* cultured in vitro. Can J Genet Cytol 14:65–70

Skoog F, Miller CO (1957) Chemical regulation of growth and organ formation in plant tissue cultured in vitro. Symp Soc Exp Biol 11:118–140

Sree Ramulu K, Devreux M, Ancora G, Laneri U (1976) Chimerism in *Lycopersicum peruvianum* plants regenerated from in vitro cultures of anthers and stem internodes. Z Pflanzenzücht 76:299–319

Sunderland N (1977) Nuclear cytology. In: Street HE (ed) Plant tissue and cell culture. Blackwell, Oxford, pp 177–205

Vig BK (1970) Sub-chromatid aberrations in *Haworthia attenuata*. Can J Genet Cytol 12:181–186

Yamabe M, Yamada T (1973) Studies on differentiation in cultured cells. 2. Chromosomes of *Haworthia* callus and of the plants grown from the callus. Kromosoma 94:2923–2931

Zagorska NA, Shamina ZB, Butenko RG (1974) The relationship of morphogenetic potency of tobacco tissue culture and its cytogenetic features. Biol Plant 16:262–274

IV.5 In Vitro Variation in *Weigela*

M. DURON and L. DECOURTYE[1]

1 Introduction

Weigela are hardy and easily grown shrubs, growing to an average height of 2 m. They have decorative flowers in June, and vary from white through pink to red. This small genus belongs to the family Caprifoliaceae. Present cultivars result from crossing between species native from Eastern Asia such as *W. florida*, *W. coraensis*, *W. praecox*, *W. hortensis*, and *W. japonica*, introduced into Europe from the middle of the 19th century, so that original species are no longer in cultivation and the exact origin of most cultivars is difficult to assess.

The name of the genus was given in memory of Christian Ehrenfried von Weigel (1748–1831), Professor in Greifswald, North Germany. Sometimes referred to in the past as Weigelia, most botanists agree today on the *Weigela* spelling.

In the past, *Weigela* was included within the genus *Diervilla*, which itself has three species, all of them native to North America. Besides this first difference in area of origin, *Diervilla* differs from *Weigela* in flower shape (irregular with two lips) flower color (always yellow), and blossoming in the current season's growth shoot.

From the 12 species of *Weigela* originally described, Rheder (1940) restricted this number to eight, and Bean (1980) differentiated only five:

W. coraensis (Thunb): a native of Japan, not of Korea, introduced to Europe about the middle of the 19th century and now the parent of many garden hybrids.

W. floribunda (Sieb. et Zucc.) K. Koch: native of the mountains of Japan, introduced to Europe in about 1860, and also a parent of garden hybrids.

W. florida (Bge.) A. DC.: native of North China and Korea, introduced to Europe in 1844. Cv. *Foliis purpureis* and var. *venusta* are closely related, as well as *W. praecox* (Lemoine) widely used by this French nurseryman in breeding.

W. japonica (Thunb) D.C., native of Japan, introduced to the USA in 1982. It is allied to *W. floribunda*.

W. hortensis (Sieb. et Zucc.) K. Koch reached Europe by the 1870's and is closely related to *W. japonica*.

W. middendorffiana (Trautv. and Mey) K. Koch: native of Japan, North China, Korea, and far East USSR. It is less vigorous than the former species (1 to 1.50 m

[1] Station d'Amélioration des Espèces Fruitières et Ornementales, Laboratoire des Arbustes d'Ornement, INRA, 49000 Angers, France

Biotechnology in Agriculture and Forestry, Vol. 11
Somaclonal Variation in Crop Improvement I (ed. by Y.P.S. Bajaj)

high), with yellow flowers, more susceptible to late spring frost; introduced to Europe in 1850.

W. maximowiczii (S. Moore) Rehd. is a similar species, but inferior as an ornamental.

Soon after the introduction into Europe, hybridizations were made between the different species. A French nurseryman, Victor Lemoine in Nancy, started in 1865 by crossing first *W. coraensis* × *W. floribunda*, extending progressively to the other species. A tremendous amount of hybrids have been named for their garden value, several of them still being under cultivation (Table 1). All the plants which have been used in the breeding program can be related to the first four species described by Bean.

No difficulty in crossing has been recorded, and the genetic meaning of these four species is questionable. All of them are diploid, with $2x = 2n = 36$ chromosomes, as well as the many cultivars we have counted in root tips (Fig. 1). Observation of meiosis in many cultivars has always shown regular figures with 18 bivalents. Analysis of pigments by chromatography has always given one anthocyan: cyanidine.

The morphological characters, when submitted to botanical classification, could be under quite simple genetical control. From a genetical point of view, the first four species described by Bean seem to belong to one group. This fits also with the results of two other breeders, Poszwinska (1961) in Poland and Weigle and Beck (1974) in the USA.

On the other hand, Weigle and Beck (1974) report the failure of crosses with *W. middendorfiana*, which thus appears as a more remote species, but Poszwinska (1961) crossed *W. maximowiczii* successfully with plants of the first group.

Table 1. *Weigela* cultivars bred by Lemoine nurseries from 1880 to 1915

[a]Abel Carrière	1876	Espérance	1906	Lacepède	1886
André Thouin	1882	[a]Eva Rathke	1892	[a]Le Printemps	1901
Auguste Wilhelm	1881	[a]Féerie	1926	Majestueux	1930
Avalanche	1909	Fleur de Mai	1899	Mémoire de Madame van Houtte	1884
Avant Garde	1906	[a]Floréal	1901		
Bayard	1893	Fraicheur	1904	Messager	1911
[a]Béranger	1881	Gavarnie	1884	Mont Blanc	1898
Bouquet Rose	1899	[a]Girondin	1913	Montesquieu	1886
Buisson Fleuri	1911	[a]Gloire des Bosquets	1881	[a]Othello	1882
Congo	1886	Glorieux	1904	Pascal	1891
Conquérant	1904	Gracieux	1904	Pavillon Blanc	1901
[a]Conquête	1896	Grandiflora Flore Alba	1882	Perle	1902
Dame Blanche	1902	Gratissima	1881	Profusion	1915
Daubenton	1886	[a]Heroine	1896	Seduction	1908
De Jussieu	1882	Ideal	1926	Superba	1894
[a]Descartes	1891	Jean Mace	1882	Teniers	1884
Diderot	1886	John Witter	1881	Vestale	1912
[a]Emile Galle	1881	Juvenal	1893	Voltaire	1882

[a]These cultivars are still being propagated in some nurseries. (Decourtye 1982)

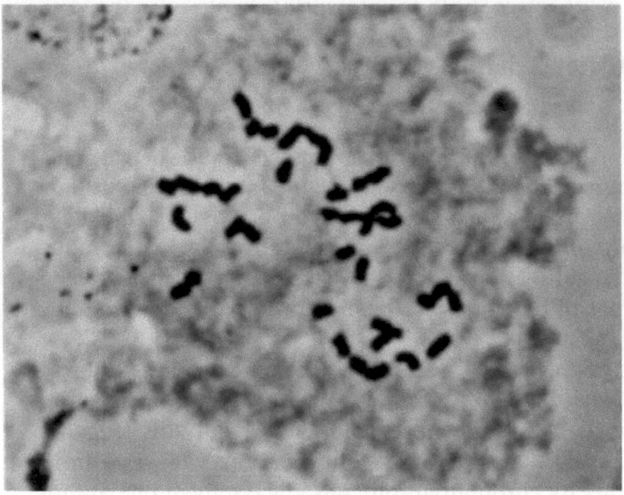

Fig. 1. Metaphase plate of New Port Red root tip mitosis (2x = 2n = 36 chromosomes)

Weigela belongs to the honeysuckle family (Caprifoliacea) close to *Abelia* and *Kolkwitzia*, both small genera with a diploid chromosome number of 2n = 32, and close to the more variable genus *Lonicera*, which has polyploid series of 2n = 18, 36, or 54 chromosomes (Darlington and Janaki Ammal 1945).

Weigela is well adapted to temperate climatic zones. Buds have dormancy in winter, to overcome which they need cold for some weeks. On the other hand, *Weigela* is not very hardy, and can suffer damage from very cold winter temperatures, as in Poland. Besides their native areas (China, Japan), *Weigela* hybrids are widely used in gardens and parks of Western European countries, in the United States and Canada, with some limitations again for frost, and in the equivalent countries of the Southern Hemisphere, like New Zealand, Weigela is available and sold in most nurseries and garden centers. The total number of plants sold in France every year reaches one million. This number can be estimated at 2.5 million in Europe and nearly 4 million for Europe and North America.

2 Conventional Practices

2.1 Culture

Weigela thrives in many different kinds of soils. It benefits from some irrigation during summer drought. Best flowering is obtained in the open field, but it can withstand some shade. It can be weeded with most chemical weedicides and shows no special sensitivity to any of them; no particular disease has been reported in this genus.

There are no more species under cultivation, having been superseded by a great number of cultivars. These have been reviewed and tested in comparative trials by different authors. The most complete review has been given by Howard (1965), who recognized at least 100 distinct cultivars among nearly 350 names, the others being synonymous or mispelled.

After a comparative trial among 27 of them, Grootendorst (1968) recommends 14 hybrids with different levels of merit symbolized by stars. The selections of the Boskoop Research Station (The Netherlands), released in the 1960's, Eva Suprême, Fiesta, Rosabella, and Ballet, are among the winners. Special mention is made for frost-resistance in favor of Eva Rathke.

More recently, Hieke (1978) published the results of observations on a large assortment from 1972 to 1977 in Czechoslovakia. From the 42 cultivars identified, described, and evaluated, four were classified as first class (Ballet, Floreal Venusta, and Bristol Ruby), 12 as very good, while 26 were classified as inferior or quite unsatisfactory in their properties. The author emphasizes the need for improvement in cold resistance. This was the aim of Weigle, who released two cultivars suitable for the conditions of Northern states in the USA, Pink Princess (1974) and later Red Prince (1979). At the Ottawa Research Station (Canada), Svejda selected Minuet (1982) and Rumba (1985) mainly for hardiness.

Private nurseries contribute also to new selections, for example, Geers in the Netherlands, who released Evita (Van de Laar 1981) and some other sister seedlings, and in France, Briant nurseries (1986) with a leaf-variegated sport Olympiade Rubidor.

2.2 Pruning and Flowering Control

Flowers do not appear in summer on current year growth, but will bloom the following year. The best flowers are on young wood. Pruning consists of removal of older shoots just after blooming; this enhances the development of new shoots and reduces the competition of the seeds for nutrients.

Flowering in *Weigela* is under long-day control (Nitsch 1957). Short days with high temperature will yield a continuous elongation, like a climber, while continuous lighting will induce a compact plant with intense blooming. In a growth chamber experiment, Lepage (1981) obtained a more complicated pattern, depending widely on the genotypes under trial, i.e., Eva Suprême and Bristol Ruby. As a rule, vegetative growth stops when daylight is shorter than 9 h, whatever the temperature is. Flowering is enhanced by long days (20 h daylight) but can also occur in short days (9 h) after a longer time (9 months) when plants are grown under high temperature (22°C). She concluded that the initial stimulus of meristem differentiation for floral development could occur in short days as well as in long days, but that the differentiation and development of flowers is long-day dependent. Maloupa-Ikonomou and Jacques (1986), using scanning electron microscopy, studied the first changes in the meristem following long-day exposures. After 9 long days, the meristem started to enlarge and after 12 days, early axillary buds were initiated in the leaf axils. The different steps of floral development were completed about 20 days after the beginning of floral induction.

2.3 Propagation

Propagation can be achieved through hardwood cuttings during winter, but good results are also obtained with soft wood cuttings in summer, with root-promoting substances under mist. Vegetative propagation is the only way to keep the true type of the cultivar. Seed propagation is equally easy; it needs some cold stratification (60 to 90 days at +4°C) to obtain quick and abundant germination. The juvenile phase is not very long and first blooming will appear on the third year growth. Due to their genetical origin, cultivars of *Weigela* are highly heterozygous; they are mostly self-sterile, so that seedlings show a wide range of variation: this has been the origin of many new cultivars.

2.4 Improvement of Weigela Through Mutagenesis

Since 1972, *Weigela* has been chosen in our laboratory as a model plant to improve the methodology of mutagenesis on vegetatively propagated plants. The choice of the best part of the plant to be irradiated and the method of propagation in the successive steps to isolate the mutation were the main purposes.

The high level of heterozygosity is a favorable condition for the expression of a mutated gene. In a short breeding period, the genus showed a great capacity for diversification. For laboratory purposes, a method of propagation of one internode with two opposite buds has been successfully developed. Practical results could also be expected from such a program, mainly toward reduced size, giving better suitability for small gardens, but also any possible change in habit of growth, flower color, or variegated leaf.

The first step was the irradiation of 1-year rooted cuttings in the dormant stage (Decourtye 1978a). Eleven cultivars (Abel Carrière, Ballet, Bristol Ruby, Eva Rathke, Eva Suprême, Féerie, Gloire des Bosquets, Le Printemps, Newport Red, Rosabella and Snowflake) have been irradiated with gamma rays (^{60}Cobalt) at a dose ranging from 30 to 70 grays. Each irradiated bud has been propagated into five cuttings; the terminal shoot and the four lower shoots developing from the two nodes beneath the apex. 700 V2 plants were screened, giving nearly 12% mutants, with a great difference between cultivars, with Le Printemps, Bristol Ruby, and Eva Suprême yielding the highest percentage. Most mutants expressed as chimera and wide heterogeneity was shown by the cuttings, due to their different ages.

To reduce these problems, the second step was the irradiation of 3-year-old plants (cv. Eva Suprême), severely pruned and disbudded so that after irradiation the collar zone produced a large population of cuttings as neoformation (Decourtye 1978b). This method did not yield a better mutation rate than the first one (4.25 mutants for 100 cuttings), most of which were again in periclinal condition. Nevertheless, this was an improvement in easier handling and labeling of a great number of cuttings of the same size.

As bud neoformation on the collar zone does not yield solid mutants (as can be expected if only one cell provides the neoformation), we tried to use neoformation of shoots from roots. The method needs large enough roots (diameter > 5 mm) and only few cultivars are capable of such neoformation: Le Printemps shows the most ability, followed by Bristol Ruby and to a very small extent Abel Carrière. Eva Rathke, Féerie, and Newport Red do not develop more than very friable white callus. Absorption of kinetin at 40 ppm for 9 h enhances the number of shoots per root, as does a preirradiation of 5 grays, 24 h before the main irradiation at 60 grays (unpubl.). Thirty different mutants have been obtained from 450 neoformed shoots, again, most of them in periclinal condition. Thus it appears that neoformation of shoots from the root does not yield solid mutants and is not better than the preceding method.

The success in in vitro culture of *Weigela* (Duron 1975) has opened new possibilities which will be discussed in the following section. From this former work, practical results have nevertheless been achieved for nursery purposes: true new cultivars have been released: Courtatom Couleur d'Automne, which is a variegated form of Le Printemps with nice autumn color; Courtavif Rubivif, which is an improvement in stable red color of the flower from Bristol Ruby, and Courtanin, which is a very floriferous dwarf mutant of Eva Rathke (Decourtye 1986).

3 In Vitro Approaches

Little work has been published on tissue culture of *Weigela*. It is summarized in Table 2.

Table 2. In vitro culture of *Weigela*

Cultivar	Explant source	Result	Reference
Five cultivars[a]	Meristem	Rooted plantlets	Duron (1975)
Red Prince	Buds	Multiplication	Calvert and Stephens (1986)
Bristol Ruby Eva Rathke Eva Suprême Le Printemps	Stem internodes	Bud neoformation	Duron (1981)

[a] Bristol Ruby, Eva Suprême, Eva Rathke, Mme Ballet, Le Printemps.

3.1 Sterilization of the Explant

For meristem culture the explant was picked either in January or after flowering in late spring or summer. The shoots were cut in single node parts and surface sterilized by dipping into alcohol 70° for 5 min without rinsing.

Dissection is easier after flowering for, at that time, all buds are vegetative and it is not necessary to distinguish between vegetative and floral buds. The explants were isolated without requiring the sterile environment of an air-flow cabinet. Nevertheless, nearly all the cultured explants remained generally uncontaminated.

The internode explants were stem pieces of plants already grown in vitro, so sterilization was not necessary.

3.2 In Vitro Culture

The culture media are based on the mineral solution of Murashige and Skoog (1962) (MS) used at full or half strength. The main differences depend on the aim of the culture. The hormonal balance is different if it is necessary to promote the development of meristems or to induce the differentiation of new meristems (Table 2).

3.2.1 Meristem Culture

Meristems were used as starting material to initiate in vitro culture (Duron 1975). Two media were used successively. The first one released the activity of the meristem, which evolved into a small shoot, the other allowed elongation and rooting.

The meristems separated from the upper part of the buds were laid on the surface of 2.5 ml of culture medium (Table 3) contained in hemolyse tubes. The size of the explants (the meristem plus one or two leaf primordia) was about 0.2–0.4 mm.

The explants developed well, whatever the season, and the meristems were removed. Three to 4 weeks later, they had one or two pairs of well-developed leaves and had to be transferred into a new medium deprived of cytokinin (Table 4) favorable for rooting and elongation.

Table 3. Meristem, bud, and stem internode culture media

	Explants		
	Meristems	Buds	Internodes
Reference	Duron (1975)	Calvert and Stephens (1986)	Duron (1981)
Macronutrients	MS	MS	MS
Micronutrients	MS	MS	MS
Vitamins (mg/l)			
Myoinositol	100	100	100
Thiamine. HCl	1	NS	30
Pyridoxin. HCl	1	"	1
Ca-pantothenate	1	"	–
Nicotinic acid		"	10
Biotin	0.01	"	–
Growth regulators (mg/l)			
Kinetin	0.01		
BAP		1.12	1
GA_3	1		
IBA			0.1
NAA	0.1		
Agar (g/l)	6	NS	7
Sucrose (g/l)	20	NS	30
pH	5.5	NS	5.5

NS = Not specified.

Table 4. Media used for rooting and elongation

	A	B	
Macronutrients	MS 1/2	Knop 1/2	see Gautheret (1959)
Micronutrients	MS	Berthelot	
Vitamins (mg/l)			
Myoinositol	100	100	
Thiamine. HCl	1	1	
Pyridoxin HCl	1	1	
Ca-pantothenate	1	1	
Nicotinic acid	1	1	
Biotin	0,01	0,01	
Growth regulators (mg/l)			
GA_3	0.5	0.5	
IBA	0.3	0	
NAA	0	0.1	
Sucrose (g/l)	10	5	
Agar (g/l)	6	6	
pH	5.5	5.5	

Five cultivars were grown in the test tube in the first step (Fig. 2) of the same procedure. For one of them, Eva Rathke, it was necessary to use a different second medium (B). On half-strength MS medium (A) the plants had a marked tendency to produce vitrous or flower-bearing plants (Fig. 3). Flowering is exceptional when the culture medium has a low salt composition such as Knop's macroelements associated with Berthelot's microelements (Gautheret 1959) and a reduced amount of sugar (5 g/l) (Table 4 – B medium).

On A or B medium, cuttings began to elongate while rooting took place. They can stay on this medium for 3–4 months. Generally the elongation is limited to three or four internodes. Except for the most vigorous cultivar, Le Printemps, the plants stopped elongating before reaching the top of the test tube.

The slow rate of multiplication by apical and nodal cuttings provided us with enough plants for further use. All our cultivars have been maintained in vitro by subcultures twice a year.

3.2.2 Internode Explants

Internode culture was performed in order to try to induce bud neoformation, as a way to obtain somaclonal variation or to take advantage of adventitious buds in induced mutagenesis (Broertjes 1982).

It was possible to induce bud neoformation from stem internodes of *Weigela* on MS salt medium (Duron 1981). The main difficulty was to find the right hormonal balance which, among cell division, will trigger off the differentiation of one or a few cells toward the organization of meristems. On a given medium, bud neoformation

Fig. 2. Two month-old apical cutting of Bristol Ruby on A medium

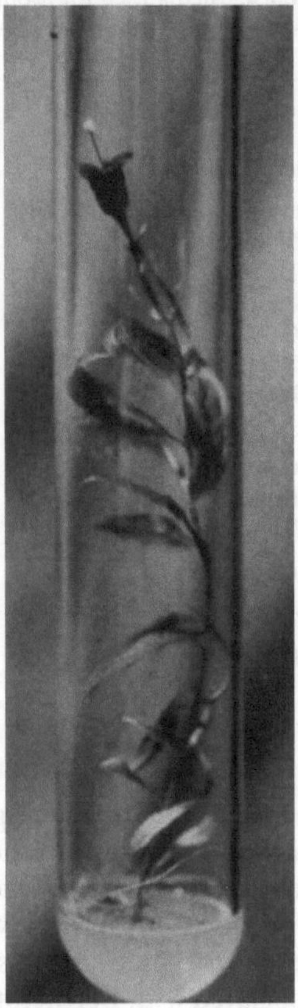

Fig. 3. Flowering in vitro of an apical cutting of Eva Rathke three months old on A medium

was genotype-dependent. Similar results were obtained trying to regenerate plants from callus culture of *Nicotiana* (Branchard 1984).

We found that with the hormonal balance (mg/l) BAP 1/IBA 0.1 among the four cultivars tested the percentage of explants able to give adventitious buds ranged from 100 for Eva Rathke to zero for Le Printemps. Buds appeared earlier (2–3 weeks after the beginning of culture) and in greater number on Eva Rathke internodes (Fig. 4). The stems produced by this technique elongated and rooted on the same medium as the cultivar from which they were obtained.

3.2.3 *Bud Culture*

Vegetative buds were removed from 30–50 cm stems. On MS medium, the number of days for bud break is related to the position on the stem. Bud breaking is fastest

Fig. 4. Adventitious buds on stem internode of Eva Rathke 1 month after the beginning of culture on bud-inducing medium

for the buds nearer to the apex. It depends also on the concentration of BAP in the culture medium: 12 days are necessary without BAP, only 9 days with 10 μM BAP.

The main shoot number increased from two without BAP to four with BAP. The shoots were then rooted in vitro with 5 μ IAA (Calvert and Stephens 1986).

3.3 In Vitro Propagation

Weigela cultivars are very easy to propagate by cuttings using traditional methods. At present, in vitro propagation is not the most economical way. It might be used to provide nurserymen quickly with plants of a new cultivar that will be then exploited as maternal stock plants in traditional propagation.

4 In Vitro-Induced Variation

4.1 Somaclonal Variation

Before using regenerated plants from stem internodes in mutagen treatments, it was necessary to assess if these plants were true to type. One hundred and fifty of them produced by adventitious buds on internodes of Eva Rathke were observed for 2 years. Only one presented minimal flower modification (Fig. 5). Flowers are smaller than those of the control, as are the leaves of the flowering branches. After flowering, this plant presented no other modifications. Thus in this case almost all the plants regenerated were not distinguishable from the control, which might be related to the earliness of bud formation on the explants, which limits the callus phase. From the pioneer results of Lutz (1966) with tobacco cells and Sibi (1976) with lettuce callus, it has been established that the best way to obtain somaclonal variation is to

Fig. 5. Flowering branches of Eva Rathke (W_3) and a variant induced after adventitious bud formation

regenerate plants from cells or callus that, if possible, have been subcultured. Such work should be performed with a genotype of good regenerating potentiality such as Eva Rathke.

4.2 Effect of Mutagenic Agents

In vitro-grown *Weigela* plants are good material for experiments in mutagen treatments. Meristems of in vitro plants are smaller (reduction of 10^{-1} in one dimension) than those of open-air plants (Duron unpubl.). Watelet-Genot and Favre (1981) established that meristems of *Dahlia variabilis* in vitro had a reduced number of cells. If, in the same way, the smaller size of *Weigela* meristems may result from a reduction of cell number, this should allow reduced competition between mutated cells following mutagen treatment. They represent an intermediate step toward the single-cell technique. Furthermore, the number of leaf primordia is reduced and they are not tightly folded above the apical dome (Fig. 6a). This situation should allow better penetration by the chemicals than treating buds of open air-grown plants (Fig. 6b). In a first set of experiments, whole plants were used in treatments with colchicine or gamma rays. In the second one, we combined the mutagenic treatment with the adventitious bud technique.

4.2.1 Colchicine Treatment

Colchicine solution was applied following the technique of Maia et al. (1973). The opposite leaves of the apical bud are a good support to maintain a drop of colchicine solution and to allow it to reach the meristem. Three concentrations (0.1–0.2–0.4%) of colchicine were assayed. The most efficient in obtaining tetraploids were the two

A 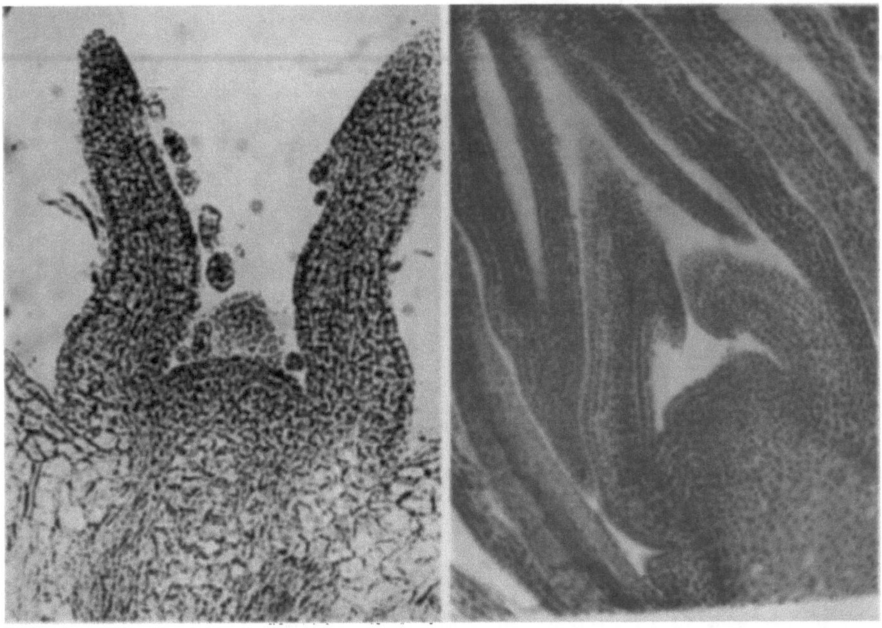 B

Fig. 6A,B. Anatomical structure of an apex of *Weigela*. **A** Grown in vitro on A medium (Table 4). **B** Grown in the field

lower ones; 0.4% was frequently toxic. A drop of colchicine solution, applied to 12 apical cuttings of five cultivars: Eva Rathke, Eva Suprême, Bristol Ruby, Mme Ballet, New Port Red, resulted in the formation of at least one solid tetraploid from the 12 treated cuttings with 0.1% or 0.2% of colchicine.

The characteristics of tetraploids are common to all tetraploid plants. They have reduced vigor, thicker and broader leaves. All the flower parts are also thicker and are easily distinguishable from the diploid ones (Fig. 7). The floriferousness of the tetraploid *Weigela* is less developed and the color is often less attractive. None of them presents improved characters to make a new ornamental variety.

Pollination of five tetraploids by eight diploid cultivars, or conversely, allowed us to observe 30 hybrid combinations and a total of 1200 triploid plants, which was an unknown level of ploidy in the genus *Weigela*. Triploids exhibit a wide range of vigor, some of them being less vigorous than the tetraploid parents. The leaves and the flowers are often larger than those of the parents (Figs. 8 and 9).

Tetraploids of Eva Suprême, New Port Red, and Bristol Ruby provided the greater number of preselections. Three triploid selections have already been released: Courtared Lucifer, a vigorous deep-red large-flowered shrub, Courtalor Carnaval with large white and pink flowers. Next is Courtamon, with salmon-colored trumpet-like flowers and a drooping habit.

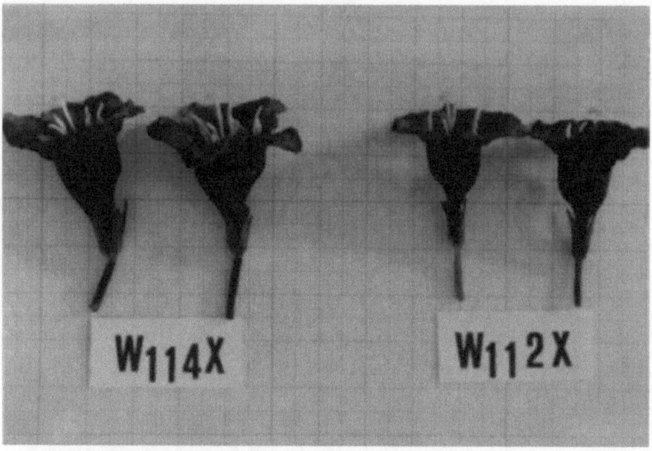

Fig. 7. Comparison of flowers of Bristol Ruby (W_{11}) diploid (2X) and tetraploid (4X)

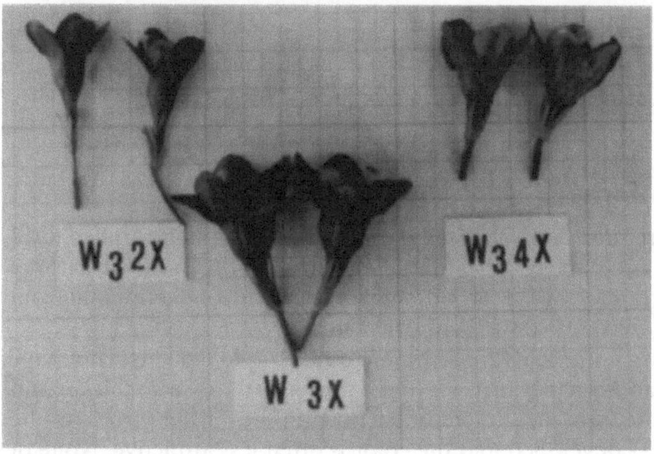

Fig. 8. Comparison of the flower size of Eva Rathke (W_3) di- and tetraploid with a triploid

4.2.2 Irradiation of in Vitro Plantlets

The small size of the shrubs in the test tube permitted treatments with a gamma ray apparatus used in radiotherapy. The experimental device and the effects of the irradiation have been described (Duron and Decourtye 1986). The efficiency of the treatment was assessed by the observation of the plants (cv. Bristol Ruby) originating from each irradiated bud. After irradiation, the rooted plantlets were separated in three parts, the apical and two nodal cuttings immediately below, and transferred on the same fresh medium.

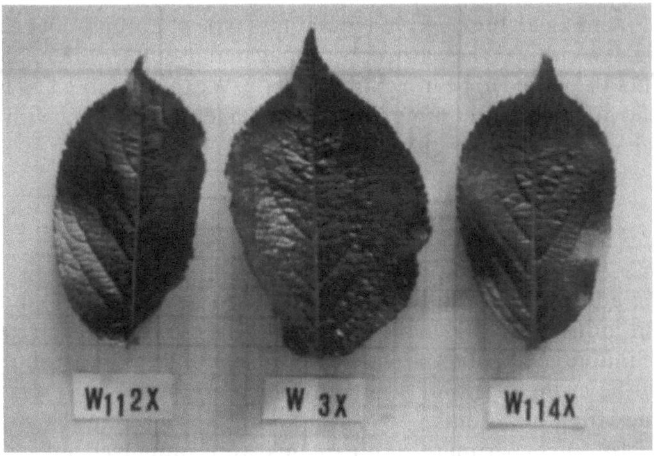

Fig. 9. Comparison of the leaf size of Bristol Ruby di- and tetraploid with a triploid

Bud survival exhibits a sharp decrease between 40 and 50 Gy[2] (Table 5); 60 Gy is almost a lethal dose. From only two replications we cannot give conclusive results, nevertheless, it seems that no clearcut difference exists in radiation susceptibility between apical and nodal cuttings.

The first mutants were noticed in vitro, especially those characterized by a reduced vigor and/or a modification of the leaf shape. Most of them were observed after the transfer to soil of the cuttings developed from each irradiated bud. Following this process, they resulted either in normal or in whole mutated plants. As expected, the number of mutants was related to the dose. From a practical point of view, it seems that 30 and 40 Gy are the most suitable doses to obtain interesting plants. A 50-Gy exposure dose induced too high a number of polymutated phenotypes, which makes it difficult to obtain, in one and the same plant, a good ornamental aspect without additional unfavorable characters.

Table 5. Survival of apical and axillary buds of Weigela after increasing doses of gamma rays. Fourteen plants were treated in each irradiation (IR_1 and IR_2)

Dose (Gy)	20		30		40		50		60	
Buds	IR_1	IR_2	IR_1	IR_2	IR_1	IR_2	IR_1	IR_2	IR_1	IR_2
Apical bud	14	14	8	13	8	11	7	3	5	1
Axillary bud no. 1	28	28	25	23	22	19	11	3	7	1
Axillary bud no. 2	28	28	22	26	16	14	6	5	4	1

Axillary buds were separated according to their position below the apex. The no. 1 is just beneath the apex.

[2] 1Gy = 100 rad

Among all the observed mutants, a type of ground-covering plant was selected as a new cultivar (Fig. 10), now propagated by nurserymen. Later, about half of the mutants proved to be in a chimeral state. We thus established the best way to obtain solid mutants: the combination of a mutagenic treatment with the adventitious bud technique (Broertjes 1982).

4.2.3 Mutagenesis and Adventitious Bud Formation

The adventitious bud technique was used by Broertjes and coworkers in irradiation-induced mutagenesis (Broertjes et al. 1976; Van Harten et al. 1981). This technique is also suitable to test the efficiency of chemical mutagens. The cells which will be at the origin of a new meristem are more easily reached than those of a preexisting meristems.

In our experiment, the explants were immersed into a 0.5% EMS solution containing 3% dimethyl sulfoxide for 90 min. They were rinsed three times in sterilized water and put (one in each test tube) at the surface of the culture medium (Table 2) where they stayed for 60 days. (Duron unpubl.).

The treatment did not modify the bud-forming ability of the explants (internodes on in vitro grown plantlets cv. Bristol Ruby). It lasted for 5 months between the beginning of the culture and the transfer of the plantlets to soil. The number of shoots per explant ranged from 1 to 23.

Fig. 10. A ground-covering mutant of Bristol Ruby

Out of 380 still living plants produced by 36 treated explants, two different mutants were first observed. They have a dwarf growth habit and look like solid mutants. From three plants having a normal phenotype during 3 to 4 years, new shoots appeared with different phenotypes. (Fig. 11). The last one noticed in the summer of 1987 has two kinds of new shoots, each with a different flower color, both different from the control. The other two presented only one new phenotype, which was not the same in each case. It is too early to know if one of these mutants will be of commercial interest. All the remaining plants have been severely pruned to facilitate the expression of other mutated cell clusters that might be present in the inner tissues, to better estimate the efficiency of the mutagen treatment. Until now, the percentage of mutants is less than that observed after a 20-Gy treatment of buds by gamma rays. The only interest of the method is that these mutants look like solid mutants.

The last three observed mutants raise the question of the origin of adventitious buds. Apparently they were of multicellular origin. For two of them, two kinds of genetically different cells have contributed to the formation of the meristems. There were three for the last observed mutants. In our observations, two out of five mutant plants were built from only one cell, in contrast to the situation observed by Broertjes with Saintpaulia.

Fig. 11. A mutated shoot (*arrow*) emerging from a "normal" EMS-treated plant 3 years after planting in the field (notice the larger leaves)

5 Conclusion

In vitro techniques may be a useful tool in different ways. For *Weigela*, micropropagation could be practiced, but the conventional cuttings seems now more economical. The method could be used to provide nurserymen with mother plants of a new variety. Meristem culture is very easy to perform and could also help to solve health problems.

In vitro plantlets have mainly been used to take advantage of the small size of the meristems and the anatomical character of the apices in mutagen treatments. Colchicine applied to apical buds easily resulted in solid tetraploids which allowed the creation of triploids, three of which became commercial cultivars.

It is possible by the observation of plantlets originating from irradiated buds to isolate many mutants, about half of them being solid mutants. One of these, noticed for its ground-covering growth habit and its flowering, is being propagated by nurserymen.

An original treatment was performed associating EMS treatment and adventitious bud techniques on stem internodes of Bristol Ruby. Five different phenotypes, looking like solid mutants, were observed. EMS will have had a very low mutagenic effect in our experimental conditions if no more mutants appear during the next growing season.

References

Bean WJ (1980) Trees and shrubs hardy in the British Isles, vol 2. Murray, London, pp 743–749

Branchard M (1984) Application de vitrométhodes à la mise en oeuvre de programmes de sélection de plantes résistantes à des maladies. Agronomie 4(9):905–911

Briant A (1986) Weigela Olympiade Rubidor. Catalogue André Briant (Jeunes Plants). Automne 1986, p 19

Broertjes C (1982) Significance of in vitro adventitious bud technique for mutation breeding of vegetatively propagated crops. In: Induced mutations in vegetatively propagated plants, vol 2. IAEA, Vienna, pp 1–10

Broertjes C, Roest S, Bokelman GS (1976) Mutation breeding of *Chrysanthemum morifolium Ram.*, using in vivo and in vitro adventitious bud techniques. Euphytica 25:11–19

Calvert LK, Stephens LC (1986) In vitro propagation of *Weigela florida* 'Red Prince'. Hortsci 21(3): (Abstr) 815

Darlington C, Janaki Ammal EK (1945) Chromosome atlas of cultivated plants. Allen & Unwin, London, 397 pp

Decourtye L (1978a) Utilisation de la mutagénèse pour l'obtention de nouvelles variétés d'arbustes ornementaux. C R Acad Agric Fr 64:664–669

Decourtye L (1978b) Amélioration du *Weigela* par mutagénèse: recherche d'une méthode. Ann Amelior Plantes 28(6):697–707

Decourtye L (1982) Urgent tasks in breeding of woody ornamentals. In: 21st Int Hortic Congr, Hamburg, FRG, vol 2, pp 799–810

Decourtye L (1986) Différents facteurs déterminant l'évolution de la gamme variétale en pépinière. C R Acad Agric Fr 72:809–816

Duron M (1975) La culture in vitro de méristèmes de quelques cultivars de *Weigela* Thunb hybrides. C R Acad Sci Paris Ser D 281:865–868

Duron M (1981) Induction de néoformations caulinaires chez des *Weigela* Thunb. hybrides cultivés in vitro. Agronomie, 1(10):865–868

Duron M, Decourtye L (1986) Effets biologiques des rayons gamma appliqués à des plantes de *Weigela* cv. Bristol Ruby cultivés in vitro. In: Nuclear techniques and in vitro culture for plant improvement. IAEA, Vienna, pp 103–111
Gautheret RJ (1959) La culture des tissus végétaux. Masson, Paris, 863 pp
Grootendorst HJ (1968) Weigela. Dendroflora 5:56–60
Hieke K (1978) Pruhonicky sortiment rodu *Weigela* Thunb. v Letech 1972–1977. Acta Pruhoniciana 39:1–91
Hieke K (1983) Evaluation of the size and habit of *Weigela* species and varieties. Sb UVTIZ Zahradninctvi 9:127–138
Howard RA (1965) A check list of cultivar names in *Weigela*. Arnoldia 25(9–11):49–69
Lepage I (1981) Approche du déterminisme de la floraison chez un arbuste ornemental: le *Weigela*. Rapp Stage DEA Amélioration et développement des végétaux. Paris XI, 11 pp
Lutz A (1966) Obtention de plantes de tabac à partir de cultures unicellulaires provenant d'une souche anergiée. C R Acad Sci 262:1856–1858
Maia E, Bettachini B, Beck D, Venard P, Maia N (1973) Contribution à l'amélioration de létat sanitaire du Lavandin clone 'Abrial'. Ann Phytopathol 5(2):115–124
Maloupa-Ikonomou E, Jacques M (1986) Development of meristems of *Weigela florida* variety Bristol Ruby following reproductive induction in long days. Ann Bot (London) 58:887–895
Morel G, Wetmore RH (1951) Tissue culture of monocotyledons. Am J Bot 38:141–143
Murashige F, Skoog T (1962) A revised medium for rapid growth and bioassay with tobacco tissue cultures. Physiol Plant 15:473–497
Nitsch JP (1957) Photoperiodism in woody plants. Proc Am Soc Hortic Sci 70:526–544
Poszwinska J (1961) Preliminary progeny analysis of some *Weigela* crosses. Arbor Korn Rocznik 6:143–167
Rehder A (1940) *Weigela*. In: Manual of cultivated trees and shrubs, 2nd edn. MacMillan, New York, pp 849–852
Sibi M (1976) La notion de programme génétique chez les végétaux supérieurs. II. Aspect expérimental: obtention de variants par culture de tissus in vitro sur *Lactuca sativa* L.; apparition de vigueur chez les croisements. Ann Amelior Plantes 26(4):523–547
Svejda F (1982) Minuet *Weigela*. Can J Plant Sci 62(1):249–250
Svejda F (1985) Rumba *Weigela*. HortSci 20(1):149
Van de Laar H (1981) *Weigela* Evita. Dendroflora 18:73
Van Harten AM, Bouter H, Broertjes C (1981) In vitro adventitious bud techniques for vegetative propagation and mutation breeding of potato *Solanum tuberum* L. II. Significance of mutation breeding. Euphytica 30:1–8
Von Schmalscheidt W (1983) Wertzeugnisse für Gehölze. *Weigela* Evita – Wertzeugnis (Getuigschrift van Verdienste). Dtsch Baumschule 4:164
Watelet-Gonod MC, Favre JM (1981) Miniaturisation et rajeunissement chez *Dahlia variabilis* (variété Télévision) cultivé in vitro. Ann Sci Nat Bot Paris, 13ème Ser 2/3:51–67
Weigle J, Beck AR (1974) Pink Princess *Weigela*. HortSci 9(6):606

IV.6 Somaclonal Variation in *Nicotiana sylvestris*

D. PRAT[1], R. DE PAEPE[2], and X.Q. LI[3]

1 Introduction

Nicotiana sylvestris is a diploid species (2n = 24) from the Solanaceae family, native of the Andean foothills of northwestern Argentina, where it grows from 500 to 1600 m, generally in moist woods, but also on red silt and sandy banks. In 1899, Spegazzini and Comes described it as "a robust plant of the aspect of cultivated tobacco" (cited by Goodspeed 1954).

General characteristics of the species were given by Goodspeed and Avery (1939) and Goodspeed (1954) in his taxonomy of the genus *Nicotiana*. Adult plants are 1 to 2 m high, with auricled, sometimes decurrent, broad leaves. The white flowers, about 85 mm long, are grouped in panicles. At the dehiscence stage, the anthers are in contact with the stigma, resulting in spontaneous self-fertilization. The obvious morphological similarities of *N. sylvestris* with cultivated tobacco, *N. tabacum* – a tetraploid species (2n = 48) – led Goodspeed and Clausen (1928) to suggest that *N. sylvestris* and *N. tomentosa* – or *N. tomentosiformis* –, were the progenitor species of *N. tabacum*. Kostoff (1938a) formulated the same hypothesis on the basis of crossabilities of artificial hybrids. Analysis of the small subunit of ribulose biphosphate carboxylase definitely proved that *N. tabacum* arose from the cross of *N. sylvestris* as female with *N. tomentosiformis* as male (Kung 1976).

Very little variability exists between *N. sylvestris* collections from different laboratories, possibly (as suggested by Goodspeed 1954), because most seedlots originate from the botanical gardens of the Faculté de Médecine de Lyon (France). However, the same author mentioned a cultivar with shorter flowers (65–75 mm). Another line, slightly different from Comes' description, and from more recent samplings in the natural range of the species, showed complete absence of variation over more than 20 years of self-pollination (Goodspeed and Avery 1939). In other cases, however, variant forms were obtained from lines maintained by self-pollination: in the Phytotron Greenhouses at Gif-sur-Yvette (France), three mutant forms appeared over a 12-year period (De Paepe 1983): two are under nuclear

Laboratoire de Génétique et Physiologie du Développement des Plantes, CNRS, 91190 Gif-sur-Yvette, France. Present addresses: [1]Laboratoire de Génétique des Populations d'Arbres Forestiers, ENGREF, 54042 Nancy Cedex, France
[2]Laboratoire de Génétique Moléculaire des Plantes, UA 115, Université Paris-Sud, 91405 Orsay Cedex, France
[3]University of Beijing, Beijing, China

Biotechnology in Agriculture and Forestry, Vol. 11
Somaclonal Variation in Crop Improvement I (ed. by Y.P.S. Bajaj)

control – a light green line and one with slightly reduced growth and smaller flowers –; the last line has slow development and leaf abnormalities that are maternally inherited (De Paepe 1983).

N. sylvestris is highly responsive to chromosomal imbalance. All possible trisomics have been obtained that may be distinguished by their morphology (Goodspeed and Avery 1939). A spontaneous haploid plant had slower growth, reduced leaf and flower dimensions, and was completely sterile (Kostoff 1934, 1938b). As in *N. tabacum*, various types of in vitro cultures – tissue, protoplast, and pollen cultures – are successful in *N. sylvestris* and numerous plants may be regenerated. The diploid level facilitates genetic analysis of the variability, and this species can be considered as a model for the study of somaclonal variation. Moreover, useful new characteristics could be transferred to *N. tabacum* or other Solanaceae species by sexual or somatic hybridization.

2 In Vitro Techniques

A general survey of in vitro studies performed in *N. sylvestris* is given in Table 1 and a detailed description of some media in Table 2.

2.1 Tissue Culture

First in vitro studies in *N. sylvestris* were performed by Dix and Street (1974), followed by Malepszy et al. (1977) and Maliga et al. (1979). Calli were induced from leaves of greenhouse plants, or from in vitro-cultured shoots. After sterilization, haploid or diploid plant material was cultured on Linsmaier and Skoog's (1965) basal medium, with added auxins (IAA or 2,4-D) and cytokinins (BAP or Kin). The obtained calli were subcultured or used for initiation of cell suspensions. Cell cultures maintained for a year in liquid nitrogen were still able to grow when put in favorable conditions (Maddox et al. 1983). Direct regeneration of shoots from leaf epidermis was obtained on Murashige and Skoog's (1962) medium with IAA and BAP (Li unpubl.).

Restivo and Tassi (1986) studied the inhibitory effect of 5-fluorouracil on morphogenesis. The effect depended on time of addition of the substance into the culture.

2.2 Protoplast Culture

The ability of protoplasts to regenerate plants was first demonstrated by Nagata and Takebe (1971) in *N. tabacum*. In *N. sylvestris*, plants were regenerated from leaf mesophyll protoplasts simultaneously by Banks and Evans (1976), Bourgin et al. (1976), and Nagy and Maliga (1976). Methods of protoplast isolation and culture are summarized in Table 2. Typical features of protoplast development are shown in Fig. 1.

Table 1. In vitro techniques in *N. sylvestris*

Plant material and peculiarities	Reference
Androgenesis	
Anther culture, just before or after pollen mitosis	Bourgin and Nitsch (1967)
Anther culture, microspores	Rashid and Street (1974)
Anther culture	Tomes and Collins (1976)
Pollen culture, 2 days after pollen mitosis	De Paepe et al. (1977)
Anther culture, immediately after pollen mitosis	McComb and McComb (1977)
Pollen culture, consecutive androgenetic cycles	De Paepe et al. (1981)
Anther culture from polyploid and aneuploid plants	Niizeki et al. (1982)
Anther culture	Misoo et al. (1984)
Anther culture from trisomic plants	Niizeki et al. (1984)
Tissue and cell cultures	
Haploid and diploid petioles of greenhouse plants	Dix and Street (1974)
Selection for NaCl tolerance	Dix and Street (1975)
Leaf strips of axenic haploid shoots	Dix et al. (1977)
Young leaves of greenhouse haploid plants	Malepszy et al. (1977)
Selection for kanamycin and streptomycin tolerance	Maliga et al. (1979)
Selection for NaCl tolerance	Dix and Pearce (1981)
Haploid leaf sections, selection for streptomycin tolerance	Maliga (1981)
Selection for TMV tolerance	Murakishi and Carlson (1982)
Preservation of diploid cells in liquid nitrogen	Maddox et al. (1983)
Morphogenetic effects of 5-fluorouracil on leaf calli	Restivo and Tassi (1986)
Leaf explants and epidermis	Li (unpublished)
Protoplast Culture	
Leaves of greenhouse diploid plants	Banks and Evans (1976)
Leaves of greenhouse diploid and haploid plants	Bourgin et al. (1976)
Leaves of greenhouse diploid plants	Nagy and Maliga (1976)
Leaves of greenhouse diploid plants	Bourgin et al. (1979)
Haploid in vitro-grown shoots	Durand (1979)
Haploid in vitro-grown shoots	Facciotti and Pilet (1979)
Diploid in vitro-grown shoots	Magnien et al. (1980)
Haploid and diploid in vitro-grown shoots	Negrutiu and Mousseau (1980)
Haploid and diploid in vitro-grown shoots	Van Slogteren et al. (1980)
Haploid in vitro-grown shoots, mutant selection	Negrutiu and Muller (1981)
In vitro-grown shoots from dihaploid lines	Prat (1983)
Haploid in vitro-grown shoots, selection for auxotrophic mutants	Tsala et al. (1983)
Haploid in vitro-grown shoots, selection for biochemical traits	Durand (1984)
Diploid in vitro-grown shoots, selection for lysine-analogue tolerance	Negrutiu et al. (1984)
Diploid in vitro-grown shoots from protoplast-derived lines	Li et al. (1985)
Somatic Hybridization	
Kanamycin-tolerant *N. sylvestris* + noninducible shoot *N. knightiana*	Maliga et al. (1977)
N. sylvestris + cms *N. tabacum* (*suaveolens*-like cytoplasm)	Zelcer et al. (1978)
N. sylvestris + *N. sylvestris*, amino acid analogue-tolerant lines	White and Vasil (1979)
N. sylvestris + cytoplasmic streptomycin-tolerant *N. tabacum*	Medgyesy et al. (1980)
Cytoplasmic lincomycin-tolerant *N. sylvestris* + *N. plumbaginifolia*	Csepló et al. (1983)
N. sylvestris + albino Su/Su *N. tabacum*	Evans et al. (1983)
cms *N. sylvestris* (*bigelovii*-like cytoplasm) + cms *N. tabacum* (*undulata*-like cytoplasm)	Fluhr et al. (1983)
N. sylvestris + *N. tabacum* identified plastomes and chondriomes	Aviv et al. (1984a)
N. sylvestris + *N. rustica* identified plastomes and chondriomes	Aviv et al. (1984b)

cms: cytoplasmic male sterility

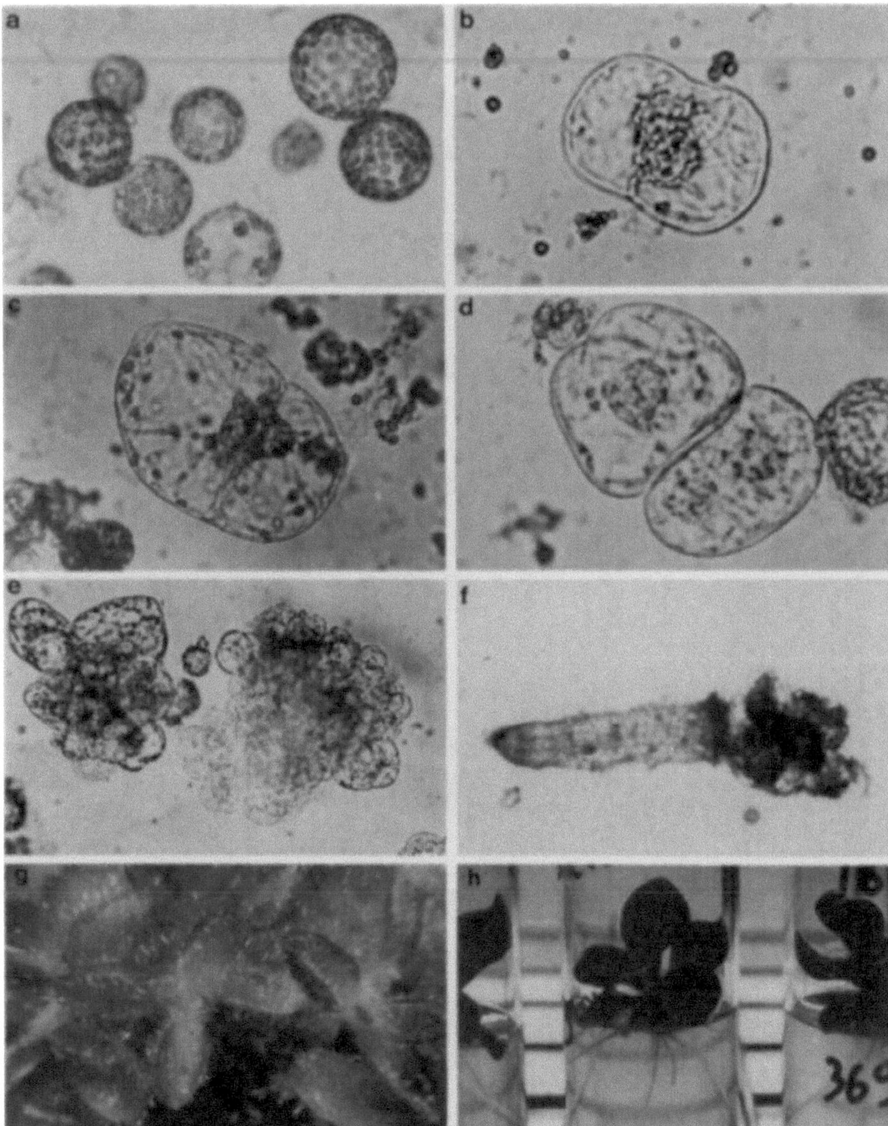

Fig. 1a-h. Development of *N. sylvestris* protoplasts (culture conditions: Prat 1983). **a** Freshly isolated protoplasts (×350); **b,c** first division (5 days after isolation, ×515); **d** second division (6 days after isolation, ×540); **e** protoplast-derived callus (10 days after protoplast isolation, ×60); **f** protoplast-derived embryoid (1 month after protoplast isolation, ×30); **g** morphogenetic callus (3 months after protoplast isolation, ×5.4); **h** regenerated plantlet (5 months after protoplast isolation, ×1.1)

Table 2. Tissue and protoplast cultures in *N. sylvestris*: culture medium composition and culture conditions

Constituent (mg/l)	Tissue culture	Protoplast isolation					Protoplast culture				
	Malepszy et al. (1977)	Banks and Evans (1976)	Bourgin et al. (1976)	Nagy and Maliga (1976)	Durand (1979)	Facciotti and Pilet (1979)	Banks and Evans (1976)	Bourgin et al. (1976)	Nagy and Maliga (1976)	Durand (1979)	Facciotti and Pilet (1979)
Enzymes											
Cellulase R10			1,000	20,000	600	3,000					
Driselase (Kyowo Hakko)			500		25	300					
Macerozyme (Yakult)				5,000							
Macerozyme (Japan Biochemicals)		5,000									
Macerozyme R10			200								
Meicelase (Meiji Seika Kaisha)		50,000									
Pectinol fest					200	1,500					
Macronutrients											
KNO_3	1,900	101	950	2,500		1,900	263	950	2,500	950	2,850
NH_4NO_3	1,650		825	250		1,650	206	825	250	825	2,475
$(NH_4)_2SO_4$				134					134		
$CaCl_2$, $2H_2O$	440	1,480	220	900		440	425	220	900	1,220	2,660
$CaHPO_4$				50					50		
KH_2PO_4	170	27.2	85		136	170	177	85		85	255
NaH_2PO_4				150					150		
$MgSO_4$, $7H_2O$	370	246	185	250	1,000	370	370	185	250	185	555
KCl					2,500						
Micronutrients											
$FeSO_4$, $7H_2O$	27.8		27.8	27.8		27.8	7.0	27.8	27.8	27.8	27.8
Na_2 EDTA	37.3		37.3	37.3		37.3	9.4	37.3	37.3	37.3	37.3
$ZnSO_4$, $7H_2O$	8.6		1	2		8.6	2.2	1	2	1	8.6
H_3BO_3	6.2		1	3		6.2	1.5	1	3	1	6.2
$MnSO_4$, $4H_2O$	22.3		0.1	10		22.3	5.6	0.1	10	0.1	22.3
$CuSO_4$, $5H_2O$	0.025	0.025	0.03	0.039		0.025	0.012	0.03	0.039	0.03	0.025
$AlCl_3$			0.03					0.03		0.03	
$NiCl_2$			0.03					0.03		0.03	
KI	0.83		0.01	0.75		0.83	0.25	0.01	0.75	0.01	0.83
Na_2MoO_4, $2H_2O$	0.25	0.16		0.25		0.25	0.062		0.25		0.25
$CoCl_2$	0.025			0.025		0.025	0.007		0.025		0.025

Organic addenda											
Inositol			100	100		100	50	100	100	250	100
Calcium panthotenate			1					1		2.5	
Nicotinic acid			1	1		0.5		1	1	2.5	0.5
Pyridoxin HCl			1	1		0.5		1	1	2.5	0.5
Thiamine HCl			1	10		0.1	0.5	1	10	2.5	0.1
Biotin			0.01					0.01		0.025	2
Glycin						2	0.5				
Conditioned medium (ml)										20	
Growth regulators											
Benzylaminopurine			1	0.2			0.5	1	0.2	1	
Kinetin	0.15										0.6
Indole acetic acid	2.0										
Naphthalene acetic acid			3	1			0.75	3	1	3	0.6
2.4 Dichlorophenoxy-acetic acid	0.08			0.1			0.25		0.1		0.2
Organic supplement											
Sucrose	15,000			137,000			5,000	20,000	137,000	20,000	
Mannitol		90,000	80,000				9,000	80,000		80,000	
Sorbitol					73,000	60,000					32,000
Xylose				250					250		
Agar							6,000				
Culture conditions											
pH	5.8	5.8	5.5	5.6	5.5	6.0	5.8	5.5	5.6	5.5	6.0
Temperature		25°C	22°C	28°C	24°C	23°C					
Duration		18 h	14–16 h	3–4 h	16 h	6–8 h					
Light		dark	dark	dark	dark	dark					
first days of culture (lx)							700	800	300	Dark	Dark
culture (lx)							700	2,000	1,500	1,000	Dark
Cell density (m/l)							10^5	$1–5\times10^4$	$1–4\times10^4$	$5–10\times10^4$	1.8×10^3

Protoplast Isolation. In first studies, mother plants were grown in greenhouse conditions and protoplasts isolated from young leaves. The lower epidermis was peeled off and the excised leaf pieces floated on the isolation medium, containing cellulases and pectinases. After digestion, the protoplasts were centrifuged at low speed (100 *g*), and washed several times with the culture medium.

Several improvements to this method were described. Better protoplast yields, plating efficiencies, and colony formation were obtained when the enzymatic digestion was performed from shoots grown in sterile conditions, either haploid (Durand 1979; Facciotti and Pilet 1979; Negrutiu and Mousseau 1980, 1981) or diploid (Prat 1983). Magnien et al. (1980) purified the protoplast suspension on a discontinuous iso-osmotic gradient. Viable protoplasts were in the lower density phases.

Protoplast Culture. Protoplast suspensions, at a density of about 50,000 cells/ml are placed in the dark or under low light for a few days at about 23°C. All media used have auxin/cytokinin ratios higher than 1. Sorbitol always gave better results than mannitol either in protoplast isolation, colony development (Zelcer et al. 1978), or callus formation (Table 3). The conditioned medium added by Durand (1979) – a filtrate of exponential *N. tabacum* cell culture – was successfully replaced by either four organic acids, pyruvic acid (20 mg/l), malic acid (40 mg/l), citric acid (40 mg/l) and fumaric acid (40 mg/l) (Negrutiu and Mousseau 1980), or by two amino acids, glutamine (200 mg/l) and asparagine (40 mg/l) (Prat 1983). Van Slogteren et al. (1980) suggested adding 200 mg/l casein hydrolysate, on the basis of thymidine uptake by *N. sylvestris* protoplasts.

Table 3. Number of calli per dish in 20-day-old protoplast culture: effect of sorbitol and mannitol (Prat unpubl.)

Culture medium	Isolation medium	
	0.4 M Sorbitol	0.4 M Mannitol
0.4 M Sorbitol	96[a]	3[c]
0.4 M Mannitol	19[b]	2[c]

[a,b,c] Groups of nonsignificantly different means at 1% level.

In general, callus formation took place, but sometimes embryos developed directly (Facciotti and Pilet 1979; Prat 1982).

Plant Regeneration. Shooting was generally induced in the basal medium used for protoplast culture and solidified with agar, by diminishing the auxin/cytokinin ratio. Rooting was induced in low salt medium, either without growth substances (Bourgin et al. 1976) or with NAA only (Banks and Evans 1976). Regenerated plants may be transplanted to the greenhouse about 4–6 months after the beginning of culture.

2.3 Anther and Pollen Culture

Production of *N. sylvestris* androgenetic plants was first described by Bourgin and Nitsch (1967). The anthers were excised before the first pollen mitosis, and after sterilization (ethanol 70%, followed by immersion in a 7% calcium hypochlorite solution) were planted on a mineral medium containing iron, vitamins, 2% sucrose, and 0.8% agar. The regenerated plantlets were haploid. Nitsch (1969) discussed the role of different substances in the culture medium and that of the development stage of the anthers, and found that the optimal stage coincides with the occurrence of mitosis in the microspore. Rashid and Street (1974) also started the culture with uninucleated microspores, in contrast with McComb and McComb (1977), who obtained best results with binucleated grains. Successful culture of isolated pollen cells was reported by De Paepe et al. (1977). The method was similar to that used for *Datura* (Nitsch and Norreel 1973) and *N. tabacum* (Nitsch 1974), with some modifications concerning pollen stage and temperature. Further improvements were described by De Paepe et al. (1981), allowing the regeneration of several hundred plantlets per flower put into culture. Details of this protocol are described here: floral buds were collected after the anthers had reached the top level on the sepals, i.e., 1–3 days after the first pollen mitosis. The anthers were excised, sterilized with 7% calcium hypochlorite, and placed into Nitsch's (1974) sterile liquid culture medium (macronutrients of Lin and Staba 1961, Fe EDTA 37 mg/l, sucrose 20 g/l) for 3 days at 17 °C and 3 days at 23 °C in the dark. The pollen was then pressed out of the anthers, filtered through 100-μm nylon sieves to eliminate the anther somatic cells, and centrifugated at 500 *g* for 5 min. The pollen pellet was then resuspended in the same liquid medium supplemented with 5 g/l inositol, 0.5 g/l glutamine and 0.03 g/l serine; 2 ml of the pollen suspension are put into 5-cm diameter Petri dishes, at 22–25 °C, 2000 lx of fluorescent light. After about 3 weeks of culture, young regenerated plantlets were transferred on a solid medium (macronutrients of Murashige and Skoog 1962, micronutrients and vitamin solution of Nitsch 1969, sucrose 10 g/l, agar 7 g/l). After rooting, the plantlets were transferred to the greenhouse, and maintained at 24 °C day, 17 °C night, with a 16-h photoperiod.

Pollen Development. In the most favorable conditions (preculture of the anthers at 17 °C for a few days) about 50% of the binucleated pollen grains started dividing after 3–5 days at 22 °C. Apparently, the generative nucleus did not divide more than once, as further divisions could only be observed in the vegetative cell (Fig. 2). After 10 days, 10% of the grains had formed globular proembryos, always surrounded by the exine. The embryos went through the "heart" and "torpedo" stages (of Nitsch 1969), turned green, and developed into well-differentiated plantlets in 3 weeks. About 1% of the pollen grains put into culture regenerated into plants.

Factors Affecting Androgenesis. Influence of pollen development stage and temperature during anther culture prior to the isolated pollen culture are shown in Table 4. Only buds picked 1 to 2 days after first pollen mitosis produced an appreciable number of plantlets. On the other hand, it was clear that low temperatures at the beginning of culture improved yield in viable plantlets.

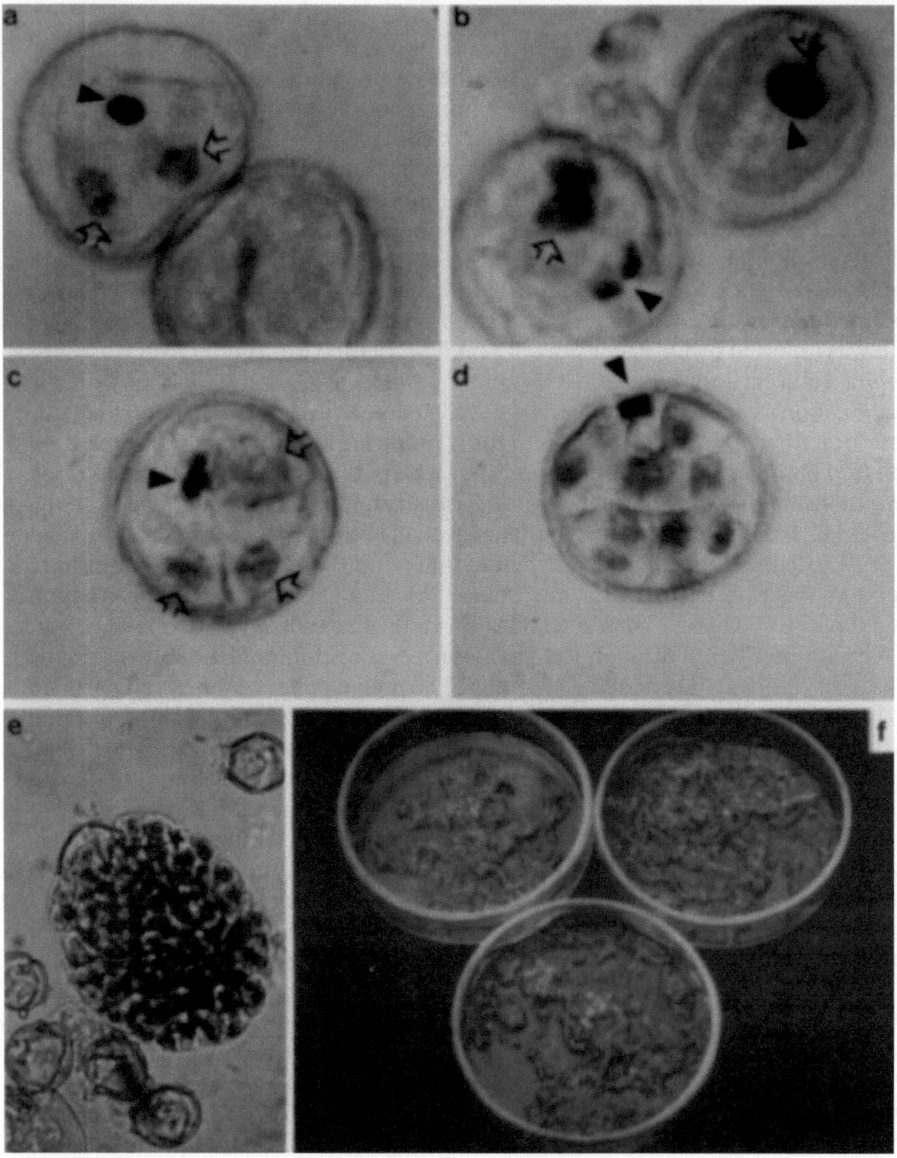

Fig. 2a-f. Isolated pollen culture at 22 °C (De Paepe et al. 1981). **a** Division of the vegetative nucleus only (4 days after pollen isolation, × 775); **b** division of both generative and vegetative nuclei (*left*), fusion of the vegetative and generative nuclei (*right*), after 4 days of culture (× 780); **c,d** proembryos where the generative has not divided, after 8 days of culture (× 590); **e** embryo after 10 days of culture (× 220); **f** regenerated plantlets after 3 weeks of culture (× 0.8). *Filled arrowheads* ▶ generative nucleus; *open arrowheads* ▷ vegetative nucleus

Table 4. Influence of flower bud stage (A) and temperature (B) on pollen culture. (After De Paepe et al. 1983)

A

Days from mitosis	− 1	0	+ 1	+ 2	+ 3
Bud length mm	12	15	20	27	40
No. of plantlets regenerated per flower	0	0	150	120	0

Temperature conditions during anther preculture: 3 days at 17°C and 7 days at 22°C

B

Temperature conditions	No. of days of anther culture before pollen isolation					
	(6) 17°C	(6–7) 22°C	(10–13) 22°C	(3) 17°C + (2) 22°C	(3) 17°C + (4) 22°C	(5–7) 27°C
No. of plantlets regenerated per flower	1–2	50	62	98	127	0

Floral buds were 25 mm long

Ploidy of the Regenerated Plants. With the isolated pollen culture method described above, doubling of the haploid genome occurs spontaneously, either by fusion between generative and vegetative nuclei (McComb and McComb 1977), or more probably, by endomitosis in the vegetative nucleus or its derivatives. About 15% of the regenerated plants were diploid (called doubled-haploids, DH), a few percent triploid, tetraploid, or even octoploid, either by anther (McComb and McComb 1977), or by pollen culture (De Paepe and Pernès 1978). Aneuploids were extremely rare. The doubling procedure described by Nitsch (1974) increased the proportion of DH regenerated plants up to 30%: the flower buds were placed overnight in a vacuum flask containing the sterile basal medium supplemented with 2% dimethylsulfoxide and 0.05% colchicine.

Ploidy could be determined in root tip mitoses, but this method led to some imprecision, some of the regenerated plants being mixoploids. Chromosomal counts in male gametophytes at the meiotic stage were more reliable, as the meiocytes of one single plant always had the same ploidy. To test efficiently a great number of regenerated plants, it was possible to rely only on morphology: indeed, haploid plants always have leaf length/width ratios higher than 2, small flowers (about 7.5 cm instead of 9 cm for diploids), and were completely sterile.

Cycles of Androgenesis. Consecutive androgeneses were carried out as follows: plants regenerated from pollen of original lines were called H_1 (haploids) and DH_1 (diploids). The pollens of some DH_1 chosen at random were put into culture separately and the regenerated plants called H_2 and DH_2. In this way, ten consecutive cycles of androgenesis were performed. The number of regenerated

Table 5. Number of plantlets regenerated per flower put into culture (N) and proportion of diploids during consecutive androgeneses (% 2 n). (De Paepe 1983)

	Androgenetic cycle				
	1	2	3	4	5–8
N	127	66	52		—18—
% 2 n	18	30	38	54	53

plantlets fell dramatically after a few cycles of culture, when the proportion of diploids increased up to 50% (Table 5). Progeny of diploid plants was compared to parental lines in statistical assays.

3 Variation by Androgenesis

Androgenetic *N. sylvestris* doubled-haploid plants were all somewhat changed as compared to parental lines. This new variability can be expressed at the phenotypic, quantitative, cellular, physiological, and chromosomal levels. Unless clearly mentioned, this section will deal with only true diploid regenerated plants and offspring, called DH, and presumed to be completely homozygous.

3.1 Phenotypic Variability

The vast majority of the DH_1 presented typical alterations in leaf morphology (Fig. 3A), that was called "crumpling", in contrast with the flat aspect of the parental leaves (De Paepe and Pernès 1978). These crumpled plants appeared whether or not colchicine treatment had been applied to the anthers put into culture, although their proportion was higher in the first case (De Paepe et al. 1981). The crumpling is transmitted through selfing. Other regenerated DH_1 plants, apparently normal, also gave crumpled plants in their selfed progenies. After three cycles of androgenesis, all the regenerated plants were crumpled, even if colchicine had never been applied (De Paepe et al. 1983). Genetic studies showed that crumpling is governed by a few recessive nuclear genes, the number of which increases with the number of androgenetic cycles. Other types of abnormalities appeared after several cycles, such as epistyly, foliar, and ovary teratomes.

3.2 Quantitative Variability

After the first cycle of androgenesis, regenerated DH_1 plants and their progenies did not show any significant changes in adult dimensions, but they generally grew at slower rates than parental lines. Growing abilities were partly restored

Fig. 3a,b. Morphological characteristics of androgenetic plants. **a** Original line plant (*left*) and crumpled DH_1 plant (*right*). **b** decrease in adult dimensions in DH plants regenerated from consecutive pollen cultures: original line T (*left*). DH_1, DH_2, DH_3, DH_4, and DH_5 lines (*right*). (**a** from De Paepe et al. 1983)

Table 6. Growth and adult characteristics of androgenetic plants. (De Paepe et al. 1983)

Character		T	DH_1	DH_2	DH_3	DH_4	DH_5	$(S_1 S_2)$
Rosette stage								
Leaf length	mm	197^{a}	184^{b}	183^{b}	168^{c}	170^{c}	168^{c}	0.69**
Leaf width	mm	105^{a}	$95^{a,b}$	$101^{a,b}$	86^{c}	$90^{b,c}$	88^{c}	0.68**
Adult stage								
Time of flowering	days	124^{a}	121^{a}	116^{a}	123^{a}	117^{a}	120^{a}	0.10
Stem height	mm	610^{a}	560^{a}	550^{a}	370^{b}	380^{b}	370^{b}	0.70**
Leaf length	mm	350^{a}	$312^{a,b}$	$302^{b,c}$	$261^{c,d}$	234^{d}	254^{d}	0.67**
Leaf width	mm	169^{a}	151^{a}	$150^{a,b}$	$148^{a,b}$	135^{b}	141^{b}	0.44**
Coralla length	mm	87^{a}	87^{a}	84^{a}	73^{b}	70^{b}	73^{b}	0.98**
Corolla radius	mm	37^{a}	37^{a}	36^{a}	32^{b}	31^{b}	32^{b}	0.92**
Number of capsules		34^{a}	27^{a}	28^{a}	8^{b}	4^{b}	7^{b}	0.80**
Weight of seeds per capsule	mg	112^{a}	75^{b}	80^{b}	$55^{c,d}$	50^{d}	$53^{c,d}$	0.68**

T: original line; DH_1-DH_5: second generation of selfing of diploid androgenetic plants obtained by consecutive cycles of pollen culture.
$(S_1 S_2)$: coefficient of correlation between the first and the second generation of selfing.
^{a}, ^{b}, ^{c}, ^{d}, as in Table 3.
**Significant at the 1% level.

by cross-breeding (De Paepe et al. 1977). Decrease in growth rates as well as in organ dimensions – stem, leaves, and flowers – continued during consecutive androgenesis up to the 5th cycle, when a plateau was reached: leaf and flower sizes of DH_{4-5} plants are reduced to the two-thirds of those in original line plants (Fig. 3B). DH dimensions are transmitted to the second generation of selfing (Table 6).

3.3 Cellular and Physiological Variability

In spite of their smaller dimensions, DH plants had greater foliar cells, with increased water content and changes in chlorophyll a/b levels (Prat 1982). Cell disorganization in parenchymatic tissue of a DH_7 plant is clearly visible in Fig. 4. In addition, all the DH plants so far analyzed, even normal plants, showed higher peroxidase activities, at all stages of development. The increase concerned both acid and basic peroxidases; electrophoretic patterns differed quantitatively between DH lines (De Paepe 1983), but new bands were not detected (Prat et al. 1983).

3.4 Chromosomal and DNA Organization Changes

Number and aspect of mitotic chromosomes of DH plants were not changed as compared to original line plants, but unusual configurations were sometimes observed in meiotic prophases of DH. Diakinesis in *N. sylvestris* usually shows 11

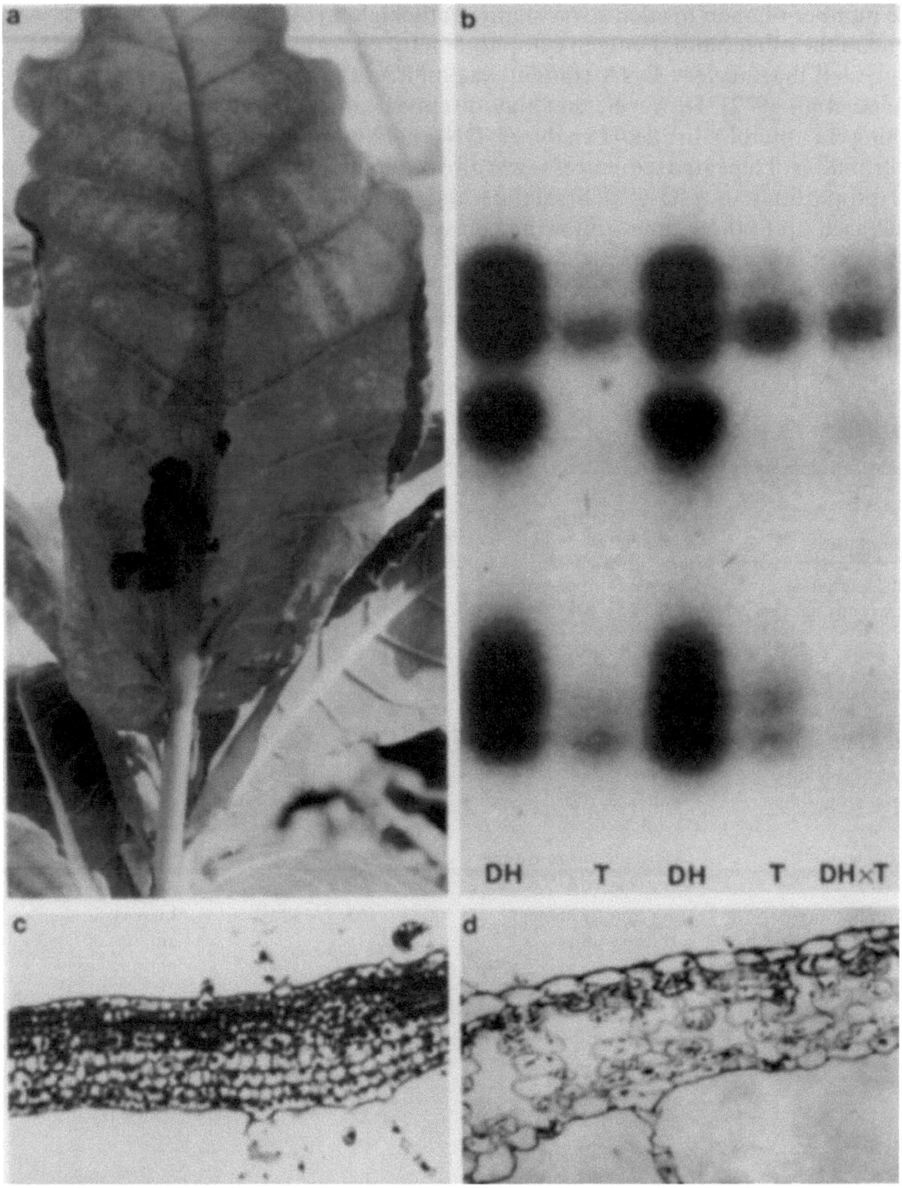

Fig. 4a-d. Characteristic features of DH plants. **a** Foliar teratomes in a DH_7 plant; **b** acidic peroxidases of original line (T), two DH_7 plants (DH)and their hybrid (DH × T); **c** leaf transversal section in an original line plant (× 170); **d** leaf transversal section in a DH_8 plant (× 170)

ring and one open bivalents (Kostoff 1938b; Goodspeed 1954). In most DH plants, the number of open bivalents was significantly higher (Fig. 5).

On the other hand, Feulgen cytophotometry performed on squashed root tips suggested that nuclear DNA content was on average higher for DH plants (De Paepe et al. 1982). However, this may disappear after the second generation of selfing (Li unpubl.). In all DH analyzed, DNA genome organization – number and distribution of repeated sequences – was also different from original (T) lines (Table 7): proportions of a G-C rich satellite and of rapidly reassociating sequences increased up to three times following consecutive androgeneses. Taken together, these results suggest that some classes of repeated sequences, including inversed repeats, are either amplified or reorganized in the genome of *N. sylvestris* androgenetic plants. So far, no differences were found concerning the ribosomal cistrons.

Table 7. Summary of the changes observed in DH DNA organization. (De Paepe et al. 1982)

Technique	Discriminating characters	T	DH_1	DH_4-DH_8	Interpretation
CsCl analytical ultracentrifugation	Heavy component 1.703%	1.0	1.6	2.7	Increase in G-C rich sequences or methylation
Thermal denaturation in 0.12 M PB	Sequences denaturing below 85°C (%)	28	33	36	Increase in A-T-rich sequences
Hydroxylapatite chromatography	Zero-binding sequences (%)	5	10	15	Increase in highly repeated sequences
S_1 nuclease	Resistance of zero-time reassociated DNA (%)	1.3	2.5	–	Increase in inverted repeats
Cytophotometry 560 nm (Feulgen)	Lng $(\text{transmittance})^{-1}$ arbitrary units	100	105	115	Increase in the amount of DNA per nucleus

T: original line; DH_1-DH_8: second generation of selfing of diploid androgenetic plants obtained by consecutive cycles of pollen culture. PB: phosphate buffer (0.5 M NaH_2PO_4 + 0.5 M Na_2HPO_4).

Fig. 5a-f. Male gametogenesis in androgenetic plants. **a** Typical diacinesis in an original line plant showing one open bivalent (*open arrowhead*) and the nucleolus (*filled arrowhead*); **b** characteristic diacinesis in a DH_5 plant, showing nine ring and three open bivalents; **c,d** meiosis in plants with supernumerary chromosomes. Diacinesis in a Su_2 plant: supernumerary chromosome (*arrow*) association with the nucleole (**c**); metaphasis I in a Su_1 plant: free Su_1 (**d**); **e,f** further abnormal development of the gametophyte in a Su_2 plant (*arrow* micronucleus): tetrad stage (**e**) and diad (**f**)

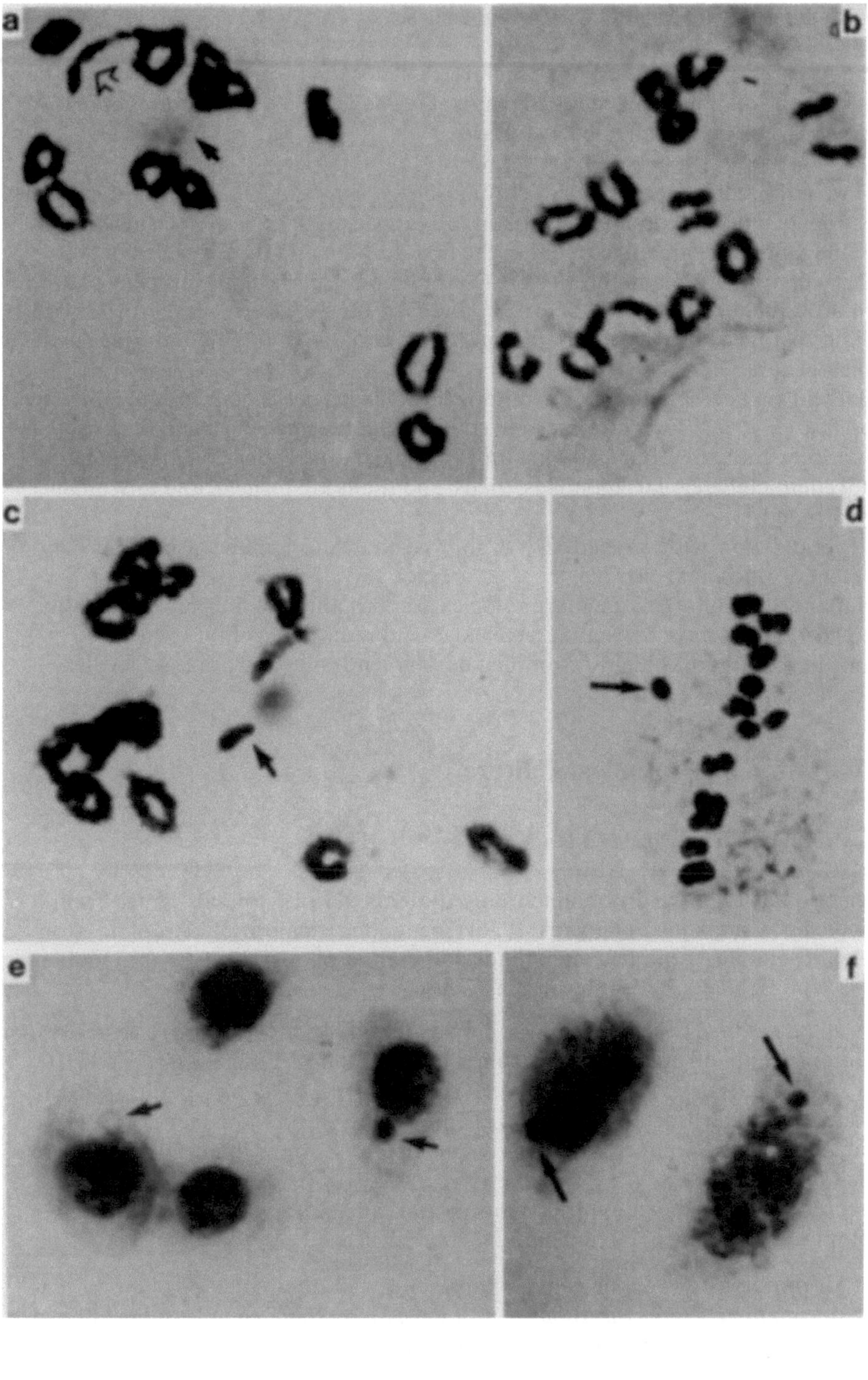
a
b
c
d
e
f

3.5 Plants with Supernumerary Chromosomes

In a few cases, plants with supernumerary chromosomal fragments were found in the offspring of diploid regenerated plants (De Paepe 1983). Key features of one of these cases can be summarized as follows:

– Two different supernumerary chromosomes, Su_1 and Su_2, appeared in the offspring of a normal diploid DH_1 plant: Su_1 appeared once in the first generation of selfing, Su_2 appeared five times, in the first, second, and third generation.
– Once present in a plant, the Su's are transmitted by selfing to the progeny plants, with a frequency of about 30% for Su_1 and 50% for Su_2. Reciprocal crosses with diploid plants showed that both Su's are partially eliminated during male gametogenesis.
– Both Su's are heterochromatic in mitotic (Lespinasse, pers. commun.) and meiotic prophase (Fig. 5c). Su_2 is often seen in contact with the nucleole. At diakinesis, the Su's are generally free, but may sometimes be linked to a bivalent. Tetrad formation is not completely regular; monoads, diads, and triads are formed, and micronuclei are extruded.
– Hybridization with labeled rRNA showed that Su_2 contains a haploid complement of ribosomal cistrons.
– Su_1 and Su_2 confer unique morphological abnormalities to the plants possessing them: Su_1 plants have few light green leaves and reduced fertility; in contrast, Su_2 plants have numerous highly crumpled dark green leaves and are fully fertile.

4 Variation by Protoplast Culture

4.1 Protoplast-Regenerated Plants

Facciotti and Pilet (1979) obtained only diploids out of 50 plants issued from a haploid line, and Banks and Evans (1976) regenerated a majority of tetraploids from diploid protoplasts, and few aneuploids. Different diploid lines were used by Prat (1983) and Li et al. (1985) as protoplast source:

– A botanical line, called T, that was already used for pollen cultures (see above).
– Doubled-haploid lines, obtained from successive pollen cultures, and called DH_1, DH_2 ... DH_5, as described above (cycles of androgenesis). Pooled plants from the second generation of selfing of such 15 DH lines (three per androgenetic cycle) were used as protoplast source. Protoplast plants obtained from T and DH lines were called PR_1.
– A protoplast line, phenotypically normal and called PR_{1a}, issued from the T line. Protoplast plants regenerated from PR_{1a} were called PR_2.

All protoplast source lines had similar effects on the ploidy of plants regenerated from them (Table 8). About 50% of the calli regenerated at least one diploid plant. Calli from which two plants at least were regenerated were of three types: those giving only diploids, those giving only tetraploids, and those giving both

Table 8. Ploidy and morphological abnormalities of protoplast-derived plants and their selfed progenies. (De Paepe unpubl)

Protoplast source lines	T	DH_1-DH_5		PR_{1a}
Calli producing at least one diploid plant (%)	49	58	42	65
Diploid regenerated plants (%)	42	60	37	60
Abnormal diploid regenerated plants (%)	0	0	0	22
Number of selfed progenies tested	22	13	64	18
Mutant progenies (%)	68	77	39	94

T, DH_1-DH_5 as in Table 6.

ploidies. These three types occurred at about the same frequency. On the average, about half the regenerated plants were diploid, the other being essentially tetraploid (Fig. 6). No aneuploid plants were identified.

Tetraploid plants were recognizable by their morphology: wider and thicker leaves, longer flowers, lower fertility. PR_1 diploid plants were not distinguishable from their parental line: in particular, protoplast plants derived from androgenetic lines had the typical phenotype of androgenetic plants. In contrast, 18 of the 92 diploid regenerated PR_2 plants showed morphological abnormalities, mainly leaf shape and color, and sterility.

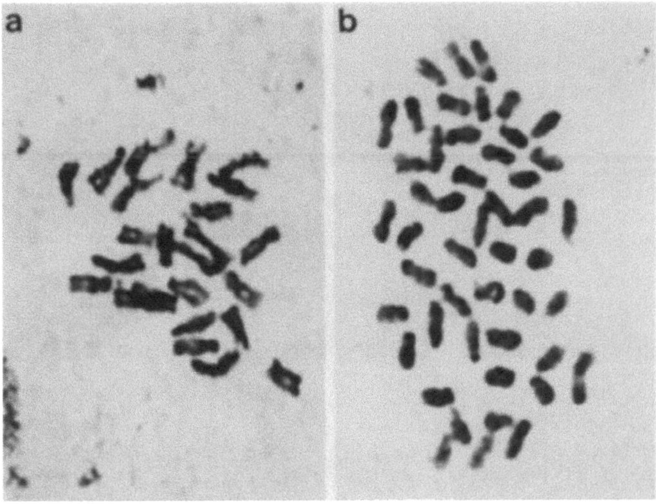

Fig. 6a,b. Root tip metaphases in plants obtained from diploid protoplast culture. **a** Diploid regenerated plant (2n = 24); **b** tetraploid regenerated plant (4n = 49). (Prat 1983)

4.2 Genetic Studies

Phenotypic Variability. Eighty-nine PR_1 plants and 18 PR_2 plants were analyzed in selfing or crosses. A great proportion of the PR progenies showed mutant phenotypes, generally in segregation. The frequency of mutant progenies differs according to the protoplast source line: in PR_1 progenies, it is higher from the T line (68%) than from the DH lines (39%, $X^2 = 5.7^*$). The proportion is still higher in PR_2 progenies (90%). Mutant phenotypes were observed at all developmental stages:

- germination – albinism, lethal necrosis of cotyledons and young leaves (only 10% of plantlets survive), callus-like seedlings,
- young plants – alteration in leaf color and shape, variegation, dwarfism (Fig. 7),
- adult plants – late flowering, abnormal flowers.

Plants regenerated from the same callus often possess the same mutation, some plants are doubled mutants. These mutations, never observed in parental lines, are generally under nuclear recessive control, but some are semi-dominant or maternally inherited, namely lethal necrosis and male sterility. The latter can apparently depend on different genetic mechanisms. Indeed, sterility, always associated with similar leaf and flower abnormalities (Fig. 8), was observed in several PR plants obtained independently:

- first some sterile plants were found in selfed progenies of PR_1 plants originated from one T, one DH_2, and one DH_5 lines, respectively,
- second, two sterile PR_2 plants were regenerated from the same PR_{1a} callus.

Fig. 7a,b. Two nuclear recessive mutations obtained by protoplast culture. **a** Dwarfism – aspect of the plant (**a$_1$**), transversal section of the leaf midrib (abnormal metaxylum) (**a$_2$**); root tip metaphase (2n = 24) (**a$_3$**); **b** variegation – aspect of the plant (**b$_1$**), leaf transversal section (irregular distribution of the chloroplasts) (**b$_2$**), metaphase II (2n = 24) (**b$_3$**)

Fig. 8a,b. Morphology of male-sterile plants. **a** Male sterile plant in segregating in selfed progeny – plant morphology (**a**$_1$) flower details showing abnormal anthers (**a**$_2$), pollinic mitosis (n = 12) (**a**$_3$); **b** cytoplasmic male sterile plant regenerated from second-cycle protoplast culture – plant morphology (**b**$_1$), flower details showing abnormal anthers (**b**$_2$), metaphase II plate (n = 12) (**b**$_3$)

In all cases, the sterile plants showed normal seed production when crossed as females with fertile plants as males, and were thus male sterile plants (ms). The crossed progenies of all three PR_1 ms were morphologically normal and fertile, and the sterility is probably controlled by several recessive nuclear genes. In contrast, all the progeny plants of both PR_2 ms were male sterile with unchanged morphological abnormalities, which proved that the mutation is cytoplasmic (cms). Molecular analyses showed that rearrangements in mitochondrial DNA are probably at the origin of the cms (Li et al. 1988).

Quantitative Variation. Selfed progenies of PR_1 plants – regenerated from the T and DH_5 lines – and PR_2 plants were compared to their respective parental lines. Only normal plants – without obvious abnormalities – were taken into account in the statistical tests (Table 9). On the average, the PR_1 plants showed a slight decrease in growth and adult characteristics. The depression effect was emphasized in the PR_2 plants (Li 1987).

Table 9. Quantitative variation in normal selfed progenies of normal diploid plants regenerated from protoplasts. (Prat 1983)

Character measured		T line		DH_5 line	
		Source line progenies	Regenerated-plant progenies	Source line progenies	Regenerated-plant progenies
Rosette plants					
Leaf length	mm	144[a]	131[b]	102[c]	95[d]
Leaf width	mm	79[a]	73[b]	57[c]	54[c]
Flowering plants					
Leaf length	mm	345[a]	336[a]	306[b]	300[b]
Leaf width	mm	157[a]	157[a]	132[b]	128[b]
Height	cm	75[a]	73[a]	59[b]	60[b]
Corolla length	mm	89.4[a]	89.5[a]	80.7[b]	79.7[c]
Corolla radius	mm	20.4[a]	20.6[a]	19.2[b]	18.6[c]
Seed-bearing plants					
Leaf length	mm	361[a]	345[a]	310[b]	297[b]
Leaf width	mm	171[a]	163[a]	150[b]	142[b]

T, DH_5, [a,b,c,d] as in Table 6

5 Variation by Tissue Culture

Malepszy et al. (1977) initiated tissue culture from *N. sylvestris* haploid lines with and without mutagenic treatment (EMS). The proportion of haploids decreased with the amount of EMS added, from 25% in nontreated series to 15% in treated ones. Most of the other regenerated plants were diploid, with 5% tetraploids and less than 1% aneuploids. No mutant phenotype was observed among the nontreated regenerated plants, whether haploid or diploid. From EMS-treated series, 20 to 60% of the haploid plants had new genotypes, most commonly albinism. Other mutations were dwarfing, leaf curving, and reduced number of petals (four instead of five). Among ten abnormal diploid plants, six were shown to be homozygous for Mendelian recessive mutations, the four others were apparently heterozygous. It can be concluded that mutations occur either before or after the doubling of the genome during culture.

6 Induction of New Characteristics and Transfer by Somatic Hybridization

Mutants in cell culture are identified by their unusual phenotype such as the ability to survive in the presence of a toxic drug, or to form green pigment when normal cultures are bleaching. Some procedures are developed (Negrutiu et al. 1984a) to produce mutants, generally obtained after mutagenesis. Regenerated resistant plants often showed morphological abnormalities and somatic hybridization by protoplast fusion was then carried out to combine desirable parental properties. In *N. sylvestris*, cells tolerant to salts, antibiotics, and amino acid analogs were recovered.

6.1 Salts

Tolerance to sodium chloride was induced in both haploid and diploid cell lines (Dix and Street 1975). The resistance was still expressed after several subcultures carried out in the absence of salt. No plants were regenerated and the genetic basis of the resistance is not known.

6.2 Amino Acid Analogs

White and Vasil (1979) obtained cell lines resistant to 5 (2-Aminoethyl)-L-cystein and found that the resistance was expressed as a dominant trait in fused cells. Negrutiu et al. (1984b) selected lines tolerant to a lysine analog, after UV or EMS treatment. The mutation conferred insensitivity to the lysine pathway feedback, leading to an accumulation of lysine 10 to 20 fold higher in resistant lines than in controls; it was inherited as a dominant nuclear gene. Similarly, Durand (1984, 1987) obtained lines resistant to a phenylalanine analog.

6.3 Antibiotics

Tolerant kanamycin cells were selected by adding kanamycin sulfate during the first 21 days of culture of protoplasts from haploid calli (Dix et al. 1977). One cell line was also resistant to streptomycin and unable to differentiate shoots. Restoration of morphogenic potential was induced by somatic hybridization with a *N. knightiana* cell line, also deficient in shoot-inducible formation (Maliga et al. 1977). Streptomycin-resistant lines were also obtained from calli cultured in presence of streptomycin sulfate. A first line, originated from a diploid callus, regenerated a diploid plant with an altered morphology and was unable to flower (Maliga et al. 1979). A translocation was detected in its karyotype. Other streptomycin-resistant lines were obtained from haploid calli, the resistance was controlled by a recessive nuclear mutation (Maliga 1981). Durand (1984) obtained lines resistant to oligomycin; in one of these lines, isolated mitochondria also were resistant to the antibiotic. Finally, lincomycin-resistant lines were selected on the basis of greening (Cseplò et al. 1983). The resistance was transferred from a *N. sylvestris* resistant line, unable to flower, into *N. plumbaginifolia*. Regenerated *N. plumbaginifolia* plants were lincomycin-resistant and the character was maternally inherited.

6.4 Other Somatic Hybridization Experiments

Evans et al. (1983) carried out fusion between *N. sylvestris* (green) and *N. tabacum* (albinos) protoplasts. Some of the regenerated plants showed a characteristic leaf pigmentation – high frequency of dark green and yellow spots – that was transmissible. Protoplast fusion between a *N. tabacum* streptomycin-tolerant line, with maternal inheritance, and *N. sylvestris* produced somatic hybrids and some resistant plants morphologically identical to *N. sylvestris* (Medgyesy et al. 1980), that were presumed to possess *N. sylvestris* nuclear genes and fused cytoplasms (cybrids). The mtDNA restriction profiles of the cybrids did not show any differences with those of the parents – *N. sylvestris* and *N. tabacum* mtDNA's are not distinguishable – and the resistance was confirmed to be chloroplastic (Nagy et al. 1983). In another fusion experiment between a *N. tabacum* streptomycin-resistant line and *N. sylvestris*, chloroplasts and mitochondria were independently assorted in cybrids (Fluhr et al. 1983); in some of them a heterogeneous chloroplast population was maintained at least up to the second sexual generation.

N. sylvestris was used in several fusion experiments related to male sterility. In *Nicotiana* species, male sterility is under nucleo-cytoplasmic control, and is probably mitochondrially encoded (review by Hanson and Conde 1985). Zelcer et al. (1978) transferred cytoplasmic male sterility from *N. tabacum* (*N. suaveolens* cytoplasm) into *N. sylvestris*. About one quarter of the regenerated plants were cybrids (Aviv and Galun 1980), with a *N. sylvestris* nucleus and a variable part of the *N. tabacum* cytoplasm: some of them had *N. sylvestris* chloroplasts and *N. tabacum* mitochondria. Fertility and the chloroplast genome were not correlated (Galun et al. 1982). Finally, cybrids derived from fusion between protoplasts of *N. sylvestris* and *N. tabacum* cytoplasmic male sterile lines showed increasing restoration of fertility through successive generations (Aviv and Galun 1986), associated with changes in mtDNA. Stabilization appeared only in third generation offspring (Aviv and Galun 1987).

7 Discussion

In *N. sylvestris*, variation after in vitro culture has been evidenced at the chromosomal, morphological, and biochemical levels. The changes will be discussed, in relation to other *Nicotiana* species, namely *N. tabacum*.

7.1 Chromosomal Changes

Depending on the parental genotype, about 15% to 50% of pollen-derived plants are nonhaploids (McComb and McComb 1977; De Paepe and Pernès 1978; De Paepe et al. 1983). Haploid somatic cells regenerated 75% of diploid plants by tissue culture (Malepszy et al. 1977) and near 100% by protoplast culture (Facciotti and Pilet 1979). From diploid protoplasts, 40 to 60% of the regenerated plants were tetraploid (Prat 1983; Li et al. 1985). Clearly, the extent of polyploidization increases with duration of culture, presence of growth factors, and strength of mutagenic treatment (Malepszy et al. 1977). Colchicine added at the beginning of anther culture also increases the proportion of diploid androgenetic plants. In all cases, very few aneuploids were obtained. Chromosomal number changes in the regenerated plants occur principally by genome doubling. Such extensive polyploidization was not observed in other *Nicotiana* species, whatever their chromosomal number (Evans 1979). The proportion of polyploid cells in *N. tabacum* calli gradually increased during culture (Berlyn 1983; Hayashi and Nakajima 1984), but most regenerated plants had the normal ploidy level of the species. The propensity of *N. sylvestris* genome to double its chromosomal stock cannot be attributed only to its diploid level, as *N. langsdorffii* (2n = 18), is much more stable (Evans 1979).

Other types of chromosomal changes have occasionally been described. A translocation was detected in a streptomycin-resistant line (Maliga et al. 1979), and some protoplast lines probably carry chromosomal rearrangements (Prat and De Paepe unpubl.). Another interesting feature is the presence of supernumerary chromosomes in several selfed progenies of diploid androgenetic plants (De Paepe 1983). These transmissible chromosomes are heterochromatic and different from normal chromosomes, and differ also from one another, one of them carrying a nucleolar organizer. Their occurrence shows that nuclear DNA of *N. sylvestris* DH plants could be in some instable configuration. This is confirmed by changes observed in DH genome organization, that suggest an increase in the proportion and/or localization of some highly repeated sequences (De Paepe et al. 1982). Chromosomal breaks could occur at the heterochromatin accumulation points, resulting in the release of chromosomal fragments. Similar changes in DNA organization have been described in androgenetic plants of *N. tabacum* (Dhillon et al. 1983).

7.2 Phenotypic Changes

Pollen cultures and somatic cell cultures, either protoplast or tissue culture, can be distinguished by the morphological modifications in the regenerated plants. Androgenesis induces characteristic "soft" modifications of leaf morphology in nearly all DH plants, that was called "crumpling" (De Paepe and Pernès 1978). Successive cycles of pollen culture always emphasize the changes in the same way, in association with decreases in adult dimensions and growing abilities. All these parameters are under the control of nuclear recessive or semi-dominant genes, the number of which increases with the number of androgenetic cycles. In contrast, protoplast and tissue cultures induce various types of "hard" phenotypic changes, such as dwarfism, albinism, variegation, and male sterility. The number of possible mutations seems to be limited, as similar phenotypes were recorded several times. The proportion of plants mutated in nuclear recessive genes was about 50% after a first cycle of protoplast culture (Prat 1983), and was found to be even higher in second-cycle plants, where semi-dominant mutants were also obtained (Li et al. 1985). Some mutations can be under cytoplasmic gene control, such as seedling necrosis and male sterility. In this last case, specific mitochondrial DNA and protein changes were observed (Li et al. 1988). Nonmutated protoplast plants show a slow decrease in growth rates and some adult plant dimensions. In several cases lines selected for resistance to salts or antibiotics had lost their shooting abilities, or the regenerated plants had an abnormal morphology.

7.3 Origin of the Genetic Variation

The question is, why do pollen cultures and somatic cell cultures induce different sorts of variability? The term "somaclonal variation" was first used by Larkin and Scowcroft (1981) for any type of variation in plants issued from in vitro cultures, especially sexually inherited ones (Scowcroft 1984). Evans et al. (1984) proposed distinguishing soma- from gametoclonal variation for three reasons: (1) mutations are expressed directly in haploid plants; (2) recombination events would result from meiotic crossing-over; (3) colchicine used for doubling chromosomes may induce mutations. It is clear that, in the case of *N. sylvestris*, these reasons alone are not sufficient to explain the observed differences: firstly, androgenesis was carried out from pure lines, and the variability increased after subsequent cycles of culture performed on homozygous plants; secondly, spontaneous diploid androgenetic plants may be "crumpled" or contain crumpled plants in their progeny (De Paepe et al. 1981); thirdly, the spectrum of variability is much larger by protoplast culture than by androgenesis. In our opinion, the differences between soma- and gametoclonal variation may have at least two origins:

1. The Culture Conditions. The pollen develops directly into an embryo in a few weeks on hormone-free medium minimizing the risk of mutational events, while in somatic cell cultures plants are formed after a long disorganized period on a hormone-rich medium. In *N. sylvestris*, as well as in *N. tabacum* or in other plant species, several reports have shown that culture duration and strength of mutagenic treatments increase both the proportions of polyploid cells and the mutation frequency (Berlyn 1983; Scowcroft 1984; D'Amato 1985).

2. The Type of Cell from which the Regenerated Plant Comes. In *N. sylvestris*, the pollen grain put into culture contains two cells, the generative cell, that ensures fertilization in vivo, and the vegetative cell, that generally develops in in vitro conditions (De Paepe et al. 1977). We previously suggested that changes in vegetative nuclear DNA are at the origin of DH characteristics (De Paepe and Pernès 1978; De Paepe et al. 1981); such differences between macro- and micro-DNA's are well known in Ciliates (review by Steinbrück 1986). On the other hand, several types of cells may divide during tissue or protoplast culture – epidermal, palissadal, or spongy mesophyll cells. Besides that, they may be in different states of DNA differentiation, and these cells may be nonprotected by efficient DNA repair mechanisms, as they are not involved in the reproductive pathway.

7.4 Specificity of Somaclonal Variation in *N. sylvestris*

In *N. sylvestris*, nearly all the androgenetic plants and most of the plants regenerated from protoplasts are genetically different from the maternal line. Consecutive cycles of culture only produce abnormal plants. Does in vitro culture also induce a similar extent of variability in other plant species? Abnormal phenotypes and quantitative variation were reported in most species, but data do not, generally, give a clear estimation of variability:

– hybrid plants often used do not allow distinction between somaclonal variation and recombination events,
– in the presence of a mutagen, spontaneous mutational events cannot be evidenced,
– in crop plants only advantageous characters are selected for, and the abnormal regenerated plants were frequently eliminated,
– finally, two cycles of protoplast culture and several consecutive androgenetic cycles were performed only on *N. sylvestris*. In tobacco (Brown et al. 1983; Deaton et al. 1986) and wheat (Picard 1984), the second androgenetic cycle also induces a new depression effect. It is likely that similar decreases through consecutive cycles of culture would be evidenced in other species.

However, it appears from the literature that the extent of somaclonal variation may vary widely from one species to another, and may be especially high in *N. sylvestris*. In *N. tabacum*, the variation induced by androgenesis is similar to that of its progenitor species. Genetic studies are unfortunately lacking concerning protoplast and somatic cell cultures.

7.5 Utilization and Prospects

Although *N. sylvestris* is not a cultivated species, interesting characters could be transferred to *N. tabacum* by sexual or somatic hybridization. Most of the variability obtained after in vitro culture is not useful, but some traits – salt and antibiotic resistance, cytoplasmic male sterility – are of interest. It must be noted that *N. sylvestris* is the only species where cms plants were regenerated without somatic hybridization. Androgenetic plants show depression effects, but good hybrid lines can be obtained by crossing them (De Paepe 1983). A similar breeding program could be applied to *N. tabacum* HD's, previously selected for characters such as disease resistance or low alkaloid content.

References

Aviv D, Galun E (1980) Restoration of fertility in cytoplasmic male sterile (CMS) *Nicotiana sylvestris* by fusion with X-irradiated *N. tabacum* protoplasts. Theor Appl Genet 58:121–128

Aviv D, Galun E (1986) Restoration of male fertile *Nicotiana* by fusion of protoplasts derived from two different cytoplasmic male-sterile cybrids. Plant Mol Biol 7:411–417

Aviv D, Galun E (1987) Chondriome analysis in sexual progenies of *Nicotiana* cybrids. Theor Appl Genet 73:821–826

Aviv D, Arzee-Gonen P, Bleichman S, Galun E (1984a) Novel alloplasmic *Nicotiana* plants by "donor-recipient" protoplast fusion: cybrids having *N. tabacum* or *N. sylvestris* nuclear genomes and either or both plastomes and chondriomes from alien species. Mol Gen Genet 196:244–253

Aviv D, Bleichman S, Arzee-Gonen P, Galun E (1984b) Intersectional cytoplasmic hybrids in *Nicotiana*: identification of plastomes and chondriomes in *N. sylvestris* + *N. rustica* cybrids having *N. sylvestris* nuclear genomes. Theor Appl Genet 67:499–504

Banks MS, Evans PK (1976) A comparison of the isolation and culture of mesophyll protoplasts from several *Nicotiana* species and their hybrids. Plant Sci Lett 7:409–419

Berlyn MB (1983) Patterns of variability in DNA content and nuclear volume in regenerating cultures of *Nicotiana tabacum*. Can J Genet Cytol 25:354–360

Bourgin JP, Nitsch JP (1967) Obtention de *Nicotiana* haploïdes à partir d'étamines cultivées in vitro. Ann Physiol Vég 9:377–382

Bourgin JP, Missonier C, Chupeau Y (1976) Culture de protoplastes de mésophylle de *Nicotiana sylvestris* Spegazzini et Comes haploïde et diploïde. CR Acad Sci Paris Ser D 282:1853–1856

Bourgin JP, Chupeau Y, Missonier C (1979) Plant regeneration from mesophyll protoplasts of several *Nicotiana* species. Physiol Plant 45:288–292

Brown JS, Wernsman EA, Schnell RJ (1983) Effect of a second cycle of anther culture on flue-cured lines of tobacco. Crop Sci 23:729–733

Csepló A, Nagy F, Maliga P (1983) Rescue of the cytoplasmic lincomycin resistance factor from *Nicotiana sylvestris* into *Nicotiana plumbaginifolia* by protoplast fusion. In: Potrykus I, Harms CT, Hinnen A, Hütter R, King PJ, Shillito RD (eds) Protoplasts 1983 – poster proceedings. Birkhäuser, Basel, pp 126–127

D'Amato F (1985) Cytogenetics of plant cell and tissue cultures and their regenerates. CRC Crit Rev Plant Sci 3:73–112

Deaton WR, Collins GB, Nielsen MT (1986) Vigor and variation expressed by anther-derived doubled haploids of burley tobacco (*Nicotiana tabacum* L.). II. Evaluation of first- and second-cycle doubled haploids. Euphytica 35:41–48

De Paepe R (1983) Etude génétique des plantes haploïdes-doublées (HD) obtenues par culture de pollen isolé chez *Nicotiana sylvestris*: aspects morphogénétiques, cellulaires et chromosomiques. Thése D d'Etat, Univ Paris XI-Orsay

De Paepe R, Pernès J (1978) Exemples de variations à hérédité mendélienne induites au cours du développement des plantes. Physiol Vég 16:195–204

De Paepe R, Nitsch C, Godard M, Pernès J (1977) Potential from haploid and possible use in agriculture. In: Barz W, Reinhard E (eds) Plant tissue culture and its bio-technological application. Springer, Berlin Heidelberg New York, pp 341–352

De Paepe R, Bleton D, Gnangbe F (1981) Basis and extent of genetic variability among doubled haploid plants obtained by pollen culture in *Nicotiana sylvestris*. Theor Appl Genet 59:177–184

De Paepe R, Prat D, Huguet T (1982) Heritable nuclear DNA changes in doubled haploid plants obtained by pollen culture of *Nicotiana sylvestris*. Plant Sci Lett 28:11–28

De Paepe R, Prat D, Knight J (1983) Effects of consecutive androgeneses on morphology and fertility in *Nicotiana sylvestris*. Can J Bot 61:2038–2046

Dhillon SJ, Wernsman EA, Miksche JP (1983) Evaluation of nuclear DNA content and heterochromatin changes in anther-derived dihaploids of tobacco (*Nicotiana tabacum*) cv. Coker 139. Can J Genet Cytol 25:169–173

Dix PJ, Pearce RS (1981) Proline accumulation in NaCl resistant and sensitive cell lines of *Nicotiana sylvestris*. Z Pflanzenphysiol 102:243–248

Dix PJ, Street HE (1974) Effect of p-fluorophenylalanine (PFP) on the growth of cell lines differing in ploidy and derived from *Nicotiana sylvestris*. Plant Sci Lett 3:283–288

Dix PJ, Street HE (1975) Sodium chloride-resistant cultured cell lines from *Nicotiana sylvestris* and *Capsicum annuum*. Plant Sci Lett 5:231–237

Dix PJ, Joo F, Maliga P (1977) A cell line of *Nicotiana sylvestris* with resistance to kanamycin and streptomycin. Mol Gen Genet 157:285–290

Durand J (1979) High and reproducible plating efficiencies of protoplasts isolated from in vitro grown haploid *Nicotiana sylvestris* Spegaz. et Comes. Z Pflanzenphysiol 93:283–295

Durand J (1984) Mutations de résistance en culture de cellules chez une plante supérieure. *Nicotiana sylvestris*; isolement de mutations nucléaires et recherche de mutations du génôme mitochondrial. Thèse D d'Etat, Univ Paris XI-Orsay

Durand J (1987) Isolation of antibiotic resistant variants in a higher plant, *Nicotiana sylvestris*. Plant Sci 51:113–118

Evans DA (1979) Chromosome stability of plants regenerated from mesophyll protoplasts of *Nicotiana* species. Z Pflanzenphysiol 95:459–463

Evans DA, Bravo JE, Kut SA, Flick CE (1983) Genetic behavior of somatic hybrids in the genus *Nicotiana*: *N. otophora* + *N. tabacum* and *N. sylvestris* + *N. tabacum*. Theor Appl Genet 65:93–101

Evans DA, Sharp WR, Medina-Filho HP (1984) Somaclonal and gametoclonal variation. Am J Bot 71:759–774

Facciotti D, Pilet PE (1979) Plants and embryoids from haploid *Nicotiana sylvestris* protoplasts. Plant Sci Lett 15:1–6

Fluhr R, Aviv D, Edelman M, Galun E (1983) Cybrids containing mixed and sorted-out chloroplasts following interspecific somatic fusions in *Nicotiana*. Theor Appl Genet 65:289–294

Galun E, Arzee-Gonen P, Fluhr R, Edelman M, Aviv D (1982) Cytoplasmic hybridization in *Nicotiana*: mitochondrial DNA analysis in progenies resulting from fusion between protoplasts having different organelle constitutions. Mol Gen Genet 186:50–56

Goodspeed TH (1954) The genus *Nicotiana*. Chronica botanica, Waltham, Mass, pp 1–536

Goodspeed TH, Avery P (1939) Trisomic and other types in *Nicotiana sylvestris*. J Genet 38:381–467

Goodspeed TH, Clausen RE (1928) Interspecific hybridization in *Nicotiana*. VIII. The *sylvestris-tomentosa-tabacum* hybrid triangle and its bearing on the origin of *tabacum*. Univ Cal Publ Bot 11:245–256

Hanson MR, Conde MF (1985) Functioning and variation of cytoplasmic genomes: lessons from cytoplasmic-nuclear interactions affecting male-sterility in plants. Int Rev Cytol 94:214–267

Hayashi M, Nakajima T (1984) Genetic stability in regenerated plants derived from tabacco mesophyll protoplasts through three-step culture. Jpn J Breed 34:409–415

Kostoff D (1934) A haploid plant of *Nicotiana sylvestris*. Nature (London) 23:949–950

Kostoff D (1938a) Studies on polyploid plants. XVIII. Cytogenetics studies on *Nicotiana sylvestris* × *N. tomentosiformis* hybrids and amphidiploids and their bearings on the problem of the origin of *N. tabacum*. C R Acad Sci USSR 18:459

Kostoff D (1938b) The problem of haploidy – cytogenetic studies on *Nicotiana* haploids and their bearings to some other cytogenetic problems. Bibliogr Genet 13:1–147

Kung S (1976) Tobacco fraction I protein – a unique genetic marker. Science 191:429–434

Larkin PJ, Scowcroft WR (1981) Somaclonal variation – a novel source of variability from cell cultures for plant improvement. Theor Appl Genet 60:197–214

Li XQ (1987) Variabilité par culture in vitro: étude génétique à partir de culture de protoplastes chez *Nicotiana sylvestris*; description des plantes régénérées chez *Medicago lupulina*. Thèse D d'Etat, Univ Paris XI-Orsay

Li XQ, Prat D, De Paepe R, Pernès J (1985) Variability induced in *Nicotiana sylvestris* by two successive cycles of protoplast culture. Genet Manipul Crops Newslett 1:81–84

Li XQ, Chetrit P, Mathieu C, Vedel F, De Paepe R, Remy R, Ambard-Breteville F (1988) Regeneration of cytoplasmic male sterile protoclones of *Nicotiana sylvestris* with mitochondrial variation. Curr Genet 13:261–266

Lin ML, Staba S (1961) Peppermint and spearmint tissue culture. Callus formation in submerged culture. Lloydia 24:139–145

Linsmaier EM, Skoog F (1965) Organic growth factor requirements of tobacco tissue culture. Physiol Plant 18:100–127

Maddox AD, Gonsalves F, Shields R (1983) Successful preservation of suspension cultures of three *Nicotiana* species at the temperature of liquid nitrogen. Plant Sci Lett 28:157–162

Magnien E, Dalschaert X, Roumengous M, Devreux M (1980) Improvement of protoplast isolation and çulture technique from axenic plantlets of wild *Nicotiana* species. Acta Genet Sin 7:231–242

Malepszy S, Grunewaldt J, Maluszynski M (1977) Über die Selektion von Mutanten in Zellkulturen aus haploider *Nicotiana sylvestris* Spegazz. et Comes. Z Pflanzenzücht 79:160–166

Maliga P (1981) Streptomycin resistance is inherited as a recessive Mendelian trait in a *Nicotiana sylvestris* line. Theor Appl Genet 60:1–3

Maliga P, Lazar G, Joo F, Nagy AH, Menczel L (1977) Restoration of morphogenetic potential in *Nicotiana* by somatic hybridization. Mol Gen Genet 157:291–296

Maliga P, Kiss ZR, Dix PJ, Lazar G (1979) A streptomycin-resistant line of *Nicotiana sylvestris* unable to flower. Mol Gen Genet 172:13–15

McComb JA, McComb AJ (1977) The cytology of plantlets derived from cultured anthers of *Nicotiana sylvestris*. New Phytol 79:679–688

Medgyesy P, Menczel L, Maliga P (1980) The use of cytoplasmic streptomycin resistance – chloroplast transfer from *Nicotiana tabacum* into *Nicotiana sylvestris*, and isolation of their somatic hybrids. Mol Gen Genet 179:693–698

Misoo S, Takeuchi H, Fukuhara K, Matsubayashi M (1984) Studies on the mechanisms of pollen embryogenesis. V. Effects of genome constitution, ploidy level and cytoplasm on anther culture in tobacco and its ancestral species. Sci Rep Fac Agric Kobe Univ 16:35

Murakishi HH, Carlson PS (1982) In vitro selection of *Nicotiana sylvestris* variant with limited resistance to TMV. Plant Cell Rep 1:94–97

Murashige T, Skoog F (1962) A revised medium for rapid growth and bioassay with tobacco tissue cultures. Physiol Plant 15:473–497

Nagata T, Takebe I (1971) Plating of isolated tobacco mesophyll protoplasts on agar medium. Planta 99:12–20

Nagy JI, Maliga P (1976) Callus induction and plant regeneration from mesophyll protoplasts of *Nicotiana sylvestris*. Z Pflanzenphysiol 78:453–455

Nagy F, Lazar G, Menczel L, Maliga P (1983) A heteroplasmic state induced by protoplast fusion is a necessary condition for detecting rearrangements in *Nicotiana* mitochondrial DNA. Theor Appl Genet 66:203–207

Negrutiu I, Mousseau J (1980) Protoplast culture from in vitro grown plants of *Nicotiana sylvestris* Spegg. and Comes. Z Pflanzenphysiol 100:373–376

Negrutiu I, Mousseau J (1981) Culture sur grande échelle de protoplastes de *Nicotiana sylvestris* Spegazzini et Comes à partir de plantes cultivées in vitro. Z Pflanzenphysiol 103:367–372

Negrutiu I, Muller JF (1981) Culture conditions of protoplast-derived cells of *Nicotiana sylvestris* for mutant selection. Plant Cell Rep 1:14–17

Negrutiu I, Jacobs M, Caboche M (1984a) Advances in somatic cell genetics of higher plants – the protoplast approach in basic studies on mutagenesis and isolation of biochemical mutants. Theor Appl Genet 67:289–304

Negrutiu I, Cattoir-Reynearts A, Verbruggen I, Jacobs M (1984b) Lysine overproducer mutants with an altered dihydrodipicolinate synthase from protoplast culture of *Nicotiana sylvestris* (Spegazzini and Comes). Theor Appl Genet 68:11–20

Niizeki M, Saito K, Kita F (1982) Studies on plant cell and tissue culture. XII. Anther culture of various polyploid and aneuploid plants in *Nicotiana sylvestris*. Bull Fac Agric Hirosaki Univ 37:20–28

Niizeki M, Hayashi H, Saito K (1984) Production of disomic haploids by anther culture of a series of trisomic plants in *Nicotiana sylvestris*. Jpn J Breed 34:1–8

Nitsch C (1974) Pollen culture – a new technique for mass production of haploids and homozygous plants. In: Kasha KJ (ed) Haploids in higher plants – advances and potential. Univ Press, Guelph, pp 123–135

Nitsch C, Norreel B (1973) Effet d'un choc thermique sur le pouvoir embryogène du pollen de *Datura innoxia* cultivé dans l'anthère et isolé de l'anthère. CR Acad Sci Paris Ser D 276:303–306

Nitsch JP (1969) Experimental androgenesis in *Nicotiana*. Phytomorphology 19:389–401

Picard E (1984) Contribution à l'étude de l'hérédité et de l'utilisation en sélection de l'haplodiploïdisation par androgenèse chez une céréale autogame: *Triticum aestivum* L. Thèse D d'Etat, Univ Paris XI-Orsay

Prat D (1982) Etude de la variabilité génétique induite au cours de cycles successifs d'androgenèse, et par culture de protoplastes chez *Nicotiana sylvestris* Spegaz. et Comes. Thèse 3ème Cycle, Univ Paris XI-Orsay

Prat D (1983) Genetic variability induced in *Nicotiana sylvestris* by protoplast culture. Theor Appl Genet 64:223–230

Prat D, Gnangbe F, De Paepe R (1983) Alterations of leaf characteristics and peroxidase activity in diploid androgenetic lines of *Nicotiana sylvestris* Spegaz. and Comes. Plant Sci Lett 31:19–34

Rashid A, Street HE (1974) Segmentation in microspores of *Nicotiana sylvestris* and *Nicotiana tabacum* with lead to embryoid formation in anther culture. Protoplasma 80:323–334

Restivo FM, Tassi F (1986) Effect of 5-fluorouracil on growth and morphogenesis of tissue culture of *Nicotiana sylvestris*. Plant Cell Physiol 27:785–790

Scowcroft WR (1984) Genetic variability in tissue culture: impact on germplasm conservation and utilization. Int Bd Plant Genet Resourc. Rome. pp 1–41

Steinbrück G (1986) Molecular reorganisation during nuclear differentiation in ciliates. In: Hennig W (ed) Germ line – soma differentiation. Springer. Berlin Heidelberg New York Tokyo. pp 105–174

Tomes DT. Collins GB (1976) Factors affecting haploid plant production from in vitro anther cultures of *Nicotiana* species. Crop Sci 16:837–840

Tsala G. Brettell R. Zryd JP (1983) Attempt to isolate auxotrophic mutants. using protoplasts of *Nicotiana sylvestris*. In: Potrykus I. Harms CT. Hinnen A. Hütter R. King PS. Shillito RD (eds) Protoplasts 1983 – poster proceedings. Birkhäuser. Basel. pp 160–161

Van Slogteren GMS. Planque K. Lekkerkerk J (1980) Evaluation of parameters affecting the initiation of division of protoplasts of haploid and diploid *Nicotiana sylvestris* and *Nicotiana tabacum*. Plant Sci Lett 20:35–45

White DWR. Vasil IK (1979) Use of amino acid analogue-resistant cell lines for selection of *Nicotiana sylvestris* somatic cell hybrids. Theor Appl Genet 55:107–112

Zelcer A. Aviv D. Galun E (1978) Interspecific transfer of cytoplasmic male sterility by fusion between protoplasts of normal *Nicotiana sylvestris* and X-ray irradiated protoplasts of male-sterile *N. tabacum*. Z Pflanzenphysiol 90:397–407

IV.7 Somaclonal Variation in Alfalfa (*Medicago sativa* L.)

L.B. JOHNSON and M.R. THOMAS[1]

1 Introduction

1.1 Importance and Distribution of Alfalfa

Alfalfa (*Medicago sativa* L.), or lucerne as it is also called, is generally considered to be the world's most important forage crop, although precise quantitative data on its agronomic utilization are difficult to obtain. The crop is widely adapted within the temperate zones, wherever it is not limited by winterkill or lack of moisture. Michaud et al. (1988) estimated worldwide alfalfa production at over 32 million ha, with more than 41% of the world acreage occurring in North America. In the United States, it is currently the fourth most important crop in terms of acreage; hay production alone occupied 10.7 million ha in 1984 and accounted for 60% of the total hay tonnage produced (US Department of Agriculture 1986).

Alfalfa is variously harvested for hay, silage, production of dehydrated protein supplements for livestock, and seed. It is also used as pasture. Use for food by humans is limited, although it is a potential source of protein for direct human consumption (Edwards et al. 1975).

Alfalfa is a deep-rooted, perennial legume that is capable of frequent regrowth from crown buds following cutting. Leaves are typically trifoliolate. Flowers possess a tripping mechanism that helps ensure cross-pollination, typically by bees (Lesins and Lesins 1979). An important attribute of alfalfa is its ability to interact symbiotically with the bacterium *Rhizobium meliloti* in the biological fixation of atmospheric nitrogen. Given its adaptability and high forage yield capability, the crop's future is bright.

1.2 Objectives for Improvement

Alfalfa is cross-pollinated, largely self-sterile, and highly heterozygous, and most commercial cultivars are autotetraploids ($2n = 4x = 32$). The difficulties of varietal improvement, where literally every plant in a population is genetically different, where five allelic states are possible at each locus (nulliplex through quadriplex), and where from one to four multiple alleles can theoretically occur at any locus, are

[1]Department of Plant Pathology, Throckmorton Hall, Kansas State University, Manhattan, KS 66506, USA

Biotechnology in Agriculture and Forestry, Vol. 11
Somaclonal Variation in Crop Improvement I (ed. by Y.P.S. Bajaj)

further confounded by the fact that maximum vigor is typically achieved through maximum heterozygosity. Special breeding techniques are being developed (Bingham 1983).

Much effort in alfalfa breeding programs has been expended on improving insect and disease resistance (Barnes et al. 1977). The importance of pest resistance in alfalfa results in part from its perennial nature, which provides an extended period for buildup of insect populations and pathogen inoculum. Alfalfa scientists have estimated that the combined annual losses from all pests equal almost 40% of the national production in the United States (Elgin et al. 1984).

Graham et al. (1979) list 14 foliar diseases, 17 crown and root diseases, 12 virus diseases, and one disease apparently caused by a mycoplasma-like organism, as being important on alfalfa. Three important plant parasitic nematodes were also listed as inflicting damage, as was dodder, a parasitic higher plant. While crown and root diseases are often more spectacular, defoliation by foliar pathogens may reduce yields by 20% or more (Willis et al. 1969). Not only does this disproportionately reduce nutrient quality, but it also stimulates the plant to accumulate estrogenic compounds (e.g., coumestrol), which can cause reproductive problems in livestock (Elliott et al. 1972). Insects also inflict significant losses in alfalfa production. Twenty-seven insects are of economic importance on the crop in the midwestern United States alone, with 14 of them categorized as major pests (Edwards et al. 1980).

Both insect and disease losses are best reduced through the use of pest-resistant cultivars (Elgin et al. 1984). The number of potentially important pests, the difficulties of working with autotetraploid genetics, and in some cases, the lack of good resistance sources (Graham et al. 1979) clearly complicate alfalfa improvement by standard breeding techniques.

A major factor limiting the utilization of alfalfa for pasture of ruminants is its propensity to cause bloat, a potentially lethal condition resulting from the accumulation of microbially induced foaming in the rumen. Efforts to modify initial rate of digestion (Kudo et al. 1985) or to initiate tannin synthesis in foliar plant parts (Lees 1984) are in progress, as scientists seek to reduce alfalfa's bloat-inducing potential.

Other traits desired in alfalfa varieties include greater efficiency of water use; increased nitrogen fixation potential; increased tolerance to low soil pH, alkaline soils, and heaving; and better adaptation for pastures and rangelands (Barnes et al. 1977).

1.3 Available Genetic Variability

Genetic variability in alfalfa cannot be discussed without briefly considering taxonomy, since there is much disagreement as to what constitutes a species. Two recent classification systems are given below. Gunn et al. (1978) divided the perennial alfalfa species complex, *Medicago sativa* L., into nine subspecies, including *M. sativa* subsp. *sativa*, a purple-flowered form with coiled seedpods selected over time by mankind during improvement of agronomic traits, and *M. sativa* subsp. *falcata*, a yellow-flowered, naturally occurring form, with falcated seedpods and a more northern distribution. These and seven other subspecies were regarded as a polymorphic, interbreeding species complex. By contrast, Lesins and Lesins (1979) divided the section Falcago into a subsection, Falcatae, which contained five species, including *M. sativa* and *M. falcata*.

For the present review, all *Medicago* taxa described as either *M. sativa* subspecies (Gunn et al. 1978) or as members of subsection *Falcatae* (Lesins and Lesins 1979) will be considered alfalfa, because all can readily hybridize, and most, if not all, have already contributed to alfalfa cultivar development. Given the present uncertainty of classification, we will simply retain the designations as used in the papers cited.

Largely through the activities of mankind, these species and subspecies have been variously mixed, transported, and allowed to adapt over time. There are now considered to be nine distinct germplasm sources available to alfalfa breeders in the United States: *M. falcata, M. varia*, Turkistan, Ladak, Flemish, Chilean, Peruvian, Indian, and African (Barnes et al. 1977). Clearly, much genetic diversity is available, yet as mentioned earlier, many traits are lacking, including resistance to specific pests and being bloat-safe.

2 Somaclonal Variation

2.1 Brief Review of In Vitro Studies on Alfalfa

Much research has transpired on alfalfa tissue culture since the first report of plant regeneration from various explant-derived calli by Saunders and Bingham (1972). Even from this initial report, evidence of somaclonal variation was documented. Because recent reviews on somaclonal variation in alfalfa (Bingham and McCoy 1986) and alfalfa tissue culture (McCoy and Walker 1984; Mroginski and Kartha 1984; Bingham et al. 1988; see also Arcioni et al. 1989) are available, we will emphasize variation in systems involving alfalfa regeneration from protoplasts and recent papers not available to earlier authors.

The term somaclonal variation was first used by Larkin and Scowcroft (1981). They defined it as variation displayed among plants derived from any form of cell culture. We will also use the term protoclonal variation to describe variation in plants regenerated from protoplast-derived calli. Clearly, to be agronomically useful, such variation must be heritable, i.e., not of an epigenetic nature. Variation may be directed by selection in culture through modifications of the medium or environment, or may represent a random spectrum of variants arising without any deliberate attempt to bias regeneration of a specific genotype. In most of the papers reviewed here, no deliberate effort was made to induce mutations in cells or calli before plant regeneration. Mutagens are variously effective with alfalfa cell cultures (Lo Schiavo et al. 1980; Bingham and McCoy 1986), and as reviewed by Maliga (1980), differences would be expected depending on the system.

The ability to regenerate from culture is heritable in alfalfa. Starting with one DuPuits and four Saranac clones from a population averaging 12% regenerability from hypocotyl callus, Bingham et al. (1975) increased the frequency of regenerable plants in the population to 67% with two cycles of recurrent selection. Seed derived from this population and named Regen S was made available to others, and has been the source of regenerable genotypes for many laboratories, including our own.

Studies by Reisch and Bingham (1980) with cell suspensions of a diploid line (HG2) of alfalfa exhibiting a high frequency of plant regeneration indicated that two dominant genes, designated Rn_1 and Rn_2, controlled regenerability. When both dominant alleles were present, regenerability was greater than 75%, whereas plants possessing only one of the two dominant genes yielded calli of lesser regenerability.

Wan et al. (1988) examined plant regenerability in several tetraploid alfalfas. Two dominant, complementary genes controlling regeneration, designated Rn_3 and Rn_4, were identified. Both dominant genes were required for regeneration from petiole-derived calli on the media used, and evidence for a gene dosage effect was obtained.

Studies of Regen S/Saranac and Rangelander genotypes suggest that they differ in the genes controlling plant regeneration (Seitz Kris and Bingham 1988). It is thus clear from this work and that of Reisch and Bingham (1980) that regenerability is heritable, but it is more complex than was initially thought, and needs further study. Two groups have also reported a cytoplasmic influence on alfalfa regeneration (Walton and Brown 1988; Wan et al. 1988).

Since the original work by Saunders and Bingham (1972), numerous studies of regenerative potential of various sources of alfalfa germplasm have been made. Mitten et al. (1984) examined somatic embryo formation by 35 tetraploid alfalfas representing all of the nine US germplasm sources designated by Barnes et al. (1977). Ladak, Norseman, Turkistan, and Nomad were considered superior. Brown and Atanassov (1985) tested 76 cultivars of *M. sativa, M. falcata*, and *M. varia* for embryo production from callus of hypocotyl and cotyledon explants. Certain creeping-rooted cultivars with strong genetic backgrounds from Ladak and *M. falcata* were best, with Rangelander being superior. Nagarajan et al. (1986) compared ten cultivars and breeding lines from *M. sativa* and *M. media* for regenerability. Two *M. media* lines, although intermediate in the percentage of embryogenic calli formed from explants, yielded the highest number of regenerated plants. Chen and Marowitch (1987) identified 10 of 17 *M. falcata* accessions as being regenerative. Bianchi et al. (1988), in a study involving 12 different accessions, detected high regenerative potential in a Chinese alfalfa population (Mestnaya) and in cultivar Europe.

Nineteen diploid accessions of *M. sativa* of four subspecies, *sativa, caerula, falcata,* and *xvaria* were also tested for somatic embryo production from hypocotyl-derived calli by Meijer and Brown (1985). Diploids generally showed poor regenerability, and high regeneration frequencies did not correlate with any germplasm source.

Several observations can be made about the aforementioned research. First, the question must be raised as to whether plant regenerability itself is being studied, or whether untried medium modifications can convert low or nonregenerators into high regenerators, since many studies were done with a single medium sequence. Efforts to answer this question have produced variable results. Kao and Michayluk (1981), using nine cell suspension lines from single plants of Canadian No. 1 or 2, noted that only through hormone or salt modifications were they able to obtain embryo formation from some lines. Using various hormonally modified MS media, Oelck and Schieder (1983) obtained redifferentiation from callus on three of four alfalfa cultivars, and one of these responded only when 6-benzylaminopurine was the sole cytokinin. After screening 76 cultivars for embryogenesis, Brown and Atanassov (1985) studied two high- and two low-regenerating cultivars on five different media protocols. They tested not only two variations of the widely used low cytokinin, high auxin, three-step induction method, but also the Blaydes medium-based two-step sequence and two variations of a high cytokinin, low auxin ratio induction step. Embryo formation was then allowed to occur on hormone-free BOi2Y medium for all protocols. They found that, although absolute scores were sometimes influenced by the medium, high- and low-regenerating types were still clearly defined. Chen et al. (1987a) sampled 50 plants each of three alfalfa cultivars for embryogenesis from callus. They noted that highly productive genotypes produced embryos over a range of medium protocols or with various explants. Poorer genotypes exhibited medium-specific or explant-specific embryogenesis. Nagarajan et al. (1986) also observed significant genotype × medium interactions, but explant differences were not significant. Meijer and Brown (1987a,b) recently demonstrated medium requirement differences between Regen S genotypes and certain genotypes of *M. sativa* subsp. *falcata* background. Results suggested two different pathways of somatic embryogenesis, with Regen S requiring an extended dedifferentiation period unneeded by the rapidly developing *falcata*. It should be noted, however, that Skotnicki (1986) reported rapid regeneration of Regen S plantlets in 6 weeks, although the frequency with which it occurred was not stated. In comparison of four elite regenerating genotypes from Regen S/Saranac and Rangelander on two culture protocols, highly significant genotype × protocol interactions were found for embryo formation (Seitz Kris and Bingham 1988). Type of culture vessel also influenced embryogenesis. Regardless of medium protocol or vessel, however, all genotypes showed some embryo formation.

A second observation is that a high frequency of somatic embryo formation does not necessarily translate into a high frequency of plantlet conversion from embryos. Factors such as ammonium ion concentration (Walker and Sato 1981; Meijer and Brown 1987b); organic nitrogen source, in particular millimolar amounts of L-proline, L-glutamine, L-arginine, or L-alanine (Mezentsev 1981; Stuart and Strickland 1984a,b; Skokut et al. 1985; Stuart et al. 1985a,b); 2,4-D concentration and length of exposure (Saunders and Bingham 1975; Walker et al. 1978, 1979; Stuart et al. 1985b); and carbohydrate source (Strickland et al. 1987) all affect both the quantity and quality of embryos obtained. The proline effect may be mediated through maintaining a high average intracellular pH (Schaefer 1985). Some of these factors may be genotype-specific (Brown and Atanassov 1985; Chen et al. 1987a; Meijer and Brown 1987a; Seitz Kris and Bingham 1988). Nagarajan et al. (1986) noted that the cultivar Heinrichs produced numerous somatic embryos, but that many failed to form plantlets. Studies that compare efficiencies of regeneration must ultimately extend beyond embryogenesis to plantlet conversion.

It is obvious that alfalfa possesses wide variation between and within cultivars in regenerability. Cultivars possessing high frequency regeneration are uncommon, although it has been suggested that regenerators can be recovered from all cultivars, if enough plants are screened (Chen et al. 1987a). There is also currently no evidence as to which plant developmental traits are controlled by genes conditioning regenerability in culture. Although high regenerability is frequent in creeping-rooted alfalfas, there are clear exceptions (Brown and Atanassov 1985). Since the trait seems uncommon in field populations, it would be of interest to know if introduction of this trait into cultivars for use in varietal improvement would ultimately result in possible negative effects.

A third and final point is that considerable confusion exists in the literature concerning the developmental pathway taken by alfalfa during plant regeneration in culture, i.e., organogenesis (shoot formation, followed by root initiation) vs. embryogenesis. Reports of both are common. The frequent abnormal appearance of the developing structures made this determination difficult. Histological studies by Santos et al. (1983) now provide strong evidence for an embryogenic pathway, as do studies by Stuart et al. (1985b, 1988) documenting the appearance of embryo-specific seed proteins in the somatic embryos. This is not to imply that all alfalfa differentiation in cultures must be embryogenic, regardless of genotype or medium protocol. For example, histological studies by Bianchi et al. (1988) document genotype-dependent organogenic and embryogenic regenerative pathways.

2.2 In Vitro Regeneration of Plants

2.2.1 Various Explant Sources

Successful callus initiation and plant regeneration have been accomplished with many different explants, and it is beyond the scope of this review to do more than cite a few examples. Saunders and Bingham (1972) obtained embryogenic callus and plants from interlocular connective tissue of immature anthers, internode sections, seedling hypocotyls, and immature ovaries. Kao and Michayluk (1981) were similarly successful with shoot tips, as were Walker et al. (1978) with stem and petiole tissue. Santos et al. (1980) regenerated plants using leaf explants. Brown and Atanassov (1985) obtained somatic embryos from cotyledons, as did Nagarajan et al. (1986) from roots. Lupotto (1986) derived recurrent embryogenic callus from epicotyls. Bingham et al. (1988) list petals and sepals as regenerable tissues. Efficiency of embryo formation from callus may be explant-dependent (Novák and Konečná 1982; Nagarajan et al. 1986; Chen and Marowitch 1987), and genotype × medium interactions may occur (Nagarajan et al. 1986; Chen and Marowitch 1987).

Direct somatic embryogenesis was reported with alfalfa by Maheswaran and Williams (1984). They obtained somatic embryos from immature zygotic embryos.

2.2.2 Callus and Suspension Cultures

As documented above, regenerable alfalfa callus cultures are attainable from many explant sources. Media utilized for callus isolation and maintenance have included variously modified Blaydes (Saunders and Bingham 1972), Schenk-Hildebrandt (Walker et al. 1978), Murashige and Skoog (Lupotto 1983), Uchimiya and Murashige (Santos et al. 1980), B5 (Atanassov and Brown 1984), or PC-L2 (Phillips and Collins 1979). Although callus growth may be maintained with low (25 μM) concentrations of naphthaleneacetic acid (NAA) as an auxin (Walker et al. 1979), embryogenesis is commonly induced with 2,4-dichlorophenoxyacetic acid (2,4-D) (Saunders and Bingham 1975; Walker et al. 1978). High (100 μM) NAA concentrations (Walker et al. 1979) or Picloram (Phillips and Collins 1979) are also capable of induction. Induced calli are typically transferred to any of a number of hormone-free media to facilitate embryo development, as first reported by Saunders and Bingham (1972). Stavarek et al. (1980) reported using a high cytokinin-low auxin ratio for restoring plant regenerability to long-term cultures of alfalfa callus, but because this was accomplished on high NaCl-adapted cultures, its general utility is unclear.

Regenerable cell suspensions of alfalfa have been reported by many investigators (McCoy and Bingham 1977; Kao and Michayluk 1981; Mezentsev 1981; Novák and Konečná 1982), although suspensions typically are more difficult to obtain than callus growth. Ability to form soft callus is required, and additional genetic factors beyond those needed for regeneration from explants may be necessary (Reisch and Bingham 1980).

2.2.3 Anther and Pollen Culture

As noted earlier, Saunders and Bingham (1972) derived callus and plants from interlocular connective tissue of alfalfa anthers, but plants were not haploid. Zagorska et al. (1984) have obtained mixoploids and near haploids (n = 16–20) from alfalfa anther cultures. Xu (1979) reported obtaining haploid pollen-derived plants from *Medicago denticulata*, now considered synonymous with *M. polymorpha* (Lesins and Lesins 1979). However, the species that they used is unclear, since their pollen donor had 32 chromosomes, while 2n equals 14 in *M. polymorpha* (Lesins and Lesins 1979).

2.2.4 Embryo Rescue and Pod Culture

Ovule-embryo culture has been used to rescue interspecific hybrid embryos of crosses between *M. sativa* and other perennial *Medicago* species (McCoy 1985; McCoy and Smith 1984, 1986). At 14–20 days after pollination, fertilized ovules were removed and cultured for 4 to 5 days, after which embryos were removed and germinated on fresh medium. This technique has potential for recovering other interspecific hybrids in which postzygotic breakdown occurs. Krause (1980) and Bauchan (1987) obtained plants from embryos of alfalfa and several annual *Medicago* species at 8 to 12 and 4 to 8 days after selfing, respectively.

As an alternative approach to embryo rescue, pod culture of alfalfa and several annual *Medicago* species has also been demonstrated (Wang et al. 1984).

2.2.5 *Protoplasts*

Since the early 1980's, a number of papers have been published reporting regeneration of alfalfa plants from mesophyll-derived (Kao and Michayluk 1980; Santos et al. 1980; Mezentsev 1981; Johnson et al. 1981; Lu et al. 1983; Atanassov and Brown 1984; Pezzotti et al. 1984), cell suspension-derived (Mezentsev 1981; Arcioni et al. 1982; Atanassov and Brown 1984; Pezzotti et al. 1984), cotyledon-derived (Lu et al. 1982, 1983; Gilmour et al. 1987), or root-derived (Xu et al. 1982; Lu et al. 1983; Pezzotti et al. 1984) protoplasts of alfalfa and its immediate relatives. Kao and Michayluk (1980, 1981) were the first to report direct somatic embryogenesis with little or no intervening callus, both from protoplasts and cell suspensions of their Canadian alfalfa. Direct embryogenesis was later observed at various frequencies by Lu et al. (1983), Pezzotti et al. (1984), and Gilmour et al. (1987). Most other reports describe alfalfa embryogenesis through an intervening callus stage.

Dijak and Brown (1987) reported that the type of embryogenic response by mesophyll protoplasts is cultivar-specific. Protoplasts of Rangelander frequently responded directly, while single Rambler and Regen S clones required an intermediate callus phase. We also observed only indirect embryogenesis by Regen S (Johnson et al. 1981). Direct embryogenesis was typically characterized by a high frequency of cell clumping followed by extensive random cell death. Brief exposure of Rangelander cultures to low voltage (0.02–0.05 V) electrical fields consistently caused aggregation, high cell death, and direct embryogenesis in all cultures, whereas this occurred in only 40% of the cultures of untreated protoplasts (Dijak et al. 1986). Relative to controls, electrically stimulated protoplasts showed more disorganized microtubules, asymmetric first divisions, and more rapid early divisions with little expansive growth (Dijak and Simmonds 1988). Interestingly, Regen S exhibited a similar direct embryogenic response following exposure to an electrical field, but the globular structures eventually callused and did not form plants (Dijak et al. 1986). As a result of the elimination of callus-induced variation (D'Amato 1977), plants formed from direct embryogenesis might be anticipated to show little or no somaclonal variation, although this has not been clearly demonstrated. It should be pointed out that mutations, mitotic aberrations, or changes in ploidy have been reported for protoplasts of alfalfa (Meijer et al. 1988) and other species within the first few divisions (Barbier and Dulieu 1983; Sree Ramulu et al. 1984). Early alfalfa mesophyll protoplast abnormalities observed included multinucleate cells and mitotic malfunctionings, the latter being frequent during protoplast budding (Meijer et al. 1988).

Given that our results were the first to extensively document chromosomal and morphological changes in alfalfa plants regenerated from protoplasts (Johnson et al. 1984), there follows a brief description of the procedures that were used. All protoplasts were obtained from one of two donor plants, designated RS-K1 and RS-K2, that were selected from Regen S (Bingham et al. 1975) based on their ability to regenerate plants both from petiole sections and mesophyll protoplasts. Following selection, these plants have been clonally propagated by cuttings and maintained in a controlled environment to provide a source of leaf material for protoplast isolation.

The enzyme isolation medium used was essentially that of Kao and Wetter (1977), except that 1% Macerozyme R-10 replaced Pectinase and $CaCl_2$ was omitted (Johnson et al. 1981). Macerozyme and Cellulase R10 were required for isolation, but omission of Rhozyme HP-150 merely delayed protoplast release.

Dark-pretreated leaves were incubated overnight in the isolation medium. Protoplasts were washed by centrifugation, resuspended, diluted with KWM medium to 0.2 to 1×10^5 protoplasts/ml, and incubated in 100 μl drops in the dark for 6 to 10 days. KWM is a slightly modified medium of Kao and Wetter (1977). Drops were then spread onto the surface of agarose-solidified KM3:1 medium, and the resulting calli were maintained on KM3:1 plates for about 2 months (Johnson et al. 1981).

Although the calli occasionally formed embryos when transferred directly to hormone-free BI2Y medium (Saunders and Bingham 1975), transfer of calli to an SH induction medium containing 50 μM 2,4-D and 5 μM kinetin (after Walker et al. 1978) was typically required to obtain significant embryogenesis. A 3- to 4-week induction period was used rather than the 3 to 4 days used by Walker et al. (1978) because shorter induction periods with the material gave reduced embryo numbers. In these early studies, RS-K1 and RS-K2 protoclones averaged 6.6 and 6.2 months from protoplast plating until plantlets were moved to vermiculite. Through various procedural changes, this time has been significantly shortened. Stages in our protoclonal regeneration procedure are shown in Fig. 1.

Regeneration of plants from protoplasts is apparently more complex than regeneration from explant callus. Single Arc and Riley plants selected for regenerability from petiole-derived calli, as was done for RS-K1 and RS-K2, have not yielded protoplasts that divide well or regenerate plants on media used for the Regen S clones (Johnson unpubl.). Whether this failure is due to physiologic differences in the Arc and Riley clones under our growth conditions or to some other genetic factor limiting protoplast division in the media has not been determined.

2.3 Genomic Changes in Regenerated Alfalfa Plants

2.3.1 Chromosomal Changes

The first of many reports of chromosomal changes in alfalfa somaclones derived from explant tissues was of increased ploidy (Saunders and Bingham 1972). Increased ploidy has occurred in regenerates from both diploid (McCoy and Bingham 1977; Reisch and Bingham 1981) and tetraploid (Saunders and Bingham 1972; Bingham and Saunders 1974; Groose and Bingham 1984; Hartman et al. 1984; Nagarajan and Walton 1987) donors. Tetraploid Regen S cell suspensions gave rise to 8x and 16x cells over time, with some chromosome loss at all ploidy levels (Atanassov and Brown 1984). We know of no reports of plants regenerated at greater than the octoploid level.

Both hyper- and hypoaneuploidy were also observed by various investigators in explant and cell suspension-derived regenerates. Reisch and Bingham (1981) found $2n=31$ and $2n=33$ aneuploids from cell suspension regenerates of a $2n=16$ line. Zagorska et al. (1984) found some mixoploid regenerates ($2n=16$ to 20) from

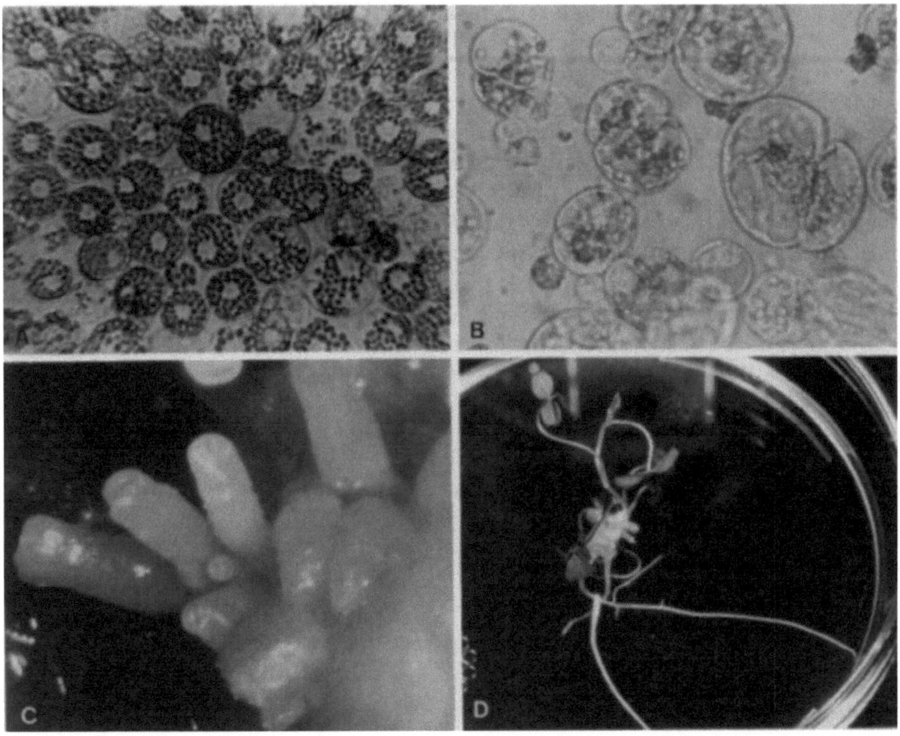

Fig. 1A-D. Stages in the regeneration of *Medicago sativa* L. Regen S plantlets from protoplasts. **A** Protoplasts several hours after isolation. Wall formation, as indicated by change in shape, is conspicuous after overnight incubation. **B** Early cell divisions. Protoplast-derived cells begin division 2 to 5 days after protoplast isolation. **C** Somatic embryos developing from a callus on hormone-free medium about 4 weeks after a 3-week embryo induction treatment on a medium containing 50 μM 2,4-D and 5 μM kinetin. **D** Protoplast-derived plantlet just before transplanting. (Johnson et al. 1981)

alfalfa anther culture. To date, we are unaware of any alfalfa regenerates with 15 or fewer chromosomes. Plants with 28, 30, and 31 chromosomes have been regenerated from a 2n = 32 alfalfa genotype (Groose and Bingham 1984). Tarczy et al. (1986) reported aneuploidy in somatic embryos, while callus culture cells had both increased aneuploidy and high ploidy levels. Both aneuploids and heteroploids were obtained by Nagarajan and Walton (1987), who noted that cytological instability was genotypically determined.

Feher et al. (1989) observed mitotic aberrations in culture, and documented both aneuploidy and ploidy increases in regenerates of Rambler alfalfa, a *M. varia* cultivar. They suggest, however, that considerable preexisting variation also occurs, and that alfalfa is a polysomatic/aneusomatic species. This suggestion was based on the high frequency of mixoploidy and aneuploidy observed in their seedlings, rooted cuttings from plants, and somaclones. Given that such mosaicism has not been reported by most workers for other alfalfa genotypes, this generalization needs further study.

It should be noted that not all reports suggest that alfalfa cultures exhibit high karyotypic variation. Mezentsev (1981) and Binarová and Doležel (1989) observed karyotypically normal regenerates and embryogenic cell suspensions with stable nuclear DNA contents, respectively.

Early studies with alfalfa plants regenerated from protoplasts did not address chromosome changes. However, significant changes in chromosomes were observed in protoclones from both of our 32-chromosome RS-K1 and RS-K2 protoplast doners (Johnson et al. 1984). Changes observed included increased polyploidy, aneuploidy, and translocations (Table 1). Over half the protoclones analyzed differed in chromosome number from their donor plant. Octoploidy or near octoploidy was the most frequent change, and occurred in 40% of the protoclones. Aneuploids (2n = 31 or 33) and those with translocations were 19 and 10% of the total protoclones, respectively. The karyotype of an aneuploid RS-K2 protoclone with two translocations is shown in Fig. 2.

Table 1. Chromosome number frequencies in root tips of RS-K1 and RS-K2 protoclones. (Johnson et al. 1984)

2n Chromosome no.	RS-K1		RS-K2	
	No. of protoclones	Percent of total	No. of protoclones	Percent of total
31, no obvious translocation	3	9	6	8
31, translocation	1	3	5	7
32, no obvious translocation	10	30	34	45
32, translocation	0	0	1	1
33, no obvious translocation	0	0	4	5
33, translocation	0	0	1	1
52–64, no obvious translocation	18	55	22	29
52–64, translocation	1	3	2	3
Total	33	100	75	100

Latunde-Dada and Lucas (1983) cytologically examined five randomly selected alfalfa protoclones of variant phenotype. All had increased ploidy, with chromosome counts suggesting hexaploid, heptaploid, and octoploid plants. Karyotypes did not reveal whether variants were euploids or aneuploids.

Our frequency of observed chromosomal and phenotypic modifications was higher than that found in other somaclonal and protoclonal studies with alfalfa, and this was the only study to document translocations (Johnson et al. 1984; Schlarbaum et al. 1988). Reasons are unclear, but it may prove to be that these differences will ultimately be shown to result from factors such as genotype (Nagarajan and Walton 1987), culture medium (Keyes and Bingham 1979), time in culture (McCoy and Bingham 1977; Hartman et al. 1984), type of embryogenesis (Dijak and Brown 1987), or tissue source used for protoplast isolation.

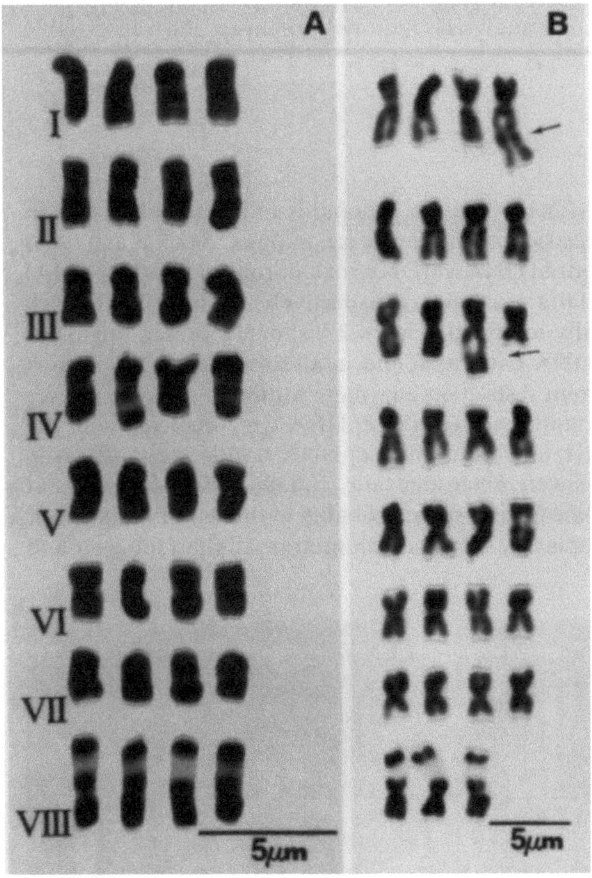

Fig. 2A,B. Karyotypes of protoplast donor and protoclone (Schlarbaum et al. 1988). **A** Karyotype of RS-K2. **B** Karyotype of protoclone K2-74S, a 31-chromosome aneuploid with two translocations (*arrows*). Note the missing satellite chromosome

2.3.2 Possible Involvement of a Transposable Element

There is evidence to suggest that a transposable element may be involved in at least part of the somaclonal variation seen in alfalfa. An unstable anthocyanin mutant was regenerated from a tetraploid purple-flowered alfalfa by Groose and Bingham (1986a,b). Reversion of this recessive white mutant to a stable purple state occurred both in shoots of rooted cuttings and at a higher frequency in plants from an additional tissue culture cycle. Results suggested that the mutation arose as the result of culture-induced migration of a transposable element into a functional C2 flower color allele that was present in the simplex condition in the donor. The occurrence of populations of cells during reculture that would not revert and yield purple-flowered regenerates gave credence to the hypothesis, suggesting imprecise excision of the element. Molecular analysis is required to confirm the transposable element hypothesis.

2.3.3 Possible Changes in Organelle DNA

It is of interest that instability in the nuclear genome is independent of that in cytoplasmic genomes. Using restriction endonuclease analysis, Rose et al. (1986) found no evidence of mitochondrial DNA (mtDNA) or chloroplast DNA (cpDNA) rearrangement in 20 and 23 alfalfa protoclones, respectively. All but eight of each population were chromosomally modified. Figure 3A shows representatives exhibiting the characteristic mtDNA patterns found in all protoclones and donors after digestion with four different restriction enzymes. Although no evidence for cpDNA rearrangement was found, two types of cpDNA were detected with the enzymes XbaI (Fig. 3B), HpaII, and MboI. The cpDNA results are most easily explained if early donor plants were heteroplastidic, and with plastid sorting out (Birky 1983; Smith et al. 1986) occurring early, probably in the donor plants prior to protoplast isolation. Heteroplasmy is not uncommon in alfalfa (Johnson and

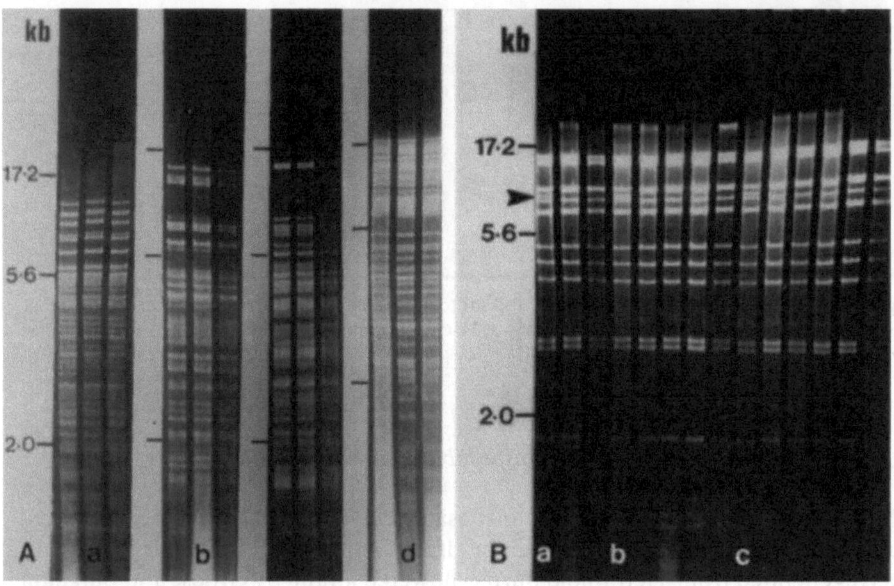

Fig. 3A,B. Ethidium bromide stained gels of DNA restriction fragments following electrophoretic separation. The location of three fragment size markers is shown for all gels (Rose et al. 1986). **A** mtDNA patterns representative of both donor plants and protoclones following digestion with BamHI (*a*), BglII (*b*), Eco RI (*c*), or XhoI (*d*) and separation on a 0.85% agarose gel. **B** cpDNA pattern of RS-K2 (*a*), RS-K1 (*b*) protoclone K2-66S (*c*), and 11 other protoclones following digestion with XbaI and separation on a 1.0% agarose gel. *Arrowhead* shows the location of a 7.8 kb band (refined size estimate) found only in lanes *a*, *b*, and *c*

Palmer 1989) and results from biparental plastid transmission (Smith et al. 1986), even with crosses between plants possessing normal green plastids (Masoud 1989).

Smith et al. (1986) obtained a sectoring yellow-green variant from tissue culture. They reported that it was a plastid mutant and demonstrated biparental inheritance. The sectoring trait must have resulted from the sorting out (Birky 1983) of either a pre-existing or a culture-derived cpDNA mutation.

Thus there is no clear evidence to date for either mtDNA or cpDNA recombination in alfalfa somaclones. However, mtDNA but not cpDNA restriction patterns of *Medicago sativa* × *M. falcata* somatic hybrids indicated recombination (D'hont et al. 1987).

2.4 Examples of Somaclonal Variants in Alfalfa

2.4.1 Variation Observed in Regenerates Without Deliberate Selection in Culture

Reisch and Bingham (1981) were the first to extensively document unselected somaclonal variation in traits such as dry matter yield and plant morphology in alfalfa, using regenerates from explant-derived cell suspensions from a single diploid plant. However, initial reports gave little indication of variation among alfalfa protoclones. For example, Kao and Michayluk (1980) noted that although embryo cotyledons were often deformed, normal plantlets typically developed. Of the few grown in the field, most were reported normal and fertile. It should be noted, however, that other early reports typically involved regeneration from explants of populations of seedlings. Thus, direct comparisons were not possible between protoclones and the source plants, given the intracultivar variation in alfalfa.

Our first protoclones, derived from cloned RS-K1 and RS-K2 protoplast donor plants, initially appeared normal (Johnson et al. 1981), although at that stage the only possible comparison of these young plants was with the mature donor plants. As protoclones matured and as more became available, considerable variation in phenotype became obvious, particularly in leaf morphology, internode length, and growth habit (Fig. 4A-C). These and other available protoclones were cloned by rooting of shoot cuttings, which were then analyzed for chromosomal deviations, and used as a source of material for field studies (Johnson et al. 1984; Schlarbaum et al. 1988).

The aim of our field studies was to have four replications of each protoclone and 12 of each donor plant in a randomized complete block design. Winter damage was excessive in protoclones, and by the first cutting the following year, only 12 RS-K1 and 51 RS-K2 protoclones had three or more surviving replicates. Materials with fewer than three replications were not analyzed in order to decrease the likelihood that differences were due to random variation.

Of 63 protoclones, only K1-13A ($2n = 32$) and K1-12A ($2n = 31$) were statistically superior ($P = 0.05$) to their donor plants (Table 2) in the first cutting. In the second cutting, no protoclones were superior. K1-13A, K1-34S, and K2-52S were superior at cutting three, prior to which plants were extremely stressed by a potato leafhopper [*Empoasca fabae* (Harris)] infestation. The extreme chlorosis and

Fig. 4A-C. Somaclonal variants of *Medicago sativa* L. and comparisons to the protoplast donor plant. **A** Cutting-derived RS-K1 protoplast donor plant as photographed in the field just prior to the first cutting. A board marked with 30×30 cm squares behind the plant permits an estimation of size. **B** K1-13A, a protoclone derived from RS-K1, as photographed the same day. K1-13A (2n = 32) was one of two protoclones exhibiting significantly greater dry weight forage yield at cutting one. Note the difference in growth habit. **C** Trifoliolate leaves of RS-K2 (*left*) and protoclone K2-11B (*right*) as obtained from greenhouse-grown cuttings. K2-11B is a 33-chromosome aneuploid possessing a translocation

tipburn shown by these three suggested that any possible resistance was of little value.

No protoclones were significantly superior in forage yield to their donor plant in cutting four. However, between the third and fourth cutting, a crown rot complex caused by *Fusarium* and *Rhizoctonia* killed nearly 60% of the remaining protoclones. Although eight of 11 replications of the RS-K2 donor plant were killed, all four ramets of protoclone K2-60S (2n = 32) survived and outyielded RS-K2 by nearly 60%. This probably represents an example of tissue culture-derived disease resistance, although further studies are needed. The extreme crown rot susceptibility of our materials has precluded further field trials. Latunde-Dada and Lucas (1983) obtained unselected *Verticillium*-tolerant protoclones of alfalfa, but suggested that a gene dosage effect because of increased ploidy was involved.

Reisch (1983) described somaclonal variation as a form of mutation breeding. The significant, negative aspects of this type of nondirected acquisition of variants in our protoclonal system was obvious, given the extremely high frequency of chromosomal abnormalities (Johnson et al. 1984). The poorest performers in the field were those with translocations, all of which were statistically inferior to RS-K2 on the first cutting. Excluding those with translocations, five of 11 octoploids and three of nine aneuploids were also statistically inferior to their donor plants on that

Table 2. Dry weight forage yields of second year RS-K1 and RS-K2 donor plants and selected protoclones, listed after the donor plant in order of decreasing mean first-cutting yield[a]

Donor or protoclone	2n chromosome no.[b]	Mean grams dry weight per cutting			
		1st	2nd	3rd	4th
RS-K1		102	77	17	30
K1-13A	32	149[c]	97	28[c]	–[d]
K1-12A	31	149[c]	89	22	–
K1-34S	32	132	93	28[c]	–
K1-1S	64	104	68	17	–
K1-23S	64	90	63	15	–
K1-16S	31	79	63	16	–
K1-5B	32	68	63	11	–
RS-K2		140	101	24	27
KS-86S	33	163	121	25	–
K2-9S	32	161	108	28	–
K2-52S	32	161	114	33[c]	–
K2-24C	64	148	110	32	–
K2-60S	32	114	95	25	43
K2-72S	31	107	72[c]	14	–
K2-13A	64, tr	90[c]	77	21	–
K2-24B	32, tr	56[c]	54[c]	16	–
K2-48S	32	37[c]	36[c]	11[c]	–
K2-26S	64	10[c]	15[c]	–	–
K2-90A	31, tr	9[c]	–	–	–

[a] From Tables 4 and 5, Johnson et al. (1984).
[b] From counts of five or more root tip cells. 64 = near octoploid, as counts varied between cells. tr = translocation.
[c] Significantly different (P = 0.05) from donor for that cutting, using Duncan's multiple range test adjusted for unequal sample size (Kramer 1956).
[d] Means not shown, as two or fewer replications were surviving.

cutting. Plants in all these categories would be of little agronomic importance. Poor growth also occurred in regenerates with 32 chromosomes, where 10 of 36 were significantly inferior. The aforementioned populations also excluded all inferior protoclones not surviving the first summer and winter.

To date, we have made no effort to study inheritance of variation seen in 32-chromosome protoclones. There is no reason to doubt that many of these traits are heritable, however, based on similar studies with plants from explant-derived alfalfa callus. For example, several culture-derived mutants from diploid alfalfa cultures have shown single gene segregation (Bingham and McCoy 1986; Bingham et al. 1988). Examples include a recessive abnormal flower mutant, a recessive mutant for simple leaves and rudimentary flowers, and a dominant branching mutant. Several of these came from cultures subjected to mutagens.

In addition, Pfeiffer and Bingham (1984) demonstrated the sexual transmission of increased herbage yield with their best somaclonal variant through a set of tester crosses. Herbage yield is considered a multigenic trait. Within the limitations of their experiment, however, they concluded that improvement in this multigenic trait by tissue culture was inefficient compared to sexual procedures.

We are unaware of any efforts to identify modified enzymes in regenerated alfalfa mutants. However, Baertlein and McDaniel (1987), using discriminant analysis of electrophoretically separated total proteins of alfalfa somaclones, reported variants at an overall frequency of 7.3%. However, breeding and chromosomal studies are required in order to establish whether these represented genetic changes.

2.4.2 Selection in Culture for Amino Acid Overproduction in Regenerates

Efforts to apply selection pressures in culture to favor the regeneration of specific somaclonal variants have also been applied to alfalfa. Reisch et al. (1981), seeking to select methionine overproducers in diploid alfalfa, cultured mutagenized cell suspensions in the presence of ethionine, a methionine analog. Cell lines with up to a tenfold increase in soluble methionine were obtained. Calli from seven of 25 regenerated plants of these lines were ethionine-resistant, but no plants had enhanced foliar levels of soluble or protein-bound methionine (Bingham and McCoy 1986; Bingham et al. 1988). Although these efforts were ineffective in improving herbage nutrient quality, experiments with other genotypes and selection strategies are required before discarding this approach toward alfalfa improvement (Bingham and McCoy 1986).

2.4.3 Selection in Culture for Disease Resistance in Regenerates

If a pathogen-produced toxin is involved in disease development, then selection in culture using the toxin should result in regeneration of plants with enhanced disease resistance. Hartman et al. (1984) selected cell lines resistant to toxin-containing culture filtrates of *Fusarium oxysporum* f. sp. *medicaginis*. Most regenerated plants obtained after a 7-month selection cycle were chromosomally normal and resistant when inoculated with this pathogen. Calli derived from them were also filtrate-resistant. Longer-term selection cycles also yielded resistant plants, but ploidy levels were elevated. Preliminary results suggested the involvement of a dominant gene mutation (Hartman and Knous 1984).

In later work, Arcioni et al. (1987) confirmed that plants resistant to *Fusarium* wilt could be selected during culture of susceptible lines of alfalfa. Calli of two genotypes yielded disease-resistant plants following selection on a medium containing a culture filtrate of *F. oxysporum* f. sp. *medicaginis*. Detailed genetic studies are still required before the utility of screening with pathogen filtrates to obtain resistant alfalfa plants is established. The approach would appear to have promise for any disease in which a pathogen-produced toxin is a factor, if the toxin is active at the cellular level.

However, Latunde-Dada and Lucus (1988), using a toxic fraction from *Verticillium albo-atrum* culture filtrates, found little benefit in its use for increasing the frequency of *Verticillium* wilt-resistant regenerates from tissue culture. They suggested that the technique offered little advantage over conventional recurrent selection in alfalfa.

2.4.4 *Selection in Culture for Salt Tolerance in Regenerates*

Cultures of NaCl-tolerant alfalfa cells have been obtained (Croughan et al. 1978, 1981; Smith and McComb 1983; Stavarek and Rains 1984), but to date we are unaware of reports of NaCl-tolerant regenerates. Reasons for this failure are not totally clear, and further studies would seem in order. However, it should be noted that McCoy (1987) has demonstrated distinct differences between in vitro and whole plant tolerances to NaCl in *Medicago*. Genotypes where calli were tolerant to 170 mM NaCl were identified in *M. dzhawakhetica, M. rhodopea, M. rupestris*, and *M. sativa*, but they were salt-sensitive at the whole plant level. Conversely, the calli of *M. marina*, a species exhibiting whole plant salt tolerance, were extremely NaCl-sensitive. Thus it is not obvious whether selection for NaCl tolerance in alfalfa at the cellular level can result in salt-tolerant plants.

2.4.5 *Selection for Miscellaneous Traits Without Subsequent Plant Regeneration*

Without plant regeneration and genetic studies, it is difficult to distinguish genetic from epigenetic changes, and the selection of cell variant lines per se falls by definition outside the topic of somaclonal variation. Such studies, however, may indicate potential for future selection-regeneration studies, and are thus briefly mentioned here. Selection for resistance in alfalfa cells to the herbicide L-phosphinothricin has been reported, with amplification of a glutamine synthetase gene as the suggested mechanism (Donn et al. 1984). Ploidy levels were not determined. Both positive (Giuliano et al. 1986) and negative (Hamill and Mariotti 1987) results have been reported in efforts to select nitrate reductase minus variants in alfalfa cells and calli, respectively, using the toxic nitrate analog chlorate. Methylammonium-resistant variants have also been selected (Giuliano et al. 1986). These latter examples will more likely facilitate studies on nitrogen metabolism.

2.5 Usefulness of Somaclonal Variation in Alfalfa

As might be expected, there are both positive and negative aspects to the occurrence of somaclonal variation in alfalfa. From a positive standpoint, culture-derived hypoaneuploids could be used in mapping genes to specific chromosomes, as suggested by Bingham and McCoy (1986). We are unaware of any gene mapping in alfalfa, for which even karyotyping is difficult because of its small chromosomes (Schlarbaum et al. 1988). The only published report of a somatic hybrid to date in *Medicago* was between *M. falcata* and *M. sativa* (Téoulé 1983; Téoulé and Dattée 1987), which are sexually compatible. However, a somatic hybrid plantlet of *M. sativa* and *M. intertexta*, an annual species with which it is sexually incompatible, has recently been obtained, although it did not survive to transplanting (M. Thomas, L. Johnson, and F. White, unpublished). In genetically wider fusions such as this, infertility is possible. Further tissue culture cycles could facilitate loss of specific chromosomes, thus restoring fertility. Bingham and McCoy (1986) stated that culture-induced gene introgression will be tested with *Medicago* interspecific hybrids, in efforts to obtain chromosome interchanges where they might otherwise be unlikely. The development of near-isogenic lines for physiologic studies, e.g., for disease resistance and susceptibility, might be accomplished using somaclonal mutants from a single donor, although examination of mutants for more extensive genetic change would be required. Chromosomal doubling in diploids was useful in bringing genotypes to the tetraploid level (Pfeiffer and Bingham 1984). Clearly, where culturally applied screening techniques exist for traits expressed in plants (Hartman et al. 1984; Arcioni et al. 1987), somaclonal variation has potential for alfalfa improvement.

We are less positive about programs seeking to improve alfalfa by the examination of large numbers of unselected somaclonal variants for any one character, but especially when quantitatively inherited traits are involved. Given the significant effort involved in regeneration from culture and the high frequency of deleterious mutants, improvement through standard breeding programs would probably be more productive in a species possessing the genetic diversity of alfalfa. When desirable traits are unavailable in accessible germplasms, a program initiated at the diploid level, and possibly involving selfing and intercrossing of vigorous R_0 regenerates and screening of R_1 progeny (Evans et al. 1984), would appear most productive. Thus true gene mutations could be identified, and all infertile materials could be discarded.

It is also obvious that there are many negative aspects to somaclonal variation. Transformation followed by plant regeneration has been accomplished in a number of laboratories with alfalfa (Schreier et al. 1985; Deak et al. 1986; Reich et al. 1986; Shahin et al. 1986; Spanò et al. 1987; Sukhapinda et al. 1987; Beach and Gresshoff 1988; Chabaud et al. 1988). To date, the only introduced foreign genes are those found on the *Agrobacterium* Ti or Ri plasmids or bacterial antibiotic resistance genes. However, as work extends beyond model systems to the introduction of agronomically useful genes, desirable cultivars will be used, and genomic stability will be important. Likewise, work with alfalfa on the production of artificial seed, in which somatic embryos are encapsulated for direct seeding (Redenbaugh et al. 1986, 1987; Stuart et al. 1987; Chen et al. 1987b), would necessitate genetic stability. In somatic hybridization, when true amphiploids are desired and culture-induced sterility is unacceptable, genetic stability in culture is needed.

As with most plant species, somaclonal variation in alfalfa can be either a blessing or a curse. Obviously, basic studies on its mechanisms and control are needed, so that its presence or absence is ultimately under the control of the investigator.

Acknowledgments. The technical assistance of R.K. Higgins, D.Z. Skinner, H.L. Douglas, and K.D. Oakleaf is acknowledged, as are the collaborative efforts of Drs. D.L. Stuteville, S.E. Schlarbaum, R.J. Rose, and R.J. Kemble. Research was supported in part by USDA-CSRS Research Agreement No. 5901-0410-9-0336-0 and 85-CRCR-1-1768 and National Science Foundation grant PCM 8022556. Any opinions, findings, conclusions, or recommendations expressed in this publication are those of the authors and do not necessarily reflect the view of either the US Department of Agriculture or the National Science Foundation. Contribution no. 87-498-B, Kansas Agricultural Experiment Station, Kansas State University, Manhattan, KS 66506.

References

Arcioni S, Davey MR, Santos AVP dos, Cocking EC (1982) Somatic embryogenesis in tissues from mesophyll and cell suspension protoplasts of *Medicago coerulea* and *M. glutinosa*. Z Pflanzenphysiol 106:105–110

Arcioni S, Pezzotti M, Damiani F (1987) In vitro selection of alfalfa plants resistant to *Fusarium oxysporum* f. sp. *medicaginis*. Theor Appl Genet 74:700–705

Arcioni S, Damiani F, Pezzotti M, Lupotto E (1990) Alfalfa, lucerne (*Medicago* species). In: Bajaj YPS (ed) Biotechnology in agriculture and forestry, vol 10. Legumes and oilseed crops I. Springer, Berlin Heidelberg New York Tokyo, pp 197–241

Atanassov A, Brown DCW (1984) Plant regeneration from suspension culture and mesophyll protoplasts of *Medicago sativa* L. Plant Cell Tissue Org Cult 3:149–162

Baertlein DA, McDaniel RG (1987) Molecular divergence of alfalfa somaclones. Theor Appl Genet 73:575–580

Barbier M, Dulieu H (1983) Early occurrence of genetic variants in protoplast cultures. Plant Sci Lett 29:201–206

Barnes DK, Bingham ET, Murphy RP, Hunt OJ, Beard DF, Skrdla WH, Teuber LR (1977) Alfalfa germplasm in the United States: genetic vulnerability, use, improvement, and maintenance. US Dep Agric Tech Bull 1571

Bauchan GR (1987) Embryo culture of *Medicago scutellata* and *M. sativa*. Plant Cell Tissue Org Cult 10:21–29

Beach KH, Gresshoff PM (1988) Characterization and culture of *Agrobacterium rhizogenes* transformed roots of forage legumes. Plant Sci 57:73–81

Bianchi S, Flament P, Dattée Y (1988) Embryogenèse somatique et organogenèse in vitro chez la luzerne: évaluation des potentialités de divers génotypes. Agronomie 8:121–126

Binarová P, Doležel J (1988) Alfalfa embryogenic cell suspension culture: growth and ploidy level stability. J Plant Physiol 133:561–566

Bingham ET (1983) Maximizing hybrid vigour in autotetraploid alfalfa. In: Better crops for food. Ciba Found Symp 97. Pitman, London, pp 130–143

Bingham ET, McCoy TJ (1986) Somaclonal variation in alfalfa. In: Janick J (ed) Plant breeding reviews, vol 4. AVI, Westport, Conn, pp 123–152

Bingham ET, Saunders JW (1974) Chromosome manipulations in alfalfa: scaling the cultivated tetraploid to seven ploidy levels. Crop Sci 14:474–477

Bingham ET, Hurley LV, Kaatz DM, Saunders JW (1975) Breeding alfalfa which regenerates from callus tissue in culture. Crop Sci 15:719–721

Bingham ET, McCoy TJ, Walker KA (1988) Alfalfa tissue culture. In: Hanson AA, Barnes DK, Hill RR Jr (eds) Alfalfa and alfalfa improvement, agronomy monograph 29. Am Soc Agron, Crop Sci Soc Am, Soil Sci Soc Am, Madison, Wisc, pp 903–929

Birky CW Jr (1983) Relaxed cellular controls and organelle heredity. Science 222:468–475

Brown DCW, Atanassov A (1985) Role of genetic background in somatic embryogenesis in *Medicago*. Plant Cell Tissue Org Cult 4:111–122

Chabaud M, Passiatore JE, Cannon F, Buchanan-Wollaston V (1988) Parameters affecting the frequency of kanamycin resistant alfalfa obtained by *Agrobacterium tumefaciens* mediated transformation. Plant Cell Rep 7:512–516

Chen THH, Marowitch J (1987) Screening of *Medicago falcata* germplasm for in vitro regeneration. J Plant Physiol 128:271–277

Chen THH, Marowitch J, Thompson BG (1987a) Genotypic effects on somatic embryogenesis and plant regeneration from callus cultures of alfalfa. Plant Cell Tissue Org Cult 8:73–81

Chen THH, Thompson BG, Gerson DF (1987b) In vitro production of alfalfa somatic embryos in fermentation systems. J Ferment Technol 65:353–357

Croughan TP, Stavarek SJ, Rains DW (1978) Selection of a NaCl tolerant line of cultured alfalfa cells. Crop Sci 18:959–963

Croughan TP, Stavarek SJ, Rains DW (1981) In vitro development of salt resistant plants. Environ Exp Bot 21:317–324

D'Amato F (1977) Cytogenetics of differentiation in tissue and cell cultures. In: Reinert J, Bajaj YPS (eds) Applied and fundamental aspects of plant cell, tissue, and organ culture. Springer, Berlin Heidelberg New York, pp 343–357

Deak M, Kiss GB, Koncz C, Dudits D (1986) Transformation of *Medicago* by *Agrobacterium* mediated gene transfer. Plant Cell Rep 5:97–100

D'hont A, Quetier F, Teoule E, Dattee Y (1987) Mitochondrial and chloroplast DNA analysis of interspecific somatic hybrids of a Leguminosae: *Medicago* (alfalfa). Plant Sci 53:237–242

Dijak M, Brown DCW (1987) Patterns of direct and indirect embryogenesis from mesophyll protoplasts of *Medicago sativa*. Plant Cell Tissue Org Cult 9:121–130

Dijak M, Simmonds DH (1988) Microtubule organization during early direct embryogenesis from mesophyll protoplasts of *Medicago sativa* L. Plant Sci 58:183–191

Dijak M, Smith DL, Wilson TJ, Brown DCW (1986) Stimulation of direct embryogenesis from mesophyll protoplasts of *Medicago sativa*. Plant Cell Rep 5:468–470

Donn G, Tischer E, Smith JA, Goodman HM (1984) Herbicide-resistant alfalfa cells: an example of gene amplification in plants. J Mol Appl Genet 2:621–635

Edwards CR, Abrams RI, Anderson MJ, Blair BD, Christensen CM, Evans JK, Ferris JM, Hankins BJ, Jordan TN, Meyer RW, Shurtleff MC, Stuckey RE, Wedberg JL, Witt WW (1980) Alfalfa: a guide to production and integrated pest management in the Midwest. N Cent Reg Ext Publ 113, Purdue Univ, Lafayette, IN

Edwards RH, Miller RE, Defremery DD, Knuckles BE, Bickoff EM, Kohler GO (1975) Pilot plant production of an edible white fraction leaf protein concentrate from alfalfa. J Agric Food Chem 23:620–626

Elgin JH Jr, Barnes DK, Ratcliffe RH, Frosheiser FI, Nielson MW, Leath KT, Sorensen EL, Lehman WF, Ostazeski SA, Stuteville DL, Kehr WR, Peaden RN, Rumbaugh MD, Manglitz GR, McMurtrey JE III, Hill RR Jr, Thyr BD, Hartman BJ (1984) Standard tests to characterize pest resistance in alfalfa cultivars. US Dep Agric Misc Publ 1434

Elliott FC, Johnson IJ, Schonhorst MH (1972) Breeding for forage yield and quality. In: Hanson CH (ed) Alfalfa science and technology. Am Soc Agron, Madison, Wisc, pp 319–334

Evans DA, Sharp WR, Medina-Filho HP (1984) Somaclonal and gametoclonal variation. Am J Bot 71:759–774

Feher F, Hangyel Tarczy M, Bocsa I, Dudits D (1989) Somaclonal chromosome variation in tetraploid alfalfa. Plant Sci 60:91–99

Gilmour DM, Davey MR, Cocking EC (1987) Plant regeneration from cotyledon protoplasts of wild *Medicago* species. Plant Sci 48:107–112

Giuliano G, Mita G, Indiogine SEP, Lo Schiavo F, Terzi M (1986) Medicago cell variants showing altered nitrogen utilization. Plant Cell Rep 5:325–328

Graham JH, Frosheiser FI, Stuteville DL, Erwin DC (1979) A compendium of alfalfa diseases. Am Phytopathol Soc, St Paul, Minn

Groose RW, Bingham ET (1984) Variation in plants regenerated from tissue culture of tetraploid alfalfa heterozygous for several traits. Crop Sci 24:655–658

Groose RW, Bingham ET (1986a) An unstable anthocyanin mutation recovered from tissue culture of alfalfa (*Medicago sativa*). 1. High frequency of reversion upon reculture. Plant Cell Rep 5:104–107

Groose RW, Bingham ET (1986b) An unstable anthocyanin mutation recovered from tissue culture of alfalfa (*Medicago sativa*). 2. Stable nonrevertants derived from reculture. Plant Cell Rep 5:108–110

Gunn CR, Skrdla WH, Spencer HC (1978) Classification of *Medicago sativa* L. using legume characters and flower colors. US Dep Agric Tech Bull 1574

Hamill JD, Mariotti D (1987) Factors affecting chlorate sensitivity in callus cultures of *Medicago sativa*. J Plant Physiol 128:425–432

Hartman CL, Knous TR (1984) Field testing and preliminary progeny evaluation of alfalfa regenerated from cell lines resistant to the toxins produced by *Fusarium oxysporum* f. sp. *medicaginis* (Abstr). Phytopathology 74:818

Hartman CL, McCoy TJ, Knous TR (1984) Selection of alfalfa (*Medicago sativa*) cell lines and regeneration of plants resistant to the toxin(s) produced by *Fusarium oxysporum* f. sp. *medicaginis*. Plant Sci Lett 34:183–194

Johnson LB, Palmer JD (1989) Heteroplasmy of chloroplast DNA in *Medicago*. Plant Mol Biol 12:3–11

Johnson LB, Stuteville DL, Higgins RK, Skinner DZ (1981) Regeneration of alfalfa plants from protoplasts of selected Regen S clones. Plant Sci Lett 20:297–304

Johnson LB, Stuteville DL, Schlarbaum SE, Skinner DZ (1984) Variation in phenotype and chromosome number in alfalfa protoclones regenerated from nonmutagenized calli. Crop Sci 24:948–951

Kao KN, Michayluk MR (1980) Plant regeneration from mesophyll protoplasts of alfalfa. Z Pflanzenphysiol 96:135–141
Kao KN, Michayluk MR (1981) Embryoid formation in alfalfa cell suspension cultures from different plants. In Vitro 17:645–648
Kao KN, Wetter LR (1977) Advances in techniques of plant protoplast fusion and culture of heterokaryocytes. In: Brinkley BR, Porter K (eds) International cell biology 1976–1977. Rockefeller Univ Press, New York, pp 216–224
Keyes GJ, Bingham ET (1979) Heterosis and ploidy effects on the growth of alfalfa callus. Crop Sci 19:473–476
Kramer CY (1956) Extension of multiple range tests to group means with unequal numbers of replication. Biometrics 12:307–310
Krause JW (1980) Embryo culture and interspecific hybridization of perennial and annual *Medicago* species. MS Thesis, Kansas State Univ, Manhattan
Kudo H, Cheng K-J, Hanna MR, Howarth RE, Goplen BP, Costerton JW (1985) Ruminal digestion of alfalfa strains selected for slow and fast initial rates of digestion. Can J Anim Sci 65:157–161
Larkin PJ, Scowcroft WR (1981) Somaclonal variation – a novel source of variability from cell cultures for plant improvement. Theor Appl Genet 60:197–214
Latunde-Dada AO, Lucas JA (1983) Somaclonal variation and reaction to *Verticillium* wilt in *Medicago sativa* L. plants regenerated from protoplasts. Plant Sci Lett 32:205–211
Latunde-Dada AO, Lucas JA (1988) Somaclonal variation and resistance to *Verticillium* wilt in lucerne, *Medicago sativa* L., plants regenerated from callus. Plant Sci 58:111–119
Lees GL (1984) Tannins and legume pasture bloat. In: Rep 29th Alfalfa Impr Conf, Lethbridge, Alberta, Can, p 68
Lesins KA, Lesins I (1979) Genus *Medicago* (Leguminosae), a taxogenetic study. Junk, The Hague Boston London
Lo Schiavo F, Nuti Ronchi V, Terzi M (1980) Genetic effects of griseofulvin on plant cell cultures. Theor Appl Genet 58:43–47
Lu DY, Pental D, Cocking EC (1982) Plant regeneration from seedling cotyledon protoplasts. Z Pflanzenphysiol 107:59–63
Lu DY, Davey MR, Cocking EC (1983) A comparison of the cultural behaviour of protoplasts from leaves, cotyledons and roots of *Medicago sativa*. Plant Sci Lett 31:87–99
Lupotto E (1983) Propagation of an embryogenic culture of *Medicago sativa* L. Z Pflanzenphysiol 111:95–104
Lupotto E (1986) The use of single somatic embryo culture in propagating and regenerating lucerne (*Medicago sativa* L.). Ann Bot (London) 57:19–24
Maheswaran G, Williams EG (1984) Direct somatic embryoid formation on immature embryos of *Trifolium repens*, *T. pratense* and *Medicago sativa*, and rapid clonal propagation of *T. repens*. Ann Bot (London) 54:201–211
Maliga P (1980) Isolation, characterization, and utilization of mutant cell lines in higher plants. In: Vasil IK (ed) International review of cytology. Perspectives in plant cell and tissue culture, Suppl 11A. Academic Press, New York London, pp 225–250
Masoud SA (1989) High transmission of paternal plastid DNA in alfalfa plants as shown by restriction polymorphic analysis. MS Thesis, Kansas State Univ, Manhattan
McCoy TJ (1985) Interspecific hybridization of *Medicago sativa* L. and *M. rupestris* M.B. using ovule-embryo culture. Can J Genet Cytol 27:238–245
McCoy TJ (1987) Tissue culture evaluation of NaCl tolerance in *Medicago* species: cellular versus whole plant response. Plant Cell Rep 6:31–34
McCoy TJ, Bingham ET (1977) Regeneration of diploid alfalfa plants from cells grown in suspension culture. Plant Sci Lett 10:59–66
McCoy TJ, Smith LY (1984) Uneven ploidy levels and a reproductive mutant required for interspecific hybridization of *Medicago sativa* L. × *Medicago dzhawakhetica* Bordz. Can J Genet Cytol 26:511–518
McCoy TJ, Smith LY (1986) Interspecific hybridization of perennial *Medicago* species using ovule-embryo culture. Theor Appl Genet 71:772–783
McCoy T, Walker K (1984) Alfalfa. In: Ammirato PV, Evans DA, Sharp WR, Yamada Y (eds) Handbook of plant cell culture, vol 3. Crop species. Macmillan, New York, pp 171–192
Meijer EGM, Brown DCW (1985) Screening of diploid *Medicago sativa* germplasm for somatic embryogenesis. Plant Cell Rep 4:285–288

Meijer EGM, Brown DCW (1987a) A novel system for rapid high frequency somatic embryogenesis in *Medicago sativa* L. Physiol Plant 69:591–596

Meijer EGM, Brown DCW (1987b) Role of exogenous reduced nitrogen and sucrose in rapid high frequency somatic embryogenesis in *Medicago sativa*. Plant Cell Tissue Org Cult 10:11–19

Meijer EGM, Keller WA, Simmonds DH (1988) Cytological abnormalities and aberrant microtubule organization during early divisions in mesophyll protoplast cultures of *Medicago sativa* and *Nicotiana tabacum*. Physiol Plant 74:233–239

Mezentsev AV (1981) Mass regeneration of plants from the cells and protoplasts of lucerne. Dokl Vses Akad SKh Nauk Im. V I Lenina 4:22–23 (in Russian)

Michaud R, Lehman WF, Rumbaugh MD (1988) World distribution and historical development. In: Hanson AA, Barnes DK, Hill RR Jr (eds) Alfalfa and alfalfa improvement, agronomy monograph 29. Am Soc Agron, Crop Sci Soc Am, Soil Sci Soc Am, Madison, Wisc, pp 25–91

Mitten DH, Sato SJ, Skokut TA (1984) In vitro regenerative potential of alfalfa germplasm sources. Crop Sci 24:943–945

Mroginski LA, Kartha KK (1984) Tissue culture of legumes for crop improvement. In: Janick J (ed) Plant breeding reviews, vol 2. AVI, Westport, Conn, pp 215–264

Nagarajan P, Walton PD (1987) A comparison of somatic chromosomal instability in tissue culture regenerants from *Medicago media* Pers. Plant Cell Rep 6:109–113

Nagarajan P, McKenzie JS, Walton PD (1986) Embryogenesis and plant regeneration of *Medicago* spp. in tissue culture. Plant Cell Rep 5:77–80

Novák FJ, Konečná D (1982) Somatic embryogenesis in callus and cell suspension cultures of alfalfa (*Medicago sativa* L.). Z Pflanzenphysiol 105:279–284

Oelck MM, Schieder O (1983) Genotypic differences in some legume species affecting the redifferentiation ability from callus to plants. Z Pflanzenzücht 91:312–321

Pezzotti M, Arcioni S, Mariotti D (1984) Plant regeneration from mesophyll, root and cell suspension protoplasts of *Medicago sativa* cv. Adriana. Genet Agr 38:195–208

Pfeiffer TW, Bingham ET (1984) Comparisons of alfalfa somaclonal and sexual derivatives from the same genetic source. Theor Appl Genet 67:263–266

Phillips GC, Collins GB (1979) In vitro tissue culture of selected legumes and plant regeneration from callus cultures of red clover. Crop Sci 19:59–64

Redenbaugh K, Paasch BD, Nichol JW, Kossler ME, Viss PR, Walker KA (1986) Somatic seeds: encapsulation of asexual plant embryos. Bio/Technology 4:797–801

Redenbaugh K, Slade D, Viss P, Fujii JA (1987) Encapsulation of somatic embryos in synthetic seed coats. HortSci 22:803–809

Reich TJ, Iyer VN, Miki BL (1986) Efficient transformation of alfalfa protoplasts by the intranuclear microinjection of Ti plasmids. Bio/Technology 4:1001–1004

Reisch B (1983) Genetic variability in regenerated plants. In: Evans DA, Sharp WR, Ammirato PV, Yamada Y (eds) Handbook of plant cell culture, vol 1, techniques for propagation and breeding. Macmillan, New York, pp 748–769

Reisch B, Bingham ET (1980) The genetic control of bud formation from callus cultures of diploid alfalfa. Plant Sci Lett 20:71–77

Reisch B, Bingham ET (1981) Plants from ethionine-resistant alfalfa tissue cultures: variation in growth and morphological characteristics. Crop Sci 21:783–788

Reisch B, Duke SH, Bingham ET (1981) Selection and characterization of ethionine-resistant alfalfa (*Medicago sativa* L.) cell lines. Theor Appl Genet 59:89–94

Rose RJ, Johnson LB, Kemble RJ (1986) Restriction endonuclease studies on the chloroplast and mitochondrial DNAs of alfalfa (*Medicago sativa* L.) protoclones. Plant Mol Biol 6:331–338

Santos AVP dos, Outka DE, Cocking EC, Davey MR (1980) Organogenesis and somatic embryogenesis in tissues derived from leaf protoplasts and leaf explants of *Medicago sativa*. Z Pflanzenphysiol 99:261–270

Santos AVP dos, Cutter EG, Davey MR (1983) Origin and development of somatic embryos in *Medicago sativa* L. (alfalfa). Protoplasma 117:107–115

Saunders JW, Bingham ET (1972) Production of alfalfa plants from callus tissue. Crop Sci 12:804–808

Saunders JW, Bingham ET (1975) Growth regulator effects on bud initiation in callus cultures of *Medicago sativa*. Am J Bot 62:850–855

Schaefer J (1985) Regeneration in alfalfa tissue culture: characterization of intracellular pH during somatic embryo production by solid-state P-31 NMR. Plant Physiol 79:584–589

Schlarbaum SE, Johnson LB, Stuteville DL (1988) Characterization of somatic chromosome morphology in alfalfa, *Medicago sativa* L.: comparison of donor plant with regenerated protoclone. Cytologia 53:499–507

Schreier PH, Kuntz M, Lipphardt S, Lörz H, Baker B, Simons A, de Bruijn F, Schell J, Bohnert HJ, Reiss B, Wasmann CC (1985) New developments in plant transformation technology: its application to cellular organelles, cereals, and dicotyledonous crop plants. In: Zaitlin M, Day P, Hollaender A (eds) Biotechnology in plant science. Relevance to agriculture in the eighties. Academic Press, New York London, pp 237–246

Seitz Kris MH, Bingham ET (1988) Interactions of highly regenerative genotypes of alfalfa (*Medicago sativa*) and tissue culture protocols. In Vitro Cell Dev Biol 24:1047–1052

Shahin EA, Spielmann A, Sukhapinda K, Simpson RB, Yashar M (1986) Transformation of cultivated alfalfa using disarmed *Agrobacterium tumefaciens*. Crop Sci 26:1235–1239

Skokut TA, Manchester J, Schaefer J (1985) Regeneration in alfalfa tissue culture: stimulation of somatic embryo production by amino acids and N-15 NMR determination of nitrogen utilization. Plant Physiol 79:579–583

Skotnicki ML (1986) Rapid regeneration of alfalfa plants from tissue culture. Cytobios 46:189–192

Smith MK, McComb JA (1983) Selection for NaCl tolerance in cell cultures of *Medicago sativa* and recovery of plants from a NaCl-tolerant cell line. Plant Cell Rep 2:126–128

Smith SE, Bingham ET, Fulton RW (1986) Transmission of chlorophyll deficiencies in *Medicago sativa*. J Hered 77:35–38

Spanò L, Mariotti D, Pezzotti M, Damiani F, Arcioni S (1987) Hairy root transformation in alfalfa (*Medicago sativa* L.). Theor Appl Genet 73:523–530

Sree Ramulu K, Dijkhuis P, Roest S, Bokelmann GS, De Groot B (1984) Early occurrence of genetic instability in protoplast cultures of potato. Plant Sci Lett 36:79–86

Stavarek SJ, Rains DW (1984) Cell culture techniques: selection and physiological studies of salt tolerance. In: Staples RC, Toenniessen GH (eds) Salinity tolerance in plants: strategies for crop improvement. John Wiley & Sons, New York, pp 321–334

Stavarek SJ, Croughan TP, Rains DW (1980) Regeneration of plants from long-term cultures of alfalfa cells. Plant Sci Lett 19:253–261

Strickland SG, Nichol JW, McCall CM, Stuart DA (1987) Effect of carbohydrate source on alfalfa somatic embryogenesis. Plant Sci 48:113–121

Stuart DA, Strickland SG (1984a) Somatic embryogenesis from cell cultures of *Medicago sativa* L. I. The role of amino acid additions to the regeneration medium. Plant Sci Lett 34:165–174

Stuart DA, Strickland SG (1984b) Somatic embryogenesis from cell cultures of *Medicago sativa* L. II. The interaction of amino acids with ammonium. Plant Sci Lett 34:175–181

Stuart DA, Nelsen J, McCall CM, Strickland SG, Walker KA (1985a) Physiology of the development of somatic embryos in cell cultures of alfalfa and celery. In: Zaitlin M, Day P, Hollaender A (eds) Biotechnology in plant science. Relevance to agriculture in the eighties. Academic Press, New York London, pp 35–47

Stuart DA, Nelsen J, Strickland SG, Nichol JW (1985b) Factors affecting developmental processes in alfalfa cell cultures. In: Henke RR, Hughes KW, Constantin MJ, Hollaender A (eds) Tissue culture in forestry and agriculture. Plenum, New York London, pp 59–73

Stuart DA, Strickland SG, Walker KA (1987) Bioreactor production of alfalfa somatic embryos. HortSci 22:800–803

Stuart DA, Nelsen J, Nichol JW (1988) Expression of 7S and 11S alfalfa seed storage proteins in somatic embryos. J Plant Physiol 132:134–139

Sukhapinda K, Spivey R, Shahin EA (1987) Ri-plasmid as a helper for introducing vector DNA into alfalfa plants. Plant Mol Biol 8:209–216

Tárczy MH, Fehér F, Deák M (1986) Chromosome variation of somaclones in tetraploid lucerne. Novenytermeles 35:281–286 (in Hungarian, English Abstr)

Téoulé E (1983) Somatic hybridization between *Medicago sativa* L. and *Medicago falcata* L. C R Acad Sci Paris 297:13–16

Téoulé E, Dattée Y (1987) Recherche d'une méthode fiable de culture de protoplastes, d'hybridation somatique et de régénération chez *Medicago*. Agronomie 7:575–584

US Department of Agriculture - USDA (ed) (1986) Agricultural statistics 1986. US Gov Print Off. Washington DC
Walker KA, Sato SJ (1981) Morphogenesis in callus tissue of *Medicago sativa*: the role of ammonium ion in somatic embryogenesis. Plant Cell Tissue Org Cult 1:109–121
Walker KA, Yu PC, Sato SJ, Jaworski EG (1978) The hormonal control of organ formation in callus of *Medicago sativa* L. cultured in vitro. Am J Bot 65:654–659
Walker KA, Wendeln ML, Jaworski EG (1979) Organogenesis in callus tissue of *Medicago sativa*. The temporal separation of induction processes from differentiation processes. Plant Sci Lett 16:23–30
Walton PD, Brown DCW (1988) Screening of *Medicago* wild species for callus formation and the genetics of somatic embryogenesis. J Genet 67:95–100
Wan Y, Sorensen EL, Liang GH (1988) Genetic control of in vitro regeneration in alfalfa (*Medicago sativa* L.). Euphytica 39:3–9
Wang JW, Sorensen EL, Liang GH (1984) In vitro culture of pods from annual and perennial *Medicago* species. Plant Cell Rep 3:146–148
Willis WG, Stuteville DL, Sorensen EL (1969) Effects of leaf and stem diseases on yield and quality of alfalfa forage. Crop Sci 9:637–640
Xu S (1979) Success in induction of alfalfa pollen plants. Plant Mag 6:5 (in Chinese)
Xu Z-H, Davey MR, Cocking EC (1982) Organogenesis from root protoplasts of the forage legumes *Medicago sativa* and *Trigonella foenum-graecum*. Z Pflanzenphysiol 107:231–235
Zagorska N, Robeva P, Dimitrov B, Shtereva R, Gancheva V (1984) Induction of regeneration in anther cultures in *Medicago sativa* L. C R Bulg Acad Sci 37:1099–1102

Subject Index

MIX
Papier aus verantwortungsvollen Quellen
Paper from responsible sources
FSC® C105338

If you have any concerns about our products,
you can contact us on
ProductSafety@springernature.com

In case Publisher is established outside the EU,
the EU authorized representative is:
Springer Nature Customer Service Center GmbH
Europaplatz 3, 69115 Heidelberg, Germany

Printed by Libri Plureos GmbH
in Hamburg, Germany